Digital
Control System
Design

SECOND EDITION

Digital
Control System
Design

SECOND EDITION

MOHAMMED S. SANTINA
The Aerospace Corporation

ALLEN R. STUBBERUD
University of California, Irvine

GENE H. HOSTETTER
Late of University of California, Irvine

SAUNDERS COLLEGE PUBLISHING
Harcourt Brace College Publishers

Fort Worth   Philadelphia   San Diego
New York   Orlando   Austin
San Antonio   Toronto   Montreal
London   Sydney   Tokyo

Text Typeface: Times Roman
Compositor: Bi-Comp, Inc.
Acquisitions Editor: Emily Barrosse
Managing Editor: Carol Field
Project Editor: Laura Shur
Copy Editor: Linda Davoli
Manager of Art and Design: Carol Bleistine
Associate Art Director: Jennifer Dunn
Cover Designer: Lawrence R. DiDona
Text Artwork: Grafacon
Director of EDP: Tim Frelick
Production Manager: Joanne Cassetti
Marketing Manager: Monica Wilson

Printed in the United States of America

Digital Control System Design, 2/e

ISBN: 0-03-076012-7

Library of Congress Catalog Card Number: 93-087483

3456 118 987654321

To our families:
Dalia, Michael, Gene, and Elham
May, Peter and Laura, and Stephen
Donna, Colleen, and Kristen

Contents

Preface   xi

1 Introduction   1

1.1   Preview 1
1.2   Control System Terminology 1
1.3   An Overview of Classical Control 5
1.4   Precision Temperature System 9
1.5   Summary 14
       References 15

2 Discrete-Time Systems and Z-Transformation   16

2.1   Preview 16
2.2   Discrete-Time Signals 16
2.3   Discrete-Time Systems 31
2.4   Sampling and Reconstruction 55
2.5   Analysis of Hybrid Systems 76
2.6   Design of a Videotape Drive Control 88
2.7   Summary 96
       References 98
       Problems 99

3 State Space Description of Dynamic Systems   109

3.1   Preview 109
3.2   Continuous-Time State Equations 110
3.3   Discrete-Time State Equations and System Response 129
3.4   Sampled Continuous-Time Systems 143
3.5   Canonical Forms 149

3.6   Uncoupling State Equations 158
3.7   Observability and Controllability 170
3.8   A Monorail System 182
3.9   Summary 185
        References 191
        Problems 192

4   Discrete-Time Observation, Control, and Feedback   205

4.1   Preview 205
4.2   Observability and State Observation 206
4.3   Controllability and State Control 221
4.4   State Feedback 235
4.5   Output Feedback 253
4.6   Pole Placement with Feedback Compensation 261
4.7   Quadratic Optimal Regulation 267
4.8   Closed-Form Solution for Optimal Gain 282
4.9   Summary 297
        References 299
        Problems 301

5   Digital Observers and Regulator Design   312

5.1   Preview 312
5.2   Full-Order State Observers 312
5.3   More About Observers 325
5.4   Observer Design 333
5.5   Lower-Order Observers 344
5.6   Eigenvalue Placement with Observer Feedback 355
5.7   Step-Varying Observers 367
5.8   Control of Flexible Spacecrafts 380
5.9   Summary 391
        References 395
        Problems 396

6   Digital Tracking System Design   405

6.1   Preview 405
6.2   Ideal Tracking System Design 406

6.3   Response Model Tracking System Design 421
6.4   Reference Model Tracking System Design 438
6.5   Disturbance Rejection 455
6.6   A Digital Phase-Locked Loop 465
6.7   Tracking System Design for Step-Varying Systems 472
6.8   Summary 478
       References 480
       Problems 481

7   Digital Control of Continuous-Time Systems   490

7.1   Preview 490
7.2   Digitizing Analog Controllers 491
7.3   Sampled Continuous-Time Systems 512
7.4   Designing Between-Sample Response 520
7.5   Digital Hardware for Control 528
7.6   Computer Software for Control 543
7.7   Design Examples 551
7.8   Summary 563
       References 565
       Problems 566

8   Stochastic Systems and Recursive Estimation   580

8.1   Preview 580
8.2   Response of Linear Systems to Random Inputs 582
8.3   State Variable Representation of Linear Systems Driven
       by Random Inputs 596
8.4   Least Squares Estimation 605
8.5   Linear Minimum Mean Square Estimation 617
8.6   The Discrete-Time Kalman Filter 625
8.7   Extensions 642
8.8   Stochastic Optimal Control 653
8.9   Summary 658
       References 660
       Problems 663

Appendix A:
Elements of Linear Algebra   680

Appendix B:
Review of Selected Topics From the Theory of
Probability and Stochastic Processes   746

Appendix C:
Basic Results for Linear Minimum Mean Square
Error Estimation   785

Preface

This text is a revised and expanded edition of *Digital Control System Design*, which was originally written by the late Gene H. Hostetter and published by Holt, Rinehart and Winston in 1988.

The dramatic development of small, inexpensive, and highly capable digital computers is changing the nature of most control system design from analog-based to digital. At the same time, powerful digital computation tools for control system design are increasingly accessible, making sophisticated design methods cost-effective. These advances have already had a tremendous effect in such areas as the aerospace industry. For most industries, however, the flourishing of digital control is really just beginning, and today's engineering and scientific graduates will be responsible for its design.

Intended Audience

This is a design-oriented digital control systems text intended for use in two academic courses at the advanced undergraduate and beginning graduate levels and for reference by practicing engineers in industry. The text is suitable for both electrical and mechanical engineering programs. It is intended to cover those general concepts that are most effectively taught in the classroom, in intensive short courses, and for individual study. Of course, much more, especially experience and common sense, is expected of a seasoned designer. Our orientation is that of state space but with an appreciation for classical viewpoints and methods. We expect that most readers have backgrounds in continuous classical control concepts and an understanding of basic linear algebra. For those students, the material in Appendix A serves as a concise review and a convenient way of initiating study. For the student who lacks the initial preparation, the material in Appendix A is sufficiently detailed to serve as a tutorial. The material in Chapter 8 requires random variables and processes. These concepts are concisely reviewed in Appendix B or whenever necessary in Chapter 8.

Style

The style in this edition is similar to that of the first edition and is strongly oriented on design. A general background is given throughout for the use of digital computation in the design process. Several design examples are provided in each chapter, and at least one major practical problem is explained and solved toward the end of the chapter. This format, too, is similar to the first edition.

Approach

The text is written to be read by the student. In short, this is a text suitable for self-study, giving the instructor time to interpret and augment the material if desired. The approach tends toward the applied rather than the abstract, rich but not overwhelming in new concepts and terms. We strove for a presentation that demonstrates relevance while exploring new horizons. The "definition-lemma-theorem," although suitable for some purposes, is poorly suited to this approach.

Organization of the Text

The text is organized into eight chapters and three appendices. The areas covered in each chapter are detailed in the following sections.

Chapter 1: Introduction

Overview of methods and goals of digital and modern control theory and implementation. Control system terminology, emphasis, and viewpoint.

Chapter 2: Discrete-Time Systems and Z-Transformation

Z-transformation and properties. Discrete-time signals and systems, Jury stability criterion. Discrete frequency response. Sampling and reconstruction. Discrete-time equivalents of continuous-time systems. Analysis of sampled data systems. Overview of classical root locus design methods.

Chapter 3: State Space Description of Dynamic Systems

Continuous, discrete, and sampled system models and response. Time- and step-varying state equations and solutions. Canonical forms. Decoupling state equations. Observability and controllability. Transformation to controllable and observable forms.

Chapter 4: Discrete-Time Observation, Control, and Feedback

Observability and state observation. Parameter identification and state estimation. Observability, state determination, controllability, and state control of step-varying systems. State feedback, eigenvalue placement, Ackermann's formula, state feedback for step-varying systems. Pole placement with feedback compensation. Quadratic optimal control.

Chapter 5: Digital Observers and Regulator Design

Observer concepts and properties. Design of state and observer feedback for acceptable transient response. Step-varying observers.

Chapter 6: Digital Tracking System Design

Design for acceptable reference input tracking via inverse filtering, response model, and reference model tracking system design methods. Introduction to disturbance rejection. Tracking system controller design for step-varying systems.

Chapter 7: Digital Control of Continuous-Time Systems

Digitizing analog controllers. Sampled continuous-time systems. Design of between-sample response when a continuous-time plant is controlled digitally. Introduction to digital hardware and software considerations.

Chapter 8: Stochastic Systems and Recursive Estimation

Response of linear systems to random inputs. Stochastic state equations and solutions. Least squares and recursive least squares state estimation. Kalman filtering techniques and extensions. Stochastic control.

Changes in the Second Edition

In undertaking this revision, we have made it our primary objective to retain the thrust of Gene Hostetter's basic ideas. We updated the main body of the text to conform to current technological conditions and advancements. We also included stochastic concepts to provide the reader with greater insight into the design process. Three appendices were also added to the second edition.

Based on comments and feedback from students, faculty, and reviewers, we introduced many minor clarifications to explain the material better. Numerical errors in the examples were also corrected. A large number of the problems at the ends of the chapters are new.

Chapter 1 of the second edition is a brief introductory chapter that provides motivation and perspective for the subject of digital control systems in general and our treatment of it in particular. As stated earlier, its function is to establish terminology, emphasis, and viewpoint for the material to follow in later chapters. Section 1.3 of Chapter 1 of the first edition was retained in Appendix A of the second edition, and the material of Sections 1.4 and 1.6 of the first edition was included (not necessarily in this order) in Chapter 3 of the second edition.

Chapter 2 of the first edition is significantly expanded in this edition. The discussion of z-transformation of Section 2.2 is now better organized, and many additional examples are presented to illustrate the ideas involved. Section 2.3 of the second edition contains additional material on stability testing of discrete-time systems, such as Jury stability criteria. The material of Section 2.4 of the second edition on signal recovery by filtering is new. We believe that this material provides a clear understanding of the concepts of sampling and reconstruction. The material of Section 2.5 of the second edition on block diagram manipulation of sampled data systems with samplers at various locations in the system is also new.

Chapter 3 of the second edition contains material from Chapters 1, 3, and 7 of the first edition; however, due to the importance of the topics covered in this chapter, the presentation of the material is more formal than that in the first edition. The material in Section 3.2 on linear time-varying state equations and solutions is new. The material on state equations and response of step-varying systems in Chapter 8 of the first edition is now in Section 3.3 of the second edition. Some of the material on discretizing continuous-time systems in Section 7.2 of the first edition is now in Section 3.4 of the second edition. Other discretization methods not presented in the first edition are also discussed along with several examples. Section 3.5 of the second edition generally follows Section 3.2 of the first edition. The characteristic value problem in Section 3.4 of the first edition is now in Appendix A of the second edition. Section 3.5 of the first edition is now in Section 3.6 of the second edition and contains new material on alternative representations of systems with complex eigenvalues. The material in Section 3.6 of the first edition is now in Section 3.7 of the second edition. The monorail system discussed in Section 1.6 of the first edition is in Section 3.8 of the second edition.

Chapter 4 of the second edition contains material from Chapters 4 and 8 of the first edition. Additional material, such as minimal realization of a system with a prescribed transfer function matrix, is discussed in Subsection 4.3.4. Pole placement with feedback compensation and other methods of placing closed-loop system eigenvalues, such as Ackermann's formula, are presented. A spacecraft example that ties many of these concepts together is presented. The material on optimal control in the first edition is greatly expanded in the second edition. Illustrative examples are also provided.

Chapter 5 of the second edition contains material from Chapters 5 and 8 of the first edition. The contents are similar, with the exception of the control of flexible spacecraft design example presented in Section 5.8 of the second edition.

Chapter 6 of the new edition contains material from Chapters 6 and 8 of the first edition. The contents are similar but improved in presentation.

Most of the material presented in Chapter 7 of the second edition is an update of the material presented in Chapter 7 of the first edition. The material in Section 7.2 of the second edition on digitizing analog controllers is treated differently from that in Section 2.5 of the first edition. In the second edition, our treatment is standard. For example, the approach of presenting the transfer function of a controller in Subsection 2.5.1 of the first edition with state variables, numerically approximating each of the analog integration operations and then determining the discrete transfer function of the controller, is modified. Each s-variable is directly replaced with a desired numerical approximation, and subsequently the transfer function of the digital controller is formed. Additional methods of digitizing analog controllers are presented in the second edition. The elevator controller design example in Section 7.6 of the first edition is revised and updated.

Chapter 8 of the second edition deals with stochastic concepts. The chapter best serves as a reference and should provide a solid foundation for the use of Kalman filtering. Response of linear systems to random inputs is discussed in detail in Sections 8.2 and 8.3. Section 8.4 develops the fundamental ideas of least squares estimation. The basic linear minimum mean squares estimation problem is formulated in Section 8.5. Section 8.6 contains detailed derivation of the Kalman filtering equations and a block diagram of the filter, which is seen to include a replica of the state variable model for the measurements. A computer program for demonstrating Kalman filtering implementations is given.

Section 8.7 introduces extensions of the Kalman filter to situations involving noise coupling matrices, deterministic inputs to the model, nonzero mean values, known initial conditions, correlated noises, bias estimation, and nonlinear estimation. Section 8.8 is concerned with stochastic optimal control using Kalman filtering techniques. A number of examples are given to illustrate key concepts.

Appendix A in the second edition, Elements of Linear Algebra, is very detailed and comprehensive and contains material from Chapters 1, 3, and 8 of the first edition. Several new topics are also covered in the appendix.

Appendix B, Review of Selected Topics from the Theory of Probability and Stochastic Processes, is lengthy and serves as a precise review of topics relevant to Chapter 8.

Finally, Appendix C, Basic Results for Linear Minimum Mean Square Error Estimation, contains proofs of the theorems that are necessary for the derivation of the Kalman filter equations.

Objectives

In this book, four design problems of digital tracking feedback controllers for linear plants are carefully formulated and solved. Basically, the design of digital tracking feedback controllers for linear plants is composed of two fundamental parts:

1. Design for acceptable feedback system zero-input (or transient) response
2. Design for acceptable feedback system zero-state response for a relevant or representative class of reference inputs to be tracked.

If the plant to be controlled is continuous time, it is also necessary to

3. Design for acceptable between-sample response of the plant. Additionally, if the plant has stochastic disturbance inputs or outputs or both, it is necessary to
4. Design the feedback controller so as to *minimize* the effects of these disturbances on system response.

This is our agenda. At the appropriate time, we also consider some aspects of the digital hardware and software that are unlikely to be obsolete in the near future.

Computation

The desirability of connecting a text such as this too closely with specific computational programs is inadvisable. The computer-aided design (CAD) support available at universities and in industry varies widely in its capabilities and idiosyncrasies. Those with good and easily accessed digital computer support for modern control system design should simply apply the tools in place, with what are probably already-familiar formats. For those who are just beginning, computational packages such as Matlab/Simulink, Matrix-x, and EASY5 are recommended. This text, however, *does not* depend on any particular software.

Acknowledgments

A large number of students and colleagues enthusiastically contributed to this work. We are especially indebted to Jose B. Cruz, Jr., Jeffrey Burl, Tim Jordanides, Claude Lindquist, Leonard Ferrari, James H. Mulligan, Jr., Raymond Stefani, Mike Borrello, Jim Mullins, Jeff Finley, Patrick Wang, Joe Anselmi, Dominic Camillone, Kenneth Hagen, Kyle N. Vaught, Jerry Marke, Bruce E. Gardner, and Charlie Gray. A distinguished group of reviewers also contributed greatly to this effort. They are Professors Enrique Barbieri, Tulane University; John Bennet, Clemson University; Louis J. Galbiati, Jr., State University of New York Institute of Technology at Utica-Rome; Kenneth Unklesbay, University of Missouri-Columbia; Richard Christiansen, Brigham Young University; Robert E. Fenton, Ohio State University; Ralph Hippenstiel, US Naval Postgraduate School; W. Richard Kolk, Hartford Graduate Center; Lal Tummala, Michigan State University; Harris McClamrock, University of Michigan; Robert Mulholland, University of Oklahoma; Don Pierre, University of Montana; Vijay Raman, Florida International University; Michael Rekoff, University of Alabama; Charles Thompson, University of Lowell.

Finally, we would like to extend a special note of appreciation to Connie Murphy and to the professional staff at Saunders College Publishing, especially Emily Barrosse, Executive Editor, Laura Shur, Project Editor, Jeff Lyons, Editorial Assistant, and Linda Davoli, Copy Editor.

M. S. Santina
A. R. Stubberud
November 1993

Gene H. Hostetter
(1939–1988)

Dr. Gene Huber Hostetter, former professor of electrical engineering at the University of California, Irvine, passed away on July 30, 1988, after a long illness. Gene was born on September 14, 1939, in Spokane, Washington. He received the B. S. and M. S. degrees in electrical engineering from the University of Washington, Seattle, and the Ph.D. degree in engineering from the University of California, Irvine, in 1973.

Gene came to UCI after 10 years of industrial experience at the Seattle Broadcasting Company and the Boeing Company. He was also a consultant to several organizations, including Battelle Institute, Meditek, Sundstrand Corporation, and Western Digital Corporation.

Gene's research interests were in the areas of microprocessor-based digital control and signal processing, digital and linear electronic system design, and electrooptics.

He was elected as a Fellow of the Institute of Electrical and Electronics Engineers (IEEE) for his "contribution to the theory and algorithms for observers of complex control systems, and to engineering education." He also received the IEEE Centennial Medal.

Gene authored four electrical engineering textbooks, more than a hundred scholarly technical papers, and numerous reviews and review articles.

As a teacher, Gene was consistently judged to be outstanding. As his students progress in the community, they will continue to praise the lifetime of analytical tools and the experience in critical thinking he gave them.

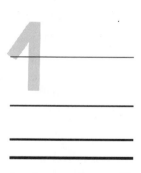

Introduction

1.1 Preview

This is a book about fundamental concepts in modern digital control system design. The rapid development of digital technology has radically changed the boundaries of practical control system design options. It is now routinely feasible to employ very complicated high-order digital controllers and to carry out the extensive calculations required for their design. These advances in implementation and design capability can be achieved at low cost because of the widespread availability of inexpensive, powerful digital computers and related devices.

The focus of our study is the modern, state space-based design of digital control systems. To begin, we introduce basic control system terminology and present an overview of the classical approach to continuous-time tracking system design. The key concepts, involving specification of transient and steady state response requirements are also much a part of the modern approach. Although our concern is with digital control, digital controllers are most often used to control continuous-time plants. It is important then to have a good understanding of the plant to be controlled as well as of the controller and its interfaces with the plant.

1.2 Control System Terminology

A *control system* is an interconnection of components to provide a desired function. The portion of the system to be controlled is called the *plant*, and the part doing the controlling is the *controller*. Often, a control system designer has little or no design freedom with the plant; it is fixed. The designer's task is, therefore, to develop a controller that will control the given plant acceptably. When measurements of plant response are available to the controller (which in turn generates signals affecting the plant), the configuration is a *feedback control system*.

A *digital control system* uses digital hardware, usually in the form of a programmed digital computer, as the heart of the controller. In contrast, the controller in an *analog control system* is composed of analog hardware, typically analog electronic, mechanical, electro-mechanical, and hydraulic devices. Digital controllers normally have analog elements at their periphery to interface with the plant; it is the internal workings of the controller that distinguish digital from analog control.

Digital control systems offer many advantages over their analog counterparts. Among these advantages are the following:

1. Low susceptibility to environmental conditions, such as temperature, humidity, and component aging
2. The cost reduction and interference rejection associated with digital signal transmission
3. Zero "drift" of parameters
4. High potential reliability
5. The ability to perform highly complex tasks at low cost
6. The potential flexibility of easily making changes in software
7. Relatively simple interfaces with other digital systems, such as those for accounting, forecasting, and data collection

To be sure, there are possible disadvantages also, among which are the following:

1. The introduction of errors (or "noise") due to the finite precision of digital computation and the abrupt changes due to the discrete-time nature of digital control
2. The need for more sophisticated engineering in order to take advantage of higher-performance control algorithms
3. Greater limitations on speed of operation
4. Greater potential for catastrophic failure

The signals used in the description of control systems are classified as continuous-time or discrete-time. *Continuous-time signals* are functions of a continuous variable, whereas *discrete-time signals* are defined only for discrete values of the variable, usually evenly spaced steps. Discrete-time signals and their manipulations are inherently well suited to digital computation and are used in describing the digital portions of a control system. Most often, continuous-time signals are involved in describing the plant and the interfaces between a controller and the plant it controls. Signals are further classified as being of *continuous amplitude* or *discrete amplitude*. Discrete-amplitude (or *quan-*

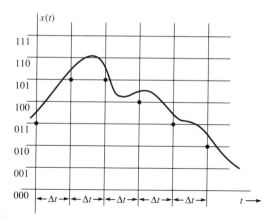

Figure 1-1 An example of a 3-bit quantized signal.

tized) signals can attain only discrete values, usually evenly spaced. For example, the continuous amplitude signal shown in Figure 1-1 is represented by a 3-bit binary code at evenly spaced time instants. In general, an n-bit binary code can represent only 2^n different values. Because of the complexity of dealing with quantized signals, digital control system design proceeds as if computer-generated signals were not of discrete amplitude. If necessary, further analysis is then done to determine if a proposed level of quantization is acceptable.

A general control system diagram appears in Figure 1-2. The plant is affected by input signals, some of which (the control inputs) are accessible to the controller, and some of which (the disturbance inputs) are not. Some of the plant signals (the tracking outputs) are to be controlled, and some of the plant signals (the measurement outputs) are available to the controller. A controller generates control inputs to the plant with the objective of having the tracking outputs closely approximate the reference inputs.

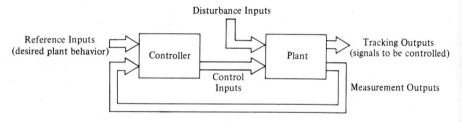

Figure 1-2 A control system.

Systems and system components are classified as to the nature of their mathematical model and termed continuous time or discrete time according to the type of signals they involve. They are classified as being *linear* if signal components in them can be superimposed. Any linear combination of signal components applied to an input produces the same linear combination of corresponding output components; otherwise, a system is *nonlinear*. A continuous-time system or component is *time-invariant* (or *constant-parameter*) if its properties do not change with time. Any time shift of the inputs produces an equal time shift of every corresponding signal. If a continuous-time system is not time-invariant, then it is *time-varying*. On the other hand, if the properties of a discrete-time system do not change with step, then it is called *step-invariant*. And, if the discrete-time system is not step-invariant, then it is *step-varying*.

An example of a digital control system for a continuous-time plant is shown in Figure 1-3. The system has two reference inputs and five outputs, only two of which are measured by the two sensors. The *analog-to-digital converters* (A/D) perform sampling of the sensor signals and produce binary representations of these sensor signals. The digital controller *algorithm* then modifies the sensor signals and generates digital control inputs $u_1(k)$ and $u_2(k)$. These control inputs are then converted to analog signals via *digital-to-analog converters* (D/A). This process of transforming digital codes to analog signals begins by converting the digital codes to signal samples and then producing *step reconstruction* from the signal samples by transforming the binary-coded digital inputs to voltages. These voltages are held constant for a *sampling period T* until the next samples are available. The process of holding each of the samples to perform step reconstruction is termed

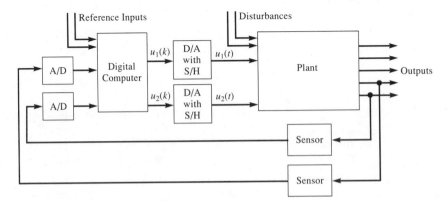

Figure 1-3 A digital control system controlling a continuous-time plant.

sample and hold. Not shown in the figure is a real time clock that synchronizes the actions of the A/D and D/A and the *shift registers.* The analog signals $u_1(t)$ and $u_2(t)$ are applied to the plant actuators or control elements to control the plant's behavior.

There are many variations on this theme, including situations in which the sampling period is not fixed, in which the A/D and D/A are not synchronized, in which the system has many controllers with different sampling periods, and in which the sensors produce digital signals and the actuators accept digital commands.

Two important classes of control systems are the *regulator* and the *tracking system* (or *servosystem*). In the former, the objective is to bring the system-tracking outputs near to zero in an acceptable manner, often in the face of disturbances. For example, a regulator might be used to keep a motor-driven satellite dish antenna on a moving vehicle accurately pointed in a fixed direction, even when the antenna base is moving and vibrating and when the antenna itself is buffeted by winds. In a tracking system, the objective is for system-tracking outputs, as nearly as possible, to equal (or ''track'' or ''follow'') an equal number of reference input signals. Regulation is a special case of tracking, in which the desired system-tracking output is zero.

1.3 An Overview of Classical Control

The starting point of most beginning study of classical and state space control is with a linear, time-invariant model of the plant to be controlled. But, in fact, most plants have nonlinearities, and many have significantly time-varying parameters.

Practical, tractable design methods exist only for very limited classes of nonlinear systems, so it is usually expedient to approximate the plant with a linear model. To obtain a linear plant model to which linear control system design methods apply, a linear approximation about the nominal plant operating conditions (the *operating point*) is made. The nominal conditions are supplied by separate *set point* signals, and the linear controller adds to, or subtracts from, these to control relatively small excursions of the plant from the operating point. If it is necessary to change the operating point, as when control objectives are changed or when the plant parameters change appreciably, the set point signals, plant model, and controller change.

The tools of classical linear control system design are the Laplace transform, stability testing, root locus, and frequency response. Laplace transformation is used to convert system descriptions in terms of integrodifferential equations to equivalent algebraic relations involving rational functions. These are conveniently manipulated in the form of transfer functions with block diagrams and signal flow graphs. Through superposition, the *zero-input response,* that due to initial conditions, and the individual *zero-state response* contributions of each input can be dealt with separately. Alternatively, one can deal with *natural* (or *transient*) and *forced* (or *steady state*) response components.

When the transfer functions describing a linear, time-invariant system are rational, the locations of denominator roots, the *poles,* indicate whether or not the system is input–output stable, according to whether or not all transfer function poles are in the left half of the complex plane. Stability is tested by factoring the transfer function denominator polynomials or with a simpler method, such as the Routh-Hurwitz test. A pole-zero plot is a plot, on the complex plane, of the poles (denominator roots) and the zeros (numerator roots) of a transfer function.

Simple feedback systems can be described by their *open-loop* transfer functions, their transfer functions when the feedback is not connected, or their overall *closed-loop* transfer functions. Usually it is easier for the designer to deal directly with open-loop quantities than with closed-loop ones because the open-loop poles and zeros are known or can be easily found. A root locus plot consists of a pole-zero plot of the open-loop transfer function of a feedback system, upon which is superimposed the locus of the poles of the closed-loop transfer function as some parameter is varied.

Frequency response characterizations of systems have long been popular because of the ease and practicality of steady state sinusoidal response measurements. These methods apply to systems with irrational transfer functions, such as those involving time delays. They do not require explicit knowledge of system transfer function models. A stability test, the Nyquist criterion, is available, but its principal measures of zero-input system performance, gain margin, and phase margin are difficult to relate closely to the character of the feedback system's response. Frequency response methods are most useful in developing models from experimental data, in verifying the performance of a system designed by other methods, and in dealing with those systems and situations in which rational transfer function models are not adequate.

The classical approach to designing a digital controller directly, which has many variations, parallels the classical approach to analog controller design. One begins with simple discrete-time controllers, increasing their complexity until the performance requirements can be met. The controller poles, zeros, and multiplying constants are selected to give feedback system pole locations that result in acceptable zero-input response. At the same time, the parameters are constrained so that the resulting system has adequate zero-state (or steady state) response to important inputs, such as steps and ramps. From today's perspective, the classical discrete-time design procedures have the same limitations as their classical counterparts in continuous-time control. The concepts and tools are, nonetheless, necessary and useful in understanding and applying the state-variable methods to follow.

Tracking System Design

A typical classical control system design problem is to determine and specify the transfer function $G_c(s)$ of a cascade compensator that results in a unity feedback tracking system with prescribed performance requirements. This is only a part of complete control system design, of course. It applies after a suitable model has been found and when the performance requirements are quantified. In learning solution methods for idealized problems such as these, we separate general design principles from the highly specialized details of a particular application.

The basic system configuration for this problem is shown in Figure 1-4. There are, of course, many variations on this theme, including situations in which the system structure is more involved, in which it is not a tracking system, in which there are disturbance inputs to be considered, and in which the designer specifies portions of the plant.

Tracking system design has the following two basic concerns:

1. Obtaining acceptable zero-input system response
2. Obtaining acceptable zero-state system response to reference inputs

The character of a system's zero-input response is determined by its pole locations, so the first concern of tracking system design is met by choosing a compensator $G_c(s)$ that results in acceptable pole locations for the overall transfer function

$$T(s) = \left.\frac{Y(s)}{R(s)}\right|_{\text{Zero initial conditions}} = \frac{G_c(s)G_p(s)}{1 + G_c(s)G_p(s)}$$

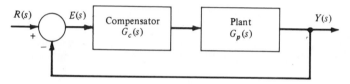

Figure 1-4 Cascade compensation of a unity feedback system.

Root locus is an important design tool because, with it, the effects on closed-loop system pole locations of varying design parameters are quickly and easily visualized.

If only the rate of decay of the plant's zero-input response is of concern, then for the zero-input response to decay at least as fast as $\exp(-\sigma t)$, all poles of $T(s)$ must be to the left of $s = -\sigma$ on the complex plane. If, in addition, all oscillatory zero-input response terms are to have damping ratios larger than some specific ratio ζ, then all complex conjugate pairs of poles of $T(s)$ must lie between angles $180° \pm$ arc cos ζ in the left half of the complex plane.

The second concern of tracking system design is obtaining acceptable closed-loop zero-state response. Zero-state performance is particularly simple to deal with if it can be expressed as a maximum steady state error to a power-of-time input. The transfer function that relates the error between the plant output and input

$$E(s) = R(s) - Y(s)$$

is

$$T_E(s) = \frac{E(s)}{R(s)}\bigg|_{\text{Zero initial conditions}} = 1 - T(s)$$

If we denote the open-loop transfer function numerator and denominator polynomials by

$$G_c(s)G_p(s) = \frac{n(s)}{d(s)}$$

then

$$T(s) = \frac{n(s)}{d(s) + n(s)} \qquad T_E(s) = \frac{d(s)}{d(s) + n(s)}$$

The plant and error transfer functions have the same poles.

Assuming the plant is stable, the steady state error to a power-of-time input of the form

$$r(t) = \frac{1}{i!} t^i u(t) \qquad R(s) = \frac{1}{s^{i+1}} \tag{1-1}$$

where $u(t)$ is the unit step function, is given by the final value theorem

$$\lim_{t \to \infty} e(t) = \lim_{s \to 0} [sE(s)] = \lim_{s \to 0} [sT_E(s)R(s)] = \lim_{s \to 0} \left\{ \frac{d(s)}{s^i[n(s) + d(s)]} \right\}$$

The steady state error is finite if the open-loop transfer function has i poles (the roots of $d(s)$) at $s = 0$. If the open-loop transfer function has more than i poles at $s = 0$, the steady state error to an input [equation (1-1)] is zero; if there are fewer than i open-loop poles at $s = 0$, the steady state error is infinite.

In practice, if necessary, compensator poles are placed at or near $s = 0$ to achieve the required steady state performance. The compensator gain and its zero locations are selected so that the resulting closed-loop system pole locations result in an acceptable zero-input response. One generally begins with a simple compensator and analyzes the possibilities, increasing the compensator order until the performance requirements are met.

The extension of methods such as these to the control of complicated feedback structures involving many loops, each of which might include a compensator, is not easy. Put another way, nowadays we want to design compensators having multiple inputs and multiple outputs, sometimes dozens of each. Design *is* iterative, and it does involve trial and error. When there are many design variables, however, it is important to deal efficiently with those design decisions that need not be iterative. The powerful methods of *state space* offer insights about what is possible and what is not. They also provide an excellent framework for general methods of approaching and accomplishing our objectives.

1.4 Precision Temperature System

One of the most complicated problems that confronts control engineers is obtaining an accurate model of the plant to be controlled. In this final section of this chapter, we present a simple example for the determination of a plant model of a chamber with an electric heater in which the temperature needs to be controlled precisely. The physical plant to be controlled is nonlinear, and a linear model, adequate for sufficiently small deviations from a nominal operating point, is derived.

1.4.1 Finding the Plant Model

Figure 1-5 shows a schematic diagram of a chamber (or oven) for precise temperature control, possibly one used for cell culture growth for the production of a medicine. A simple model of this plant is described as follows, assuming good mixing of the heated air in the oven and with all quantities expressed in SI units. The rate of change of the

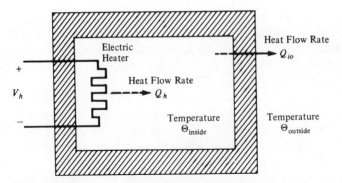

Figure 1-5 Schematic diagram of the oven.

temperature difference in the heat flow Q_h supplied by the heater, and the flow rate through the oven insulation, Q_{io} are related by

$$c \frac{d}{dt} (\Theta_{in} - \Theta_{out}) = Q_h - Q_{io}$$

For constant (or very slowly varying) outside temperature, this relation becomes

$$c \frac{d\Theta_{in}}{dt} = Q_h - Q_{io} \tag{1-2}$$

The constant of proportionality c is the thermal capacity of the oven (in joules per degree kelvin). Q_h and Q_{io} are heat flow rates (in joules per second = watts), and Θ_{in} and Θ_{out} are temperatures (in degrees kelvin).

The heat flow rate supplied by the heater is

$$Q_h = \frac{V_h^2}{R} \text{ joules/second} = \text{watts} \tag{1-3}$$

where V_h is the heater voltage (in volts), and R is the electric resistance of the heater (in ohms). The rate of heat loss through the insulation is proportional to the temperature difference across the insulation

$$Q_{io} = \frac{\Theta_{in} - \Theta_{out}}{r} \tag{1-4}$$

where r is the thermal resistance of the insulation (in kelvin-seconds per joule = kelvins per watt).

Substituting from equations (1-3) and (1-4) into equation (1-2) gives

$$\frac{d\Theta_{out}}{dt} + \frac{1}{rc} \Theta_{in} = \frac{1}{rc} \Theta_{out} + \frac{1}{Rc} V_h^2 \tag{1-5}$$

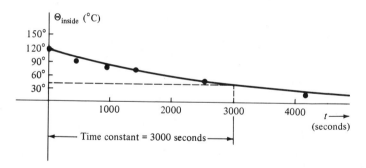

Figure 1-6 Plot of oven temperature decay data.

Because the derivative removes any reference level from Θ_{out} and the other temperature terms involve a temperature difference, degrees celsius can be used for temperatures instead of degrees kelvin, if desired.

1.4.2 Identifying Model Parameters

Numerical values of the parameters rc and Rc in equation (1-5) could be estimated by the designer from the material properties and dimensions of the oven. For an oven already in existence, as we suppose this one to be, measurements can be used to determine the parameters. Table 1-1 gives temperature data for the oven when the heater has been turned off. An exponential decay curve has been fitted to the data, in Figure 1-6. The exponential curve has the time constant

$rc = 3000$ sec

The outside temperature of about 30°C can also be inferred from this data, but this is of no direct use in determining rc.

Table 1-1 Oven Temperature Decay Data

Time	Temperature (°C)
14:23:10	120
14:31:00	108
14:39:30	92
14:48:35	80.5
15:05:00	63
15:34:00	40.5

Data that allow determination of the constant Rc are given in Table 1-2. From equation (1-5), the steady state (or forced) oven temperature for constant heater voltage V_h is given by

$$\Theta_{in} = \Theta_{out} + \left(\frac{r}{R}\right) V_h^2 \qquad (1\text{-}6)$$

Table 1-2 Steady State Oven
Temperature Data

Heater Voltage (volts)	Steady State Oven Temperature (°C)
9.8	34.1
20.0	50.5
29.5	74.0
40.0	110.0

In Figure 1-7, the steady state oven temperature data are plotted versus the heater voltage, along with a curve of the form of equation (1-6). The outside temperature is again found to be about 30°C and

$$\frac{r}{R} = 5 \times 10^{-2}$$

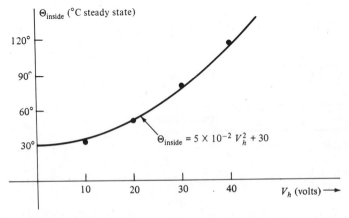

Figure 1-7 Plot of steady state oven temperature data.

giving

$$Rc = (rc)\left(\frac{R}{r}\right) = (3000)\left(\frac{1}{5 \times 10^{-2}}\right) = 60,000 \text{ sec}$$

The oven model is thus established to be approximately

$$\frac{d\Theta_{in}}{dt} + \frac{1}{3000}\,\Theta_{in} = \frac{1}{3000}\,\Theta_{out} + \frac{1}{60,000}\,V_h^2 \qquad \textbf{(1-7)}$$

1.4.3 Linearization about the Operating Point

For sufficiently small changes about an operating point, equation (1-7) can be linearized by expressing each of the signals involved as the sum of a constant nominal value plus a deviation from the nominal:

$$V_h(t) = \overline{V}_h + v_h(t)$$
$$\Theta_{in}(t) = \overline{\Theta}_{in} + \theta_{in}(t)$$
$$\Theta_{out}(t) = \overline{\Theta}_{out} + \theta_{out}(t) \qquad \textbf{(1-8)}$$

The symbols with overbars are the constant nominal values, and the lowercase symbols represent deviations from the nominal.

Suppose that the nominal operation of the oven is to be with

$$\overline{\Theta}_{in} = 50°C$$
$$\overline{\Theta}_{out} = 30°C$$

With these inside and outside temperatures, the steady state solution of equation (1-7) for the heater voltage, its nominal value, is

$$\overline{V}_h = 20 \text{ V}$$

The collection of nominal conditions is called the *operating point* (or *set point*) of the plant.

Substituting

$$V_h(t) = 20 + v_h(t)$$
$$\Theta_{in}(t) = 50 + \theta_{in}(t)$$
$$\Theta_{out}(t) = 30 + \theta_{out}(t)$$

in equation (1-7) gives

$$\frac{d\theta_{in}}{dt} + \frac{1}{3000}\,\theta_{in}(t) = \frac{1}{3000}\,\theta_{out}(t) + \frac{40}{60,000}\,v_h(t) + \frac{1}{60,000}\,v_h^2(t) \qquad \textbf{(1-9)}$$

For sufficiently small perturbations $v_h(t)$ of the heater voltage from the nominal value, the v_h^2 term in equation (1-9) can be ignored compared with the v_h term. The linearized equation, valid for small deviations from the operating point, is

$$\frac{d\theta_{in}}{dt} + \frac{1}{3000} \theta_{in}(t) = \frac{1}{3000} \theta_{out}(t) + \frac{40}{60,000} v_h(t) \tag{1-10}$$

The ignored v_h^2 term is less than 20% of the linear v_h term for v_h less than 8 V. The effect of changing the operating point is to change the coefficient of v_h in equation (1-10).

This plant model is of the first order and has a linearized state equation

$$\dot{\theta}_{in}(t) = - \frac{1}{3000} \theta_{in}(t) + \frac{1}{3000} \theta_{out}(t) + \frac{1}{1500} v_h(t)$$

where $\theta_{out}(t)$ is a disturbance input and $v_h(t)$ is a control input. In practice, the nominal constant heater voltage $\overline{V}_h$ is summed with $v_h(t)$ and applied to the heater. Necessary changes in the operating point, as when Θ_{out} changes drastically or when the oven door has been left open, are made very slowly or in concert with changes in the opposite sense in $v_h(t)$.

1.5 Summary

In this first chapter, some control terminology was discussed, and a brief survey of classical continuous-time feedback control system design was given. In the classical approach, whether in the discrete domain or continuous domain, the designer begins with a low-order controller and raises the controller order as necessary to meet the feedback system performance requirements. The controller poles, zeros, and multiplying constant are chosen so that the overall system has poles at locations for which the character of its zero-input response is acceptable. Simultaneously, it is required that the zero-state (or steady state) response to a representative class of inputs, such as steps and ramps, be adequate.

In the final section of this chapter, an interesting example system was examined. A chamber for precision temperature control involves a simple but significant nonlinearity. Although the form of the equations describing this system is known from physical principles, it would be difficult to predict accurately the numerical values of some of its parameters. The parameters were instead identified by curve fitting experimental data. The resulting model was then linearized about an operating point.

References

The references at the end of each chapter do not constitute a bibliography. They are intended to help the reader to find additional, compatible material on the chapter's subjects quickly.

Classical continuous-time control system design was taught to generations of engineers and scientists with such early texts as the following:

J. G. Truxal, *Control System Synthesis.* New York: McGraw-Hill, 1955;

C. J. Savant, Jr., *Basic Feedback Control System Design.* New York: McGraw-Hill, 1958;

R. N. Clark, *Introduction to Automatic Control Systems.* New York: Wiley, 1962.

In later years, other texts, such as

R. C. Dorf, *Modern Control Systems,* 6th ed. Reading, MA: Addison-Wesley, 1992;

B. C. Kuo, *Automatic Control Systems,* 6th ed. Englewood Cliffs, NJ: Prentice-Hall, 1991;

J. J. D'Azzo and C. H. Houpis, *Linear Control System Analysis and Design.* New York: McGraw-Hill, 1981;

and their later editions were popular.
We are a bit partial toward

J. J. DiStefano III, A. R. Stubberud, and I. J. Williams, *Feedback and Control Systems (Schaum's Outline),* 2/e. New York: McGraw-Hill, 1990;

G. H. Hostetter, C. J. Savant, Jr., and R. T. Stefani, *Design of Feedback Control Systems.* New York: Holt, Rinehart and Winston, 1989.

Physical system modeling is covered well and in detail in

R. H. Cannon, Jr., *Dynamics of Physical Systems.* New York: McGraw-Hill, 1967;

W. R. Perkins and J. B. Cruz, Jr., *Engineering of Dynamic Systems.* New York: Wiley, 1969;

C. M. Close and D. K. Frederick, *Modeling and Analysis of Dynamic Systems.* Boston: Houghton Mifflin, 1978.

Discrete-Time Systems and Z-Transformation

2.1 Preview

This chapter covers fundamental concepts relating to digital control. We first introduce the idea of a discrete-time sequence and the z-transformation. In Section 2.3, discrete-time systems are described in terms of difference equations and z-transfer functions. Then, the decomposition of system response into zero-input/zero-state is discussed carefully. The stability of a linear, step-invariant, discrete-time system is defined, and numerical procedures for stability testing in terms of the system's characteristic polynomial are presented and discussed.

In Section 2.4, the sampling of continuous-time signals and their reconstruction from samples are considered. Then, in Section 2.5, discrete-time equivalents of continuous-time systems and block diagram reduction of sampled data systems are covered and illustrated.

In the final section of the chapter, there is an overview of direct digital design methods paralleling those of classical continuous-time control. From today's perspective, the classical discrete-time design procedures have the same limitations as their classical counterparts in continuous-time control. Nonetheless, the concepts and tools are necessary and useful in understanding and applying the state-variable methods to follow.

2.2 Discrete-Time Signals

A discrete-time signal $f(k)$ is a function of a discrete variable k. It is a sequence of numbers, called *samples*, indexed by the sample number k. We now examine common and important sequences, most of which consist of evenly spaced samples of familiar continuous-time functions. The z-transformation is introduced and is used to simplify sequence manipulation. Transforms of basic sequences and important z-transforms properties are shown, and transform inversion is discussed.

2.2.1 Representing Sequences

Some fundamental sequences, all having samples that are zero prior to $k = 0$, are shown in Figure 2-1. The unit pulse sequence, Figure 2-1(a), has a unit sample for $k = 0$, and all subsequent samples are zero. The unit step sequence, Figure 2-1(b), has samples that are unity for $k = 0$ and thereafter. The unit ramp sequence, Figure 2-1(c), is defined by

$$f(k) = ku(k)$$

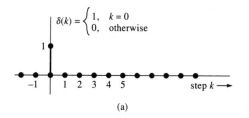

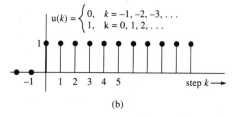

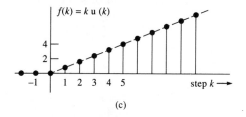

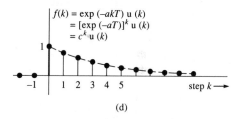

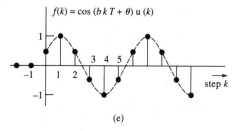

Figure 2-1 Some fundamental sequences. (a) Unit pulse. (b) Unit step. (c) Unit ramp. (d) Geometric (or exponential). (e) Sinusoidal.

where $u(k)$ is the unit step sequence. The step and ramp sequences consist of samples that are values of the corresponding continuous-time function at evenly spaced points in time. The unit pulse sequence and the unit impulse function are not related in this way because the impulse is infinite at $t = 0$ and the pulse has a unit sample at $k = 0$.

Evenly spaced samples of an exponential function

$$f(k) = e^{-at}|_{t=kT} = e^{-akT} = (e^{-aT})^k = c^k$$

as in Figure 2-1(d), result in a geometric (or exponential) sequence. Geometric sequences have samples that are progressive powers of the number

$$c = e^{-aT}$$

Similarly, evenly spaced samples of a sinusoidal continuous-time function

$$f(k) = \cos(\omega t + \theta)|_{t=kT} = \cos(\omega kT + \theta)$$

as in Figure 2-1(e), constitute a sinusoidal sequence. When a sequence is derived by sampling a continuous-time function, the constant T, which is the time interval between samples, is termed the *sampling interval* or *sampling period*. The use of the symbol T for both transfer functions and sampling interval is common and should not be confusing.

An arbitrary sequence can be expressed as the sum of scaled and shifted pulses. The sequence

$$f(k) = \begin{cases} 4, & k = 0 \\ -3, & k = 1 \\ 2, & k = 2 \\ 1, & k = 3 \\ 0, & \text{otherwise} \end{cases}$$

for example, is, in terms of pulses:

$$f(k) = 4\delta(k) - 3\delta(k - 1) + 2\delta(k - 2) + \delta(k - 3)$$

2.2.2 Z-Transformation

The *one-sided* z-transform of a sequence $f(k)$ is defined by the following equation:

$$\mathcal{L}[f(k)] = F(z) = \sum_{k=0}^{\infty} f(k)z^{-k}$$

The z-transform is indicated by $\mathscr{L}[\ \ ]$ or by the uppercase symbol for the sequence function, for example, $F(z)$. In the one-sided z-transform, samples before step zero are not included in the transform.

The z-transform of a sequence plays much the same role in the description of discrete-time signals as the Laplace transform does with continuous-time signals. When written out, the transform is

$$F(z) = f(0)z^{-0} + f(1)z^{-1} + f(2)z^{-2} + f(3)z^{-3} + \cdots$$

Each sample of the sequence, starting with step zero, is the coefficient of an inverse power of the variable z. The utility of the transform is that simple expressions can be obtained for important sequences and that the transform converts linear, step-invariant difference equations to equivalent linear algebraic equations.

Fundamental Transforms

Table 2-1 lists some important sequences and their z-transforms. The z-transform of the unit pulse is

$$\mathscr{L}[\delta(k)] = \sum_{k=0}^{\infty} \delta(k)z^{-k} = z^{-0} = 1$$

Table 2-1 Z-Transform Pairs

$f(k)$	$F(z)$
$\delta(k)$, unit pulse	1
$u(k)$, unit step	$\dfrac{z}{z-1}$
$ku(k)$	$\dfrac{z}{(z-1)^2}$
$c^k u(k)$	$\dfrac{z}{z-c}$
$kc^k u(k)$	$\dfrac{cz}{(z-c)^2}$
$u(k)\sin \Omega k$	$\dfrac{z \sin \Omega}{z^2 - 2z \cos \Omega + 1}$
$u(k)\cos \Omega k$	$\dfrac{z(z - \cos \Omega)}{z^2 - 2z \cos \Omega + 1}$
$u(k)c^k \sin \Omega k$	$\dfrac{z(c \sin \Omega)}{z^2 - (2c \cos \Omega)z + c^2}$
$u(k)c^k \cos \Omega k$	$\dfrac{z(z - c \cos \Omega)}{z^2 - (2c \cos \Omega)z + c^2}$

The unit step sequence has a z-transform given by

$$\mathcal{Z}[u(k)] = \sum_{k=0}^{\infty} z^{-k}$$

Using the geometric series formula

$$\sum_{k=0}^{\infty} x^k = \frac{1}{1-x} \qquad |x| < 1$$

this becomes

$$\mathcal{Z}[u(k)] = \sum_{k=0}^{\infty} z^{-k} = \sum_{k=0}^{\infty} \left(\frac{1}{z}\right)^k = \frac{1}{1-(1/z)} = \frac{z}{z-1} \qquad \left|\frac{1}{z}\right| < 1$$

The ability to recover a sequence uniquely from its z-transform rests on there being at least one value of z for which the infinite series converges. The unit step sequence converges for any complex value of z with

$$|z| > 1$$

and there are similar regions of convergence on the z-plane for the z-transforms of other sequences.*

A geometric sequence

$$f(k) = c^k$$

---

* The inverse z-transform can be expressed as

$$f(k) = \frac{1}{2\pi j} \oint F(z)z^{k-1}\, dz$$

in which the integration is performed in a counterclockwise direction along a closed contour on the complex plane within the region of convergence of $F(z)$, similar to the situation with the inverse Laplace transformation. It is usually much easier to deal with the infinite series $F(z)$ directly, however. Like the Laplace transform, the *two-sided* z-transform

$$\bar{F}(z) = \sum_{k=-\infty}^{\infty} f(k)z^{-k}$$

which will be used later in Chapter 8, is not generally unique. That is, two different sequences can have the same two-sided z-transform, each with a different region of convergence. It is then necessary to specify the region of convergence in order to specify the sequence.

has the z-transform

$$\mathcal{Z}[f(k)] = \sum_{k=0}^{\infty} c^k z^{-k} = \sum_{k=0}^{\infty} \left(\frac{c}{z}\right)^k = \frac{1}{1 - (c/z)} = \frac{z}{z - c}$$

The transform converges for any complex value of z for which

$$|z| > |c|$$

A sampled exponential function

$$f(k) = e^{-akT} = (e^{-aT})^k$$

whether decaying or expanding, is of this form in which the constant c is chosen to be

$$c = e^{-aT}$$

2.2.3 Z-Transform Properties

Important properties of the z-transformation which are listed in Table 2-2 are now derived.

Multiplication by a Constant

The z-transform of a constant c times a sequence $f(k)$ is the constant times the transform of the sequence

$$\mathcal{Z}[cf(k)] = \sum_{k=0}^{\infty} cf(k)z^{-k}$$

$$= cF(z)$$

The z-transform of a product of two sequences is generally *not* the product of the individual transforms.

Linearity

The z-transform is linear in that the transform of a linear combination of sequences is the linear combination of the individual transforms, that is,

$$\mathcal{Z}[c_1 f_1(k) \pm c_2 f_2(k)] = \sum_{k=0}^{\infty} c_1 f_1(k)z^{-k} \pm \sum_{k=0}^{\infty} c_2 f_2(k)z^{-k}$$

$$= c_1 F_1(z) \pm c_2 F_2(z)$$

where c_1 and c_2 are any scalars.

As a numerical example, the z-transform of a sampled sinusoid as given in Table 2-1 can be found by expanding the sequence into a linear combination of complex exponentials and applying the linearity property. For the sampled cosine

$$\mathcal{L}[\cos \Omega k] = \frac{1}{2}\mathcal{L}[e^{j\Omega k}] + \frac{1}{2}\mathcal{L}[e^{-j\Omega k}] = \frac{1}{2}\frac{z}{z - e^{j\Omega}} + \frac{1}{2}\frac{z}{z - e^{-j\Omega}}$$

$$= \frac{\frac{1}{2}z(z - e^{-j\Omega}) + \frac{1}{2}z(z - e^{j\Omega})}{(z - e^{j\Omega})(z - e^{-j\Omega})} = \frac{z^2 - \frac{1}{2}(e^{j\Omega} + e^{-j\Omega})z}{z^2 - (e^{j\Omega} + e^{-j\Omega})z + 1}$$

$$= \frac{z^2 - z \cos \Omega}{z^2 - 2z \cos \Omega + 1}$$

Multiplication by a Geometric Sequence c^k

A sequence $f(k)$ that is weighted by a geometric sequence c^k has the z-transform that is simply related to the transform of the sequence

$$\mathcal{L}[c^k f(k)] = \sum_{k=0}^{\infty} c^k f(k) z^{-k} = \sum_{k=0}^{\infty} f(k) \left(\frac{z}{c}\right)^{-k}$$

$$= F\left(\frac{z}{c}\right)$$

Table 2-2 Z-Transform Properties

$$\mathcal{L}[f(k)] = \sum_{k=0}^{\infty} f(k)z^{-k} = F(z)$$

$\mathcal{L}[cf(k)] = cF(z)$ c a constant

$\mathcal{L}[f(k) + g(k)] = F(z) + G(z)$

$$\mathcal{L}[kf(k)] = -z\frac{dF(z)}{dz}$$

$\mathcal{L}[c^k f(k)] = F(z/c)$ c a constant

$\mathcal{L}[f(k - 1)] = f(-1) + z^{-1}F(z)$

$\mathcal{L}[f(k - 2)] = f(-2) + z^{-1}f(-1) + z^{-2}F(z)$

$\mathcal{L}[f(k - n)] = f(-n) + z^{-1}f(1 - n) + z^{-2}f(2 - n) + \cdots + z^{1-n}f(-1) + z^{-n}F(z)$

$\mathcal{L}[f(k + 1)] = zF(z) - zf(0)$

$\mathcal{L}[f(k + 2)] = z^2F(z) - z^2f(0) - zf(1)$

$\mathcal{L}[f(k + n)] = z^n F(z) - z^n f(0) - z^{n-1}f(1) - \cdots - z^2 f(n - 2) - zf(n - 1)$

$f(0) = \lim_{z \to \infty} F(z)$

Table 2-2 *Cont.*

If $\lim_{k \to \infty} f(k)$ exists and is finite, $\lim_{k \to \infty} f(k) = \lim_{z \to 1} \left[\dfrac{z-1}{z} F(z) \right]$

$\mathscr{L} \left[\displaystyle\sum_{i=0}^{k} f_1(k-i)f_2(i) \right] = F_1(z)F_2(z)$

Applying this result to the unit ramp sequence gives

$$\mathscr{L}[c^k k u(k)] = \left. \frac{z}{(z-1)^2} \right|_{z=(z/c)} = \frac{cz}{(z-c)^2}$$

Applying it to the sampled cosine sequence gives

$$\mathscr{L}[c^k u(k)\cos \Omega k] = \left. \frac{z(z - \cos \Omega)}{z^2 - 2z \cos \Omega + 1} \right|_{z=(z/c)}$$

$$= \frac{z(z - c \cos \Omega)}{z^2 - (2c \cos \Omega)z + c^2}$$

as in Table 2-1.

Differentiation

A sequence weighted by the step index k has the z-transform

$$\mathscr{L}[kf(k)] = \sum_{k=0}^{\infty} kf(k)z^{-k} = \sum_{k=0}^{\infty} f(k)(kz^{-k})$$

$$= \sum_{k=0}^{\infty} \left\{ f(k) \left[-z \frac{d}{dz} z^{-k} \right] \right\} = -z \frac{d}{dz} \left[\sum_{k=0}^{\infty} f(k)z^{-k} \right]$$

$$= -z \frac{d}{dz} F(z)$$

Applying this result to the unit step sequence gives the transform of the unit ramp sequence

$$\mathscr{L}[ku(k)] = -z \frac{d}{dz} \left(\frac{z}{z-1} \right) = -z \left[\frac{z-1-z}{(z-1)^2} \right]$$

$$= \frac{z}{(z-1)^2}$$

Applying the result to the geometric sequence gives

$$\mathscr{L}[kc^k u(k)] = -z \frac{d}{dz} \left(\frac{z}{z-c} \right) = -z \left[\frac{z-c-z}{(z-c)^2} \right]$$

$$= \frac{cz}{(z-c)^2}$$

as listed in Table 2-1.

In general, z-transforms for sequences weighted by higher powers of k can be evaluated by repeatedly applying the z-transform as follows:

$$\mathscr{L}[k^2 f(k)] = \mathscr{L}[k\{kf(k)\}] = -z \frac{d}{dz} \{\mathscr{L}[kf(k)]\}$$

$$= -z \frac{d}{dz} \left\{ -z \frac{dF(z)}{dz} \right\}$$

For example,

$$\mathscr{L}[k^2 u(k)] = \mathscr{L}[k\{ku(k)\}] = -z \frac{d}{dz} \left\{ \frac{z}{(z-1)^2} \right\}$$

$$= z \left[\frac{z+1}{(z-1)^3} \right]$$

Step-Shifted Relations

A sequence that is delayed one step has the z-transform

$$\mathscr{L}[f(k-1)] = \sum_{k=0}^{\infty} f(k-1)z^{-k} = \sum_{k=-1}^{\infty} f(k)z^{-(k+1)}$$

$$= f(-1) + z^{-1} \sum_{k=0}^{\infty} f(k)z^{-k} = f(-1) + z^{-1} F(z)$$

The initial condition term $f(-1)$ occurs because the z-transform of the shifted sequence is affected by an additional sample that has no effect on the z-transform of the original sequence. Applying this result repeatedly gives

$$\mathscr{L}[f(k-2)] = f(-2) + z^{-1}[f(-1) + z^{-1}F(z)]$$
$$= f(-2) + z^{-1}f(-1) + z^{-2}F(z)$$
$$\mathscr{L}[f(k-3)] = f(-3) + z^{-1}[f(-2) + z^{-1}f(-1) + z^{-2}F(z)]$$
$$= f(-3) + z^{-1}f(-2) + z^{-2}f(-1) + z^{-3}F(z)$$

and so on.

A one-step advance of a sequence has the z-transform

$$\mathscr{L}[f(k+1)] = \sum_{k=0}^{\infty} f(k+1)z^{-k} = \sum_{k=1}^{\infty} f(k)z^{-k+1}$$

$$= z \sum_{k=1}^{\infty} f(k)z^{-k} = zF(z) - zf(0)$$

The advance of a sequence also involves an initial condition term, in this case $f(0)$. Applying this result repeatedly

$$\mathscr{L}[f(k+2)] = z[zF(z) - zf(0)] - zf(1)$$

$$= z^2F(z) - z^2f(0) - zf(1)$$

$$\mathscr{L}[f(k+3)] = z[z^2F(z) - z^2f(0) - zf(1)] - zf(2)$$

$$= z^3F(z) - z^3f(0) - z^2f(1) - zf(2)$$

Initial Value Theorem

The initial value of a sequence $f(k)$ is the value of the sequence at $k = 0$. A sequence's initial value may be found from the z-transform of the sequence by the relation

$$f(0) = \lim_{z \to \infty} [f(0) + f(1)z^{-1} + f(2)z^{-2} + f(3)z^{-3} + \cdots]$$

$$= \lim_{z \to \infty} F(z)$$

For example, the initial value of the sequence that has the following z-transform

$$F(z) = \frac{3z^2 + 5z + 1}{2z^2 + z + 1}$$

is

$$f(0) = \lim_{z \to \infty} F(z) = \lim_{z \to \infty} \frac{3 + (5/z + 1/z^2)}{2 + (1/z + 1/z^2)} = \frac{3}{2}$$

Final Value Theorem

Provided that the sequence $f(k)$ approaches a finite limit, the final value theorem for sequences is

$$\lim_{k \to \infty} f(k) = f_\infty = \lim_{z \to 1} \left[\frac{z-1}{z} F(z) \right]$$

For the final value theorem to apply, the transform $F(z)$ must have all poles within the unit circle on the complex plane except possibly a single pole at $z = 1$. To show this result, consider the two finite summations

$$\sum_{k=0}^{n} f(k)z^{-k} = f(0) + f(1)z^{-1} + f(2)z^{-2} + \cdots + f(n)z^{-n}$$

and

$$\sum_{k=1}^{n} f(k-1)z^{-k} = f(0)z^{-1} + f(1)z^{-2} + \cdots + f(n-1)z^{-n}$$

$$= z^{-1} \sum_{k=0}^{n-1} f(k)z^{-k}$$

In the limit, as z approaches unity, the difference between these two summations is

$$\lim_{z \to 1} \left[\sum_{k=0}^{n} f(k)z^{-k} - z^{-1} \sum_{k=0}^{n-1} f(k)z^{-k} \right] = f(n)$$

The limit of this equality as n becomes very large is

$$\lim_{n \to \infty} \left\{ \lim_{z \to 1} \left[\sum_{k=0}^{n} f(k)z^{-k} - z^{-1} \sum_{k=0}^{n-1} f(k)z^{-k} \right] \right\} = f_\infty$$

when the limit exists. Interchanging the order of the limits

$$\lim_{z \to 1} \left\{ \lim_{n \to \infty} \left[\sum_{k=0}^{n} f(k)z^{-k} - z^{-1} \sum_{k=0}^{n-1} f(k)z^{-k} \right] \right\} = f_\infty$$

or

$$\lim_{z \to 1} [(1 - z^{-1})F(z)] = f_\infty$$

For example, the sequence $f(k)$ corresponding to the transform

$$F(z) = \frac{z}{z^2 - 1.5z + 0.5} = \frac{z}{(z - 1)(z - 0.5)}$$

has a final value given by

$$f(\infty) = \lim_{z \to 1} \left\{ \left(\frac{z - 1}{z} \right) \left[\frac{z}{(z - 1)(z - 0.5)} \right] \right\} = 2$$

Unfortunately, the final value theorem gives reasonable-looking answers even when a final value does not exist. For the function with the z-transform

$$F(z) = \frac{z^2}{(z-1)(z-2)}$$

for example,

$$\lim_{z \to 1}\left\{\left(\frac{z-1}{z}\right)F(z)\right\} = -1$$

However, the final value theorem does not apply because the transform $F(z)$ has a pole outside the unit circle at $z = 2$.

2.2.4 Z-Transform Inversion

In the solution of problems involving systems and the difference equations that describe them, we often wish to invert a z-transformed signal; that is, we wish to find the inverse transform of a given function of z. Two methods for obtaining a sequence of samples for a given z-transformed function are discussed below. A third method is given in Problem 2-7.

(1) Long-Division Method

The sequence of samples represented by a rational z-transform can be obtained by long division, expanding the transform into a power series in z^{-1} with coefficients that are the samples:

$$F(z) = f(0)z^{-0} + f(1)z^{-1} + f(2)z^{-2} + f(3)z^{-3} + \cdots$$

For the z-transform

$$F(z) = \frac{2z^3 + 2z^2 + 3z - 4}{z^3 + 0.4z^2 - 0.6z + 0.3}$$

for example, long division gives

$$
\begin{array}{r}
2 + 1.2z^{-1} + 3.72z^{-2} - 5.368z^{-3} + \cdots \\
\hline
z^3 + 0.4z^2 - 0.6z + 0.3 \,\big|\, 2z^3 + 2z^2 + 3z - 4 \\
2z^3 + 0.8z^2 - 1.2z + 0.6 \\
\hline
1.2z^2 + 4.2z - 4.6 \\
1.2z^2 + 0.48z - 0.72 + 0.36z^{-1} \\
\hline
3.72z - 3.88 - 0.36z^{-1} \\
3.72z + 1.488 - 2.232z^{-1} + 1.116z^{-2} \\
\hline
-5.368 + 1.872z^{-1} - 1.116z^{-2}
\end{array}
$$

The first several samples of the sequence are given by

$$f(k) = 2\delta(k) + 1.2\delta(k - 1) + 3.72\delta(k - 2) - 5.368\delta(k - 3) + \cdots$$

or

$$f(0) = 2$$
$$f(1) = 1.2$$
$$f(2) = 3.72$$
$$f(3) = -5.368$$
$$\vdots$$

In principle, as many terms in the sequence as desired can be found in this way, but a closed-form expression for the sequence does not generally result.

The inverse one-sided z-transform gives no information on values of the sequence prior to $k = 0$.

(2) Partial-Fraction Expansion Method

To find formulas for the sequence of samples represented by a rational z-transform, a partial fraction expansion can be used. Rather than expanding a z-transform $F(z)$ directly in partial fractions, the function $F(z)/z$ is expanded so that terms with a z in the numerator result. For example, for the z-transform

$$F(z) = \frac{-2z^2 + 11z}{z^2 - z - 6}$$

$$\frac{F(z)}{z} = \frac{-2z + 11}{z^2 - z - 6} = \frac{k_1}{z + 2} + \frac{k_2}{z - 3} = \frac{(k_1 + k_2)z - 3k_1 + 2k_2}{z^2 - z - 6}$$

equating coefficients gives

$$F(z) = \frac{-3z}{z + 2} + \frac{z}{z - 3}$$

$$f(k) = -3(-2)^k + 3^k \qquad k = 0, 1, 2, 3, \ldots$$

In general, for rational functions with nonrepeated denominator roots,

$$\frac{F(z)}{z} = \frac{\text{Numerator polynomial}}{(z + z_1)(z + z_2) \ldots (z + z_n)}$$

$$= \frac{k_1}{z + z_1} + \frac{k_2}{z + z_2} + \cdots + \frac{k_n}{z + z_n}$$

then the k's are determined as follows:

$$k_i = (z + z_i) \left. \frac{F(z)}{z} \right|_{z = -z_i} \qquad i = 1, 2, \ldots, n$$

Another example is the following:

$$F(z) = \frac{2z - 3}{(z - 0.5)(z + 0.3)}$$

$$\frac{F(z)}{z} = \frac{2z - 3}{z(z - 0.5)(z + 0.3)} = \frac{k_1}{z} + \frac{k_2}{z - 0.5} + \frac{k_3}{z + 0.3}$$

$$k_1 = \left. \frac{2z - 3}{(z - 0.5)(z + 0.3)} \right|_{z=0} = 20$$

$$k_2 = \left. \frac{2z - 3}{z(z + 0.3)} \right|_{z=0.5} = -5$$

$$k_3 = \left. \frac{2z - 3}{z(z - 0.5)} \right|_{z=-0.3} = -15$$

so that

$$F(z) = 20 - \frac{5z}{z - 0.5} - \frac{15z}{z + 0.3}$$

$$f(k) = 20\delta(k) - 5(0.5)^k - 15(-0.3)^k \qquad k = 0, 1, 2, 3, \ldots$$

Complex Conjugate Pairs

When complex conjugate pairs of terms occur, they can be combined into a single term with a quadratic denominator, then separated into terms of the form of the last two entries in Table 2-1. For example, for the z-transform

$$F(z) = \frac{-6z^2 + z}{(z - 1)(z - \frac{1}{2} + j\frac{1}{4})(z - \frac{1}{2} - j\frac{1}{4})}$$

$$\frac{F(z)}{z} = \frac{-6z + 1}{(z - 1)(z - \frac{1}{2} + j\frac{1}{4})(z - \frac{1}{2} - j\frac{1}{4})}$$

$$= \frac{-16}{z - 1} + \frac{8 + j4}{z - \frac{1}{2} + j\frac{1}{4}} + \frac{8 - j4}{z - \frac{1}{2} - j\frac{1}{4}} = \frac{-16}{z - 1} + \frac{16z - 6}{z^2 - z + (5/16)}$$

$$F(z) = \frac{-16z}{z - 1} + \frac{z(16z - 6)}{z^2 - z + (5/16)}$$

Equating

$$\frac{z(16z - 6)}{z^2 - z + (5/16)} = \frac{K_1 z(c \sin \Omega)}{z^2 - (2c \cos \Omega)z + c^2} + \frac{K_2 z(z - c \cos \Omega)}{z^2 - (2c \cos \Omega)z + c^2}$$

where c, Ω, K_1, and K_2 are determined as follows:

$$c^2 = \frac{5}{16} \qquad c = 0.56$$

$$2c \cos \Omega = 1 \qquad \cos \Omega = 0.893 \qquad \Omega = 0.468 \text{ rad}$$

$$c \sin \Omega = 0.56 \sin (0.468) = 0.252$$

Then

$$F(z) = \frac{-16z}{z - 1} + \frac{(7.9365)z(0.252)}{z^2 - z + (5/16)} + \frac{(16)z(z - 0.5)}{z^2 - z + (5/16)}$$

$$f(k) = -16 + 7.9365c^k \sin(\Omega k) + 16c^k \cos(\Omega k)$$

$$= -16 + (0.56)^k[7.9365 \sin(0.468k) + 16 \cos(0.468k)]$$

$$k = 0, 1, 2, \ldots$$

Repeated Roots

When the denominator polynomial of a rational function of z has repeated roots z_1, the corresponding partial fraction expansion terms are of the form

$$\frac{F(z)}{z} = \frac{\text{Numerator polynomial}}{(z + z_1)^r(z + z_2) \ldots (z + z_n)}$$

$$= \frac{k_1}{z + z_1} + \frac{k_2}{(z + z_1)^2} + \cdots + \frac{k_r}{(z + z_1)^r}$$

$$+ \text{ terms for other different roots}$$

The repeated root coefficients may be found by evaluation according to

$$k_i = \frac{1}{(r - i)!} \left\{ \frac{d^{r-i}}{dz^{r-i}} [(z + z_1)^r F(z)] \right\} \Bigg|_{z=-z_1}$$

For example, for the z-transform

$$F(z) = \frac{2z - 3}{z(z - 0.5)}$$

we have

$$\frac{F(z)}{z} = \frac{2z - 3}{z^2(z - 0.5)} = \frac{k_1}{z} + \frac{k_2}{z^2} + \frac{k_3}{z - 0.5}$$

$$k_1 = \frac{1}{(2-1)!}\left\{\frac{d^{2-1}}{dz^{2-1}}\left[z^2\frac{F(z)}{z}\right]\right\}\Bigg|_{z=0} = 8$$

$$k_2 = \frac{1}{(2-2)!}\left\{\frac{d^{2-2}}{dz^{2-2}}\left[z^2\frac{F(z)}{z}\right]\right\}\Bigg|_{z=0} = 6$$

$$k_3 = \frac{2z-3}{z^2}\Bigg|_{z=0.5} = -8$$

so

$$f(k) = 8\delta(k) + 6\delta(k-1) - 8(0.5)^k \qquad k = 0, 1, 2, 3, \ldots$$

2.3 Discrete-Time Systems

The input–output behavior of discrete-time systems is described by difference equations, analogous to the differential equations that describe continuous-time systems. In this section, we examine the properties of difference equations and apply z-transform methods to find their solutions. Discrete-frequency response, an important technique for finding z-transfer function models experimentally, is also introduced.

2.3.1 Difference Equations

Discrete-time systems are described by difference equations, which in the linear, step-invariant case have the form

$$\begin{aligned}
y(k+n) &+ a_{n-1}y(k+n-1) + a_{n-2}y(k+n-2) + \cdots \\
&+ a_1y(k+1) + a_0y(k) \\
&= b_mr(k+m) + b_{m-1}r(k+m-1) + \cdots \\
&\qquad\qquad + b_1r(k+1) + b_0r(k) \quad (2\text{-}1)
\end{aligned}$$

where y is the output and r is an input. When expressed this way, the coefficient of $y(k+n)$ is made equal to unity. The *order* of a difference equation is n, the number of past output steps that are involved in calculating the present output:

$$y(k+n) =$$

$$\underbrace{-a_{n-1}y(k+n-1) - a_{n-2}y(k+n-2) - \cdots - a_1y(k+1) - a_0y(k)}_{n \text{ terms}}$$

$$\underbrace{+ b_mr(k+m) + b_{m-1}r(k+m-1) + \cdots + b_1r(k+1) + b_0r(k)}_{m \text{ terms}}$$

A discrete-time system is termed *causal* if $m \leqslant n$ so that only present and past inputs (not future ones) are involved in the calculation of the present output. If a system has more than a single output, there is a difference equation for each output. If a discrete-time system has more than a single input, there are terms in the equation(s) for each input.

Because a difference equation holds for each integer value of the index, k, there are many equivalent forms for the equation. Replacing k by $k - n$ in equation (2-1), for example, gives

$$y(k) + a_{n-1}y(k - 1) + a_{n-2}y(k - 2) + \cdots$$
$$+ a_1y(k - n + 1) + a_0y(k - n)$$

$$= b_mr(k - n + m) + b_{m-1}r(k - n + m - 1) + \cdots$$
$$+ b_1r(k - n + 1) + b_0r(k - n)$$

Given the input $r(k)$ for $k = -n, -n + 1, \ldots, 0, 1, 2, \ldots$ and the initial conditions $y(-1), y(-2), y(-3), \ldots, y(-n)$, the solution $y(k)$ to an nth order causal difference equation can be calculated for $k = 0, 1, 2, \ldots$ by repeatedly using the equation. Consider the difference equation

$$y(k + 2) = 2y(k + 1) - 3y(k) + 4r(k + 1) - r(k)$$

with input

$$r(k) = (-1)^k$$

and initial conditions

$$y(-1) = 5 \qquad y(-2) = -6$$

Using $k = -2$ and substituting into the difference equation gives

$$y(0) = 2y(-1) - 3y(-2) + 4r(-1) - r(-2)$$
$$= 10 + 18 - 4 - 1 = 23$$

Using $k = -1$ and substituting

$$y(1) = 2y(0) - 3y(-1) + 4r(0) - r(-1)$$
$$= 46 - 15 + 4 + 1 = 36$$

Similarly,

$$y(2) = 2y(1) - 3y(0) + 4r(1) - r(0) = 72 - 69 - 4 - 1 = -2$$
$$y(3) = 2y(2) - 3y(1) + 4r(2) - r(1)$$
$$= -4 - 108 + 4 + 1 = -107$$

and so on. This is known as a *recursive* solution of the equation, using the equation and solutions at past steps to calculate the solution at the next step.

A difference equation can be constructed using a computer by programming its recursive solution. For example, the second-order difference equation

$$y(k + 2) + \frac{1}{2} y(k + 1) - \frac{1}{3} y(k) = r(k + 2) - 4r(k)$$

or

$$y(k) = -\frac{1}{2} y(k - 1) + \frac{1}{3} y(k - 2) + r(k) - 4r(k - 2)$$

is realized by the digital computer program in BASIC given in Table 2-3. The initial conditions incorporated were chosen arbitrarily. A digital hardware realization of this difference equation can also be constructed by coding the signals as binary words, storing present and past values of the input and output in registers, and using binary arithmetic devices to multiply the signals by the equation coefficients and add them to form the output.

2.3.2 Z-Transfer Function Methods

Formulas for the responses of step-invariant, discrete-time systems can be found using z-transformation. For example, consider the single-input/single-output system

$$y(k + 2) + \frac{1}{4} y(k + 1) - \frac{1}{8} y(k) = 3r(k + 1) - r(k)$$

with input

$$r(k) = (-1)^k u(k)$$

and initial conditions

$$y(-1) = 5 \qquad y(-2) = -6$$

The difference equation is first converted to the equivalent form

$$y(k) + \frac{1}{4} y(k - 1) - \frac{1}{8} y(k - 2) = 3r(k - 1) - r(k - 2)$$

Then taking the z-transform of both sides gives

$$Y(z) + \frac{1}{4} [z^{-1} Y(z) + y(-1)] - \frac{1}{8} [z^{-2} Y(z) + z^{-1} y(-1) + y(-2)]$$

$$= 3[z^{-1} R(z) + r(-1)] - [z^{-2} R(z) + z^{-1} r(-1) + r(-2)]$$

Table 2-3 Digital Computer Program in BASIC to Realize a Difference Equation

```
100   REM DIFFERENCE EQUATION SOLUTION
110   REM
120   REM Y IS PRESENT OUTPUT
130   REM Y1 IS OUTPUT DELAYED ONE STEP
140   REM Y2 IS OUTPUT DELAYED TWO STEPS
150   REM R IS PRESENT INPUT
160   REM R1 IS INPUT DELAYED ONE STEP
170   REM R2 IS INPUT DELAYED TWO STEPS
180   REM K IS STEP NUMBER
190   REM
200   REM ASSIGN INITIAL CONDITIONS
210   Y2 = 3
220   Y1 = -1
230   R1 = -2
240   R2 = 4
250   REM START AT STEP ZERO
260   K = 0
270   REM ENTER INPUT R(K)
280   PRINT "ENTER STEP"; K;"INPUT:";
290   INPUT R
300   REM COMPUTE AND PRINT Y(K)
310   Y = -(1/2)*Y1 +(1/3)*Y2 + R - 4*R2
320   PRINT "STEP"; K;"OUTPUT IS"; Y
330   REM INCREMENT STEP INDEX
350   K = K + 1
360   Y2 = Y1
370   Y1 = Y
380   R2 = R1
390   R1 = R
400   REM LOOP TO COMPUTE NEXT OUTPUT
410   GOTO 270
420   END
```

Because

$$r(-1) = r(-2) = 0$$

$$\left(1 + \frac{1}{4} z^{-1} - \frac{1}{8} z^{-2}]\right) Y(z) = (3z^{-1} - z^{-2}) R(z) + \frac{5}{8} z^{-1} - 2$$

$$\left(z^2 + \frac{1}{4} z - \frac{1}{8}\right) Y(z) = (3z - 1)R(z) + \frac{5}{8} z - 2z^2$$

$$Y(z) = \frac{3z - 1}{z^2 + \frac{1}{4}z - \frac{1}{8}} R(z) + \frac{-2z^2 + \frac{5}{8}z}{z^2 + \frac{1}{4}z - \frac{1}{8}}$$

Zero-state component   Zero-input component

For

$$R(z) = \mathcal{Z}[(-1)^k] = \frac{z}{z + 1}$$

then

$$Y(z) = \frac{-2z^3 + (13/8)z^2 - \frac{3}{8}z}{(z + \frac{1}{2})(z - \frac{1}{4})(z + 1)}$$

and $Y(z)$ is inverted to obtain $y(k)$.

Expanding $Y(z)/z$ into partial fractions

$$\frac{Y(z)}{z} = \frac{-2z^2 + (13/8)z - \frac{3}{8}}{(z + \frac{1}{2})(z - \frac{1}{4})(z + 1)} = \frac{\frac{9}{2}}{z + \frac{1}{2}} + \frac{-1/10}{z - \frac{1}{4}} + \frac{-32/5}{z + 1}$$

then

$$Y(z) = \frac{(\frac{9}{2})z}{z + \frac{1}{2}} + \frac{(-1/10)z}{z - \frac{1}{4}} + \frac{-(32/5)z}{z + 1}$$

so that

$$y(k) = \left[\left(\frac{9}{2}\right)\left(-\frac{1}{2}\right)^k - \left(\frac{1}{10}\right)\left(\frac{1}{4}\right)^k - \left(\frac{32}{5}\right)(-1)^k\right] \quad k \geq 0$$

In general, an nth-order causal single-input system with an input–output difference equation of

$$y(k + n) + a_{n-1} y(k + n - 1) + \cdots + a_1 y(k + 1) + a_0 y(k)$$
$$= b_m r(k + m) + \cdots + b_1 r(k + 1) + b_0 r(k)$$

or

$$y(k) + a_{n-1} y(k - 1) + \cdots + a_1 y(k - n + 1) + a_0 y(k - n)$$
$$= b_m r(k + m - n) + \cdots + b_1 r(k - n + 1) + b_0 r(k - n)$$

has a z-transformed output given by

$$Y(z) + a_{n-1}[z^{-1} Y(z) + y(-1)] + \cdots$$
$$+ a_1[z^{-n+1} Y(z) + z^{-n+2} y(-1) + \cdots + y(-n + 1)]$$
$$+ a_0[z^{-n} Y(z) + z^{-n+1} y(-1) + \cdots + y(-n)]$$
$$= b_m[z^{-n+m}R(z) + z^{-n+m-1}r(-1) + \cdots + r(-n + m - 1)] + \cdots$$
$$+ b_0[z^{-n}R(z) + z^{-n+1}r(-1) + \cdots + r(-n)]$$

$$Y(z) = \frac{b_m z^m + b_{m-1}z^{m-1} + \cdots + b_1 z + b_0}{z^n + a_{n-1}z^{n-1} + \cdots + a_1 z + a_0} R(z)$$

Zero-state component

$$+ \frac{\text{(Polynomial in } z \text{ of degree } n \text{ or less with coefficients}}{z^n + a_{n-1}z^{n-1} + \cdots + a_1 z + a_0}$$

Zero-input component

The zero-input response component is zero if all the initial conditions are zero. The zero-state response component is the product of the system *z-transfer function*

$$T(z) = \frac{b_m z^m + b_{m-1}z^{m-1} + \cdots + b_1 z + b_0}{z^n + a_{n-1}z^{n-1} + \cdots + a_1 z + a_0}$$

and the z-transform of the input:

$$Y_{\text{zero state}}(z) = T(z)R(z)$$

The transfer function is the ratio of z-transformed output to z-transformed input when all initial conditions are zero.

It is also common practice to separate system response into two other components, *natural* (or *transient*) and *forced* (or *steady state*). Both the zero-state and the zero-input response components share the system's characteristic polynomial

$$q(z) = z^n + a_{n-1}z^{n-1} + \cdots + a_1 z + a_0 = (z - z_1)(z - z_2)\ldots(z - z_n)$$

and each of these potentially contributes terms of the form

$$\frac{k_1 z}{z - z_1}, \quad \frac{k_2 z}{z - z_2}, \quad \ldots, \quad \frac{k_n z}{z - z_n}$$

(or the equivalent if the characteristic equation has repeated roots) to the output transform. The sum of all these is the natural response component, and the remainder of the response, which has a form dependent on the specific input, is the forced response component.

Stability and Response Terms

As indicated in Figure 2-2, the unit pulse response of a linear, step-invariant system, the response when the input is the unit pulse $\delta(k)$ and all initial conditions are zero, is given by

$$Y_{\text{pulse}}(z) = R(z)T(z) = T(z)$$

Its z-transform is equal to the transfer function. A single-input/single-output system is said to be *input–output stable* if its unit pulse re-

$$R(z) = 1 \quad \boxed{\quad T(z) \quad} \quad Y_{\text{pulse}}(z) = R(z)\,T(z) = T(z)$$

$$r(k) = \delta(k) \qquad\qquad\qquad y_{\text{pulse}}(k)$$

Figure 2-2 Unit pulse response of a discrete-time system.

sponse decays asymptotically to zero. This occurs if and only if the denominator polynomial of the transfer function has all of its roots inside the unit circle on the complex plane.

Table 2-4 shows, for rational z-transforms, the types of sequences corresponding to various pole (denominator root) locations. For poles inside the unit circle on the complex plane, these sequences decay with step. For nonrepeated poles on the unit circle, the sequences neither expand nor decay with step. Poles outside the unit circle and repeated poles on the unit circle represent sequences that expand with step.

Discrete Convolution

The last entry in the listing of z-transform properties, Table 2-2, is that the inverse z-transform of the product of two transforms, $F_1(z)$ and $F_2(z)$, is the discrete convolution of the two sequences $f_1(k)$ and $f_2(k)$. Using

$$\mathcal{Z}\left[\sum_{i=0}^{k} f_1(k - i) f_2(i)\right] = F_1(z) F_2(z)$$

and inverse transforming both sides,

$$\mathcal{Z}^{-1}[F_1(z) F_2(z)] = \sum_{i=0}^{k} f_1(k - i) f_2(i) = \text{convolution}[f_1(k), f_2(k)]$$

$$= \text{convolution}[f_2(k), f_1(k)] = \sum_{i=0}^{k} f_2(k - i) f_1(i)$$

The convolution relation holds for values of z for which both $F_1(z)$ and $F_2(z)$ converge. This is the form of the zero-state component of a single-input system's output

$$y_{\text{zero-state}}(k) = \mathcal{Z}^{-1}[T(z) R(z)] = \text{convolution}[h(k), r(k)]$$

where $T(z)$ is the transfer function and $R(z)$ is the input. Using the relation, the zero-state output is seen to be the convolution of the system's unit pulse response

$$h(k) = \mathcal{Z}^{-1}[T(z)]$$

with the input $r(k)$.

Table 2-4 Sequences Corresponding to Various Z-Transform Pole Locations

Pole location(s) on the complex plane	Sequence

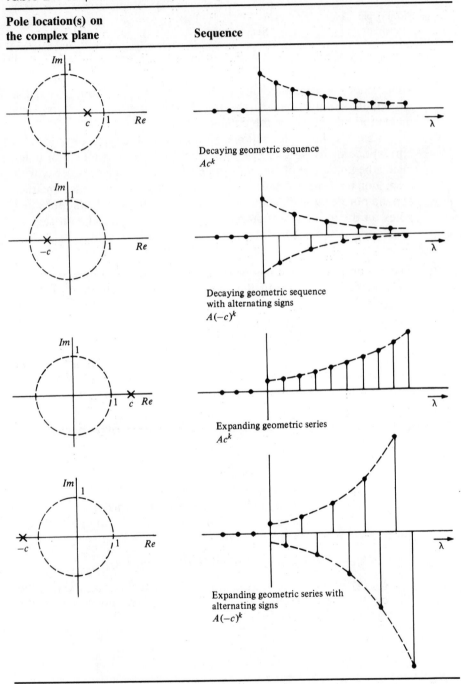

Decaying geometric sequence
Ac^k

Decaying geometric sequence
with alternating signs
$A(-c)^k$

Expanding geometric series
Ac^k

Expanding geometric series with
alternating signs
$A(-c)^k$

Table 2-4 (cont.)

Pole location(s) on the complex plane	Sequence

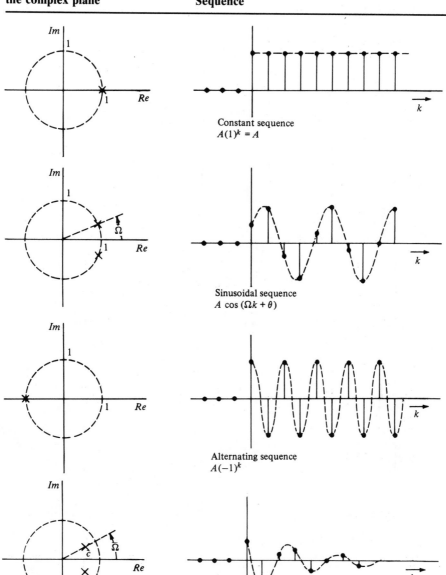

Constant sequence
$A(1)^k = A$

Sinusoidal sequence
$A \cos (\Omega k + \theta)$

Alternating sequence
$A(-1)^k$

Damped sinusoidal sequence
$Ac^k \cos (\Omega k + \theta)$

Table 2-4 (cont.)

Pole location(s) on the complex plane	Sequence

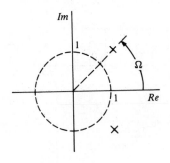

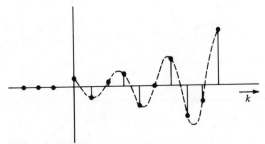

Exponentially expanding sinusoidal sequence
$A^k \cos (\Omega k + \theta)$

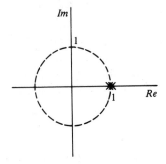

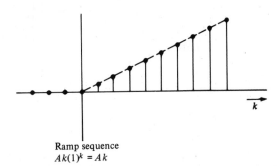

Ramp sequence
$Ak(1)^k = Ak$

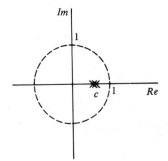

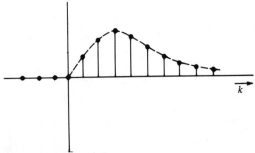

Ramp-weighted geometric sequence
Akc^k

The convolution relation is derived as follows. For convergent transforms and sequences that are zero for negative step:

$$F_1(z)F_2(z) = \left[\sum_{i=0}^{\infty} f_2(i)z^{-i}\right]\left[\sum_{k=0}^{\infty} f_1(k)z^{-k}\right] = \left[\sum_{i=0}^{\infty} f_2(i)\left[z^{-i}\sum_{k=0}^{\infty} f_1(k)z^{-k}\right]\right]$$

Using the delayed sequence property, with $f_2(k)$ zero prior to $k = 0$ so that the initial condition terms are all zero

$$z^{-i}\sum_{k=0}^{\infty} f_1(k)z^{-k} = \mathscr{L}[f_1(k-i)] = \sum_{k=0}^{\infty} f_1(k-i)z^{-k}$$

and

$$F_1(z)F_2(z) = \sum_{i=0}^{\infty} f_2(i)\left[\sum_{k=0}^{\infty} f_1(k-i)z^{-k}\right] = \sum_{k=0}^{\infty}\left[\sum_{i=0}^{\infty} f_1(k-i)f_2(i)\right]z^{-k}$$

Now, because f_1 is zero for negative arguments

$$F_1(z)F_2(z) = \sum_{k=0}^{\infty}\left[\sum_{i=0}^{k} f_1(k-i)f_2(i)\right]z^{-k} = \mathscr{L}\left[\sum_{i=0}^{k} f_1(k-i)f_2(i)\right]$$

Figure 2-3 shows the process involved in discrete convolution. In Figure 2-3(a), two sequences, $f_1(k)$ and $f_2(k)$, are shown. They are plotted offset slightly for clarity. In Figure 2-3(b), the step index is changed to i, and $f_1(k-i)$ and $f_2(i)$ are plotted. Changing the argument of a sequence from i to $k-i$ (versus step i) reverses the direction of the sequence and delays it k steps, as shown. For a specific step k, the products $f_1(k-i)f_2(i)$ are plotted versus i in Figure 2-3(c), and the sum of these is the convolution for that step k. The sum of products as a function of k, which is the convolution of the two sequences, is shown in Figure 2-3(d).

As a numerical example of convolution, let

$$f_1(k) = \left(\frac{1}{4}\right)^k \qquad f_2(k) = \left(\frac{1}{2}\right)^k \qquad k \geq 0$$

Then

$$g(k) = \text{convolution}[f_1(k), f_2(k)] = \sum_{i=0}^{k} f_1(k-i)f_2(i) = \sum_{i=0}^{k}\left(\frac{1}{4}\right)^{k-i}\left(\frac{1}{2}\right)^i$$

The first several values of the convolution are as follows:

$$g(0) = \left(\frac{1}{4}\right)^0\left(\frac{1}{2}\right)^0 = 1$$

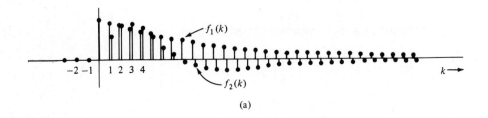

(a)

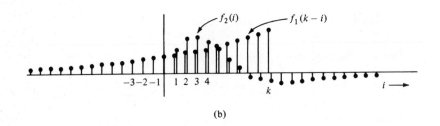

(b)

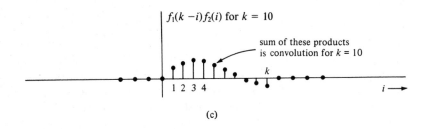

(c)

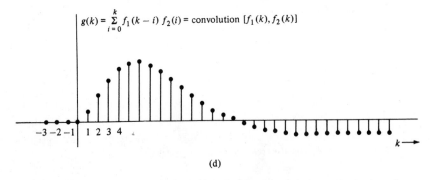

(d)

Figure 2-3 Discrete convolution of two sequences that are zero prior to *k* = 0.
(a) Functions to be convolved set to zero prior to *k* = 0. (b) Step index changed to
i and one function reversed in step. (c) Products of overlapping samples. (d) Sums
of products of overlapping samples as a function of the shift *k*.

$$g(1) = \left(\frac{1}{4}\right)^1\left(\frac{1}{4}\right)^0 + \left(\frac{1}{4}\right)^0\left(\frac{1}{2}\right)^1 = \frac{1}{4} + \frac{1}{2} = \frac{3}{4}$$

$$g(2) = \left(\frac{1}{4}\right)^2\left(\frac{1}{2}\right)^0 + \left(\frac{1}{4}\right)^1\left(\frac{1}{2}\right)^1 + \left(\frac{1}{4}\right)^0\left(\frac{1}{2}\right)^2 = \frac{1}{16} + \frac{1}{8} + \frac{1}{4} = \frac{7}{16}$$

$$g(3) = \left(\frac{1}{4}\right)^3\left(\frac{1}{2}\right)^0 + \left(\frac{1}{4}\right)^2\left(\frac{1}{2}\right)^1 + \left(\frac{1}{4}\right)^1\left(\frac{1}{2}\right)^2 + \left(\frac{1}{4}\right)^0\left(\frac{1}{2}\right)^3$$

$$= \frac{1}{64} + \frac{1}{32} + \frac{1}{16} + \frac{1}{8} = \frac{15}{64}$$

These values compare with the transform solution

$$G(z) = \left(\frac{z}{z - \frac{1}{4}}\right)\left(\frac{z}{z - \frac{1}{2}}\right) = \frac{-z}{z - \frac{1}{4}} + \frac{2z}{z - \frac{1}{2}}$$

$$g(k) = \left[-\left(\frac{1}{4}\right)^k + 2\left(\frac{1}{2}\right)^k\right] u(k)$$

which gives

$$g(0) = -\left(\frac{1}{4}\right)^0 + 2\left(\frac{1}{2}\right)^0 = 1$$

$$g(1) = -\left(\frac{1}{4}\right)^1 + 2\left(\frac{1}{2}\right)^1 = -\frac{1}{4} + 1 = \frac{3}{4}$$

$$g(2) = -\left(\frac{1}{4}\right)^2 + 2\left(\frac{1}{2}\right)^2 = -\frac{1}{16} + \frac{2}{4} = \frac{7}{16}$$

$$g(3) = -\left(\frac{1}{4}\right)^3 + 2\left(\frac{1}{2}\right)^3 = -\frac{1}{64} + \frac{2}{8} = \frac{15}{64}$$

Z-transfer functions for discrete-time systems and system components are manipulated in the same way as transfer functions for continuous-time systems. For example, a system block diagram is shown in Figure 2-4. Using the block diagram equivalences as for continuous-time systems, the overall z-transfer function is found to be

$$T(z) = \frac{\left(\frac{1}{z + \frac{1}{2}}\right)\left(\frac{1}{z^2 - \frac{1}{2}z + 2}\right)}{1 + \left(\frac{1}{z + \frac{1}{2}}\right)\left(\frac{1}{z^2 - \frac{1}{2}z + 2}\right)\left(z - \frac{1}{2}\right)} = \frac{1}{z^3 + \frac{11}{4}z + \frac{1}{2}}$$

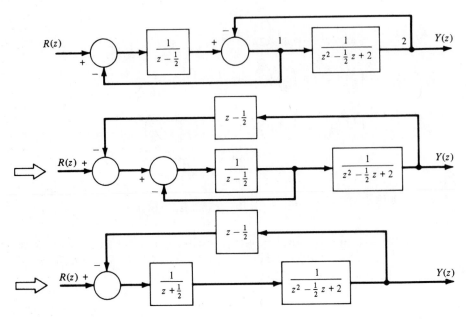

Figure 2-4 Manipulating z-transforms.

Another way of simplifying this block diagram is to move the pick-off point labeled 1 in the figure in front of the gain block to node 2 and proceed to manipulate the blocks. The answer is the same, of course.

As with continuous-time systems, when a discrete-time system has multiple inputs or multiple outputs or both, there is a z-transfer function relating each one of the inputs to each one of the outputs, with all other inputs zero

$$T_{ij} = \frac{Y_i(z)}{R_j(z)} \Bigg|$$
When all initial conditions
are zero and when all inputs
except R_j are zero

In general, when the system's initial conditions are zero, the system's outputs are given by

$$Y_1(z) = T_{11}(z)R_1(z) + T_{12}(z)R_2(z) + T_{13}(z)R_3(z) + \cdots$$
$$Y_2(z) = T_{21}(z)R_1(z) + T_{22}(z)R_2(z) + T_{23}(z)R_3(z) + \cdots$$
$$Y_3(z) = T_{31}(z)R_1(z) + T_{32}(z)R_2(z) + T_{33}(z)R_3(z) + \cdots$$
$$\vdots$$

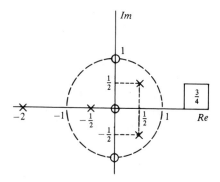

Figure 2-5 An example.

A multiple-input/multiple-output system is input–output stable if and only if all of the poles of all of its z-transfer functions are inside the unit circle on the complex plane.

A pole-zero plot of a z-transfer function consists of ×'s denoting poles and ○'s denoting zeros on the complex plane. It is helpful also to place the multiplying constant of the function in a box at the right of the plot. The z-transfer function

$$T(z) = \frac{3z + 3z^3}{4z^4 + 6z^3 - 4z^2 + z + 2}$$

$$= \left(\frac{3}{4}\right) \frac{z(z + j)(z - j)}{(z + 2)(z + \frac{1}{2})(z - \frac{1}{2} + j\frac{1}{2})(z - \frac{1}{2} - j\frac{1}{2})}$$

has the pole-zero plot given in Figure 2-5. It represents an unstable discrete-time system because there is a pole (at $z = -2$) outside the unit circle. The usual graphical evaluation and root locus techniques can also be applied without change in the discrete-time case, although it is the unit circle rather than the imaginary axis that is the stability boundary.

2.3.3. Jury Stability Test and the Bilinear Transformation

A linear, step-invariant, discrete-time system is said to be *stable* if and only if its unit pulse response decays asymptomatically to zero with time. As the z-transfer function is the z-transform of the unit pulse response, a system is stable if and only if all of its characteristic roots lie inside the unit circle on the complex plane. If the discrete-time system has one or more characteristic roots outside the unit circle or any repeated roots on the unit circle, its pulse response expands with time and the system is said to be unstable. Characteristic roots on the unit circle, if not repeated, give a pulse response that neither expands nor decays with time and the system is *marginally stable*.

We now discuss two methods for testing the location of the roots of a characteristic polynomial on the complex z-plane without factoring the polynomial. The first method is the Jury test, after E. I. Jury who developed it, and the second method is the bilinear transformation.

Jury Test

Analogous to the Routh-Hurwitz test, the Jury test is a numerical procedure for testing the location of the roots on the complex plane without solving for the roots explicitly.

For the characteristic polynomial

$$q(z) = a_n z^n + a_{n-1} z^{n-1} + a_{n-2} z^{n-2} + \cdots + a_1 z + a_0 \ (a_n > 0)$$

the initial part of the array is formed by entering the coefficients of the characteristic polynomial, starting with the coefficient of the lowest power in the polynomial a_0 until a_n, in the first row of the array. The entries in the second row of the array are simply the entries of the first row but in reverse order as shown below:

Row						
1	a_0	a_1	a_2	$\cdots$	a_{n-1}	a_n
2	a_n	a_{n-1}	a_{n-2}	$\cdots$	a_1	a_0
3	b_0	b_1	b_2	$\cdots$	b_{n-1}	
4	b_{n-1}	b_{n-2}	b_{n-3}	$\cdots$	b_0	
5	c_0	c_1	c_2	$\cdots c_{n-2}$		
$\vdots$	$\vdots$	$\vdots$	$\vdots$			
$2n-5$	d_0	d_1	d_2	d_3		
$2n-4$	d_3	d_2	d_1	d_0		
$2n-3$	e_0	e_1	e_2			

The array is completed by proceeding two rows at a time, calculating the elements of the next row. Each element calculated is derived from four elements in the above two rows as follows:

$$b_j = \begin{vmatrix} a_0 & a_{n-j} \\ a_n & a_j \end{vmatrix} \qquad c_j = \begin{vmatrix} b_0 & b_{n-1-j} \\ b_{n-1} & b_j \end{vmatrix}$$

$$e_0 = \begin{vmatrix} d_0 & d_3 \\ d_3 & d_0 \end{vmatrix} \qquad e_1 = \begin{vmatrix} d_0 & d_2 \\ d_3 & d_1 \end{vmatrix}$$

and

$$e_2 = \begin{vmatrix} d_0 & d_1 \\ d_3 & d_2 \end{vmatrix}$$

In each case, the calculated element is the determinant of the four elements in the two rows above it. This process is continued until row $2n - 3$, which has three elements. The procedure is then terminated.

Jury Stability Criteria

The necessary and sufficient conditions for the polynomial $q(z)$ to have all its roots inside the unit circle on the z-plane are

1. $q(z = 1) > 0$

2. $q(z = -1) = \begin{cases} > 0 & \text{for } n \text{ even} \\ < 0 & \text{for } n \text{ odd} \end{cases}$

3. $\begin{cases} |a_0| < a_n \\ |b_0| > |b_{n-1}| \\ |c_0| > |c_{n-2}| \\ \vdots \\ |d_0| > |d_3| \\ |e_0| > |e_2| \end{cases}$

That is, all three conditions should be satisfied simultaneously.
For example

$$q(z) = z^4 - \frac{1}{2} z^2 + \frac{1}{16}$$

$$= \left(z - \frac{1}{2}\right)^2 \left(z + \frac{1}{2}\right)^2$$

has all its roots inside the unit circle on the complex z-plane. The conditions

$$q(1) = \frac{9}{16} > 0$$

and

$$q(-1) = \frac{9}{16} > 0$$

are both satisfied. Then, it is necessary to construct the Jury array. Note, however, that if either condition 1 or condition 2 is not satisfied, the system is unstable, and it is not necessary to form the array.

Form the Jury array as follows:

Row					
1	$\dfrac{1}{16}$	0	$-\dfrac{1}{2}$	0	1
2	1	0	$-\dfrac{1}{2}$	0	$\dfrac{1}{16}$
3	$-\dfrac{255}{256}$	0	$\dfrac{15}{32}$	0	
4	0	$\dfrac{15}{32}$	0	$-\dfrac{255}{256}$	
5	$\dfrac{65{,}025}{65{,}536}$	0	$-\dfrac{3825}{8192}$		

where

$$b_0 = \begin{vmatrix} \frac{1}{16} & 1 \\ 1 & \frac{1}{16} \end{vmatrix} = -\frac{255}{256}$$

$$b_1 = \begin{vmatrix} \frac{1}{16} & 0 \\ 1 & 0 \end{vmatrix} = 0$$

$$b_2 = \begin{vmatrix} \frac{1}{16} & -\frac{1}{2} \\ 1 & -\frac{1}{2} \end{vmatrix} = \frac{15}{32}$$

$$b_3 = 0$$

and

$$c_0 = \begin{vmatrix} -\dfrac{255}{256} & 0 \\ 0 & -\dfrac{255}{256} \end{vmatrix} = \frac{65{,}025}{65{,}536}$$

$$c_1 = 0$$

$$c_2 = \begin{vmatrix} -\dfrac{255}{256} & 0 \\ 0 & \dfrac{15}{32} \end{vmatrix} = -\frac{3825}{8192}$$

Because *all* the conditions in 3 are satisfied, all the roots of the characteristic equation are inside the unit circle; the system is therefore stable.

As another example, consider the polynomial

$$q(z) = z^2 - \frac{3}{2}z - 1 = (z - 2)\left(z + \frac{1}{2}\right)$$

Condition 1 of the Jury stability criteria is not satisfied because

$$q(1) = -\frac{3}{2}$$

Therefore, the system is not stable.

When designing control systems, it is often desirable to know the range of an adjustable parameter k that results in a stable system. For example, the closed-loop transfer function, in terms of k, of the system shown in Figure 2-6 is

$$T(z) = \frac{kG(z)}{1 + kG(z)}$$

$$= \frac{1.33k(z + 0.75)}{z^2 + (1.33k - 1.72)z + 0.72 + 0.9975k}$$

Applying Jury's test, conditions 1 and 2 give

$$q(1) = 2.3275k > 0$$
$$q(-1) = 3.44 - 0.3325k > 0$$

or

$$k > 0$$

and

$$k < 10.346$$

Also, the array that consists of one row only is

Row			
1	$0.72 + 0.9975k$	$1.33k - 1.72$	1

Hence, the third condition for stability is

$$0.72 + 0.9975k < 1$$

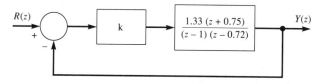

R(z)   k $\dfrac{1.33\,(z + 0.75)}{(z - 1)\,(z - 0.72)}$ Y(z)

Figure 2-6 Poles and zeros of a z-transfer function.

or

k < 0.2807

Therefore, for the system to be stable, the three conditions above should be satisfied simultaneously. That is,

0 < k < 0.2807

If

k = 0

the closed-loop poles of the system are at $z_1 = 1$ and $z_2 = 0.72$, and the system is marginally stable.

   If

k = 0.2807

the closed-loop poles are located at

$z_{1,2} = 0.673 \pm j0.739$

and the system is also marginally stable because the poles z_1 and z_2 are located on the unit circle as determined by the magnitude of z_1 and z_2.

Bilinear Transformation

Another method for determining the stability of a discrete-time system that does not require factoring the polynomial is to apply the bilinear transformation followed by a Routh-Hurwitz test.

   The bilinear transformation

$$z = \frac{1 + w}{1 - w} \qquad w = \frac{z - 1}{z + 1}$$

maps the unit circle on the z-plane to the imaginary axis on the w-plane. The interior of the unit circle in z maps to the left half-plane (LHP) in w, and the exterior of the unit circle on the z-plane maps to the right half-plane (RHP) on the w-plane. If this bilinear transformation is applied to the z-transfer function of a discrete-time system, the transfer function as a function of w has RHP poles if and only if the z-transfer function has poles outside the unit circle. The stability of a discrete-time system can thus be investigated by applying the bilinear transformation to the closed-loop z-transfer function followed by a Routh-Hurwitz test of the resulting characteristic polynomial as for continuous-time systems.

   When the bilinear transformation is applied to discrete-time frequency response data, Bode plot methods can be applied to find an approximate system model.

For example, consider the z-transfer function

$$T(z) = \frac{3z^4 + 2z^3 - z^2 + 4z + 5}{z^4 + 0.5z^3 - 0.2z^2 + z + 0.4}$$

When the bilinear change of variables is made on the denominator polynomial,

$$q(z) = z^4 + 0.5z^3 - 0.2z^2 + z + 0.4$$

there results

$$q(w) = \left(\frac{1+w}{1-w}\right)^4 + 0.5 \left(\frac{1+w}{1-w}\right)^3 - 0.2 \left(\frac{1+w}{1-w}\right)^2 + \left(\frac{1+w}{1-w}\right) + 0.4$$

$$= \frac{-0.3w^4 + 3.4w^3 + 8.8w^2 + 1.4w + 2.7}{(1-w)^4}$$

A Routh-Hurwitz testing of the zeros of $q(w)$

$$
\begin{array}{c|ccc}
w^4 & -0.3 & 8.8 & 2.7 \\
w^3 & 3.4 & 1.4 & \\
w^2 & 8.92 & 2.7 & \\
w^1 & 0.37 & & \\
w^0 & 2.7 & &
\end{array}
$$

shows that $q(w)$ has one RHP root. Therefore, $T(z)$ has one pole outside the unit circle.

2.3.4 Discrete-Frequency Response

When the input to a linear, step-invariant, discrete-time system is a geometric (or sampled exponential) sequence

$$r(k) = Ac^k$$

where A and c are constants, the corresponding forced output is also a geometric sequence:

$$y_{\text{forced}}(k) = Bc^k$$

with the same geometric constant c. This is similar to the result for the forced exponential response of a continuous-time system.

If the discrete-time system's difference equation is

$$y(k + n) + a_{n-1}y(k + n - 1) + \cdots + a_1 y(k + 1) + a_0 y(k)$$
$$= b_m r(k + m) + b_{m-1}r(k + m - 1) + \cdots + b_0 r(k)$$

these signals satisfy

$$(c^n + a_{n-1}c^{n-1} + \cdots + a_1c + a_0)Bc^k$$
$$= (b_mc^m + b_{m-1}c^{m-1} + \cdots + b_0)Ac^k$$

The ratio of forced output to geometric input is

$$\frac{y_{forced}(k) = Bc^k}{r(k) = Ac^k} = \frac{b_mc^m + b_{m-1}c^{m-1} + \cdots + b_0}{c^n + a_{n-1}c^{n-1} + \cdots + a_1c + a_0}$$

which is the z-transfer function relating the output and input, evaluated at $z = c$. As an example, if the input to a system with the z-transfer function

$$T(z) = \frac{2z - 1}{z^2 + \frac{1}{2}z}$$

is

$$r(k) = 3\left(\frac{1}{3}\right)^k$$

then the forced output is

$$y_{forced}(k) = T\left(z = \frac{1}{3}\right)r(k) = -\frac{18}{5}\left(\frac{1}{3}\right)^k$$

The forced output for a complex exponential input

$$r(k) = A\cos(\Omega k + \theta) + jA\sin(\Omega k + \theta) = Ae^{j(\Omega k + \theta)} = (Ae^{j\theta})(e^{j\Omega})^k$$

where A, Ω, and θ are constants is then

$$y_{forced}(k) = T(z = e^{j\Omega})r(k)$$

The sinusoidal sequence

$$r(k) = A\cos(\Omega k + \theta)$$

which is the real part of the complex exponential input, has a forced response that is the real part of the complex exponential forced response

$$y_{forced}(k) = Re[T(z = e^{j\Omega})Ae^{j\theta}(e^{j\Omega})^k] = B\cos(\Omega k + \phi)$$

where, for positive A

$$B = |T(z = e^{j\Omega})|A$$
$$\phi = \theta + \underline{/T(z = e^{j\Omega})}$$

Hence, when the input to a linear, step-invariant, discrete-time system is a sinusoidal sequence, the system's forced output is another sinusoidal sequence with the same frequency Ω, but generally with a different amplitude B and phase shift ϕ. Figure 2-7 illustrates the idea.

Samples of the sinusoidal continuous-time function

$$r(t) = A \cos(\omega t + \theta)$$

at sampling interval T are

$$r(k) = A \cos(\omega kT + \theta) = A \cos(\Omega k + \theta)$$

where

$$\omega T = \Omega$$

The discrete-time system's forced response

$$y_{forced}(k) = B \cos(\omega kT + \phi) = B \cos(\Omega k + \phi)$$

consists of samples of another sinusoidal function of the same frequency,

$$y(t) = B \cos(\omega t + \phi)$$

or of any frequency ω_i for which

$$\omega_i = \omega + i\frac{2\pi}{T} \qquad i = \pm 1, \pm 2, \ldots$$

The magnitude of the z-transfer function, evaluated at $z = \exp(j\Omega)$, is the ratio of forced sinusoidal output sequence amplitude to sinusoi-

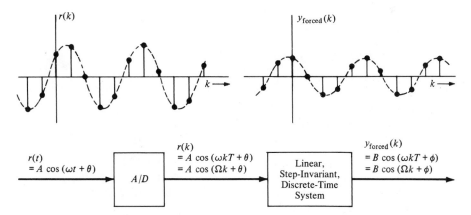

Figure 2-7 Discrete-time system with a sinuosidal input sequence and the sinusoidal forced output sequence.

dal input amplitude:

$$\frac{B}{A} = |T(z = e^{j\Omega})|$$

The angle of the z-transfer function, evaluated at $z = \exp(j\Omega)$, is the phase shift from input to output

$$\phi - \theta = \underline{/T(z = e^{j\Omega})}$$

This result is similar to the counterpart for continuous-time systems, in which the transfer function $T(s)$ is evaluated at $s = j\omega$.

Suppose that a system with z-transfer function

$$T(z) = \frac{2z}{z - \frac{1}{3}}$$

is driven with the sinusoidal input sequence

$$r(k) = 10 \cos\left(\frac{1}{2}k + \frac{\pi}{8}\right) = A \cos(\Omega k + \theta)$$

where arguments of all trigonometric functions are in radians. The z-transfer function, evaluated at $z = \exp(j\Omega)$ is

$$T(z = e^{j/2}) = T(z = 0.877 + j0.48) = \frac{2(0.877 + j0.48)}{(0.877 + j0.48) - \frac{1}{3}}$$

$$= \frac{2e^{j/2}}{0.544 + j0.48} = \frac{2e^{j/2}}{0.725e^{j0.72}} = 2.76e^{-j0.22}$$

The amplitude of the forced sinusoidal output sequence is then

$$B = A|T(z = e^{j/2})| = 10(2.76) = 27.6$$

and the output sequence phase is

$$\phi = \theta + \underline{/T(z = e^{j/2})} = \frac{\pi}{8} - 0.22 = 0.173 \text{ rad}$$

so that

$$y_{\text{forced}}(k) = 27.6 \cos\left(\frac{1}{2} + 0.173\right) \qquad k = 0, 1, 2, \ldots$$

Frequency response plots for a linear, step-invariant, discrete-time system consist of plots of the magnitude and angle of the z-transfer function, evaluated at $z = \exp(j\Omega)$, versus Ω. The magnitude plot is the ratio of forced output amplitude to input amplitude as a function of frequency. The angle plot is the phase shift between the forced output and input as a function of frequency.

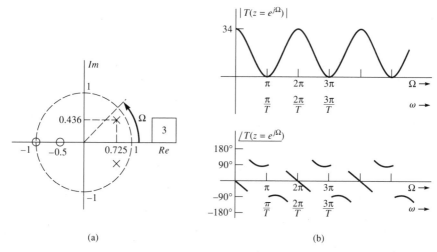

Figure 2-8 Periodicity of the frequency response of a discrete-time system. (a) Pole-zero plot of a transfer function $T(z)$. (b) Frequency response plots for $T(z)$.

Evaluation of a transfer function at $z = \exp(j\Omega)$ is the evaluation at a point z at angle Ω on the unit circle, as indicated on the pole-zero plot for a z-transfer function in Figure 2-8(a). As $\exp(j\Omega)$ is periodic in Ω with period 2π, frequency response plots for discrete-time systems are periodic as in the example in Figure 2-8(b), which gives the frequency response for the z-transfer function in the accompanying pole-zero plot. The frequency response has been deliberately graphed over a wide range of Ω to emphasize this periodicity. Frequency response plots for discrete-time systems are also symmetric about $\Omega = \pi$, as shown. The amplitude ratio is even-symmetric about $\Omega = \pi$, and the phase shift is odd-symmetric. Consequently, the frequency range of Ω from 0 to π is adequate to specify a discrete-time system's frequency response completely.

2.4 Sampling and Reconstruction

Sampling is the process of deriving a discrete-time sequence from a continuous-time function. Usually, but not always, the samples are evenly spaced in time. *Reconstruction* is the reverse; it is the formation of a continuous-time function from a sequence of samples. Many different continuous-time functions can have the same set of samples, so a reconstruction is not unique.

In this section, we discuss sampling and determine how the z-

transform of a sampled continuous-time function $f(t)$ is related to the Laplace transform $F(s)$. We then consider piecewise-constant and other reconstructions of continuous-time functions from evenly spaced samples. The sampling theorem, although it does not apply to most control system signals because they are not adequately bandlimited, is of some guidance in deciding how to process sensor signals before sampling.

2.4.1 Sampling and A/D Conversion

Samples of a continuous-time signal form a discrete-time sequence, as shown in Figure 2-9(a). Here, the samples are always spaced evenly in time and, unless otherwise stated, the sampling is synchronized so that a sample is taken at time $t = 0$, as shown. Devices for performing sampling are called *analog-to-digital* (A/D) *converters*. An electronic A/D converter produces a binary representation (typically using from 6 to 16 bits) of the applied input signal at each sample time.

The symbol for an A/D converter is given in Figure 2-9(b). An incoming continuous-time signal $f(t)$ is sampled to produce the discrete-time sequence $f(k)$. The sampling interval T is generally known and is indicated on the diagram or elsewhere. As is common, the same symbol (in this case, f) is used to represent both a continuous-time signal $f(t)$ and its related sequence of samples $f(k)$. Which is meant is inferred from the argument of the function.

Using a finite number of bits to represent a signal sample generally results in *quantization errors* in the A/D process. The maximum quantization error in 16-bit A/D conversion is $2^{-16} = 0.0015\%$, which is

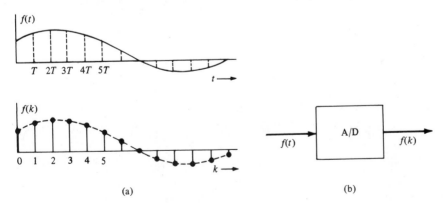

(a)

(b)

Figure 2-9 Analog-to-digital (A/D) conversion. (a) A uniformly sampled continuous-time signal. (b) Symbol for an A/D converter.

quite low compared with typical errors in analog sensors. This error, if taken to be "noise," gives a signal-to-noise ratio of $20 \log_{10} (2^{-16}) = 96.3$ dB, which is much better than that of most high-fidelity audio systems. The designer must ensure that enough bits are used to give the desired system accuracy. Beyond this, the digital computations used must maintain adequate accuracy. Study of the effects of roundoff or truncation errors in digital computation is beyond our scope in this book, but it is important to use adequate word lengths in fixed or floating-point computations. Years ago, digital hardware was very expensive, so minimizing word lengths was much more important than it is today.

When a continuous-time signal $f(t)$ is sampled to form the sequence $f(k)$, there is a close relationship between the Laplace transform of $f(t)$ and the z-transform of $f(k)$. If a rational Laplace transform is expanded into partial-fraction terms, the corresponding continuous-time signal components in the time domain are powers of time, exponentials, sinusoids, and so on. Uniform samples of these elementary signal components have, in turn, simple z-transforms that can be summed to give the z-transform of the entire sampled signal. Table 2-5 lists Laplace transform terms and the resulting z-transforms when the corresponding time functions are sampled uniformly. The z-transform terms generally involve the length T of the sampling interval.

As an example, consider the continuous-time function with the Laplace transform

$$F(s) = \frac{3s^2 + 2s + 3}{s^3 + 4s^2 + 3s} = \frac{1}{s} + \frac{-2}{s + 1} + \frac{4}{s + 3}$$

The z-transform of the sampled signal with a sampling interval $T = 0.2$ is

$$F(z) = \frac{z}{z - 1} + \frac{-2z}{z - e^{-0.2}} + \frac{4z}{z - e^{-0.6}}$$

$$= \frac{3z^3 - 5.55z^2 + 2.63z}{(z - 1)(z - 0.82)(z - 0.55)}$$

If the continuous-time signal involves delays that are multiples of the sampling interval T, the delays are simply z-transformed, each Laplace transform term of the form $\exp(-sT)$ becoming a z-transform term z^{-1} using the step-shifted relation. For example, when the continuous-time signal with the Laplace transform

$$F(s) = \frac{3e^{-0.2s} + 3}{s(s + 3)} = (e^{-0.2s} + 1)\left(\frac{1}{s} + \frac{-1}{s + 3}\right)$$

Table 2-5 Laplace and Z-Transform Pairs

$f(t)$	$F(s)$	$f(k)$	$F(z)$
$u(t)$, unit step	$\dfrac{1}{s}$	$u(k)$, unit step	$\dfrac{z}{z-1}$
$tu(t)$	$\dfrac{1}{s^2}$	$kTu(k)$	$\dfrac{Tz}{(z-1)^2}$
$e^{-at}u(t)$	$\dfrac{1}{s+a}$	$(e^{-aT})^k u(k) = c^k u(k)$ where $c = e^{-aT}$	$\dfrac{z}{z-e^{-aT}} = \dfrac{z}{z-c}$
$te^{-at}u(t)$	$\dfrac{1}{(s+a)^2}$	$kT(e^{-aT})^k u(k) = kTc^k u(k)$	$\dfrac{Tze^{-aT}}{(z-e^{-aT})^2} = \dfrac{Tcz}{(z-c)^2}$
$(\sin \omega t)u(t)$	$\dfrac{\omega}{s^2+\omega^2}$	$(\sin k\omega T)u(k) = \sin \Omega k$ where $\Omega = \omega T$	$\dfrac{z \sin \Omega}{z^2 - 2z \cos \Omega + 1}$
$(\cos \omega t)u(t)$	$\dfrac{s}{s^2+\omega^2}$	$(\cos k\omega T)u(k) = \cos \Omega k$	$\dfrac{z(z - \cos \Omega)}{z^2 - 2z \cos \Omega + 1}$
$e^{-at}(\sin \omega t)u(t)$	$\dfrac{\omega}{(s+a)^2+\omega^2}$	$(e^{-aT})^k(\sin k\omega T)u(k) = c^k(\sin \Omega k)u(k)$	$\dfrac{z(e^{-aT} \sin \Omega)}{(z - e^{(-a+j\omega)T})(z - e^{(-a-j\omega)T})}$ $= \dfrac{zc \sin \Omega}{z^2 - (2c \cos \Omega)z + c^2}$
$e^{-at}(\cos \omega t)u(t)$	$\dfrac{s+a}{(s+a)^2+\omega^2}$	$(e^{-aT})^k(\cos k\omega T)u(t) = c^k(\cos \Omega k)u(k)$	$\dfrac{z(z - e^{-aT} \cos \Omega)}{(z - e^{(-a+j\omega)T})(z - e^{(-a-j\omega)T})}$ $= \dfrac{z(z - c \cos \Omega)}{z^2 - (2c \cos \Omega)z + c^2}$

is sampled with $T = 0.1$, it has the z-transform

$$F(z) = (z^{-2} + 1)\left(\frac{z}{z - 1} + \frac{-z}{z - e^{-0.3}}\right) = \frac{0.26z(z^2 + 1)}{z^2(z - 1)(z - 0.74)}$$

We denote this process of conversion of a continuous-time function's Laplace transform $F(s)$ to the z-transform $F(z)$ of its sequence of uniform samples at interval T by

$$F(z) = \underset{\text{at } T}{\text{sample}}\left[F(s)\right]$$

If the above signal is, instead, sampled at an interval that is not a multiple of the 0.2-sec delay, say at $T = 0.3$, then inverse powers of z cannot simply be substituted for delays. With this sampling rate

$$F(s) = (e^{-0.2s} + 1)\left(\frac{1}{s} + \frac{-1}{s + 3}\right)$$

and

$$f(t) = u(t) + u(t - 0.2) - e^{-3t}u(t) - e^{-3(t-0.2)}u(t - 0.2)$$

The two delayed step functions are zero at step 0 so that

$$f(0) = 1 + 0 - 1 - 0 = 0$$

At step 1 (at time $t = T = 0.3$) and thereafter, the step functions have a value of unity, and

$$f(k) = 2 - e^{-0.9k} - e^{-0.9k+0.6} = 2 - (1 + e^{0.6})e^{-0.9k}$$
$$= 2 - 2.82e^{-0.9k} = 2 - 2.82(0.41)^k$$

Then

$$f(k) = 0.82\delta(k) + 2 - 2.82(0.41)^k \qquad k = 0, 1, 2, \ldots$$

so that

$$F(z) = 0.82 + \frac{2z}{z - 1} - \frac{2.82z}{z - 0.41} = \frac{0.84z + 0.336}{(z - 1)(z - 0.41)}$$

Mathematically, the process of sampling may be viewed as passing a continuous-time signal through an impulse sampler as shown in Figure 2-10. The sampler closes for a very short time compared with the sampling period. The output of the sampler is a series of impulses, timed at intervals T, with strength equal to the sampled value of the continuous-time signal at the corresponding sampling instants.

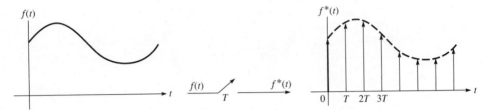

Figure 2-10 The impulse sampler.

The sampled output may be expressed as

$$f^*(t) = \sum_{k=0}^{\infty} f(t)\delta(t - kT)$$

or

$$f^*(t) = \sum_{k=0}^{\infty} f(kT)\delta(t - kT) \tag{2-2}$$

Taking the Laplace transform of (2-2) gives

$$F^*(s) = \int_0^{\infty} \left[\sum_{k=0}^{\infty} f(kT)\delta(t - kT) \right] e^{-st} \, dt$$

Using the sifting property of the impulse,

$$F^*(s) = \sum_{k=0}^{\infty} f(kT)e^{-skT}$$

and letting

$$e^{sT} = z$$

then

$$F^*(s) = \sum_{k=0}^{\infty} f(kT)z^{-k} \Big|_{z=e^{sT}}$$

which is the z-transform of the sample $f(kT)$. That is,

$$F(z) = F^*(s) \Big|_{z=e^{sT}}$$

2.4.2 Reconstruction and D/A Conversion

Reconstruction is the process of converting a discrete-time sequence into a continuous-time signal. Figure 2-11 shows a continuous-time signal $f(t)$, the sequence $f(k)$ resulting from uniform sampling, and a piecewise-constant approximation to $f(t)$ constructed from the samples $f(k)$. This latter continuous-time waveform is termed the *step reconstruction* of $f(t)$. For a sampling interval T, the step reconstruction is related to the sample sequence $f(k)$ by

$$f^0(t) = \sum_{k=0}^{\infty} f(k)\{u(t - kT) - u[t - (k + 1)T]\}$$

Although very often the continuous-time signal reconstructed from a sequence is a sampled-and-held waveform, that is not always the case.

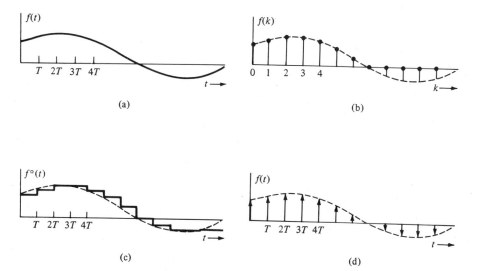

(a)

(b)

(c)

(d)

Figure 2-11 Continuous-time signal sampling and reconstruction. (a) Continuous-time signal. (b) Sequence of samples of the continuous-time signal $f(k) = f(t = kT)$.

(c) Step reconstruction of $f(t)$: $f^0(t) = \sum_{k=0}^{\infty} f(k)\{u(t - kT) - u[t - (k + 1)T]\}$

(d) Impulse train for $f(t)$: $f^*(t) = \sum_{k=0}^{\infty} f(k)\, \delta(t - kT)$

A more fundamental continuous-time signal related to a sequence of samples is a series of impulses, timed at intervals T, with strengths equal to the corresponding samples, as in Figure 2-11(d). This *impulse train* corresponding to a sequence $f(k)$ and a sampling interval T is denoted by $f^*(t)$ and is given by

$$f^*(t) = \sum_{k=0}^{\infty} f(k)\delta(t - kT)$$

Devices for performing reconstruction are called *digital-to-analog (D/A) converters*. Electronic D/A converters typically produce a step reconstruction from incoming signal samples by converting the binary-coded digital input to a voltage, transferring the voltage to the output, and holding the output voltage constant until the next sample is available. The symbol for a D/A converter that generates the step reconstruction $f^0(t)$ from signal samples $f(k)$ is given in Figure 2-12(a). D/A refers to the process of converting digital codes to signal samples. *Sample and hold* (S/H) is the operation of holding each of these samples for a sampling interval T to form the step reconstruction. The step reconstruction of a continuous-time signal from samples can be represented as the conversion of the sequence $f(k)$ to its corresponding impulse train $f^*(t)$, then conversion of the impulse train to the step

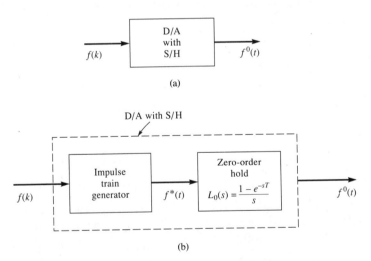

(a)

(b)

Figure 2-12 Digital-to-analog (D/A) conversion with sample-and-hold (S/H). (a) Symbol for D/A conversion with S/H. (b) Representation of a D/A converter with S/H as an impulse train generator driving a zero-order hold.

reconstruction, as in Figure 2-12(b). This viewpoint neatly separates conversion of the discrete sequence to a continuous-time waveform and the details of the shape of the reconstructed waveform.

The continuous-time transfer function that converts an impulse train with sampling interval T to a step reconstruction, termed the *zero-order hold,* has the impulse response pictured in Figure 2-13(a). Each

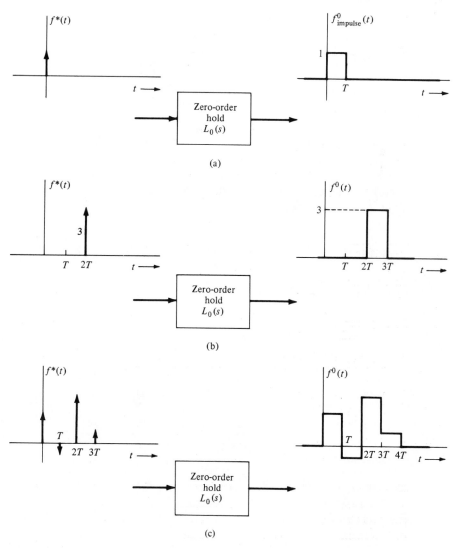

Figure 2-13 Responses to the zero-order hold. (a) Unit impulse response. (b) Response to a scaled, shifted impulse. (c) Response to an impulse train.

incoming impulse produces a rectangular pulse of duration T and of height equal to the impulse strength. An impulse train input produces a train of such pulses, which sum to give the step reconstruction, as in Figure 2-13(b) and (c). The zero-order hold has a unit impulse response given by

$$f^0_{\text{impulse}}(t) = u(t) - u(t - T)$$

or

$$F^0_{\text{impulse}}(s) = \frac{1}{s}(1 - e^{-sT})$$

so that

$$L_0(s) = \frac{1}{s}(1 - e^{-sT})$$

Other reconstructions can also be represented in a similar way. For example, practical D/A converters can exhibit substantial "rise time" compared with perfect step reconstruction, as indicated in Figure 2-14(a). A model that includes this imperfection (which, by the way, might not be detrimental to system performance at all) would simply

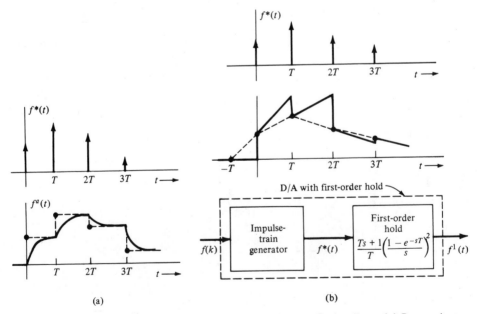

Figure 2-14 Using the impulse train in modeling other reconstructions. (a) Reconstruction with significant rise time. (b) First-order hold reconstruction.

have a different transfer function in place of that for the zero-order hold.

To improve on the accuracy of the reconstruction, higher-order holds, using more than a single sample at a time for reconstruction, can be used. A *first-order hold* uses the previous two samples to construct a straight line approximation during each sampling interval, as shown in Figure 2-14(b). Connecting the sample points with straight lines, what is called a *linear point connector*, is a possibility also, but this requires that the future sample be known in advance of the time the reconstruction is begun.

2.4.3 The Sampling Theorem

Figure 2-15 shows two different continuous-time signals that have the same samples, illustrating how, except in highly restricted circumstances, a sampled function is not uniquely determined by its samples. One important situation for which samples of a continuous-time function are unique occurs when the function is *bandlimited*. A signal $g(t)$ and its Fourier transform $G(\omega)$ are generally related by

$$G(\omega) = \int_{-\infty}^{\infty} g(t)e^{-j\omega t}\, dt$$

$$g(t) = \frac{1}{2\pi} \int_{-\infty}^{\infty} G(\omega)e^{j\omega t}\, d\omega \tag{2-3}$$

This relationship is similar to Laplace transformation with $s = j\omega$, but the transform integral of equation (2-3) extends over all time rather

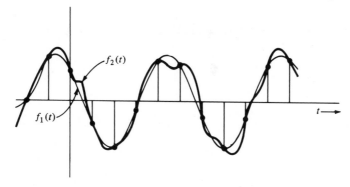

Figure 2-15 Two different continuous-time signals with the same samples.

than from $t = 0^-$ on. The Fourier transform $G(\omega)$ is termed the *spectrum* of $g(t)$. A signal is bandlimited at (hertz) frequency f_B if

$$G(\omega) = 0 \quad \text{for} \quad |\omega| > 2\pi f_B = \omega_B$$

as shown in Figure 2-16(a). Equation (2-3) becomes

$$g(t) = \frac{1}{2\pi} \int_{-\omega_B}^{\omega_B} G(\omega) e^{j\omega t} \, d\omega$$

If a signal $g(t)$ is uniformly sampled with sampling interval T to form the sequence

$$g(k) = g(t = kT)$$

then the corresponding impulse train that extends both ways in time

$$g^*(t) = \sum_{k=-\infty}^{\infty} g(kT)\delta(t - kT)$$

has the Fourier transform

$$G^*(\omega) = \frac{1}{T} \sum_{n=-\infty}^{\infty} G(\omega - n\omega_s) \tag{2-4}$$

where

$$\omega_S = 2\pi f_s = \frac{2\pi}{T}$$

To prove this result, the periodic function

$$s(t) = \sum_{k=0}^{\infty} \delta(t - kT)$$

is represented by an exponential Fourier series of the form

$$s(t) = \sum_{n=-\infty}^{\infty} d_n e^{(jn2\pi/T)t}$$

where

$$d_n = \frac{1}{T} \int_{-T/2}^{T/2} \sum_{k=-\infty}^{\infty} \delta(t - kT) e^{-(jn2\pi/T)t} \, dt$$

$$= \frac{1}{T}$$

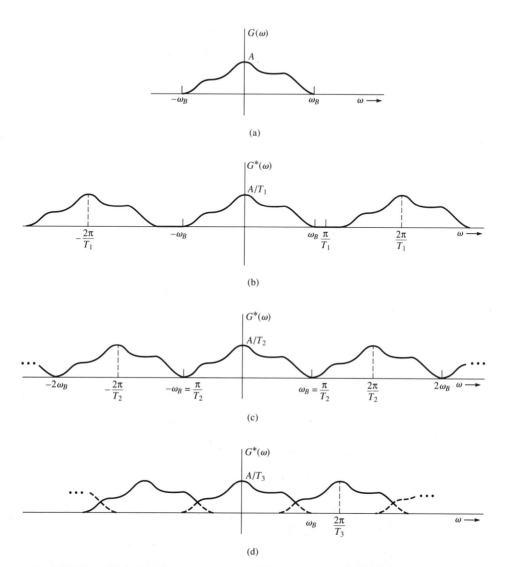

Figure 2-16 Frequency spectra of a signal sampled at various frequencies. (a) Frequency spectrum of an analog bandlimited signal $g(t)$. (b) Frequency spectrum of a sampled signal $g^*(t)$ with $f_{s1} > 2f_B(f_{s1} = 1/T_1)$. (c) Frequency spectrum of a sampled signal $g^*(t)$ with $f_{s2} = 2f_B(f_{s2} = 1/T_2)$. (d) Frequency spectrum of a sampled signal $g^*(t) < 2f_B$ $(f_{s3} = 1/T_3)$.

Hence,

$$s(t) = \frac{1}{T} \sum_{n=-\infty}^{\infty} e^{(jn2\pi/T)t}$$

Substituting this result into the impulse train

$$g^*(t) = \sum_{k=-\infty}^{\infty} g(kT)\delta(t - kT)$$

gives

$$g^*(t) = \frac{1}{T} g(t) \sum_{n=-\infty}^{\infty} e^{(jn2\pi/T)t}$$

and taking the Fourier transform yields

$$G^*(\omega) = \frac{1}{T} \sum_{n=-\infty}^{\infty} \int_{-\infty}^{\infty} g(t)e^{(jn2\pi/T)t}e^{-j\omega t} \, dt$$

and therefore,

$$G^*(\omega) = \frac{1}{T} \sum_{n=-\infty}^{\infty} G\left(\omega - n\frac{2\pi}{T}\right)$$

$$= \frac{1}{T} \sum_{n=-\infty}^{\infty} G(\omega - n\omega_s)$$

which completes the proof.

The function $G^*(\omega)$ in equation (2-4) is periodic, and each individual term in the series has the same form as the original $G(\omega)$, with the exception that each term is centered at

$$\omega = n\frac{2\pi}{T} \qquad n = \ldots -2, -1, 0, 1, 2, \ldots$$

If the sampling frequency f_S is more than twice the bandlimit frequency f_B, the individual terms in equation (2-4) do not overlap as shown in Figure 2-16(b), and $G(\omega)$ and thus $g(t)$ can be determined from $G^*(\omega)$, which, in turn, is determined from the samples $g(k)$. If the sampling frequency f_s is twice the bandlimited frequency f_B, the individual terms in equation (2-4) do not overlap as shown in Figure 2-16(c).

Note that, in terms of the sampling period,

$$\omega_s = 2\omega_B$$

and

$$T = \frac{2\pi}{\omega_s}$$

Then

$$T = \frac{\pi}{\omega_B} = \frac{1}{2f_B}$$

which relates the sampling period to the highest frequency f_B in the signal.

Consider, for example, the time function $g(t)$, which has a frequency spectrum as shown in Figure 2-17(a). If this signal is sampled at a rate of 10^6 samples per second, that is,

$$T = 10^{-6} \text{ sec}$$

then the sampling frequency

$$\omega_s = \frac{2\pi}{T} = 2\pi \times 10^6 \text{ rad/sec}$$

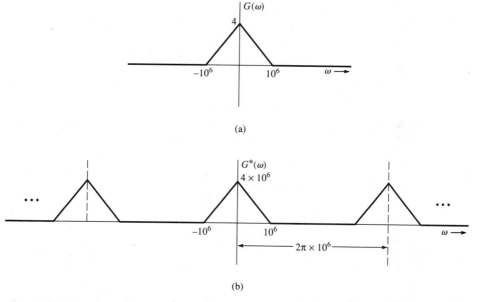

(a)

(b)

Figure 2-17 An example of the frequency spectrum of a sampled signal. (a) Frequency spectrum of the analog signal $g(t)$. (b) Frequency spectrum of the corresponding sampled signal $g^*(t)$.

being greater than twice the bandlimited frequency

$$\omega_B = 10^6 \text{ rad/sec}$$

indicates that the individual terms in the frequency spectrum of the sampled signal $g^*(t)$ do not overlap as shown in Figure 2-17(b).

A statement of the sampling theorem is:

> **The uniform samples of a signal $g(t)$ that is bandlimited above (hertz) frequency f_B are unique if and only if the sampling frequency is higher than $2f_B$.**

The frequency $2f_B$ is termed the *Nyquist frequency* for a bandlimited signal. As shown in Figure 2-16(d), if the sampling frequency does not exceed the Nyquist frequency, the individual terms in equation (2-4) overlap, a phenomenon called *aliasing* (or *foldover*).

2.4.4 Signal Recovery by Filtering

Another statement of the sampling theorem is: If a bandlimited signal $g(t)$ is uniformly sampled at a rate greater than twice the highest frequency in $g(t)$, then the signal $g(t)$ may be recovered from its samples by passing the sampled signal $g^*(t)$ through an ideal low-pass filter with the frequency response

$$H_{LP}(s = j\omega) = \begin{cases} T; & \dfrac{-\omega_s}{2} \le \omega \le \dfrac{\omega_s}{2} \\ 0; & \text{otherwise} \end{cases}$$

as shown in Figure 2-18.

The output of the low-pass filter becomes

$$G(j\omega) = G^*(j\omega)H_{LP}(j\omega)$$

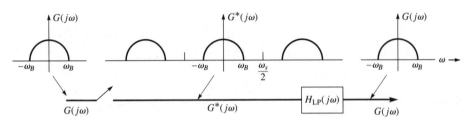

Figure 2-18 Signal recovery by passing a sampled signal through an ideal low-pass filter.

which is the Fourier transform of the original signal $g(t)$. Recalling that multiplication in the frequency domain is equivalent to convolution in the time domain, the original signal $g(t)$ may be determined by convolving $g^*(t)$ with the impulse response of the low-pass filter

$$g(t) = \text{convolution}[g^*(t), h_{LP}(t)]$$

The impulse response of the ideal low-pass filter is given by

$$h_{LP}(t) = \frac{1}{2\pi} \int_{-\omega_s/2}^{\omega_s/2} T e^{j\omega t} \, d\omega$$

$$= \frac{T}{2\pi jt} e^{j\omega t} \Big|_{-\omega_s/2}^{\omega_s/2} = T \frac{e^{j\omega_s t/2} - e^{-j\omega_s t/2}}{2\pi jt}$$

$$= T \frac{\sin(\omega_s/2)t}{\pi t} = \frac{T\omega_s}{2\pi} \frac{\sin(\omega_s/2)t}{(\omega_s/2)t}$$

as shown in Figure 2-19. Therefore,

$$g(t) = \int_{-\infty}^{\infty} g(\tau) \sum_{k=-\infty}^{\infty} \delta(\tau - kT)T \frac{\omega_s}{2\pi} \frac{\sin[(\omega_s/2)(t - \tau)]}{[(\omega_s/2)(t - \tau)]} \, d\tau$$

Interchanging the summation and integration signs and using the sifting property of the impulse function, then

$$g(t) = T \frac{\omega_s}{2\pi} \sum_{k=-\infty}^{\infty} g(kT) \frac{[\sin(\omega_s/2)(t - kT)]}{[(\omega_s/2)(t - kT)]}$$

with

$$T = \frac{2\pi}{\omega_s}$$

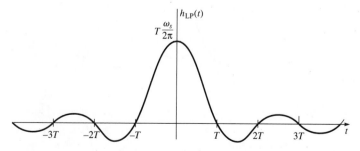

Figure 2-19 Impulse response of the ideal low-pass filter.

then the function $g(t)$ may be recovered from its samples $g(kT)$ as follows:

$$g(t) = \sum_{k=-\infty}^{\infty} g(kT) \frac{\sin[(\omega_s/2)(t - kT)]}{[(\omega_s/2)(t - kT)]} \qquad \textbf{(2-5)}$$

This result may be interpreted graphically as shown in Figure 2-20. Recognizing that the sinc function is zero at all sampling instants except at the kth sample point, we conclude that individual terms in equation (2-5) do not contribute to previous and future ones at the kth sample point.

Unfortunately, the ideal low-pass filter is not physically realizable because its impulse response starts before the input impulse arrives as shown in Figure 2-19. One way to remedy this situation is to approxi-

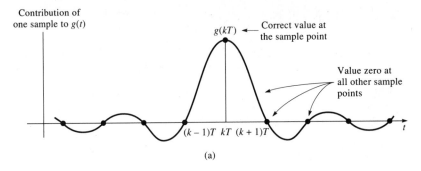

(a)

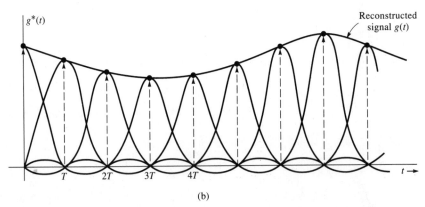

(b)

Figure 2-20 Recovering a signal $g(t)$ from the impulse train $g^*(t)$ using an ideal low-pass filter. (a) Contribution of one sample to the signal $g(t)$. (b) Signal recovery using an ideal low-pass filter.

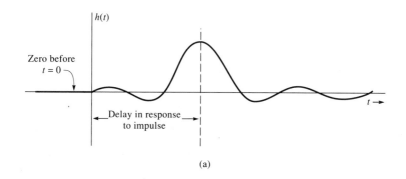

(a)

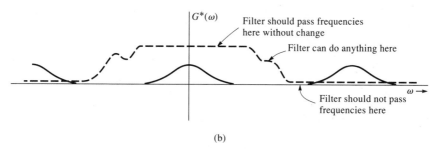

(b)

Figure 2-21 Two approaches for practical postfiltering. (a) Time delay of a low-pass filter impulse response. (b) High sampling rate.

mate the ideal low-pass filter with time delay as shown in Figure 2-21(a), which, with sufficient delay, may be done with arbitrary accuracy and complexity.**

Another approach is to use a sampling rate considerably higher than the Nyquist rate; then $g^*(t)$ has a periodic spectrum with wide gaps between the repetitions as shown in Figure 2-21(b). The low-pass filter then has far less stringent requirements.

The third approach is to use a hold device so that high frequencies in the spectrum are adequately attenuated. For example, the zero-order hold given by the transfer function

$$L_0(s) = \frac{1 - e^{-sT}}{s}$$

** In feedback control systems, large time delays usually cause system instability and should be avoided.

has a frequency response determined by replacing s with $j\omega$ in its transfer function

$$L_0(j\omega) = Te^{-j\omega T/2} \frac{\sin(\omega T/2)}{\omega T/2}$$

Its magnitude and phase plots are shown in Figure 2-22. Comparing the magnitude plot of the zero-order hold with the magnitude plot of the ideal low-pass filter shows that the zero-order hold roughly approximates the behavior of the ideal low-pass filter in the range

$$\frac{-\pi}{T} \le \omega \le \frac{\pi}{T}$$

The undesirable frequency components outside this range may be attenuated by using a low-pass filter following the zero-order hold to produce a smoothed signal.

The first-order hold, on the other hand, which is described by the transfer function

$$L_1(s) = \frac{1 + sT}{T} \left(\frac{1 - e^{-sT}}{s} \right)^2$$

has a frequency response of the form

$$L_1(j\omega) = T[1 + (\omega T)^2]^{1/2} \sin c^2 \frac{\omega T}{2} e^{-j(\omega T - \tan^{-1} \omega T)}$$

The magnitude and phase plots of the first-order hold are shown in Figure 2-22. As shown in the figure, for frequencies well below π/T, the first-order hold has less phase lag than the zero-order hold. However, at frequencies above π/T the zero-order hold may exhibit better results. Higher-order holds may have better frequency responses but with greatly increased complexity.

The sampling theorem is important to control system design because when A/D conversion is done on noisy signals that have significant frequency components above half the sampling frequency, the high frequencies produce errors in the sampling that are *indistinguishable* from the presence of lower-frequency errors. For this reason, low-pass filters, termed *prefilters*, or *antialiasing filters*, are often used to reduce high frequencies present in sensor signals before their A/D conversion.

The following is a summary of a constructive statement of how to recover a suitably bandlimited signal from its samples:

To recover a signal $g(t)$ that is bandlimited above frequency f_B from its samples $g(k)$, form the impulse train $g^*(t)$ with the sample frequency $f_S =$

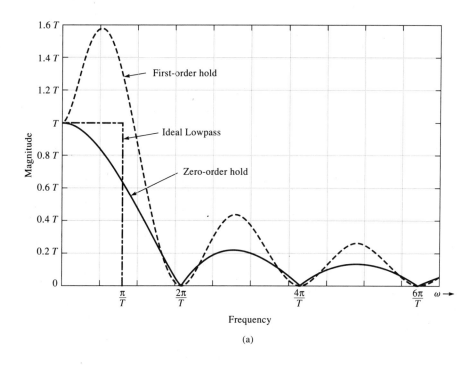

(a)

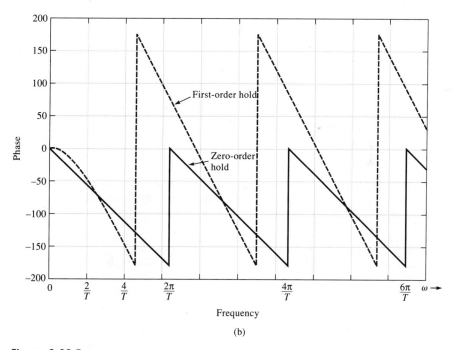

(b)

Figure 2-22 Frequency response plots of the zero- and first-order holds. (a) Magnitude responses of the zero- and first-order holds. (b) Phase responses of the zero- and first-order holds.

$1/T > 2f_B$. **Pass it through a low-pass filter that removes all frequencies in** $g^*(t)$ **above** $\frac{1}{2}f_S$ **and passes, unchanged, all frequencies below** f_B.

A great deal is implied here that is seldom evident at first. Perfect reconstruction requires that an infinite number of samples, dating from $t = -\infty$, be processed. If, instead, we begin the processing at a finite time, $t = 0$, the bandlimited signal will *never* be perfectly reconstructed, although as time goes on the reconstruction can be increasingly accurate. After all, how much can be determined about the spectrum of a signal from only, say, the first three samples? Practical bandlimited signal reconstruction devices have a response that consists of the desired perfect reconstruction plus a significant and usually long-lasting transient error.

2.5 Analysis of Hybrid Systems

In this section, procedures for analyzing hybrid systems containing discrete-time and continuous-time components are developed and demonstrated.

2.5.1 Discrete-Time Equivalents

It often happens that we wish to find the transfer function of a system or subsystem with discrete-time input and discrete-time output but which contains continuous-time components. Figure 2-23(a) shows a general situation in which a discrete-time system or subsystem has an intervening continuous-time transfer function $G(s)$. It is desired to find the overall z-transfer function $H(z)$ of the arrangement, and this can be done by finding its pulse response. For a unit pulse input of

$$f(k) = \delta(k)$$

the sampled-and-held continuous-time signal that is the input to $G(s)$ is given by

$$f^0(t) = u(t) - u(t - T)$$

or

$$F^0(s) = \frac{1 - e^{-sT}}{s}$$

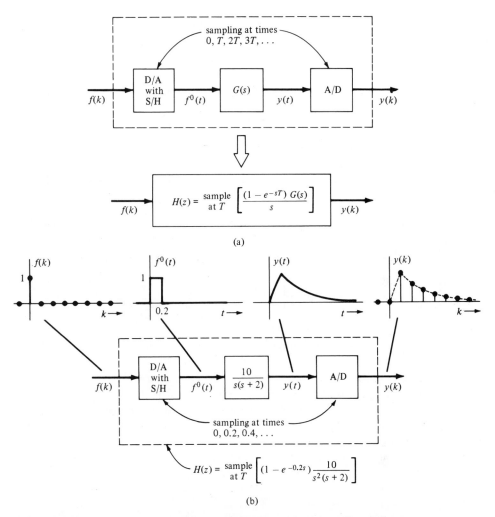

Figure 2-23 Finding the discrete-time equivalent of a system with continuous-time components. (a) General situation. (b) An example.

where T is the sampling interval. Then

$$Y(s) = F^0(s)G(s) = \frac{1 - e^{-sT}}{s} G(s)$$

and

$$H(z) = \frac{\text{sample}}{\text{at } T} \left[Y(s) \right] = \frac{\text{sample}}{\text{at } T} \left[\frac{1 - e^{-sT}}{s} G(s) \right]$$

is found by the usual substitution for each delay and partial-fraction expansion term of $Y(s)$.

For the numerical example of Figure 2-23(b) in which

$$G(s) = \frac{10}{s(s + 2)}$$

and the sampling interval is $T = 0.2$, a unit pulse input produces a continuous-time signal to be converted that has the Laplace transform

$$Y(s) = F^0(s)G(s) = \frac{1 - e^{-0.2s}}{s}\left[\frac{10}{s(s + 2)}\right] = (1 - e^{-0.2s})\left[\frac{10}{s^2(s + 2)}\right]$$

$$= (1 - e^{-0.2s})\left[\frac{-\frac{5}{2}}{s} + \frac{5}{s^2} + \frac{\frac{5}{2}}{s + 2}\right] \qquad (2\text{-}6)$$

Samples of $y(t)$ at the interval $T = 0.2$ have the z-transform

$$Y(z) = H(z) = (1 - z^{-1})\left[\frac{-(\frac{5}{2})z}{z - 1} + \frac{5(0.2)z}{(z - 1)^2} + \frac{(\frac{5}{2})z}{z - e^{-0.4}}\right]$$

$$= \frac{(z - 1)(0.175z + 0.155)}{(z - 1)^2(z - 0.67)} = \frac{0.175z + 0.155}{(z - 1)(z - 0.67)}$$

As another example, consider the closed-loop system shown in Figure 2-24(a) in which the sampling interval is $T = 0.2$ sec. A unit pulse input to the forward transmittance produces a continuous-time signal as shown in Figure 2-24(b). Then

$$Y(s) = (1 - e^{-0.2s})\left[\frac{1}{s(s + 2)}\right]$$

$$= (1 - e^{-0.2s})\left[\frac{k_1}{s} + \frac{k_2}{(s + 2)}\right]$$

$$= (1 - e^{-0.2s})\left[\frac{0.5}{s} + \frac{-0.5}{(s + 2)}\right]$$

Using Table 2-5, samples of the output $y(t)$ of the forward transmittance at the sampling interval $T = 0.2$ have the z-transform

$$Y(z) = H(z) = (1 - z^{-1})\left[\frac{0.5z}{z - 1} + \frac{-0.5z}{z - e^{-0.4}}\right]$$

$$H(z) = \frac{1}{2}\left(1 - \frac{z - 1}{z - e^{-0.4}}\right) = \frac{0.165}{z - 0.67}$$

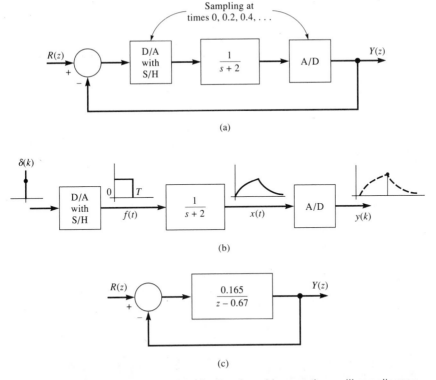

Figure 2-24 Discrete-time equivalent of a closed-loop system with continuous-time components. (a) A closed-loop, sampled data system. (b) Pulse response of the forward transmittance. (c) Equivalent closed-loop discrete-time system.

Hence, the overall discrete transfer function of the closed-loop system is

$$T(z) = \frac{H(z)}{1 + H(z)} = \frac{0.165}{z - 0.505}$$

A similar solution process also applies to situations in which the A/D and D/A converters are not synchronized. For the example with the Laplace transform given by Equation (2-6), suppose that the A/D samples are taken at the times 0.1, 0.3, 0.5, . . . instead of the times 0, 0.2, 0.4, . . . when the D/A samples are taken. Then because

$$y(t) = \left(-\frac{5}{2} + 5t + \frac{5}{2} e^{-2t}\right) u(t)$$

$$- \left[-\frac{5}{2} + 5(t - 0.2) + \frac{5}{2} e^{-2(t-0.2)}\right] u(t - 0.2)$$

and A/D samples are, as a function of step $k = 0, 1, \ldots$

$$y(t = 0.1 + 0.2k) = \left[-\frac{5}{2} + 5(0.1 + 0.2k) + \frac{5}{2} e^{-2(0.1 + 0.2k)} \right]$$

$$- \left[-\frac{5}{2} + 5(0.2k - 0.1) + \frac{5}{2} e^{-2(0.2k - 0.1)} \right] u(0.2k - 0.1)$$

$$= (-2.5 + 0.5 + k + 2.5e^{-0.2}e^{-0.4k})$$
$$- (-2.5 + k - 0.5 + 2.5e^{0.2}e^{-0.4k})u(0.2k - 0.1)$$
$$= (-2 + k + 2.05e^{-0.4k})$$
$$- (-3 + k + 3.05e^{-0.4k})u(0.2k - 0.1)$$
$$= \begin{cases} 0.05 & k = 0 \\ 1 - e^{-0.4k} & k = 1, 2, 3, \ldots \end{cases}$$

The z-transform of these samples, which is the system's z-transfer function is

$$H(z) = Y(z) = \frac{z}{z - 1} - \frac{z}{z - 0.67} - 0.95 = \frac{-0.95z^2 + 1.92z - 0.64}{(z - 1)(z - 0.67)}$$

2.5.2 Analyzing Sampled Data Systems

In general, the signals used in the description of a sampled data system may be a combination of continuous- and discrete-time signals. Another method for determining discrete-time equivalents of sampled data systems, other than the one presented in the previous section, utilizes the concept of impulse sampling presented in Subsection 2.5.1. Figure 2-25(a) shows an impulse sampler followed by a continuous-time transfer function. The continuous-time output of this system is given by

$$Y(s) = R^*(s)G(s)$$

We assume for the sake of analysis that a fictitious sampler is introduced at the output to produce the sampled output $y^*(t)$. Also, unless otherwise stated, all samplers present in a system are assumed to be synchronized and have the same sampling interval.

The Laplace transform of the sampled output $y^*(t)$ is

$$Y^*(s) = [R^*(s)G(s)]^* = R^*(s)G^*(s) \tag{2-7}$$

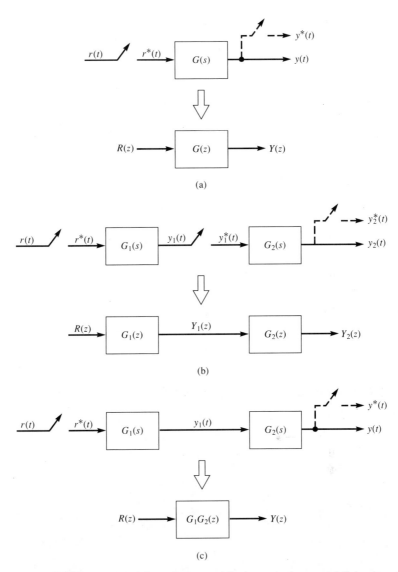

Figure 2-25 Sampled data systems and their equivalence. (a) Pulse transfer function. (b) Transfer functions with a sampler between them. (c) Transfer functions in cascade.

Equation (2-7) is important for analyzing sampled data systems. To prove it, we apply equation (2-4) to the sampled output $y*(t)$ as follows:

$$Y*(s) = \frac{1}{T} \sum_{k=-\infty}^{\infty} Y(s - jk\omega_s)$$

$$= \frac{1}{T} \sum_{k=-\infty}^{\infty} R*(s - jk\omega_s) G(s - jk\omega_s) \tag{2-8}$$

But, we know that

$$R*(s) = \frac{1}{T} \sum_{n=-\infty}^{\infty} R*(s - jn\omega_s)$$

Then

$$R*(s - jk\omega_s) = \frac{1}{T} \sum_{n=-\infty}^{\infty} R(s - jn\omega_s - jk\omega_s)$$

Letting

$$m = n + k$$

then,

$$R*(s - jk\omega_s) = \frac{1}{T} \sum_{m=-\infty}^{\infty} R(s - jm\omega_s)$$

$$= R*(s) \tag{2-9}$$

that is, $R*(s)$ is periodic. Therefore, substituting equation (2-9) in equation (2-8), we arrive at

$$Y*(s) = \frac{1}{T} R*(s) \sum_{k=-\infty}^{\infty} G(s - jk\omega_s)$$

and therefore,

$$Y*(s) = R*(s)G*(s)$$

the stated result.

In z-transform notation, this relationship becomes

$$Y(z) = R(z)G(z) \tag{2-10}$$

Another way of proving equation (2-7) is to convolve the impulse train $r*(t)$ given by equation (2-2) with the impulse response $g(t)$ of the system described by the transfer function $G(s)$. (See Problem 2-25).

Now consider the two transmittances, with a sampler between them, as shown in Figure 2-25(b). The overall transfer function for this system is determined by manipulating the variables as follows:

$$Y_1(s) = R^*(s)G_1(s)$$

then, using equation (2-7)

$$Y_1^*(s) = R^*(s)G_1^*(s) \qquad\qquad \text{(2-11)}$$

Also,

$$Y_2(s) = Y_1^*(s)G_2(s)$$

and, similarly, using equation (2-7) gives

$$Y_2^*(s) = Y_1^*(s)G_2^*(s) \qquad\qquad \text{(2-12)}$$

Therefore, substituting equation (2-11) into equation (2-12) to eliminate $Y_1^*(s)$ yields

$$Y_2^*(s) = R^*(s)G_1^*(s)G_2^*(s)$$

In z-transform terminology, this equation becomes

$$Y_2(z) = R(z)G_1(z)G_2(z)$$

If for example,

$$G_1(s) = \frac{1}{s + 2}$$

and

$$G_2(s) = \frac{1}{s}$$

then for a sampling interval $T = 0.1$ sec, the corresponding z-transforms of $G_1(s)$ and $G_2(s)$ are

$$G_1(z) = \frac{z}{z - e^{-0.2}}$$

$$G_2(z) = \frac{z}{z - 1}$$

respectively. Hence,

$$G_1(z)G_2(z) = \frac{z^2}{(z - 1)(z - 0.818)} \qquad\qquad \text{(2-13)}$$

On the other hand, the discrete-time transfer function of the sampled-data system shown in Figure 2-25(c) is determined using

$$Y^*(s) = [R^*(s)G_1(s)G_2(s)]^*$$
$$= R^*(s)[G_1(s)G_2(s)]^*$$

and, hence,

$$Y(z) = R(z)\mathscr{Z}[G_1(s)G_2(s)]$$

For convenience, we introduce the notation

$$G_1G_2(z) \triangleq \mathscr{Z}[G_1(s)G_2(s)]$$

where the symbol $\triangleq$ means "equals by definition." It is important to point out the fact that

$$G_1(z)G_2(z) \neq G_1G_2(z)$$

That is, the product of two functions of z that correspond to two functions of the complex variable s is *not* equal to the z-transform of the product of the two functions of s.

For example, for the functions discussed in the previous example,

$$G_1(s) = \frac{1}{s + 2}$$

and

$$G_2(s) = \frac{1}{s}$$

$$G_1(s)G_2(s) = \frac{1}{s(s + 2)} = \frac{0.5}{s} + \frac{-0.5}{s + 2}$$

Then for the sampling interval $T = 0.1$ sec,

$$\mathscr{Z}[G_1(s)G_2(s)] = \frac{0.5z}{z - 1} + \frac{-0.5z}{z - e^{-0.2}}$$

$$= \frac{0.096z}{(z - 1)(z - 0.818)}$$

which is different from the result given by equation (2-13).

Consider the closed loop, unity feedback system shown in Figure 2-26(a). The error signal $E(s)$ is

$$E(s) = R(s) - Y(s)$$

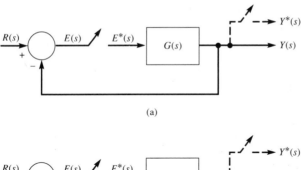

(a)

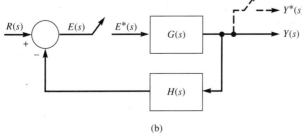

(b)

Figure 2-26 Sampled data, closed-loop systems.
(a) Sampled data, unity feedback arrangement. (b) Non-unity feedback arrangement.

which is the difference between the input and the output of the system. The sampled error signal is easily formed as

$$E^*(s) = R^*(s) - Y^*(s) \tag{2-14}$$

and the output is

$$Y(s) = E^*(s)G(s)$$

Hence, using equation (2-7),

$$Y^*(s) = [E^*(s)G(s)]^*$$
$$= E^*(s)G^*(s) \tag{2-15}$$

Substituting equation (2-14) into equation (2-15) gives

$$Y^*(s) = [R^*(s) - Y^*(s)]G^*(s)$$
$$Y^*(s)[1 + G^*(s)] = R^*(s)G^*(s)$$

Then

$$\frac{Y(z)}{R(z)} = \frac{G(z)}{1 + G(z)}$$

For the nonunity feedback system shown in Figure 2-26(b), the error signal is

$$E(s) = R(s) - H(s)Y(s) \qquad\qquad \textbf{(2-16)}$$

and the output, in terms of the sampled error, is

$$Y(s) = E^*(s)G(s) \qquad\qquad \textbf{(2-17)}$$

or

$$Y^*(s) = E^*(s)G^*(s)$$

substituting equation (2-17) into equation (2-16), then

$$E(s) = R(s) - H(s)E^*(s)G(s)$$

or

$$E^*(s) = R^*(s) - E^*(s)[G(s)H(s)]^*$$

Collecting terms,

$$E^*(s)\{1 + [G(s)H(s)]^*\} = R^*(s)$$

and

$$E^*(s) = \frac{R^*(s)}{1 + [G(s)H(s)]^*} \qquad\qquad \textbf{(2-18)}$$

Multiplying both sides of equation (2-18) by $G^*(s)$, we obtain

$$Y^*(s) = E^*(s)G^*(s)$$

$$= \frac{R^*(s)G^*(s)}{1 + [G(s)H(s)]^*}$$

and therefore,

$$\frac{Y(z)}{R(z)} = \frac{G(z)}{1 + GH(z)}$$

Not all sampled data systems have equivalent discrete-time transfer functions. Consider, for example, the system shown in Figure 2-27. Manipulating the variables

$$Y_1(s) = R(s)G_1(s) \qquad\qquad \textbf{(2-19)}$$

$$Y(s) = Y_1^*(s)G_2(s)$$

Hence,

$$Y^*(s) = Y_1^*(s)G_2^*(s) \qquad\qquad \textbf{(2-20)}$$

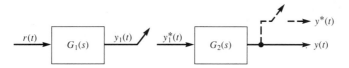

Figure 2-27 An example where an overall transfer function does not exist.

Substituting equation (2-19) into equation (2-20) results in

$$Y^*(s) = [R(s)G_1(s)]^*G_2^*(s)$$

and the input $R(s)$ cannot be separated from $G_1(s)$ to form the transfer function. However, the output of the system can be formed by first multiplying $R(s)$ by $G_1(s)$, then transforming the product to the z-domain, and finally multiplying the result by the z-transform of the transfer function $G_2(s)$.

The rules developed in this subsection may now be used to analyze digital systems containing A/D and D/A converters. Every A/D converter is replaced with an impulse sampler and every D/A converter is replaced with an impulse-train generator followed by a zero-order hold as shown in Figure 2-12. For simplicity, the block termed impulse-train generator need not be used but should be implied.

For example, the system shown in Figure 2-28(a) has an equivalent

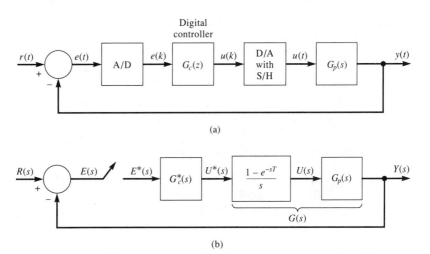

(a)

(b)

Figure 2-28 A digital control system and its equivalence. (a) Digital control system. (b) Equivalent configuration.

configuration shown in Figure 2-28(b), then

$$G(s) = \frac{1 - e^{-sT}}{s} G_p(s)$$

$$Y(s) = U^*(s)G(s)$$
$$= E^*(s)G_c^*(s)G(s)$$

or, using equation (2-7),

$$Y^*(s) = [E^*(s)G_c^*(s)G(s)]^*$$
$$= E^*(s)G_c^*(s)G^*(s)$$

But

$$E^*(s) = [R(s) - Y(s)]^*$$
$$= R^*(s) - Y^*(s)$$

Hence,

$$Y^*(s) = [R^*(s) - Y^*(s)]G_c^*(s)G^*(s)$$

Collecting like terms and simplifying then,

$$\frac{Y^*(s)}{R^*(s)} = \frac{G_c^*(s)G^*(s)}{1 + G_c^*(s)G^*(s)}$$

In the z-domain, this equation becomes

$$\frac{Y(z)}{R(z)} = \frac{G_c(z)G(z)}{1 + G_c(z)G(z)}$$

giving a closed-loop transfer function of the system in terms of the compensator transfer function and the plant's transfer function.

2.6 Design of a Videotape Drive Control

In this section we present an overview of classical discrete-time control system design using an example. There are two basic ways of approaching classical discrete-time control. In the sampled data approach, discrete-time signals are represented by continuous-time impulse trains so that all signals in a plant and controller model are continuous-time ones. This was appealing in the early days of digital control when digital concepts were new and most designers had backgrounds that were solidly in continuous-time control. There is little to recommend this complexity nowadays, though.‡ In the conventional approach, which is used here, discrete-time signals are represented as sequences.

Similar to classical continuous-time control system design, acceptable discrete-time, zero-input response is obtained by choosing a controller that results in acceptable feedback system pole locations in the z-plane. At the same time, requirements are placed on the overall system's zero-state response component for representative discrete-time input signals, such as steps or ramps. Any resulting continuous-time signals of interest are then examined and if their between-sample response is not acceptable, the design is modified, perhaps by increasing the sampling rate.

2.6.1 Plant Model and Controller Structure

The configuration of a commercial broadcast videotape-positioning system is shown in Figure 2-29(a). The relationship between the applied voltage for the drive motor armature and the tape speed at the recording and playback heads is approximated by the transfer function $G(s)$. The delay term accounts for the propagation of speed changes along the tape over the distance of physical separation of the tape drive mechanism and the recording and playback heads. The pole term in $G(s)$ represents the dynamics of the motor and tape drive capstan. Tape position is sensed by a recorded signal on the tape itself.

The digital controller that is to be designed should result in zero steady state error to any step change in desired tape position. It should have a zero-input (or transient) response that decays to no more than 10% of any initial value within a 1/30-sec interval. One-thirtieth of a second is the video frame rate.

The sampling interval of the controller is chosen to be $T = 1/120$ based on the desirability of synchronizing the tape motion control with the 1/60-sec field rate (each frame consists of two fields) of the recorded video. In Figure 2-29(b), the diagram of Fig. 2-29(a) has been rearranged to emphasize the discrete-time input $R(z)$ and the discrete-time samples $P(z)$ of the tape position. The indicated open-loop z-transfer

---

‡Among the difficulties is the possibility (indeed the likelihood) of having to deal with impulse trains derived from signals containing steps and impulses that occur at the sampling times. Offsetting sampler timing gives a "fix" of the problem, but the result is not very satisfying, particularly when feedback loops are involved.

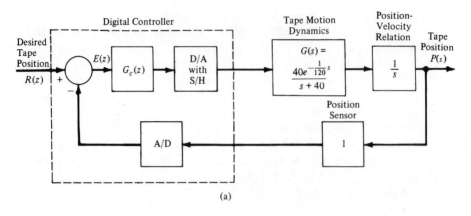

(a)

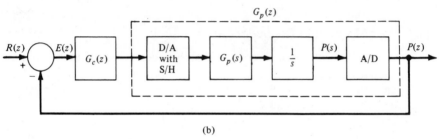

(b)

Figure 2-29 Videotape-positioning system. (a) Block diagram. (b) Relation between discrete-time signals.

function is

$$G_p(z) = \begin{array}{c} \text{sample at} \\ T = 1/120 \end{array} \left[\left(\frac{1 - e^{-(1/120)s}}{s} \right) \left(\frac{40e^{-(1/120)s}}{s + 40} \right) \left(\frac{1}{s} \right) \right]$$

$$= \begin{array}{c} \text{sample at} \\ T = 1/120 \end{array} \left\{ [1 - e^{-(1/120)s}] e^{-(1/120)s} \left[\frac{-(1/40)}{s} + \frac{1}{s^2} + \frac{1/40}{s + 40} \right] \right\}$$

$$= (1 - z^{-1})z^{-1} \left[\frac{-z/40}{z - 1} + \frac{z/120}{(z - 1)^2} + \frac{z/40}{z - 0.72} \right]$$

$$= \frac{0.00133(z + 0.75)}{z(z - 1)(z - 0.72)} \tag{2-21}$$

In terms of the compensator's z-transfer function $G_c(z)$, the position error signal is given by

$$E(z) = R(z) - Y(z)$$

$$= \left[1 - \frac{G_c(z)G_p(z)}{1 + G_c(z)G_p(z)} \right] R(z) = \frac{1}{1 + G_c(z)G_p(z)} R(z)$$

For a unit step input sequence

$$E(z) = \frac{1}{1 + G_c(z)G_p(z)} \left(\frac{z}{z - 1}\right)$$

assuming that the feedback system is stable

$$\lim_{k \to \infty} e(k) = \lim_{z \to 1} \left[\frac{z - 1}{z} E(z)\right] = \lim_{z \to 1} \left[\frac{1}{1 + G_c(z)G_p(z)}\right]$$

In view of the pole at $z = 1$ in $G_p(z)$, equation (2-21), the steady state error to a step input is zero, provided that the compensator does not have a zero at $z = 1$.

For the feedback system transient response to decay at least by a factor of 1/10 within 1/30 sec, the closed-loop poles must be located such that a decay of at least this amount occurs every four 1/120-sec steps. This is to say that all closed-loop poles must be within a radius c of the origin on the complex plane, where

$$c^4 = \frac{1}{10} \qquad c = 0.56$$

2.6.2 Root Locus Compensator Design

A compensator consisting of only a gain K is inadequate, as is shown in Figure 2-30. The feedback system is stable for

$$0 < K < 95$$

but, regardless of K, there are always poles at distances from the origin greater than the required $c = 0.56$. For negative K, the feedback system is unstable.

In Figure 2-31, the controller with the z-transfer function

$$G_c(z) = \frac{K(z - 0.72)}{z}$$

which cancels the plant pole at $z = 0.72$, is shown. As before, the feedback system is unstable for negative values of the adjustable constant K. For positive K, the root locus is as shown in Figure 2-31(b). The closed-loop system is stable for

$$0 < K < 370$$

but there is always a pole or two at radial distance greater than $c = 0.56$. For $K = 90$, this design is close to meeting the requirements, but it is not quite good enough.

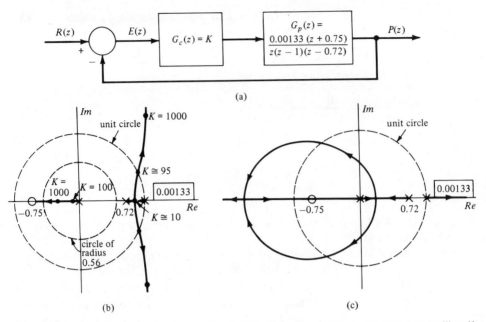

(a)

(b)   (c)

Figure 2-30 Constant-gain compensator. (a) Block diagram. (b) Root locus for positive K. (c) Root locus for negative K.

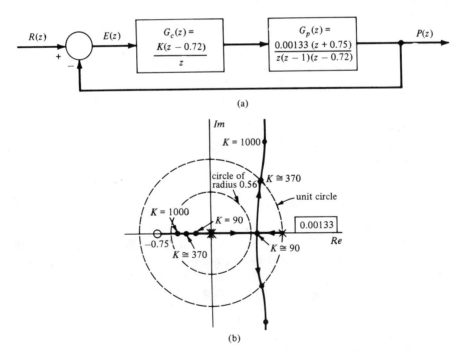

(a)

(b)

Figure 2-31 Compensator with zero at $z = 0.72$ and pole at $z = 0$. (a) Block diagram. (b) Root locus for positive K.

When the compensator pole is moved from the origin to the left, as shown in Figure 2-32, the root locus is pulled to the left and the performance requirements can be met. For the compensator with the z-transfer function

$$G_c(z) = \frac{150(z - 0.72)}{z + 0.4} \tag{2-22}$$

the feedback system z-transfer function is

$$T(z) = \frac{G_c(z)G_p(z)}{1 + G_c(z)G_p(z)} = \frac{0.2(z + 0.75)}{z^3 - 0.6z^2 - 0.2z + 0.15}$$

$$= \frac{0.2(z + 0.75)}{(z - 0.539 - j0.155)(z - 0.539 + j0.155)(z + 0.477)}$$

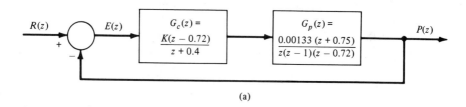

(a)

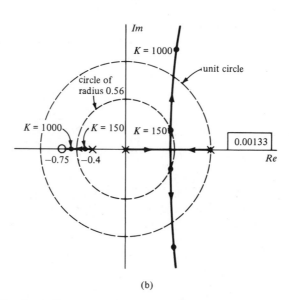

(b)

Figure 2-32 Compensator with zero at $z = 0.72$ and pole at $z = -0.4$. (a) Block diagram. (b) Root locus for positive K.

As expected, the steady state error to a step input sequence is zero

$$\lim_{z \to 1} \left\{ \left(\frac{z-1}{z} \right) [1 - T(z)] \left(\frac{z}{z-1} \right) \right\} = \lim_{z \to 1} \frac{z^3 - 0.6z^2 - 0.4z}{z^3 - 0.6z^2 - 0.2z + 0.15}$$

$$= 0$$

The steady state error to a unit ramp input sequence is

$$\lim_{z \to 1} \left\{ \left(\frac{z-1}{z} \right) [1 - T(z)] \left[\frac{Tz}{(z-1)^2} \right] \right\} = \lim_{z \to 1} \frac{\dfrac{1}{120}(z^2 + 0.4z)}{z^3 - 0.6z^2 - 0.2z + 0.15}$$

$$= \frac{1}{30}$$

For a compensator with a z-transfer function of the form

$$G_c(z) = \frac{150(z - 0.72)}{z + a}$$

the feedback system has the z-transfer function

$$T(z) = \frac{G_c(z)G_p(z)}{1 + G_c(z)G_p(z)} = \frac{0.2(z + 0.75)}{(z + a)(z^2 - z) + 0.2(z + 0.75)}$$

$$= \frac{0.2(z + 0.75)}{z^3 - z^2 + 0.2z + 0.15 + a(z^2 - z)}$$

$$= \frac{0.2(z + 0.75)/[z^3 - z^2 + 0.2z + 0.15]}{1 + az(z - 1)/[z^3 - z^2 + 0.2z + 0.15]}$$

$$= \frac{\text{numerator}}{1 + az(z - 1)/[(z - 0.637 - j0.378)}$$
$$(z - 0.637 + j0.378)(z + 0.274)]$$

A pole-zero plot in terms of positive a is shown in Figure 2-33, from which it is seen that choices of a between 0.4 and about 0.5 give a controller that meets the performance requirements.

Classical pole-zero, discrete-time control system design is much the same as its continuous-time counterpart. Increasingly complicated controllers are tried until both steady state error and transient performance requirements are met. In this iterative process, root locus is an important tool because it easily indicates qualitative closed-loop system pole locations as a function of a parameter. Once feasible control-

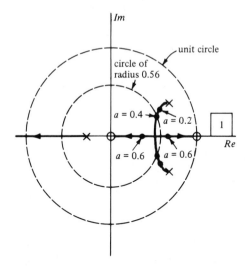

Figure 2-33 Root locus plot as a function of the compensator pole location.

lers are selected, root locus plots are refined to show quantitative results.

2.6.3 Simulation Studies

Another important system design tool is *simulation,* computer modeling of the plant and controller to verify the properties of a preliminary design and to test its performance under conditions (e.g., noise, disturbances, parameter variations, and nonlinearities) that might be difficult or cumbersome to study analytically.

When a digital controller is used to control a continuous-time plant, proper discrete-time design ensures that *samples* of continuous-time plant response behave acceptably, as is the case with this example. However, it is often additionally important that continuous-time plant signals be well behaved between the sampling times also. A simulation of the unit step response of the videotape-positioning system with a controller described by equation (2-22) is shown in Figure 2-34. The tape position $y(t)$ is seen to vary smoothly between sampling times, so the between-sample response of this design is probably quite acceptable. In Chapter 7, we will develop and apply methods for im-

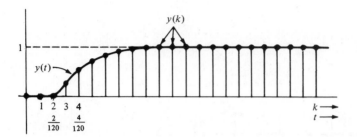

Figure 2-34 Continuous-time response of the videotape drive control.

proving the between-sample response of digitally controlled continu-ous-time plants.

2.7 Summary

A discrete-time signal or sequence is a function of a discrete variable k, termed the *step index*. The z-transform of a sequence $f(k)$ is

$$\mathcal{Z}[f(k)] = F(z) = \sum_{k=0}^{\infty} f(k)z^{-k} = f(0)z^{-0} + f(1)z^{-1} + f(2)z^{-2} + \cdots$$

Important z-transform pairs are listed in Table 2-1 and properties of the z-transform are summarized in Table 2-2. For rational $F(z)$, sequence samples can be obtained from $F(z)$ by long division. The sequence as a function of step is found by expanding $F(z)/z$ in partial fractions and using Table 2-1. Rational z-transforms with denominator roots inside the unit circle on the complex plane represent sequences that decay with step. If there are nonrepeated denominator roots on the unit cir-cle, the corresponding sequence neither decays nor expands with step. If there is a repeated unit circle root or any root outside the unit circle, the sequence is an expanding one.

Linear discrete-time systems are described by linear difference equations. If the difference equation coefficients are constant, the sys-tem is step-invariant. Given the system initial conditions and inputs, a difference equation can be applied repeatedly to calculate the system outputs recursively. Any signal in a linear discrete-time system can be expressed as the sum of a zero-input component that is due to the initial conditions but not the inputs, plus a zero-state component that is due to the inputs but not the initial conditions. The part of the zero-state response due to each input is the discrete convolution of the input and the unit pulse response for that input.

The z-transfer function of a single-input/single-output system is the z-transform of its unit pulse response when the initial conditions are zero. If there are multiple inputs and outputs, the system has a z-transfer function for each combination of input and output, each calculated with all other inputs zero. Z-transfer functions for discrete-time systems are manipulated in the same way as continuous-time system transfer functions. A linear, step-invariant, discrete-time system is input–output stable if and only if its unit pulse responses decay asymptotically to zero, which occurs only when all denominator roots (poles) of all its z-transfer functions lie inside the unit circle on the complex plane.

The Jury test is a numerical procedure for testing the location of the roots on the complex plane without solving for the roots explicitly.

The bilinear transformation

$$z = \frac{1 + w}{1 - w} \qquad w = \frac{z - 1}{z + 1}$$

maps the inside of the unit circle on the complex plane to the left half-plane. When applied to a z-transfer function, it converts a discrete-time stability problem to an equivalent continuous-time one. Like its continuous-time counterpart, discrete-frequency response is useful for the experimental determination of z-transfer functions.

Discrete-time sequences are often derived from continuous-time functions by sampling. Devices that perform sampling are called *analog-to-digital* (A/D) converters. For evenly spaced samples at sampling interval T

$$t = kT \quad k = 0, 1, 2, \ldots$$

the Laplace transform $F(s)$ of a continuous-time function and the z-transform $F(z)$ of its sequence of samples are related by

$$F(z) = \frac{\text{sample}}{\text{at } T} \left[F(s) \right]$$

where the "sample at T" operation involves expanding $F(s)$ in partial fractions and replacing each term by the corresponding z-transform term of Table 2-5. Each time delay of interval T is replaced by z^{-1}.

Reconstruction is the process of forming a continuous-time function from samples. Because many different continuous-time functions can have the same samples, reconstruction is not unique. The most common reconstruction device is the digital-to-analog (D/A) converter with sample-and-hold (S/H). It produces a stepwise-constant reconstruction that can be considered to be the production of a train of impulses $f^*(t)$ with strengths equal to the samples followed by a zero-order hold with the transfer function

$$L_0(s) = \frac{1 - e^{-sT}}{s}$$

The sampling theorem states that when a bandlimited continuous-time signal is sampled at a rate higher than twice the bandlimit frequency, the samples can be used to reconstruct uniquely the original continuous-time signal. Although the sampling theorem is not applicable to most discrete-time control systems because the signals (e.g., steps and ramps) are not bandlimited and because good reconstruction requires long time delays, it does provide some guidance in deciding how best to filter sensor signals before sampling them.

A system or subsystem with discrete-time input and output but with intervening D/A conversion with S/H, a transfer function $G(s)$, then A/D conversion, has the z-transfer function

$$H(z) = \frac{\text{sample}}{\text{at } T} \left[\frac{1 - e^{-sT}}{s} G(s) \right]$$

Finally, block diagram manipulations of sampled data systems containing discrete- and continuous-time components are carried out using the important relationship given by equation (2-7). This relationship is very useful for the simplification of system models involving both discrete-time and continuous-time elements.

The classical approach to designing a digital controller directly, which has many variations, parallels the classical approach to analog controller design. We begin with simple discrete-time controllers, increasing their complexity until the performance requirements can be met. The controller poles, zeros, and multiplying constant are selected to give feedback system pole locations that result in acceptable zero-input response. At the same time, the parameters are constrained so that the resulting system has adequate zero-state (or ''steady state'') response to important inputs, such as steps and ramps. We illustrated the classical approach with the design of a commercial videotape drive unit.

References

The period from about 1955 to 1965 was an especially exciting and productive one for the advancement of control system theory. At Columbia University, John Ragazzini and his students, including Eli Jury, Gene Franklin, Jack Sklansky, John Bertram, Bernard Friedland, Lotfi Zadeh, and Rudolf Kalman, led the development of digital control methods that paralleled those of classical continuous-time control. (Bertram and Kalman also contributed heavily to the application of state space concepts, and Kalman made fundamental advances in the design of systems for stochastic estimation and control.) Many others, at leading universities and in industry, were also highly involved, of course. Much of the original theory of sampled data control was given in the following classical texts:

J. R. Ragazzini and G. F. Franklin, *Sampled-Data Control Systems*. New York: McGraw-Hill, 1958;

E. I. Jury, *Sampled-Data Control Systems*. New York: Wiley, 1958.

A comprehensive treatment of classical discrete-time control system design is also given in

B. C. Kuo, *Analysis and Synthesis of Sampled-Data Control Systems*. Englewood Cliffs, NJ: Prentice-Hall, 1963.

Conventional classical digital control is covered in depth in such texts as

B. C. Kuo, *Digital Control Systems,* 2nd edition. Philadelphia: Saunders College Publishing, 1992;

and

C. L. Phillips and H. T. Nagle, Jr., *Digital Control System Analysis and Design*. Englewood Cliffs, NJ: Prentice-Hall, 1990;

K. Ogata, *Discrete-Time Control Systems*. Englewood Cliffs, NJ: Prentice-Hall, 1987.

A delightful mixture of classic and some modern digital control system design is found in

G. F. Franklin, J. D. Powell, and M. L. Workman, *Digital Control of Dynamic Systems,* 2nd edition. Reading, MA: Addison-Wesley, 1990.

Chapter Two Problems

2-1. Sketch the following sequences for $k = 0, 1, 2, \ldots$

    a. $f_1(k) = (0.5)^k$

    b. $f_2(k) = 2\delta(k - 1) + e^{-0.2k}[u(k - 5) - u(k - 8)]$

    c. $f_3(k) = 3 \cos 0.4k + 4 \sin 0.4k$

2-2. A bank account pays 6% annual interest, compounded monthly. Initially, a deposit of $2000 is made. Thereafter, $50 is deposited into the account each month. Describe the monthly bank balance as a function of the month k after the initial deposit.

2-3. Find more compact mathematical expressions for the following sequences for $k \geq 0$:

$$
\textbf{a. } f_1(k) = \begin{cases} e^{-k} \cos 3k & k = 0,\ 1,\ 2,\ 3 \\ 0 & \text{otherwise} \end{cases}
$$

$$
\textbf{b. } f_2(k) = \begin{cases} -6 & k \text{ odd} \\ 0 & \text{otherwise} \end{cases}
$$

$$
\textbf{c. } f_3(k) = \begin{cases} 5k & k \text{ odd} \\ -5k & k \text{ even} \end{cases}
$$

2-4. Find the z-transforms of the following sequences:

a. $f_1(k) = (0.2)^k - 3(0.1)^k$

b. $f_2(k) = 4e^{-k} - 3e^{-2k}$

$$
\textbf{c. } f_3(k) = \begin{cases} (-1)^k & k = 4,\ 5,\ 6,\ \ldots \\ 0 & \text{otherwise} \end{cases}
$$

d. $f_4(k) = \cos[2k + (\pi/6)]$

$$
\textbf{e. } f_5(k) = \begin{cases} 1 & k = 1 \\ -3 & k = 4 \\ 6 & k = 7 \\ 0 & \text{otherwise} \end{cases}
$$

f. $f_6(k) = k^3 - 6k^2 + 2k + 4$

$$
\textbf{g. } f_7(k) = \begin{cases} 4e^{-2k} & k = 1,\ 2,\ 3,\ 4 \\ 0 & \text{otherwise} \end{cases}
$$

h. $f_8(k) = ke^{-0.2k} \sin 2k$

i. $f_9(k) = \dfrac{1}{k!}$

2-5. Use long division to show that

$$
\textbf{a. } \mathscr{L}^{-1}\left[\frac{2z^2}{z^2 - \frac{1}{4}}\right] = \left(\frac{1}{2}\right)^k + \left(-\frac{1}{2}\right)^k \quad k = 0,\ 1,\ 2,\ 3,\ \ldots
$$

$$
\textbf{b. } \mathscr{L}^{-1}\left[\frac{8z}{(z - 1)^2}\right] = 8k \quad k = 0,\ 1,\ 2,\ \ldots
$$

$$
\textbf{c. } \mathscr{L}^{-1}\left[\frac{z(z - \frac{1}{2})}{z^2 - z + 1}\right] = \cos\left(\frac{1}{2}\right)k \quad k = 0,\ 1,\ 2,\ 3,\ \ldots
$$

2-6. Find closed form expressions for $k \geq 0$ for the inverse z-transforms of the following:

$$
\textbf{a. } F_1(z) = \frac{2z^2 - 3}{z^2 - 7z + 12}
$$

b. $F_2(z) = \dfrac{10}{z^3 - 7z^2 + 12z}$

c. $F_3(z) = \dfrac{4z^2 - 3z}{z^2 + z + \frac{1}{4}}$

d. $F_4(z) = \dfrac{4z^2 - 3z + 2}{z^2 + z + \frac{1}{4}}$

e. $F_5(z) = \dfrac{2z^2 - 3z}{z^2 + z + \frac{5}{4}}$

f. $F_6(z) = \dfrac{-6z + 1}{z^{50}(z^2 + 3z + 2)}$

g. $F_7(z) = \dfrac{3z^3}{(z^2 + \frac{1}{4})(z - \frac{1}{2})}$

h. $F_8(z) = \dfrac{z^2 + 4z - 5}{(z - \frac{1}{2} + j\frac{1}{4})(z - \frac{1}{2} - j\frac{1}{4})(z - \frac{1}{2})}$

i. $F_9(z) = \dfrac{4z(z - 1)(z + 1)}{(z^2 + 1)^2}$

2-7. Another method of determining the inverse z-transform of a rational function of z that is well suited to digital computation is to construct a difference equation from the rational function and then solve the difference equation recursively. Using this procedure, find the inverse z-transform of the following:

a. $Y_1(z) = \dfrac{5z^2 + 4z}{z^2 + \frac{3}{2}z + \frac{1}{2}}$

b. $Y_2(z) = \dfrac{3z + 2}{z^2 + 2z + 1}$

c. $Y_3(z) = \dfrac{2z + 6}{z^2 + 2z + 2}$

d. $Y_4(z) = \dfrac{z - 1}{z^2 - 5z + 4}$

e. $Y_5(z) = \dfrac{z^2 + 3z + 2}{z^3}$

2-8. Show that

a. $\mathscr{L}\left[\dfrac{f(k)}{k}\right] = \displaystyle\int_z^\infty \dfrac{F(z)}{z}\, dz$

b. $\mathscr{L}\left[\dfrac{f(k)}{k + n}\right] = z^n \displaystyle\int_z^\infty \dfrac{F(z)}{z^{n+1}}\, dz$

2-9. Find the z-transform of the following:

a. $f_1(k) = \dfrac{1}{k + 1}$

b. $f_2(k) = \dfrac{c^k}{k + 1}$

2-10. For the difference equation

$$y(k + 2) + 4y(k + 1) - y(k) = r(k + 1) - 3r(k)$$

if

$y(-1) = -2$ and $y(-2) = 1$

and $r(k) = 2$, for all k,

use the difference equation to calculate $y(0)$, $y(1)$, $y(2)$, and $y(3)$.

2-11. Use z-transformation to find the solutions of the following difference equations with the given initial conditions and inputs:

a. $y(k + 2) = \left(\dfrac{1}{3}\right) y(k + 1) + \left(\dfrac{1}{3}\right) y(k) + r(k + 1) - r(k)$

$y(-1) = 0 \qquad y(-2) = -1$

$r(k) = \left[3 + 4 \left(\dfrac{1}{4}\right)^k\right] u(k)$

b. $y(k + 3) - 2y(k + 2) - 2y(k + 1) = 2r(k + 2) - r(k)$
$y(-1) = 1 \qquad y(-2) = -2 \qquad y(-3) = 0$
$r(k) = 3\delta(k - 1)$

c. $y(k + 2) = -y(k + 1) - \left(\dfrac{1}{2}\right) y(k) + 4r(k + 2) - 3r(k + 1)$

$y(-1) = 2 \qquad y(-2) = 0$

$r(k) = \left(\dfrac{1}{2}\right)^k u(k)$

d. $ky(k + 1) + y(k + 1) = y(k)$
$y(0) = 1$

2-12. Use discrete convolution to find
$\mathscr{Z}^{-1}[F_1(z)F_2(z)]$
by convolving

$f_1(k) = \left(\dfrac{1}{2}\right)^k u(k) \qquad$ and $\qquad f_2(k) = 2[u(k) - u(k - 2)]$

2-13. Find the z-transfer functions of the following discrete-time systems. Express as rational functions of z.

a. $y(k + 2) + 2y(k + 1) - 4y(k) = 8r(k + 1) - 6r(k)$

b. $y(k + 2) - y(k + 1) + y(k) = r(k + 3) - r(k + 2) + 4r(k + 1) + r(k)$

c. $y(k + 3) = r(k + 3) - 2r(k + 2) + 4r(k + 1) - 2r(k)$

2-14. For discrete-time systems with the following z-transfer functions and input sequences, find the output sequences for $k \geq 0$ if the initial conditions are zero:

a. $T(z) = \dfrac{10}{z^2 - \frac{1}{4}z + (2/64)}$

$r(k) = 4u(k)$

b. $T(z) = \dfrac{z^2 + 2}{(z - 1)^2}$

$r(k) = \left(\dfrac{1}{2}\right)^k u(k)$

c. $T(z) = \dfrac{z}{z - \frac{1}{3}}$

$r(k) = \sin(k)u(k)$

2-15. Find the sequences $y(k)$ if all initial conditions are zero:

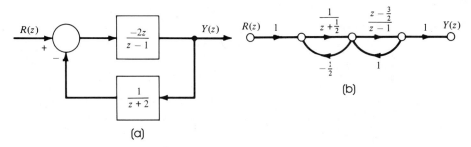

(a)

(b)

a. $r(k) = \delta(k)$, the unit pulse

b. $r(k) = 6$

2-16. For the discrete-time system with z-transfer function

$$T(z) = \dfrac{z^2 + 1}{z^2 - \frac{1}{4}}$$

input sequence

$$r(k) = \left(\dfrac{1}{4}\right)^k u(k)$$

for all k (including negative k), and initial conditions

$$y(0) = 2$$
$$y(1) = -6$$

find $y(k)$.

2-17. Determine whether each of the following discrete-time systems is stable:

a. $T(z) = \dfrac{z^3 + 4z^2 - 9z + 1}{z^3}$

b. $y(k + 2) = 3y(k + 1) - 2y(k) + 10r(k + 2) - 3r(k + 1) + 4r(k)$

c.

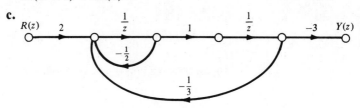

2-18. Use the Jury test to determine whether the systems with the following transfer functions are stable:

a. $T_1(z) = \dfrac{z + 1}{z^2 - \frac{3}{4}z + \frac{1}{8}}$

b. $T_2(z) = \dfrac{z - 2}{z^3 - \frac{5}{6}z^2 - \frac{1}{3}z + \frac{1}{6}}$

c. $T_3(z) = \dfrac{z + \frac{1}{3}}{z^3 - 2z^2 - \frac{1}{4}z + \frac{1}{2}}$

2-19. Use the bilinear transformations and a Routh-Hurwitz test to determine whether the systems with the following transfer functions are stable:

a. $T_1(z) = \dfrac{3z - 2}{z^2 + 0.5z + 0.9}$

b. $T_2(z) = \dfrac{1.2z^3 - 0.6z^2}{z^3 - 0.8z^2 + 1.2z - 1}$

c. $T_3(z) = \dfrac{2z^4 - 3z^3 + z^2 - 3z + 4}{z^4 + 0.6z^3 - 0.7z^2 + 0.8z - 0.9}$

2-20. Find and carefully sketch the frequency response of the following systems:

a. $T(z) = 1 + z^{-1}$

b. $y(k + 1) + \dfrac{1}{4} y(k) = r(k + 1)$

c. $T(z) = \dfrac{2}{z(z - 1)}$

d.

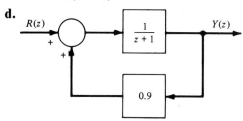

e. $T(z) = 1 + 3z^{-1} + 4.5z^{-2} + 4z^{-3} + 2.25z^{-4} + 0.75z^{-5} + 0.125z^{-6}$

2-21. For continuous-time signals with the following Laplace transforms, find the sequences of samples $f(kT)$, $k = 0, 1, 2, \ldots$ with the given sampling interval T.

a. $F_1(s) = \dfrac{3s - 2}{s^2 + 5s + 6}$ $T = 0.3$

b. $F_2(s) = \dfrac{15}{s^3 + 6s^2 + 9s}$ $T = 0.1$

c. $F_3(s) = \dfrac{2s - 6e^{-0.4s}}{(s + 2)(s^2 + 9)}$ $T = 0.2$

d. $F_4(s) = \dfrac{2 - e^{-0.2s}}{(s + 1)(s + 2)}$ $T = 0.3$

e. $F_5(s) = \dfrac{3 + se^{-0.3s}}{s(s + 2)}$ $T = 0.2$

2-22. Find the overall z-transfer functions. The sampling interval for each system is $T = 0.2$ sec.

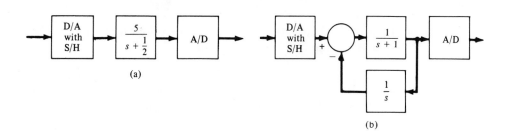

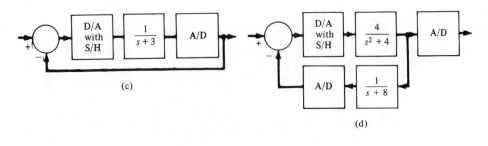

(c)

(d)

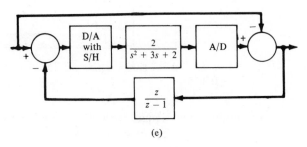

(e)

2-23. The linear point connector is a D/A reconstruction device that, for input samples, generates an analog waveform consisting of straight line segments connecting the samples. Show that the transfer function of the linear point connector is

$$G(s) = \frac{(1 - e^{sT})^2}{Ts^2}$$

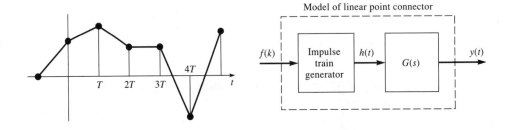

Model of linear point connector

2-24. Find the response $y(k)$ for each of the following systems. The sampling interval for each system is $T = 0.1$ sec.

(a)

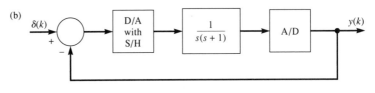

(b)

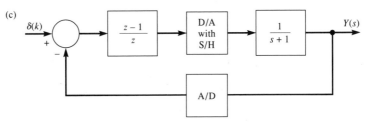

(c)

2-25. Prove the relationship in equation (2-7) using convolution.

2-26. Find the transfer function for each of the following systems. The sampling interval for each system is $T = 0.2$ sec.

  a. The system in Figure 2-25(a) with

  $$G(s) = \frac{1}{s + 1}$$

  b. The system in Figure 2-25(a) with

  $$G(s) = \left(\frac{1 - e^{-sT}}{s}\right)\left(\frac{1}{s + 1}\right)$$

  c. The system in Figure 2-25(b) with

  $$G_1(s) = \frac{1}{s + 1}$$

  $$G_2(s) = \frac{1}{s + 3}$$

  d. The system in Figure 2-25(c) with

  $$G_1(s) = \frac{1}{s + 1}$$

  $$G_2(s) = \frac{1}{s + 3}$$

2-27. Find the transfer function of the following system. The sampling interval is $T = 0.4$ sec.

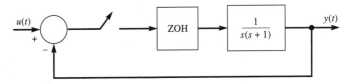

2-28. Find the transfer function for each of the following systems. Assume all samplers are synchronized.

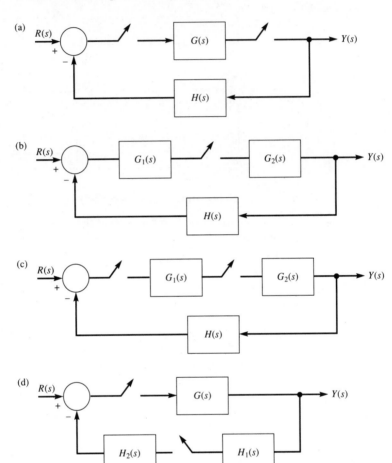

State Space Description of Dynamic Systems

3.1 Preview

The extension of the classical control methods to the control of complicated feedback structures involving many loops and many inputs and outputs is not easy. Put another way, we now want to design controllers having multiple inputs and multiple outputs. Design *is* iterative, and it involves trial and error. But when there are many design variables, it is important to deal efficiently with those design decisions that need not be iterative. The powerful methods of state variables offer insights about what is possible and what is not. State variable (or *state space*) descriptions of systems provide an important means of efficiently describing and dealing with complex systems. They give standard representations that offer economy of notation, direct applicability of matrix algebra, and ease of entry for computer-aided design. And they provide an excellent framework for general methods of approaching and accomplishing our objectives.

Most of the design methods presented in this book are carried out using state variable formulation. Unlike classical control methods, which are applicable only to linear time-invariant systems, state variable methods are also well suited as extensions to time-varying, adaptive, and nonlinear systems.

We now make the transition from classical system description and design to state variable methods. Linear, time-invariant, continuous-time systems are introduced through synthesis of systems with specified transfer functions. Once the form of the equations is established, system response is described in terms of the state transition matrix and convolution, and the matrix of transfer functions describing multiple-input/multiple-output systems is found in terms of the state equations.

Linear, time-varying, continuous-time systems are then briefly discussed, and the solution for the state is presented. As our main objective is digital control, the emphasis is on those concepts with direct bearing on digital control system design. Much more can be said about the use of state variables in continuous-time control; when the topics here have been carried over to discrete-time systems, it will be apparent how similar methods also apply to strictly continuous-time control.

Next, state variable models are developed for discrete-time systems, paralleling then extending considerably the analogous topics for continuous-time systems. The form of the equations is established, then system response is expressed in terms of discrete convolution and in terms of z-transforms. Z-transfer function matrices of multiple-input/multiple-output systems are found in terms of the state equations. State equations and the response of step-varying systems are then discussed. Next, the relationship between continuous-time state variable plant models and discrete-time models of plant signal samples are examined.

The controllable and observable forms of the state equations are discussed. Using them, it is especially easy to synthesize systems having desired z-transfer functions. The characteristic value problem, which is considered in some detail and with a number of illustrative examples in Appendix A, is fundamental to many topics to follow. A nonsingular transformation of state variables changes the internal description of a system but leaves its input–output relations unchanged. The change of variables involved in the characteristic value problem decouples the state equations from one another, simplifying their solution and giving important insight into system structure and behavior. Transformations to block real number representations in the case of complex eigenvalues, and transformations to block Jordan form in the case of repeated eigenvalues are also developed.

There is the possibility that certain systems cannot be completely controlled from their inputs, or it may be impossible to determine their state completely from the available outputs. These topics, controllability and observability, which pertain to the question of whether or not a given system *can* be controlled, are developed toward the end of this chapter.

3.2 Continuous-Time State Equations

In this section, we review the fundamentals of state variables for linear, time-invariant, continuous-time systems with an eye toward carrying these over to the discrete-time case in Section 3.3 and later to using these results in designing digital controllers for discrete- or continuous-time systems. Before we proceed, it is important to mention that the material in Appendix A is fundamental to many topics to follow and should be reviewed right away.

3.2.1 Integration Diagrams and State Equations

Continuous-time, state variable descriptions are mathematical models of systems in terms of coupled first-order differential equations. The signals involved are the system inputs, the system outputs, and internal signals. The internal signals are the state variables. For linear, time-invariant, continuous-time systems, state variable models involve the operations of summation, multiplication by a constant, and integration. A block diagram, or signal flow graph, representing a state variable model is called an *integration diagram* (or a *simulation diagram*). The output of each integrator is a state variable, and the order of the system is the number of integrators, which is the number of state variables. The diagram is arranged so that the input to each integrator is a linear combination of the system inputs and the state variables. Each system output is also a linear combination of the system inputs and the state variables.

Figure 3-1 shows an example of an integration diagram, in signal flow graph form, in terms of Laplace-transformed signals for a single-input/single-output system. For this system, there are three integrators, so the system represented is third-order. The internal system signals, the state variables, are the integrator outputs, and the integra-

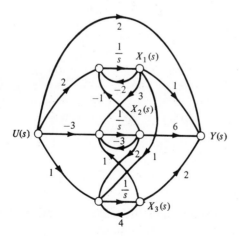

Figure 3-1 Integration diagram in terms of Laplace transforms.

tor inputs are each a linear combination of the state variables and the external input $U(s)$. From the diagram, the three Laplace-transformed *state equations* for this system are

$$\begin{cases} X_1(s) = \dfrac{1}{s}[-2X_1(s) - X_2(s) + 2U(s)] \\[2mm] X_2(s) = \dfrac{1}{s}[3X_1(s) - 3X_2(s) + X_3(s) - 3U(s)] \\[2mm] X_3(s) = \dfrac{1}{s}[X_1(s) + 2X_2(s) + 4X_3(s) + U(s)] \end{cases}$$

which, as functions of time, are

$$\begin{cases} \dfrac{dx_1}{dt} = -2x_1(t) - x_2(t) + 2u(t) \\[2mm] \dfrac{dx_2}{dt} = 3x_1(t) - 3x_2(t) + x_3(t) - 3u(t) \\[2mm] \dfrac{dx_3}{dt} = x_1(t) + 2x_2(t) + 4x_3(t) + u(t) \end{cases}$$

In matrix form:

$$\begin{bmatrix} \dot{x}_1(t) \\ \dot{x}_2(t) \\ \dot{x}_3(t) \end{bmatrix} = \begin{bmatrix} -2 & -1 & 0 \\ 3 & -3 & 1 \\ 1 & 2 & 4 \end{bmatrix} \begin{bmatrix} x_1(t) \\ x_2(t) \\ x_3(t) \end{bmatrix} + \begin{bmatrix} 2 \\ -3 \\ 1 \end{bmatrix} u(t)$$

where the dot over the symbol for a function denotes differentiation with respect to time t. The system's output is a linear combination of the state variables and the input

$$Y(s) = X_1(s) + 6X_2(s) + 2X_3(s) + 2U(s)$$

or

$$y(t) = x_1(t) + 6x_2(t) + 2x_3(t) + 2u(t)$$

In matrix form, the *output equation* for this system is

$$y(t) = \begin{bmatrix} 1 & 6 & 2 \end{bmatrix} \begin{bmatrix} x_1(t) \\ x_2(t) \\ x_3(t) \end{bmatrix} + 2u(t)$$

An integration diagram shows how the system can be constructed in hardware, with operational amplifiers connected to form scaled summations and integrals of signals, and in software, where the integration operations represent numerical integrations. Alternatively, integration diagrams can relate signals in the time domain rather than the transform domain, as in the block diagram of Figure 3-2. The collection of integrator outputs at time $t = 0$ is the initial state, $\mathbf{x}(0)$.

Even though the symbol u is sometimes used to denote the unit step function, it is common practice also to use u to represent inputs in state variable models. The ambiguity is easily resolved by context and is preferable to using nonstandard notation.

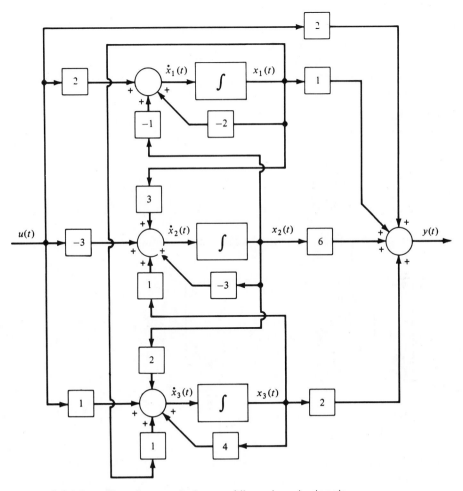

Figure 3-2 Integration diagram in terms of time domain signals.

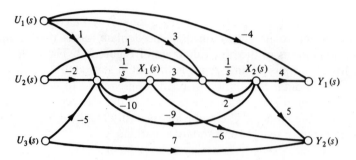

Figure 3-3 An integration diagram with multiple inputs and outputs.

Another integration diagram, a second-order one with multiple inputs and outputs, is shown in Figure 3-3. For this system, the state and output equations are

$$\begin{bmatrix} \dot{x}_1(t) \\ \dot{x}_2(t) \end{bmatrix} = \begin{bmatrix} -10 & -9 \\ 3 & 2 \end{bmatrix}\begin{bmatrix} x_1(t) \\ x_2(t) \end{bmatrix} + \begin{bmatrix} 1 & -2 & -5 \\ 3 & 1 & 0 \end{bmatrix}\begin{bmatrix} u_1(t) \\ u_2(t) \\ u_3(t) \end{bmatrix}$$

$$\begin{bmatrix} y_1(t) \\ y_2(t) \end{bmatrix} = \begin{bmatrix} 0 & 4 \\ -6 & 5 \end{bmatrix}\begin{bmatrix} x_1(t) \\ x_2(t) \end{bmatrix} + \begin{bmatrix} -4 & 0 & 0 \\ 0 & 0 & 7 \end{bmatrix}\begin{bmatrix} u_1(t) \\ u_2(t) \\ u_3(t) \end{bmatrix}$$

In general, linear, constant-coefficient, continuous-time state equations are of the form

$$\begin{bmatrix} \dot{x}_1(t) \\ \dot{x}_2(t) \\ \vdots \\ \dot{x}_n(t) \end{bmatrix} = \begin{bmatrix} a_{11} & a_{12} & \cdots & a_{1n} \\ a_{21} & a_{22} & \cdots & a_{2n} \\ \vdots & & & \\ a_{n1} & a_{n2} & \cdots & a_{nn} \end{bmatrix}\begin{bmatrix} x_1(t) \\ x_2(t) \\ \vdots \\ x_n(t) \end{bmatrix}$$

$$+ \begin{bmatrix} b_{11} & b_{12} & \cdots & b_{1r} \\ b_{21} & b_{22} & \cdots & b_{2r} \\ \vdots & & & \\ b_{n1} & b_{n2} & \cdots & b_{nr} \end{bmatrix}\begin{bmatrix} u_1(t) \\ u_2(t) \\ \vdots \\ u_r(t) \end{bmatrix} \quad \text{(3-1)}$$

or

$$\dot{\mathbf{x}}(t) = \mathbf{A}\mathbf{x}(t) + \mathbf{B}\mathbf{u}(t) \quad \text{(3-2)}$$

where the *state vector* $\mathbf{x}(t)$ is an *n*-vector, the *state coupling matrix* $\mathbf{A}$ is $n \times n$, the *input vector* $\mathbf{u}(t)$ is an *r*-vector of input signals, and the *input coupling matrix* $\mathbf{B}$ is $n \times r$. The order of the system is the dimension of the state vector *n*. The general form of the output equations is

$$
\begin{bmatrix} y_1(t) \\ y_2(t) \\ \vdots \\ y_m(t) \end{bmatrix} = \begin{bmatrix} c_{11} & c_{12} & \cdots & c_{1n} \\ c_{21} & c_{22} & \cdots & c_{2n} \\ \vdots & & & \\ c_{m1} & c_{m2} & \cdots & c_{mn} \end{bmatrix} \begin{bmatrix} x_1(t) \\ x_2(t) \\ \vdots \\ x_n(t) \end{bmatrix}
$$

$$
+ \begin{bmatrix} d_{11} & d_{12} & \cdots & d_{1r} \\ d_{21} & d_{22} & \cdots & d_{2r} \\ \vdots & & & \\ d_{m1} & d_{m2} & \cdots & d_{mr} \end{bmatrix} \begin{bmatrix} u_1(t) \\ u_2(t) \\ \vdots \\ u_r(t) \end{bmatrix} \qquad \textbf{(3-3)}
$$

or

$$\mathbf{y}(t) = \mathbf{C}\mathbf{x}(t) + \mathbf{D}\mathbf{u}(t) \qquad \textbf{(3-4)}$$

where the *output vector* $\mathbf{y}(t)$ is an *m*-vector of output signals, the *output coupling matrix* $\mathbf{C}$ is $m \times n$, and the *input-to-output coupling matrix* $\mathbf{D}$ is $m \times r$.

A block diagram that shows how the input, output, and state vectors are related in general in a continuous-time state variable system is given in Figure 3-4. The wide arrows in that diagram represent vectors of signals.

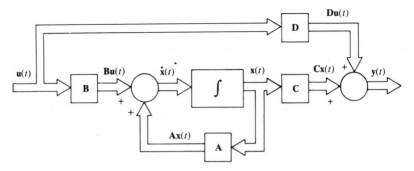

Figure 3-4 Block diagram showing the relations between signal vectors in a continuous-time, state variable model.

3.2.2 Laplace-Transformed Signals

The emphasis of state variable methods is on the time domain rather than a transform domain. However, for linear, time-invariant systems, the Laplace transform helps relate state variables to classical concepts and gives easy derivations of important time domain results. When Laplace-transformed, the state and output equations of a state variable system model, equations (3-1) and (3-3) or equations (3-2) and (3-4), become

$$
\begin{bmatrix} sX_1(s) \\ sX_2(s) \\ \vdots \\ sX_n(s) \end{bmatrix} - \begin{bmatrix} x_1(0^-) \\ x_2(0^-) \\ \vdots \\ x_n(0^-) \end{bmatrix} = \begin{bmatrix} a_{11} & a_{12} & \cdots & a_{1n} \\ a_{21} & a_{22} & \cdots & a_{2n} \\ \vdots & & & \\ a_{n1} & a_{n2} & \cdots & a_{nn} \end{bmatrix} \begin{bmatrix} X_1(s) \\ X_2(s) \\ \vdots \\ X_n(s) \end{bmatrix}
$$

$$
+ \begin{bmatrix} b_{11} & b_{12} & \cdots & b_{1r} \\ b_{21} & b_{22} & \cdots & b_{2r} \\ \vdots & & & \\ b_{n1} & b_{n2} & \cdots & b_{nr} \end{bmatrix} \begin{bmatrix} U_1(s) \\ U_2(s) \\ \vdots \\ U_r(s) \end{bmatrix}
$$

$$
\begin{bmatrix} Y_1(s) \\ Y_2(s) \\ \vdots \\ Y_m(s) \end{bmatrix} = \begin{bmatrix} c_{11} & c_{12} & \cdots & c_{1n} \\ c_{21} & c_{22} & \cdots & c_{2n} \\ \vdots & & & \\ c_{m1} & c_{m2} & \cdots & c_{mn} \end{bmatrix} \begin{bmatrix} X_1(s) \\ X_2(s) \\ \vdots \\ X_n(s) \end{bmatrix}
$$

$$
+ \begin{bmatrix} d_{11} & d_{12} & \cdots & d_{1r} \\ d_{21} & d_{22} & \cdots & d_{2r} \\ \vdots & & & \\ d_{m1} & d_{m2} & \cdots & d_{mr} \end{bmatrix} \begin{bmatrix} U_1(s) \\ U_2(s) \\ \vdots \\ U_r(s) \end{bmatrix}
$$

or

$$ sX(s) - x(0^-) = AX(s) + BU(s) $$

$$ Y(s) = CX(s) + DU(s) $$

Here, the initial conditions $\mathbf{x}(0^-)$ have been included. Solving for the transform of the state vector gives

$$(s\mathbf{I} - \mathbf{A})\mathbf{X}(s) = \mathbf{x}(0^-) + \mathbf{B}\mathbf{U}(s)$$

$$\mathbf{X}(s) = \underset{\text{Zero-Input Component of State Vector}}{(s\mathbf{I} - \mathbf{A})^{-1}\mathbf{x}(0^-)} + \underset{\text{Zero-State Component of State Vector}}{(s\mathbf{I} - \mathbf{A})^{-1}\mathbf{B}\mathbf{U}(s)}$$

where the zero-input and zero-state components of the state vector are evident. The zero-input component is the solution when the input is zero, and the zero-state component is the solution when the initial conditions are zero. The transform of the system output is then

$$\mathbf{Y}(s) = \mathbf{C}\mathbf{X}(s) + \mathbf{D}\mathbf{U}(s)$$

$$= \underset{\text{Zero-Input Component of Output}}{\mathbf{C}(s\mathbf{I} - \mathbf{A})^{-1}\mathbf{x}(0^-)} + \underset{\text{Zero-State Component of Output}}{[\mathbf{C}(s\mathbf{I} - \mathbf{A})^{-1}\mathbf{B} + \mathbf{D}]\mathbf{U}(s)} \qquad (3\text{-}5)$$

As a numerical example of finding signals in state variable models using Laplace transformation, consider the second-order, single-input/single-output system

$$\begin{bmatrix} \dot{x}_1(t) \\ \dot{x}_2(t) \end{bmatrix} = \begin{bmatrix} -7 & 1 \\ -12 & 0 \end{bmatrix} \begin{bmatrix} x_1(t) \\ x_2(t) \end{bmatrix} + \begin{bmatrix} 2 \\ -1 \end{bmatrix} u(t) = \mathbf{A}\mathbf{x}(t) + \mathbf{b}u(t)$$

$$y(t) = \begin{bmatrix} 3 & -4 \end{bmatrix} \begin{bmatrix} x_1(t) \\ x_2(t) \end{bmatrix} - 2u(t) = \mathbf{c}^\dagger\mathbf{x}(t) + du(t)$$

where the dagger symbol, $\dagger$, denotes the transpose of the quantity. The output coupling matrix $\mathbf{C}$ is, in this case, a single row, which is represented as the transpose of a column vector $\mathbf{c}$. The integration diagram for this system is shown in Figure 3-5(a). Let the system input be

$$u(t) = 3e^{-t} \quad t \geq 0$$

and let the initial system state be

$$\begin{bmatrix} x_1(0^-) \\ x_2(0^-) \end{bmatrix} = \begin{bmatrix} -6 \\ 1 \end{bmatrix}$$

The Laplace-transformed state equations are

$$\begin{cases} sX_1(s) - x_1(0^-) = -7X_1(s) + X_2(s) + 2U(s) \\ sX_2(s) - x_2(0^-) = -12X_1(s) - U(s) \end{cases}$$

If desired, the initial conditions can be included as part of the integration diagram, as in Figure 3-5(b). In matrix form,

$$\begin{bmatrix} (s + 7) & -1 \\ 12 & s \end{bmatrix} \begin{bmatrix} X_1(s) \\ X_2(s) \end{bmatrix} = \begin{bmatrix} x_1(0^-) \\ x_2(0^-) \end{bmatrix} + \begin{bmatrix} 2 \\ -1 \end{bmatrix} U(s)$$

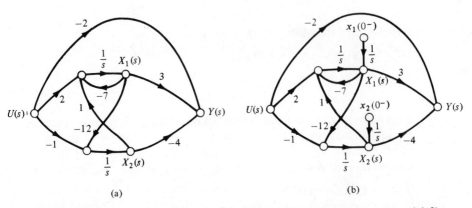

(a)   (b)

Figure 3-5 Finding the response of a continuous-time, state variable system. (a) Simulation diagram. (b) Simulation diagram including initial conditions.

or

$$(s\mathbf{I} - \mathbf{A})\mathbf{X}(s) = \mathbf{x}(0^-) + \mathbf{b}U(s)$$

The solution for the transform of the state vector is given by

$$\begin{bmatrix} X_1(s) \\ X_2(s) \end{bmatrix} = \begin{bmatrix} (s+7) & -1 \\ 12 & s \end{bmatrix}^{-1} \begin{bmatrix} x_1(0^-) \\ x_2(0^-) \end{bmatrix}$$

Zero-Input Component of State Vector

$$+ \begin{bmatrix} (s+7) & -1 \\ 12 & s \end{bmatrix}^{-1} \begin{bmatrix} 2 \\ -1 \end{bmatrix} U(s)$$

Zero-State Component of State Vector

$$= \begin{bmatrix} \dfrac{-6s^2 + s - 2}{(s+1)(s+3)(s+4)} \\[3mm] \dfrac{s^2 + 77s - 14}{(s+1)(s+3)(s+4)} \end{bmatrix}$$

which can be inverted to find $x_1(t)$ and $x_2(t)$. The system output has the transform

$$Y(s) = \mathbf{c}^t\mathbf{X}(s) + dU(s) = \frac{-22s^2 - 305s + 50}{(s+1)(s+3)(s+4)} - \frac{6}{s+1}$$

$$= \frac{297/6}{s+1} + \frac{-767/2}{s+3} + \frac{918/3}{s+4}$$

so that

$$y(t) = \frac{297}{6} e^{-t} + \frac{-767}{2} e^{-3t} + \frac{918}{3} e^{-4t} \quad t \geq 0$$

The transfer functions of a state variable system are found by expressing the Laplace transform of the output vector in terms of the Laplace transform of the input vector when the initial conditions are zero. Setting

$$\mathbf{x}(0^-) = \mathbf{0}$$

in equation (3-5) gives

$$\mathbf{Y}(s) = [\mathbf{C}(s\mathbf{I} - \mathbf{A})^{-1}\mathbf{B} + \mathbf{D}]\mathbf{U}(s)$$

The quantity

$$\mathbf{T}(s) = \mathbf{C}(s\mathbf{I} - \mathbf{A})^{-1}\mathbf{B} + \mathbf{D}$$

is an $m \times r$ matrix, where m is the number of outputs in $\mathbf{y}(t)$ and r is the number of inputs in $\mathbf{u}(t)$. The elements of $\mathbf{T}(s)$ are functions of the variable s, and the element in the ith row and jth column of $\mathbf{T}(s)$ is the transfer function relating the ith output to the jth input:

$$T_{ij}(s) = \frac{Y_i(s)}{U_j(s)} \Big|_{\substack{\text{Zero initial conditions and} \\ \text{all other inputs zero}}}$$

Because

$$(s\mathbf{I} - \mathbf{A})^{-1} = \frac{\text{adj}(s\mathbf{I} - \mathbf{A})}{|s\mathbf{I} - \mathbf{A}|}$$

the nth-degree *characteristic polynomial*

$$q(s) = |s\mathbf{I} - \mathbf{A}|$$

always occurs as the denominator polynomial of each individual transfer function. The roots of the *characteristic equation*

$$|s\mathbf{I} - \mathbf{A}| = 0 \tag{3-6}$$

are thus the poles of each of the system transfer functions. The system's poles are also called its *eigenvalues* because the solutions of equation (3-6) are the eigenvalues of the matrix $\mathbf{A}$. (Detailed discussion of the characteristic value problem is presented in Appendix A.) The system is stable if and only if all the eigenvalues of $\mathbf{A}$ are in the left half of the complex plane (LHP). This definition of stability of a linear, time-invariant system is more restrictive than the input–output stability that is usually considered in classical control because it requires that all *possible* transfer functions have all their poles in the LHP.

The three-input/two-output, second-order system

$$\begin{bmatrix} \dot{x}_1(t) \\ \dot{x}_2(t) \end{bmatrix} = \begin{bmatrix} 0 & 1 \\ -5 & -2 \end{bmatrix} \begin{bmatrix} x_1(t) \\ x_2(t) \end{bmatrix} + \begin{bmatrix} 1 & 0 & -1 \\ 0 & -2 & 3 \end{bmatrix} \begin{bmatrix} u_1(t) \\ u_2(t) \\ u_3(t) \end{bmatrix}$$

$$= \mathbf{A}\mathbf{x}(t) + \mathbf{B}\mathbf{u}(t)$$

$$\begin{bmatrix} y_1(t) \\ y_2(t) \end{bmatrix} = \begin{bmatrix} 0 & -3 \\ 2 & -4 \end{bmatrix} \begin{bmatrix} x_1(t) \\ x_2(t) \end{bmatrix} + \begin{bmatrix} 1 & -2 & 0 \\ 0 & 0 & 0 \end{bmatrix} \begin{bmatrix} u_1(t) \\ u_2(t) \\ u_3(t) \end{bmatrix}$$

$$= \mathbf{C}\mathbf{x}(t) + \mathbf{D}\mathbf{u}(t)$$

has the integration diagram of Figure 3-6 and has a transfer function matrix given by

$$\mathbf{T}(s) = \mathbf{C}(s\mathbf{I} - \mathbf{A})^{-1}\mathbf{B} + \mathbf{D}$$

$$= \begin{bmatrix} \dfrac{s^2 + 2s + 20}{s^2 + 2s + 5} & \dfrac{-2s^2 + 2s - 10}{s^2 + 2s + 5} & \dfrac{-9s - 15}{s^2 + 2s + 5} \\ \dfrac{2s + 24}{s^2 + 2s + 5} & \dfrac{8s - 4}{s^2 + 2s + 5} & \dfrac{-14s - 18}{s^2 + 2s + 5} \end{bmatrix}$$

$$= \begin{bmatrix} T_{11}(s) & T_{12}(s) & T_{13}(s) \\ T_{21}(s) & T_{22}(s) & T_{23}(s) \end{bmatrix}$$

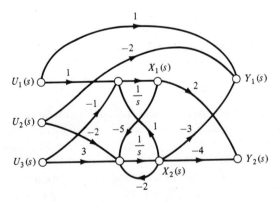

Figure 3-6 Finding a transfer function matrix.

The six individual transfer functions that are elements of this transfer function matrix are the usual ratios of Laplace-transformed signals

$$T_{11}(s) = \frac{Y_1(s)}{U_1(s)} \quad T_{12}(s) = \frac{Y_1(s)}{U_2(s)} \quad T_{13}(s) = \frac{Y_1(s)}{U_3(s)}$$

$$T_{21}(s) = \frac{Y_2(s)}{U_1(s)} \quad T_{22}(s) = \frac{Y_2(s)}{U_2(s)} \quad T_{23}(s) = \frac{Y_2(s)}{U_3(s)}$$

when the initial conditions and all other inputs are zero.

Alternatively, the transfer functions can be found from the integration diagram, using Mason's gain rule or block diagram reduction.

When the vector of outputs of one system is the vector of inputs to a second system, the overall composite system has a transfer function matrix that is the product of the individual system transfer function matrices.

3.2.3 Time Domain Solution

It is advantageous to have an expression for the solution for the state of a linear, time-invariant, continuous-time system as a function of time rather than in terms of Laplace transforms. For a first-order system with the state equation

$$\frac{dx}{dt} - ax(t) = bu(t)$$

this solution can be obtained as follows. Multiplying both sides of the equation by the "integrating factor" e^{-at} gives

$$e^{-at}\frac{dx}{dt} - ae^{-at}x(t) = e^{-at}bu(t)$$

The left side of the equation is then the derivative of a product

$$\frac{d}{dt}[e^{-at}x(t)] = e^{-at}bu(t)$$

Integrating both sides gives

$$e^{-at}x(t) = \int e^{-at}bu(t)\, dt + \text{(arbitrary constant)}$$

If the integration is begun at time $t = 0^-$, then

$$e^{-at}x(t) = \int_{0^-}^{t} e^{-a\tau}bu(\tau)\, d\tau + x(0^-) \quad t \geq 0$$

where the variable of integration has been distinguished from the time t, which occurs as a limit of integration. Solving for $x(t)$,

$$x(t) = e^{at}x(0^-) + \int_{0^-}^{t} e^{a(t-\tau)} bu(\tau) \, d\tau \quad t \geq 0$$

The integral is the *convolution* of the function e^{at} and the input term $bu(t)$.

In terms of Laplace transforms, the state of this first-order system is given by

$$sX(s) - x(0^-) = aX(s) + bU(s)$$

$$X(s) = \frac{1}{s-a} x(0^-) + \frac{1}{s-a} bU(s)$$

$$x(t) = \mathcal{L}^{-1}\left[\frac{1}{s-a} x(0^-) + \frac{1}{s-a} bU(s)\right]$$

$$= e^{at}x(0^-) + \mathcal{L}^{-1}\left[\frac{1}{s-a} bU(s)\right] \quad t \geq 0$$

so that

$$\mathcal{L}^{-1}\left[\frac{1}{s-a} bU(s)\right] = \int_{0^-}^{t} e^{a(t-\tau)}bu(\tau) \, d\tau \quad t \geq 0$$

$$= \text{convolution}[e^{at}, bu(t)]$$

For digital computation, convolution using numerical integration is generally easier than dealing with Laplace transforms.

In general, the nth-order system described by

$$\dot{\mathbf{x}}(t) = \mathbf{A}\mathbf{x}(t) + \mathbf{B}\mathbf{u}(t)$$

$$\mathbf{y}(t) = \mathbf{C}\mathbf{x}(t) + \mathbf{D}\mathbf{u}(t)$$

has the solution given by

$$\mathbf{X}(s) = (s\mathbf{I} - \mathbf{A})^{-1}\mathbf{x}(0^-) + (s\mathbf{I} - \mathbf{A})^{-1}\mathbf{B}U(s)$$

or

$$\mathbf{x}(t) = \mathcal{L}^{-1}[(s\mathbf{I} - \mathbf{A})^{-1}]\mathbf{x}(0^-) + \text{convolution}\{\mathcal{L}^{-1}[(s\mathbf{I} - \mathbf{A})^{-1}], \mathbf{B}\mathbf{u}(t)\}$$

Denoting the *state transition matrix* by

$$\mathbf{\Phi}(t) = \mathcal{L}^{-1}[(s\mathbf{I} - \mathbf{A})^{-1}] \tag{3-7}$$

then

$$\mathbf{x}(t) = \mathbf{\Phi}(t)\mathbf{x}(0^-) + \text{convolution}[\mathbf{\Phi}(t),\ \mathbf{Bu}(t)]$$

$$= \mathbf{\Phi}(t)\mathbf{x}(0^-) + \int_{0^-}^{t} \mathbf{\Phi}(t - \tau)\mathbf{Bu}(\tau)\ d\tau$$

and

$$\mathbf{y}(t) = \mathbf{C\Phi}(t)\mathbf{x}(0^-) + \int_{0^-}^{t} \mathbf{C\Phi}(t - \tau)\mathbf{Bu}(\tau)\ d\tau + \mathbf{Du}(t)$$

As discussed in Appendix A, the state transition matrix of a linear, time-invariant, continuous-time system is the matrix exponential function

$$\mathbf{\Phi}(t) = \exp(\mathbf{A}t) = \mathbf{I} + \mathbf{A}t + \frac{1}{2!}\mathbf{A}^2 t^2 + \cdots + \frac{1}{i!}\mathbf{A}^i t^i + \cdots \qquad \textbf{(3-8)}$$

Matrix exponential functions play much the same role in the solution of higher-order, linear, time-invariant, differential equations as scalar exponential functions

$$e^{at} = 1 + at + \frac{(at)^2}{2!} + \frac{(at)^3}{3!} + \cdots$$

do in the solution of first-order equations. That the function $\exp(\mathbf{A}t)$ is the state transition matrix can be shown as follows. Laplace transforming equation (3-7):

$$\mathbf{\Phi}(s) = (s\mathbf{I} - \mathbf{A})^{-1}$$

so that

$$s\mathbf{\Phi}(s) - \mathbf{I} = \mathbf{A\Phi}(s)$$

or

$$\dot{\mathbf{\Phi}}(t) = \mathbf{A\Phi}(t)$$

with

$$\mathbf{\Phi}(0) = \mathbf{I}$$

This set of differential equations and boundary conditions has the solution

$$\mathbf{\Phi}(t) = \exp(\mathbf{A}t)$$

because

$$\frac{d}{dt}[\exp(\mathbf{A}t)] = \mathbf{A}\exp(\mathbf{A}t)$$

and

$$\exp(\mathbf{0}) = \mathbf{I}$$

Some important properties of state transition matrices are summarized in Table 3-1.

One method of calculating the state transition matrix for a linear, time-invariant, continuous-time system is to use the relation

$$\boldsymbol{\Phi}(t) = \exp(\mathbf{A}t) = \mathcal{L}^{-1}[(s\mathbf{I} - \mathbf{A})^{-1}]$$

Table 3-1 Properties of State Transition Matrices

For the linear, time-invariant, continuous-time system with state equations

$$\dot{\mathbf{x}}(t) = \mathbf{A}\mathbf{x}(t) + \mathbf{B}\mathbf{u}(t)$$

$$\mathbf{x}(t) = \boldsymbol{\Phi}(t)\mathbf{x}(0^-) + \int_{0^-}^{t} \boldsymbol{\Phi}(t - \tau)\mathbf{B}\mathbf{u}(\tau) \, d\tau$$

the state transition matrix $\boldsymbol{\Phi}(t)$ has these properties:

A. $\boldsymbol{\Phi}(t) = \mathcal{L}^{-1}[(s\mathbf{I} - \mathbf{A})^{-1}]$

B. $\boldsymbol{\Phi}(t) = \exp(\mathbf{A}t) = \mathbf{I} + \mathbf{A}t + \dfrac{1}{2!}\mathbf{A}^2 t^2 + \cdots + \dfrac{1}{i!}\mathbf{A}^i t^i + \cdots$

C. $\boldsymbol{\Phi}(0) = \mathbf{I}$

D. $\boldsymbol{\Phi}(t + \tau) = \boldsymbol{\Phi}(t)\boldsymbol{\Phi}(\tau) = \boldsymbol{\Phi}(\tau)\boldsymbol{\Phi}(t)$

E. The state transition matrix is the unique $n \times n$ solution to

$$\dot{\boldsymbol{\Phi}}(t) = \mathbf{A}\boldsymbol{\Phi}(t)$$

with

$$\boldsymbol{\Phi}(0) = \mathbf{I}$$

F. $\boldsymbol{\Phi}^{-1}(t) = \boldsymbol{\Phi}(-t)$

For example, for a system with the state coupling matrix

$$\mathbf{A} = \begin{bmatrix} 0 & 1 \\ -3 & -4 \end{bmatrix}$$

the state transition matrix is

$$\Phi(t) = \mathcal{L}^{-1}\left\{\begin{bmatrix} s & -1 \\ 3 & (s+4) \end{bmatrix}^{-1}\right\} = \mathcal{L}^{-1}\begin{bmatrix} \dfrac{s+4}{s^2+4s+3} & \dfrac{1}{s^2+4s+3} \\[3mm] \dfrac{-3}{s^2+4s+3} & \dfrac{s}{s^2+4s+3} \end{bmatrix}$$

$$= \mathcal{L}^{-1}\begin{bmatrix} \dfrac{\frac{3}{2}}{s+1} + \dfrac{-\frac{1}{2}}{s+3} & \dfrac{\frac{1}{2}}{s+1} + \dfrac{-\frac{1}{2}}{s+3} \\[3mm] \dfrac{-\frac{3}{2}}{s+1} + \dfrac{-\frac{3}{2}}{s+3} & \dfrac{-\frac{1}{2}}{s+1} + \dfrac{\frac{3}{2}}{s+3} \end{bmatrix}$$

$$= \begin{bmatrix} \frac{3}{2}e^{-t} - \frac{1}{2}e^{-3t} & \frac{1}{2}e^{-t} - \frac{1}{2}e^{-3t} \\[2mm] -\frac{3}{2}e^{-t} + \frac{3}{2}e^{-3t} & -\frac{1}{2}e^{-t} + \frac{3}{2}e^{-3t} \end{bmatrix}$$

Other methods for computing the state transition matrix are given in Appendix A.

3.2.4 Linear, Time-Varying, Continuous-Time Systems

Linear, time-varying, continuous-time systems of order n are modeled by state and output equations of the form

$$\dot{x}(t) = A(t)x(t) + B(t)u(t)$$

$$y(t) = C(t)x(t) + D(t)u(t)$$

where the time t is a continuous variable, $x(t)$ is an n-vector state of the system, $u(t)$ is an r-vector of input signals, the state coupling matrix $A(t)$ is $n \times n$, and the input coupling matrix $B(t)$ is $n \times r$. The output $y(t)$ is an m-vector, the output coupling matrix $C(t)$ is $m \times n$, and the input-to-output coupling matrix $D(t)$ is $m \times r$. In general, linear, time-varying, continuous-time systems have A, B, C, and D matrices that are piecewise-continuous functions of time.

Unforced Solution

When the inputs $u(t)$ to the system are all zero, the solution of the state equation

$$\dot{x}(t) = A(t)x(t)$$

exists and is unique for all $t \geq t_0$. It is given by

$$x(t) = \Phi(t, t_0)x(t_0) \tag{3-9}$$

where the state transition matrix $\Phi(t, t_0)$ is the solution to

$$\frac{d}{dt}\,\Phi(t, t_0) = \mathbf{A}(t)\Phi(t, t_0) \tag{3-10}$$

with

$$\Phi(t_0, t_0) = \mathbf{I}$$

This solution can be easily verified by first observing from equation (3-9) that

$$\mathbf{x}(t_0) = \Phi(t_0, t_0)\mathbf{x}(t_0)$$

Then, differentiating equation (3-9) gives

$$\dot{\mathbf{x}}(t) = \frac{d}{dt}\,\Phi(t, t_0)\mathbf{x}(t_0)$$

and using equation (3-10),

$$\dot{\mathbf{x}}(t) = \mathbf{A}(t)\Phi(t, t_0)\mathbf{x}(t_0)$$

or

$$\dot{\mathbf{x}}(t) = \mathbf{A}(t)\mathbf{x}(t)$$

which completes the proof.

Before we discuss the forced solution, let us investigate some additional properties of the state transition matrix. Using equation (3-9), we have

$$\mathbf{x}(t_2) = \Phi(t_2, t_1)\mathbf{x}(t_1)$$

and

$$\mathbf{x}(t_3) = \Phi(t_3, t_2)\mathbf{x}(t_2)$$

Substituting $\mathbf{x}(t_2)$ from the first equation into the second results in

$$\mathbf{x}(t_3) = \Phi(t_3, t_2)\Phi(t_2, t_1)\mathbf{x}(t_1) \tag{3-11}$$

But equation (3-9) also gives

$$\mathbf{x}(t_3) = \Phi(t_3, t_1)\mathbf{x}(t_1) \tag{3-12}$$

Hence, comparing equation (3-11) with equation (3-12) yields the property

$$\Phi(t_3, t_1) = \Phi(t_3, t_2)\Phi(t_2, t_1) \tag{3-13}$$

Another useful property may be derived by setting $t_3 = t_1$ in equation (3-13), then

$$\Phi(t_1, t_1) = \Phi(t_1, t_2)\Phi(t_2, t_1) = \mathbf{I}$$

or

$$\Phi(t_2, t_1) = \Phi^{-1}(t_1, t_2) \tag{3-14}$$

Forced Solution

In terms of the initial state $\mathbf{x}(t_0)$ and the inputs $\mathbf{u}(t)$ at time t_0 and beyond, the solution for the state given by

$$\dot{\mathbf{x}}(t) = \mathbf{A}(t)\mathbf{x}(t) + \mathbf{B}(t)\mathbf{u}(t) \tag{3-15}$$

exists and is unique for all $t \geq t_0$. It is given by

$$\mathbf{x}(t) = \Phi(t, t_0)\mathbf{x}(t_0) + \int_{t_0}^{t} \Phi(t, \tau)\mathbf{B}(\tau)\mathbf{u}(\tau) \, d\tau \tag{3-16}$$

where the state transition matrix is still defined by equation (3-10). That equation (3-16) is the solution to equation (3-15) can be verified by direct substitution. From equation (3-16), checking the initial conditions

$$\mathbf{x}(t_0) = \Phi(t_0, t_0)\mathbf{x}(t_0) + \int_{t_0}^{t_0} \Phi(t, \tau)\mathbf{B}(\tau)\mathbf{u}(\tau) \, d\tau$$
$$= \mathbf{x}(t_0)$$

Then differentiating equation (3-16) gives

$$\dot{\mathbf{x}}(t) = \frac{\partial}{\partial t} [\Phi(t, t_0)\mathbf{x}(t_0)] + \frac{\partial}{\partial t} \int_{t_0}^{t} \Phi(t, \tau)\mathbf{B}(\tau)\mathbf{u}(\tau) \, d\tau$$

and using the Leibniz rule for partial differentiation gives

$$\dot{\mathbf{x}}(t) = \mathbf{A}(t)\Phi(t, t_0)\mathbf{x}(t_0) + \Phi(t, t)\mathbf{B}(t)\mathbf{u}(t) + \int_{t_0}^{t} \frac{\partial}{\partial t} \Phi(t, \tau)\mathbf{B}(\tau)\mathbf{u}(\tau) \, d\tau$$

$$= \mathbf{A}(t)[\Phi(t, t_0)\mathbf{x}(t_0) + \int_{t_0}^{t} \Phi(t, \tau)\mathbf{B}(\tau)\mathbf{u}(\tau) \, d\tau] + \mathbf{B}(t)\mathbf{u}(t)$$

$$= \mathbf{A}(t)\mathbf{x}(t) + \mathbf{B}(t)\mathbf{u}(t)$$

which completes the proof.

It is important to point out that the state transition matrix for a linear, time-varying, continuous-time system is *not* determined by the matrix exponential function given by equation (3-8). And except in some special cases, a *closed-form* solution does not exist in general. The most common way of determining the state transition matrix for linear, time-varying, continuous-time systems is by numerical integration of the matrix differential equation (3-10). However, if the matrix $\mathbf{A}(t)$ satisfies the following relation (commutativity):

$$\mathbf{A}(t) \int_{t_0}^{t} \mathbf{A}(\tau) \, d\tau = \left[\int_{t_0}^{t} \mathbf{A}(\tau) \, d\tau \right] \mathbf{A}(t)$$

for all t and t_0, then the state transition matrix may be determined from

$$\Phi(t, t_0) = \exp\left[\int_{t_0}^{t} A(\tau) \, d\tau\right]$$

To illustrate the ideas involved, consider the simple example

$$\begin{bmatrix} \dot{x}_1(t) \\ \dot{x}_2(t)] \end{bmatrix} = \begin{bmatrix} 0 & 0 \\ t & 0 \end{bmatrix} \begin{bmatrix} x_1(t) \\ x_2(t) \end{bmatrix}$$

with $t_0 = 0$, and the initial conditions

$$\begin{bmatrix} x_1(0) \\ x_2(0) \end{bmatrix} = \begin{bmatrix} 2 \\ 0 \end{bmatrix}$$

The elements of the 2×2 state transition matrix

$$\Phi(t, 0) = \begin{bmatrix} \phi_{11}(t, 0) & \phi_{12}(t, 0) \\ \phi_{21}(t, 0) & \phi_{22}(t, 0) \end{bmatrix}$$

are determined from equation (3-10)

$$\dot{\phi}_{11}(t, 0) = 0 \qquad\qquad \phi_{11}(0, 0) = 1$$
$$\dot{\phi}_{12}(t, 0) = 0 \qquad\qquad \phi_{12}(0, 0) = 0$$
$$\dot{\phi}_{21}(t, 0) = t\phi_{11}(t, 0) \qquad \phi_{21}(0, 0) = 0$$
$$\dot{\phi}_{22}(t, 0) = t\phi_{12}(t, 0) \qquad \phi_{22}(0, 0) = 1$$

Solving these equations using standard integration methods gives

$$\phi_{11}(t, 0) = 1$$
$$\phi_{12}(t, 0) = 0$$
$$\phi_{21}(t, 0) = \frac{t^2}{2}$$
$$\phi_{22}(t, 0) = 1$$

Therefore,

$$\Phi(t, 0) = \begin{bmatrix} 1 & 0 \\ \dfrac{t^2}{2} & 1 \end{bmatrix}$$

and

$$\begin{bmatrix} x_1(t) \\ x_2(t) \end{bmatrix} = \begin{bmatrix} 1 & 0 \\ \dfrac{t^2}{2} & 1 \end{bmatrix} \begin{bmatrix} 2 \\ 0 \end{bmatrix} = \begin{bmatrix} 2 \\ t^2 \end{bmatrix}$$

3.3 Discrete-Time State Equations and System Response

Discrete-time state variable models are now introduced. Like their continuous-time counterparts discrete-time state variables are a convenient and powerful way of describing and dealing with systems, especially those with more than a single input and a single output. The state and output signals of a discrete-time system are found from the inputs and the initial state. Repeated application of the state variable equations is used to express the state and output signals in terms of a discrete convolution. Then, the z-transform is applied to find z-transfer functions and closed-form solutions for response. Step-varying state equations, the recursive solution, and some ideas about linear discrete-time system stability for step-varying systems are introduced.

3.3.1 Delay Diagrams and State Equations

State variable descriptions of discrete-time systems are mathematical models consisting of coupled first-order difference equations. For linear, step-invariant, discrete-time systems, state variable models involve the operations of summation, multiplication by a constant, and one-step delay. A block diagram, or signal flow graph, expressed in terms of these operations is called a *delay diagram* for the system. Delay diagrams relate the system inputs, the system outputs, and the internal state variables. The output of each delay is a state variable, and the number of state variables, which is the number of delays, is the order of the system. Delay diagrams are arranged so that the input to each delay is a linear combination of the system inputs and the state variables, as is each system output.

An example of a delay diagram for a single-input/single-output system is the signal flow graph shown in Figure 3-7. In it, there are three delays, so the system represented is of third-order. The state variables are each of the delay outputs, and the inputs to the delays are each a linear combination of the system input and the state variables. From the diagram, the z-transformed state equations are

$$\begin{cases} X_1(z) = \dfrac{1}{z}\left[\dfrac{1}{2}X_1(z) + \dfrac{1}{4}X_2(z) + 2X_3(z) + U(z)\right] \\[2mm] X_2(z) = \dfrac{1}{z}\left[-\dfrac{1}{2}X_2(z) - X_3(z) - U(z)\right] \\[2mm] X_3(z) = \dfrac{1}{z}\left[3X_2(z) + \dfrac{1}{3}X_3(z) + 2U(z)\right] \end{cases}$$

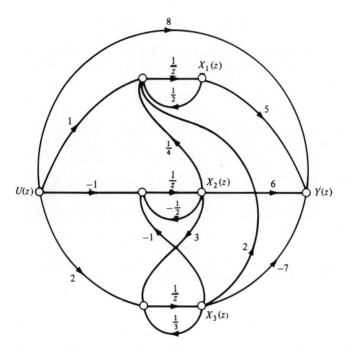

Figure 3-7 Delay diagram in terms of z-transforms.

As functions of step, these state equations are

$$\begin{cases} x_1(k+1) = \dfrac{1}{2} x_1(k) + \dfrac{1}{4} x_2(k) + 2x_3(k) + u(k) \\[2mm] x_2(k+1) = -\dfrac{1}{2} x_2(k) - x_3(k) - u(k) \\[2mm] x_3(k+1) = 3x_2(k) + \dfrac{1}{3} x_3(k) + 2u(k) \end{cases}$$

which in matrix form is

$$\begin{bmatrix} x_1(k+1) \\ x_2(k+1) \\ x_3(k+1) \end{bmatrix} = \begin{bmatrix} \frac{1}{2} & \frac{1}{4} & 2 \\ 0 & -\frac{1}{2} & -1 \\ 0 & 3 & \frac{1}{3} \end{bmatrix} \begin{bmatrix} x_1(k) \\ x_2(k) \\ x_3(k) \end{bmatrix} + \begin{bmatrix} 1 \\ -1 \\ 2 \end{bmatrix} u(k)$$

The output of the system is a linear combination of the state variables and the system input

$$Y(z) = 5X_1(z) + 6X_2(z) - 7X_3(z) + 8U(z)$$

or

$$y(k) = [5 \quad 6 \quad -7] \begin{bmatrix} x_1(k) \\ x_2(k) \\ x_3(k) \end{bmatrix} + 8u(k)$$

Delay diagrams show how a system can be constructed in hardware, with digital multipliers and adders, and with shift registers to provide the delays. It also shows how the system can be realized recursively in software, with the state and output for the next step produced during each program loop. Alternatively, a delay diagram can be expressed in terms of the signals themselves, rather than their z-transforms, as in the block diagram of Figure 3-8. The initial delay outputs are the initial system state. One can deal either in the step or the transform domain for step-invariant systems. For step-varying systems, however, z-transforms are no longer of much use.

Another delay diagram, one for a second-order system with multiple inputs and outputs, is shown in Figure 3-9. For this system, state equations are

$$\begin{bmatrix} x_1(k+1) \\ x_2(k+1) \end{bmatrix} = \begin{bmatrix} -1 & 2 \\ \frac{1}{4} & \frac{1}{2} \end{bmatrix} \begin{bmatrix} x_1(k) \\ x_2(k) \end{bmatrix} + \begin{bmatrix} 2 & 0 \\ -2 & 3 \end{bmatrix} \begin{bmatrix} u_1(k) \\ u_2(k) \end{bmatrix}$$

The output equations are

$$\begin{bmatrix} y_1(k) \\ y_2(k) \\ y_3(k) \end{bmatrix} = \begin{bmatrix} 6 & 5 \\ -3 & 0 \\ 1 & 7 \end{bmatrix} \begin{bmatrix} x_1(k) \\ x_2(k) \end{bmatrix} + \begin{bmatrix} 0 & 0 \\ 0 & 0 \\ 0 & 10 \end{bmatrix} \begin{bmatrix} u_1(k) \\ u_2(k) \end{bmatrix}$$

In general, nth-order discrete-time state equations are of the form

$$\begin{bmatrix} x_1(k+1) \\ x_2(k+1) \\ \vdots \\ x_n(k+1) \end{bmatrix} = \begin{bmatrix} a_{11} & a_{12} & \cdots & a_{1n} \\ a_{21} & a_{22} & \cdots & a_{2n} \\ \vdots & & & \\ a_{n1} & a_{n2} & \cdots & a_{nn} \end{bmatrix} \begin{bmatrix} x_1(k) \\ x_2(k) \\ \vdots \\ x_n(k) \end{bmatrix} +$$

$$\begin{bmatrix} b_{11} & b_{12} & \cdots & b_{1r} \\ b_{21} & b_{22} & \cdots & b_{2r} \\ \vdots & & & \\ b_{n1} & b_{n2} & \cdots & b_{nr} \end{bmatrix} \begin{bmatrix} u_1(k) \\ u_2(k) \\ \vdots \\ u_r(k) \end{bmatrix}$$

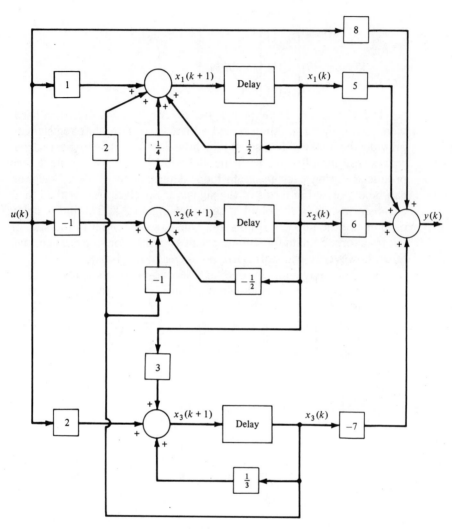

Figure 3-8 Delay diagrams in terms of time domain signals.

or

$$\mathbf{x}(k + 1) = \mathbf{A}\mathbf{x}(k) + \mathbf{B}\mathbf{u}(k)$$

where the state vector $\mathbf{x}(k)$ is an n-vector, the state coupling matrix $\mathbf{A}$ is $n \times n$, the input vector $\mathbf{u}(k)$ is an r-vector of input signals, and the

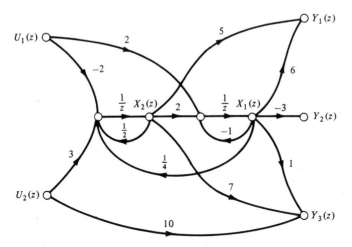

Figure 3-9 A delay diagram with multiple inputs and outputs.

input coupling matrix $\mathbf{B}$ is $n \times r$. The general form of the output equations is

$$
\begin{bmatrix} y_1(k) \\ y_2(k) \\ \vdots \\ y_m(k) \end{bmatrix} = \begin{bmatrix} c_{11} & c_{12} & \cdots & c_{1n} \\ c_{21} & c_{22} & \cdots & c_{2n} \\ \vdots & & & \\ c_{m1} & c_{m2} & \cdots & c_{mn} \end{bmatrix} \begin{bmatrix} x_1(k) \\ x_2(k) \\ \vdots \\ x_n(k) \end{bmatrix}
$$

$$
+ \begin{bmatrix} d_{11} & d_{12} & \cdots & d_{1r} \\ d_{21} & d_{22} & \cdots & d_{2r} \\ \vdots & & & \\ d_{m1} & d_{m2} & \cdots & d_{mr} \end{bmatrix} \begin{bmatrix} u_1(k) \\ u_2(k) \\ \vdots \\ u_r(k) \end{bmatrix}
$$

or

$$\mathbf{y}(k) = \mathbf{C}\mathbf{x}(k) + \mathbf{D}\mathbf{u}(k)$$

where the output vector $\mathbf{y}(k)$ is an m-vector of output signals, the output coupling matrix $\mathbf{C}$ is $m \times n$, and the input-to-output coupling matrix $\mathbf{D}$ is $m \times r$. A block diagram showing how the vectors $\mathbf{u}(k)$, $\mathbf{x}(k)$, and $\mathbf{y}(k)$ are related is given in Figure 3-10. In this diagram, the wide arrows represent vectors of signals.

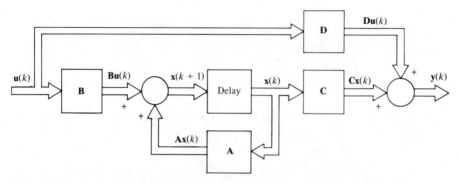

Figure 3-10 Block diagrams showing the relations between signal vectors in a discrete-time, state variable model.

As is common in practice, we have elected to use the same symbols, $\mathbf{x}$, $\mathbf{u}$, $\mathbf{y}$, $\mathbf{A}$, $\mathbf{B}$, $\mathbf{C}$, $\mathbf{D}$, to describe discrete-time state equations as were used earlier for continuous-time state equations. The quantities are not the same, of course, but real conflict occurs only when discrete-time and continuous-time systems are considered simultaneously. When that situation occurs, as it will later in connection with discrete-time models of continuous-time systems, script symbols will be used for the continuous-time, state variable matrices.

3.3.2 Response in Terms of Discrete Convolution

The response of a discrete-time system can be calculated recursively, starting with an initial state and repeatedly using the state equation

$$\mathbf{x}(k + 1) = \mathbf{A}\mathbf{x}(k) + \mathbf{B}\mathbf{u}(k)$$

From $\mathbf{x}(0)$ and $\mathbf{u}(0)$, $\mathbf{x}(1)$ can be calculated

$$\mathbf{x}(1) = \mathbf{A}\mathbf{x}(0) + \mathbf{B}\mathbf{u}(0)$$

Then, using $\mathbf{x}(1)$ and $\mathbf{u}(1)$

$$\mathbf{x}(2) = \mathbf{A}\mathbf{x}(1) + \mathbf{B}\mathbf{u}(1) = \mathbf{A}^2\mathbf{x}(0) + \mathbf{A}\mathbf{B}\mathbf{u}(0) + \mathbf{B}\mathbf{u}(1)$$

From $\mathbf{x}(2)$ and $\mathbf{u}(2)$

$$\mathbf{x}(3) = \mathbf{A}\mathbf{x}(2) + \mathbf{B}\mathbf{u}(2) = \mathbf{A}^3\mathbf{x}(0) + \mathbf{A}^2\mathbf{B}\mathbf{u}(0) + \mathbf{A}\mathbf{B}\mathbf{u}(1) + \mathbf{B}\mathbf{u}(2)$$

and in general

$$\mathbf{x}(k) = \mathbf{A}^k\mathbf{x}(0) + \mathbf{A}^{k-1}\mathbf{B}\mathbf{u}(0) + \mathbf{A}^{k-2}\mathbf{B}\mathbf{u}(1) + \cdots$$
$$+ \mathbf{A}\mathbf{B}\mathbf{u}(k - 2) + \mathbf{B}\mathbf{u}(k - 1)$$

$$= \mathbf{A}^k\mathbf{x}(0) \quad + \sum_{i=0}^{k-1} \mathbf{A}^{k-1-i}\mathbf{B}u(i)$$

zero-input component   zero-state component

$$= \mathbf{A}^k\mathbf{x}(0) + \text{convolution}[\mathbf{F}(k), \mathbf{B}u(k)] \tag{3-17}$$

The term that depends on the initial state, but not the inputs, is the *zero-input component*. The *zero-state component* does not depend on the initial conditions and is the discrete convolution of the function $\mathbf{B}u(k)$ with the function

$$\mathbf{F}(k) = \begin{cases} \mathbf{A}^{k-1} & k = 1, 2, 3, \ldots \\ \mathbf{0} & k = 0, -1, -2, \ldots \end{cases}$$

Recall that the convolution summation has an upper limit k, not $k - 1$,

$$\text{convolution } [f_1(k), f_2(k)] = \sum_{i=0}^{k} f_2(k - i)f_1(i)$$

so $\mathbf{F}(k)$ must be zero for $k = 0$.

As a numerical example, consider the system

$$\begin{bmatrix} x_1(k + 1) \\ x_2(k + 1) \end{bmatrix} = \begin{bmatrix} \frac{1}{2} & 1 \\ -\frac{1}{2} & 0 \end{bmatrix}\begin{bmatrix} x_1(k) \\ x_2(k) \end{bmatrix} + \begin{bmatrix} 0 \\ 4 \end{bmatrix} u(k) = \mathbf{A}\mathbf{x}(k) + \mathbf{b}u(k)$$

$$y(k) = [2 \quad -3]\begin{bmatrix} x_1(k) \\ x_2(k) \end{bmatrix} = \mathbf{c}^\dagger\mathbf{x}(k)$$

with

$$\begin{bmatrix} x_1(0) \\ x_2(0) \end{bmatrix} = \begin{bmatrix} -5 \\ 6 \end{bmatrix}$$

and

$$u(k) = 1 \quad \text{for all } k \geq 0.$$

At the zeroth step

$$y(0) = [2 \quad -3]\begin{bmatrix} -5 \\ 6 \end{bmatrix} = -28$$

At step one

$$\begin{bmatrix} x_1(1) \\ x_2(1) \end{bmatrix} = \begin{bmatrix} \frac{1}{2} & 1 \\ -\frac{1}{2} & 0 \end{bmatrix}\begin{bmatrix} -5 \\ 6 \end{bmatrix} + \begin{bmatrix} 0 \\ 4 \end{bmatrix} (1) = \begin{bmatrix} \frac{7}{2} \\ \frac{13}{2} \end{bmatrix}$$

$$y(1) = [2 \quad -3]\begin{bmatrix} \frac{7}{2} \\ \frac{13}{2} \end{bmatrix} = -\frac{25}{2}$$

At step two

$$\begin{bmatrix} x_1(2) \\ x_2(2) \end{bmatrix} = \begin{bmatrix} \frac{1}{2} & 1 \\ -\frac{1}{2} & 0 \end{bmatrix} \begin{bmatrix} \frac{7}{2} \\ \frac{13}{2} \end{bmatrix} + \begin{bmatrix} 0 \\ 4 \end{bmatrix} (1) = \begin{bmatrix} \frac{33}{4} \\ \frac{9}{4} \end{bmatrix}$$

$$y(2) = [2 \quad -3] \begin{bmatrix} \frac{33}{4} \\ \frac{9}{4} \end{bmatrix} = \frac{39}{4}$$

At step three

$$\begin{bmatrix} x_1(3) \\ x_2(3) \end{bmatrix} = \begin{bmatrix} \frac{1}{2} & 1 \\ -\frac{1}{2} & 0 \end{bmatrix} \begin{bmatrix} \frac{33}{4} \\ \frac{9}{4} \end{bmatrix} + \begin{bmatrix} 0 \\ 4 \end{bmatrix} (1) = \begin{bmatrix} \frac{51}{8} \\ -\frac{1}{8} \end{bmatrix}$$

$$y(3) = [2 \quad -3] \begin{bmatrix} \frac{51}{8} \\ -\frac{1}{8} \end{bmatrix} = \frac{105}{8}$$

and so on.

In terms of the convolution formula, the system state at step three is given by

$$\mathbf{x}(3) = \mathbf{A}^3 \mathbf{x}(0) + \mathbf{A}^2 \mathbf{b} u(0) + \mathbf{A} \mathbf{b} u(1) + \mathbf{b} u(2)$$

$$= \begin{bmatrix} -\frac{3}{8} & -\frac{1}{4} \\ \frac{1}{8} & -\frac{1}{4} \end{bmatrix} \begin{bmatrix} -5 \\ 6 \end{bmatrix} + \begin{bmatrix} -\frac{1}{4} & \frac{1}{2} \\ -\frac{1}{4} & -\frac{1}{2} \end{bmatrix} \begin{bmatrix} 0 \\ 4 \end{bmatrix} (1)$$

$$+ \begin{bmatrix} \frac{1}{2} & 1 \\ -\frac{1}{2} & 0 \end{bmatrix} \begin{bmatrix} 0 \\ 4 \end{bmatrix} (1) + \begin{bmatrix} 0 \\ 4 \end{bmatrix} (1)$$

$$= \begin{bmatrix} \frac{3}{8} \\ -\frac{17}{8} \end{bmatrix} + \begin{bmatrix} 2 \\ -2 \end{bmatrix} + \begin{bmatrix} 4 \\ 0 \end{bmatrix} + \begin{bmatrix} 0 \\ 4 \end{bmatrix} = \begin{bmatrix} \frac{51}{8} \\ -\frac{1}{8} \end{bmatrix}$$

$$\underbrace{\phantom{\begin{bmatrix}\frac{3}{8}\\-\frac{17}{8}\end{bmatrix}}}_{\text{Zero-input component}} \qquad \underbrace{\phantom{\begin{bmatrix} 2 \\ -2 \end{bmatrix} + \begin{bmatrix} 4 \\ 0 \end{bmatrix} + \begin{bmatrix} 0 \\ 4 \end{bmatrix}}}_{\text{Zero-state component}}$$

3.3.3 Z-Transformed Signals

Closed-form expressions for the signals in a linear, step-invariant, state variable model can be found by z-transforming the equations, solving for the transform of the signals of interest, then inverting the transforms to obtain the signals themselves. For the specific system of Figure 3-11(a) for example, the state variable equations are

$$\begin{bmatrix} x_1(k+1) \\ x_2(k+1) \end{bmatrix} = \begin{bmatrix} \frac{3}{2} & -1 \\ 1 & -1 \end{bmatrix} \begin{bmatrix} x_1(k) \\ x_2(k) \end{bmatrix} + \begin{bmatrix} 3 \\ 2 \end{bmatrix} u(k) = \mathbf{A}\mathbf{x}(k) + \mathbf{b}u(k)$$

$$\begin{bmatrix} y_1(k) \\ y_2(k) \end{bmatrix} = \begin{bmatrix} -3 & 4 \\ -1 & 1 \end{bmatrix} \begin{bmatrix} x_1(k) \\ x_2(k) \end{bmatrix} + \begin{bmatrix} -2 \\ 0 \end{bmatrix} u(k) = \mathbf{C}\mathbf{x}(k) + \mathbf{d}u(k)$$

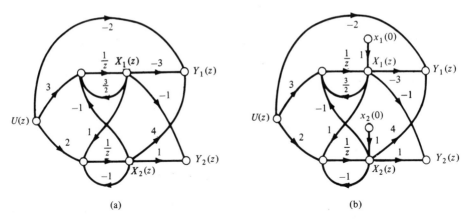

Figure 3-11 Finding the response of a discrete-time system using z-transforms. (a) Delay diagram. (b) Delay diagram including initial conditions.

Let the system input be

$$u(k) = \left(\frac{1}{2}\right)^k \quad k \geqslant 0$$

and let the initial conditions be

$$\begin{bmatrix} x_1(0) \\ x_2(0) \end{bmatrix} = \begin{bmatrix} -5 \\ 1 \end{bmatrix}$$

The z-transformed state equations are

$$\begin{cases} \left(z - \dfrac{3}{2}\right) X_1(z) + X_2(z) = -5z + \dfrac{3z}{z - \frac{1}{2}} = \dfrac{-5z^2 + (11/2)z}{z - \frac{1}{2}} \\[4mm] -X_1(z) + (z + 1)X_2(z) = z + \dfrac{2z}{z - \frac{1}{2}} = \dfrac{z^2 + \frac{3}{2}z}{z - \frac{1}{2}} \end{cases}$$

These include the initial conditions for the state variables, which can be added to the delay diagram as in Figure 3-11(b), if desired.

Solving for $X_1(z)$ and $X_2(z)$ and expanding into partial fractions (in the form suitable for the inverse z-transformation)

$$\begin{cases} X_1(z) = \dfrac{z(-5z^2 - \frac{1}{2}z + 4)}{(z - \frac{1}{2})(z^2 - \frac{1}{2}z - \frac{1}{2})} = \dfrac{-5z}{z - \frac{1}{2}} + \dfrac{2z}{z + \frac{1}{2}} + \dfrac{-2z}{z - 1} \\[4mm] X_2(z) = \dfrac{z[z^2 - 5z + (13/4)]}{(z - \frac{1}{2})(z + \frac{1}{2})(z - 1)} = \dfrac{-2z}{z - \frac{1}{2}} + \dfrac{4z}{z + \frac{1}{2}} + \dfrac{-z}{z - 1} \end{cases}$$

there results

$$\begin{cases} x_1(k) = -5\left(\dfrac{1}{2}\right)^k + 2\left(-\dfrac{1}{2}\right)^k - 2 \\ x_2(k) = -2\left(\dfrac{1}{2}\right)^k + 4\left(-\dfrac{1}{2}\right)^k - 1 \quad k = 0, 1, 2, \ldots \end{cases}$$

The system outputs are then

$$y_1(k) = -3x_1(k) + 4x_2(k) - 2u(k)$$
$$= 5\left(\frac{1}{2}\right)^k + 10\left(-\frac{1}{2}\right)^k + 2 \quad k = 0, 1, 2, \ldots$$

$$y_2(k) = -x_1(k) + x_2(k) = 3\left(\frac{1}{2}\right)^k + 2\left(-\frac{1}{2}\right)^k + 1, \quad k = 0, 1, 2, \ldots$$

In general, closed-form expressions for the signals in a linear, step-invariant, state variable model can be found by z-transforming the state equations

$$\mathbf{x}(k + 1) = \mathbf{A}\mathbf{x}(k) + \mathbf{B}u(k)$$

That is,

$$z\mathbf{X}(z) - z\mathbf{x}(0) = \mathbf{A}\mathbf{X}(z) + \mathbf{B}\mathbf{U}(z)$$

or

$$(z\mathbf{I} - \mathbf{A})\mathbf{X}(z) = z\mathbf{x}(0) + \mathbf{B}\mathbf{U}(z)$$

Hence,

$$\mathbf{X}(z) = z(z\mathbf{I} - \mathbf{A})^{-1}\mathbf{x}(0) + (z\mathbf{I} - \mathbf{A})^{-1}\mathbf{B}\mathbf{U}(z)$$

The solution for the state is then

$$\mathbf{x}(k) = \mathcal{Z}^{-1}[z(z\mathbf{I} - \mathbf{A})^{-1}]\mathbf{x}(0) + \mathcal{Z}^{-1}[(z\mathbf{I} - \mathbf{A})^{-1}\mathbf{B}\mathbf{U}(z)]$$

Comparing this result with equation (3-17) shows that

$$\mathbf{A}^k = \mathcal{Z}^{-1}[z(z\mathbf{I} - \mathbf{A})^{-1}]$$

which is analogous to the continuous-state transition matrix. It is important to point out that the methods (diagonalization, Cayley-Hamilton, Sylvester, Faddeev-Leverrier) presented in Appendix A for computing the continuous-state transition matrix can be used to determine the discrete-state transition matrix $\mathbf{A}^k$. (For further details, see problems 3-18, 3-19, and 3-20.)

The z-transfer function matrix of a system is found by z-transforming the state equations with zero initial conditions. Solving for the state:

$$z\mathbf{X}(z) = \mathbf{A}\mathbf{X}(z) + \mathbf{B}\mathbf{U}(z)$$

$$(z\mathbf{I} - \mathbf{A})\mathbf{X}(z) = \mathbf{B}\mathbf{U}(z)$$

$$\mathbf{X}(z) = (z\mathbf{I} - \mathbf{A})^{-1}\mathbf{B}\mathbf{U}(z)$$

and then finding the transform of the output vector in terms of the input vector transform gives

$$\mathbf{Y}(z) = \mathbf{C}\mathbf{X}(z) + \mathbf{D}\mathbf{U}(z) = [\mathbf{C}(z\mathbf{I} - \mathbf{A})^{-1}\mathbf{B} + \mathbf{D}]\mathbf{U}(z)$$

The $m \times r$ transfer matrix, where m is the number of outputs and r is the number of inputs, is

$$\mathbf{T}(z) = \mathbf{C}(z\mathbf{I} - \mathbf{A})^{-1}\mathbf{B} + \mathbf{D}$$

The element of $\mathbf{T}(z)$ in the ith row and jth column is the transfer function that relates the ith output to the jth input

$$T_{ij}(z) = \frac{Y_i(z)}{U_j(z)} \begin{array}{l} \text{Zero initial conditions} \\ \text{and all other inputs zero} \end{array}$$

The elements of a transfer function matrix are ratios of polynomials in the transform variable z. For an $n \times n$ matrix $\mathbf{A}$,

$$(z\mathbf{I} - \mathbf{A})^{-1} = \frac{\text{adj}(z\mathbf{I} - \mathbf{A})}{|z\mathbf{I} - \mathbf{A}|}$$

is composed of the $n \times n$ adjugate matrix with elements that are polynomials of maximum degree $n - 1$, each of which is divided by the determinant, which is an nth-degree polynomial in z. Because each of the individual z-transfer functions shares the denominator polynomial

$$q(z) = |z\mathbf{I} - \mathbf{A}|$$

each z-transfer function has the same poles, although there may be pole-zero cancellations. The z-transfer function poles (or eigenvalues) are the solutions to the characteristic equation for the system

$$q(z) = |z\mathbf{I} - \mathbf{A}| = 0$$

Stability requires that the system poles (or eigenvalues) be within the unit circle on the complex plane.

The second-order three-input/two-output system with discrete-time, state variable equations

$$\begin{bmatrix} x_1(k+1) \\ x_2(k+1) \end{bmatrix} = \begin{bmatrix} 2 & -5 \\ \frac{1}{2} & -1 \end{bmatrix}\begin{bmatrix} x_1(k) \\ x_2(k) \end{bmatrix} + \begin{bmatrix} 1 & -2 & 0 \\ 0 & 1 & 3 \end{bmatrix}\begin{bmatrix} u_1(k) \\ u_2(k) \\ u_3(k) \end{bmatrix}$$

$$= \mathbf{Ax}(k) + \mathbf{Bu}(k)$$

$$\begin{bmatrix} y_1(k) \\ y_2(k) \end{bmatrix} = \begin{bmatrix} 2 & 0 \\ 1 & -1 \end{bmatrix}\begin{bmatrix} x_1(k) \\ x_2(k) \end{bmatrix} + \begin{bmatrix} 0 & 4 & 0 \\ 0 & 0 & -2 \end{bmatrix}\begin{bmatrix} u_1(k) \\ u_2(k) \\ u_3(k) \end{bmatrix}$$

$$= \mathbf{Cx}(k) + \mathbf{Du}(k)$$

for example, has the characteristic equation

$$|z\mathbf{I} - \mathbf{A}| = \begin{vmatrix} (z-2) & 5 \\ -\frac{1}{2} & (z+1) \end{vmatrix} = z^2 - z + \frac{1}{2}$$

$$= \left(z - \frac{1}{2} - j\frac{1}{2}\right)\left(z - \frac{1}{2} + j\frac{1}{2}\right) = 0$$

Its six z-transfer functions all share the poles

$$z_1 = \frac{1}{2} + j\frac{1}{2} \qquad z_2 = \frac{1}{2} - j\frac{1}{2}$$

The transfer function matrix for the system, which is stable, is given by

$$\mathbf{T}(z) = \mathbf{C}(z\mathbf{I} - \mathbf{A})^{-1}\mathbf{B} + \mathbf{D}$$

$$= \begin{bmatrix} 2 & 0 \\ 1 & -1 \end{bmatrix}\begin{bmatrix} (z-2) & 5 \\ -\frac{1}{2} & (z+1) \end{bmatrix}^{-1}\begin{bmatrix} 1 & -2 & 0 \\ 0 & 1 & 3 \end{bmatrix}$$

$$+ \begin{bmatrix} 0 & 4 & 0 \\ 0 & 0 & -2 \end{bmatrix}$$

$$= \begin{bmatrix} \dfrac{2z+2}{z^2 - z + \frac{1}{2}} & 4 + \dfrac{-4z - 14}{z^2 - z + \frac{1}{2}} & \dfrac{-30}{z^2 - z + \frac{1}{2}} \\ \dfrac{z + \frac{1}{2}}{z^2 - z + \frac{1}{2}} & \dfrac{-3z - 4}{z^2 - z + \frac{1}{2}} & -2 + \dfrac{-3z - 9}{z^2 - z + \frac{1}{2}} \end{bmatrix}$$

The transfer functions of a causal system described by rational functions of z must have numerator polynomials of an order less than or equal to that of the denominator polynomial. Only causal systems can be represented by the standard state variable model.

3.3.4 State Equations and Response of Step-Varying Systems

A step-varying, discrete-time system has state equations of the form

$$\mathbf{x}(k + 1) = \mathbf{A}(k)\mathbf{x}(k) + \mathbf{B}(k)\mathbf{u}(k)$$
$$\mathbf{y}(k) = \mathbf{C}(k)\mathbf{x}(k) + \mathbf{D}(k)\mathbf{u}(k) \tag{3-18}$$

The **A, B, C,** and **D** matrices each might vary with step, as is indicated in the vector signal delay diagram of Figure 3-12. Most of the ideas and results for step-invariant systems have counterparts for step-varying systems. The z-transform is generally of no help in finding the response of a step-varying system, though, and matrix eigenvalues do not have the same significance for step-varying systems as they do for step-invariant ones. The response of a step-varying, discrete-time system can be calculated recursively from the state equations (3-18), the initial state $\mathbf{x}(0)$, and the input $\mathbf{u}(k)$ as follows:

$$\mathbf{x}(1) = \mathbf{A}(0)\mathbf{x}(0) + \mathbf{B}(0)\mathbf{u}(0)$$
$$\mathbf{x}(2) = \mathbf{A}(1)\mathbf{x}(1) + \mathbf{B}(1)\mathbf{u}(1)$$
$$= \mathbf{A}(1)\mathbf{A}(0)\mathbf{x}(0) + \mathbf{A}(1)\mathbf{B}(0)\mathbf{u}(0) + \mathbf{B}(1)\mathbf{u}(1)$$
$$\mathbf{x}(3) = \mathbf{A}(2)\mathbf{x}(2) + \mathbf{B}(2)\mathbf{u}(2)$$
$$= \mathbf{A}(2)\mathbf{A}(1)\mathbf{A}(0)\mathbf{x}(0) + \mathbf{A}(2)\mathbf{A}(1)\mathbf{B}(0)\mathbf{u}(0) + \mathbf{A}(2)\mathbf{B}(1)\mathbf{u}(1)$$
$$+ \mathbf{B}(2)\mathbf{u}(2)$$
$$\vdots$$

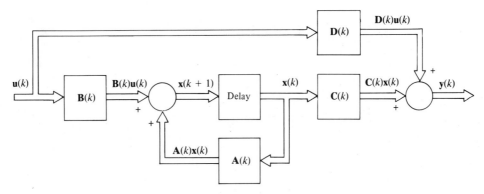

Figure 3-12 Delay diagram representation of a step-varying system. The signals are vectors.

$$\begin{aligned}
\mathbf{x}(k) = {} & \mathbf{A}(k-1)\mathbf{A}(k-2)\cdots\mathbf{A}(0)\mathbf{x}(0) + \mathbf{A}(k-1)\cdots\mathbf{A}(1)\mathbf{B}(0)\mathbf{u}(0) \\
& + \mathbf{A}(k-1)\cdots\mathbf{A}(2)\mathbf{B}(1)\mathbf{u}(1) + \cdots \\
& + \mathbf{A}(k-1)\mathbf{B}(k-2)\mathbf{u}(k-2) + \mathbf{B}(k-1)\mathbf{u}(k-1)
\end{aligned}$$

$$\vdots$$

A product of state coupling matrices

$$\mathbf{\Phi}(j, i) = \mathbf{A}(j-1)\mathbf{A}(j-2)\cdots\mathbf{A}(i)$$

is the state transition matrix of the system.

The definitions of stability for step-varying, linear, discrete-time systems are many and varied. Perhaps the best general definition is the following: A system with state equations

$$\mathbf{x}(k+1) = \mathbf{A}(k)\mathbf{x}(k) + \mathbf{B}(k)\mathbf{u}(k)$$

is *zero-input stable* if and only if for every set of finite initial conditions $\mathbf{x}_{\text{zero-input}}(0)$, the zero-input component of the state, governed by

$$\mathbf{x}_{\text{zero-input}}(k+1) = \mathbf{A}(k)\mathbf{x}_{\text{zero-input}}(k)$$

approaches zero with step. That is,

$$\lim_{k\to\infty} \|\mathbf{x}_{\text{zero-input}}(k)\| = 0$$

where the symbol $\|.\|$ denotes the Euclidean norm of the quantity.

The system is *zero-state stable* if and only if for zero initial conditions and every bounded input

$$\|\mathbf{u}(k)\| < \delta \quad k = 0, 1, 2, \ldots$$

the zero-state component of the state, governed by

$$\begin{cases}
\mathbf{x}_{\text{zero-state}}(k+1) = \mathbf{A}(k)\mathbf{x}_{\text{zero-state}}(k) + \mathbf{B}(k)\mathbf{u}(k) \\
\mathbf{x}_{\text{zero-state}}(0) = \mathbf{0}
\end{cases}$$

is bounded:

$$\|\mathbf{x}_{\text{zero-state}}(k)\| < \sigma \quad k = 0, 1, 2, \ldots$$

A linear, discrete-time system is *stable* if it is both zero-input stable and zero-state stable.

The system with state equations

$$\begin{bmatrix} x_1(k+1) \\ x_2(k+1) \end{bmatrix} = \begin{bmatrix} k & 0 \\ 0 & \frac{1}{2} \end{bmatrix} \begin{bmatrix} x_1(k) \\ x_2(k) \end{bmatrix} + \begin{bmatrix} 0 \\ 1 \end{bmatrix} u(k)$$

is zero-state stable but not zero-input stable. The system

$$\begin{bmatrix} x_1(k+1) \\ x_2(k+1) \end{bmatrix} = \begin{bmatrix} \frac{1}{3} & 0 \\ 0 & \frac{1}{2} \end{bmatrix} \begin{bmatrix} x_1(k) \\ x_2(k) \end{bmatrix} + \begin{bmatrix} k! \\ 1 \end{bmatrix} u(k)$$

is zero-input stable but not zero-state stable. Although these examples are somewhat contrived, they serve to illustrate the ideas involved. Having defined stability, it may not be an easy matter to determine whether a given step-varying system is stable. There is the possibility, though, whether the system is stable or not, of using feedback to obtain an acceptable response.

3.4 Sampled Continuous-Time Systems

The relationship between continuous-time, state variable plant models and discrete-time models of plant signal samples is now examined. Other methods for the computation of system response are also discussed.

3.4.1 Discrete-Time Models of Continuous-Time Systems

Consider a continuous-time system in state variable form

$$\dot{\mathbf{x}}(t) = \mathcal{A}\mathbf{x}(t) + \mathcal{B}\mathbf{u}(t)$$
$$\mathbf{y}(t) = \mathbf{C}\mathbf{x}(t) + \mathbf{D}\mathbf{u}(t) \tag{3-19}$$

Script symbols are now used for the state and input coupling matrices to distinguish between these and the corresponding matrices in discrete-time models. The state as a function of the initial conditions and the input in equation (3-19) is

$$\mathbf{x}(t) = e^{\mathcal{A}t}\mathbf{x}(0) + \int_0^t e^{\mathcal{A}(t-\tau)}\mathcal{B}\mathbf{u}(\tau)\, d\tau$$

At the sample times kT, $k = 0, 1, 2, \ldots$ the state is

$$\mathbf{x}(kT) = e^{\mathcal{A}kT}\mathbf{x}(0) + \int_0^{kT} e^{\mathcal{A}(kT-\tau)}\mathcal{B}\mathbf{u}(\tau)\, d\tau$$

The state at the $(k + 1)$th step can be expressed in terms of the state at the kth step as follows:

$$\mathbf{x}(kT + T) = e^{\mathcal{A}(kT+T)}\mathbf{x}(0) + \int_0^{kT+T} e^{\mathcal{A}(kT+T-\tau)}\mathcal{B}\mathbf{u}(\tau)\, d\tau$$

$$= e^{\mathscr{A}T} e^{\mathscr{A}kT} \mathbf{x}(0) + \int_0^{kT} e^{\mathscr{A}(kT+T-\tau)} \mathscr{B} \mathbf{u}(\tau) \, d\tau$$

$$+ \int_{kT}^{kT+T} e^{\mathscr{A}(kT+T-\tau)} \mathscr{B} \mathbf{u}(\tau) \, d\tau$$

$$= e^{\mathscr{A}T} \left[e^{\mathscr{A}kT} \mathbf{x}(0) + \int_0^{kT} e^{\mathscr{A}(kT-\tau)} \mathscr{B} \mathbf{u}(\tau) \, d\tau \right]$$

$$+ \int_{kT}^{kT+T} e^{\mathscr{A}(kT+T-\tau)} \mathscr{B} \mathbf{u}(\tau) \, d\tau$$

$$= e^{\mathscr{A}T} \mathbf{x}(kT) + (\text{input term})$$

The input term, involving as it does a weighted integral of $\mathbf{u}(t)$, is not generally proportional to samples of the input. When the input is constant during each sampling interval, however, as it is when it is driven by sample-and-hold signals, the input term is

$$\int_{kT}^{kT+T} e^{\mathscr{A}(kT+T-\tau)} \mathscr{B} \mathbf{u}(\tau) \, d\tau = \left[\int_{kT}^{kT+T} e^{\mathscr{A}(kT+T-\tau)} \, d\tau \right] \mathscr{B} \mathbf{u}(kT)$$

which is proportional to the input samples, the discrete-time input coupling matrix being

$$\mathbf{B} = \left[\int_{kT}^{kT+T} e^{\mathscr{A}(kT+T-\tau)} \, d\tau \right] \mathscr{B}$$

Letting

$$\gamma = kT + T - \tau \qquad d\gamma = -d\tau$$

then

$$\mathbf{B} = \left[\int_0^T e^{\mathscr{A}\gamma} \, d\gamma \right] \mathscr{B}$$

Expanding the integrand into a power series

$$e^{\mathscr{A}\gamma} = \mathbf{I} + \frac{\mathscr{A}\gamma}{1!} + \frac{\mathscr{A}^2\gamma^2}{2!} + \cdots + \frac{\mathscr{A}^i\gamma^i}{i!} + \cdots$$

and integrating term by term results in

$$\mathbf{B} = \left\{ \int_0^T \left[\mathbf{I} + \frac{\mathscr{A}\gamma}{1!} + \frac{\mathscr{A}^2\gamma^2}{2!} + \cdots + \frac{\mathscr{A}^i\gamma^i}{i!} + \cdots \right] d\gamma \right\} \mathscr{B}$$

$$= \left[\mathbf{I}T + \frac{\mathscr{A}T^2}{2!} + \frac{\mathscr{A}^2T^3}{3!} + \cdots + \frac{\mathscr{A}^iT^{i+1}}{(i+1)!} + \cdots \right] \mathscr{B}$$

Because

$$\mathscr{A}\mathbf{B} = \left[\frac{\mathscr{A}T}{1!} + \frac{\mathscr{A}^2T^2}{2!} + \cdots + \frac{\mathscr{A}^{i+1}T^{i+1}}{(i+1)!} + \cdots \right]\mathscr{B} = (e^{\mathscr{A}T} - \mathbf{I})\mathscr{B}$$

and if $\mathscr{A}$ is nonsingular, then

$$\mathbf{B} = \mathscr{A}^{-1}(e^{\mathscr{A}T} - \mathbf{I})\mathscr{B} = (e^{\mathscr{A}T} - \mathbf{I})\mathscr{A}^{-1}\mathscr{B}$$

The discrete-time model of equation (3-19) is then

$$\mathbf{x}[(k + 1)T] = \mathbf{A}\mathbf{x}(kT) + \mathbf{B}\mathbf{u}(kT)$$
$$\mathbf{y}(kT) = \mathbf{C}\mathbf{x}(kT) + \mathbf{D}\mathbf{u}(kT)$$

or

$$\mathbf{x}(k + 1) = \mathbf{A}\mathbf{x}(k) + \mathbf{B}\mathbf{u}(k)$$
$$\mathbf{y}(k) = \mathbf{C}\mathbf{x}(k) + \mathbf{D}\mathbf{u}(k)$$

where

$$\mathbf{A} = e^{\mathscr{A}T} = \mathbf{I} + \frac{\mathscr{A}T}{1!} + \frac{\mathscr{A}^2T^2}{2!} + \cdots + \frac{\mathscr{A}^iT^i}{i!} + \cdots \qquad \text{(3-20)}$$

$$\mathbf{B} = \left[\mathbf{I}T + \frac{\mathscr{A}T^2}{2!} + \frac{\mathscr{A}^2T^3}{3!} + \cdots + \frac{\mathscr{A}^iT^{i+1}}{(i+1)!} + \cdots \right]\mathscr{B} \qquad \text{(3-21a)}$$

and where

$$\mathbf{B} = \mathscr{A}^{-1}[e^{\mathscr{A}T} - \mathbf{I}]\mathscr{B} = [e^{\mathscr{A}T} - \mathbf{I}]\mathscr{A}^{-1}\mathscr{B} \qquad \text{(3-21b)}$$

when $\mathscr{A}$ is nonsingular.

As a numerical example, consider the continuous-time system

$$\begin{bmatrix} \dot{x}_1(t) \\ \dot{x}_2(t) \end{bmatrix} = \begin{bmatrix} -2 & 2 \\ 1 & -3 \end{bmatrix}\begin{bmatrix} x_1(t) \\ x_2(t) \end{bmatrix} + \begin{bmatrix} -1 \\ 5 \end{bmatrix} u(t)$$

$$y(t) = [2 \quad -4]\begin{bmatrix} x_1(t) \\ x_2(t) \end{bmatrix} + 6u(t)$$

with a sampling interval $T = 0.2$. The matrix exponential is

$$e^{\mathscr{A}T} = e^{0.2\mathscr{A}} = \begin{bmatrix} 0.696 & 0.246 \\ 0.123 & 0.572 \end{bmatrix}$$

which can be calculated by truncating the power series

$$e^{\mathscr{A}T} \cong \mathbf{I} + \mathscr{A}T + \frac{(\mathscr{A}T)^2}{2!} + \frac{(\mathscr{A}T)^3}{3!} + \cdots + \frac{(\mathscr{A}T)^i}{i!}$$

By examining the finite series as more and more terms are added, it can be decided when to truncate the series. It is good to bear in mind, however, that there are pathologic matrices for which the series converges slowly, for which the series seems to converge first to one matrix then to another, and for which numerical rounding can give misleading results.

Continuing with the example, if the input $u(t)$ is constant in each interval from kT to $kT + T$, the input term in the discrete-time model is proportional to the input samples and has the form

$$\mathbf{b} = [e^{\mathscr{A}T} - \mathbf{I}]\mathscr{A}^{-1}\mathscr{b}$$

$$= \begin{bmatrix} -0.304 & 0.246 \\ 0.123 & -0.428 \end{bmatrix} \begin{bmatrix} -\frac{3}{4} & -\frac{1}{2} \\ -\frac{1}{4} & -\frac{1}{2} \end{bmatrix} \begin{bmatrix} -1 \\ 5 \end{bmatrix}$$

$$= \begin{bmatrix} -0.021 \\ 0.747 \end{bmatrix}$$

which could also have been found using a truncated series in equation (3-20). The discrete-time model of the continuous-time system is then

$$\begin{bmatrix} x_1(kT + T) \\ x_2(kT + T) \end{bmatrix} = \begin{bmatrix} 0.696 & 0.246 \\ 0.123 & 0.572 \end{bmatrix} \begin{bmatrix} x_1(kT) \\ x_2(kT) \end{bmatrix} + \begin{bmatrix} -0.021 \\ 0.747 \end{bmatrix} u(kT) \qquad \textbf{(3-22)}$$

$$y(kT) = \begin{bmatrix} 2 & -4 \end{bmatrix} \begin{bmatrix} x_1(kT) \\ x_2(kT) \end{bmatrix} + 6u(kT)$$

3.4.2 Approximation Methods

Discrete-time approximations of continuous-time systems described by state variable equations can be obtained by integrating equation (3-19) as follows:

$$\mathbf{x}(t) = \mathbf{x}(t_0) + \int_{t_0}^{t} [\mathscr{A}\mathbf{x}(t) + \mathscr{B}\mathbf{u}(t)] \, dt$$

For evenly spaced samples, at $t = kT$, $k = 0, 1, 2, \ldots$

$$\mathbf{x}(kT + T) = \mathbf{x}(kT) + \int_{kT}^{kT+T} [\mathscr{A}\mathbf{x}(t) + \mathscr{B}\mathbf{u}(t)] \, dt \qquad \textbf{(3-23a)}$$

Applying Euler's forward rectangular approximation of the integral gives

$$\mathbf{x}(kT + T) \cong \mathbf{x}(kT) + [\mathscr{A}\mathbf{x}(kT) + \mathscr{B}\mathbf{u}(kT)]T$$

or

$$\mathbf{x}(kT + T) \cong [\mathbf{I} + \mathscr{A}T]\mathbf{x}(kT) + \mathscr{B}T\mathbf{u}(kT) \qquad \textbf{(3-23b)}$$

Beginning with the known initial conditions $\mathbf{x}(0)$, then

$$\mathbf{x}(T) \cong [\mathbf{I} + \mathcal{A}T]\mathbf{x}(0) + \mathcal{B}T\mathbf{u}(0)$$
$$\mathbf{x}(2T) \cong [\mathbf{I} + \mathcal{A}T]\mathbf{x}(T) + \mathcal{B}T\mathbf{u}(T)$$
$$\mathbf{x}(3T) \cong [\mathbf{I} + \mathcal{A}T]\mathbf{x}(2T) + \mathcal{B}T\mathbf{u}(2T)$$
$$\vdots$$

and so on.

For example, consider the continuous-time system discussed in the previous subsection

$$\begin{bmatrix} \dot{x}_1(t) \\ \dot{x}_2(t) \end{bmatrix} = \begin{bmatrix} -2 & 2 \\ 1 & -3 \end{bmatrix}\begin{bmatrix} x_1(t) \\ x_2(t) \end{bmatrix} + \begin{bmatrix} -1 \\ 5 \end{bmatrix}u(t)$$

$$y(t) = \begin{bmatrix} 2 & -4 \end{bmatrix}\begin{bmatrix} x_1(t) \\ x_2(t) \end{bmatrix} + 6u(t)$$

with a sampling interval $T = 0.2$ sec. Using Euler's forward rectangular rule gives

$$\begin{bmatrix} x_1(kT + T) \\ x_2(kT + T) \end{bmatrix} \cong \left\{ \begin{bmatrix} 1 & 0 \\ 0 & 1 \end{bmatrix} + \begin{bmatrix} -2 & 2 \\ 1 & -3 \end{bmatrix}(0.2) \right\}\begin{bmatrix} x_1(kT) \\ x_2(kT) \end{bmatrix} + \begin{bmatrix} -1 \\ 5 \end{bmatrix}(0.2)u(kT)$$

or

$$\begin{bmatrix} x_1(kT + T) \\ x_2(kT + T) \end{bmatrix} \cong \begin{bmatrix} 0.6 & 0.4 \\ 0.2 & 0.4 \end{bmatrix}\begin{bmatrix} x_1(kT) \\ x_2(kT) \end{bmatrix} + \begin{bmatrix} -0.2 \\ 1 \end{bmatrix}u(kT)$$

and

$$y(kT) = \begin{bmatrix} 2 & -4 \end{bmatrix}\begin{bmatrix} x_1(kT) \\ x_2(kT) \end{bmatrix} + 6u(kT)$$

which does not match well with the result given by equation (3-22). However, reducing the sampling interval to $T = 0.01$ sec, Euler's approximation gives

$$\begin{bmatrix} x_1(kT + T) \\ x_2(kT + T) \end{bmatrix} \cong \begin{bmatrix} 0.98 & 0.02 \\ 0.01 & 0.97 \end{bmatrix}\begin{bmatrix} x_1(kT) \\ x_2(kT) \end{bmatrix} + \begin{bmatrix} -0.01 \\ 0.05 \end{bmatrix}u(kT)$$

$$y(kT) = \begin{bmatrix} 2 & -4 \end{bmatrix}\begin{bmatrix} x_1(kT) \\ x_2(kT) \end{bmatrix} + 6u(kT)$$

For $T = 0.01$ sec, equations (3-20) and (3-21b) result in

$$\begin{bmatrix} x_1(kT + T) \\ x_2(kT + T) \end{bmatrix} \cong \begin{bmatrix} 0.98 & 0.0195 \\ 0.0097 & 0.97 \end{bmatrix}\begin{bmatrix} x_1(kT) \\ x_2(kT) \end{bmatrix} + \begin{bmatrix} -0.0082 \\ 0.0505 \end{bmatrix}u(kT)$$

which is in close agreement with Euler's result. The comparison is more evident by letting $\tau' = \tau - kT$ in the solution of the state equation

$$\mathbf{x}(kT + T) = e^{\mathcal{A}T}\mathbf{x}(kT) + \int_{kT}^{kT+T} e^{\mathcal{A}(kT+T-\tau)}\mathcal{B}\mathbf{u}(\tau) \, d\tau$$

then

$$\mathbf{x}(kT + T) = e^{\mathcal{A}T}\mathbf{x}(kT) + \int_{0}^{T} e^{\mathcal{A}(T-\tau')}\mathcal{B}\mathbf{u}(kT + \tau') \, d\tau'$$

$$= e^{\mathcal{A}T}[\mathbf{x}(kT) + \int_{0}^{T} e^{-\mathcal{A}\tau'}\mathcal{B}\mathbf{u}(kT + \tau') \, d\tau']$$

Expressing the matrix exponential in series form,

$$e^{\mathcal{A}T} = \mathbf{I} + \mathcal{A}T + \frac{1}{2!} \mathcal{A}^2T^2 + \frac{1}{3!} \mathcal{A}^3T^3 + \cdots$$

it is seen how the exact solution

$$\mathbf{x}(kT + T) = \left[\mathbf{I} + \mathcal{A}T + \frac{1}{2!} \mathcal{A}^2T^2 + \frac{1}{3!} \mathcal{A}^3T^3 + \cdots\right]\mathbf{x}(kT)$$

$$+ \int_{0}^{T} \left[\mathbf{I} + \mathcal{A}(T - \tau') + \frac{1}{2!} \mathcal{A}^2(T - \tau')^2\right.$$

$$\left. + \frac{1}{3!} \mathcal{A}^3(T - \tau')^3 + \cdots\right]\mathcal{B}\mathbf{u}(kT + \tau') \, d\tau'$$

compares with Euler's forward rectangular approximation

$$\mathbf{x}(kT + T) \cong \mathbf{x}(kT) + [\mathcal{A}\mathbf{x}(kT) + \mathcal{B}\mathbf{u}(kT)]T$$

On the other hand, applying Euler's backward rectangular approximation of the integral in equation (3-23a) gives

$$\mathbf{x}(kT + T) \cong \mathbf{x}(kT) + [\mathcal{A}\mathbf{x}(kT + T) + \mathcal{B}\mathbf{u}(kT + T)]T$$

or

$$[\mathbf{I} - \mathcal{A}T]\mathbf{x}(kT + T) \cong \mathbf{x}(kT) + \mathcal{B}T\mathbf{u}(kT + T)$$

Hence

$$\mathbf{x}(kT + T) \cong [\mathbf{I} - \mathcal{A}T]^{-1}\mathbf{x}(kT) + [\mathbf{I} - \mathcal{A}T]^{-1}\mathcal{B}T\mathbf{u}(kT + T)$$

Letting

$$\mathbf{x}'(kT + T) = \mathbf{x}(kT)$$

then

$$[\mathbf{I} - \mathcal{A}T]\mathbf{x}(kT + T) \cong \mathbf{x}'(kT + T) + \mathcal{B}T\mathbf{u}(kT + T)$$

or

$$[\mathbf{I} - \mathcal{A}T]\mathbf{x}(kT) \cong \mathbf{x}'(kT) + \mathcal{B}T\mathbf{u}(kT)$$

Hence

$$\mathbf{x}(kT) \cong [\mathbf{I} - \mathcal{A}T]^{-1}\mathbf{x}'(kT) + [\mathbf{I} - \mathcal{A}T]^{-1}\mathcal{B}T\mathbf{u}(kT)$$

and therefore,

$$\mathbf{x}'(kT + T) \cong [\mathbf{I} - \mathcal{A}T]^{-1}\mathbf{x}'(kT) + [\mathbf{I} - \mathcal{A}T]^{-1}\mathcal{B}T\mathbf{u}(kT)$$

The output equation

$$\mathbf{y}(kT) = \mathbf{C}\mathbf{x}(kT) + \mathbf{D}\mathbf{u}(kT)$$

in terms of the new variable $\mathbf{x}'$, becomes

$$\mathbf{y}(kT) \cong \mathbf{C}[\mathbf{I} - \mathcal{A}T]^{-1}\mathbf{x}'(kT) + \mathbf{C}[\mathbf{I} - \mathcal{A}T]^{-1}\mathcal{B}T\mathbf{u}(kT) + \mathbf{D}\mathbf{u}(kT)$$

or

$$\mathbf{y}(kT) \cong \mathbf{C}[\mathbf{I} - \mathcal{A}T]^{-1}\mathbf{x}'(kT) + \{\mathbf{C}[\mathbf{I} - \mathcal{A}T]^{-1}\mathcal{B}T + \mathbf{D}\}\mathbf{u}(kT)$$

Derivative approximations, such as those given in Table 7-2 in Chapter 7, are also possibilities for discretizing continuous-time state equations.

Also, improved accuracy and a reduced sampling interval may result from using more involved approximations, such as predictor–correctors, or Runge-Kutta methods.

3.5 Canonical Forms

In this section, two special canonical forms for state equations—controllable form and observable form—are discussed. Using them, it is especially easy to synthesize systems having desired z-transfer functions. Any nonsingular change of state variables is shown to result in new state variable equations with no change in the z-transfer functions relating the system's outputs and inputs.

3.5.1 Controllable Form

A particularly convenient arrangement for obtaining state equations for a single-input system with specified transfer functions is the *controllable* (or *phase variable*) *form*, shown with a single output in Figure 3-13.

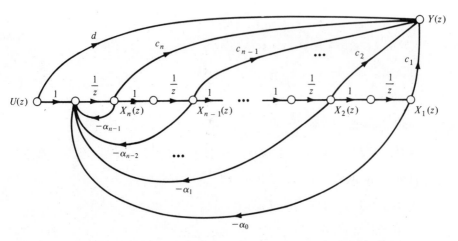

Figure 3-13 Controllable form of a single-input/single-output, discrete-time system.

Table 3-2 Signal Flow Graph Definitions and Mason's Gain Rule

Path:	A succession of branches, from input to output, in the direction of the arrows, which does not pass any node more than once.
Path Gain:	Product of the transmittances of the branches of the path. For the ith path, the path gain is denoted by P_i.
Loop:	A closed succession of branches, in the direction of the arrows, which does not pass any node more than once.
Loop Gain:	Product of the transmittances of the branches of the loop.
Touching:	Loops with one or more nodes in common are termed *touching*. A loop and a path are touching if they have a common node.
Determinant:	The determinant of a signal flow graph is $\Delta = 1 -$ (sum of all loop gains) + (sum of products of gains of all combinations of 2 nontouching loops) − (sum of products of gains of all combinations of 3 nontouching loops) + · · ·
Cofactor:	The cofactor of the ith path, denoted by Δ_i, is the determinant of the signal flow graph formed by deleting all loops touching path i.
Mason's Gain Rule:	$T(z) = \dfrac{P_1\Delta_1 + P_2\Delta_2 + \cdots}{\Delta}$

Applying Mason's gain rule (see Table 3-2 for a summary), the system z-transfer function is

$$T(z) = \frac{d\Delta + P_1\Delta_1 + P_2\Delta_2 + P_3\Delta_3 + \cdots}{\Delta} = d + \frac{P_1 + P_2 + P_3 + \cdots}{\Delta}$$

$$= d + \frac{(c_n/z) + (c_{n-1}/z^2) + \cdots + (c_2/z^{n-1}) + (c_1/z^n)}{1 + (\alpha_{n-1}/z) + (\alpha_{n-2}/z^2) + \cdots + (\alpha_1/z^{n-1}) + (\alpha_0/z^n)}$$

$$= d + \frac{c_n z^{n-1} + c_{n-1}z^{n-2} + \cdots + c_2 z + c_1}{z^n + \alpha_{n-1}z^{n-1} + \alpha_{n-2}z^{n-2} + \cdots + \alpha_1 z + \alpha_0}$$

Except for the path d, each path and each loop share the node at the input to the leftmost delay, so all of the path cofactors are unity, and there are no terms for the product of loop gain in the determinant of the signal flow graph.

The state and output equations for the system are, from the delay diagram,

$$\begin{bmatrix} x_1(k+1) \\ x_2(k+1) \\ x_3(k+1) \\ \vdots \\ x_{n-1}(k+1) \\ x_n(k+1) \end{bmatrix} = \begin{bmatrix} 0 & 1 & 0 & \cdots & 0 & 0 \\ 0 & 0 & 1 & \cdots & 0 & 0 \\ 0 & 0 & 0 & \cdots & 0 & 0 \\ \vdots & & & & & \\ 0 & 0 & 0 & \cdots & 0 & 1 \\ -\alpha_0 & -\alpha_1 & -\alpha_2 & \cdots & -\alpha_{n-2} & -\alpha_{n-1} \end{bmatrix} \begin{bmatrix} x_1(k) \\ x_2(k) \\ x_3(k) \\ \vdots \\ x_{n-1}(k) \\ x_n(k) \end{bmatrix} + \begin{bmatrix} 0 \\ 0 \\ 0 \\ \vdots \\ 0 \\ 1 \end{bmatrix} u(k)$$

$$y(k) = [c_1 \quad c_2 \quad \cdots \quad c_{n-1} \quad c_n] \begin{bmatrix} x_1(k) \\ x_2(k) \\ \vdots \\ x_{n-1}(k) \\ x_n(k) \end{bmatrix} + du(k)$$

so it is a simple matter to find state and output equations of this form with a given z-transfer function.

Single-input/multiple-output systems with specified z-transfer functions, provided that the transfer function numerator degrees are no higher than that of the denominator, can be synthesized by simply appending additional output equations to the state variable description.

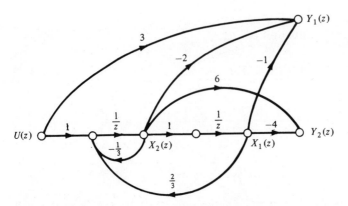

Figure 3-14 A multiple-output, discrete-time system in controllable form.

For example, a single-input/two-output system with z-transfer functions

$$T_1(z) = \frac{Y_1(z)}{U(z)}\bigg|_{\text{zero initial conditions}} = \frac{3z^2 - z - 3}{z^2 + \frac{1}{3}z - \frac{2}{3}} = 3 + \frac{-(2/z) - (1/z^2)}{1 + (\frac{1}{3}/z) - (\frac{2}{3}/z^2)}$$

$$T_2(z) = \frac{Y_2(z)}{U(z)}\bigg|_{\text{zero initial conditions}} = \frac{6}{z+1} = \frac{6(z - \frac{2}{3})}{(z+1)(z - \frac{2}{3})}$$

$$= \frac{6z - 4}{z^2 + \frac{1}{3}z - \frac{2}{3}} = \frac{(6/z) - (4/z^2)}{1 + (\frac{1}{3}/z) - (\frac{2}{3}/z^2)}$$

has the delay diagram given in Figure 3-14. Its controllable form state and output equations are

$$\begin{bmatrix} x_1(k+1) \\ x_2(k+1) \end{bmatrix} = \begin{bmatrix} 0 & 1 \\ \frac{2}{3} & -\frac{1}{3} \end{bmatrix} \begin{bmatrix} x_1(k) \\ x_2(k) \end{bmatrix} + \begin{bmatrix} 0 \\ 1 \end{bmatrix} u(k)$$

$$\begin{bmatrix} y_1(k) \\ y_2(k) \end{bmatrix} = \begin{bmatrix} -1 & -2 \\ -4 & 6 \end{bmatrix} \begin{bmatrix} x_1(k) \\ x_2(k) \end{bmatrix} + \begin{bmatrix} 3 \\ 0 \end{bmatrix} u(k)$$

3.5.2 Observable Form

The structure for the delay diagram of a single-input/single-output system in *observable form* is shown in Figure 3-15. In applying Mason's gain rule to find the system's transfer function, every path, except the uppermost one, and every loop pass through the X_1 node. Because

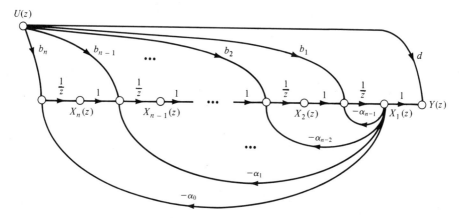

Figure 3-15 Observable form of a single-input/single-output, discrete-time system.

every one of the loops touches each of these paths, their cofactors are all unity. Because each of the loops touches one another, the determinant of the signal flow graph is simply unity minus the sum of the loop transmittances. The cofactor of the upper path is the same as the signal flow graph determinant because none of the loops touch it. The transfer function is then

$$T(s) = \frac{d\Delta + P_1\Delta_1 + P_2\Delta_2 + P_3\Delta_3 + \cdots}{\Delta} = d + \frac{P_1 + P_2 + P_3 + \cdots}{\Delta}$$

$$= d + \frac{(b_1/z) + (b_2/z^2) + \cdots + (b_{n-1}/z^{n-1}) + (b_n/z^n)}{1 + (\alpha_{n-1}/z) + (\alpha_{n-2}/z^2) + \cdots + (\alpha_1/z^{n-1}) + (\alpha_0/z^n)}$$

$$= d + \frac{b_1 z^{n-1} + b_2 z^{n-1} + \cdots + b_{n-1}z + b_n}{z^n + \alpha_{n-1}z^{n-1} + \alpha_{n-2}z^{n-2} + \cdots + \alpha_1 z + \alpha_0}$$

which has coefficients that are the gains in the signal flow graph. State variable equations for this system in observable form are

$$\begin{bmatrix} x_1(k+1) \\ x_2(k+1) \\ x_3(k+1) \\ \vdots \\ x_{n-1}(k+1) \\ x_n(k+1) \end{bmatrix} = \begin{bmatrix} -\alpha_{n-1} & 1 & 0 & \cdots & 0 & 0 \\ -\alpha_{n-2} & 0 & 1 & \cdots & 0 & 0 \\ -\alpha_{n-3} & 0 & 0 & \cdots & 0 & 0 \\ \vdots & & & & & \\ -\alpha_1 & 0 & 0 & \cdots & 0 & 1 \\ -\alpha_0 & 0 & 0 & \cdots & 0 & 0 \end{bmatrix} \begin{bmatrix} x_1(k) \\ x_2(k) \\ x_3(k) \\ \vdots \\ x_{n-1}(k) \\ x_n(k) \end{bmatrix} + \begin{bmatrix} b_1 \\ b_2 \\ b_3 \\ \vdots \\ b_{n-1} \\ b_n \end{bmatrix} u(k)$$

$$y(k) = [1 \quad 0 \quad 0 \quad \cdots \quad 0 \quad 0] \begin{bmatrix} x_1(k) \\ x_2(k) \\ x_3(k) \\ \vdots \\ x_{n-1}(k) \\ x_n(k) \end{bmatrix} + du(k)$$

To accommodate multiple inputs, the state variable equations are

$$\begin{bmatrix} x_1(k+1) \\ x_2(k+1) \\ \vdots \\ x_n(k+1) \end{bmatrix} = \begin{bmatrix} -\alpha_{n-1} & 1 & \cdots & 0 \\ -\alpha_{n-2} & 0 & \cdots & 0 \\ \vdots & & & \\ -\alpha_0 & 0 & \cdots & 0 \end{bmatrix} \begin{bmatrix} x_1(k) \\ x_2(k) \\ \vdots \\ x_n(k) \end{bmatrix}$$

$$+ \begin{bmatrix} b_{11} & b_{12} & \cdots & b_{1r} \\ b_{21} & b_{22} & \cdots & b_{2r} \\ \vdots & & & \\ b_{n1} & b_{n2} & \cdots & b_{nr} \end{bmatrix} \begin{bmatrix} u_1(k) \\ u_2(k) \\ \vdots \\ u_r(k) \end{bmatrix}$$

$$y(k) = [1 \quad 0 \quad \cdots \quad 0] \begin{bmatrix} x_1(k) \\ x_2(k) \\ \vdots \\ x_n(k) \end{bmatrix} + [d_1 \quad d_2 \quad \cdots \quad d_r] \begin{bmatrix} u_1(k) \\ u_2(k) \\ \vdots \\ u_r(k) \end{bmatrix}$$

and the z-transfer functions are

$$T_1(z) = \frac{Y(z)}{U_1(z)} \bigg|_{\substack{\text{zero initial} \\ \text{conditions}}} = d_1 + \frac{b_{11}z^{n-1} + b_{21}z^{n-2} + \cdots + b_{n1}}{z^n + \alpha_{n-1}z^{n-1} + \cdots + \alpha_1 z + \alpha_0}$$

$$T_2(z) = \frac{Y(z)}{U_2(z)} \bigg|_{\substack{\text{zero initial} \\ \text{conditions}}} = d_2 + \frac{b_{12}z^{n-1} + b_{22}z^{n-2} + \cdots + b_{n2}}{z^n + \alpha_{n-1}z^{n-1} + \cdots + \alpha_1 z + \alpha_0}$$

$$\vdots$$

$$T_r(z) = \frac{Y(z)}{U_r(z)} \bigg|_{\substack{\text{zero initial} \\ \text{conditions}}} = d_r + \frac{b_{1r}z^{n-1} + b_{2r}z^{n-2} + \cdots + b_{nr}}{z^n + \alpha_{n-1}z^{n-1} + \cdots + \alpha_1 z + \alpha_0}$$

For example, a single-input/single-output system with the z-transfer function

$$T(z) = \frac{-2z^3 + 2z^2 - z + 2}{z^3 + z^2 - z - \frac{3}{4}} = -2 + \frac{4z^2 - 3z + \frac{1}{2}}{z^3 + z^2 - z - \frac{3}{4}}$$

in observable form has the delay diagram of Figure 3-16(a) and has state variable equations

$$\begin{bmatrix} x_1(k+1) \\ x_2(k+1) \\ x_3(k+1) \end{bmatrix} = \begin{bmatrix} -1 & 1 & 0 \\ 1 & 0 & 1 \\ \frac{3}{4} & 0 & 0 \end{bmatrix} \begin{bmatrix} x_1(k) \\ x_2(k) \\ x_3(k) \end{bmatrix} + \begin{bmatrix} 4 \\ -3 \\ \frac{1}{2} \end{bmatrix} u(k)$$

$$y(k) = \begin{bmatrix} 1 & 0 & 0 \end{bmatrix} \begin{bmatrix} x_1(k) \\ x_2(k) \\ x_3(k) \end{bmatrix} - 2u(k)$$

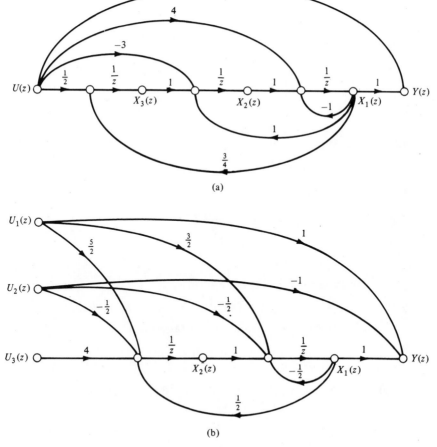

(a)

(b)

Figure 3-16 Examples of discrete-time systems in observable form. (a) A single-input system. (b) A multiple-input system.

The three-input/single-output system with the z-transfer functions

$$T_1(z) = \frac{Y(z)}{U_1(z)}\bigg|_{\substack{\text{zero initial}\\\text{conditions}}} = \frac{z^2 + 2z + 2}{z^2 + \frac{1}{2}z - \frac{1}{2}} = 1 + \frac{\frac{3}{2}z + \frac{5}{2}}{z^2 + \frac{1}{2}z - \frac{1}{2}}$$

$$T_2(z) = \frac{Y(z)}{U_2(z)}\bigg|_{\substack{\text{zero initial}\\\text{conditions}}} = \frac{-z}{z - \frac{1}{2}} = \frac{-z(z + 1)}{(z - \frac{1}{2})(z + 1)}$$

$$= -1 + \frac{-\frac{1}{2}z - \frac{1}{2}}{z^2 + \frac{1}{2}z - \frac{1}{2}}$$

$$T_3(z) = \frac{Y(z)}{U_3(z)}\bigg|_{\substack{\text{zero initial}\\\text{conditions}}} = \frac{4}{z^2 + \frac{1}{2}z - \frac{1}{2}}$$

has the observable form delay diagram of Figure 3-16(b) and the following state variable equations:

$$\begin{bmatrix} x_1(k + 1) \\ x_2(k + 1) \end{bmatrix} = \begin{bmatrix} -\frac{1}{2} & 1 \\ \frac{1}{2} & 0 \end{bmatrix} \begin{bmatrix} x_1(k) \\ x_2(k) \end{bmatrix} + \begin{bmatrix} \frac{3}{2} & -\frac{1}{2} & 0 \\ \frac{5}{2} & -\frac{1}{2} & 4 \end{bmatrix} \begin{bmatrix} u_1(k) \\ u_2(k) \\ u_3(k) \end{bmatrix}$$

$$y(k) = [1 \quad 0] \begin{bmatrix} x_1(k) \\ x_2(k) \end{bmatrix} + [1 \quad -1 \quad 0] \begin{bmatrix} u_1(k) \\ u_2(k) \\ u_3(k) \end{bmatrix}$$

3.5.3 Change of State Variables

A nonsingular change of state variables

$$\mathbf{x}(k) = \mathbf{P}\mathbf{x}'(k) \qquad \mathbf{x}'(k) = \mathbf{P}^{-1}\mathbf{x}(k)$$

in discrete-time state variable equations

$$\mathbf{x}(k + 1) = \mathbf{A}\mathbf{x}(k) + \mathbf{B}\mathbf{u}(k)$$
$$\mathbf{y}(k) = \mathbf{C}\mathbf{x}(k) + \mathbf{D}\mathbf{u}(k)$$

gives new equations of the same form

$$\mathbf{x}'(k + 1) = (\mathbf{P}^{-1}\mathbf{A}\mathbf{P})\mathbf{x}'(k) + (\mathbf{P}^{-1}\mathbf{B})\mathbf{u}(k) = \mathbf{A}'\mathbf{x}'(k) + \mathbf{B}'\mathbf{u}(k)$$
$$\mathbf{y}(k) = (\mathbf{C}\mathbf{P})\mathbf{x}'(k) + \mathbf{D}\mathbf{u}(k) = \mathbf{C}'\mathbf{x}'(k) + \mathbf{D}\mathbf{u}(k)$$

The system transfer function matrix is unchanged by a nonsingular change of state variables

$$\mathbf{T}'(z) = \mathbf{C}'(z\mathbf{I} - \mathbf{A}')^{-1}\mathbf{B}' + \mathbf{D} = \mathbf{C}\mathbf{P}(z\mathbf{P}^{-1}\mathbf{P} - \mathbf{P}^{-1}\mathbf{A}\mathbf{P})^{-1}\mathbf{P}^{-1}\mathbf{B} + \mathbf{D}$$
$$= \mathbf{C}\mathbf{P}[\mathbf{P}^{-1}(z\mathbf{I} - \mathbf{A})\mathbf{P}]^{-1}\mathbf{P}^{-1}\mathbf{B} + \mathbf{D}$$
$$= \mathbf{C}\mathbf{P}\mathbf{P}^{-1}(z\mathbf{I} - \mathbf{A})^{-1}\mathbf{P}\mathbf{P}^{-1}\mathbf{B} + \mathbf{D}$$
$$= \mathbf{C}(z\mathbf{I} - \mathbf{A})^{-1}\mathbf{B} + \mathbf{D} = \mathbf{T}(z)$$

Hence, the input–output relations of such a system have many realizations, a different one for each choice of state variables. The transformation of the state coupling matrix

$$\mathbf{A}' = \mathbf{P}^{-1}\mathbf{A}\mathbf{P}$$

is called a *similarity transformation,* which is discussed in detail in Appendix A.

When a nonsingular change of state variables, possibly step-varying,

$$\mathbf{x}(k) = \mathbf{P}(k)\mathbf{x}'(k) \qquad \mathbf{x}'(k) = \mathbf{P}^{-1}(k)\mathbf{x}(k)$$

is made for a step-varying system [equation (3-18)], the result is

$$\mathbf{P}(k + 1)\mathbf{x}'(k + 1) = \mathbf{A}(k)\mathbf{P}(k)\mathbf{x}'(k) + \mathbf{B}(k)\mathbf{u}(k)$$

or

$$\begin{aligned}
\mathbf{x}'(k + 1) &= [\mathbf{P}^{-1}(k + 1)\mathbf{A}(k)\mathbf{P}(k)]\mathbf{x}'(k) + [\mathbf{P}^{-1}(k + 1)\mathbf{B}(k)]\mathbf{u}(k) \\
&= \mathbf{A}'(k)\mathbf{x}'(k) + \mathbf{B}'(k)\mathbf{u}(k) \\
\mathbf{y}(k) &= [\mathbf{C}(k)\mathbf{P}(k)]\mathbf{x}'(k) + \mathbf{D}(k)\mathbf{u}(k) = \mathbf{C}'(k)\mathbf{x}'(k) + \mathbf{D}'(k)\mathbf{u}(k)
\end{aligned}$$

which is also in the state variable form. Unlike the step-invariant case, the new state coupling matrix

$$\mathbf{A}'(k) = \mathbf{P}^{-1}(k + 1)\mathbf{A}(k)\mathbf{P}(k)$$

is not a similarity transformation of $\mathbf{A}(k)$ unless the transformation $\mathbf{P}$ does not change with step.

A particularly interesting change of state variables is the one for which

$$\mathbf{P}(k + 1) = \mathbf{A}(k)\mathbf{P}(k) \qquad \mathbf{P}(0) = \mathbf{I} \qquad \mathbf{P}(k) = \prod_{i=1}^{k} \mathbf{A}(k - i)$$

When $\mathbf{A}(k)$ is nonsingular, as it always is when equation (3-18) is a discrete-time model of a time-invariant plant, the new state variable equations are

$$\begin{aligned}
\mathbf{x}'(k + 1) &= \mathbf{x}'(k) + \mathbf{P}^{-1}(k + 1)\mathbf{B}(k)\mathbf{u}(k) \\
\mathbf{y}(k) &= \mathbf{C}(k)\mathbf{P}(k)\mathbf{x}'(k) + \mathbf{D}(k)\mathbf{u}(k)
\end{aligned}$$

In this form, the state equations simply accumulate the effects of the inputs, and the rest of the state dynamics have been transferred to the output equations. If the $n \times n$ state coupling matrix $\mathbf{A}(k)$ is singular at any step, then at that step, the system can be described by less than n state variables.

3.6 Uncoupling State Equations

In this section, the solution of the characteristic value problem is applied to the problem of determining changes of state variables that take a system to a realization where its state coupling matrix is diagonal. Before doing so, the existence and structure of diagonal forms, found by expanding z-transfer functions into partial fractions, are examined.

3.6.1 Transformation to Diagonal Form

A single-input/single-output system with the z-transfer function

$$T(z) = \frac{2z^2 - \frac{1}{3}z + \frac{1}{6}}{(z - \frac{1}{2})(z + \frac{1}{2})(z - \frac{1}{3})} = \frac{3}{z - \frac{1}{2}} + \frac{1}{z + \frac{1}{2}} + \frac{-2}{z - \frac{1}{3}}$$

can be considered to be three tandem first-order subsystems, as shown in Figure 3-17(a). Expressing each of the subsystems in terms of a delay diagram as in Figure 3-17(b) results in state equations that have a diagonal state coupling matrix

$$\begin{bmatrix} x_1(k+1) \\ x_2(k+1) \\ x_3(k+1) \end{bmatrix} = \begin{bmatrix} \frac{1}{2} & 0 & 0 \\ 0 & -\frac{1}{2} & 0 \\ 0 & 0 & \frac{1}{3} \end{bmatrix} \begin{bmatrix} x_1(k) \\ x_2(k) \\ x_3(k) \end{bmatrix} + \begin{bmatrix} 3 \\ 1 \\ -2 \end{bmatrix} u(k)$$

$$y(k) = \begin{bmatrix} 1 & 1 & 1 \end{bmatrix} \begin{bmatrix} x_1(k) \\ x_2(k) \\ x_3(k) \end{bmatrix}$$

(3-24)

When the state coupling matrix of a system is diagonal, the individual state equations are decoupled from one another. The third-order difference equation represented by the system in equation (3-24) is in the form of three equivalent first-order equations

$$\begin{cases} x_1(k+1) = \dfrac{1}{2} x_1(k) + 3u(k) \\[2mm] x_2(k+1) = -\dfrac{1}{2} x_2(k) + u(k) \\[2mm] x_3(k+1) = \dfrac{1}{3} x_3(k) + 2u(k) \end{cases}$$

For a system

$$\mathbf{x}(k+1) = \mathbf{A}\mathbf{x}(k) + \mathbf{B}u(k)$$
$$\mathbf{y}(k) = \mathbf{C}\mathbf{x}(k) + \mathbf{D}u(k)$$

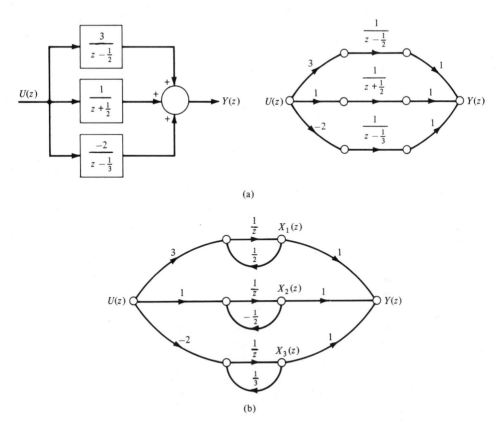

Figure 3-17 A diagonal single-input/single-output, discrete-time system. (a) Block diagram and signal flow graph of tandem first-order subsystems. (b) Subsystems in delay diagram form.

with distinct eigenvalues, finding the change of state variables

$$\mathbf{x}(k) = \mathbf{P}\mathbf{x}'(k) \qquad \mathbf{x}'(k) = \mathbf{P}^{-1}\mathbf{x}(k)$$
$$\mathbf{x}'(k + 1) = (\mathbf{P}^{-1}\mathbf{A}\mathbf{P})\mathbf{x}'(k) + (\mathbf{P}^{-1}\mathbf{B})\mathbf{u}(k)$$
$$\mathbf{y}(k) = (\mathbf{C}\mathbf{P})\mathbf{x}'(k) + \mathbf{D}\mathbf{u}(k)$$

that diagonalizes the state coupling matrix

$$\mathbf{P}^{-1}\mathbf{A}\mathbf{P} = \mathbf{\Lambda}$$

is the characteristic value problem discussed in detail in Appendix A. As a numerical example, the system

$$\begin{bmatrix} x_1(k + 1) \\ x_2(k + 1) \\ x_3(k + 1) \end{bmatrix} = \begin{bmatrix} -5 & 1 & 0 \\ -6 & 0 & 1 \\ 0 & 0 & 0 \end{bmatrix} \begin{bmatrix} x_1(k) \\ x_2(k) \\ x_3(k) \end{bmatrix} + \begin{bmatrix} 3 & 0 \\ 0 & 0 \\ 1 & -1 \end{bmatrix} \begin{bmatrix} u_1(k) \\ u_2(k) \end{bmatrix}$$

$$= \mathbf{A}x(k) + \mathbf{B}u(k)$$

$$\begin{bmatrix} y_1(k) \\ y_2(k) \end{bmatrix} = \begin{bmatrix} 1 & 2 & 0 \\ 0 & -1 & 0 \end{bmatrix} \begin{bmatrix} x_1(k) \\ x_2(k) \\ x_3(k) \end{bmatrix} + \begin{bmatrix} 2 & -1 \\ 0 & 0 \end{bmatrix} \begin{bmatrix} u_1(k) \\ u_2(k) \end{bmatrix}$$

$$= \mathbf{C}x(k) + \mathbf{D}u(k)$$

has real and distinct eigenvalues given by

$$|\lambda \mathbf{I} - \mathbf{A}| = \begin{vmatrix} (\lambda + 5) & -1 & 0 \\ 6 & \lambda & -1 \\ 0 & 0 & \lambda \end{vmatrix}$$

$$= \lambda^3 + 5\lambda^2 + 6\lambda = \lambda(\lambda + 2)(\lambda + 3) = 0$$

$$\lambda_1 = 0 \qquad \lambda_2 = -2 \qquad \lambda_3 = -3$$

The transformation $\mathbf{P}$ in

$$\mathbf{x} = \mathbf{P}\mathbf{x}' \qquad \mathbf{x}' = \mathbf{P}^{-1}\mathbf{x}$$

has columns that are the eigenvectors of $\mathbf{A}$.

The eigenvector corresponding to λ_1 satisfies

$$\begin{bmatrix} 5 & -1 & 0 \\ 6 & 0 & -1 \\ 0 & 0 & 0 \end{bmatrix} \begin{bmatrix} p_{11} \\ p_{21} \\ p_{31} \end{bmatrix} = \begin{bmatrix} 0 \\ 0 \\ 0 \end{bmatrix}$$

a nontrivial solution to which is

$$\begin{bmatrix} p_{11} \\ p_{21} \\ p_{31} \end{bmatrix} = \begin{bmatrix} 1 \\ 5 \\ 6 \end{bmatrix}$$

For λ_2

$$\begin{bmatrix} 3 & -1 & 0 \\ 6 & -2 & -1 \\ 0 & 0 & -2 \end{bmatrix} \begin{bmatrix} p_{12} \\ p_{22} \\ p_{32} \end{bmatrix} = \begin{bmatrix} 0 \\ 0 \\ 0 \end{bmatrix}; \qquad \begin{bmatrix} p_{12} \\ p_{22} \\ p_{32} \end{bmatrix} = \begin{bmatrix} 1 \\ 3 \\ 0 \end{bmatrix}$$

For λ_3

$$\begin{bmatrix} 2 & -1 & 0 \\ 6 & -3 & -1 \\ 0 & 0 & -3 \end{bmatrix} \begin{bmatrix} p_{13} \\ p_{23} \\ p_{33} \end{bmatrix} = \begin{bmatrix} 0 \\ 0 \\ 0 \end{bmatrix}; \qquad \begin{bmatrix} p_{13} \\ p_{23} \\ p_{33} \end{bmatrix} = \begin{bmatrix} 1 \\ 2 \\ 0 \end{bmatrix}$$

Then

$$\mathbf{P} = \begin{bmatrix} 1 & 1 & 1 \\ 5 & 3 & 2 \\ 6 & 0 & 0 \end{bmatrix}; \qquad \mathbf{P}^{-1} = \frac{1}{6} \begin{bmatrix} 0 & 0 & 1 \\ -12 & 6 & -3 \\ 18 & -6 & 2 \end{bmatrix}$$

$$\mathbf{A}' = \mathbf{P}^{-1}\mathbf{AP} = \frac{1}{6} \begin{bmatrix} 0 & 0 & 1 \\ -12 & 6 & -3 \\ 18 & -6 & 2 \end{bmatrix} \begin{bmatrix} -5 & 1 & 0 \\ -6 & 0 & 1 \\ 0 & 0 & 0 \end{bmatrix} \begin{bmatrix} 1 & 1 & 1 \\ 5 & 3 & 2 \\ 6 & 0 & 0 \end{bmatrix}$$

$$= \begin{bmatrix} 0 & 0 & 0 \\ 0 & -2 & 0 \\ 0 & 0 & -3 \end{bmatrix}$$

$$\mathbf{B}' = \mathbf{P}^{-1}\mathbf{B} = \frac{1}{6} \begin{bmatrix} 0 & 0 & 1 \\ -12 & 6 & -3 \\ 18 & -6 & 2 \end{bmatrix} \begin{bmatrix} 3 & 0 \\ 0 & 0 \\ 1 & -1 \end{bmatrix} = \frac{1}{6} \begin{bmatrix} 1 & -1 \\ -39 & 3 \\ 56 & -2 \end{bmatrix}$$

$$\mathbf{C}' = \mathbf{CP} = \begin{bmatrix} 1 & 2 & 0 \\ 0 & -1 & 0 \end{bmatrix} \begin{bmatrix} 1 & 1 & 1 \\ 5 & 3 & 2 \\ 6 & 0 & 0 \end{bmatrix} = \begin{bmatrix} 11 & 7 & 5 \\ -5 & -3 & -2 \end{bmatrix}$$

$$\mathbf{D}' = \mathbf{D} = \begin{bmatrix} 2 & -1 \\ 0 & 0 \end{bmatrix}$$

so that the system description in terms of the new state variables is

$$\begin{bmatrix} x_1'(k+1) \\ x_2'(k+1) \\ x_3'(k+1) \end{bmatrix} = \begin{bmatrix} 0 & 0 & 0 \\ 0 & -2 & 0 \\ 0 & 0 & -3 \end{bmatrix} \begin{bmatrix} x_1'(k) \\ x_2'(k) \\ x_3'(k) \end{bmatrix} + \begin{bmatrix} \frac{1}{6} & -\frac{1}{6} \\ -\frac{39}{6} & \frac{3}{6} \\ \frac{56}{6} & -\frac{2}{6} \end{bmatrix} \begin{bmatrix} u_1(k) \\ u_2(k) \end{bmatrix}$$

$$\begin{bmatrix} y_1(k) \\ y_2(k) \end{bmatrix} = \begin{bmatrix} 11 & 7 & 5 \\ -5 & -3 & -2 \end{bmatrix} \begin{bmatrix} x_1'(k) \\ x_2'(k) \\ x_3'(k) \end{bmatrix} + \begin{bmatrix} 2 & -1 \\ 0 & 0 \end{bmatrix} \begin{bmatrix} u_1(k) \\ u_2(k) \end{bmatrix}$$

and the state equations are decoupled from one another. A delay diagram for this system is shown in Figure 3-18.

3.6.2 Complex Eigenvalues

When the system eigenvalues are complex, diagonal state equations involve complex coefficients. It is often more convenient to combine each pair of nonrepeated, first-order, complex conjugate subsystems

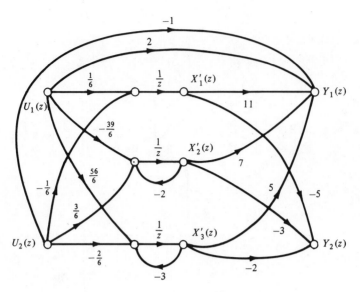

Figure 3-18 Delay diagram in diagonal form of a system with real, distinct eigenvalues.

into an equivalent second-order subsystem. The result is a state coupling matrix with 2×2 blocks along the diagonal for each complex eigenvalue pair.

For example, the single-input/single-output system with the z-transfer function

$$T(z) = \frac{4z^2 + 3z - \frac{3}{8}}{z^3 + \frac{1}{2}z^2 - (3z/16) - (5/32)}$$

$$= \frac{2}{z - \frac{1}{2}} + \frac{1 + j2}{z + \frac{1}{2} + j\frac{1}{4}} + \frac{1 - j2}{z + \frac{1}{2} - j\frac{1}{4}}$$

involves complex eigenvalues. A delay diagram of a diagonal system having this z-transfer function is shown in Figure 3-19(a). The corresponding state and output equations are

$$\begin{bmatrix} x_1(k+1) \\ x_2(k+1) \\ x_3(k+1) \end{bmatrix} = \begin{bmatrix} \frac{1}{2} & 0 & 0 \\ 0 & (-\frac{1}{2} - j\frac{1}{4}) & 0 \\ 0 & 0 & (-\frac{1}{2} + j\frac{1}{4}) \end{bmatrix} \begin{bmatrix} x_1(k) \\ x_2(k) \\ x_3(k) \end{bmatrix} + \begin{bmatrix} 1 \\ 1 \\ 1 \end{bmatrix} u(k)$$

$$y(k) = [2 \quad (1 + j2) \quad (1 - j2)] \begin{bmatrix} x_1(k) \\ x_2(k) \\ x_3(k) \end{bmatrix}$$

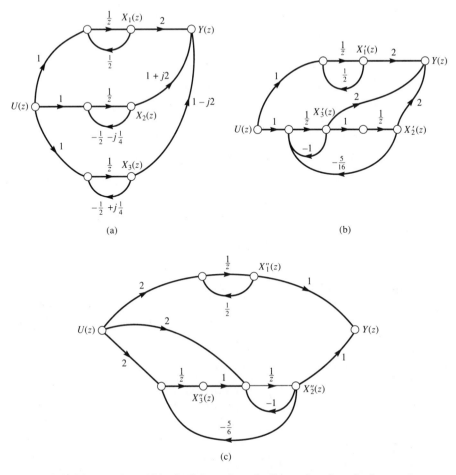

Figure 3-19 Diagonal and block diagonal realizations of a discrete-time system with complex eigenvalues. (a) System in diagonal form. (b) System in block diagonal controllable form. (c) Block diagonal representation of complex eigenvalues using observable form.

If the complex conjugate terms in the z-transfer function are combined, the individual terms each involve only real numbers

$$T(z) = \frac{2}{z - \frac{1}{2}} + \frac{2z + 2}{z^2 + z + (5/16)}$$

The second-order subsystem can be realized with real number coefficients in a variety of ways with a resulting 2×2 block along the diagonal of the system's state coupling matrix. Expanding this second-order subsystem in controllable form

$$\frac{2z + 2}{z^2 + z + (5/16)} = \frac{(2/z) + (2/z^2)}{1 + (1/z) + [(5/16)/z^2]}$$

gives the realization of Figure 3-19(b) for which the state and output equations are

$$\begin{bmatrix} x_1'(k + 1) \\ x_2'(k + 1) \\ x_3'(k + 1) \end{bmatrix} = \begin{bmatrix} \frac{1}{2} & 0 & 0 \\ 0 & 0 & 1 \\ 0 & -\frac{5}{16} & -1 \end{bmatrix} \begin{bmatrix} x_1'(k) \\ x_2'(k) \\ x_3'(k) \end{bmatrix} + \begin{bmatrix} 1 \\ 0 \\ 1 \end{bmatrix} u(k)$$

$$y(k) = [2 \quad 2 \quad 2] \begin{bmatrix} x_1'(k) \\ x_2'(k) \\ x_3'(k) \end{bmatrix}$$

Alternatively, a 2×2 block representing a pair of complex eigenvalues can have a state coupling like that in the observable form, as the matrix shown in the delay diagram of Figure 3-19(c), for which the block diagonal state and output equations are

$$\begin{bmatrix} x_1''(k + 1) \\ x_2''(k + 1) \\ x_3''(k + 1) \end{bmatrix} = \begin{bmatrix} \frac{1}{2} & 0 & 0 \\ 0 & -1 & 1 \\ 0 & -\frac{5}{6} & 0 \end{bmatrix} \begin{bmatrix} x_1''(k) \\ x_2''(k) \\ x_3''(k) \end{bmatrix} + \begin{bmatrix} 2 \\ 2 \\ 2 \end{bmatrix} u(k)$$

$$y(k) = [1 \quad 1 \quad 0] \begin{bmatrix} x_1''(k) \\ x_2''(k) \\ x_3''(k) \end{bmatrix}$$

Yet another convenient block diagonal form for complex eigenvalues is the *normal form*, in which each second-order subsystem has the structure shown in Figure 3-20. Using Mason's gain rule, the second-order system has the transmittance

$$G(z) = \frac{(b_1c_1/z)[1 - (\alpha/z)] + (b_2c_2/z)[1 - (\alpha/z)]}{1 - (\alpha/z) - (\alpha/z) + (\beta^2/z^2) + (\alpha^2/z^2)}$$

$$= \frac{(b_1c_1 + b_2c_2)(z - \alpha) + \beta(b_2c_1 - b_1c_2)}{(z - \alpha)^2 + \beta^2}$$

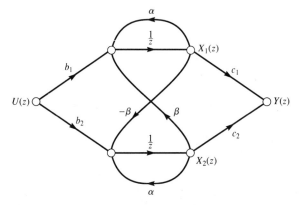

Figure 3-20 Normal form representation of complex eigenvalues.

Its complex conjugate eigenvalues are $\alpha \pm j\beta$.

· The system

$$
\begin{bmatrix} x_1(k+1) \\ x_2(k+1) \\ x_3(k+1) \\ x_4(k+1) \\ x_5(k+1) \end{bmatrix} = \begin{bmatrix} -\frac{1}{2} & \frac{1}{4} & 0 & 0 & 0 \\ -\frac{1}{4} & -\frac{1}{2} & 0 & 0 & 0 \\ 0 & 0 & -\frac{1}{3} & \frac{1}{5} & 0 \\ 0 & 0 & -\frac{1}{5} & -\frac{1}{3} & 0 \\ 0 & 0 & 0 & 0 & \frac{1}{6} \end{bmatrix} \begin{bmatrix} x_1(k) \\ x_2(k) \\ x_3(k) \\ x_4(k) \\ x_5(k) \end{bmatrix} + \begin{bmatrix} 2 \\ -1 \\ 3 \\ 5 \\ 2 \end{bmatrix} u(k)
$$

$$
y(k) = \begin{bmatrix} 1 & 0 & -3 & 2 & -1 \end{bmatrix} \begin{bmatrix} x_1(k) \\ x_2(k) \\ x_3(k) \\ x_4(k) \\ x_5(k) \end{bmatrix} + u(k)
$$

for example, is in normal form. It has the transfer function

$$
T(z) = 1 + \frac{2(z + \frac{1}{2}) + \frac{1}{4}(-1)}{(z + \frac{1}{2})^2 + (\frac{1}{4})^2} + \frac{1(z + \frac{1}{3}) + \frac{1}{5}(-21)}{(z + \frac{1}{3})^2 + (\frac{1}{5})^2} + \frac{-2}{z - \frac{1}{6}}
$$

3.6.3 Block Jordan Forms

A z-transfer function with repeated poles generally cannot be realized by a diagonal system because its partial-fraction expansion is not the sum of first-order subsystems. For example, a single-input/single-output system with the z-transfer function

$$T(z) = \frac{3z^2 - 4z + 6}{(z - \frac{1}{3})^3} = \frac{3}{z - \frac{1}{3}} + \frac{-2}{(z - \frac{1}{3})^2} + \frac{5}{(z - \frac{1}{3})^3}$$

can be considered as the tandem (parallel) connection of subsystems shown in Figure 3-21(a). The system shown, however, is of sixth order. By interleaving the three identical first-order subsystems as in Figure 3-21(b), a third-order realization results. Expanding this into a delay diagram, Figure 3-21(c) gives the state and output equations

$$\begin{bmatrix} x_1(k + 1) \\ x_2(k + 1) \\ x_3(k + 1) \end{bmatrix} = \begin{bmatrix} \frac{1}{3} & 1 & 0 \\ 0 & \frac{1}{3} & 1 \\ 0 & 0 & \frac{1}{3} \end{bmatrix} \begin{bmatrix} x_1(k) \\ x_2(k) \\ x_3(k) \end{bmatrix} + \begin{bmatrix} 0 \\ 0 \\ 1 \end{bmatrix} u(k)$$

$$y(k) = \begin{bmatrix} 5 & -2 & 3 \end{bmatrix} \begin{bmatrix} x_1(k) \\ x_2(k) \\ x_3(k) \end{bmatrix}$$

These have a state coupling matrix with the repeated eigenvalue along the diagonal and ones just above the diagonal. When repeated eigenvalues are represented in this way, the equations are in *upper block Jordan* form. It can also be arranged so that the ones are just below the diagonal, obtaining a *lower block Jordan* form, if desired.

To transform a matrix $\mathbf{A}_0$ with all eigenvalues $\lambda = \lambda_0$ repeated to the upper block Jordan form

$$\mathbf{\Lambda}_0 = \begin{bmatrix} \lambda_0 & 1 & 0 & 0 & \cdots & 0 & 0 \\ 0 & \lambda_0 & 1 & 0 & \cdots & & \\ \vdots & & & & & & \\ 0 & 0 & 0 & 0 & \cdots & \lambda_0 & 1 \\ 0 & 0 & 0 & 0 & \cdots & 0 & \lambda_0 \end{bmatrix}$$

the transformation matrix $\mathbf{P}$ must satisfy

$$\mathbf{P}^{-1}\mathbf{A}_0\mathbf{P} = \mathbf{\Lambda}_0$$

or

$$\mathbf{A}_0\mathbf{P} = \mathbf{P}\mathbf{\Lambda}_0$$

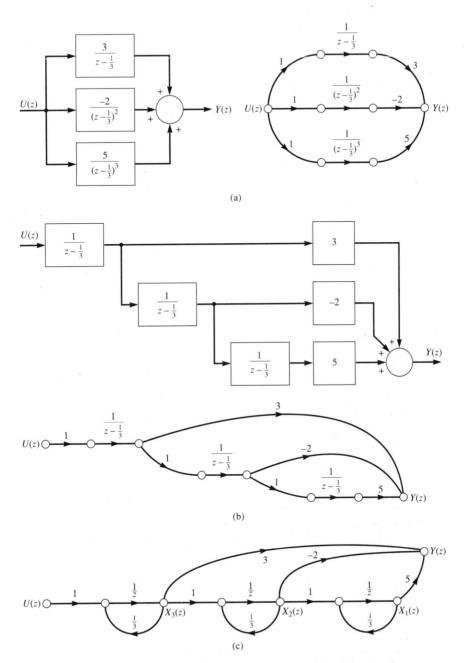

Figure 3-21 System with repeated eigenvalues in upper block Jordan form.
(a) Tandem subsystems. (b) Interleaved first-order subsystems. (c) Delay diagram.

Partitioning $\mathbf{P}$ into columns,

$$\mathbf{A}_0[\mathbf{p}^1 \mid \mathbf{p}^2 \mid \cdots \mid \mathbf{p}^n]$$

$$= [\mathbf{p}^1 \mid \mathbf{p}^2 \mid \cdots \mid \mathbf{p}^n] \begin{bmatrix} \lambda_0 & 1 & 0 & \cdots & 0 & 0 \\ 0 & \lambda_0 & 1 & \cdots & 0 & 0 \\ \vdots & & & & & \\ 0 & 0 & 0 & \cdots & \lambda_0 & 1 \\ 0 & 0 & 0 & \cdots & 0 & \lambda_0 \end{bmatrix}$$

and equating the columns results in

$$\begin{cases} \mathbf{A}_0\mathbf{p}^1 = \lambda_0\mathbf{p}^1 \\ \mathbf{A}_0\mathbf{p}^2 = \lambda_0\mathbf{p}^2 + \mathbf{p}^1 \\ \mathbf{A}_0\mathbf{p}^3 = \lambda_0\mathbf{p}^3 + \mathbf{p}^2 \\ \quad\vdots \\ \mathbf{A}_0\mathbf{p}^n = \lambda_0\mathbf{p}^n + \mathbf{p}^{n-1} \end{cases}$$

or

$$\begin{cases} (\lambda_0\mathbf{I} - \mathbf{A}_0)\mathbf{p}^1 = \mathbf{0} \\ (\lambda_0\mathbf{I} - \mathbf{A}_0)\mathbf{p}^2 = -\mathbf{p}^1 \\ (\lambda_0\mathbf{I} - \mathbf{A}_0)\mathbf{p}^3 = -\mathbf{p}^2 \\ \quad\vdots \\ (\lambda_0\mathbf{I} - \mathbf{A}_0)\mathbf{p}^n = -\mathbf{p}^{n-1} \end{cases}$$

When an $n \times n$ matrix $\mathbf{A}$ contains a mixture of eigenvalues, the eigenvectors satisfying

$$(\lambda_i\mathbf{I} - \mathbf{A})\mathbf{p}^i = \mathbf{0}$$

are found until an eigenvalue repetition is encountered. If, as is usually the case, no additional linearly independent eigenvectors exist, then the relations

$$\begin{cases} (\lambda_i\mathbf{I} - \mathbf{A})\mathbf{p}^{i+1} = -\mathbf{p}^i \\ (\lambda_i\mathbf{I} - \mathbf{A})\mathbf{p}^{i+2} = -\mathbf{p}^{i+1} \\ \quad\vdots \end{cases}$$

are used to form columns of the transformation matrix for the eigenvalue repetitions. The resulting matrix Λ has any nonrepeated eigenvalues along its diagonal and an upper Jordan block along the diagonal for the repeated eigenvalues.

The system

$$\begin{bmatrix} x_1(k+1) \\ x_2(k+1) \\ x_3(k+1) \end{bmatrix} = \begin{bmatrix} 0 & 1 & 0 \\ 0 & 0 & 1 \\ -3 & 5 & -1 \end{bmatrix} \begin{bmatrix} x_1(k) \\ x_2(k) \\ x_3(k) \end{bmatrix} + \begin{bmatrix} 0 \\ 0 \\ 1 \end{bmatrix} u(k) = \mathbf{Ax}(k) + \mathbf{b}u(k)$$

$$\begin{bmatrix} y_1(k) \\ y_2(k) \end{bmatrix} = \begin{bmatrix} 0 & 3 & 0 \\ 1 & 0 & -2 \end{bmatrix} \begin{bmatrix} x_1(k) \\ x_2(k) \\ x_3(k) \end{bmatrix} + \begin{bmatrix} 2 \\ 0 \end{bmatrix} u(k) = \mathbf{Cx}(k) + \mathbf{d}u(k)$$

for example, has eigenvalues given by

$$|\lambda \mathbf{I} - \mathbf{A}| = \begin{vmatrix} \lambda & -1 & 0 \\ 0 & \lambda & -1 \\ 3 & -5 & (\lambda + 1) \end{vmatrix}$$

$$= \lambda^3 + \lambda^2 - 5\lambda + 3 = (\lambda + 3)(\lambda - 1)^2 = 0$$

$$\lambda_1 = -3 \quad \lambda_2 = 1 \quad \lambda_3 = 1$$

which involves a repeated eigenvalue. For λ_1

$$(\lambda_1 \mathbf{I} - \mathbf{A})\mathbf{p}^1 = \mathbf{0}$$

$$\begin{bmatrix} -3 & -1 & 0 \\ 0 & -3 & -1 \\ 3 & -5 & -2 \end{bmatrix} \begin{bmatrix} p_{11} \\ p_{21} \\ p_{31} \end{bmatrix} = \begin{bmatrix} 0 \\ 0 \\ 0 \end{bmatrix}; \quad \mathbf{p}^1 = \begin{bmatrix} -1 \\ 3 \\ -9 \end{bmatrix}$$

For λ_2

$$(\lambda_2 \mathbf{I} - \mathbf{A})\mathbf{p}^2 = \mathbf{0}$$

$$\begin{bmatrix} 1 & -1 & 0 \\ 0 & 1 & -1 \\ 3 & -5 & 2 \end{bmatrix} \begin{bmatrix} p_{12} \\ p_{22} \\ p_{32} \end{bmatrix} = \begin{bmatrix} 0 \\ 0 \\ 0 \end{bmatrix}; \quad \mathbf{p}^2 = \begin{bmatrix} 1 \\ 1 \\ 1 \end{bmatrix}$$

For the repetition of the root at $\lambda = 1$, because there is not another linearly independent solution of the equations for $\mathbf{p}^2$, we solve

$$(\lambda_2 \mathbf{I} - \mathbf{A})\mathbf{p}^3 = -\mathbf{p}^2$$

$$\begin{bmatrix} 1 & -1 & 0 \\ 0 & 1 & -1 \\ 3 & -5 & 2 \end{bmatrix} \begin{bmatrix} p_{13} \\ p_{23} \\ p_{33} \end{bmatrix} = - \begin{bmatrix} 1 \\ 1 \\ 1 \end{bmatrix}$$

which has the solution

$$\mathbf{p}^3 = \begin{bmatrix} 0 \\ 1 \\ 2 \end{bmatrix}$$

A transformation to upper block Jordan form then has

$$\mathbf{P} = \begin{bmatrix} -1 & 1 & 0 \\ 3 & 1 & 1 \\ -9 & 1 & 2 \end{bmatrix}; \quad \mathbf{P}^{-1} = \frac{1}{16} \begin{bmatrix} -1 & 2 & -1 \\ 15 & 2 & -1 \\ -12 & 8 & 4 \end{bmatrix}$$

giving

$$\mathbf{A}' = \mathbf{P}^{-1}\mathbf{A}\mathbf{P}$$

$$= \frac{1}{16} \begin{bmatrix} -1 & 2 & -1 \\ 15 & 2 & -1 \\ -12 & 8 & 4 \end{bmatrix} \begin{bmatrix} 0 & 1 & 0 \\ 0 & 0 & 1 \\ -3 & 5 & -1 \end{bmatrix} \begin{bmatrix} -1 & 1 & 0 \\ 3 & 1 & 1 \\ -9 & 1 & 2 \end{bmatrix} = \begin{bmatrix} -3 & 0 & 0 \\ 0 & 1 & 1 \\ 0 & 0 & 1 \end{bmatrix}$$

$$\mathbf{b}' = \mathbf{P}^{-1}\mathbf{b} = \frac{1}{16} \begin{bmatrix} -1 & 2 & -1 \\ 15 & 2 & -1 \\ -12 & 8 & 4 \end{bmatrix} \begin{bmatrix} 0 \\ 0 \\ 1 \end{bmatrix} = \frac{1}{16} \begin{bmatrix} -1 \\ -1 \\ 4 \end{bmatrix}$$

$$\mathbf{C}' = \mathbf{C}\mathbf{P} = \begin{bmatrix} 0 & 3 & 0 \\ 1 & 0 & -2 \end{bmatrix} \begin{bmatrix} -1 & 1 & 0 \\ 3 & 1 & 1 \\ -9 & 1 & 2 \end{bmatrix} = \begin{bmatrix} 9 & 3 & 3 \\ 17 & -1 & -4 \end{bmatrix}$$

so that the system description in terms of the new state variables is

$$\begin{bmatrix} x_1'(k+1) \\ x_2'(k+1) \\ x_3'(k+1) \end{bmatrix} = \begin{bmatrix} -3 & 0 & 0 \\ 0 & 1 & 1 \\ 0 & 0 & 1 \end{bmatrix} \begin{bmatrix} x_1'(k) \\ x_2'(k) \\ x_3'(k) \end{bmatrix} + \begin{bmatrix} -\frac{1}{16} \\ -\frac{1}{16} \\ \frac{1}{4} \end{bmatrix} u(k)$$

$$\begin{bmatrix} y_1(k) \\ y_2(k) \end{bmatrix} = \begin{bmatrix} 9 & 3 & 3 \\ 17 & -1 & -4 \end{bmatrix} \begin{bmatrix} x_1'(k) \\ x_2'(k) \\ x_3'(k) \end{bmatrix} + \begin{bmatrix} 2 \\ 0 \end{bmatrix} u(k)$$

The state coupling matrix of the system in terms of the state variables $\mathbf{x}'$ is in upper block Jordan form.

3.7 Observability and Controllability

When state variable equations for a system are placed in diagonal form or, in the case of repeated eigenvalues, block Jordan form, fundamental structural aspects of the system are apparent. We now examine that structure and find whether it is possible to control a system's state from its inputs and whether it is possible to determine a system's state from its outputs.

3.7.1 Unobservable and Uncontrollable Modes

When state equations are in diagonal form, each state variable appears in only one equation so the equations are all of first order and are decoupled from one another. The form of the zero-input response of each of these first-order subsystems is termed its *mode*, or the mode associated with the corresponding state variable, and each output of the system has a zero-input response consisting of a linear combination of the individual modes.

If, when in diagonal form, any column of the output coupling matrix is zero, the corresponding state variable does not couple to any output, and the mode associated with that diagonal form state variable is termed *unobservable*. Otherwise, a mode is observable. If, in diagonal form, any row of the input coupling matrix is zero, no input can affect the corresponding first-order equation, and that equation's mode is termed *uncontrollable*; otherwise it is controllable. Each mode of a system is either observable or unobservable. And each mode of a system is either controllable or uncontrollable. If all the modes of a system are observable, the system is *completely observable*. Similarly, if all system modes are controllable, the system is *completely controllable*.

For the diagonal system

$$
\begin{bmatrix} x_1'(k+1) \\ x_2'(k+1) \\ x_3'(k+1) \\ x_4'(k+1) \end{bmatrix} = \begin{bmatrix} 1 & 0 & 0 & 0 \\ 0 & 0 & 0 & 0 \\ 0 & 0 & -2 & 0 \\ 0 & 0 & 0 & \frac{1}{3} \end{bmatrix} \begin{bmatrix} x_1'(k) \\ x_2'(k) \\ x_3'(k) \\ x_4'(k) \end{bmatrix} + \begin{bmatrix} -2 \\ 4 \\ 0 \\ 3 \end{bmatrix} u(k)
$$

$$
\begin{bmatrix} y_1(k) \\ y_2(k) \end{bmatrix} = \begin{bmatrix} -1 & 1 & 0 & 0 \\ 0 & -2 & 0 & 0 \end{bmatrix} \begin{bmatrix} x_1'(k) \\ x_2'(k) \\ x_3'(k) \\ x_4'(k) \end{bmatrix} + \begin{bmatrix} 5 \\ 0 \end{bmatrix} u(k)
$$

the individual decoupled first-order equations are

$$
\begin{cases} x_1'(k+1) = x_1'(k) - 2u(k) \\ x_2'(k+1) = 4u(k) \\ x_3'(k+1) = -2x_3'(k) \\ x_4'(k+1) = \frac{1}{3}x_4'(k) + 3u(k) \end{cases}
$$

and the corresponding modes are 1^k, 0^k, $(-2)^k$, and $(\tfrac{1}{3})^k$. The input does not couple to the $(-2)^k$ mode, so that mode is uncontrollable. The $(-2)^k$ and $(\tfrac{1}{3})^k$ modes do not couple to either output, so both these modes are unobservable. The 1^k and 0^k modes are each controllable and observable.

A delay diagram from this system is shown in Figure 3-22. The two system transfer functions are

$$T_1(z) = \frac{Y_1(z)}{U(z)} \bigg|_{\substack{\text{zero initial} \\ \text{conditions}}} = \frac{(-1)(-2)}{z-1} + \frac{(1)(4)}{z} + 5 = \frac{5z^2 + z - 4}{z(z-1)}$$

$$T_2(z) = \frac{Y_2(z)}{U(z)} \bigg|_{\substack{\text{zero initial} \\ \text{conditions}}} = \frac{(-2)(4)}{z} = -\frac{8}{z}$$

Whenever, in diagonal form, there is no coupling to a mode from an input or no coupling from a mode to an output, the transfer function relating that output and input does not have a pole (eigenvalue) corresponding to that mode. In calculating the system transfer function matrix according to

$$\mathbf{T}(z) = \mathbf{C}(z\mathbf{I} - \mathbf{A})^{-1}\mathbf{B} + \mathbf{D}$$

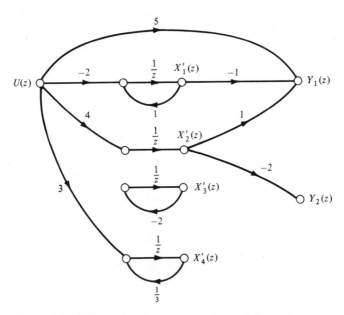

Figure 3-22 Diagonal system with unobservable and uncontrollable modes.

where every element of $\mathbf{T}(z)$ shares the denominator polynomial $|z\mathbf{I} - \mathbf{A}|$, a z-transfer function without one of these poles must have a pole-zero cancellation. If any mode is unobservable or uncontrollable, every element of the transfer function matrix has a pole-zero cancellation. In general, a system can be decomposed into four parts, one each in which the modes are

controllable and observable

controllable but not observable

not controllable but observable

not controllable and not observable

as indicated in Figure 3-23. The transfer function matrix characterizes only the controllable and observable subsystem.

The modes associated with a Jordan block in the state coupling matrix representing repeated eigenvalues a are a^k, ka^k, k^2a^k, and so on. As an example, the delay diagram for the portion of a system represented by a 4×4 Jordan block

$$\begin{bmatrix} x_i(k+1) \\ x_{i+1}(k+1) \\ x_{i+2}(k+1) \\ x_{i+3}(k+1) \end{bmatrix} = \begin{bmatrix} a & 1 & 0 & 0 \\ 0 & a & 1 & 0 \\ 0 & 0 & a & 1 \\ 0 & 0 & 0 & a \end{bmatrix} \begin{bmatrix} x_i(k) \\ x_{i+1}(k) \\ x_{i+2}(k) \\ x_{i+3}(k) \end{bmatrix} + \text{(input terms)}$$

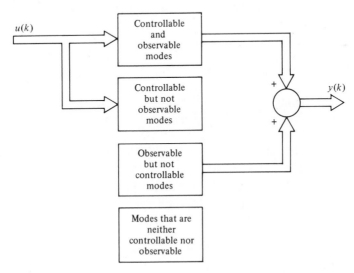

Figure 3-23 Decomposition of a system into subsystems with various combinations of mode observability and controllability.

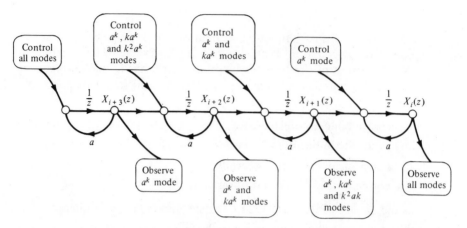

Figure 3-24 Illustration of controllability and observability of modes associated with repeated system eigenvalues.

is shown in Figure 3-24. Input coupling to the last state variable, in this case to the $x_{i+3}(k)$ equation, affects all the block's state variables. If no input couples to the last equation of the block, the highest degree mode, k^3a^k in the example, cannot be controlled. In general, controllability of all the modes associated with a Jordan block rests on whether there is input coupling to the last state variable in the block. If there is not, the highest degree mode is not controllable. Controllability of the rest of the modes depends on whether there is input coupling to the next-to-last state variable, and so on.

If the first state variable in the Jordan block couples to an output, all of the state variables of the block can affect the output, and all of the block's modes are observable. If the first state variable in the block does not couple to any output, the highest degree mode, k^3a^k in the example, cannot be observed. Observability of the k^2a^k, ka^k, and a^k modes requires coupling to an output from the second state variable, and so on.

Most of the fine points connected with observability and controllability have to do with systems with repeated eigenvalues, where the repeated eigenvalue has more than one Jordan block. For example, a system with the block diagonal state coupling matrix

$$\mathbf{A} = \begin{bmatrix} -\frac{1}{4} & 0 & 0 \\ 0 & \frac{1}{2} & 1 \\ 0 & 0 & \frac{1}{2} \end{bmatrix}$$

has the eigenvalue $\frac{1}{2}$ repeated in a single Jordan block, whereas a system with the same eigenvalues but with the diagonal state coupling matrix

$$\mathbf{A} = \begin{bmatrix} -\frac{1}{4} & 0 & 0 \\ 0 & \frac{1}{2} & 0 \\ 0 & 0 & \frac{1}{2} \end{bmatrix}$$

has two Jordan "blocks" with the same eigenvalue. For the latter system to be completely controllable, there must be two inputs, each of which couples in different ways to the repeated mode. For example,

$$\begin{bmatrix} x_1(k+1) \\ x_2(k+1) \\ x_3(k+1) \end{bmatrix} = \begin{bmatrix} -\frac{1}{4} & 0 & 0 \\ 0 & \frac{1}{2} & 0 \\ 0 & 0 & \frac{1}{2} \end{bmatrix} \begin{bmatrix} x_1(k) \\ x_2(k) \\ x_3(k) \end{bmatrix} + \begin{bmatrix} 2 & 3 \\ 1 & 0 \\ 0 & 1 \end{bmatrix} \begin{bmatrix} u_1(k) \\ u_2(k) \end{bmatrix}$$

is completely controllable, whereas

$$\begin{bmatrix} x_1(k+1) \\ x_2(k+1) \\ x_3(k+1) \end{bmatrix} = \begin{bmatrix} -\frac{1}{4} & 0 & 0 \\ 0 & \frac{1}{2} & 0 \\ 0 & 0 & \frac{1}{2} \end{bmatrix} \begin{bmatrix} x_1(k) \\ x_2(k) \\ x_3(k) \end{bmatrix} + \begin{bmatrix} 3 \\ 1 \\ 2 \end{bmatrix} u(k) \qquad \textbf{(3-25)}$$

is not. In equation (3-25) the change of state variable

$$x_3'(k) = -2x_2(k) + x_3(k)$$

gives the realization

$$\begin{bmatrix} x_1(k+1) \\ x_2(k+1) \\ x_3'(k+1) \end{bmatrix} = \begin{bmatrix} -\frac{1}{4} & 0 & 0 \\ 0 & \frac{1}{2} & 0 \\ 0 & 0 & \frac{1}{2} \end{bmatrix} \begin{bmatrix} x_1(k) \\ x_2(k) \\ x_3'(k) \end{bmatrix} + \begin{bmatrix} 3 \\ 1 \\ 0 \end{bmatrix} u(k)$$

from which it is seen that there is a part of the state, the state variable $x_3'(k)$, that is unaffected by any input. In general, controllability of a repeated mode requires that there be a number of system inputs equal to the maximum number of Jordan blocks having the same eigenvalue. Each must couple differently to Jordan blocks having the same eigenvalue.

Similarly, observability of a repeated mode having more than one Jordan block requires that the system have independent outputs from each such block. For example, the system

$$\begin{bmatrix} x_1(k+1) \\ x_2(k+1) \\ x_3(k+1) \end{bmatrix} = \begin{bmatrix} -\frac{1}{3} & 0 & 0 \\ 0 & \frac{1}{2} & 0 \\ 0 & 0 & \frac{1}{2} \end{bmatrix} \begin{bmatrix} x_1(k) \\ x_2(k) \\ x_3(k) \end{bmatrix} + \begin{bmatrix} 2 & 3 \\ 1 & 0 \\ 0 & 1 \end{bmatrix} \begin{bmatrix} u_1(k) \\ u_2(k) \end{bmatrix}$$

$$\begin{bmatrix} y_1(k) \\ y_2(k) \end{bmatrix} = \begin{bmatrix} 4 & 1 & 0 \\ -1 & 0 & 1 \end{bmatrix} \begin{bmatrix} x_1(k) \\ x_2(k) \\ x_3(k) \end{bmatrix}$$

is completely observable, but

$$
\begin{bmatrix} x_1(k+1) \\ x_2(k+1) \\ x_3(k+1) \end{bmatrix} = \begin{bmatrix} -\frac{1}{3} & 0 & 0 \\ 0 & \frac{1}{2} & 0 \\ 0 & 0 & \frac{1}{2} \end{bmatrix} \begin{bmatrix} x_1(k) \\ x_2(k) \\ x_3(k) \end{bmatrix} + \begin{bmatrix} 2 & 3 \\ 1 & 0 \\ 0 & 1 \end{bmatrix} \begin{bmatrix} u_1(k) \\ u_2(k) \end{bmatrix}
$$

$$
y(k) = \begin{bmatrix} 4 & 3 & 2 \end{bmatrix} \begin{bmatrix} x_1(k) \\ x_2(k) \\ x_3(k) \end{bmatrix}
$$

(3-26)

is not. In equation (3-26), the change of state variables where

$$
x_2(k) = x_2'(k) + 2x_3'(k) \qquad x_2'(k) = x_2(k) + \frac{2}{3} x_3(k)
$$

$$
x_3(k) = -3x_3'(k) \qquad x_3'(k) = -\frac{1}{3} x_3(k)
$$

gives the realization

$$
\begin{bmatrix} x_1(k+1) \\ x_2'(k+1) \\ x_3'(k+1) \end{bmatrix} = \begin{bmatrix} -\frac{1}{3} & 0 & 0 \\ 0 & \frac{1}{2} & 0 \\ 0 & 0 & \frac{1}{2} \end{bmatrix} \begin{bmatrix} x_1(k) \\ x_2'(k) \\ x_3'(k) \end{bmatrix} + \begin{bmatrix} 2 & 3 \\ 1 & \frac{2}{3} \\ 0 & -\frac{1}{3} \end{bmatrix} \begin{bmatrix} u_1(k) \\ u_2(k) \end{bmatrix}
$$

$$
y(k) = \begin{bmatrix} 4 & 3 & 0 \end{bmatrix} \begin{bmatrix} x_1(k) \\ x_2'(k) \\ x_3'(k) \end{bmatrix}
$$

which shows that part of the state, the variable $x_3'(k)$, does not affect the output.

3.7.2 Transformation to Controllable Form

Any completely controllable single-input system can be realized in controllable form. The needed change of state variables, which will be used again early in the next chapter, is now derived. Let the transformation matrix **P** in

$$
\mathbf{x} = \mathbf{P}\mathbf{x}' \qquad \mathbf{x}' = \mathbf{P}^{-1}\mathbf{x}
$$

take a completely controllable single-input system into the controllable form. Partition **P** into columns and let the characteristic equation of **A** be

$$\lambda^n + \alpha_{n-1}\lambda^{n-1} + \cdots + \alpha_1\lambda + \alpha_0 = 0$$

where the coefficients α_i have been found. Then

$$\mathbf{P}^{-1}\mathbf{b} = \mathbf{b}_c \qquad \mathbf{b} = \mathbf{P}\mathbf{b}_c$$

or

$$\mathbf{b} = [\mathbf{p}^1 \mid \mathbf{p}^2 \mid \cdots \mid \mathbf{p}^n] \begin{bmatrix} 0 \\ 0 \\ \vdots \\ 0 \\ 1 \end{bmatrix}$$

which gives

$$\mathbf{p}^n = \mathbf{b} \tag{3-27}$$

Using

$$\mathbf{P}^{-1}\mathbf{A}\mathbf{P} = \mathbf{A}_c \qquad \mathbf{A}\mathbf{P} = \mathbf{P}\mathbf{A}_c$$

then

$$\mathbf{A}[\mathbf{p}^1 \mid \mathbf{p}^2 \mid \cdots \mid \mathbf{p}^{n-1} \mid \mathbf{p}^n]$$

$$= [\mathbf{p}^1 \mid \mathbf{p}^2 \mid \cdots \mid \mathbf{p}^{n-1} \mid \mathbf{p}^n] \begin{bmatrix} 0 & 1 & 0 & \cdots & 0 & 0 \\ 0 & 0 & 1 & \cdots & 0 & 0 \\ \vdots & & & & & \\ 0 & 0 & 0 & \cdots & 0 & 1 \\ -\alpha_0 & -\alpha_1 & -\alpha_2 & \cdots & -\alpha_{n-2} & -\alpha_{n-1} \end{bmatrix}$$

Equating columns, from right to left gives

$$\begin{cases} \mathbf{A}\mathbf{p}^n = \mathbf{p}^{n-1} - \alpha_{n-1}\mathbf{p}^n \\ \mathbf{A}\mathbf{p}^{n-1} = \mathbf{p}^{n-2} - \alpha_{n-2}\mathbf{p}^n \\ \vdots \\ \mathbf{A}\mathbf{p}^2 = \mathbf{p}^1 - \alpha_1\mathbf{p}^n \\ \mathbf{A}\mathbf{p}^1 = -\alpha_0\mathbf{p}^n \end{cases} \tag{3-28}$$

then equations (3-27) and (3-28) form the recursive algorithm

$$
\begin{cases}
\mathbf{p}^n = \mathbf{b} \\
\mathbf{p}^{n-1} = \mathbf{A}\mathbf{p}^n + \alpha_{n-1}\mathbf{b} \\
\quad \vdots \\
\mathbf{p}^2 = \mathbf{A}\mathbf{p}^3 + \alpha_2\mathbf{b} \\
\mathbf{p}^1 = \mathbf{A}\mathbf{p}^2 + \alpha_1\mathbf{b}
\end{cases}
\tag{3-29}
$$

where the relation

$$\mathbf{A}\mathbf{p}^1 = -\alpha_0\mathbf{p}^n$$

is not used because, using the Cayley-Hamilton theorem, it can be shown to be redundant. It is, however, a good check on the calculations.

As a numerical example, for the system

$$
\begin{bmatrix} x_1(k+1) \\ x_2(k+1) \\ x_3(k+1) \end{bmatrix}
=
\begin{bmatrix} -2 & 1 & 1 \\ 0 & 2 & -1 \\ 0 & 1 & 3 \end{bmatrix}
\begin{bmatrix} x_1(k) \\ x_2(k) \\ x_3(k) \end{bmatrix}
+
\begin{bmatrix} 1 \\ 0 \\ -1 \end{bmatrix} u(k)
$$

$$
\begin{bmatrix} y_1(k) \\ y_2(k) \end{bmatrix}
=
\begin{bmatrix} 2 & 0 & 1 \\ 0 & -2 & 4 \end{bmatrix}
\begin{bmatrix} x_1(k) \\ x_2(k) \\ x_3(k) \end{bmatrix}
+
\begin{bmatrix} -3 \\ 5 \end{bmatrix} u(k)
$$

the characteristic equation is

$$
|\lambda\mathbf{I} - \mathbf{A}| =
\begin{vmatrix}
\lambda + 2 & -1 & -1 \\
0 & (\lambda - 2) & 1 \\
0 & -1 & (\lambda - 3)
\end{vmatrix}
= (\lambda + 2)(\lambda^2 - 5\lambda + 7)
$$

$$
= \lambda^3 - 3\lambda^2 - 3\lambda + 14 = 0
$$

The columns of the transformation matrix $\mathbf{P}$ are given by

$$
\mathbf{p}^3 = \mathbf{b} = \begin{bmatrix} 1 \\ 0 \\ -1 \end{bmatrix}
$$

$$
\mathbf{p}^2 = \mathbf{A}\mathbf{p}^3 - 3\mathbf{b} =
\begin{bmatrix} -2 & 1 & 1 \\ 0 & 2 & -1 \\ 0 & 1 & 3 \end{bmatrix}
\begin{bmatrix} 1 \\ 0 \\ -1 \end{bmatrix}
- 3
\begin{bmatrix} 1 \\ 0 \\ -1 \end{bmatrix}
=
\begin{bmatrix} -6 \\ 1 \\ 0 \end{bmatrix}
$$

$$\mathbf{p}^1 = \mathbf{A}\mathbf{p}^2 - 3\mathbf{b} = \begin{bmatrix} -2 & 1 & 1 \\ 0 & 2 & -1 \\ 0 & 1 & 3 \end{bmatrix} \begin{bmatrix} -6 \\ 1 \\ 0 \end{bmatrix} - 3 \begin{bmatrix} 1 \\ 0 \\ -1 \end{bmatrix} = \begin{bmatrix} 10 \\ 2 \\ 4 \end{bmatrix}$$

so that

$$\mathbf{P} = \begin{bmatrix} 10 & -6 & 1 \\ 2 & 1 & 0 \\ 4 & 0 & -1 \end{bmatrix}$$

A controllable form representation of a single-input system exists if and only if the system is completely controllable. This can be shown as follows: Using the recursive transformation relations of equation (3-29), the transformation can be expressed in the form

$$\mathbf{P} = [\mathbf{p}^1 \mid \mathbf{p}^2 \mid \cdots \mid \mathbf{p}^n] = \underbrace{[\mathbf{b} \mid \mathbf{Ab} \mid \mathbf{A}^2\mathbf{b} \mid \cdots \mid \mathbf{A}^{n-1}\mathbf{b}]}_{\mathbf{M}_c}$$

$$\times \begin{bmatrix} \alpha_1 & \alpha_2 & \cdots & \alpha_{n-2} & \alpha_{n-1} & 1 \\ \alpha_2 & \alpha_3 & \cdots & \alpha_{n-1} & 1 & 0 \\ \alpha_3 & \alpha_4 & \cdots & 1 & 0 & 0 \\ \vdots & \vdots & & & & \\ \alpha_{n-2} & \alpha_{n-1} & \cdots & 0 & 0 & 0 \\ \alpha_{n-1} & 1 & \cdots & 0 & 0 & 0 \\ 1 & 0 & \cdots & 0 & 0 & 0 \end{bmatrix}$$

where the matrix $\mathbf{M}_c$ is termed the *controllability matrix* of the system. The matrix at the right is nonsingular for any α_i's because each column, right to left, is linearly independent of the previous columns. Thus, the transformation to controllable form is nonsingular if and only if the single-input system's controllability matrix $\mathbf{M}_c$ is nonsingular. As will be demonstrated in the next chapter, a single-input system is completely controllable if and only if $\mathbf{M}_c$ is nonsingular.

3.7.3 Transformation to Observable Form

The transformation matrix $\mathbf{Q} = \mathbf{P}^{-1}$ in

$$\mathbf{x} = \mathbf{P}\mathbf{x}' = \mathbf{Q}^{-1}\mathbf{x}' \qquad \mathbf{x}' = \mathbf{P}^{-1}\mathbf{x} = \mathbf{Q}\mathbf{x}$$

that takes a completely observable single-output system into the observable form can be found as follows. Partition the rows of $\mathbf{Q}$ and let the characteristic equation of $\mathbf{A}$ be

$$\lambda^n + \alpha_{n-1}\lambda^{n-1} + \cdots + \alpha_1\lambda + \alpha_0 = 0$$

Then

$$\mathbf{c}_0^\dagger = \mathbf{c}^\dagger\mathbf{P} = \mathbf{c}^\dagger\mathbf{Q}^{-1} \qquad \mathbf{c}_0^\dagger\mathbf{Q} = \mathbf{c}^\dagger$$

or

$$[1 \quad 0 \quad \cdots \quad 0]\begin{bmatrix} \mathbf{q}_1^\dagger \\ \hline \mathbf{q}_2^\dagger \\ \hline \vdots \\ \vdots \\ \hline \mathbf{q}_n^\dagger \end{bmatrix} = \mathbf{c}^\dagger$$

which gives

$$\mathbf{q}_1^\dagger = \mathbf{c}^\dagger \tag{3-30}$$

and

$$\mathbf{P}^{-1}\mathbf{A}\mathbf{P} = \mathbf{Q}\mathbf{A}\mathbf{Q}^{-1} = \mathbf{A}_0 \qquad \mathbf{Q}\mathbf{A} = \mathbf{A}_0\mathbf{Q}$$

or

$$\begin{bmatrix} \mathbf{q}_1^\dagger \\ \hline \mathbf{q}_2^\dagger \\ \hline \vdots \\ \hline \mathbf{q}_{n-1}^\dagger \\ \hline \mathbf{q}_n^\dagger \end{bmatrix} \mathbf{A} = \begin{bmatrix} -\alpha_{n-1} & 1 & 0 & \cdots & 0 & 0 \\ -\alpha_{n-2} & 0 & 1 & \cdots & 0 & 0 \\ \vdots & & & & & \\ -\alpha_1 & 0 & 0 & \cdots & 0 & 1 \\ -\alpha_0 & 0 & 0 & \cdots & 0 & 0 \end{bmatrix}\begin{bmatrix} \mathbf{q}_1^\dagger \\ \hline \mathbf{q}_2^\dagger \\ \hline \vdots \\ \hline \mathbf{q}_{n-1}^\dagger \\ \hline \mathbf{q}_n^\dagger \end{bmatrix} \tag{3-31}$$

Equating rows from top to bottom and using equation (3-30) gives the recursive algorithm

$$\begin{cases} \mathbf{q}_1^\dagger = \mathbf{c}^\dagger \\ \mathbf{q}_2^\dagger = \mathbf{q}_1^\dagger\mathbf{A} + \alpha_{n-1}\mathbf{c}^\dagger \\ \vdots \\ \mathbf{q}_{n-1}^\dagger = \mathbf{q}_{n-2}^\dagger\mathbf{A} + \alpha_2\mathbf{c}^\dagger \\ \mathbf{q}_n^\dagger = \mathbf{q}_{n-1}^\dagger\mathbf{A} + \alpha_1\mathbf{c}^\dagger \end{cases}$$

where the last equation from the set in equation (3-31) is redundant and is not used.

For example, for the system

$$\begin{bmatrix} x_1(k+1) \\ x_2(k+1) \\ x_3(k+1) \end{bmatrix} = \begin{bmatrix} -2 & 2 & 0 \\ -1 & -3 & 0 \\ 1 & 0 & 1 \end{bmatrix} \begin{bmatrix} x_1(k) \\ x_2(k) \\ x_3(k) \end{bmatrix} + \begin{bmatrix} 3 & 1 \\ 4 & 1 \\ -2 & 0 \end{bmatrix} \begin{bmatrix} u_1(k) \\ u_2(k) \end{bmatrix}$$

$$= \mathbf{A}x(k) + \mathbf{B}u(k)$$

$$y(k) = [0 \quad -2 \quad 1] \begin{bmatrix} x_1(k) \\ x_2(k) \\ x_3(k) \end{bmatrix} + [1 \quad 0] \begin{bmatrix} u_1(k) \\ u_2(k) \end{bmatrix}$$

$$= \mathbf{c}^\dagger x(k) + \mathbf{d}^\dagger u(k)$$

the characteristic equation is

$$|\lambda\mathbf{I} - \mathbf{A}| = \begin{bmatrix} (\lambda+2) & -2 & 0 \\ 1 & (\lambda+3) & 0 \\ -1 & 0 & (\lambda-1) \end{bmatrix} = (\lambda-1)(\lambda^2 + 5\lambda + 8)$$

$$= \lambda^3 + 4\lambda^2 + 3\lambda - 8 = 0$$

The rows of the inverse transformation matrix $\mathbf{Q}$ are given by

$$\mathbf{q}_1^\dagger = \mathbf{c}^\dagger = [0 \quad -2 \quad 1]$$

$$\mathbf{q}_2^\dagger = \mathbf{q}_1^\dagger\mathbf{A} + \alpha_2\mathbf{c}^\dagger = [0 \quad -2 \quad 1] \begin{bmatrix} -2 & 2 & 0 \\ -1 & -3 & 0 \\ 1 & 0 & 1 \end{bmatrix} + 4[0 \quad -2 \quad 1]$$

$$= [3 \quad -2 \quad 5]$$

$$\mathbf{q}_3^\dagger = \mathbf{q}_2^\dagger\mathbf{A} + \alpha_1\mathbf{c}^\dagger = [3 \quad -2 \quad 5] \begin{bmatrix} -2 & 2 & 0 \\ -1 & -3 & 0 \\ 1 & 0 & 1 \end{bmatrix} + 3[0 \quad -2 \quad 1]$$

$$= [1 \quad 6 \quad 8]$$

so that

$$\mathbf{Q} = \begin{bmatrix} 0 & -2 & 1 \\ 3 & -2 & 5 \\ 1 & 6 & 8 \end{bmatrix}$$

Similar to the situation with the controllable form, the transformation matrix $\mathbf{Q}$ for the observable canonical form of a single-output system is nonsingular if and only if the system is completely observable.

3.8 A Monorail System

In this section, a detailed example that consolidates the material on continuous-time state equations is presented. A simple one-dimensional model of a two-car (plus engine) monorail train is shown in Figure 3-25(a). The engine is modeled as the mass M_1. Its air and drive system frictions are modeled by B_1, and the equivalent linear force produced by the engine's rotation of its drive wheels is represented by the applied force $f(t)$. The two cars also have masses and air and track frictions, and each is coupled to the others by devices having both spring K and damping B properties. The position of the engine from some stationary reference point is $x_1(t)$. Similarly, the positions of the cars are $x_2(t)$ and $x_3(t)$.

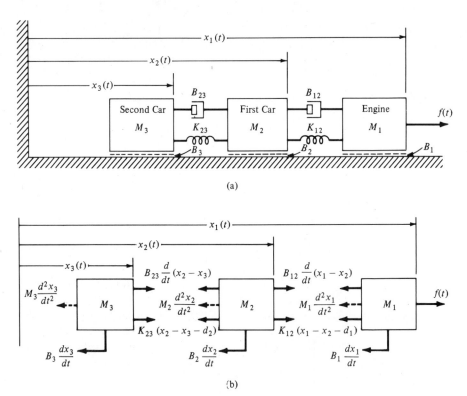

(a)

(b)

Figure 3-25 Monorail train model. (a) Spring, mass, and damper model. (b) Freebody diagram.

A freebody diagram for this model is drawn in Figure 3-25(b). The air and track frictional forces

$$B_1 \frac{dx_1}{dt}; \qquad B_2 \frac{dx_2}{dt}; \qquad B_3 \frac{dx_3}{dt}$$

are proportional to the velocity of each car. The frictional forces of the car couplings

$$B_{12} \frac{d}{dt}(x_1 - x_2); \qquad B_{23} \frac{d}{dt}(x_2 - x_3)$$

are proportional to the difference in velocities of the two cars involved. The spring forces of the car coupling

$$K_{12}(x_1 - x_2 - d_1) \qquad K_{23}(x_2 - x_3 - d_2)$$

are proportional to the relative distances between adjacent cars, where d_1 and d_2 are the equilibrium distances for which there is no spring force. If the distances x_2 and x_3 are translated by d_1 and $(d_1 + d_2)$, respectively, the constants d_1 and d_2 in the force relations above become zero. We assume this has been done. Equating the forces to the rate of change of momentum for each car gives the simultaneous differential equations

$$
\begin{cases}
M_1 \dfrac{d^2x_1}{dt^2} + (B_1 + B_{12}) \dfrac{dx_1}{dt} - B_{12} \dfrac{dx_2}{dt} + K_{12}x_1 - K_{12}x_2 = f(t) \\[2mm]
M_2 \dfrac{d^2x_2}{dt^2} + (B_2 + B_{23} + B_{12}) \dfrac{dx_2}{dt} - B_{12} \dfrac{dx_1}{dt} - B_{23} \dfrac{dx_3}{dt} \\[2mm]
\qquad\qquad + (K_{12} + K_{23})x_2 - K_{12}x_1 - K_{23}x_3 = 0 \\[2mm]
M_3 \dfrac{d^2x_3}{dt^2} + (B_3 + B_{23}) \dfrac{dx_3}{dt} - B_{23} \dfrac{dx_2}{dt} + K_{23}x_3 - K_{23}x_2 = 0 \qquad \textbf{(3-32)}
\end{cases}
$$

The equations (3-32) are not in state variable form. They involve second derivatives, and they do not equate the derivative of each state variable to a linear combination of the state variables and the input. They can be converted to a form involving only first derivatives by defining the mass velocities as

$$v_1 = \frac{dx_1}{dt}; \qquad v_2 = \frac{dx_2}{dt}; \qquad v_3 = \frac{dx_3}{dt} \qquad \textbf{(3-33)}$$

and substituting, using

$$\frac{d^2x_1}{dt^2} = \frac{dv_1}{dt}; \qquad \frac{d^2x_2}{dt^2} = \frac{dv_2}{dt}; \qquad \frac{d^2x_3}{dt^2} = \frac{dv_3}{dt}$$

so that equation (3-32) becomes

$$\begin{cases} M_1 \dfrac{dv_1}{dt} + (B_1 + B_{12})v_1 - B_{12}v_2 + K_{12}x_1 - K_{12}x_2 = f(t) \\[2ex] M_2 \dfrac{dv_2}{dt} + (B_2 + B_{23} + B_{12})v_2 - B_{12}v_1 - B_{23}v_3 \\[2ex] \qquad + (K_{12} + K_{13})x_2 - K_{12}x_1 - K_{23}x_3 = 0 \\[2ex] M_3 \dfrac{dv_3}{dt} + (B_3 + B_{23})v_3 - B_{23}v_2 + K_{23}x_3 - K_{23}x_2 = 0 \end{cases} \qquad \textbf{(3-34)}$$

The collection of the three equations of (3-33) and the three equations of (3-34) constitute a set of six equations in the six variables x_1, x_2, x_3, v_1, v_2, and v_3. Solving for the first derivative of each of these variables gives the state equations

$$\begin{cases} \dfrac{dx_1}{dt} = v_1 \\[2ex] \dfrac{dv_1}{dt} = -\dfrac{B_1 + B_{12}}{M_1}v_1 + \dfrac{B_{12}}{M_1}v_2 - \dfrac{K_{12}}{M_1}x_1 + \dfrac{K_{12}}{M_1}x_2 + f(t) \\[2ex] \dfrac{dx_2}{dt} = v_2 \\[2ex] \dfrac{dv_2}{dt} = -\dfrac{B_2 + B_{23} + B_{12}}{M_2}v_2 + \dfrac{B_{12}}{M_2}v_1 + \dfrac{B_{23}}{M_2}v_3 - \dfrac{K_{12} + K_{13}}{M_2}x_2 \\[2ex] \qquad + \dfrac{K_{12}}{M_2}x_1 + \dfrac{K_{23}}{M_2}x_3 \\[2ex] \dfrac{dx_3}{dt} = v_3 \\[2ex] \dfrac{dv_3}{dt} = -\dfrac{B_3 + B_{23}}{M_3}v_3 + \dfrac{B_{23}}{M_3}v_2 - \dfrac{K_{23}}{M_3}x_3 + \dfrac{K_{23}}{M_3}x_2 \end{cases}$$

or

$$\begin{bmatrix} \dot{x}_1(t) \\ \dot{v}_1(t) \\ \dot{x}_2(t) \\ \dot{v}_2(t) \\ \dot{x}_3(t) \\ \dot{v}_3(t) \end{bmatrix} = \begin{bmatrix} 0 & 1 & 0 & 0 & 0 & 0 \\ -\left(\dfrac{K_{12}}{M_1}\right) & -\left(\dfrac{B_1 + B_{12}}{M_1}\right) & \dfrac{K_{12}}{M_1} & \dfrac{B_{12}}{M_1} & 0 & 0 \\ 0 & 0 & 0 & 1 & 0 & 0 \\ \dfrac{K_{12}}{M_2} & \dfrac{B_{12}}{M_2} & -\left(\dfrac{K_{12} + K_{13}}{M_2}\right) & -\left(\dfrac{B_2 + B_{23} + B_{12}}{M_2}\right) & \dfrac{K_{23}}{M_2} & \dfrac{B_{23}}{M_2} \\ 0 & 0 & 0 & 0 & 0 & 1 \\ 0 & 0 & \dfrac{K_{23}}{M_3} & \dfrac{B_{23}}{M_3} & -\left(\dfrac{K_{23}}{M_3}\right) & -\left(\dfrac{B_3 + B_{23}}{M_3}\right) \end{bmatrix}$$

$$\times \begin{bmatrix} x_1(t) \\ v_1(t) \\ x_2(t) \\ v_2(t) \\ x_3(t) \\ v_3(t) \end{bmatrix} + \begin{bmatrix} 0 \\ 1 \\ 0 \\ 0 \\ 0 \\ 0 \end{bmatrix} f(t)$$

If tachometers are placed on a wheel of each car, an electrical signal proportional to each car's velocity is produced,

$$y_1(t) = \alpha v_1(t) \qquad y_2(t) = \alpha v_2(t) \qquad y_3(t) = \alpha v_3(t)$$

where α is the constant of proportionality relating tachometer voltage to car velocity. These outputs

$$\begin{bmatrix} y_1(t) \\ y_2(t) \\ y_3(t) \end{bmatrix} = \begin{bmatrix} 0 & \alpha & 0 & 0 & 0 & 0 \\ 0 & 0 & 0 & \alpha & 0 & 0 \\ 0 & 0 & 0 & 0 & 0 & \alpha \end{bmatrix} \begin{bmatrix} x_1(t) \\ v_1(t) \\ x_2(t) \\ v_2(t) \\ x_3(t) \\ v_3(t) \end{bmatrix}$$

might be used in a feedback arrangement to the engine controller to aid in giving a smooth ride.

Clearly, the model for a system of this kind could be of a very high order if more cars were involved, if the dynamics of the engine were included, and if the rotational moments of the wheels were modeled.

3.9 Summary

Linear, time-invariant systems have the form

$$\dot{\mathbf{x}}(t) = \mathbf{A}\mathbf{x}(t) + \mathbf{B}\mathbf{u}(t)$$
$$\mathbf{y}(t) = \mathbf{C}\mathbf{x}(t) + \mathbf{D}\mathbf{u}(t)$$

where $\mathbf{u}(t)$ is an r-vector of inputs, $\mathbf{y}(t)$ is an m-vector of outputs, and $\mathbf{x}(t)$ is the n-vector state. The relations between signals in a state variable system can be visualized with an integration diagram, involving only the operations of integration, multiplication by a constant, and addition.

In terms of Laplace transforms, the solution for the state and output of time-invariant systems, in terms of an initial state and the inputs, is

$$\mathbf{X}(s) = (s\mathbf{I} - \mathbf{A})^{-1}\mathbf{x}(0^-) + (s\mathbf{I} - \mathbf{A})^{-1}\mathbf{B}\mathbf{U}(s)$$

 zero-input component zero-state component

$$\mathbf{Y}(s) = \mathbf{C}\mathbf{X}(s) + \mathbf{D}\mathbf{U}(s)$$

The matrix of transfer functions, each relating an output to an input, is given by

$$\mathbf{T}(s) = \mathbf{C}(s\mathbf{I} - \mathbf{A})^{-1}\mathbf{B} + \mathbf{D}$$

Each transfer function element shares the denominator polynomial

$$q(s) = |s\mathbf{I} - \mathbf{A}|$$

the roots of which are the transfer function poles and also the eigenvalues of the matrix $\mathbf{A}$. The state variable model is stable if and only if the eigenvalues of $\mathbf{A}$ are in the left half of the complex plane.

The time domain solution for the state of a linear, time-invariant, continuous-time system is

$$\mathbf{x}(t) = \mathbf{\Phi}(t)\mathbf{x}(0^-) + \int_0^t \mathbf{\Phi}(t - \tau)\mathbf{B}\mathbf{u}(\tau)\, d\tau$$

$$= \mathbf{\Phi}(t)\mathbf{x}(0^-) + \text{convolution}[\mathbf{\Phi}(t), \mathbf{B}\mathbf{u}(t)]$$

where the state transition matrix is

$$\mathbf{\Phi}(t) = \exp(\mathbf{A}t) = \mathbf{I} + \mathbf{A}t + \frac{1}{2!}\mathbf{A}^2 t^2 + \cdots + \frac{1}{i!}\mathbf{A}^i t^i \cdots$$

$$= \mathcal{L}^{-1}[(s\mathbf{I} - \mathbf{A})^{-1}]$$

In general, linear, time-varying, continuous-time systems have $\mathbf{A}$, $\mathbf{B}$, $\mathbf{C}$, and $\mathbf{D}$ matrices that are piecewise-continuous functions of time. The solution for the state is

$$\mathbf{x}(t) = \mathbf{\Phi}(t, t_0)\mathbf{x}(t_0) + \int_{t_0}^t \mathbf{\Phi}(t, \tau)\mathbf{B}(\tau)\mathbf{u}(\tau)\, d\tau$$

where the state transition matrix satisfies

$$\frac{d}{dt}\mathbf{\Phi}(t, t_0) = \mathbf{A}(t)\mathbf{\Phi}(t, t_0)$$

with

$$\mathbf{\Phi}(t_0, t_0) = \mathbf{I}$$

Linear, step-invariant, discrete-time system models have the form

$$\mathbf{x}(k + 1) = \mathbf{A}\mathbf{x}(k) + \mathbf{B}\mathbf{u}(k)$$
$$\mathbf{y}(k) = \mathbf{C}\mathbf{x}(k) + \mathbf{D}\mathbf{u}(k)$$

A delay diagram is an easy way to visualize discrete-time systems graphically, in terms of delays, multiplications by constants, and additions.

In terms of the initial state and the inputs, the discrete-time solution of the state is

$$\mathbf{x}(k) = \underbrace{\mathbf{A}^k\mathbf{x}(0)}_{\text{Zero-input component}} + \underbrace{\sum_{i=0}^{k-1} \mathbf{A}^{k-1-i}\mathbf{B}\mathbf{u}(i)}_{\text{Zero-state component}}$$

$$\mathbf{y}(k) = \mathbf{C}\mathbf{x}(k) + \mathbf{D}\mathbf{u}(k)$$

and the matrix of system z-transfer functions is given by

$$\mathbf{T}(z) = \mathbf{C}(z\mathbf{I} - \mathbf{A})^{-1}\mathbf{B} + \mathbf{D}$$

Each element of $\mathbf{T}(z)$ is a ratio of polynomials in z that shares the denominator polynomial

$$|z\mathbf{I} - \mathbf{A}| = z^n + \alpha_{n-1}z^{n-1} + \alpha_{n-2}z^{n-2} + \cdots + \alpha_1 z + \alpha_0$$

which is the characteristic polynomial of the state coupling matrix $\mathbf{A}$. The roots of

$$|z\mathbf{I} - \mathbf{A}| = 0,$$

the eigenvalues of $\mathbf{A}$, are the poles of each of the z-transfer function elements. A state variable system is stable if and only if all the eigenvalues of $\mathbf{A}$ are within the unit circle on the complex plane.

State equations for linear, step-varying systems are of the form

$$\mathbf{x}(k + 1) = \mathbf{A}(k)\mathbf{x}(k) + \mathbf{B}(k)\mathbf{u}(k)$$
$$\mathbf{y}(k) = \mathbf{C}(k)\mathbf{x}(k) + \mathbf{D}(k)\mathbf{u}(k)$$

For any initial state $\mathbf{x}(0)$, the state thereafter is composed of the zero-input component and the zero-state component

$$\mathbf{x}(k) = \mathbf{x}_{\text{zero-input}}(k) + \mathbf{x}_{\text{zero-state}}(k)$$

where

$$\begin{cases} \mathbf{x}_{\text{zero-input}}(k + 1) = \mathbf{A}(k)\mathbf{x}_{\text{zero-input}}(k) \\ \mathbf{x}_{\text{zero-input}}(0) = \mathbf{x}(0) \end{cases}$$

and

$$\begin{cases} \mathbf{x}_{\text{zero-state}}(k + 1) = \mathbf{A}(k)\mathbf{x}_{\text{zero-state}}(k) + \mathbf{B}(k)\mathbf{u}(k) \\ \mathbf{x}_{\text{zero-state}}(0) = \mathbf{0} \end{cases}$$

A linear, step-varying system is stable if the zero-input component of its state decays to zero as k becomes large

$$\lim_{k \to \infty} \mathbf{x}_{\text{zero-input}}(k) = \mathbf{0}$$

and if the zero-state component of its state is bounded for any bounded input.

The relationship between a linear, time-invariant, continuous-time plant driven by sampled-and-held inputs

$$\dot{\mathbf{x}}(t) = \mathcal{A}\mathbf{x}(t) + \mathcal{B}\mathbf{u}(t)$$
$$\mathbf{y}(t) = \mathbf{C}\mathbf{x}(t) + \mathbf{D}\mathbf{x}(t)$$

and the linear, step-invariant, discrete-time model of its evenly spaced samples at intervals T

$$\mathbf{x}(k + 1) = \mathbf{A}\mathbf{x}(k) + \mathbf{B}\mathbf{u}(k)$$
$$\mathbf{y}(k) = \mathbf{C}\mathbf{x}(k) + \mathbf{D}\mathbf{u}(k)$$

is

$$\mathbf{A} = e^{\mathcal{A}T} = \left[\mathbf{I} + \frac{\mathcal{A}T}{1!} + \frac{\mathcal{A}^2 T^2}{2!} + \cdots + \frac{\mathcal{A}^i T^i}{i!} + \cdots\right]$$

and

$$\mathbf{B} = \left[\mathbf{I}T + \frac{\mathcal{A}T^2}{2!} + \frac{\mathcal{A}^2 T^3}{3!} + \cdots + \frac{\mathcal{A}^i T^{i+1}}{(i + 1)!} + \cdots\right]\mathcal{B}$$

Truncated power series are a convenient computational method. Computer simulation of the plant and controller is an important design tool.

For a single-input system, when the $n \times n$ state coupling matrix and the $n \times 1$ input coupling matrix have the forms

$$\mathbf{A} = \begin{bmatrix} 0 & 1 & 0 & \cdots & 0 & 0 \\ 0 & 0 & 1 & \cdots & 0 & 0 \\ 0 & 0 & 0 & \cdots & 0 & 0 \\ & \vdots & & & & \\ 0 & 0 & 0 & \cdots & 0 & 1 \\ -\alpha_0 & -\alpha_1 & -\alpha_2 & \cdots & -\alpha_{n-2} & -\alpha_{n-1} \end{bmatrix} \quad \mathbf{b} = \begin{bmatrix} 0 \\ 0 \\ 0 \\ \vdots \\ 0 \\ 1 \end{bmatrix}$$

the system is said to be in controllable form. Its z-transfer function numerator polynomials are easily related to the elements of the **C** and **d** matrices, and all z-transfer functions share the denominator polynomial

$$z^n + \alpha_{n-1}z^{n-1} + \alpha_{n-2}z^{n-2} + \cdots + \alpha_1 z + \alpha_0$$

which is the characteristic polynomial of the system. When a single-output system has

$$\mathbf{A} = \begin{bmatrix} -\alpha_{n-1} & 1 & 0 & \cdots & 0 & 0 \\ -\alpha_{n-2} & 0 & 1 & \cdots & 0 & 0 \\ -\alpha_{n-3} & 0 & 0 & \cdots & 0 & 0 \\ \vdots & & & & & \\ -\alpha_1 & 0 & 0 & \cdots & 0 & 1 \\ -\alpha_0 & 0 & 0 & \cdots & 0 & 0 \end{bmatrix} \qquad \mathbf{c}^\dagger = [1 \ \ 0 \ \ 0 \ \ \cdots \ \ 0 \ \ 0]$$

it is in observable form. Its z-transfer functions share the denominator polynomial

$$z^n + \alpha_{n-1}z^{n-1} + \alpha_{n-2}z^{n-2} + \cdots + \alpha_1 z + \alpha_0$$

and have numerator polynomials that are easily related to the elements of the **B** and **d** matrices.

A nonsingular change of state variables,

$$\mathbf{x}(k) = \mathbf{P}\mathbf{x}'(k) \qquad \mathbf{x}'(k) = \mathbf{P}^{-1}\mathbf{x}(k)$$

results in new state variable equations of the same form

$$\mathbf{x}'(k+1) = (\mathbf{P}^{-1}\mathbf{A}\mathbf{P})\mathbf{x}'(k) + (\mathbf{P}^{-1}\mathbf{B})\mathbf{u}(k) = \mathbf{A}'\mathbf{x}'(k) + \mathbf{B}'\mathbf{u}(k)$$
$$\mathbf{y}(k) = (\mathbf{C}\mathbf{P})\mathbf{x}'(k) + \mathbf{D}\mathbf{u}(k) = \mathbf{C}'\mathbf{x}'(k) + \mathbf{D}\mathbf{u}(k)$$

but no change in the z-transfer function matrix. Each different set of state variable equations having the same z-transfer function matrix is termed a realization of the z-transfer functions.

If the eigenvalues of **A** are distinct, there are changes of state variables for which

$$\Lambda = \mathbf{P}^{-1}\mathbf{A}\mathbf{P}$$

is diagonal, with the eigenvalues of **A** along the diagonal. The state equations are then decoupled from one another. Each equation involves only one state variable. Each equation's zero-input response is a geometric sequence that is termed a mode of the system. The zero-input part of every system output is a linear combination of the modes.

Finding a diagonalizing transformation is the characteristic value problem of linear algebra, which is developed in Appendix A. Also discussed in that Appendix is the related Cayley-Hamilton theorem: an $n \times n$ matrix $\mathbf{A}$ satisfies its own characteristic equation. This means that $\mathbf{A}^n$ and any higher powers of $\mathbf{A}$ can be expressed as a linear combination of $\mathbf{A}^{n-1}$, $\mathbf{A}^{n-2}$, ..., $\mathbf{A}^0 = \mathbf{I}$.

When a system's state coupling matrix $\mathbf{A}$ has repeated eigenvalues, a change of state variables places it in upper block Jordan form where, for a repeated eigenvalue λ_i, a block of the following form occurs:

$$
\mathbf{A} = \begin{bmatrix}
\ddots & & & & & \\
& \lambda_i & 1 & 0 & \cdots & \\
& 0 & \lambda_i & 1 & & \\
& 0 & 0 & \lambda_i & & \\
& & \vdots & & & \\
& & & & & \ddots
\end{bmatrix}
$$

The corresponding modes are $(\lambda_i)^k$, $k(\lambda_i)^k$, $k^2(\lambda_i)^k$, ... and so on.

Structural properties are especially apparent when a system is in diagonal or block diagonal form. If, in diagonal form, there is no coupling to one of the state equations from any input, the mode associated with that equation cannot be affected by any input and is said to be uncontrollable. Similarly, modes are unobservable if they do not couple to any output. For repeated eigenvalues, if all repetitions of an eigenvalue are in the same upper Jordan block, there must be coupling of an input to the last state variable of the block for controllability of all the block's modes. There must be coupling to an output of the first state variable of the block for observability of all the block's modes. As discussed, if the same eigenvalue occurs in more than one block, the situation can be more complicated.

The change of state variables that take a completely controllable single-input system to controllable form was derived. Then the change of variables that transforms a completely observable single-output system to observable form was found. In the concluding section of the chapter, a monorail train example for developing a state variable model of a translational mechanical system was presented.

References

A comprehensive treatment of the elements of linear system theory for both continuous-time and discrete-time systems can be found in many books, including

L. A. Zadeh and C. A. Desoer, *Linear System Theory*. New York: McGraw-Hill, 1963;

T. Kailath, *Linear Systems*. Englewood Cliffs, NJ: Prentice-Hall, 1980;

W. L. Brogan, *Modern Control Theory,* 2nd ed. Englewood Cliffs, NJ: Prentice-Hall, 1985;

C. T. Chen, *Linear System Theory and Design*. Philadelphia, PA: Saunders College Publishing, 1984.

An outstanding book on continuous-time, state variable system models is

P. M. DeRusso, R. J. Roy, and C. M. Close, *State Variables for Engineers*. New York: Wiley, 1965.

Many other texts contain several chapters on continuous-time state variables. The paper by Kalman,

R. E. Kalman, "Mathematical Description of Linear Dynamical Systems," *SIAM J. Control series A,* 1, 1963, pp. 152–192,

that did much to popularize state variable control system models is good and quite easy to read. The paper

I. C. Horowitz and U. Shaked, "Superiority of Transfer Function over State Variable Methods in Linear, Time-Invariant Feedback System Design," *IEEE Trans. Automatic Control,* Vol. AC-20, February, 1975, pp. 84–97,

compares state variable transfer function design methods.
Discrete-time state equations are discussed in many other books, including

J. A. Cadzow, *Discrete-Time Systems*. Englewood Cliffs, NJ: Prentice-Hall, 1973;

D. G. Luenberger, *Introduction to Dynamic Systems*. New York: Wiley, 1979;

and as part of texts on control system design, such as

H. F. VanLandingham, *Introduction to Digital Control Systems*. New York: Macmillan, 1985.

Key papers on observability and controllability include

E. G. Gilbert, "Controllability and Observability in Multivariable Control Systems," *J. Soc. Ind. Appl. Math.,* ser. A, Vol. 1, 1963, pp. 128–51;

R. E. Kalman, "Canonical Structure of Linear Dynamical Systems," *Proc. Nat. Acad. Sci.,* Vol. 48, no. 4, April 1962, pp. 596–600.

The transformations to controllable and observable forms used here are based on those proposed by

D. S. Rane, "A Simplified Transformation to (Phase-Variable) Canonical Form," *IEEE Trans. Automatic Control,* Vol. AC-11, July 1966, p. 608;

W. G. Tuel, Jr., "On the Transformation to (Phase-Variable) Canonical Form," *IEEE Trans. Automatic Control,* Vol. AC-11, April 1966, pp. 300–303.

Physical system modeling is covered well and in detail in

R. H. Cannon, Jr., *Dynamics of Physical Systems.* New York: McGraw-Hill, 1967;

C. M. Close and D. K. Frederick, *Modeling and Analysis of Dynamic Systems.* Boston: Houghton Mifflin, 1978;

W. R. Perkins and J. B. Cruz, Jr., *Engineering of Dynamic Systems.* New York: Wiley, 1969.

Chapter Three Problems

3-1. Find the state and output equations represented by each of the following integration diagrams.

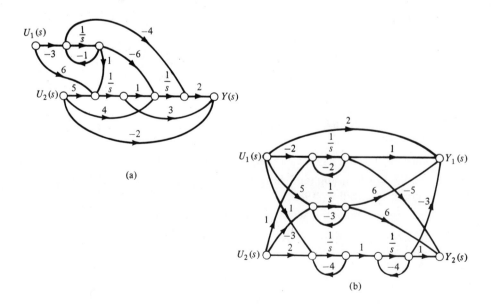

3-2. Draw integration diagrams to represent the following systems:

a.
$$\begin{bmatrix} \dot{x}_1(t) \\ \dot{x}_2(t) \end{bmatrix} = \begin{bmatrix} 1 & 3 \\ -2 & 3 \end{bmatrix} \begin{bmatrix} x_1(t) \\ x_2(t) \end{bmatrix} + \begin{bmatrix} 2 \\ -3 \end{bmatrix} u(t)$$

$$y(t) = \begin{bmatrix} 0 & 2 \end{bmatrix} \begin{bmatrix} x_1(t) \\ x_2(t) \end{bmatrix}$$

b.
$$\begin{bmatrix} \dot{x}_1(t) \\ \dot{x}_2(t) \\ \dot{x}_3(t) \end{bmatrix} = \begin{bmatrix} 3 & -1 & 2 \\ 1 & 6 & 0 \\ 1 & 0 & -1 \end{bmatrix} \begin{bmatrix} x_1(t) \\ x_2(t) \\ x_3(t) \end{bmatrix} + \begin{bmatrix} 8 & -1 \\ 3 & -3 \\ 0 & 6 \end{bmatrix} \begin{bmatrix} u_1(t) \\ u_2(t) \end{bmatrix}$$

$$\begin{bmatrix} y_1(t) \\ y_2(t) \end{bmatrix} = \begin{bmatrix} 1 & -2 & 1 \\ 0 & 3 & 1 \end{bmatrix} \begin{bmatrix} x_1(t) \\ x_2(t) \\ x_3(t) \end{bmatrix} + \begin{bmatrix} 1 & 0 \\ -1 & 1 \end{bmatrix} \begin{bmatrix} u_1(t) \\ u_2(t) \end{bmatrix}$$

c.
$$\dot{\mathbf{x}}(t) = \begin{bmatrix} 2 & 1 & 0 & 0 \\ 3 & 0 & 6 & 1 \\ -1 & 2 & -2 & 1 \\ 2 & -2 & 0 & 1 \end{bmatrix} \mathbf{x}(t) + \begin{bmatrix} 0 & 1 & 0 \\ -2 & 0 & 1 \\ 7 & 8 & 1 \\ 1 & 2 & -2 \end{bmatrix} \mathbf{u}(t)$$

$$\mathbf{y}(t) = \begin{bmatrix} 1 & 0 & 6 & 1 \\ -1 & 0 & -1 & 0 \end{bmatrix} \mathbf{x}(t)$$

3-3. Find a set of state equations for the following system. A solution to this problem is not unique.

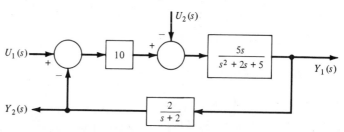

3-4. Use Laplace transformation to find the state $\mathbf{x}(t)$ and the output $\mathbf{y}(t)$ for $t \geq 0$:

a.
$$\begin{bmatrix} \dot{x}_1(t) \\ \dot{x}_2(t) \end{bmatrix} = \begin{bmatrix} -2 & -3 \\ -4 & -6 \end{bmatrix} \begin{bmatrix} x_1(t) \\ x_2(t) \end{bmatrix} + \begin{bmatrix} 1 \\ 1 \end{bmatrix} u(t)$$

$$y(t) = \begin{bmatrix} 3 & -2 \end{bmatrix} \begin{bmatrix} x_1(t) \\ x_2(t) \end{bmatrix}$$

$$u(t) = 2e^{-4t}$$

$$\begin{bmatrix} x_1(0) \\ x_2(0) \end{bmatrix} = \begin{bmatrix} 1 \\ -3 \end{bmatrix}$$

b.
$$\begin{bmatrix} \dot{x}_1(t) \\ \dot{x}_2(t) \end{bmatrix} = \begin{bmatrix} 0 & 4 \\ -4 & 0 \end{bmatrix} \begin{bmatrix} x_1(t) \\ x_2(t) \end{bmatrix} + \begin{bmatrix} 1 & 1 \\ 0 & 1 \end{bmatrix} \begin{bmatrix} u_1(t) \\ u_2(t) \end{bmatrix}$$

$$\begin{bmatrix} y_1(t) \\ y_2(t) \end{bmatrix} = \begin{bmatrix} 7 & 0 \\ -1 & 2 \end{bmatrix} \begin{bmatrix} x_1(t) \\ x_2(t) \end{bmatrix} + \begin{bmatrix} 0 & 0 \\ 2 & 0 \end{bmatrix} \begin{bmatrix} u_1(t) \\ u_2(t) \end{bmatrix}$$

$$\begin{bmatrix} u_1(t) \\ u_2(t) \end{bmatrix} = \begin{bmatrix} \delta(t) \\ 7 \end{bmatrix}$$

where $\delta(t)$ is the unit impulse, and $\mathbf{x}(0^-) = \mathbf{0}$.

c. $\dot{\mathbf{x}}(t) = \begin{bmatrix} -4 & 0 & 0 \\ 0 & -1 & 1 \\ 0 & 0 & -1 \end{bmatrix} \mathbf{x}(t) + \begin{bmatrix} 2 \\ 0 \\ 1 \end{bmatrix} u(t)$

$$\mathbf{y}(t) = \begin{bmatrix} 3 & 1 & 0 \\ 0 & 0 & -1 \end{bmatrix} \mathbf{x}(t) + \begin{bmatrix} 0 \\ 2 \end{bmatrix} u(t)$$

$$u(t) = 10e^{-3t}$$

$$\mathbf{x}(0^-) = \mathbf{0}$$

3-5. Find the transfer function matrices of the following systems, using

$$T(s) = \mathbf{C}[s\mathbf{I} - \mathbf{A}]^{-1}\mathbf{B} + \mathbf{D}$$

a.
$$\begin{bmatrix} \dot{x}_1(t) \\ \dot{x}_2(t) \end{bmatrix} = \begin{bmatrix} -2 & 4 \\ 3 & -6 \end{bmatrix} \begin{bmatrix} x_1(t) \\ x_2(t) \end{bmatrix} + \begin{bmatrix} 3 & 0 \\ 1 & 6 \end{bmatrix} \begin{bmatrix} u_1(t) \\ u_2(t) \end{bmatrix}$$

$$y(t) = \begin{bmatrix} 1 & -2 \end{bmatrix} \begin{bmatrix} x_1(t) \\ x_2(t) \end{bmatrix}$$

b. $\dot{\mathbf{x}}(t) = \begin{bmatrix} 2 & 0 & 3 \\ 3 & -1 & 0 \\ -2 & 0 & 1 \end{bmatrix} \mathbf{x}(t) + \begin{bmatrix} 0 & 0 \\ 1 & 3 \\ -1 & 2 \end{bmatrix} \mathbf{u}(t)$

$$\mathbf{y}(t) = \begin{bmatrix} 0 & 1 & 2 \\ 0 & 1 & -2 \end{bmatrix} \mathbf{x}(t) + \begin{bmatrix} 0 & 1 \\ 0 & 0 \end{bmatrix} \mathbf{u}(t)$$

3-6. Find the transfer function matrices of the following system by converting to an integration diagram and finding transfer functions from the diagram.

a.
$$\begin{bmatrix} \dot{x}_1(t) \\ \dot{x}_2(t) \end{bmatrix} = \begin{bmatrix} -2 & 1 \\ 8 & -4 \end{bmatrix} \begin{bmatrix} x_1(t) \\ x_2(t) \end{bmatrix} + \begin{bmatrix} 0 \\ 1 \end{bmatrix} u(t)$$

$$y(t) = \begin{bmatrix} 2 & 0 \end{bmatrix} \begin{bmatrix} x_1(t) \\ x_2(t) \end{bmatrix}$$

b.
$$\begin{bmatrix} \dot{x}_1(t) \\ \dot{x}_2(t) \\ \dot{x}_3(t) \end{bmatrix} = \begin{bmatrix} 1 & 0 & 2 \\ 0 & 2 & -2 \\ 1 & 1 & 0 \end{bmatrix} \begin{bmatrix} x_1(t) \\ x_2(t) \\ x_3(t) \end{bmatrix} + \begin{bmatrix} 0 & 0 \\ 8 & -1 \\ 2 & 3 \end{bmatrix} \begin{bmatrix} u_1(t) \\ u_2(t) \end{bmatrix}$$

$$y(t) = \begin{bmatrix} 1 & 0 & 3 \end{bmatrix} \begin{bmatrix} x_1(t) \\ x_2(t) \\ x_3(t) \end{bmatrix}$$

3-7. Find the state transition matrices using $\Phi(t) = \mathcal{L}^{-1}[(s\mathbf{I} - \mathbf{A})^{-1}]$ for systems with the following state coupling matrices $\mathbf{A}$:

a. $\mathbf{A} = \begin{bmatrix} -5 & 3 \\ -2 & 0 \end{bmatrix}$

b. $\mathbf{A} = \begin{bmatrix} 0 & 1 \\ -1 & -2 \end{bmatrix}$

c. $\mathbf{A} = \begin{bmatrix} -1 & 1 & -2 \\ -2 & 2 & -1 \\ 0 & 0 & 3 \end{bmatrix}$

3-8. For the precision temperature system presented in Chapter 1, find a linearized state model about the operating point in which

$$\overline{\Theta}_{out} = 35°C$$

$$\overline{\Theta}_{in} = 47°C$$

if the oven temperature and steady state data are as given in the tables.

Oven Temperature Decay Data

Time	Temperature (°C)
10:05:00 AM	135
10:08:30	125
10:13:10	110
10:21:30	90
10:31:30	75
10:37:30	60
10:53:00	50
11:10:00	40

Steady State Oven Temperature Data

Heater Voltage V_h (volts)	Steady State Oven Temperature (°C)
10	35
15	47
20	65
25	88

3-9. Find state equations for the simple monorail system having three cars instead of two. Let the cars and the couplings be identical, with

$$M_2 = M_3 = M_4 = 7000 \text{ kg} \qquad B_{12} = B_{23} = B_{34} = 380 \text{ N} \cdot \text{s/m}$$
$$B_2 = B_3 = B_4 = 1500 \text{ N} \cdot \text{s/m} \qquad K_{12} = K_{23} = K_{34} = 4600 \text{ N/m}$$

and let

$$M_1 = 10,500 \text{ kg} \qquad B_1 = 8000 \text{ N} \cdot \text{s/m}$$

3-10. Write state equations for the following translational mechanical systems, using the positions of the masses as the state variables. All quantities are in SI units: Masses are in kilograms (kg), the spring constants in newtons per meter (N/m), viscous friction constants in newton seconds per meter (N · s/m), and applied forces in newtons (N).

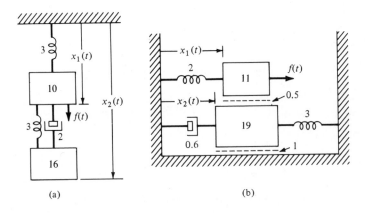

(a)   (b)

3-11. Write state equations for the following rotational mechanical systems, using the angles of rotation of the masses as the state variables. All quantities are in SI units: Moments of inertia are in kilogram square meters kg · m², rotational spring constants are in newton meters (N · m), rotational viscous friction constants are in newton meter seconds (N · m · s), and applied torques are in N · m.

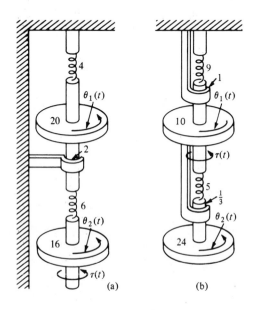

3-12. Find the state $\mathbf{x}(t)$ for $t \geq 0$

a.
$$\begin{bmatrix} \dot{x}_1(t) \\ \dot{x}_2(t) \end{bmatrix} = \begin{bmatrix} 2 & 0 \\ t & 0 \end{bmatrix} \begin{bmatrix} x_1(t) \\ x_2(t) \end{bmatrix}$$

with
$$\begin{bmatrix} x_1(0) \\ x_2(0) \end{bmatrix} = \begin{bmatrix} 1 \\ 1 \end{bmatrix}$$

b.
$$\begin{bmatrix} \dot{x}_1(t) \\ \dot{x}_2(t) \end{bmatrix} = \begin{bmatrix} -1 & 0 \\ e^t & -1 \end{bmatrix} \begin{bmatrix} x_1(t) \\ x_2(t) \end{bmatrix}$$

with
$$\begin{bmatrix} x_1(0) \\ x_2(0) \end{bmatrix} = \begin{bmatrix} -1 \\ 2 \end{bmatrix}$$

3-13. Sketch a delay diagram for the system with state equations

$$\begin{bmatrix} x_1(k+1) \\ x_2(k+1) \\ x_3(k+1) \end{bmatrix} = \begin{bmatrix} 1 & 0 & 1 \\ 0 & -3 & 6 \\ 1 & 1 & 0 \end{bmatrix} \begin{bmatrix} x_1(k) \\ x_2(k) \\ x_3(k) \end{bmatrix} + \begin{bmatrix} -2 \\ 0 \\ -2 \end{bmatrix} u(k)$$

$$y(k) = \begin{bmatrix} 9 & -3 & 0 \end{bmatrix} \begin{bmatrix} x_1(k) \\ x_2(k) \\ x_3(k) \end{bmatrix} - 2u(k)$$

Use Mason's gain rule to find the system's z-transfer function.

3-14. Draw a delay diagram for a one-input/two-output system in controllable form with the z-transfer functions

$$T_1(z) = \frac{2z^2 + 5z - 4}{z^3 + 3z^2 - 2z + 4}$$

$$T_2(z) = \frac{-3z^3}{z^3 + 3z^2 - 2z + 4}$$

Find state variable equations for this system.

3-15. Draw a delay diagram for a two-input/one-output system in observable form with the z-transfer functions

$$T_1(z) = \frac{-2z^2 + 8z - 3}{z^3 - 2z^2 + z - 3}$$

$$T_2(z) = \frac{2z^3 + 6}{z^3 - 2z^2 + z - 3}$$

Find state variable equations for this system.

3-16. Given the following block diagram for a discrete-time system, find a state variable representation for the system.

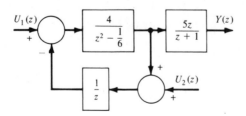

3-17. For the system

$$
\begin{bmatrix} x_1(k + 1) \\ x_2(k + 1) \\ x_3(k + 1) \end{bmatrix} = \begin{bmatrix} -1 & 0 & 0 \\ 1 & 1 & 4 \\ 1 & -5 & 2 \end{bmatrix} \begin{bmatrix} x_1(k) \\ x_2(k) \\ x_3(k) \end{bmatrix} + \begin{bmatrix} 3 \\ -1 \\ 0 \end{bmatrix} u(k)
$$

$$
y(k) = [1 \quad 4 \quad 0] \begin{bmatrix} x_1(k) \\ x_2(k) \\ x_3(k) \end{bmatrix}
$$

find the state vectors and outputs, $\mathbf{x}(1)$, $\mathbf{x}(2)$, $\mathbf{x}(3)$, $y(1)$, $y(2)$, $y(3)$ if

$$
\mathbf{x}(0) = \begin{bmatrix} -1 \\ 2 \\ 4 \end{bmatrix}
$$

and

$$
u(k) = 2 - 3\delta(k)
$$

3-18. Using the definition of z-transform, find $\mathscr{Z}[\mathbf{A}^k]$.

3-19. The methods presented in Appendix A for computing the state transition matrix of a continuous-time system are directly applicable to discrete-time systems. For the matrix

$$
\mathbf{A} = \begin{bmatrix} 0 & \frac{1}{2} \\ -\frac{1}{2} & 0 \end{bmatrix}
$$

find the discrete state transition matrix $\mathbf{A}^k$ using:

a. The diagonalization method.

b. The Cayley-Hamilton method.

c. Sylvester's method.

d. Residue matrices.

3-20. Repeat problem 3-19 for the matrix

$$
\mathbf{A} = \begin{bmatrix} -\frac{1}{4} & -\frac{1}{6} & \frac{1}{4} \\ 0 & -\frac{1}{3} & -\frac{1}{2} \\ 0 & 0 & -\frac{1}{2} \end{bmatrix}
$$

3-21. Find the z-transform of the state and output of the system in problem 3-17.

3-22. Find the z-transfer function matrix of the following system:

$$\begin{bmatrix} x_1(k+1) \\ x_2(k+1) \\ x_3(k+1) \end{bmatrix} = \begin{bmatrix} 1 & 0 & 0 \\ 0 & 6 & 1 \\ 0 & -1 & 0 \end{bmatrix} \begin{bmatrix} x_1(k) \\ x_2(k) \\ x_3(k) \end{bmatrix} + \begin{bmatrix} -2 & 0 \\ 0 & -1 \\ 1 & 0 \end{bmatrix} \begin{bmatrix} u_1(k) \\ u_2(k) \end{bmatrix}$$

$$\begin{bmatrix} y_1(k) \\ y_2(k) \end{bmatrix} = \begin{bmatrix} 0 & 0 & -2 \\ 1 & 0 & 0 \end{bmatrix} \begin{bmatrix} x_1(k) \\ x_2(k) \\ x_3(k) \end{bmatrix} + \begin{bmatrix} 2 & 3 \\ 0 & 0 \end{bmatrix} \begin{bmatrix} u_1(k) \\ u_2(k) \end{bmatrix}$$

3-23. For the system in problem 3-22, make the change of state variables

$$\mathbf{x}' = \begin{bmatrix} 3 & -1 & 0 \\ -2 & 1 & 0 \\ 0 & 0 & 2 \end{bmatrix} \mathbf{x}$$

and find the state and output equations in terms of $\mathbf{x}'$.

3-24. Find the state and output of the system

$$\begin{bmatrix} x_1(k+1) \\ x_2(k+1) \end{bmatrix} = \begin{bmatrix} (-\frac{1}{2})^k & -1 \\ 0 & 6 \end{bmatrix} \begin{bmatrix} x_1(k) \\ x_2(k) \end{bmatrix} + \begin{bmatrix} k \\ 2 \end{bmatrix} u(k)$$

$$y(k) = \begin{bmatrix} \frac{1}{k+1} & 0 \end{bmatrix} \begin{bmatrix} x_1(k) \\ x_2(k) \end{bmatrix} - u(k)$$

for steps 1, 2, and 3 if

$$\mathbf{x}(0) = \begin{bmatrix} 1 \\ -4 \end{bmatrix} \quad \text{and} \quad u(k) = 4 - k$$

3-25. Find the new state and output equations in terms of

$$\mathbf{x}'(k) = \begin{bmatrix} 1 & k \\ 0 & -1 \end{bmatrix} \mathbf{x}(k)$$

for the system in Problem 3-24.

3-26. Find discrete-time models of each of the following continuous-time systems. The inputs are all piecewise-constant during each sampling interval T.

a. $\dot{x}(t) = -3x(t) + 2u(t)$
   $y(t) = 2x(t)$
   $T = 0.01$ sec

b. $\begin{bmatrix} \dot{x}_1(t) \\ \dot{x}_2(t) \end{bmatrix} = \begin{bmatrix} -2 & 1 \\ 1 & -2 \end{bmatrix} \begin{bmatrix} x_1(t) \\ x_2(t) \end{bmatrix} + \begin{bmatrix} 1 \\ -1 \end{bmatrix} u(t)$

$\begin{bmatrix} y_1(t) \\ y_2(t) \end{bmatrix} = \begin{bmatrix} 1 & 0 \\ 2 & -3 \end{bmatrix} \begin{bmatrix} x_1(t) \\ x_2(t) \end{bmatrix} + \begin{bmatrix} 0 \\ -1 \end{bmatrix} u(t)$

$T = 0.2$ sec

c. $\begin{bmatrix} \dot{x}_1(t) \\ \dot{x}_2(t) \end{bmatrix} = \begin{bmatrix} 0 & 1 \\ -1 & 0 \end{bmatrix} \begin{bmatrix} x_1(t) \\ x_2(t) \end{bmatrix} + \begin{bmatrix} 0 \\ 1 \end{bmatrix} u(t)$

$y(t) = \begin{bmatrix} 1 & 0 \end{bmatrix} \begin{bmatrix} x_1(t) \\ x_2(t) \end{bmatrix}$

$T = 0.02$ sec

d. $\begin{bmatrix} \dot{x}_1 \\ \dot{x}_2 \\ \dot{x}_3 \end{bmatrix} = \begin{bmatrix} 0 & 1 & 0 \\ 0 & 0 & 1 \\ -6 & -11 & -6 \end{bmatrix} \begin{bmatrix} x_1 \\ x_2 \\ x_3 \end{bmatrix} + \begin{bmatrix} 0 & 0 \\ 2 & -1 \\ 1 & 1 \end{bmatrix} \begin{bmatrix} u_1(t) \\ u_2(t) \end{bmatrix}$

$y(t) = \begin{bmatrix} 1 & 0 & -1 \end{bmatrix} \begin{bmatrix} x_1 \\ x_2 \\ x_3 \end{bmatrix} + \begin{bmatrix} 0 & 1 \end{bmatrix} \begin{bmatrix} u_1(t) \\ u_2(t) \end{bmatrix}$

$T = 0.1$ sec

3-27. Find a discrete-time model of the continuous-time system

$\dot{\mathbf{x}}(t) = \mathcal{A}\mathbf{x}(t) + \mathcal{B}\mathbf{u}(t)$

$\mathbf{y}(t) = \mathbf{C}\mathbf{x}(t) + \mathbf{D}\mathbf{u}(t)$

using trapezoidal approximation of the integral in equation (3-23a).

3-28. Find diagonal state equations for a two-input/one-output system with the z-transfer functions

$T_1(z) = \dfrac{z^2 + z + 3}{z^3 - \frac{1}{4}z}$

$T_2(z) = \dfrac{3z - 1}{z^3 - \frac{1}{4}z}$

3-29. Find a transformation matrix $\mathbf{P}$ in

$\mathbf{x} = \mathbf{P}\mathbf{x}' \qquad \mathbf{x}' = \mathbf{P}^{-1}\mathbf{x}$

that takes the system

$$\begin{bmatrix} x_1(k+1) \\ x_2(k+1) \\ x_3(k+1) \end{bmatrix} = \begin{bmatrix} -4 & 1 & 0 \\ -3 & 0 & 1 \\ 0 & 0 & 0 \end{bmatrix} \begin{bmatrix} x_1(k) \\ x_2(k) \\ x_3(k) \end{bmatrix} + \begin{bmatrix} 2 & 0 \\ 0 & 0 \\ 1 & -1 \end{bmatrix} \begin{bmatrix} u_1(k) \\ u_2(k) \end{bmatrix}$$

$$\begin{bmatrix} y_1(k) \\ y_2(k) \end{bmatrix} = \begin{bmatrix} 1 & 2 & 0 \\ 0 & -1 & 0 \end{bmatrix} \begin{bmatrix} x_1(k) \\ x_2(k) \\ x_3(k) \end{bmatrix} + \begin{bmatrix} 2 & -1 \\ 0 & 0 \end{bmatrix} \begin{bmatrix} u_1(k) \\ u_2(k) \end{bmatrix}$$

into a diagonal form. Find the new state equations in terms of $\mathbf{x'}$.

3-30. Find a set of diagonal state equations for a single-input/single-output system with the z-transfer function

$$T(z) = \frac{-4z^2 - 2z + 5}{z^3 + z^2 + (13/36)z}$$

Then find an alternative block diagonal representation that does not involve complex numbers.

3-31. Find a transformation matrix $\mathbf{P}$ in

$$\mathbf{x} = \mathbf{Px'} \qquad \mathbf{x'} = \mathbf{P^{-1}x}$$

that takes the system with complex eigenvalues

$$\begin{bmatrix} x_1(k+1) \\ x_2(k+1) \end{bmatrix} = \begin{bmatrix} -1 & -1 \\ 1 & -1 \end{bmatrix} \begin{bmatrix} x_1(k) \\ x_2(k) \end{bmatrix} + \begin{bmatrix} 1 \\ -3 \end{bmatrix} u(k)$$

$$y(k) = \begin{bmatrix} 1 & 0 \end{bmatrix} \begin{bmatrix} x_1(k) \\ x_2(k) \end{bmatrix} + u(k)$$

into diagonal form. Find the new state equations in terms of $\mathbf{x'}$.

3-32. Find a set of block Jordan state equations for a single-input/two-output system with the z-transfer functions

$$T_1(z) = \frac{z^2 + 2z - 1}{z^3 - z^2 + \frac{1}{4}z}$$

$$T_2(z) = \frac{3z^3 - 3z^2 + 2}{z^3 - z^2 + \frac{1}{4}z}$$

3-33. Find a transformation matrix $\mathbf{P}$ in

$$\mathbf{x} = \mathbf{Px'} \qquad \mathbf{x'} = \mathbf{P^{-1}x}$$

that takes the system with repeated eigenvalues

$$\begin{bmatrix} x_1(k+1) \\ x_2(k+1) \\ x_3(k+1) \end{bmatrix} = \begin{bmatrix} 0 & 1 & 0 \\ 0 & 0 & 1 \\ \frac{1}{8} & -\frac{3}{4} & \frac{3}{2} \end{bmatrix} \begin{bmatrix} x_1(k) \\ x_2(k) \\ x_3(k) \end{bmatrix} + \begin{bmatrix} 1 \\ 0 \\ -1 \end{bmatrix} u(k)$$

$$y(k) = \begin{bmatrix} -1 & 1 & 1 \end{bmatrix} \begin{bmatrix} x_1(k) \\ x_2(k) \\ x_3(k) \end{bmatrix} - 2u(k)$$

into a block Jordan form.

3-34. Show that for a system with distinct eigenvalues, the eigenvectors of the state coupling matrix are the initial states for which the amplitudes of all but one of the modes in the zero-input response are zero. That is, initial conditions equal to an eigenvector excite only a single zero-input mode, the one associated with the corresponding eigenvalue. Then, for the system

$$\begin{bmatrix} x_1(k+1) \\ x_2(k+1) \\ x_3(k+1) \end{bmatrix} = \begin{bmatrix} -\frac{1}{2} & 1 & 0 \\ \frac{1}{2} & 0 & 1 \\ 0 & 0 & 0 \end{bmatrix} \begin{bmatrix} x_1(k) \\ x_2(k) \\ x_3(k) \end{bmatrix} = \mathbf{A}\mathbf{x}(k)$$

determine nonzero initial conditions $\mathbf{x}(0)$ such that only the $(\frac{1}{2})^k$ mode is excited.

3-35. Determine whether the following system is completely controllable and whether it is completely observable by transforming it into a diagonal form:

$$\begin{bmatrix} x_1(k+1) \\ x_2(k+1) \\ x_3(k+1) \end{bmatrix} = \begin{bmatrix} 1 & -1 & 0 \\ 2 & -5 & 0 \\ 0 & 0 & -1 \end{bmatrix} \begin{bmatrix} x_1(k) \\ x_2(k) \\ x_3(k) \end{bmatrix} + \begin{bmatrix} 2 & -2 \\ 1 & -1 \\ 2 & 0 \end{bmatrix} \begin{bmatrix} u_1(k) \\ u_2(k) \end{bmatrix}$$

$$y(k) = \begin{bmatrix} -3 & 2 & 1 \end{bmatrix} \begin{bmatrix} x_1(k) \\ x_2(k) \\ x_3(k) \end{bmatrix} + \begin{bmatrix} 3 & 0 \end{bmatrix} \begin{bmatrix} u_1(k) \\ u_2(k) \end{bmatrix}$$

Identify any modes that are uncontrollable, unobservable, or both.

3-36. For each of the following systems, find the transformation matrix $\mathbf{P}$ in

$$\mathbf{x} = \mathbf{P}\mathbf{x}' \qquad \mathbf{x}' = \mathbf{P}^{-1}\mathbf{x}$$

that takes the system into the controllable form. Find the new state equations in terms of $\mathbf{x}'$.

a.
$$\begin{bmatrix} x_1(k+1) \\ x_2(k+1) \end{bmatrix} = \begin{bmatrix} 4 & -1 \\ 1 & -1 \end{bmatrix} \begin{bmatrix} x_1(k) \\ x_2(k) \end{bmatrix} + \begin{bmatrix} 0 \\ 2 \end{bmatrix} u(k)$$

$$y(k) = [-2 \quad 1] \begin{bmatrix} x_1(k) \\ x_2(k) \end{bmatrix} - u(k)$$

b.
$$\begin{bmatrix} x_1(k+1) \\ x_2(k+1) \\ x_3(k+1) \end{bmatrix} = \begin{bmatrix} -4 & 1 & 1 \\ 0 & 2 & -1 \\ 0 & 1 & 3 \end{bmatrix} \begin{bmatrix} x_1(k) \\ x_2(k) \\ x_3(k) \end{bmatrix} + \begin{bmatrix} 1 \\ 0 \\ -1 \end{bmatrix} u(k)$$

$$\begin{bmatrix} y_1(k) \\ y_2(k) \end{bmatrix} = \begin{bmatrix} 2 & 0 & 1 \\ 0 & -2 & 4 \end{bmatrix} \begin{bmatrix} x_1(k) \\ x_2(k) \\ x_3(k) \end{bmatrix} + \begin{bmatrix} -3 \\ 5 \end{bmatrix} u(k)$$

3-37. For each of the following systems, find the transformation matrix $\mathbf{P}$ in

$$\mathbf{x} = \mathbf{P}\mathbf{x}' \qquad \mathbf{x}' = \mathbf{P}^{-1}\mathbf{x}$$

that takes the system into the observable form. Find the new state equations in terms of $\mathbf{x}'$.

a.
$$\begin{bmatrix} x_1(k+1) \\ x_2(k+1) \end{bmatrix} = \begin{bmatrix} 2 & -1 \\ -2 & -3 \end{bmatrix} \begin{bmatrix} x_1(k) \\ x_2(k) \end{bmatrix} + \begin{bmatrix} 2 & -1 \\ 0 & 1 \end{bmatrix} \begin{bmatrix} u_1(k) \\ u_2(k) \end{bmatrix}$$

$$y(k) = [1 \quad 1] \begin{bmatrix} x_1(k) \\ x_2(k) \end{bmatrix} + [4 \quad 0] \begin{bmatrix} u_1(k) \\ u_2(k) \end{bmatrix}$$

b.
$$\begin{bmatrix} x_1(k+1) \\ x_2(k+1) \\ x_3(k+1) \end{bmatrix} = \begin{bmatrix} -1 & 2 & 0 \\ -1 & 4 & 0 \\ 1 & 0 & 1 \end{bmatrix} \begin{bmatrix} x_1(k) \\ x_2(k) \\ x_3(k) \end{bmatrix} + \begin{bmatrix} 3 \\ 2 \\ -2 \end{bmatrix} u(k)$$

$$y(k) = [0 \quad -2 \quad 3] \begin{bmatrix} x_1(k) \\ x_2(k) \\ x_3(k) \end{bmatrix} + u(k)$$

4

Discrete-Time Observation, Control, and Feedback

4.1 Preview

Feedback control involves determining something about the plant state from plant outputs and using this information to generate control inputs that force the plant to behave in a desired way. It is the use of plant outputs that forms a *feedback* system. For a plant with a known model, the plant state summarizes everything that can currently be known about the plant, so the most information that can be derived from the outputs is the state of the plant. This is not to say that it is necessary to determine the plant state in order to perform good feedback control, but if we are able to do so, all possible relevant information about the plant is available for generating the control. Similarly, one can do no more than to control a plant's state. We may or may not need to control the entire state of a plant, but if we are able to do so, then we are able to control the plant to any lesser degree also.

In this chapter, methods are developed for the determination of the state of a completely observable plant from its outputs and inputs in a finite number of steps. Similar methods are then developed for driving a completely controllable plant through its inputs to a given desired state in a finite number of steps.

When a completely controllable plant's state is available for feedback, it is always possible to place the feedback system eigenvalues at any locations selected by the designer. This important result is derived and a number of illustrative design examples are given. Then the design of a step-varying state feedback controller is considered in some detail, particularly for the case of step-varying deadbeat feedback. When the plant state is not available for feedback, arbitrary placement of the feedback system eigenvalues using output feedback alone might not be possible. This situation is discussed in Section 4.5 and is considered further in the next chapter.

The final section of the chapter concerns quadratic optimal control. In the previous design methods, the properties of zero-input response convergence of the plant with state feedback were selected directly. In the optimal control approach, state feedback is chosen to minimize a designer-selected performance measure, indirectly resulting in a zero-input response. The optimal quadratic solution is first derived via dynamic programming and examples are given. Then a closed-form solution for the optimal gain is derived from which the numerical value of the gain can be calculated for any instant of time.

4.2 Observability and State Observation

Observability was defined in Chapter 3 in terms of the coupling of modes to plant outputs. Complete observability also means that the state of a plant can be determined from a finite number of its most recent inputs and outputs. That is the subject of this section. Determining the state of a plant from its inputs and outputs is an important capability in control system design because in some situations it is necessary (or at least convenient) to know the actual state to control that state effectively. In the process of learning how to calculate the state of a plant from input and output measurements, we develop the observability matrix rank test.

4.2.1 Observability and the Rank Test

If a discrete-time system is completely observable, its state $\mathbf{x}(k)$ at any specific step k can be determined from the system model and a finite number of steps of its inputs and outputs, starting at step k. For a step-invariant system, if it is possible to determine the state at any step, say $\mathbf{x}(0)$, then with a shift of step, the state at any other step can be determined in the same way.

For a system

$$\mathbf{x}(k + 1) = \mathbf{A}\mathbf{x}(k) + \mathbf{B}\mathbf{u}(k)$$
$$\mathbf{y}(k) = \mathbf{C}\mathbf{x}(k) + \mathbf{D}\mathbf{u}(k)$$

the initial state $\mathbf{x}(0)$ is, in terms of the outputs and inputs, given by

$$
\begin{cases}
\mathbf{y}(0) = \mathbf{Cx}(0) + \mathbf{Du}(0) \\
\mathbf{y}(1) = \mathbf{Cx}(1) + \mathbf{Du}(1) = \mathbf{CAx}(0) + \mathbf{CBu}(0) + \mathbf{Du}(1) \\
\mathbf{y}(2) = \mathbf{Cx}(2) + \mathbf{Du}(2) = \mathbf{CA}^2\mathbf{x}(0) + \mathbf{CABu}(0) + \mathbf{CBu}(1) + \mathbf{Du}(2) \\
\quad \vdots \\
\mathbf{y}(n-1) = \mathbf{Cx}(n-1) + \mathbf{Du}(n-1) \\
\qquad\quad = \mathbf{CA}^{n-1}\mathbf{x}(0) + \mathbf{CA}^{n-2}\mathbf{Bu}(0) + \mathbf{CA}^{n-3}\mathbf{Bu}(1) + \cdots \\
\qquad\qquad + \mathbf{CBu}(n-2) + \mathbf{Du}(n-1)
\end{cases}
$$

Collecting the $\mathbf{x}(0)$ terms on the left

$$
\begin{cases}
\mathbf{Cx}(0) = \mathbf{y}(0) - \mathbf{Du}(0) \\
\mathbf{CAx}(0) = \mathbf{y}(1) - \mathbf{CBu}(0) - \mathbf{Du}(1) \\
\mathbf{CA}^2\mathbf{x}(0) = \mathbf{y}(2) - \mathbf{CABu}(0) - \mathbf{CBu}(1) - \mathbf{Du}(2) \\
\quad \vdots \\
\mathbf{CA}^{n-1}\mathbf{x}(0) = \mathbf{y}(n-1) - \mathbf{CA}^{n-2}\mathbf{Bu}(0) - \cdots - \mathbf{CBu}(n-2) \\
\qquad\qquad\qquad - \mathbf{Du}(n-1)
\end{cases}
$$

This set of linear algebraic equations can be solved for $\mathbf{x}(0)$ only if the array of coefficients

$$
\mathbf{M}_0 = \begin{bmatrix} \mathbf{C} \\ \hline \mathbf{CA} \\ \hline \mathbf{CA}^2 \\ \hline \vdots \\ \hline \mathbf{CA}^{n-1} \end{bmatrix}
$$

is of full rank. Additional outputs are of no help because they yield additional equations with coefficients $\mathbf{CA}^n$, $\mathbf{CA}^{n+1}, \ldots$, which by the Cayley-Hamilton theorem can be expressed in terms of a linear combination of $\mathbf{A}^{n-1}$ and lower powers of $\mathbf{A}$. Additional equations are thus linear combinations of those already considered.*

---

* There are special cases for which the state of a system *after the initial step k* can be determined from system inputs and outputs even when $\mathbf{M}_0$ is singular. For example, if $\mathbf{A} = \mathbf{0}$, then $\mathbf{x}(k)$ might not be determinable from the plant inputs and outputs beginning with step k, but $\mathbf{x}(k + 1) = \mathbf{Bu}(k)$. That is, it might be possible to find $\mathbf{x}(k + 1)$ or the state

As an illustration, the system

$$\begin{bmatrix} x_1(k+1) \\ x_2(k+1) \end{bmatrix} = \begin{bmatrix} 0 & -6 \\ -1 & 1 \end{bmatrix} \begin{bmatrix} x_1(k) \\ x_2(k) \end{bmatrix} + \begin{bmatrix} 0 & 3 \\ -4 & 7 \end{bmatrix} \begin{bmatrix} u_1(k) \\ u_2(k) \end{bmatrix}$$

$$y(k) = \begin{bmatrix} 1 & 2 \end{bmatrix} \begin{bmatrix} x_1(k) \\ x_2(k) \end{bmatrix} + \begin{bmatrix} 4 & 0 \end{bmatrix} \begin{bmatrix} u_1(k) \\ u_2(k) \end{bmatrix}$$

has the observability matrix

$$\mathbf{M}_0 = \begin{bmatrix} 1 & 2 \\ -2 & -4 \end{bmatrix}$$

It is not completely observable because

$$\begin{vmatrix} 1 & 2 \\ -2 & -4 \end{vmatrix} = 0$$

which indicates that $\mathbf{M}_0$ is of rank 1, not the required rank 2.

The system

$$\begin{bmatrix} x_1(k+1) \\ x_2(k+1) \\ x_3(k+1) \end{bmatrix} = \begin{bmatrix} 3 & -2 & 1 \\ 1 & 0 & 0 \\ 1 & -2 & 3 \end{bmatrix} \begin{bmatrix} x_1(k) \\ x_2(k) \\ x_3(k) \end{bmatrix} + \begin{bmatrix} 4 \\ 5 \\ -6 \end{bmatrix} u(k)$$

$$\begin{bmatrix} y_1(k) \\ y_2(k) \end{bmatrix} = \begin{bmatrix} 1 & 0 & -1 \\ 2 & 0 & 0 \end{bmatrix} \begin{bmatrix} x_1(k) \\ x_2(k) \\ x_3(k) \end{bmatrix}$$

is completely observable because its observability matrix,

$$\mathbf{M}_0 = \begin{bmatrix} 1 & 0 & -1 \\ 2 & 0 & 0 \\ 2 & 0 & -2 \\ 6 & -4 & 2 \\ 4 & 0 & -4 \\ 16 & -16 & 12 \end{bmatrix}$$

at some later step using *previous* inputs and outputs. Because of this possibility, some authors use the term *observability* more loosely, allowing the use of input and output data from before step k to find $\mathbf{x}(k)$. The term *constructability* is used for the more stringent definition. System modes should then probably be classified as "constructable" and "unconstructable" instead of "observable" and "unobservable." There is no distinction between the two definitions when $\mathbf{A}$ is nonsingular.

is of full rank. Rows 1, 2, and 4, for example, are linearly independent because

$$\begin{vmatrix} 1 & 0 & -1 \\ 2 & 0 & 0 \\ 6 & -4 & 2 \end{vmatrix} \neq 0$$

For a multiple-output system, the smallest integer ν for which

$$\mathbf{M}_0(\nu) = \begin{bmatrix} \mathbf{C} \\ \hline \mathbf{CA} \\ \hline \mathbf{CA}^2 \\ \hline \vdots \\ \hline \mathbf{CA}^{\nu-1} \end{bmatrix}$$

has full rank is termed the *observability index* of the system. It is the minimum number of steps required to determine the system state.

A continuous-time system is completely observable if its state $\mathbf{x}(t)$ at any specific time t can be determined from the system model and its inputs and measurement outputs for a finite interval of time. Similar to the test for complete observability of a step-invariant discrete-time system, a time-invariant continuous-time system governed by the state and output equations

$$\dot{\mathbf{x}}(t) = \mathscr{A}\mathbf{x}(t) + \mathscr{B}\mathbf{u}(t)$$
$$\mathbf{y}(t) = \mathbf{C}\mathbf{x}(t) + \mathbf{D}\mathbf{u}(t)$$

is completely observable if and only if its observability matrix

$$\mathbf{M}_0 = \begin{bmatrix} \mathbf{C} \\ \hline \mathbf{C}\mathscr{A} \\ \hline \mathbf{C}\mathscr{A}^2 \\ \hline \vdots \\ \hline \mathbf{C}\mathscr{A}^{n-1} \end{bmatrix}$$

is of full rank.

4.2.2 Observing Single- and Multiple-Output Systems

The state of a completely observable nth-order single-output system

$$\mathbf{x}(k + 1) = \mathbf{A}\mathbf{x}(k) + \mathbf{B}\mathbf{u}(k)$$
$$y(k) = \mathbf{c}^{\dagger}\mathbf{x}(k) + \mathbf{d}^{\dagger}\mathbf{u}(k) \tag{4-1}$$

can always be found at any step i from its outputs and inputs at steps i through $i + n - 1$. The resulting n simultaneous equations in the n unknown components of the state $\mathbf{x}(i)$

$$
\begin{cases}
y(i) = \mathbf{c}^{\dagger}\mathbf{x}(i) + \mathbf{d}^{\dagger}\mathbf{u}(i) \\
y(i + 1) = \mathbf{c}^{\dagger}[\mathbf{A}\mathbf{x}(i) + \mathbf{B}\mathbf{u}(i)] + \mathbf{d}^{\dagger}\mathbf{u}(i + 1) \\
\vdots \\
y(i + n - 1) = \mathbf{c}^{\dagger}[\mathbf{A}^{n-1}\mathbf{x}(i) + \mathbf{A}^{n-2}\mathbf{B}\mathbf{u}(i) + \mathbf{A}^{n-3}\mathbf{B}\mathbf{u}(i + 1) + \cdots \\
\qquad\qquad\qquad + \mathbf{B}\mathbf{u}(i + n - 2)] + \mathbf{d}^{\dagger}\mathbf{u}(i + n - 1)
\end{cases}
$$

have a unique solution whenever the system's observability matrix (which is $n \times n$ for a single-output system) is nonsingular.

Once $\mathbf{x}(i)$ is known, the state equation (4-1) can be used to find $\mathbf{x}(i + 1), \mathbf{x}(i + 2), \ldots, \mathbf{x}(i + n)$. If the input is also known at step $i + n$ and perhaps beyond, the state equation can also be used to find $\mathbf{x}(i + n + 1)$ and so on. If the input is known at step $i - 1$ and perhaps further backward in step, the state equation can be used as

$$\mathbf{x}(k) = \mathbf{A}^{-1}[\mathbf{x}(k + 1) - \mathbf{B}\mathbf{u}(k)]$$

to solve for the state at steps prior to step i if $\mathbf{A}$ is nonsingular.

As a numerical example, consider the single-output system

$$
\begin{bmatrix} x_1(k + 1) \\ x_2(k + 1) \\ x_3(k + 1) \end{bmatrix} = \begin{bmatrix} 2 & -1 & 1 \\ 0 & 1 & 3 \\ -2 & 0 & 1 \end{bmatrix} \begin{bmatrix} x_1(k) \\ x_2(k) \\ x_3(k) \end{bmatrix} + \begin{bmatrix} 3 \\ 1 \\ 0 \end{bmatrix} u(k) = \mathbf{A}\mathbf{x}(k) + \mathbf{b}u(k)
$$

$$
y(k) = [0 \quad 2 \quad -1] \begin{bmatrix} x_1(k) \\ x_2(k) \\ x_3(k) \end{bmatrix} + 2u(k) = \mathbf{c}^{\dagger}\mathbf{x}(k) + 2u(k)
$$

Suppose that

$$y(0) = 8 \qquad y(1) = 5 \qquad y(2) = -4$$
$$u(0) = 3 \qquad u(1) = 0 \qquad u(2) = -2$$

and that it is desired to find $\mathbf{x}(0)$ from these. Then

$$y(0) = 8 = \mathbf{c}^\dagger\mathbf{x}(0) + 2u(0) = [0 \quad 2 \quad -1]\begin{bmatrix} x_1(0) \\ x_2(0) \\ x_3(0) \end{bmatrix} + 2(3)$$

$$= 2x_2(0) - x_3(0) + 6$$

$$y(1) = 5 = \mathbf{c}^\dagger\mathbf{A}\mathbf{x}(0) + \mathbf{c}^\dagger\mathbf{b}u(0) + du(1)$$

$$= [0 \quad 2 \quad -1]\begin{bmatrix} 2 & -1 & 1 \\ 0 & 1 & 3 \\ -2 & 0 & 1 \end{bmatrix}\begin{bmatrix} x_1(0) \\ x_2(0) \\ x_3(0) \end{bmatrix}$$

$$+ [0 \quad 2 \quad -1]\begin{bmatrix} 3 \\ 1 \\ 0 \end{bmatrix}(3) + 2(0)$$

$$= 2x_1(0) + 2x_2(0) + 5x_3(0) + 6$$

$$y(2) = -4 = \mathbf{c}^\dagger\mathbf{A}^2\mathbf{x}(0) + \mathbf{c}^\dagger\mathbf{A}\mathbf{b}u(0) + \mathbf{c}^\dagger\mathbf{b}u(1) + du(2)$$

$$= [2 \quad 2 \quad 5]\begin{bmatrix} 2 & -1 & 1 \\ 0 & 1 & 3 \\ -2 & 0 & 1 \end{bmatrix}\begin{bmatrix} x_1(0) \\ x_2(0) \\ x_3(0) \end{bmatrix}$$

$$+ [2 \quad 2 \quad 5]\begin{bmatrix} 3 \\ 1 \\ 0 \end{bmatrix}(3) + 2(0) + 2(-2)$$

$$= -6x_1(0) + 13x_3(0) + 20$$

Solving these equations for $\mathbf{x}(0)$

$$\begin{cases} 2x_2(0) - x_3(0) = 2 \\ 2x_1(0) + 2x_2(0) + 5x_3(0) = -1 \\ -6x_1(0) + 13x_3(0) = -24 \end{cases}$$

$$\begin{bmatrix} x_1(0) \\ x_2(0) \\ x_3(0) \end{bmatrix} = \begin{bmatrix} \frac{105}{62} \\ \frac{29}{62} \\ -\frac{66}{62} \end{bmatrix}$$

From $\mathbf{x}(0)$ and the inputs, $\mathbf{x}(1)$, $\mathbf{x}(2)$, and $\mathbf{x}(3)$ can now be found also, if desired

$$\mathbf{x}(1) = \mathbf{A}\mathbf{x}(0) + \mathbf{b}u(0) = \begin{bmatrix} \frac{673}{62} \\ \frac{17}{62} \\ -\frac{276}{62} \end{bmatrix}$$

$$\mathbf{x}(2) = \mathbf{A}\mathbf{x}(1) + \mathbf{b}u(1) = \begin{bmatrix} \frac{1053}{62} \\ -\frac{811}{62} \\ -\frac{1622}{62} \end{bmatrix}$$

$$\mathbf{x}(3) = \mathbf{A}\mathbf{x}(2) + \mathbf{b}u(2) = \begin{bmatrix} \frac{923}{62} \\ -\frac{5801}{62} \\ -\frac{3728}{62} \end{bmatrix}$$

If, instead, this system's outputs and inputs for steps 4, 5, and 6 were known, then the state at the first of these steps, $\mathbf{x}(4)$, would be found from

$$\begin{cases} y(4) = 8 = \mathbf{c}^t\mathbf{x}(4) + du(4) \\ y(5) = 5 = \mathbf{c}^t\mathbf{A}\mathbf{x}(4) + \mathbf{c}^t\mathbf{b}u(4) + du(5) \\ y(6) = -4 = \mathbf{c}^t\mathbf{A}^2\mathbf{x}(4) + \mathbf{c}^t\mathbf{A}\mathbf{b}u(4) + \mathbf{c}^t\mathbf{b}u(5) + du(6) \end{cases}$$

From $\mathbf{x}(4)$ and the inputs, $\mathbf{x}(5)$, $\mathbf{x}(6)$, and $\mathbf{x}(7)$ can be determined.

The outputs of a multiple-output system are *linearly independent* if the rows of the output coupling matrix $\mathbf{C}$ are linearly independent of one another. Linearly dependent outputs are combinations of the other outputs and inputs, and so are of no help in observing a system's state because they do not contribute any linearly independent rows to the system's observability matrix. In practice there may be other reasons, such as improved reliability, for using linearly dependent outputs. However, for the present, we assume that all outputs are linearly independent.

When multiple linearly independent outputs are available from an nth-order system, the system's initial state can be determined from its inputs and outputs in fewer than n steps because each step yields more than one equation involving the n components of the initial state. For example, in the two-output system

$$\begin{bmatrix} x_1(k+1) \\ x_2(k+1) \\ x_3(k+1) \end{bmatrix} = \begin{bmatrix} 0 & 0 & 2 \\ 1 & -1 & 0 \\ -2 & 0 & 1 \end{bmatrix} \begin{bmatrix} x_1(k) \\ x_2(k) \\ x_3(k) \end{bmatrix} + \begin{bmatrix} 1 \\ 0 \\ 4 \end{bmatrix} u(k) = \mathbf{A}\mathbf{x}(k) + \mathbf{b}u(k)$$

$$\begin{bmatrix} y_1(k) \\ y_2(k) \end{bmatrix} = \begin{bmatrix} 1 & 0 & 2 \\ -1 & 3 & 0 \end{bmatrix} \begin{bmatrix} x_1(k) \\ x_2(k) \\ x_3(k) \end{bmatrix} = \mathbf{C}\mathbf{x}(k)$$

(4-2)

suppose that

$$\mathbf{y}(0) = \begin{bmatrix} 4 \\ 5 \end{bmatrix} \qquad \mathbf{y}(1) = \begin{bmatrix} -7 \\ 2 \end{bmatrix}$$

and

$$u(k) = 1 \quad k = 0, 1, 2, \ldots$$

Then

$$\begin{cases} \mathbf{y}(0) = \mathbf{C}\mathbf{x}(0) \\ \mathbf{y}(1) = \mathbf{C}\mathbf{x}(1) = \mathbf{C}[\mathbf{A}\mathbf{x}(0) + \mathbf{b}u(0)] \end{cases}$$

or

$$\begin{cases} \begin{bmatrix} 4 \\ 5 \end{bmatrix} = \begin{bmatrix} 1 & 0 & 2 \\ -1 & 3 & 0 \end{bmatrix} \begin{bmatrix} x_1(0) \\ x_2(0) \\ x_3(0) \end{bmatrix} \\ \begin{bmatrix} -7 \\ 2 \end{bmatrix} = \begin{bmatrix} -4 & 0 & 4 \\ 3 & -3 & -2 \end{bmatrix} \begin{bmatrix} x_1(0) \\ x_2(0) \\ x_3(0) \end{bmatrix} + \begin{bmatrix} 9 \\ -1 \end{bmatrix} \end{cases}$$

The two vector outputs then yield the following four linear equations:

$$\begin{bmatrix} \mathbf{C} \\ \hline \mathbf{C}\mathbf{A} \end{bmatrix} \mathbf{x}(0) = \begin{bmatrix} 1 & 0 & 2 \\ -1 & 3 & 0 \\ -4 & 0 & 4 \\ 3 & -3 & -2 \end{bmatrix} \begin{bmatrix} x_1(0) \\ x_2(0) \\ x_3(0) \end{bmatrix} = \begin{bmatrix} 4 \\ 5 \\ -16 \\ 3 \end{bmatrix}$$

(4-3)

The first three equations represented by equation (4-3)

$$\begin{bmatrix} 1 & 0 & 2 \\ -1 & 3 & 0 \\ -4 & 0 & 4 \end{bmatrix} \begin{bmatrix} x_1(0) \\ x_2(0) \\ x_3(0) \end{bmatrix} = \begin{bmatrix} 4 \\ 5 \\ -16 \end{bmatrix}$$

have the solution

$$x_1(0) = 4 \qquad x_2(0) = 3 \qquad x_3(0) = 0$$

The fourth equation of (4-3) is automatically satisfied.

4.2.3 Observability in Diagonal Form

When the concept of complete observability was first introduced in Chapter 3, our concern was whether every system mode could be detected at the system's outputs. If not, the system state could not be determined from its inputs and outputs because an unobservable mode affects the state but not the outputs. The observability of each mode of a system is most apparent when the system is in diagonal form or, if its eigenvalues have repetitions, when it is in block Jordan form. Now we have another test for a system's complete observability, that of the rank of the observability matrix $\mathbf{M}_0$. The rank test is simpler than diagonalization, but it does not indicate which modes are unobservable. We now outline the relation between the two tests.

A nonsingular change of state variables

$$\mathbf{x} = \mathbf{P}\mathbf{x}' \qquad \mathbf{x}' = \mathbf{P}^{-1}\mathbf{x}$$
$$\mathbf{A}' = \mathbf{P}^{-1}\mathbf{A}\mathbf{P} \qquad \mathbf{C}' = \mathbf{C}\mathbf{P}$$

does not affect the rank of the observability matrix because

$$\mathbf{M}_0' = \begin{bmatrix} \mathbf{C}' \\ \hline \mathbf{C}'\mathbf{A}' \\ \hline \mathbf{C}'\mathbf{A}'^2 \\ \hline \vdots \\ \hline \mathbf{C}'\mathbf{A}'^{n-1} \end{bmatrix} = \begin{bmatrix} \mathbf{C}\mathbf{P} \\ \hline \mathbf{C}\mathbf{P}\mathbf{P}^{-1}\mathbf{A}\mathbf{P} \\ \hline \mathbf{C}\mathbf{P}\mathbf{P}^{-1}\mathbf{A}^2\mathbf{P} \\ \hline \vdots \\ \hline \mathbf{C}\mathbf{P}\mathbf{P}^{-1}\mathbf{A}^{n-1}\mathbf{P} \end{bmatrix} = \begin{bmatrix} \mathbf{C} \\ \hline \mathbf{C}\mathbf{A} \\ \hline \mathbf{C}\mathbf{A}^2 \\ \hline \vdots \\ \hline \mathbf{C}\mathbf{A}^{n-1} \end{bmatrix} \mathbf{P} = \mathbf{M}_0\mathbf{P}$$

When a system with distinct eigenvalues is placed in a diagonal form, the state equations are decoupled from one another, each having the form

$$x_i'(k + 1) = \lambda_i x_i'(k) + \mathbf{b}_i'^\dagger \mathbf{u}(k)$$

where $\mathbf{b}_i'^\dagger$ is the ith row of the input coupling matrix $\mathbf{B}'$. The state signals have solutions of the form

$$x_i'(k) = (\lambda_i)^k x_i'(0) + \sum_{p=0}^{k-1} (\lambda_i)^{k-1-p}\mathbf{b}_i'^\dagger\mathbf{u}(p)$$

If there is no coupling of x_i' to any of the system outputs, the initial condition $x_i'(0)$ cannot be determined from the system outputs and inputs. If, in diagonal form, the ith column of the output coupling matrix $\mathbf{C}'$ is zero, the ith column of the diagonal system's observability matrix

$$\mathbf{M}_0' = \begin{bmatrix} \mathbf{C}' \\ \hline \mathbf{C}'\Lambda \\ \hline \mathbf{C}'\Lambda^2 \\ \hline \vdots \\ \hline \mathbf{C}'\Lambda^{n-1} \end{bmatrix}$$

is zero, indicating that the system is not completely observable. If, in diagonal form, the output coupling matrix does not have any columns of zeros, the system is completely observable and the observability matrix is necessarily of full rank. Whatever collection of decoupled state variables connect to some output y_i, the initial states of those signals can be determined from the inputs and that single output, because (due to decoupling) that output could as well be the output of a system with only the connected modes present.

This analysis is extended to the case of repeated eigenvalues by considering the system to be in upper block Jordan form. It is straightforward but tedious to show that the observability matrix is of full rank only if all plant modes are observable.

Observability is an algebraic property of the matrices $\mathbf{A}$ and $\mathbf{C}$, independent of any state space interpretations. Thus, it applies to the same algebraic problem for continuous-time systems.

4.2.4 Parameter Identification

Measurements of the input and the output of a system can be used to determine the coefficients of the system's input–output equation(s) or, equivalently, the coefficients of the system transfer function(s). This is called *system identification*. It can be performed on all or part of an existing plant to aid in developing a plant model before controller design is done. Or, it can be performed as part of the control strategy in what is called *adaptive control*. In adaptive control, the plant is repetitively identified and the controller is changed if the measurements indicate that the plant has changed.

The input–output relations of a system are unchanged by any nonsingular change of state variables, so many state variable models can account for the same data. Only the transfer functions or some equivalent set of unique parameters can be found from the system inputs and outputs. To convey the ideas involved without getting bogged down in notation for general equations, we consider a specific system of relatively low order. It is straightforward to extend these results to higher-order systems, systems with multiple inputs and multiple outputs, and situations in which some of the parameters are known.

A causal second-order single-input/single-output system has an input–output difference equation of the form

$$y(k + 2) + a_1 y(k + 1) + a_0 y(k) = b_2 u(k + 2) + b_1 u(k + 1) + b_0 u(k)$$
(4-4)

If the initial conditions are zero so that the initial inputs $u(-1)$, $u(-2)$ and the initial outputs $y(-1)$, $y(-2)$ are zero, then

$$\begin{cases} y(0) = b_2 u(0) \\ y(1) = -a_1 y(0) + b_2 u(1) + b_1 u(0) \\ y(2) = -a_1 y(1) - a_0 y(0) + b_2 u(2) + b_1 u(1) + b_0 u(0) \\ y(3) = -a_1 y(2) - a_0 y(1) + b_2 u(3) + b_1 u(2) + b_0 u(1) \\ \vdots \end{cases}$$

Given the outputs y and the inputs u, the first five of these equations can be solved for the five unknown system parameters a_1, a_0, b_2, b_1, and b_0. Any additional equations are automatically satisfied if the data and calculations are without error. For example, suppose that the initial conditions for the system described by equation (4-4) are zero and that the first five outputs and inputs are

$$\begin{array}{ll} y(0) = 1 & u(0) = 1 \\ y(1) = 0 & u(1) = -1 \\ y(2) = 1 & u(2) = 0 \\ y(3) = 3 & u(3) = 5 \\ y(4) = -2 & u(4) = -3 \end{array}$$
(4-5)

Then

$$\begin{cases} y(0) = b_2 u(0) \\ y(1) = -a_1 y(0) + b_2 u(1) + b_1 u(0) \\ y(2) = -a_1 y(1) - a_0 y(0) + b_2 u(2) + b_1 u(1) + b_0 u(0) \\ y(3) = -a_1 y(2) - a_0 y(1) + b_2 u(3) + b_1 u(2) + b_0 u(1) \\ y(4) = -a_1 y(3) - a_0 y(2) + b_2 u(4) + b_1 u(3) + b_0 u(2) \end{cases}$$

or

$$\begin{cases} & b_2 & & = & 1 \\ -a_1 & -b_2 & +b_1 & = & 0 \\ & -a_0 & -b_1 & +b_0 & = & 1 \\ -a_1 & +5b_2 & & -b_0 & = & 3 \\ -3a_1 & -a_0 & -3b_2 & +5b_1 & = & -2 \end{cases}$$

which has the solutions

$$a_1 = -1 \qquad a_0 = 2$$
$$b_2 = 1 \qquad b_1 = 0 \qquad b_0 = 3$$

If the system's initial conditions are not known, one must begin at a step at which all the y's and u's are known. If the inputs and outputs are known beginning at step 0, the second-order example system equation (4-4) is

$$\begin{cases} y(2) = -a_1y(1) - a_0y(0) + b_2u(2) + b_1u(1) + b_0u(0) \\ y(3) = -a_1y(2) - a_0y(1) + b_2u(3) + b_1u(2) + b_0u(1) \\ y(4) = -a_1y(3) - a_0y(2) + b_2u(4) + b_1u(3) + b_0u(2) \\ \quad \vdots \end{cases}$$

$$(4\text{-}6)$$

Five such equations are enough to find the five unknown parameters, but seven successive inputs and outputs are required. Suppose that for the example system the initial conditions are not known, but the sixth and seventh outputs and inputs

$$y(5) = -1 \qquad u(5) = 3$$
$$y(6) = -2 \qquad u(6) = 4$$

are known in addition to the other five outputs and inputs listed in equations (4-5). Then the first five equations of (4-6) are

$$\begin{cases} y(2) = -a_1y(1) - a_0y(0) + b_2u(2) + b_1u(1) + b_0u(0) \\ y(3) = -a_1y(2) - a_0y(1) + b_2u(3) + b_1u(2) + b_0u(1) \\ y(4) = -a_1y(3) - a_0y(2) + b_2u(4) + b_1u(3) + b_0u(2) \\ y(5) = -a_1y(4) - a_0y(3) + b_2u(5) + b_1u(4) + b_0u(3) \\ y(6) = -a_1y(5) - a_0y(4) + b_2u(6) + b_1u(5) + b_0u(4) \end{cases}$$

or

$$
\begin{bmatrix}
0 & -1 & 0 & -1 & 1 \\
-1 & 0 & 5 & 0 & -1 \\
-3 & -1 & -3 & 5 & 0 \\
2 & -3 & 3 & -3 & 5 \\
1 & 2 & 4 & 3 & -3
\end{bmatrix}
\begin{bmatrix}
a_1 \\
a_0 \\
b_2 \\
b_1 \\
b_0
\end{bmatrix}
=
\begin{bmatrix}
1 \\
3 \\
-2 \\
-1 \\
-2
\end{bmatrix}
$$

which can be solved for a_1, a_0, b_2, b_1, and b_0.

It is possible that, for some input sequences, the equations for the system parameters are such that one or more parameters cannot be identified. The simplest examples of such sequences are ones that are identically zero or a constant, but there are usually other possible pathologic sequences that can frustrate identification too. The common terminology is that identification is hampered if the input is not *persistently exciting*. Similarly, if a transfer function to be identified has a pole-zero cancellation, the coefficients associated with the cancellation can be found only if the initial conditions are such that the associated mode appears in the zero-input response component of the system output. Because of the cancellation, the mode does not appear in the zero-state response component.

If, as is often the case, the available system output and input data have small (but significant) errors, it is appropriate to use more than the minimum number of measurements necessary for parameter identification. Least squares estimates of the parameters from overdetermined equations is an especially simple and generally useful method. It is discussed in Chapter 8.

4.2.5 Combined Parameter Identification and State Estimation

Once the independent parameters of a system have been identified, input and output measurements can also be used to determine the system's state. The parameters, once found, determine the unknown part of the state variable model; then the usual methods are applied to determine the state from input and output measurements. The same data can be used for both parameter identification and state estimation, if desired, and each can employ least squares estimation.

As a simple numerical example, consider a causal single-input/single-output second-order system with measured outputs and inputs as follows:

$$y(0) = 2 \qquad u(0) = 1$$
$$y(1) = 1 \qquad u(1) = 2$$

$$y(2) = 5 \quad u(2) = -1$$
$$y(3) = -2 \quad u(3) = 0$$
$$y(4) = 0 \quad u(4) = -2$$
$$y(5) = -6 \quad u(5) = 3$$
$$y(6) = 7 \quad u(6) = 4 \qquad \text{(4-7)}$$

The input–output equation coefficients in

$$y(k + 2) = -a_1 y(k + 1) - a_0 y(k) + b_2 u(k + 2) + b_1 u(k + 1) + b_0 u(k)$$

satisfy

$$\begin{cases} -a_1 & -2a_0 & -b_2 & +2b_1 & +b_0 = 5 \\ -5a_1 & -a_0 & & -b_1 & +2b_0 = -2 \\ 2a_1 & -5a_0 & -2b_2 & & -b_0 = 0 \\ & 2a_0 & +3b_2 & -2b_1 & = -6 \\ 6a_1 & & +4b_2 & +3b_1 & -2b_0 = 7 \end{cases}$$

which has the solutions

$$a_1 = -1 \quad a_0 = 0$$
$$b_2 = 0 \quad b_1 = 3 \quad b_0 = -2$$

The system's z-transfer function has thus been identified to be

$$T(z) = \frac{b_2 z^2 + b_1 z + b_0}{z^2 + a_1 z + a_0} = \frac{3z - 2}{z^2 - z}$$

If the state variable description of this system is in observable form, it must then be

$$\begin{bmatrix} x_1(k + 1) \\ x_2(k + 1) \end{bmatrix} = \begin{bmatrix} 1 & 1 \\ 0 & 0 \end{bmatrix} \begin{bmatrix} x_1(k) \\ x_2(k) \end{bmatrix} + \begin{bmatrix} 3 \\ -2 \end{bmatrix} u(k) = \mathbf{A}\mathbf{x}(k) + \mathbf{b}u(k)$$

$$y(k) = \begin{bmatrix} 1 & 0 \end{bmatrix} \begin{bmatrix} x_1(k) \\ x_2(k) \end{bmatrix} = \mathbf{c}^\dagger \mathbf{x}(k)$$

Using some of the same data of equation (4-7) again, we obtain

$$y(0) = \mathbf{c}^\dagger \mathbf{x}(0)$$
$$y(1) = \mathbf{c}^\dagger \mathbf{x}(1) = \mathbf{c}^\dagger \mathbf{A}\mathbf{x}(0) + \mathbf{c}^\dagger \mathbf{b}u(0)$$

so

$$\mathbf{x}(0) = \begin{bmatrix} \mathbf{c}^\dagger \\ \hline \mathbf{c}^\dagger \mathbf{A} \end{bmatrix}^{-1} \begin{bmatrix} y(0) \\ \hline [y(1) - \mathbf{c}^\dagger \mathbf{b}u(0)] \end{bmatrix} = \begin{bmatrix} 1 & 0 \\ -1 & 1 \end{bmatrix} \begin{bmatrix} 2 \\ -2 \end{bmatrix} = \begin{bmatrix} 2 \\ -4 \end{bmatrix}$$

Then

$$\mathbf{x}(1) = \mathbf{A}\mathbf{x}(0) + \mathbf{b}u(0) = \begin{bmatrix} 1 & 1 \\ 0 & 0 \end{bmatrix}\begin{bmatrix} 2 \\ -4 \end{bmatrix} + \begin{bmatrix} 3 \\ -2 \end{bmatrix}(1) = \begin{bmatrix} 1 \\ -2 \end{bmatrix}$$

and so on, through

$$\mathbf{x}(7) = \mathbf{A}\mathbf{x}(6) + \mathbf{b}u(6) = \begin{bmatrix} 7 \\ -6 \end{bmatrix}$$

4.2.6 Observability and State Determination of Step-Varying Systems

A step-varying system governed by

$$\mathbf{x}(k + 1) = \mathbf{A}(k)\mathbf{x}(k) + \mathbf{B}(k)\mathbf{u}(k)$$
$$\mathbf{y}(k) = \mathbf{C}(k)\mathbf{x}(k) + \mathbf{D}(k)\mathbf{u}(k) \qquad\qquad \textbf{(4-8)}$$

is *observable at step 0* if $\mathbf{x}(0)$ can be determined from the system outputs $\mathbf{y}(0)$, $\mathbf{y}(1)$, . . . , $\mathbf{y}(N - 1)$ and $\mathbf{u}(0)$, $\mathbf{u}(1)$, . . . , $\mathbf{u}(N - 1)$ for some finite number of steps N. From $\mathbf{x}(0)$ and the inputs and outputs, the system state at any later step can be found. A system (4-8) is *observable at step k* if $\mathbf{x}(k)$ can be determined from the outputs $\mathbf{y}(k)$, $\mathbf{y}(k + 1)$, . . . , $\mathbf{y}(k + N - 1)$ and inputs $\mathbf{u}(k)$, $\mathbf{u}(k + 1)$, . . . , $\mathbf{u}(k + N - 1)$ for some finite number of steps N. If the system is observable at every step, it is termed *uniformly observable*.

The observability of a system rests on whether or not there is a unique solution for the state given the outputs and inputs. For a general step-varying system (4-8), the state at step 0, in terms of outputs and inputs from step 0 through step $N - 1$, satisfies

$$\begin{cases}
\mathbf{y}(0) = \mathbf{C}(0)\mathbf{x}(0) + \mathbf{D}(0)\mathbf{u}(0) \\
\mathbf{y}(1) = \mathbf{C}(1)\mathbf{x}(1) + \mathbf{D}(1)\mathbf{u}(1) = \mathbf{C}(1)\mathbf{A}(0)\mathbf{x}(0) + \mathbf{C}(1)\mathbf{B}(0)\mathbf{u}(0) \\
\qquad\quad + \mathbf{D}(1)\mathbf{u}(1) \\
\mathbf{y}(2) = \mathbf{C}(2)\mathbf{A}(1)\mathbf{A}(0)\mathbf{x}(0) + \mathbf{C}(2)\mathbf{A}(1)\mathbf{B}(0)\mathbf{u}(0) + \mathbf{C}(2)\mathbf{B}(1)\mathbf{u}(1) \\
\qquad\quad + \mathbf{D}(2)\mathbf{u}(2) \\
\qquad \vdots \\
\mathbf{y}(N - 1) = \mathbf{C}(N - 1)\mathbf{A}(N - 2) \cdots \mathbf{A}(0)\mathbf{x}(0) \\
\qquad\quad + \mathbf{C}(N - 1)\mathbf{A}(N - 2) \cdots \mathbf{A}(1)\mathbf{B}(0)\mathbf{u}(0) \\
\qquad\quad + \mathbf{C}(N - 1)\mathbf{A}(N - 2) \cdots \mathbf{A}(2)\mathbf{B}(1)\mathbf{u}(1) + \cdots \\
\qquad\quad + \mathbf{C}(N - 1)\mathbf{A}(N - 2)\mathbf{B}(N - 3)\mathbf{u}(N - 3) \\
\qquad\quad + \mathbf{C}(N - 1)\mathbf{B}(N - 2)\mathbf{u}(N - 2) + \mathbf{D}(N - 1)\mathbf{u}(N - 1)
\end{cases}$$

or

$$\begin{bmatrix} \mathbf{C}(0) \\ \hline \mathbf{C}(1)\mathbf{A}(0) \\ \hline \mathbf{C}(2)\mathbf{A}(1)\mathbf{A}(0) \\ \hline \vdots \\ \hline \mathbf{C}(N-1)\mathbf{A}(N-2)\cdots\mathbf{A}(0) \end{bmatrix} \mathbf{x}(0) = \mathbf{M}_0(0, N-1)\mathbf{x}(0) = \begin{bmatrix} \text{Known} \\ \text{quantities} \end{bmatrix}$$

If, for some finite N, the observability matrix $\mathbf{M}_0(0, N-1)$ is of full rank (that is, if its n columns are linearly independent), the system is observable at step 0; otherwise, it is not. Similarly, the system is observable at step k if and only if the observability matrix

$$\mathbf{M}_0(k, k+N-1) = \begin{bmatrix} \mathbf{C}(k) \\ \hline \mathbf{C}(k+1)\mathbf{A}(k) \\ \hline \mathbf{C}(k+2)\mathbf{A}(k+1)\mathbf{A}(k) \\ \hline \vdots \\ \hline \mathbf{C}(k+N-1)\mathbf{A}(k+N-2)\cdots\mathbf{A}(k) \end{bmatrix}$$

is of full rank for some finite N.

In the step-invariant case, the matrices $\mathbf{A}$ and $\mathbf{C}$ do not vary with step. Because powers of the $n \times n$ matrix $\mathbf{A}$ above $\mathbf{A}^{n-1}$ can be expressed in terms of $\mathbf{A}^{n-1}$ and lower powers of $\mathbf{A}$, there is no need to use outputs from more than $N = n$ consecutive steps. And, in the step-invariant case, observability at any step means that the system is observable at any other step because the equations to be solved for the initial state are unchanged by a translation in step. Neither of these properties necessarily holds for a step-varying system.

4.3 Controllability and State Control

Controllability was defined in Chapter 3 in terms of input access to each plant mode. Complete controllability also means that there are plant inputs that will bring a plant from any arbitrary initial state to any desired later state in a finite number of steps. This is the subject of this section. Occasionally, this is what one wants to do for control. For example, moving a space vehicle from one orbit to another involves changing the vehicle's state from one three-dimensional position, velocity, orientation, and motion about its center of mass, to another.

In tracking systems, a more demanding result is desired: one or more plant tracking output signals are to be nearly equal to available external reference signals. For tracking, controllability is one of several concerns.

In the course of learning how to control a system's state at a later step, the controllability rank test will be derived.

4.3.1 Controllability and the Rank Test

If a discrete-time system is completely controllable, then knowing the system model and its state $x(k)$ at any initial step k, an input sequence $u(k)$, $u(k + 1)$, . . . , $u(k + i - 1)$ can be determined that takes the system to any desired later state $x(k + i)$ in a finite number of steps. For a step-invariant system, if it is possible to move the state at any step, say $x(0)$, to an arbitrary state at a later step, then it is possible to move to an arbitrary desired state starting with any beginning step.

For a system

$$x(k + 1) = Ax(k) + Bu(k)$$

$$y(k) = Cx(k) + Du(k)$$

and a desired state δ, the system state at step n, in terms of the initial state $x(0)$ and the inputs, is

$$\delta = x(n) = A^n x(0) + \sum_{i=0}^{n-1} A^{n-1-i} Bu(i)$$

or

$$Bu(n - 1) + ABu(n - 2) + \cdots + A^{n-2}Bu(1)$$
$$+ A^{n-1}Bu(0) = \underbrace{\delta - A^n x(0)}_{\text{known}}$$

For a nonzero known vector, these equations have a solution for the inputs $u(0)$, $u(1)$, . . . , $u(n - 1)$ if and only if the array of coefficients, which is the system's controllability matrix

$$M_c = [B \mid AB \mid \cdots \mid A^{n-2}B \mid A^{n-1}B]$$

is of full rank. Additional steps, giving additional equations with coefficients $A^n B$, $A^{n+1}B$, and so on, do not affect this result because, by the Cayley-Hamilton theorem, the higher powers of A can be expressed as a linear combination of A^{n-1} and lower powers of A. The resulting additional equations are linearly dependent on those already considered.

For example, the system

$$\begin{bmatrix} x_1(k+1) \\ x_2(k+1) \\ x_3(k+1) \end{bmatrix} = \begin{bmatrix} 1 & -1 & 0 \\ 4 & -2 & 0 \\ 2 & 3 & 1 \end{bmatrix} \begin{bmatrix} x_1(k) \\ x_2(k) \\ x_3(k) \end{bmatrix} + \begin{bmatrix} 2 \\ -1 \\ 0 \end{bmatrix} u(k)$$

has the controllability matrix

$$\mathbf{M}_c = \begin{bmatrix} 2 & 3 & -7 \\ -1 & 10 & -8 \\ 0 & 1 & 37 \end{bmatrix}$$

and is completely controllable because

$$\begin{vmatrix} 2 & 3 & -7 \\ -1 & 10 & -8 \\ 0 & 1 & 37 \end{vmatrix} \neq 0$$

which indicates that all three columns of $\mathbf{M}_c$ are linearly independent.
For the system with state equations

$$\begin{bmatrix} x_1(k+1) \\ x_2(k+1) \\ x_3(k+1) \end{bmatrix} = \begin{bmatrix} -10 & 0 & -18 \\ 4 & -3 & 6 \\ 6 & 0 & 11 \end{bmatrix} \begin{bmatrix} x_1(k) \\ x_2(k) \\ x_3(k) \end{bmatrix} + \begin{bmatrix} 8 & 0 \\ 5 & -1 \\ -4 & 0 \end{bmatrix} \begin{bmatrix} u_1(k) \\ u_2(k) \end{bmatrix}$$

the controllability matrix is

$$\mathbf{M}_c = [\mathbf{B} \mid \mathbf{AB} \mid \mathbf{A}^2\mathbf{B}] = \begin{bmatrix} 8 & 0 & -8 & 0 & 8 & 0 \\ 5 & -1 & -7 & 3 & 13 & -9 \\ -4 & 0 & 4 & 0 & -4 & 0 \end{bmatrix}$$

The controllability matrix has only two linearly independent columns and so is of rank 2, not 3. The system is therefore not completely controllable.

For a multiple-input system, the smallest integer η for which

$$\mathbf{M}_c(\eta) = [\mathbf{B} \mid \mathbf{AB} \mid \mathbf{A}^2\mathbf{B} \mid \cdots \mid \mathbf{A}^{\eta-1}\mathbf{B}]$$

has full rank is called the *controllability index* of the system. It is the minimum number of input steps required to control the system state.

The replacements

$$\begin{cases} \mathbf{A} \rightarrow \mathbf{A}^\dagger \\ \mathbf{B} \rightarrow \mathbf{C}^\dagger \\ \mathbf{C} \rightarrow \mathbf{B}^\dagger \end{cases}$$

create a system with a controllability matrix which is the observability matrix of the original system and an observability matrix that is the controllability matrix of the original system. Every controllability result thus has a corresponding observability result and vice versa, a concept termed *duality*. This concept generalizes to step-varying discrete-time and time-varying continuous-time systems as well.

Similar to the situation with observability, a nonsingular change of state variables does not affect the rank of the controllability matrix, because if

$$\mathbf{x} = \mathbf{P}\mathbf{x}' \qquad \mathbf{x}' = \mathbf{P}^{-1}\mathbf{x}$$

then

$$\mathbf{M}'_c = [\mathbf{P}^{-1}\mathbf{B} \mid \mathbf{P}^{-1}\mathbf{A}\mathbf{P}\mathbf{P}^{-1}\mathbf{B} \mid \mathbf{P}^{-1}\mathbf{A}^2\mathbf{P}\mathbf{P}^{-1}\mathbf{B} \mid \cdots \mid \mathbf{P}^{-1}\mathbf{A}^{n-1}\mathbf{P}\mathbf{P}^{-1}\mathbf{B}]$$

$$= \mathbf{P}^{-1}\mathbf{M}_c$$

The dual result to the rank test for complete observability is that a system is completely controllable if and only if its controllability matrix is of full rank.**

A continuous-time system is completely controllable if, by knowing the system model and its state at any specific time, $\mathbf{x}(t_0)$, inputs $\mathbf{u}(t)$ for a finite time interval $t_0 \le t \le t_1$ can be determined that takes the system to any desired state at time t_1. A time-invariant continuous-time system governed by the state equation

$$\dot{\mathbf{x}}(t) = \mathscr{A}\mathbf{x}(t) + \mathscr{B}\mathbf{u}(t)$$

is completely controllable if and only if its controllability matrix

$$\mathbf{M}_c = [\mathscr{B} \mid \mathscr{A}\mathscr{B} \mid \mathscr{A}^2\mathscr{B} \mid \cdots \mid \mathscr{A}^{n-1}\mathscr{B}]$$

is of full rank.

**The definition of controllability involves beginning with an *arbitrary* known initial state and moving to an *arbitrary* desired final state. If the desired final state is the origin, $\mathbf{x}(k + i) = \mathbf{0}$, then "controllability to the origin" might be achieved even if $\mathbf{M}_c$ is not of full rank. It could be that the system state coupling matrix has the property $\mathbf{A}^i = \mathbf{0}$, for example, so that the state decays to zero without any input at all.

Reachability is control of the state from an initial zero state to an arbitrary final state. A completely controllable system (not just "controllable to the origin") is reachable and vice versa. There are a variety of viewpoints in the literature, where "controllable to the origin" has sometimes been substituted for controllability. Because of this, some authors prefer the term *reachability*. However, one should probably then speak of "reachable" and "unreachable" system modes instead of "controllable" and "uncontrollable" ones.

4.3.2 Controlling Single- and Multiple-Input Systems

As an example of finding the control inputs needed to bring a completely controllable system to a desired state, consider the single-input system with state equations

$$\begin{bmatrix} x_1(k+1) \\ x_2(k+1) \\ x_3(k+1) \end{bmatrix} = \begin{bmatrix} 1 & 2 & 0 \\ 4 & -1 & 0 \\ 0 & 1 & 3 \end{bmatrix} \begin{bmatrix} x_1(k) \\ x_2(k) \\ x_3(k) \end{bmatrix} + \begin{bmatrix} -1 \\ 1 \\ 0 \end{bmatrix} u(k) = \mathbf{A}\mathbf{x}(k) + \mathbf{b}u(k)$$

Suppose that it is desired to take the state from

$$\begin{bmatrix} x_1(0) \\ x_2(0) \\ x_3(0) \end{bmatrix} = \begin{bmatrix} 0 \\ -1 \\ 3 \end{bmatrix} \quad \text{to} \quad \begin{bmatrix} x_1(i) \\ x_2(i) \\ x_3(i) \end{bmatrix} = \begin{bmatrix} 6 \\ -8 \\ 2 \end{bmatrix}$$

If the input sequence is begun at step 0, this can be accomplished by step $i = n = 3$ if the system is completely controllable. The desired state at step 3 is related to the system initial state and inputs by

$$\mathbf{x}(3) = \begin{bmatrix} 6 \\ -8 \\ 2 \end{bmatrix} = \mathbf{A}^3\mathbf{x}(0) + \mathbf{A}^2\mathbf{b}u(0) + \mathbf{A}\mathbf{b}u(1) + \mathbf{b}u(2)$$

or

$$\begin{bmatrix} 24 \\ -17 \\ -64 \end{bmatrix} = \begin{bmatrix} -9 \\ 9 \\ -2 \end{bmatrix} u(0) + \begin{bmatrix} 1 \\ -5 \\ 1 \end{bmatrix} u(1) + \begin{bmatrix} -1 \\ 1 \\ 0 \end{bmatrix} u(2)$$

These three simultaneous linear algebraic equations

$$\underbrace{\begin{bmatrix} -1 & 1 & -9 \\ 1 & -5 & 9 \\ 0 & 1 & -2 \end{bmatrix}}_{\mathbf{M}_c} \underbrace{\begin{bmatrix} u(2) \\ u(1) \\ u(0) \end{bmatrix}}_{\substack{\text{inputs in} \\ \text{reverse order}}} = \begin{bmatrix} 24 \\ -17 \\ -64 \end{bmatrix}$$

have the solutions

$$u(0) = \frac{249}{8} \qquad u(1) = -\frac{7}{4} \qquad u(2) = -\frac{2447}{8}$$

The state trajectory with these inputs is as follows:

$$\begin{bmatrix} x_1(1) \\ x_2(1) \\ x_3(1) \end{bmatrix} = \begin{bmatrix} 1 & 2 & 0 \\ 4 & -1 & 0 \\ 0 & 1 & 3 \end{bmatrix} \begin{bmatrix} 0 \\ -1 \\ 3 \end{bmatrix} + \begin{bmatrix} -1 \\ 1 \\ 0 \end{bmatrix} \frac{249}{8} = \begin{bmatrix} -\frac{265}{8} \\ \frac{257}{8} \\ 8 \end{bmatrix}$$

$$\begin{bmatrix} x_1(2) \\ x_2(2) \\ x_3(2) \end{bmatrix} = \begin{bmatrix} 1 & 2 & 0 \\ 4 & -1 & 0 \\ 0 & 1 & 3 \end{bmatrix} \begin{bmatrix} -\frac{265}{8} \\ \frac{257}{8} \\ 8 \end{bmatrix} + \begin{bmatrix} -1 \\ 1 \\ 0 \end{bmatrix} \left(-\frac{7}{4} \right) = \begin{bmatrix} \frac{263}{8} \\ -\frac{1331}{8} \\ \frac{449}{8} \end{bmatrix}$$

$$\begin{bmatrix} x_1(3) \\ x_2(3) \\ x_3(3) \end{bmatrix} = \begin{bmatrix} 1 & 2 & 0 \\ 4 & -1 & 0 \\ 0 & 1 & 3 \end{bmatrix} \begin{bmatrix} \frac{263}{8} \\ -\frac{1331}{8} \\ \frac{449}{8} \end{bmatrix} + \begin{bmatrix} -1 \\ 1 \\ 0 \end{bmatrix} \left(-\frac{2447}{8} \right) = \begin{bmatrix} 6 \\ -8 \\ 2 \end{bmatrix}$$

If, instead, it is desired to move this system's state from

$$\mathbf{x}(5) = \begin{bmatrix} 0 \\ -1 \\ 3 \end{bmatrix} \quad \text{to} \quad \mathbf{x}(8) = \begin{bmatrix} 6 \\ -8 \\ 2 \end{bmatrix}$$

then

$$\mathbf{x}(8) = \mathbf{A}^3\mathbf{x}(5) + \mathbf{A}^2\mathbf{b}u(5) + \mathbf{A}\mathbf{b}u(6) + \mathbf{b}u(7)$$

and the inputs required are

$$u(5) = \frac{249}{8} \qquad u(6) = -\frac{7}{4} \qquad u(7) = \frac{-2447}{8}$$

When it is desired to control a system's state at a step later than the minimum number of steps necessary to achieve control, some of the inputs can be chosen arbitrarily. For example, for the system with state equations

$$\begin{bmatrix} x_1(k + 1) \\ x_2(k + 1) \end{bmatrix} = \begin{bmatrix} 1 & 1 \\ 2 & 0 \end{bmatrix} \begin{bmatrix} x_1(k) \\ x_2(k) \end{bmatrix} + \begin{bmatrix} 2 \\ -1 \end{bmatrix} u(k)$$

if it is desired to apply inputs to bring the state from

$$\mathbf{x}(0) = \begin{bmatrix} x_1(0) \\ x_2(0) \end{bmatrix} = \begin{bmatrix} 0 \\ -4 \end{bmatrix} \quad \text{to} \quad \mathbf{x}(k) = \begin{bmatrix} x_1(k) \\ x_2(k) \end{bmatrix} = \begin{bmatrix} 6 \\ 1 \end{bmatrix}$$

this can be accomplished in two steps. That is, one can solve

$$\mathbf{x}(2) = \mathbf{A}^2\mathbf{x}(0) + \mathbf{A}\mathbf{b}u(0) + \mathbf{b}u(1)$$

for the inputs $u(0)$ and $u(1)$ that achieve the desired state at step 2. Suppose, though, that it is desired to achieve the given state instead at step 3. Then three input steps are available to the designer, because

$$\mathbf{x}(3) = \mathbf{A}^3\mathbf{x}(0) + \mathbf{A}^2\mathbf{b}u(0) + \mathbf{A}\mathbf{b}u(1) + \mathbf{b}u(2)$$

One solution is to apply any input at the first step. This results in some state $\mathbf{x}(1)$ at step 1. With this as an initial state, the system can then be brought to the desired state in the remaining two steps. Another method is to impose additional requirements on the input sequence or the state sequence or both.

The inputs to a multiple-input system are said to be *linearly independent* if the columns of the input coupling matrix $\mathbf{B}$ are linearly independent of one another. A linearly dependent input is of no additional help in controlling a system because it does not contribute any linearly independent columns to the system's controllability matrix. There may be other reasons, such as increased reliability, for including linearly dependent inputs in practice, but for the present we assume that all system inputs are linearly independent unless otherwise stated.

When multiple linearly independent inputs are available for control of an nth-order system, the system state can be moved to a desired state in fewer than n steps, because each step involves more than a single control variable. For example, in the two-input system with state equations

$$\begin{bmatrix} x_1(k+1) \\ x_2(k+1) \\ x_3(k+1) \end{bmatrix} = \begin{bmatrix} 0 & 1 & 1 \\ -1 & 0 & 2 \\ 0 & 3 & 0 \end{bmatrix} \begin{bmatrix} x_1(k) \\ x_2(k) \\ x_3(k) \end{bmatrix} + \begin{bmatrix} 1 & 0 \\ 0 & 0 \\ 1 & -1 \end{bmatrix} \begin{bmatrix} u_1(k) \\ u_2(k) \end{bmatrix}$$

$$= \mathbf{A}\mathbf{x}(k) + \mathbf{B}\mathbf{u}(k)$$

suppose that it is desired to take the state from

$$\begin{bmatrix} x_1(0) \\ x_2(0) \\ x_3(0) \end{bmatrix} = \begin{bmatrix} -1 \\ 3 \\ -2 \end{bmatrix} \quad \text{to} \quad \begin{bmatrix} x_1(i) \\ x_2(i) \\ x_3(i) \end{bmatrix} = \begin{bmatrix} 1 \\ 2 \\ 3 \end{bmatrix}$$

at the earliest later step i. In terms of the inputs, the system state at step 2 is

$$\mathbf{x}(2) = \mathbf{A}\mathbf{x}(1) + \mathbf{B}\mathbf{u}(1) = \mathbf{A}[\mathbf{A}\mathbf{x}(0) + \mathbf{B}\mathbf{u}(0)] + \mathbf{B}\mathbf{u}(1)$$

$$= \mathbf{A}^2\mathbf{x}(0) + \mathbf{A}\mathbf{B}\mathbf{u}(0) + \mathbf{B}\mathbf{u}(1)$$

Equating to the desired state

$$
\begin{bmatrix} 1 \\ 2 \\ 3 \end{bmatrix} = \begin{bmatrix} 0 & 1 & 1 \\ -1 & 0 & 2 \\ 0 & 3 & 0 \end{bmatrix} \begin{bmatrix} 0 & 1 & 1 \\ -1 & 0 & 2 \\ 0 & 3 & 0 \end{bmatrix} \begin{bmatrix} -1 \\ 3 \\ -2 \end{bmatrix}
$$

$$
+ \begin{bmatrix} 0 & 1 & 1 \\ -1 & 0 & 2 \\ 0 & 3 & 0 \end{bmatrix} \begin{bmatrix} 1 & 0 \\ 0 & 0 \\ 1 & -1 \end{bmatrix} \begin{bmatrix} u_1(0) \\ u_2(0) \end{bmatrix} + \begin{bmatrix} 1 & 0 \\ 0 & 0 \\ 1 & -1 \end{bmatrix} \begin{bmatrix} u_1(1) \\ u_2(1) \end{bmatrix}
$$

results in the following three simultaneous linear algebraic equations in four unknowns:

$$
\begin{cases} u_1(0) - u_2(0) + u_1(1) & = -5 \\ u_1(0) - 2u_2(0) & = -15 \\ u_1(1) - u_2(1) = & 12 \end{cases}
$$

Arbitrarily choosing the input $u_1(0)$ as

$$u_1(0) = 0$$

gives

$$u_2(0) = \frac{15}{2} \qquad u_1(1) = \frac{5}{2} \qquad u_2(1) = -\frac{19}{2}$$

With these inputs, the state trajectory is

$$
\begin{bmatrix} x_1(1) \\ x_2(1) \\ x_3(1) \end{bmatrix} = \begin{bmatrix} 0 & 1 & 1 \\ -1 & 0 & 2 \\ 0 & 3 & 0 \end{bmatrix} \begin{bmatrix} -1 \\ 3 \\ -2 \end{bmatrix} + \begin{bmatrix} 1 & 0 \\ 0 & 0 \\ 1 & -1 \end{bmatrix} \begin{bmatrix} 0 \\ \frac{15}{2} \end{bmatrix} = \begin{bmatrix} 1 \\ -3 \\ \frac{3}{2} \end{bmatrix}
$$

$$
\begin{bmatrix} x_1(2) \\ x_2(2) \\ x_3(2) \end{bmatrix} = \begin{bmatrix} 0 & 1 & 1 \\ -1 & 0 & 2 \\ 0 & 3 & 0 \end{bmatrix} \begin{bmatrix} 1 \\ -3 \\ \frac{3}{2} \end{bmatrix} + \begin{bmatrix} 1 & 0 \\ 0 & 0 \\ 1 & -1 \end{bmatrix} \begin{bmatrix} \frac{5}{2} \\ -\frac{19}{2} \end{bmatrix} = \begin{bmatrix} 1 \\ 2 \\ 3 \end{bmatrix}
$$

Each different choice of $u_1(0)$ (or any other of the input components, for that matter) results in a different set of inputs that brings this third-order system to the desired state at step 2.

Control, in the sense of controllability, is fundamentally different from observation. Once enough linearly independent equations have been collected, one can continue to observe a known system's state at every step. The state can be controlled, however, only every η steps or so. And the state to be reached must be known η steps in advance. Actually, one can do a little better than this if the equations for control in η steps are underdetermined, because their solution can be chosen to aid in control at one or more later steps.

4.3.3 Controllability and State Control of Step-Varying Systems

A step-varying system (4-8) is *controllable at step 0* if there exists a finite length input sequence $\mathbf{u}(0)$, $\mathbf{u}(1)$, ... , $\mathbf{u}(N - 1)$, depending on $\mathbf{x}(0)$ and $\mathbf{x}(N)$, that takes the system from any known state $\mathbf{x}(0)$ to any desired state $\mathbf{x}(N)$. A system is *controllable at step k* if a finite length input sequence $\mathbf{u}(k)$, $\mathbf{u}(k + 1)$, ... , $\mathbf{u}(k + N - 1)$ exists such that the state can be moved from any known $\mathbf{x}(k)$ to any desired $\mathbf{x}(N)$. If the system is controllable at every step, it is termed *uniformly controllable*.

For a general step-varying system (4-8), the state at step N in terms of the state at step 0 and the inputs $\mathbf{u}(0)$, $\mathbf{u}(1)$, ... , $\mathbf{u}(N - 1)$ is given by

$$\mathbf{x}(N) = \mathbf{A}(N - 1) \ldots \mathbf{A}(0)\mathbf{x}(0) + \mathbf{A}(N - 1) \ldots \mathbf{A}(1)\mathbf{B}(0)\mathbf{u}(0)$$
$$+ \mathbf{A}(N - 1) \ldots \mathbf{A}(2)\mathbf{B}(1)\mathbf{u}(1) + \cdots$$
$$+ \mathbf{A}(N - 1)\mathbf{B}(N - 2)\mathbf{u}(N - 2) + \mathbf{B}(N - 1)\mathbf{u}(N - 1)$$

Taking the initial state $\mathbf{x}(0)$ and the desired state $\mathbf{x}(N)$ to be known, these are of the form

$$[\mathbf{B}(N - 1) \vdots \mathbf{A}(N - 1)\mathbf{B}(N - 2) \vdots \cdots$$
$$\vdots \mathbf{A}(N - 1) \ldots \mathbf{A}(2)\mathbf{B}(1) \vdots \mathbf{A}(N - 1) \ldots \mathbf{A}(1)\mathbf{B}(0)]$$

$$\times \begin{bmatrix} \mathbf{u}(N - 1) \\ \hline \mathbf{u}(N - 2) \\ \hline \vdots \\ \hline \mathbf{u}(1) \\ \hline \mathbf{u}(0) \end{bmatrix} = \mathbf{M}_c(0, N - 1) \begin{bmatrix} \mathbf{u}(N - 1) \\ \vdots \\ \hline \mathbf{u}(0) \end{bmatrix} = \begin{bmatrix} \text{Known} \\ \text{quantities} \end{bmatrix}$$

If, for some finite N, the controllability matrix, $\mathbf{M}_c(0, N - 1)$ is of full rank (that is, if its n rows are linearly independent), the system is controllable at step 0; otherwise, it is not. The system is controllable at step k if and only if the controllability matrix

$$\mathbf{M}_c(k, k + N - 1) = [\mathbf{B}(k + N - 1) \vdots \mathbf{A}(k + N - 1)\mathbf{B}(k + N - 2) \vdots$$
$$\cdots \vdots \mathbf{A}(k + N - 1) \ldots \mathbf{A}(k + 1)\mathbf{B}(k)]$$

is of full rank for some finite N.

4.3.4 Minimal Realizations

A minimal realization of a system with a prescribed transfer function matrix is a state variable model that is both completely controllable and completely observable. Such a model is of the lowest possible order. Synthesizing a system with both multiple inputs and multiple outputs with a given transfer function matrix is an important general problem. One can always form each output with a multiple-input/single-output subsystem as in Figure 4-1(a). One also can always form each output from the sum of corresponding outputs of a set of single-input/multiple-output subsystems as in Figure 4-1(b). There is no guarantee, however, that either of these realizations will be of *minimal* order.

As a numerical example, consider a two-input/three-output system with the following six transfer functions:

$$T_{11}(z) = \frac{10z + 4}{z^2 + \frac{5}{6}z + \frac{1}{6}} = \frac{6}{z + \frac{1}{2}} + \frac{4}{z + \frac{1}{3}}$$

$$T_{12}(z) = \frac{3z + 3}{z^2 + \frac{5}{6}z + \frac{1}{6}} = \frac{-9}{z + \frac{1}{2}} + \frac{12}{z + \frac{1}{3}}$$

$$T_{21}(z) = \frac{1}{z + \frac{1}{3}} = \frac{0}{z + \frac{1}{2}} + \frac{1}{z + \frac{1}{3}}$$

$$T_{22}(z) = \frac{\frac{1}{2}}{z^2 + \frac{5}{6}z + \frac{1}{6}} = \frac{-3}{z + \frac{1}{2}} + \frac{3}{z + \frac{1}{3}}$$

$$T_{31}(z) = \frac{3}{z + \frac{1}{3}} = \frac{0}{z + \frac{1}{2}} + \frac{3}{z + \frac{1}{3}}$$

$$T_{32}(z) = \frac{12z + \frac{11}{2}}{z^2 + \frac{5}{6}z + \frac{1}{6}} = \frac{3}{z + \frac{1}{2}} + \frac{9}{z + \frac{1}{3}}$$

The system composed of three two-input/single-output subsystems shown in Figure 4-2(a) has the required transfer functions and is described by the following equations:

$$\begin{bmatrix} x_1(k+1) \\ x_2(k+1) \\ x_3(k+1) \\ x_4(k+1) \\ x_5(k+1) \\ x_6(k+1) \end{bmatrix} = \begin{bmatrix} -\frac{1}{2} & 0 & 0 & 0 & 0 & 0 \\ 0 & -\frac{1}{3} & 0 & 0 & 0 & 0 \\ 0 & 0 & -\frac{1}{2} & 0 & 0 & 0 \\ 0 & 0 & 0 & -\frac{1}{3} & 0 & 0 \\ 0 & 0 & 0 & 0 & -\frac{1}{2} & 0 \\ 0 & 0 & 0 & 0 & 0 & -\frac{1}{3} \end{bmatrix} \begin{bmatrix} x_1(k) \\ x_2(k) \\ x_3(k) \\ x_4(k) \\ x_5(k) \\ x_6(k) \end{bmatrix}$$

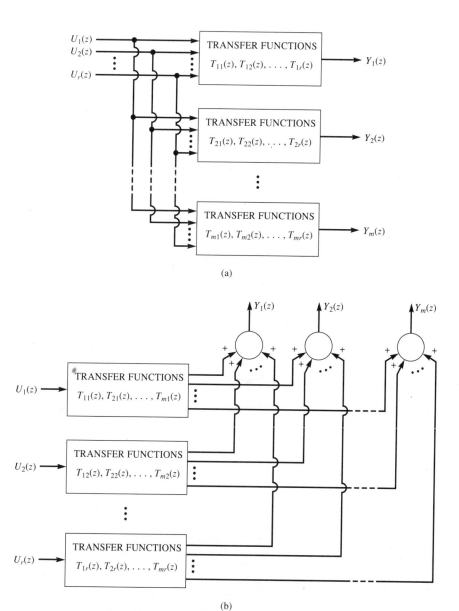

(a)

(b)

Figure 4-1 Synthesizing multiple-input/multiple-output systems using single-output and single-input subsystems. (a) Using single-output subsystems. (b) Using single-input subsystems.

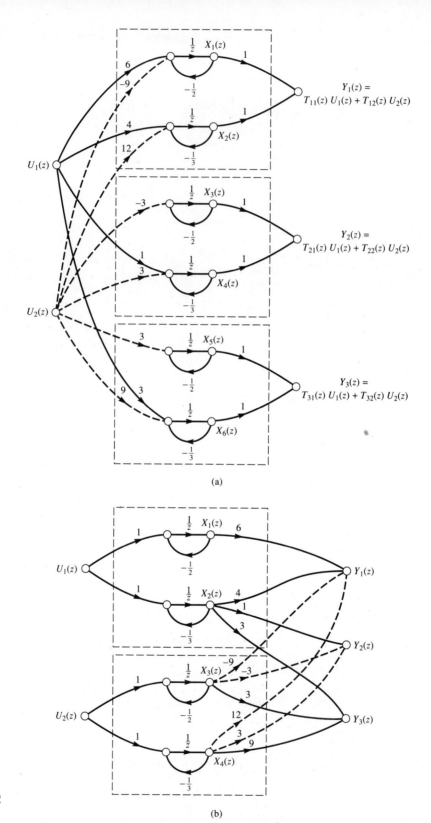

(a)

(b)

$$+ \begin{bmatrix} 6 & -9 \\ 4 & 12 \\ 0 & -3 \\ 1 & 3 \\ 0 & 3 \\ 3 & 9 \end{bmatrix} \begin{bmatrix} u_1(k) \\ u_2(k) \end{bmatrix}$$

$$\begin{bmatrix} y_1(k) \\ y_2(k) \\ y_3(k) \end{bmatrix} = \begin{bmatrix} 1 & 1 & 0 & 0 & 0 & 0 \\ 0 & 0 & 1 & 1 & 0 & 0 \\ 0 & 0 & 0 & 0 & 1 & 1 \end{bmatrix} \begin{bmatrix} x_1(k) \\ x_2(k) \\ x_3(k) \\ x_4(k) \\ x_5(k) \\ x_6(k) \end{bmatrix}$$

The system composed of two single-input/three-output subsystems in Figure 4-2(b) is of the fourth order, has the required transfer functions, and is described by

$$\begin{bmatrix} x_1(k+1) \\ x_2(k+1) \\ x_3(k+1) \\ x_4(k+1) \end{bmatrix} = \begin{bmatrix} -\frac{1}{2} & 0 & 0 & 0 \\ 0 & -\frac{1}{3} & 0 & 0 \\ 0 & 0 & -\frac{1}{2} & 0 \\ 0 & 0 & 0 & -\frac{1}{3} \end{bmatrix} \begin{bmatrix} x_1(k) \\ x_2(k) \\ x_3(k) \\ x_4(k) \end{bmatrix} + \begin{bmatrix} 1 & 0 \\ 1 & 0 \\ 0 & 1 \\ 0 & 1 \end{bmatrix} \begin{bmatrix} u_1(k) \\ u_2(k) \end{bmatrix}$$

$$= \mathbf{A}x(k) + \mathbf{B}u(k)$$

$$\begin{bmatrix} y_1(k) \\ y_2(k) \\ y_3(k) \end{bmatrix} = \begin{bmatrix} 6 & 4 & -9 & 12 \\ 0 & 1 & -3 & 3 \\ 0 & 3 & 3 & 9 \end{bmatrix} \begin{bmatrix} x_1(k) \\ x_2(k) \\ x_3(k) \\ x_4(k) \end{bmatrix} = \mathbf{C}x(k)$$

. Even this system (because the numbers are just right) is not of minimal order, however. Its controllability matrix is

$$\mathbf{M}_c = [\mathbf{b} \;\vdots\; \mathbf{Ab} \;\vdots\; \mathbf{A^2b} \;\vdots\; \mathbf{A^3b}] = \begin{bmatrix} 1 & 0 & -\frac{1}{2} & 0 & \frac{1}{4} & 0 & -\frac{1}{8} & 0 \\ 1 & 0 & -\frac{1}{3} & 0 & \frac{1}{9} & 0 & -\frac{1}{27} & 0 \\ 0 & 1 & 0 & -\frac{1}{2} & 0 & \frac{1}{4} & 0 & -\frac{1}{8} \\ 0 & 1 & 0 & -\frac{1}{3} & 0 & \frac{1}{9} & 0 & -\frac{1}{27} \end{bmatrix}$$

Figure 4-2 Synthesizing a two-input/three-output system. (a) Using three single-output subsystems. (b) Using two single-input subsystems.

which is of full rank. It must be of full rank because each of the controllable form subsystems is completely controllable. The observability matrix of this realization is

$$
M_o = \begin{bmatrix} C \\ \hdashline CA \\ \hdashline CA^2 \\ \hdashline CA^3 \end{bmatrix} = \begin{bmatrix}
6 & 4 & -9 & 12 \\
0 & 1 & -3 & 3 \\
0 & 3 & 3 & 9 \\
-3 & -\frac{4}{3} & \frac{9}{2} & -4 \\
0 & -\frac{1}{3} & \frac{3}{2} & -1 \\
0 & -1 & -\frac{3}{2} & -3 \\
\frac{3}{2} & \frac{4}{9} & -\frac{9}{4} & \frac{4}{3} \\
0 & \frac{1}{9} & -\frac{3}{4} & \frac{1}{3} \\
0 & \frac{1}{3} & \frac{3}{4} & 1 \\
-\frac{3}{4} & -\frac{4}{27} & \frac{9}{8} & -\frac{4}{9} \\
0 & -\frac{1}{27} & \frac{3}{8} & -\frac{1}{9} \\
0 & -\frac{1}{9} & -\frac{3}{8} & -1
\end{bmatrix}
$$

which has rank 3, not rank 4. Because this system is not completely observable, an unobservable mode can be deleted, resulting in a realization of lower order.

If there were a column of zeros in the output coupling matrix C for this system, the corresponding state variable could be deleted, reducing the system order. This statement is true only when the system eigenvalues are distinct. With repeated eigenvalues with more than one Jordan block for the same eigenvalue, as is the case here, there is the additional possibility that one of the blocks is redundant. In this example, the two 1×1 Jordan blocks with $z = -\frac{1}{3}$ eigenvalue are redundant. As shown in Figure 4-3(a), the portions of the six transfer functions involving the two state variables $x_2(z)$ and $x_4(z)$ can be realized with a single delay element, giving the third-order realization of Figure 4-3(b). This realization,

$$
\begin{bmatrix} x_1(k+1) \\ x_2'(k+1) \\ x_3(k+1) \end{bmatrix} = \begin{bmatrix} -\frac{1}{2} & 0 & 0 \\ 0 & -\frac{1}{3} & 0 \\ 0 & 0 & -\frac{1}{2} \end{bmatrix} \begin{bmatrix} x_1(k) \\ x_2'(k) \\ x_3(k) \end{bmatrix} + \begin{bmatrix} 1 & 0 \\ 1 & 3 \\ 0 & 1 \end{bmatrix} \begin{bmatrix} u_1(k) \\ u_2(k) \end{bmatrix}
$$

$$
\begin{bmatrix} y_1(k) \\ y_2(k) \\ y_3(k) \end{bmatrix} = \begin{bmatrix} 6 & 4 & -9 \\ 0 & 1 & -3 \\ 0 & 3 & 3 \end{bmatrix} \begin{bmatrix} x_1(k) \\ x_2'(k) \\ x_3(k) \end{bmatrix}
$$

has controllability and observability matrices of full rank and thus is a minimal realization.

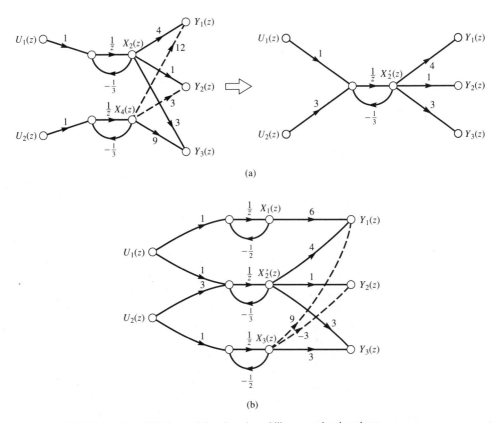

Figure 4-3 Minimal realization of the two-input/three-output system.
(a) Redundancy of one of the two 1×1 Jordan blocks with $z = -\frac{1}{3}$ eigenvalue. (b) Third-order realization.

A general theory for obtaining a minimal realization of a system is beyond our scope and therefore will not be discussed further.

4.4 State Feedback

We now examine the situation in which the plant input, instead of being some sequence dictated in part by a known initial and desired later state, is a linear transformation of the plant state at each step. For a plant with state equations

$$\mathbf{x}(k + 1) = \mathbf{A}\mathbf{x}(k) + \mathbf{B}\mathbf{u}(k)$$

we consider the state feedback

$$\mathbf{u}(k) = \mathbf{E}\mathbf{x}(k) + \boldsymbol{\rho}(k)$$

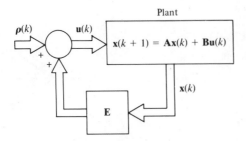

Figure 4-4 State feedback.

where $\rho(k)$ is a vector of external inputs, as shown in Figure 4-4. Provided that the plant is completely controllable, a feedback gain matrix $\mathbf{E}$ can always be chosen so that each of the eigenvalues of the feedback system

$$\mathbf{x}(k + 1) = (\mathbf{A} + \mathbf{BE})\mathbf{x}(k) + \mathbf{B}\rho(k)$$

is at an arbitrary location selected by the designer. Methods for placing feedback system eigenvalues will now be developed.

Eigenvalue placement ultimately provides a solution to controlling the zero-input component of a plant's response. By selecting the feedback system eigenvalues, the designer specifies the closed-loop zero-input response modes and thus the character of that response, the plant's transient response. Because the plant initial conditions are not usually known, this is ordinarily the best that can be done; whatever the initial conditions, the zero-input response is made to decay in a desired manner. For now, it is supposed that the plant state is available for feedback. Normally it is not, and subsequently we will extend these results to the usual case in which only the plant inputs and outputs are accessible.

4.4.1 Eigenvalue Placement for Single-Input Systems

If a single-input plant is in controllable form, finding the feedback gains for arbitrary eigenvalue placement is especially simple. For example, consider the single-input system with state equations

$$\begin{bmatrix} x_1(k + 1) \\ x_2(k + 1) \\ x_3(k + 1) \end{bmatrix} = \begin{bmatrix} 0 & 1 & 0 \\ 0 & 0 & 1 \\ -4 & -2 & -1 \end{bmatrix} \begin{bmatrix} x_1(k) \\ x_2(k) \\ x_3(k) \end{bmatrix} + \begin{bmatrix} 0 \\ 0 \\ 1 \end{bmatrix} u(k) = \mathbf{A}\mathbf{x}(k) + \mathbf{b}u(k)$$

which has a characteristic equation with coefficients given by the last row of the state coupling matrix

$$\lambda^3 + \lambda^2 + 2\lambda + 4 = 0$$

Suppose that it is desired to have the three eigenvalues of the state feedback system at

$$\lambda_1 = 0 \qquad \lambda_2 = -\frac{1}{2} + j\frac{1}{2} \qquad \lambda_3 = -\frac{1}{2} - j\frac{1}{2}$$

Then the desired characteristic equation is

$$\lambda \left(\lambda + \frac{1}{2} + j\frac{1}{2} \right)\left(\lambda + \frac{1}{2} - j\frac{1}{2} \right) = \lambda^3 + \lambda^2 + \frac{1}{2}\lambda = 0$$

The feedback

$$u(k) = \mathbf{e}^\dagger \mathbf{x}(k) + \rho(k) = [e_1 \quad e_2 \quad e_3] \begin{bmatrix} x_1(k) \\ x_2(k) \\ x_3(k) \end{bmatrix} + \rho(k)$$

$$= e_1 x_1(k) + e_2 x_2(k) + e_3 x_3(k) + \rho(k)$$

as indicated in Figure 4-5, results in a feedback system governed by

$$\begin{bmatrix} x_1(k+1) \\ x_2(k+1) \\ x_3(k+1) \end{bmatrix} = \begin{bmatrix} 0 & 1 & 0 \\ 0 & 0 & 1 \\ -4 & -2 & -1 \end{bmatrix}\begin{bmatrix} x_1(k) \\ x_2(k) \\ x_3(k) \end{bmatrix}$$

$$+ \begin{bmatrix} 0 \\ 0 \\ 1 \end{bmatrix} [e_1 \quad e_2 \quad e_3]\begin{bmatrix} x_1(k) \\ x_2(k) \\ x_3(k) \end{bmatrix} + \begin{bmatrix} 0 \\ 0 \\ 1 \end{bmatrix}\rho(k)$$

$$= \begin{bmatrix} 0 & 1 & 0 \\ 0 & 0 & 1 \\ (e_1 - 4) & (e_2 - 2) & (e_3 - 1) \end{bmatrix}\begin{bmatrix} x_1(k) \\ x_2(k) \\ x_3(k) \end{bmatrix} + \begin{bmatrix} 0 \\ 0 \\ 1 \end{bmatrix}\rho(k)$$

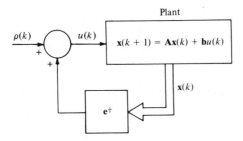

Figure 4-5 A single-input plant with state feedback.

The state feedback system has the characteristic equation

$$\lambda^3 + (1 - e_3)\lambda^2 + (2 - e_2)\lambda + (4 - e_1) = 0$$

and the feedback gains

$$\mathbf{e}^\dagger = [4 \quad \tfrac{3}{2} \quad 0]$$

result in the desired characteristic equation

$$\lambda^3 + \lambda^2 + \frac{1}{2}\lambda = 0$$

In general, however, the plant is not in controllable canonical form. One way for calculating the state feedback gain for eigenvalue placement for plants that are not in controllable canonical form is to transform the plant to controllable form, calculate the state feedback gain for the transformed system, and then transform the state feedback gain of the transformed system back to the original system. A step-by-step procedure for computing the state feedback gain vector $\mathbf{e}$ is as follows:

1. Find the characteristic polynomial of the matrix $\mathbf{A}$.

$$\Delta_0(\lambda) = |\lambda\mathbf{I} - \mathbf{A}| = \lambda^n + \alpha_{n-1}\lambda^{n-1} + \alpha_{n-2}\lambda^{n-2} + \cdots + \alpha_1\lambda + \alpha_0$$

2. Find the transformation

$$\mathbf{x} = \mathbf{P}\mathbf{x}' \qquad \mathbf{x}' = \mathbf{P}^{-1}\mathbf{x}$$

that takes the system to controllable form. The transformation matrix is

$$\mathbf{P} = [\mathbf{p}^1 \mid \mathbf{p}^2 \mid \mathbf{p}^3 \mid \cdots \mid \mathbf{p}^n]$$

where

$$\begin{cases} \mathbf{p}^n = \mathbf{b} \\ \mathbf{p}^{n-1} = \mathbf{A}\mathbf{p}^n + \alpha_{n-1}\mathbf{b} \\ \vdots \\ \mathbf{p}^2 = \mathbf{A}\mathbf{p}^3 + \alpha_2\mathbf{b} \\ \mathbf{p}^1 = \mathbf{A}\mathbf{p}^2 + \alpha_1\mathbf{b} \end{cases}$$

Then find $\mathbf{P}^{-1}$.

3. Given the desired feedback system eigenvalues, find the desired characteristic polynomial

$$\Delta_c(\lambda) = (\lambda + \lambda_1)(\lambda + \lambda_2) \cdots (\lambda + \lambda_n)$$
$$= \lambda^n + \beta_{n-1}\lambda^{n-1} + \beta_{n-2}\lambda^{n-2} + \cdots + \beta_1\lambda + \beta_0$$

4. Calculate the feedback gain vector $\mathbf{e}'^\dagger$ of the transformed system

$$\mathbf{e}'^\dagger = [e_1' \quad e_2' \quad \cdots \quad e_n']$$
$$= [(\alpha_0 - \beta_0) \quad (\alpha_1 - \beta_1) \quad \cdots \quad (\alpha_{n-1} - \beta_{n-1})]$$

5. Calculate the feedback gain vector of the original system

$$\mathbf{e}^\dagger = \mathbf{e}'^\dagger \mathbf{P}^{-1}$$

For example, the completely controllable single-input system with state equations

$$
\begin{bmatrix} x_1(k+1) \\ x_2(k+1) \\ x_3(k+1) \end{bmatrix} = \begin{bmatrix} 0 & 3 & -1 \\ -2 & 0 & 2 \\ 2 & 0 & -3 \end{bmatrix} \begin{bmatrix} x_1(k) \\ x_2(k) \\ x_3(k) \end{bmatrix} + \begin{bmatrix} 4 \\ 0 \\ 0 \end{bmatrix} u(k)
$$

$$= \mathbf{A}\mathbf{x}(k) + \mathbf{b}u(k) \tag{4-9}$$

has the characteristic equation

$$|\lambda\mathbf{I} - \mathbf{A}| = \begin{vmatrix} \lambda & -3 & 1 \\ 2 & \lambda & -2 \\ -2 & 0 & (\lambda+3) \end{vmatrix} = \lambda^3 + 3\lambda^2 + 8\lambda + 6$$

$$= \lambda^3 + \alpha_2\lambda^2 + \alpha_1\lambda + \alpha_0 = 0$$

The transformation

$$\mathbf{x} = \mathbf{P}\mathbf{x}' \qquad \mathbf{x}' = \mathbf{P}^{-1}\mathbf{x}$$

that takes this system to controllable form

$$\mathbf{P} = [\mathbf{p}^1 \mid \mathbf{p}^2 \mid \mathbf{p}^3]$$

is given (as shown in Section 3.7) by

$$\mathbf{p}^3 = \mathbf{b} = \begin{bmatrix} 4 \\ 0 \\ 0 \end{bmatrix}$$

$$\mathbf{p}^2 = \mathbf{A}\mathbf{p}^3 + \alpha_2\mathbf{b} = \begin{bmatrix} 12 \\ -8 \\ 8 \end{bmatrix}$$

$$\mathbf{p}^1 = \mathbf{A}\mathbf{p}^2 + \alpha_1\mathbf{b} = \begin{bmatrix} 0 \\ -8 \\ 0 \end{bmatrix}$$

so that

$$\mathbf{P} = \begin{bmatrix} 0 & 12 & 4 \\ -8 & -8 & 0 \\ 0 & 8 & 0 \end{bmatrix} \qquad \mathbf{P}^{-1} = \frac{1}{8} \begin{bmatrix} 0 & -1 & -1 \\ 0 & 0 & 1 \\ 2 & 0 & -3 \end{bmatrix}$$

In terms of the new state variables, the plant is described by the state equations

$$\begin{bmatrix} x_1'(k+1) \\ x_2'(k+1) \\ x_3'(k+1) \end{bmatrix} = \begin{bmatrix} 0 & 1 & 0 \\ 0 & 0 & 1 \\ -6 & -8 & -3 \end{bmatrix} \begin{bmatrix} x_1'(k) \\ x_2'(k) \\ x_3'(k) \end{bmatrix} + \begin{bmatrix} 0 \\ 0 \\ 1 \end{bmatrix} u(k)$$

$$= \mathbf{A}'\mathbf{x}'(k) + \mathbf{b}'u(k)$$

For the desired feedback system eigenvalues

$$\lambda = \frac{1}{2}, -\frac{1}{4}, -1$$

the desired characteristic equation is

$$\left(\lambda - \frac{1}{2}\right)\left(\lambda + \frac{1}{4}\right)(\lambda + 1) = \lambda^3 + \frac{3}{4}\lambda^2 - \frac{3}{8}\lambda - \frac{1}{8} = 0$$

If the *primed* state is fed back according to

$$u(k) = \mathbf{e}'^{\mathsf{t}}\mathbf{x}'(k) + \rho(k) = e_1'x_1'(k) + e_2'x_2'(k) + e_3'x_3'(k) + \rho(k)$$

the composite system is described by

$$\mathbf{x}'(k+1) = \mathbf{A}'\mathbf{x}'(k) + \mathbf{b}'\mathbf{e}'^{\mathsf{t}}\mathbf{x}'(k) + \mathbf{b}'\rho(k)$$

$$= (\mathbf{A}' + \mathbf{b}'\mathbf{e}'^{\mathsf{t}})\mathbf{x}'(k) + \mathbf{b}'\rho(k)$$

$$= \begin{bmatrix} 0 & 1 & 0 \\ 0 & 0 & 1 \\ (e_1' - 6) & (e_2' - 8) & (e_3' - 3) \end{bmatrix} \begin{bmatrix} x_1'(k) \\ x_2'(k) \\ x_3'(k) \end{bmatrix} + \begin{bmatrix} 0 \\ 0 \\ 1 \end{bmatrix} \rho(k)$$

which has the characteristic equation

$$\lambda^3 + (3 - e_3')\lambda^2 + (8 - e_2')\lambda + (6 - e_1') = 0$$

Choosing

$$\mathbf{e}'^{\mathsf{t}} = \begin{bmatrix} \frac{49}{8} & \frac{67}{8} & \frac{9}{4} \end{bmatrix}$$

gives the desired characteristic equation.

But it is the original state $\mathbf{x}(k)$, not $\mathbf{x}'(k)$, that is assumed available for feedback. The two are related by

$$\mathbf{x}'(k) = \mathbf{P}^{-1}\mathbf{x}(k)$$

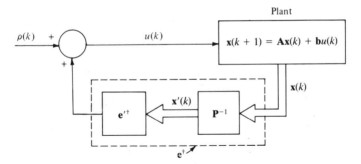

Figure 4-6 Use of controllable form to calculate state feedback gains.

so that the feedback

$$\mathbf{e'^{\dagger}P^{-1}x}(k) = \mathbf{e^{\dagger}x}(k)$$

where

$$\mathbf{e^{\dagger}} = \mathbf{e'^{\dagger}P^{-1}} = [\tfrac{9}{16} \quad -\tfrac{49}{64} \quad -\tfrac{9}{16}] \tag{4-10}$$

results in the desired eigenvalue placement, as indicated in Figure 4-6. The original system has these same eigenvalues because the eigenvalues of a system are unchanged by a change in state variables.

4.4.2 Other Methods for Eigenvalue Placement

There are a number of other methods for finding the state feedback gain vector of single-input plants. Two of these are now discussed and a third method, based on optimal control, is considered later in the chapter.

One way for finding the state feedback gain vector that does not require transforming the system to controllable canonical form is to recognize that the transformation matrix $\mathbf{P}$ is expressed in the form

$$\mathbf{P} = \underbrace{[\mathbf{b} \; \vdots \; \mathbf{Ab} \; \vdots \; \mathbf{A^2b} \; \vdots \; \cdots \; \vdots \; \mathbf{A^{n-1}b}]}_{\mathbf{M}_c} \begin{bmatrix} \alpha_1 & \alpha_2 & \cdots & \alpha_{n-2} & \alpha_{n-1} & 1 \\ \alpha_2 & \alpha_3 & \cdots & \alpha_{n-1} & 1 & 0 \\ \alpha_3 & \alpha_4 & \cdots & 1 & 0 & 0 \\ \vdots & & & \vdots & \vdots & \vdots \\ \alpha_{n-2} & \alpha_{n-1} & \cdots & 0 & 0 & 0 \\ \alpha_{n-1} & 1 & \cdots & 0 & 0 & 0 \\ 1 & 0 & \cdots & 0 & 0 & 0 \end{bmatrix}$$

Hence, the state feedback gain vector can be evaluated directly as

$$\mathbf{e}^{\dagger} = \mathbf{e}'^{\dagger}\mathbf{P}^{-1}$$

where $\mathbf{e}'^{\dagger}$ is determined as in step 4 of Subsection 4.4.1, that is,

$$\mathbf{e}'^{\dagger} = [(\alpha_0 - \beta_0) \quad (\alpha_1 - \beta_1) \quad \cdots \quad (\alpha_n - \beta_n)]$$

where the α's are the coefficients of the characteristic polynomial of matrix $\mathbf{A}$, and the β's are the coefficients of the desired characteristic equation of the feedback system.

Ackermann's Formula

Another method for calculating the state feedback gain vector is due to J. Ackermann. Ackermann showed that the state feedback gain vector is given by (Ackermann's formula)

$$\mathbf{e}^{\dagger} = -\mathbf{j}_n^{\dagger}\mathbf{M}_c^{-1}\,\Delta_c(\mathbf{A}) \tag{4-11}$$

where $\mathbf{j}_n^{\dagger}$ is the transpose of the nth-unit coordinate vector

$$\mathbf{j}_n^{\dagger} = [0 \quad 0 \quad \cdots \quad 0 \quad 1]$$

$\mathbf{M}_c$ is the controllability matrix of the system, and $\Delta_c(\mathbf{A})$ is the desired characteristic equation with the matrix $\mathbf{A}$ substituted for the variable z.

To prove Ackermann's formula, consider the system

$$\mathbf{x}(k + 1) = \mathbf{A}\mathbf{x}(k) + \mathbf{b}u(k)$$

with the characteristic equation

$$\Delta_p(z) = |z\mathbf{I} - \mathbf{A}| = z^n + \alpha_{n-1}z^{n-1}$$
$$+ \alpha_{n-2}z^{n-2} + \cdots + \alpha_1 z + \alpha_0 \tag{4-12}$$

We showed earlier that the transformation matrix $\mathbf{P}$, which takes the system to controllable canonical form

$$\mathbf{x}'(k + 1) = \mathbf{A}'\mathbf{x}'(k) + \mathbf{b}'u(k)$$
$$\mathbf{A}' = \mathbf{P}^{-1}\mathbf{A}\mathbf{P}$$
$$\mathbf{b}' = \mathbf{P}^{-1}\mathbf{b}$$

results in no change in the z-transfer function matrix.

For a given set of desired eigenvalues for the feedback system, the desired characteristic equation is

$$\Delta_c(z) = z^n + \beta_{n-1}z^{n-1} + \beta_{n-2}z^{n-2} + \cdots + \beta_1 z + \beta_0 \tag{4-13}$$

Furthermore, the desired characteristic equation of the transformed system in terms of the feedback gain terms is

$$\Delta_c(z) = z^n + (\alpha_{n-1} - e'_{n-1})z^{n-1} + (\alpha_{n-2} - e'_{n-2})z^{n-2}$$
$$+ \cdots + (\alpha_1 - e'_1)z + (\alpha_0 - e'_0)$$

Hence,

$$\alpha_{n-1} - e'_{n-1} = \beta_{n-1}$$
$$\alpha_{n-2} - e'_{n-2} = \beta_{n-2}$$
$$\vdots$$
$$\alpha_1 - e'_1 = \beta_1$$
$$\alpha_0 - e'_0 = \beta_0$$

or, in vector form

$$\boldsymbol{\alpha} - \mathbf{e}' = \boldsymbol{\beta}$$

Substituting the matrix $\mathbf{A}'$ in equation (4-12) and recalling the Cayley-Hamilton theorem, which states that every matrix satisfies its own characteristic equation, gives

$$\Delta_p(\mathbf{A}') = \mathbf{A}'^n + \alpha_{n-1}\mathbf{A}'^{n-1} + \cdots + \alpha_1\mathbf{A}' + \alpha_0\mathbf{I} = \mathbf{0} \qquad \textbf{(4-14)}$$

Also, substituting $\mathbf{A}'$ into equation (4-13) gives

$$\Delta_c(\mathbf{A}') = \mathbf{A}'^n + \beta_{n-1}\mathbf{A}'^{n-1} + \cdots + \beta_1\mathbf{A}' + \beta_0\mathbf{I} \qquad \textbf{(4-15)}$$

Subtracting equation (4-14) from equation (4-15) yields

$$\Delta_c(\mathbf{A}') = (\beta_{n-1} - \alpha_{n-1})\mathbf{A}'^{n-1} + \cdots$$
$$+ (\beta_1 - \alpha_1)\mathbf{A}' + (\beta_0 - \alpha_0)\mathbf{I} \qquad \textbf{(4-16)}$$

Define the basis vectors

$$\mathbf{j}_1 = \begin{bmatrix} 1 \\ 0 \\ 0 \\ \vdots \\ 0 \end{bmatrix} \quad \mathbf{j}_2 = \begin{bmatrix} 0 \\ 1 \\ 0 \\ \vdots \\ 0 \end{bmatrix} \quad \cdots \quad \mathbf{j}_n = \begin{bmatrix} 0 \\ 0 \\ 0 \\ \vdots \\ 1 \end{bmatrix}$$

Then premultiplying the matrix $\mathbf{A}'$ with $\mathbf{j}_1^\dagger$ gives

$$\mathbf{j}_1^\dagger\mathbf{A}' = [1 \quad 0 \quad 0 \quad \cdots \quad 0]\begin{bmatrix} 0 & 1 & 0 & \cdots & 0 \\ 0 & 0 & 1 & \cdots & 0 \\ \vdots & & & & \\ -\alpha_0 & -\alpha_1 & -\alpha_2 & \cdots & -\alpha_{n-1} \end{bmatrix}$$

$$= [0 \quad 1 \quad 0 \quad \cdots \quad 0] = \mathbf{j}_2^\dagger$$

and

$$(\mathbf{j}_1^\dagger\mathbf{A}')\mathbf{A}' = \mathbf{j}_2^\dagger\mathbf{A}' = [0 \quad 0 \quad 1 \quad \cdots \quad 0] = \mathbf{j}_3^\dagger$$

Also,

$$[(\mathbf{j}_1^\dagger\mathbf{A}')\mathbf{A}']\mathbf{A}' = (\mathbf{j}_2^\dagger\mathbf{A}')\mathbf{A}' = \mathbf{j}_3^\dagger\mathbf{A}' = \mathbf{j}_4^\dagger$$

and so on until

$$\mathbf{j}_{n-1}^\dagger\mathbf{A}' = \mathbf{j}_n^\dagger$$

Premultiplying equation (4-16) with $\mathbf{j}_1^\dagger$ gives

$$\mathbf{j}_1^\dagger \, \Delta_c(\mathbf{A}') = (\beta_{n-1} - \alpha_{n-1})\mathbf{j}_n^\dagger + \cdots + (\beta_1 - \alpha_1)[0 \quad 1 \quad 0 \quad \cdots \quad 0]$$
$$+ (\beta_0 - \alpha_0)[1 \quad 0 \quad 0 \quad \cdots \quad 0]$$

or

$$\mathbf{j}_1^\dagger \, \Delta_c(\mathbf{A}') = [(\beta_0 - \alpha_0) \quad (\beta_1 - \alpha_1) \cdots (\beta_{n-1} - \alpha_{n-1})]$$
$$= -[e_0' \quad e_1' \quad e_2' \quad \cdots \quad e_{n-1}'] = -\mathbf{e}'^\dagger \qquad \textbf{(4-17)}$$

But

$$\mathbf{e}^\dagger = \mathbf{e}'^\dagger\mathbf{P}^{-1} \qquad \textbf{(4-18)}$$

then substituting equation (4-17) in equation (4-18) gives

$$\mathbf{e}^\dagger = -\mathbf{j}_1^\dagger \, \Delta_c(\mathbf{A}')\mathbf{P}^{-1}$$

or

$$\mathbf{e}^\dagger = -\mathbf{j}_1^\dagger \, \Delta_c(\mathbf{P}^{-1}\mathbf{A}\mathbf{P})\mathbf{P}^{-1}$$
$$= -\mathbf{j}_1^\dagger\mathbf{P}^{-1} \, \Delta_c(\mathbf{A}) \qquad \textbf{(4-19)}$$

because, using the results in section A.4.3 of Appendix A,

$$(\mathbf{P}^{-1}\mathbf{A}\mathbf{P})^k = \mathbf{P}^{-1}\mathbf{A}^k\mathbf{P}$$

In addition, we showed earlier in the chapter that the transformation matrix $\mathbf{P}$ is related to the controllability matrix of the original system

$\mathbf{M}_c$ and the controllability matrix of the transformed system $\mathbf{M}'_c$ as follows:

$$\mathbf{P}^{-1} = \mathbf{M}'_c \mathbf{M}_c^{-1} \tag{4-20}$$

and substituting equation (4-20) into equation (4-19) gives

$$\mathbf{e}^\dagger = -\mathbf{j}_1^\dagger \mathbf{M}'_c \mathbf{M}_c^{-1} \Delta_c(\mathbf{A})$$

But

$$\mathbf{j}_1^\dagger \mathbf{M}'_c = [1 \quad 0 \quad 0 \quad \cdots \quad 0] \begin{bmatrix} 0 & 0 & 0 & \cdots & 1 \\ 0 & 0 & 0 & \cdots & xxx \\ \vdots & & & & \\ 0 & 1 & xxx & \cdots & xxx \\ 1 & xxx & xxx & \cdots & xxx \end{bmatrix} = \mathbf{j}_n^\dagger$$

where xxx indicates locations of terms that are functions of the α's. Therefore,

$$\mathbf{e}^\dagger = -\mathbf{j}_n^\dagger \mathbf{M}_c^{-1} \Delta_c(\mathbf{A})$$

which completes the proof.

As a numerical example, consider the system

$$\begin{bmatrix} x_1(k+1) \\ x_2(k+1) \\ x_3(k+1) \end{bmatrix} = \begin{bmatrix} 0 & 3 & -1 \\ -2 & 0 & 2 \\ 2 & 0 & -3 \end{bmatrix} \begin{bmatrix} x_1(k) \\ x_2(k) \\ x_3(k) \end{bmatrix} + \begin{bmatrix} 4 \\ 0 \\ 0 \end{bmatrix} u(k)$$

which is given in Subsection 4.4.1. Applying Ackermann's formula (4-11) gives

$$\mathbf{e}^\dagger = -[0 \quad 0 \quad 1] \begin{bmatrix} 4 & 0 & -32 \\ 0 & -8 & 16 \\ 0 & 8 & -24 \end{bmatrix}^{-1} \{\mathbf{A}^3 + \tfrac{3}{4}\mathbf{A}^2 - \tfrac{3}{8}\mathbf{A} - \tfrac{1}{8}\mathbf{I}\}$$

$$= (\tfrac{1}{24})(\tfrac{1}{8})[0 \quad 0 \quad 1] \begin{bmatrix} -6 & 24 & 24 \\ 0 & 9 & 6 \\ 0 & 3 & 3 \end{bmatrix} \begin{bmatrix} 95 & -201 & -95 \\ 62 & 59 & -102 \\ -26 & -108 & 26 \end{bmatrix}$$

$$= \tfrac{1}{192}[108 \quad -147 \quad -76]$$

$$= [\tfrac{9}{16} \quad -\tfrac{49}{64} \quad -\tfrac{9}{16}]$$

which is the same result as that obtained in the previous example.

4.4.3 Eigenvalue Placement with Multiple Inputs

If the plant for eigenvalue placement has multiple inputs and if it is completely controllable from one of the inputs, then that one input alone could be used for the feedback. For example, consider the system with state equations

$$
\begin{bmatrix} x_1(k+1) \\ x_2(k+1) \\ x_3(k+1) \end{bmatrix} = \begin{bmatrix} 0 & 3 & -1 \\ -2 & 0 & 2 \\ 2 & 0 & -3 \end{bmatrix} \begin{bmatrix} x_1(k) \\ x_2(k) \\ x_3(k) \end{bmatrix} + \begin{bmatrix} 4 & 3 \\ 0 & -1 \\ 0 & 1 \end{bmatrix} \begin{bmatrix} u_1(k) \\ u_2(k) \end{bmatrix}
$$

This is system (4-9) with an additional input. To place the eigenvalues so that the characteristic equation of this system is, as in the earlier example

$$
\lambda^3 + \frac{3}{4}\lambda^2 - \frac{3}{8}\lambda - \frac{1}{8} = 0
$$

the second input can be ignored so far as feedback is concerned, with the state fed back through the gain of equation (4-10) found earlier

$$
\mathbf{e}^\dagger = \begin{bmatrix} \frac{9}{16} & -\frac{49}{64} & -\frac{9}{16} \end{bmatrix}
$$

to the first input. Adding additional external inputs $\rho_1(k)$ and $\rho_2(k)$,

$$
u_1(k) = \mathbf{e}^\dagger \mathbf{x}(k) + \rho_1(k)
$$
$$
u_2(k) = \rho_2(k)
$$

This arrangement is shown in Figure 4-7.

    If the plant is not completely controllable from a single input, a single input can usually be distributed to the multiple ones in such a way that the plant is completely controllable from the single input. For

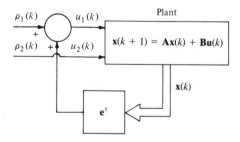

Figure 4-7 Use of a single plant input for state feedback.

example, the two-input system with diagonal state equations

$$\begin{bmatrix} x_1(k + 1) \\ x_2(k + 1) \\ x_3(k + 1) \end{bmatrix} = \begin{bmatrix} -\frac{1}{2} & 0 & 0 \\ 0 & \frac{1}{2} & 0 \\ 0 & 0 & 0 \end{bmatrix} \begin{bmatrix} x_1(k) \\ x_2(k) \\ x_3(k) \end{bmatrix} + \begin{bmatrix} 1 & 0 \\ 0 & -2 \\ -1 & 1 \end{bmatrix} \begin{bmatrix} u_1(k) \\ u_2(k) \end{bmatrix}$$

$$= \mathbf{A}\mathbf{x}(k) + \mathbf{B}\mathbf{u}(k)$$

is obviously not completely controllable from either input $u_1(k)$ or input $u_2(k)$ alone. Distributing the single input $\mu(k)$ to $u_1(k)$ and $u_2(k)$ in almost any way, for example as

$$u_1(k) = 3\mu(k)$$
$$u_2(k) = \mu(k)$$

results in a single-input system that is completely controllable from $\mu(k)$

$$\begin{bmatrix} x_1(k + 1) \\ x_2(k + 1) \\ x_3(k + 1) \end{bmatrix} = \begin{bmatrix} -\frac{1}{2} & 0 & 0 \\ 0 & \frac{1}{2} & 0 \\ 0 & 0 & 0 \end{bmatrix} \begin{bmatrix} x_1(k) \\ x_2(k) \\ x_3(k) \end{bmatrix} + \begin{bmatrix} 3 \\ -2 \\ -2 \end{bmatrix} \mu(k) = \mathbf{A}\mathbf{x}(k) + \mathbf{h}\mu(k)$$

This distribution of a single input to multiple plant inputs is shown in Figure 4-8.

With state feedback to the single input $\mu(k)$, the feedback system eigenvalues can be placed at any desired locations. The characteristic equation of the plant is

$$\left(\lambda + \frac{1}{2}\right)\left(\lambda - \frac{1}{2}\right)(\lambda) = \lambda^3 - \frac{1}{4}\lambda = 0$$

The change of state variables

$$\mathbf{x} = \mathbf{P}\mathbf{x}' \qquad \mathbf{x}' = \mathbf{P}^{-1}\mathbf{x}$$

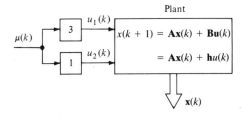

Figure 4-8 Distribution of a single input to multiple plant inputs for state feedback.

that takes the system with input $\mu(k)$ into the controllable form is

$$\mathbf{P} = \begin{bmatrix} 0 & -\frac{3}{2} & 3 \\ 0 & -1 & -2 \\ \frac{1}{2} & 0 & -2 \end{bmatrix} \qquad \mathbf{P}^{-1} = \frac{1}{12}\begin{bmatrix} 8 & -12 & 24 \\ -4 & -6 & 0 \\ 2 & -3 & 0 \end{bmatrix}$$

and feedback of the primed state variables, of the form

$$\mu(k) = \mathbf{e}'^{\dagger}\mathbf{x}'(k)$$

results in the composite system equations

$$\begin{bmatrix} x_1'(k+1) \\ x_2'(k+1) \\ x_3'(k+1) \end{bmatrix} = \left\{ \begin{bmatrix} 0 & 1 & 0 \\ 0 & 0 & 1 \\ 0 & \frac{1}{4} & 0 \end{bmatrix} + \begin{bmatrix} 0 \\ 0 \\ 1 \end{bmatrix}[e_1' \quad e_2' \quad e_3'] \right\}\begin{bmatrix} x_1'(k) \\ x_2'(k) \\ x_3'(k) \end{bmatrix}$$

$$= \begin{bmatrix} 0 & 1 & 0 \\ 0 & 0 & 1 \\ e_1' & \frac{1}{4}+e_2' & e_3' \end{bmatrix}\begin{bmatrix} x_1'(k) \\ x_2'(k) \\ x_3'(k) \end{bmatrix}$$

If it is desired that all three eigenvalues of the feedback system be at $\lambda = 0$, so that the desired characteristic equation is

$$\lambda^3 = 0$$

then

$$\mathbf{e}'^{\dagger} = [0 \quad -\tfrac{1}{4} \quad 0]$$

In terms of the original state variables

$$\mathbf{e}^{\dagger} = \mathbf{e}'^{\dagger}\mathbf{P}^{-1} = [0 \quad -\tfrac{1}{4} \quad 0]\begin{bmatrix} 8 & -12 & 24 \\ -4 & -6 & 0 \\ 2 & -3 & 0 \end{bmatrix}\left(\frac{1}{12}\right) = [\tfrac{1}{12} \quad \tfrac{1}{8} \quad 0]$$

Adding external inputs to each of the original two plant inputs, the resulting feedback system design is

$$\begin{bmatrix} x_1(k+1) \\ x_2(k+1) \\ x_3(k+1) \end{bmatrix} = \begin{bmatrix} -\frac{1}{2} & 0 & 0 \\ 0 & \frac{1}{2} & 0 \\ 0 & 0 & 0 \end{bmatrix}\begin{bmatrix} x_1(k) \\ x_2(k) \\ x_3(k) \end{bmatrix} + \begin{bmatrix} 1 & 0 \\ 0 & -2 \\ -1 & 1 \end{bmatrix}\begin{bmatrix} u_1(k) \\ u_2(k) \end{bmatrix}$$

where

$$u_1(k) = 3\mu(k) + \rho_1(k) = 3\mathbf{e}^{\dagger}\mathbf{x}(k) + \rho_1(k)$$
$$u_2(k) = \mu(k) + \rho_2(k) = \mathbf{e}^{\dagger}\mathbf{x}(k) + \rho_2(k)$$

or

$$
\begin{bmatrix} u_1(k) \\ u_2(k) \end{bmatrix} = \begin{bmatrix} 3\mathbf{e}^\dagger \\ \hline \mathbf{e}^\dagger \end{bmatrix} \begin{bmatrix} x_1(k) \\ x_2(k) \\ x_3(k) \end{bmatrix} + \begin{bmatrix} \rho_1(k) \\ \rho_2(k) \end{bmatrix} = \begin{bmatrix} \frac{1}{4} & \frac{3}{8} & 0 \\ \frac{1}{12} & \frac{1}{8} & 0 \end{bmatrix} \begin{bmatrix} x_1(k) \\ x_2(k) \\ x_3(k) \end{bmatrix} + \begin{bmatrix} \rho_1(k) \\ \rho_2(k) \end{bmatrix}
$$

$$= \mathbf{E}\mathbf{x}(k) + \boldsymbol{\rho}(k)$$

The structure of this system is shown in Figure 4-9.

When a feedback system is designed in this way, the feedback gain matrix $\mathbf{E}$ is of rank 1 and is not unique because the single input $\mu(k)$ can be distributed to the original plant inputs in many different ways. The feedback need not be designed in this way at all; distribution of a single input is simply a convenient design option because we already know how to place eigenvalues in a single-input system. This option is not always available, though, because it is not always possible to distribute a single input to a completely controllable multiple-input plant in such a way that the plant is completely controllable from the single input. If, in a block diagonal plant realization, there is more than one Jordan block involving the same eigenvalue and if each of two such blocks couple to different inputs (as they must for complete controllability), then making those inputs proportional destroys controllability of the individual modes.

One may wish to feed the state back to more than one input even when the plant is completely controllable from a single input. Using

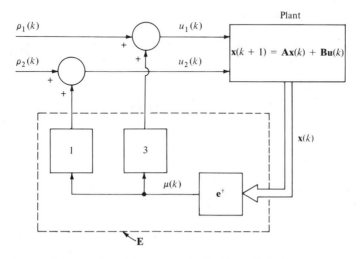

Figure 4-9 State feedback to a plant with multiple inputs.

more than one input can be more reliable in that all feedback is not lost if one input is damaged and becomes inactive. Also, the use of more than one input might make the overall system less susceptible to noise.

4.4.4 State Feedback for Step-Varying Systems

Deadbeat or perhaps some other shaping of a step-varying plant's zero-input response with state feedback is now discussed. For a single-input nth-order plant with state equations

$$\mathbf{x}(k + 1) = \mathbf{A}(k)\mathbf{x}(k) + \mathbf{b}(k)u(k)$$

the state feedback

$$u(k) = \mathbf{e}^\dagger(k)\mathbf{x}(k) + \rho(k)$$

where $\rho(k)$ is an external input, gives feedback system state equations

$$\mathbf{x}(k + 1) = [\mathbf{A}(k) + \mathbf{b}(k)\mathbf{e}^\dagger(k)]\mathbf{x}(k) + \mathbf{b}(k)\rho(k) \tag{4-21}$$

This arrangement is shown in Figure 4-10. The zero-input response of the feedback system (4-21) is

$$\mathbf{x}_{\text{zero-input}}(k) = \mathbf{\Phi}(k - 1)\mathbf{\Phi}(k - 2) \cdots \mathbf{\Phi}(0)\mathbf{x}(0)$$

where

$$\mathbf{\Phi}(i) = \mathbf{A}(i) + \mathbf{b}(i)\mathbf{e}^\dagger(i)$$

Let

$$\mathbf{j}_1, \mathbf{j}_2, \ldots, \mathbf{j}_n$$

be any linearly independent set of n-vectors. These are called *basis vectors* because they span the n-dimensional space. Requiring that

$$\mathbf{\Phi}(n - 1)\mathbf{\Phi}(n - 2) \cdots \mathbf{\Phi}(0) = \mathbf{0} \tag{4-22}$$

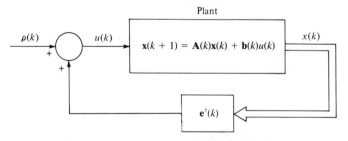

Figure 4-10 Step-varying state feedback for a single-input plant.

so that the zero-input response of the state feedback system is zero at step n and beyond is equivalent to requiring that

$$\Phi^\dagger(0) \cdots \Phi^\dagger(n - 2)\Phi^\dagger(n - 1) = 0$$

or

$$\begin{cases} \Phi^\dagger(0) \cdots \Phi^\dagger(n - 2)\Phi^\dagger(n - 1)\mathbf{j}_1 = 0 \\ \Phi^\dagger(0) \cdots \Phi^\dagger(n - 2)\Phi^\dagger(n - 1)\mathbf{j}_2 = 0 \\ \quad \vdots \\ \Phi^\dagger(0) \cdots \Phi^\dagger(n - 2)\Phi^\dagger(n - 1)\mathbf{j}_n = 0 \end{cases} \tag{4-23}$$

In turn, the following requirements assure equation (4-23) and thus equation (4-22)

$$\begin{cases} \Phi^\dagger(n - 1)\mathbf{j}_1 = 0 \\ \Phi^\dagger(n - 2)\Phi^\dagger(n - 1)\mathbf{j}_2 = 0 \\ \quad \vdots \\ \Phi^\dagger(0) \cdots \Phi^\dagger(n - 2)\Phi^\dagger(n - 1)\mathbf{j}_n = 0 \end{cases} \tag{4-24}$$

The first equation of (4-24) is, in terms of the feedback gain $\mathbf{e}(n - 1)$,

$$\mathbf{A}^\dagger(n - 1)\mathbf{j}_1 + \mathbf{e}(n - 1)\mathbf{b}^\dagger(n - 1)\mathbf{j}_1 = 0$$

or

$$\mathbf{e}(n - 1) = - \frac{\mathbf{A}^\dagger(n - 1)\mathbf{j}_1}{\mathbf{b}^\dagger(n - 1)\mathbf{j}_1}$$

Similarly, the second equation of (4-24) gives

$$\mathbf{e}(n - 2) = - \frac{\mathbf{A}^\dagger(n - 2)\Phi^\dagger(n - 1)\mathbf{j}_2}{\mathbf{b}^\dagger(n - 2)\Phi^\dagger(n - 1)\mathbf{j}_2}$$

where

$$\Phi(n - 1) = \mathbf{A}(n - 1) + \mathbf{b}(n - 1)\mathbf{e}^\dagger(n - 1)$$

and in general

$$\mathbf{e}(k) = - \frac{\mathbf{A}^\dagger(k)\Phi^\dagger(k + 1) \cdots \Phi^\dagger(n - 1)\mathbf{j}_{n-k}}{\mathbf{b}^\dagger(k)\Phi^\dagger(k + 1) \cdots \Phi^\dagger(n - 1)\mathbf{j}_{n-k}} \tag{4-25}$$

where

$$\Phi(i) = \mathbf{A}(i) + \mathbf{b}(i)\mathbf{e}^\dagger(i) \tag{4-26}$$

The computational procedure is summarized in Table 4-1. The calculations are inherently backward in step.

Table 4-1 Deadbeat State Feedback for Single-Input Plants

Plant with State Feedback

$$u(k) = \mathbf{e}^\dagger(k)\mathbf{x}(k) + \rho(k)$$

in

$$\mathbf{x}(k + 1) = \mathbf{A}(k)\mathbf{x}(k) + \mathbf{b}(k)u(k) = [\mathbf{A}(k) + \mathbf{b}(k)\mathbf{e}^\dagger(k)]\mathbf{x}(k) + \mathbf{b}(k)\rho(k)$$

Feedback Gain Calculation

Choose any set of linearly independent n-vectors (basis vectors) $\mathbf{j}_1, \mathbf{j}_2, \ldots,$ $\mathbf{j}_n$. Then

$$\begin{cases} \mathbf{e}(n - 1) = -\dfrac{\mathbf{A}^\dagger(n - 1)\mathbf{j}_1}{\mathbf{b}^\dagger(n - 1)\mathbf{j}_1} \\ \boldsymbol{\Phi}(n - 1) = \mathbf{A}(n - 1) + \mathbf{b}(n - 1)\mathbf{e}^\dagger(n - 1) \end{cases}$$

$$\begin{cases} \mathbf{e}(n - 2) = -\dfrac{\mathbf{A}^\dagger(n - 2)\boldsymbol{\Phi}^\dagger(n - 1)\mathbf{j}_2}{\mathbf{b}^\dagger(n - 2)\boldsymbol{\Phi}^\dagger(n - 1)\mathbf{j}_2} \\ \boldsymbol{\Phi}(n - 1) = \mathbf{A}(n - 2) + \mathbf{b}(n - 2)\mathbf{e}^\dagger(n - 2) \end{cases}$$

$$\vdots$$

$$\begin{cases} \mathbf{e}(k) = -\dfrac{\mathbf{A}^\dagger(k)\boldsymbol{\Phi}^\dagger(k + 1) \ldots \boldsymbol{\Phi}^\dagger(n - 1)\mathbf{j}_{n-k}}{\mathbf{b}^\dagger(k)\boldsymbol{\Phi}^\dagger(k + 1) \ldots \boldsymbol{\Phi}^\dagger(n - 1)\mathbf{j}_{n-k}} \\ \boldsymbol{\Phi}(k) = \mathbf{A}(k) + \mathbf{b}(k)\mathbf{e}^\dagger(k) \end{cases}$$

$$\vdots$$

$$\begin{cases} \mathbf{e}(0) = -\dfrac{\mathbf{A}^\dagger(0)\boldsymbol{\Phi}^\dagger(1) \ldots \boldsymbol{\Phi}^\dagger(n - 1)\mathbf{j}_n}{\mathbf{b}^\dagger(0)\boldsymbol{\Phi}^\dagger(1) \ldots \boldsymbol{\Phi}^\dagger(n - 1)\mathbf{j}_n} \\ \boldsymbol{\Phi}(0) = \mathbf{A}(0) + \mathbf{b}(0)\mathbf{e}^\dagger(0) \end{cases}$$

If for any gain calculation $\mathbf{e}(k)$, the scalar divisor involved is zero

$$\mathbf{b}^\dagger(k)\boldsymbol{\Phi}^\dagger(k + 1) \ldots \boldsymbol{\Phi}^\dagger(n - 1)\mathbf{j}_{n-k} = 0$$

a special situation has occurred for which the state of the plant cannot be controlled in the $\mathbf{j}_{n-k}$ direction at that step. Reordering the *remaining* basis vectors gives a solution whenever one exists. If all remaining basis vectors give a zero scalar divisor, the plant state cannot be changed by the input as desired at this step.

Consider the plant

$$\begin{bmatrix} x_1(k + 1) \\ x_2(k + 1) \end{bmatrix} = \begin{bmatrix} k & 2 \\ 1 & -1 \end{bmatrix} \begin{bmatrix} x_1(k) \\ x_2(k) \end{bmatrix} + \begin{bmatrix} -1 \\ \left(\dfrac{1}{k + 1}\right) \end{bmatrix} u(k)$$

$$= \mathbf{A}(k)\mathbf{x}(k) + \mathbf{b}(k)u(k)$$

with state feedback

$$u(k) = [e_1(k) \quad e_2(k)]\begin{bmatrix} x_1(k) \\ x_2(k) \end{bmatrix} + \rho(k) = \mathbf{e}^\dagger(k)\mathbf{x}(k) + \rho(k)$$

Relations (4-25) and (4-26) and the choices of basis vectors $\mathbf{j}_1$ and $\mathbf{j}_2$ as unit coordinate vectors

$$\mathbf{j}_1^\dagger = [1 \quad 0] \qquad \mathbf{j}_2^\dagger = [0 \quad 1]$$

give the following feedback gains for deadbeat zero-input response:

$$\mathbf{e}(1) = -\frac{\mathbf{A}^\dagger(1)\mathbf{j}_1}{\mathbf{b}^\dagger(1)\mathbf{j}_1} = \begin{bmatrix} 1 \\ 2 \end{bmatrix}$$

$$\boldsymbol{\Phi}(1) = \mathbf{A}(1) + \mathbf{b}(1)\mathbf{e}^\dagger(1) = \begin{bmatrix} 0 & 0 \\ \frac{3}{2} & 0 \end{bmatrix}$$

$$\mathbf{e}(0) = -\frac{\mathbf{A}^\dagger(0)\boldsymbol{\Phi}^\dagger(1)\mathbf{j}_2}{\mathbf{b}^\dagger(0)\boldsymbol{\Phi}^\dagger(1)\mathbf{j}_2} = \begin{bmatrix} 0 \\ 2 \end{bmatrix}$$

For continued deadbeat state feedback, this computational procedure can simply be repeated to obtain $\mathbf{e}(3)$ and $\mathbf{e}(2)$, then $\mathbf{e}(5)$ and $\mathbf{e}(4)$, and so on. Different pairs of gain vectors generally result because $\mathbf{A}(k)$ and $\mathbf{b}(k)$ can change with step. The next two gains calculated in this way are

$$\mathbf{e}(3) = -\frac{\mathbf{A}^\dagger(3)\mathbf{j}_1}{\mathbf{b}^\dagger(3)\mathbf{j}_1} = \begin{bmatrix} 3 \\ 2 \end{bmatrix}$$

$$\boldsymbol{\Phi}(3) = \mathbf{A}(3) + \mathbf{b}(3)\mathbf{e}^\dagger(3) = \begin{bmatrix} 0 & 0 \\ \frac{7}{4} & -\frac{1}{2} \end{bmatrix}$$

$$\mathbf{e}(2) = -\frac{\mathbf{A}^\dagger(2)\boldsymbol{\Phi}^\dagger(3)\mathbf{j}_2}{\mathbf{b}^\dagger(2)\boldsymbol{\Phi}^\dagger(3)\mathbf{j}_2} = \begin{bmatrix} -3 \\ 3 \end{bmatrix}$$

If the plant has multiple inputs, a scalar feedback signal can be distributed to the inputs and the feedback designed as if the plant were single input. Better, the state feedback to each of the inputs can be designed to drive the state vector to zero in more than one direction at each step.

4.5 Output Feedback

It is the measurement output vector of a plant, not the state vector, that is available for feedback. In this section, we consider what eigenvalue placement can be performed with output feedback alone. With enough linearly independent outputs, the plant state can be formed from the outputs and inputs and the state feedback results applied. With a single plant input and only a few outputs, the designer's options for placing feedback system eigenvalues could be (and often are) severely limited. Multiple plant inputs can also be used to advantage for eigenvalue placement with output feedback, but it still may not be possible to achieve an acceptable design.

4.5.1 State Feedback Derived from Outputs

If an nth-order plant has n linearly independent outputs, then the plant state can be recovered from the outputs and (if there is direct input-to-output coupling) the plant inputs. For example, for the system with state equations (4-9)

$$\begin{bmatrix} x_1(k+1) \\ x_2(k+1) \\ x_3(k+1) \end{bmatrix} = \begin{bmatrix} 0 & 3 & -1 \\ -2 & 0 & 2 \\ 2 & 0 & -3 \end{bmatrix} \begin{bmatrix} x_1(k) \\ x_2(k) \\ x_3(k) \end{bmatrix} + \begin{bmatrix} 4 \\ 0 \\ 0 \end{bmatrix} u(k)$$

where it was found that the state feedback gain of equation (4-10)

$$\mathbf{e}^\dagger = \begin{bmatrix} \frac{9}{16} & -\frac{49}{64} & -\frac{9}{16} \end{bmatrix}$$

resulted in the desired eigenvalue placement, suppose that the state is not directly available for feedback but that the three linearly independent outputs

$$\begin{bmatrix} y_1(k) \\ y_2(k) \\ y_3(k) \end{bmatrix} = \begin{bmatrix} 2 & 0 & 1 \\ -1 & 3 & 1 \\ 0 & -2 & 4 \end{bmatrix} \begin{bmatrix} x_1(k) \\ x_2(k) \\ x_3(k) \end{bmatrix} + \begin{bmatrix} 3 \\ 0 \\ -1 \end{bmatrix} u(k) = \mathbf{C}\mathbf{x}(k) + \mathbf{d}u(k)$$

are available. Then the state is related to the outputs and input by

$$\mathbf{x}(k) = \mathbf{C}^{-1}[\mathbf{y}(k) - \mathbf{d}u(k)] = \mathbf{C}^{-1}\mathbf{y}(k) - \mathbf{C}^{-1}\mathbf{d}u(k)$$

so that

$$u(k) = \mathbf{e}^\dagger\mathbf{x}(k) + \rho(k) = \mathbf{e}^\dagger\mathbf{C}^{-1}\mathbf{y}(k) - \mathbf{e}^\dagger\mathbf{C}^{-1}\mathbf{d}u(k) + \rho(k)$$

as shown in Figure 4-11(a), where

$$\mathbf{e}^\dagger\mathbf{C}^{-1} = \begin{bmatrix} \frac{9}{16} & -\frac{49}{64} & -\frac{9}{16} \end{bmatrix} \frac{1}{30} \begin{bmatrix} 14 & -2 & -3 \\ 4 & 8 & -3 \\ 2 & 4 & 6 \end{bmatrix} = \begin{bmatrix} \frac{59}{480} & -\frac{152}{480} & -\frac{177}{1920} \end{bmatrix}$$

and

$$\mathbf{e}^\dagger\mathbf{C}^{-1}\mathbf{d} = \frac{177}{384}$$

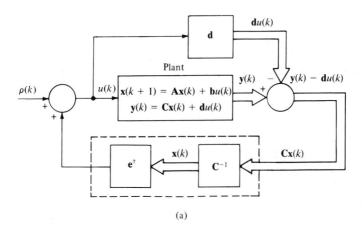

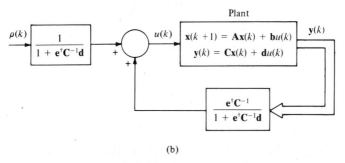

Figure 4-11 Use of output feedback for eigenvalue placement. (a) Recovering the plant state from the input and output, then feeding it back. (b) State feedback expressed directly as output feedback.

Feedback to the plant input can be expressed simply as a linear combination of plant outputs and external input, as in Figure 4-11(b), if desired, according to

$$u(k) = \mathbf{e}^\dagger \mathbf{C}^{-1}\mathbf{y}(k) - \mathbf{e}^\dagger \mathbf{C}^{-1}\mathbf{d}u(k) + \rho(k)$$

$$= \frac{\mathbf{e}^\dagger \mathbf{C}^{-1}}{1 + \mathbf{e}^\dagger \mathbf{C}^{-1}\mathbf{d}} \mathbf{y}(k) + \frac{1}{1 + \mathbf{e}^\dagger \mathbf{C}^{-1}\mathbf{d}} \rho(k)$$

The external input $\rho(k)$ could be added directly to $u(k)$ instead of being scaled by the factor $1/(1 + \mathbf{e}^\dagger \mathbf{C}^{-1}\mathbf{d})$, of course.

The special case in which

$$\mathbf{e}^\dagger \mathbf{C}^{-1}\mathbf{d} = -1$$

occurs when the state $\mathbf{x}(k)$ cannot be determined directly from the system outputs when there is feedback. Under these conditions, the feedback has caused the feedback system outputs to be no longer linearly independent of one another so that the state cannot be determined directly from the output, as was originally assumed. A small change in the feedback gains $\mathbf{e}^\dagger$, corresponding to small changes in the desired feedback system eigenvalue locations, eliminates the singularity.

If an nth-order system has more than n outputs, only n of these can be linearly independent, so excess linearly dependent output equations can simply be ignored when recovering a system's state from its output. For example, if a third-order system has outputs

$$
\begin{bmatrix} y_1(k) \\ y_2(k) \\ y_3(k) \\ y_4(k) \\ y_5(k) \end{bmatrix} = \begin{bmatrix} 0 & 1 & 0 \\ 2 & -1 & 3 \\ -2 & 0 & -3 \\ 1 & 1 & 0 \\ -1 & 4 & 2 \end{bmatrix} \begin{bmatrix} x_1(k) \\ x_2(k) \\ x_3(k) \end{bmatrix} + \begin{bmatrix} 3 \\ -1 \\ 0 \\ 4 \\ -5 \end{bmatrix} u(k) = \mathbf{C}\mathbf{x}(k) + \mathbf{d}u(k)
$$

deleting the third equation (it is linearly dependent on the first two) and the fifth equation leaves

$$
\begin{bmatrix} y_1(k) \\ y_2(k) \\ y_4(k) \end{bmatrix} = \begin{bmatrix} 0 & 1 & 0 \\ 2 & -1 & 3 \\ 1 & 1 & 0 \end{bmatrix} \begin{bmatrix} x_1(k) \\ x_2(k) \\ x_3(k) \end{bmatrix} + \begin{bmatrix} 3 \\ -1 \\ 4 \end{bmatrix} u(k) = \tilde{\mathbf{C}}\mathbf{x}(k) + \tilde{\mathbf{d}}u(k)
$$

The state can be recovered directly from these outputs via

$$
\mathbf{x}(k) = \tilde{\mathbf{C}}^{-1} \left(\begin{bmatrix} y_1(k) \\ y_2(k) \\ y_4(k) \end{bmatrix} - \begin{bmatrix} 3 \\ -1 \\ 4 \end{bmatrix} u(k) \right)
$$

so

$$
\begin{bmatrix} x_1(k) \\ x_2(k) \\ x_3(k) \end{bmatrix} = \begin{bmatrix} -1 & 0 & 0 & 1 & 0 \\ 1 & 0 & 0 & 0 & 0 \\ 1 & \frac{1}{3} & 0 & -\frac{2}{3} & 0 \end{bmatrix} \begin{bmatrix} y_1(k) \\ y_2(k) \\ y_3(k) \\ y_4(k) \\ y_5(k) \end{bmatrix} - \begin{bmatrix} 1 \\ 3 \\ 0 \end{bmatrix} u(k)
$$

To improve feedback system reliability and its performance in the presence of noise, one may wish instead to combine linearly dependent outputs with other of the outputs rather than to ignore them.

4.5.2 Output Feedback with a Single-Input Plant

When a single-input plant does not have enough linearly independent outputs for the state to be recovered directly from it, output feedback does not allow arbitrary feedback system eigenvalue placement. For example, consider the system with state equations

$$\begin{bmatrix} x_1(k+1) \\ x_2(k+1) \\ x_3(k+1) \end{bmatrix} = \begin{bmatrix} 0 & 1 & 0 \\ 0 & 0 & 1 \\ 0 & 2 & -1 \end{bmatrix} \begin{bmatrix} x_1(k) \\ x_2(k) \\ x_3(k) \end{bmatrix} + \begin{bmatrix} 0 \\ 0 \\ 1 \end{bmatrix} u(k) = \mathbf{A}\mathbf{x}(k) + \mathbf{b}u(k)$$

For convenience, these state equations are in controllable form. If there is a single output

$$y(k) = [4 \quad 0 \quad 1] \begin{bmatrix} x_1(k) \\ x_2(k) \\ x_3(k) \end{bmatrix} + 4u(k) = \mathbf{c}^\dagger\mathbf{x}(k) + du(k)$$

output feedback of the form

$$u(k) = f[y(k) - 4u(k)] + \rho(k) = [4f \quad 0 \quad f] \begin{bmatrix} x_1(k) \\ x_2(k) \\ x_3(k) \end{bmatrix} + \rho(k)$$

where f is a scalar, gives the following state equations for the feedback system:

$$\begin{bmatrix} x_1(k+1) \\ x_2(k+1) \\ x_3(k+1) \end{bmatrix} = \begin{bmatrix} 0 & 1 & 0 \\ 0 & 0 & 1 \\ 4f & 2 & (f-1) \end{bmatrix} \begin{bmatrix} x_1(k) \\ x_2(k) \\ x_3(k) \end{bmatrix} + \begin{bmatrix} 0 \\ 0 \\ 1 \end{bmatrix} \rho(k)$$

This feedback system, shown in Figure 4-12(a), has the following characteristic equation, in terms of f:

$$\lambda^3 + (1 - f)\lambda^2 - 2\lambda - 4f = 0 \tag{4-27}$$

The characteristic equations that can be achieved with output feedback are of this form, for some value of f.

Collecting terms and arranging equation (4-27) as a root locus problem results in

$$\lambda^3 + \lambda^2 - 2\lambda - f(\lambda^2 + 4) = 0$$

$$1 - f\frac{(\lambda + j2)(\lambda - j2)}{\lambda(\lambda - 1)(\lambda + 2)} = 0$$

Root locus plots for negative and positive values of the adjustable constant f are shown in Figure 4-12(b). In this case, no value of f results in even a stable system. Of course, this system could also be examined using z-transfer functions and classical root locus methods, where the same plots would result.

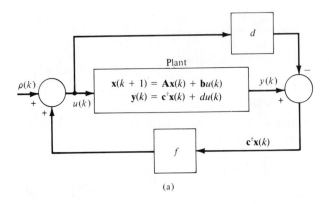

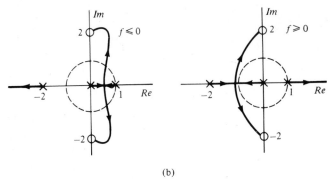

(b)

Figure 4-12 A third-order single-output system with output feedback. (a) System block diagram. (b) Root locus for the system, as a function of the output feedback gain f.

If there are instead two outputs, say

$$\begin{bmatrix} y_1(k) \\ y_2(k) \end{bmatrix} = \begin{bmatrix} 1 & 0 & -1 \\ -2 & 1 & 1 \end{bmatrix} \begin{bmatrix} x_1(k) \\ x_2(k) \\ x_3(k) \end{bmatrix} + \begin{bmatrix} 0 \\ 2 \end{bmatrix} u(k) = \mathbf{C}\mathbf{x}(k) + \mathbf{d}u(k)$$

then output feedback of the form

$$u(k) = \mathbf{f}^\dagger[\mathbf{y}(k) - \mathbf{d}u(k)] + \rho(k) = [f_1 \quad f_2] \begin{bmatrix} 1 & 0 & -1 \\ -2 & 1 & 1 \end{bmatrix} \begin{bmatrix} x_1(k) \\ x_2(k) \\ x_3(k) \end{bmatrix} + \rho(k)$$

$$= [(f_1 - 2f_2) \quad f_2 \quad (-f_1 + f_2)] \begin{bmatrix} x_1(k) \\ x_2(k) \\ x_3(k) \end{bmatrix} + \rho(k)$$

gives a feedback system governed by

$$
\begin{bmatrix} x_1(k+1) \\ x_2(k+1) \\ x_3(k+1) \end{bmatrix}
$$

$$
= \begin{bmatrix} 0 & 1 & 0 \\ 0 & 0 & 1 \\ (f_1 - 2f_2) & (2 + f_2) & (-1 - f_1 + f_2) \end{bmatrix} \begin{bmatrix} x_1(k) \\ x_2(k) \\ x_3(k) \end{bmatrix} + \begin{bmatrix} 0 \\ 0 \\ 1 \end{bmatrix} \rho(k)
$$

Characteristic polynomials of the form

$$
\lambda^3 + (1 + f_1 - f_2)\lambda^2 + (-2 - f_2)\lambda + (-f_1 + 2f_2) = 0 \tag{4-28}
$$

for any f_1 and f_2 can be achieved with output feedback in this system. As the three roots of equation (4-28) are constrained, it may or may not be possible to choose the constants f_1 and f_2 to achieve acceptable eigenvalue locations for the feedback system.

4.5.3 General Output Feedback

The dual result of placing the eigenvalue of a single-output system with state feedback is that the eigenvalues of a completely observable single-output system having independent inputs to each state equation

$$
\mathbf{x}(k+1) = \mathbf{A}\mathbf{x}(k) + \mathbf{u}(k)
$$
$$
y(k) = \mathbf{c}^\dagger\mathbf{x}(k) \tag{4-29}
$$

can be placed arbitrarily by appropriate choice of the feedback gains $\mathbf{f}$ in

$$
\mathbf{u}(k) = \mathbf{f}y(k) \tag{4-30}
$$

In equation (4-29), a separate component of the n-vector input $\mathbf{u}(k)$ couples to each of the n state equations. This result also applies for a system

$$
\mathbf{x}(k+1) = \mathbf{A}\mathbf{x}(k) + \mathbf{B}\mathbf{u}(k)
$$
$$
y(k) = \mathbf{c}^\dagger\mathbf{x}(k) \tag{4-31}
$$

where $\mathbf{B}$ is of rank n. If the input coupling matrix is square and nonsingular, for example, the choice

$$
\mathbf{u}(k) = \mathbf{B}^{-1}\mathbf{f}y(k) = \mathbf{f}'y(k)
$$

gives the same feedback system

$$
\mathbf{x}(k+1) = (\mathbf{A} + \mathbf{f}\mathbf{c}^\dagger)\mathbf{x}(k)
$$

for equation (4-31) as for equation (4-30).

When such a system is in observable form, the feedback gains are especially easy to determine. For example, the system

$$\begin{bmatrix} x_1(k+1) \\ x_2(k+1) \\ x_3(k+1) \end{bmatrix} = \begin{bmatrix} 2 & 1 & 0 \\ 1 & 0 & 1 \\ 3 & 0 & 0 \end{bmatrix} \begin{bmatrix} x_1(k) \\ x_2(k) \\ x_3(k) \end{bmatrix} + \begin{bmatrix} 2 & -1 & 0 \\ -1 & 1 & 0 \\ 0 & 0 & 1 \end{bmatrix} \begin{bmatrix} u_1(k) \\ u_2(k) \\ u_3(k) \end{bmatrix}$$

$$= \mathbf{A}\mathbf{x}(k) + \mathbf{B}\mathbf{u}(k)$$

$$y(k) = \begin{bmatrix} 1 & 0 & 0 \end{bmatrix} \begin{bmatrix} x_1(k) \\ x_2(k) \\ x_3(k) \end{bmatrix} = \mathbf{c}^\dagger \mathbf{x}(k)$$

with feedback

$$\begin{bmatrix} u_1(k) \\ u_2(k) \\ u_3(k) \end{bmatrix} = \begin{bmatrix} f_1 \\ f_2 \\ f_3 \end{bmatrix} y(k) = \mathbf{B}^{-1} \begin{bmatrix} f_1' \\ f_2' \\ f_3' \end{bmatrix} y(k)$$

is described by

$$\begin{bmatrix} x_1(k+1) \\ x_2(k+1) \\ x_3(k+1) \end{bmatrix} = \begin{bmatrix} (2+f_1') & 1 & 0 \\ (1+f_2') & 0 & 1 \\ (3+f_3') & 0 & 0 \end{bmatrix} \begin{bmatrix} x_1(k) \\ x_2(k) \\ x_3(k) \end{bmatrix} = (\mathbf{A} + \mathbf{f}'\mathbf{c}^\dagger)\mathbf{x}(k)$$

which has the characteristic equation

$$\lambda^3 - (2+f_1')\lambda^2 - (1+f_2')\lambda - (3+f_3') = 0$$

Choosing

$$\mathbf{f}' = \begin{bmatrix} -2 \\ -\frac{3}{4} \\ -3 \end{bmatrix} \qquad \mathbf{f} = \mathbf{B}^{-1}\mathbf{f}' = \begin{bmatrix} -\frac{11}{4} \\ -\frac{7}{2} \\ -3 \end{bmatrix}$$

gives the characteristic equation

$$\lambda \left(\lambda + \frac{1}{2} \right) \left(\lambda - \frac{1}{2} \right) = 0$$

If an nth-order single-output plant has multiple inputs, but not n linearly independent ones, the characteristic equations that can be achieved are constrained, as they are in the case of a multiple-output/single-input plant.

When a completely observable and completely controllable plant has both multiple outputs and multiple inputs, but not n of either, it may or may not be possible to place the feedback system eigenvalues arbitrarily. Certainly, the number of gains to be chosen must be at least as large as the system (and characteristic equation) order. Except in the single-input and the single-output cases, the coefficients of the feedback system characteristic equation are nonlinear functions of the feedback gains.

4.6 Pole Placement with Feedback Compensation

In the previous sections, eigenvalue placement with state and output feedback was discussed in detail. In this section, we present another viewpoint for placing the feedback system poles using a transfer function approach. Although our discussion is limited to single-input and single-output plants, the results can be generalized to the case of plants with multiple inputs and multiple outputs.

4.6.1 Pole Placement Design

Similar to output feedback, the transfer function approach to pole placement design assumes that the measurement outputs of a plant, not the state vector, are available for feedback.

For an nth-order, linear, step-invariant, discrete-time system described by the transfer function $G(z)$, arbitrary placement of the feedback system poles can be accomplished with an mth-order feedback compensator as shown in Figure 4-13. Denoting the numerator and denominator polynomials of the plant transfer function $G(z)$ by $N_p(z)$ and $D_p(z)$, respectively, and denoting the feedback compensator trans-

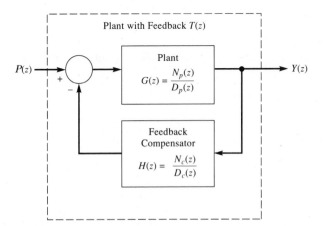

Figure 4-13 Pole placement with feedback compensation.

fer function $H(z)$ by $N_c(z)$ and $D_c(z)$, then the overall transfer function of the system is

$$T(z) = \frac{G(z)}{1 + G(z)H(z)} = \frac{N_p(z)/D_p(z)}{1 + [N_p(z)/D_p(z)][N_c(z)/D_c(z)]}$$

$$= \frac{N_p(z)D_c(z)}{D_p(z)D_c(z) + N_p(z)N_c(z)} = \frac{P(z)}{Q(z)}$$

The overall transfer function $T(z)$ has closed-loop zeros in $P(z)$ that are those of the plant, in $N_p(z)$, together with zeros that are the poles of the feedback compensator, in $D_c(z)$. For a desired set of poles of $T(z)$, given with a multiplicative constant by the polynomial $Q(z)$, then

$$D_p(z)D_c(z) + N_p(z)N_c(z) = Q(z) \qquad \text{(4-32)}$$

Equating the coefficients in equation (4-32) results in $n + m + 1$ linear algebraic equations in $2m + 1$ unknown coefficients of the feedback compensator $H(z)$ and the one multiplying constant of $Q(z)$. In general, for a solution to exist there must be at least as many unknowns as equations

$$n + m + 1 \leq 2m + 2$$

or

$$m \geq n - 1 \qquad \text{(4-33)}$$

where n is the order of the plant and m is the order of the compensator. Equation (4-33) states that the order of the feedback compensator is at least one less than the plant order. If the plant transfer function has *coprime* numerator and denominator polynomials (that is, plant pole-zero cancellations have been made), then a solution is guaranteed to exist.

For example, consider a second-order plant described by the following transfer function:

$$G(z) = \frac{(z + 1)(z + 0.5)}{z(z - 1)} = \frac{N_p(z)}{D_p(z)}$$

Then, according to equation (4-33), a first-order feedback compensator of the form

$$H(z) = \frac{\alpha_1 z + \alpha_2}{z + \alpha_3} = \frac{N_c(z)}{D_c(z)}$$

places the three poles of the plant with feedback at any desired location in the z-plane by appropriate choice of the compensator coefficients α_1, α_2, and α_3.

Suppose it is desired to have all the poles of the plant with feedback at $z = 0.1$, then the desired characteristic equation is

$$Q(z) = \alpha_0(z - 0.1)^3 = \alpha_0(z^3 - 0.3z^2 + 0.03z - 0.001) \qquad \textbf{(4-34)}$$

In terms of the compensator coefficients, the characteristic equation of the plant with feedback is

$$
\begin{aligned}
D_p(z)&D_c(z) + N_p(z)N_c(z) \\
&= z(z - 1)(z + \alpha_3) + (z + 1)(z + 0.5)(\alpha_1 z + \alpha_2) \\
&= (\alpha_1 + 1)z^3 + (1.5\alpha_1 + \alpha_2 + \alpha_3 - 1)z^2 \\
&\quad + (0.5\alpha_1 + 1.5\alpha_2 - \alpha_3)z + 0.5\alpha_2 \qquad \textbf{(4-35)}
\end{aligned}
$$

Equating coefficients in equations (4-34) and (4-35) gives the following set of four linear algebraic equations in four unknowns:

$$
\begin{aligned}
\alpha_1 && + 1 &= \alpha_0 \\
1.5\alpha_1 + && \alpha_2 + \alpha_3 - 1 &= -0.3\alpha_0 \\
0.5\alpha_1 + 1.5\alpha_2 - \alpha_3 && &= 0.03\alpha_0 \\
0.5\alpha_2 && &= -0.001\alpha_0
\end{aligned}
$$

which has a unique solution of

$$\alpha_0 = 1.325 \qquad \alpha_1 = 0.325 \qquad \alpha_2 = -0.00265 \qquad \alpha_3 = 0.1185$$

The plant and resulting feedback compensator are shown in Figure 4-14, and the overall transfer function is

$$
\begin{aligned}
T(z) &= \frac{[(z + 1)(z + 0.5)/z(z - 1)]}{1 + [(z + 1)(z + 0.5)/z(z - 1)][(0.325z - 0.00265)/(z + 0.1185)]} \\
&= \frac{(z + 1)(z + 0.5)(z + 0.1185)}{1.325(z^3 - 0.3z^2 + 0.03z - 0.001)}
\end{aligned}
$$

as desired.

As another example, consider a plant described by the following transfer function

$$G(z) = \frac{z + 2}{(z + \tfrac{1}{2})(z - 2)}$$

This plant is unstable and has a root outside the unit circle. According to equation (4-33), because the plant is second-order, a feedback compensator of order $n - 1 = 1$ of the form

$$H(z) = \frac{\alpha_1 z + \alpha_2}{z + \alpha_3} = \frac{N_c(z)}{D_c(z)}$$

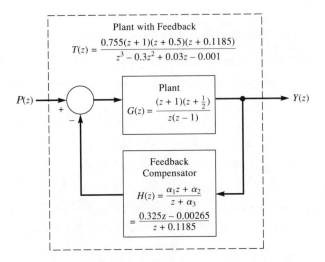

Figure 4-14 Arbitrary pole placement design example.

arbitrarily places the feedback system poles. If it is desired to place all the poles of the plant with feedback at $z = -0.5$, then

$$Q(z) = \alpha_0(z^3 + 1.5z^2 + 0.75z + 0.125)$$
$$= D_p(z)D_c(z) + N_p(z)N_c(z)$$
$$= \left(z + \frac{1}{2}\right)(z - 2)(z + \alpha_3) + (z + 2)(\alpha_1 z + \alpha_2)$$
$$= z^3 + (\alpha_1 + \alpha_3 - 1.5)z^2 + (2\alpha_1 + \alpha_2 - 1.5\alpha_3 - 1)z$$
$$+ 2\alpha_2 - \alpha_3$$

Equating coefficients and letting $\alpha_0 = 1$ results in the following set of three linear algebraic equations in three unknowns:

$$\begin{cases} \alpha_1 \qquad + \quad \alpha_3 - 1.5 = 1.5 \\ 2\alpha_1 + \ \alpha_2 - 1.5\alpha_3 - 1 \ = 0.75 \\ \qquad 2\alpha_2 - \quad \alpha_3 \qquad = 0.125 \end{cases}$$

which has the unique solution

$$\alpha_1 = 1.5625 \qquad \alpha_2 = 0.78125 \qquad \alpha_3 = 1.4375$$

Hence, the resulting feedback compensator has the unstable transfer function

$$H(z) = \frac{1.5625z + 0.78125}{z + 1.4375}$$

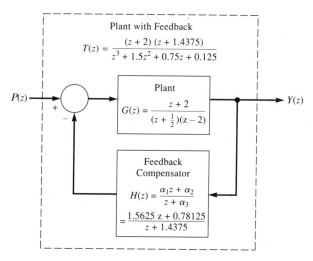

Figure 4-15 Arbitrary pole placement design example.

Here, although the compensator and the plant are both unstable, the combination of the plant and feedback, as shown in Figure 4-15, is stable. There are no problems in using or implementing an unstable feedback compensator provided that pole-zero cancellation of unstable roots is not involved. That the overall transfer function has a zero outside the unit circle requires special consideration later in the tracking portion of the design. This will be discussed in detail in Chapter 6.

4.6.2 Constrained Pole Placement Design

In the previous section, we showed that as far as feedback system pole placement is concerned a feedback compensator of order $n - 1$, where n is the order of the plant, can *always* be designed. It is possible, however, that a lower-order feedback compensator may give acceptable feedback pole locations even though those locations are constrained and not completely arbitrary. This is the thrust of classical control system design, in which increasingly higher order controllers are tested until satisfactory results are obtained.

For the previous example, where the plant transfer function is

$$G(z) = \frac{(z + 1)(z + 0.5)}{z(z - 1)}$$

a zero-order feedback compensator of the form

$$H(z) = K$$

where K is a constant, gives the plant with feedback transfer function

$$T(z) = \frac{G(z)}{1 + G(z)H(z)}$$

$$= \frac{(z + 1)(z + \frac{1}{2})}{(1 + K)z^2 + (1.5K - 1)z + 0.5K}$$

as shown in Figure 4-16. For example, if

$$K = \frac{1}{6}$$

then

$$T(z) = \frac{\frac{6}{7}(z + 1)(z + \frac{1}{2})}{z^2 - (9/14)z + (1/14)}$$

and the overall transfer function has poles at $z = 0.1428$ and $z = 0.5$, which might be an adequate pole placement design. Obviously, arbitrary pole placement cannot be achieved because the number of equations exceeds the number of unknowns in the denominator polynomial.

For plants with multiple inputs and outputs, feedback compensation is more conveniently accomplished in a different *observer* feedback form. This is the subject of Chapter 5, where we show that if there are r linearly independent plant outputs, the feedback compensator order need only be $n - r$, and if there are multiple inputs, the compensator order may usually be reduced still further.

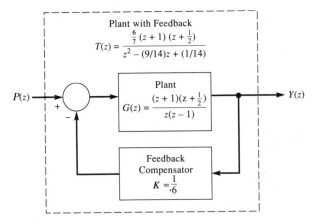

Figure 4-16 Constrained pole placement design example.

4.7 Quadratic Optimal Regulation

We showed in previous sections that, provided the plant is completely controllable, a feedback gain matrix $\mathbf{E}$ can always be chosen so that all of the eigenvalues of the feedback system are at arbitrary locations selected by the designer. Using the controllable canonical form, it can be proved that for nth-order, single-input systems specifying all the n eigenvalues of the feedback system *uniquely* determines the n elements of the feedback gain vector. This uniqueness, however, is destroyed when dealing with multiple-input systems because there are many feedback gain matrices that lead to the same set of feedback eigenvalues. The process of choosing an *optimum* feedback gain matrix from among the many possible gain matrices is developed in this section.

4.7.1 Optimal Control

To elucidate the concept of optimal control, consider the example of a space satellite. The state $\mathbf{x}$ might represent the satellite's orbital position and velocity, and the input $\mathbf{u}$ might be thrusts from positioning rockets. Although any number of $\mathbf{u}$ sequences could bring the satellite to a desired orbital state, one that does so with a minimum expenditure of energy or fuel is clearly desirable.

As another simple numerical example, consider the system

$$\begin{bmatrix} x_1(k+1) \\ x_2(k+1) \end{bmatrix} = \begin{bmatrix} 1 & 1 \\ 2 & 0 \end{bmatrix} \begin{bmatrix} x_1(k) \\ x_2(k) \end{bmatrix} + \begin{bmatrix} 2 \\ -1 \end{bmatrix} u(k)$$

with

$$\mathbf{x}(0) = \begin{bmatrix} 0 \\ -4 \end{bmatrix}$$

and with the state at step 3 required to be

$$\mathbf{x}(3) = \begin{bmatrix} 6 \\ 1 \end{bmatrix}$$

The inputs $u(0)$, $u(1)$, and $u(2)$ in

$$\mathbf{x}(3) = \mathbf{A}^3\mathbf{x}(0) + \mathbf{A}^2\mathbf{b}u(0) + \mathbf{A}\mathbf{b}u(1) + \mathbf{b}u(2)$$

or

$$\begin{bmatrix} 6 \\ 1 \end{bmatrix} = \begin{bmatrix} -12 \\ -8 \end{bmatrix} + \begin{bmatrix} 5 \\ 2 \end{bmatrix} u(0) + \begin{bmatrix} 1 \\ 4 \end{bmatrix} u(1) + \begin{bmatrix} 2 \\ -1 \end{bmatrix} u(2)$$

are to be selected. Rather than arbitrarily choosing $u(0)$ and then solving for the resulting necessary $u(1)$ and $u(2)$, suppose that we seek the solution for which

$$J = u^2(0) + u^2(1) + u^2(2)$$

is minimum. The solution in terms of $u(0)$ is

$$\begin{cases} u(1) + 2u(2) = 18 - 5u(0) \\ 4u(1) - u(2) = 9 - 2u(0) \end{cases}$$

or

$$u(1) = \frac{\begin{vmatrix} 18 - 5u(0) & 2 \\ 9 - 2u(0) & -1 \end{vmatrix}}{\begin{vmatrix} 1 & 2 \\ 4 & -1 \end{vmatrix}} = \frac{-36 + 9u(0)}{-9} = 4 - u(0)$$

$$u(2) = \frac{\begin{vmatrix} 1 & 18 - 5u(0) \\ 4 & 9 - 2u(0) \end{vmatrix}}{-9} = \frac{-63 + 18u(0)}{-9} = 7 - 2u(0)$$

In terms of $u(0)$, the performance index is

$$J = u^2(0) + u^2(1) + u^2(2) = u^2(0) + [4 - u(0)]^2 + [7 - 2u(0)]^2$$
$$= 6u^2(0) - 36u(0) + 65$$

and its minimum with respect to $u(0)$ is given by

$$\frac{\partial J}{\partial u(0)} = 12u(0) - 36 = 0 \qquad u(0) = 3$$

Then

$$u(1) = 4 - u(0) = 1$$
$$u(2) = 7 - 2u(0) = 1$$

There are many variations on this theme of *optimal control*, including problems in which the performance measure J includes terms involving the error between desired and actual states and in which the final step number is not fixed. Next, we derive and discuss the solution to a highly useful general optimal control problem.

4.7.2 Principle of Optimality

Another approach to the state feedback portion of tracking system design is that of *optimal regulation,* in which the plant feedback is chosen to minimize a scalar performance measure that weights the control input and the error from zero of the plant state at each step. Instead of choosing state feedback gains to give zero-input response convergence properties directly, the gains are selected to minimize a performance measure, indirectly resulting in whatever convergence results from the minimization. The designer trades selecting the nature of the decay of the feedback system's zero-input response at each step for selecting the step-varying weighting matrices in the performance measure.

The discrete-time, linear-quadratic, optimal control problem is to find the inputs $\mathbf{u}(0), \mathbf{u}(1), \ldots, \mathbf{u}(N-1)$ to the plant with linear state equations

$$\mathbf{x}(k+1) = \mathbf{A}(k)\mathbf{x}(k) + \mathbf{B}(k)\mathbf{u}(k)$$

such that a scalar quadratic performance measure (or *cost function*)

$$J = \mathbf{x}^\dagger(N)\mathbf{P}(N)\mathbf{x}(N) + \sum_{k=0}^{N-1} [\mathbf{x}^\dagger(k)\mathbf{Q}(k)\mathbf{x}(k) + \mathbf{u}^\dagger(k)\mathbf{R}(k)\mathbf{u}(k)] \qquad \textbf{(4-36)}$$

is minimized. The matrix $\mathbf{P}(N)$, the matrices $\mathbf{Q}(0), \mathbf{Q}(1), \ldots,$ $\mathbf{Q}(N-1)$, and the matrices $\mathbf{R}(0), \mathbf{R}(1), \ldots, \mathbf{R}(N-1)$ are each taken to be symmetric because each defines a quadratic form. Each is assumed to be positive semidefinite, which means that contributions to J by each of the individual terms are never negative.

Solution of the linear-quadratic optimal control problem is obtained by applying the *principle of optimality,* a technique developed by Richard Bellman in the 1950s in connection with his invention of *dynamic programming*. This method also offers insight into the backward-in-step nature of the solution. To solve the problem, we use the partial performance measures $J(i, j)$, which are the costs of control from step i to a later step j

$$J(i, j) = \sum_{k=i}^{j-1} [\mathbf{x}^\dagger(k)\mathbf{Q}(k)\mathbf{x}(k) + \mathbf{u}^\dagger(k)\mathbf{R}(k)\mathbf{u}(k)]$$

when step j is not the last step N, and

$$J(i, N) = \mathbf{x}^\dagger(N)\mathbf{P}(N)\mathbf{x}(N) + \sum_{k=i}^{N-1} [\mathbf{x}^\dagger(k)\mathbf{Q}(k)\mathbf{x}(k) + \mathbf{u}^\dagger(k)\mathbf{R}(k)\mathbf{u}(k)]$$

when the last step is N.

The problem at hand is to find, as a function of the initial state $\mathbf{x}(0)$, the sequence of inputs $\mathbf{u}(0), \mathbf{u}(1), \ldots, \mathbf{u}(N-1)$ that results in minimal cost from the zeroth step to the Nth step.

$$\min_{\mathbf{u}(0), \ldots, \mathbf{u}(N-1)} \left\{ J(0, N) \right\}$$

$$= \min_{\mathbf{u}(0), \ldots, \mathbf{u}(N-1)}$$

$$\left\{ \mathbf{x}^t(N)\mathbf{P}(N)\mathbf{x}(N) + \sum_{k=0}^{N-1} [\mathbf{x}^t(k)\mathbf{Q}(k)\mathbf{x}(k) + \mathbf{u}^t(k)\mathbf{R}(k)\mathbf{u}(k)] \right\}$$

Separating the cost of the last control $\mathbf{u}(N-1)$ from the rest of the cost

$$\min_{\mathbf{u}(0), \ldots, \mathbf{u}(N-1)} \left\{ J(0, N) \right\}$$

$$= \min_{\mathbf{u}(0), \ldots, \mathbf{u}(N-1)} \left\{ J(0, N-1) + J(N-1, N) \right\}$$

we note that $\mathbf{u}(N-1)$ does not affect $J(0, N-1)$ so that

$$\min_{\mathbf{u}(0), \ldots, \mathbf{u}(N-1)} \left\{ J(0, N) \right\}$$

$$= \min_{\mathbf{u}(0), \ldots, \mathbf{u}(N-2)} \left\{ J(0, N-1) \right\}$$

$$+ \min_{\mathbf{u}(0), \ldots, \mathbf{u}(N-1)} \left\{ J(N-1, N) \right\}$$

$$= \min_{\mathbf{u}(0), \ldots, \mathbf{u}(N-2)}$$

$$\left\{ J(0, N-1) + \min_{\mathbf{u}(N-1)} [J(N-1, N)] \right\}$$

In general

$$\min_{\mathbf{u}(0), \ldots, \mathbf{u}(N-1)} \left\{ J(0, N) \right\}$$

$$= \min_{\mathbf{u}(0), \ldots, \mathbf{u}(i-1)}$$

$$\left\{ J(0, i) + \min_{\mathbf{u}(i), \ldots, \mathbf{u}(N-1)} [J(i, N)] \right\} \tag{4-37}$$

The term involving $J(i, N)$ is a minimization problem of the same type. It begins at step i with the initial state $\mathbf{x}(i)$. Relation (4-37) is a statement of the principle of optimality, which in words is: whatever the initial state $\mathbf{x}(0)$ of the system and the initial control sequence $\mathbf{u}(0)$, $\ldots$, $\mathbf{u}(i - 1)$, the remaining control sequence $\mathbf{u}(i)$, $\ldots$, $\mathbf{u}(N - 1)$ is optimal with regard to the state $\mathbf{x}(i)$ that results from the initial control sequence.

To apply the principle of optimality, one begins at the next-to-last step $N - 1$ and finds the last input $\mathbf{u}(N - 1)$ that minimizes the cost of control from step $N - 1$ to step N, $J(N - 1, N)$, as a function of the beginning state for that step $\mathbf{x}(N - 1)$. Then, the input $\mathbf{u}(N - 2)$ is found that minimizes $J(N - 2, N)$ when $\mathbf{u}(N - 1)$ is as previously determined. One proceeds in this manner finding one control vector at a time, from the last to the first, as a function of the system's state. In consequence of the quadratic nature of the cost function, the optimal input at each step turns out to be a linear transformation of the system state at that step, so the result is in the form of state feedback. The transformation is generally not the same at each step, so the optimal control solution is step-varying feedback, whether or not the controlled system is step-invariant.

At the next-to-last step, substituting

$$\mathbf{x}(N) = \mathbf{A}(N - 1)\mathbf{x}(N - 1) + \mathbf{B}(N - 1)\mathbf{u}(N - 1)$$

the cost of control to the last step is

$$
\begin{aligned}
J(N - 1, N) &= \mathbf{x}^t(N)\mathbf{P}(N)\mathbf{x}(N) + \mathbf{x}^t(N - 1)\mathbf{Q}(N - 1)\mathbf{x}(N - 1) \\
&\quad + \mathbf{u}^t(N - 1)\mathbf{R}(N - 1)\mathbf{u}(N - 1) \\
&= \mathbf{x}^t(N - 1)[\mathbf{A}^t(N - 1)\mathbf{P}(N)\mathbf{A}(N - 1) \\
&\quad + \mathbf{Q}(N - 1)]\mathbf{x}(N - 1) \\
&\quad + \mathbf{x}^t(N - 1)\mathbf{A}^t(N - 1)\mathbf{P}(N)\mathbf{B}(N - 1)\mathbf{u}(N - 1) \\
&\quad + \mathbf{u}^t(N - 1)\mathbf{B}^t(N - 1)\mathbf{P}(N)\mathbf{A}(N - 1)\mathbf{x}(N - 1) \\
&\quad + \mathbf{u}^t(N - 1)[\mathbf{B}^t(N - 1)\mathbf{P}(N)\mathbf{B}(N - 1) \\
&\quad + \mathbf{R}(N - 1)]\mathbf{u}(N - 1)
\end{aligned}
$$

Taking the partial derivatives of this expression with respect to each of the r scalar components of $\mathbf{u}(N - 1)$ gives

$$
\left[\frac{\partial J}{\partial \mathbf{u}(N - 1)}\right]^t =
\begin{bmatrix}
\dfrac{\partial J}{\partial u_1(N - 1)} \\[2ex]
\dfrac{\partial J}{\partial u_2(N - 1)} \\[2ex]
\vdots \\[2ex]
\dfrac{\partial J}{\partial u_r(N - 1)}
\end{bmatrix}
$$

$$
\begin{aligned}
= &\ \mathbf{0} + \mathbf{B}^{\dagger}(N - 1)\mathbf{P}^{\dagger}(N)\mathbf{A}(N - 1)\mathbf{x}(N - 1) \\
&+ \mathbf{B}^{\dagger}(N - 1)\mathbf{P}(N)\mathbf{A}(N - 1)\mathbf{x}(N - 1) \\
&+ [\mathbf{B}^{\dagger}(N - 1)\mathbf{P}(N)\mathbf{B}(N - 1) \\
&+ \mathbf{R}(N - 1)]\mathbf{u}(N - 1) \\
&+ [\mathbf{B}^{\dagger}(N - 1)\mathbf{P}^{\dagger}(N)\mathbf{B}(N - 1) \\
&+ \mathbf{R}^{\dagger}(N - 1)]\mathbf{u}(N - 1) \\
= &\ 2\mathbf{B}^{\dagger}(N - 1)\mathbf{P}(N)\mathbf{A}(N - 1)\mathbf{x}(N - 1) \\
&+ 2[\mathbf{B}^{\dagger}(N - 1)\mathbf{P}(N)\mathbf{B}(N - 1) \\
&+ \mathbf{R}(N - 1)]\mathbf{u}(N - 1)
\end{aligned}
$$

using partial derivative identities from Table A-1 of Appendix A and

$$\mathbf{P}^{\dagger}(N) = \mathbf{P}(N) \qquad \mathbf{R}^{\dagger}(N - 1) = \mathbf{R}(N - 1)$$

because of the symmetry of $\mathbf{P}$ and $\mathbf{R}$. Equating this result to zero,

$$
\begin{aligned}
\mathbf{u}(N - 1) &= -[\mathbf{B}^{\dagger}(N - 1)\mathbf{P}(N)\mathbf{B}(N - 1) + \mathbf{R}(N - 1)]^{-1} \\
&\quad \times \mathbf{B}^{\dagger}(N - 1)\mathbf{P}(N)\mathbf{A}(N - 1)\mathbf{x}(N - 1) \\
&= \mathbf{E}(N - 1)\mathbf{x}(N - 1)
\end{aligned}
$$

where

$$
\begin{aligned}
\mathbf{E}(N - 1) &= -[\mathbf{B}^{\dagger}(N - 1)\mathbf{P}(N)\mathbf{B}(N - 1) + \mathbf{R}(N - 1)]^{-1} \\
&\quad \times \mathbf{B}^{\dagger}(N - 1)\mathbf{P}(N)\mathbf{A}(N - 1) \tag{4-38}
\end{aligned}
$$

Because of the quadratic nature of the performance measure, the solution is a global minimum.

The optimal last-step input $\mathbf{u}(N - 1)$ is a linear transformation of the state $\mathbf{x}(N - 1)$. The transformation (or gain) $\mathbf{E}(N - 1)$ of the state feedback is given by the complicated but straightforward expression (4-38). The matrix inversion involved is of an $r \times r$ matrix, where r is the number of plant inputs. The number r is usually not as large as n, the plant order. Existence of the inverse in equation (4-38) is guaranteed if $\mathbf{R}(N - 1)$ is positive definite rather than the assumed positive semidefinite; that is, when any nonzero control input contributes to the performance measure. Its existence is also guaranteed under other conditions, such as when $\mathbf{P}(N)$ is positive definite and the columns of $\mathbf{B}(N - 1)$ are linearly independent. If the inverse does not exist, it means that the solution for the optimal control input is not unique.

When the optimal last-step input $\mathbf{u}(N - 1)$ is used, the cost of control from the next-to-last step to the last step is

$$\min_{\mathbf{u}(N - 1)} \left\{ J(N - 1), N \right\}$$

$$
\begin{aligned}
&= \mathbf{x}^\dagger(N)\mathbf{P}(N)\mathbf{x}(N) + \mathbf{x}^\dagger(N - 1)\mathbf{Q}(N - 1)\mathbf{x}(N - 1) \\
&\quad + \mathbf{u}^\dagger(N - 1)\mathbf{R}(N - 1)\mathbf{u}(N - 1) \\
&= \mathbf{x}^\dagger(N - 1)[\mathbf{A}^\dagger(N - 1)\mathbf{P}(N)\mathbf{A}(N - 1) \\
&\quad + \mathbf{Q}(N - 1)]\mathbf{x}(N - 1) \\
&\quad + \mathbf{x}^\dagger(N - 1)\mathbf{A}^\dagger(N - 1)\mathbf{P}(N)\mathbf{B}(N - 1)\mathbf{E}(N - 1)\mathbf{x}(N - 1) \\
&\quad + \mathbf{x}^\dagger(N - 1)\mathbf{E}^\dagger(N - 1)\mathbf{B}^\dagger(N - 1)\mathbf{P}(N)\mathbf{A}(N - 1)\mathbf{x}(N - 1) \\
&\quad + \mathbf{x}^\dagger(N - 1)\mathbf{E}^\dagger(N - 1)[\mathbf{B}^\dagger(N - 1)\mathbf{P}(N)\mathbf{B}(N - 1) \\
&\quad + \mathbf{R}(N - 1)]\mathbf{E}(N - 1)\mathbf{x}(N - 1) \\
&= \mathbf{x}^\dagger(N - 1)\{[\mathbf{A}(N - 1) + \mathbf{B}(N - 1)\mathbf{E}(N - 1)]^\dagger\mathbf{P}(N) \\
&\quad\quad \times [\mathbf{A}(N - 1) + \mathbf{B}(N - 1)\mathbf{E}(N - 1)] \\
&\quad\quad + \mathbf{E}^\dagger(N - 1)\mathbf{R}(N - 1)\mathbf{E}(N - 1) \\
&\quad\quad + \mathbf{Q}(N - 1)\}\mathbf{x}(N - 1) \\
&= \mathbf{x}^\dagger(N - 1)\mathbf{P}(N - 1)\mathbf{x}(N - 1)
\end{aligned}
$$

where

$$
\begin{aligned}
\mathbf{P}(N - 1) &= [\mathbf{A}(N - 1) + \mathbf{B}(N - 1)\mathbf{E}(N - 1)]^\dagger \\
&\quad \times \mathbf{P}(N)[\mathbf{A}(N - 1) + \mathbf{B}(N - 1)\mathbf{E}(N - 1)] \\
&\quad + \mathbf{E}^\dagger(N - 1)\mathbf{R}(N - 1)\mathbf{E}(N - 1) + \mathbf{Q}(N - 1)
\end{aligned}
$$

It is significant that the minimum cost is quadratic in $\mathbf{x}(N - 1)$ because finding the optimal input $\mathbf{u}(N - 2)$ is another problem of exactly the same form as the last-step problem.

Applying the principle of optimality, the optimum input $\mathbf{u}(N - 2)$ is given by

$$
\begin{aligned}
&\min_{\mathbf{u}(N - 2),\, \mathbf{u}(N - 1)} \left\{ J(N - 2), N \right\} \\
&= \min_{\mathbf{u}(N - 2)} \left\{ J(N - 2, N - 1) + \min_{\mathbf{u}(N - 1)} [J(N - 1), N] \right\} \\
&= \min_{\mathbf{u}(N - 2)} \left\{ \mathbf{x}^\dagger(N - 2)\mathbf{Q}(N - 2)\mathbf{x}(N - 2) + \mathbf{u}^\dagger(N - 2)\mathbf{u}(N - 2) \right. \\
&\quad\quad \left. + \mathbf{x}^\dagger(N - 1)\mathbf{P}(N - 1)\mathbf{x}(N - 1) \right\}
\end{aligned}
$$

which is of the form of the previous problem, with the indices of each of the quantities moved back one step to $N - 2$. Using the previous results, the optimal input is given by

$$
\mathbf{u}(N - 2) = \mathbf{E}(N - 2)\mathbf{x}(N - 2)
$$

where

$$\mathbf{E}(N - 2) = -[\mathbf{B}^\dagger(N - 2)\mathbf{P}(N - 1)\mathbf{B}(N - 2) + \mathbf{R}(N - 2)]^{-1}$$
$$\times \mathbf{B}^\dagger(N - 2)\mathbf{P}(N - 1)\mathbf{A}(N - 2)$$

and where $\mathbf{P}(N - 1)$ is as given above. The cost of using the optimal controls $\mathbf{u}(N - 2)$ and $\mathbf{u}(N - 1)$ is

$$\min_{\mathbf{u}(N - 2), \, \mathbf{u}(N - 1)} \left\{ J(N - 2), N \right\} = \mathbf{x}^\dagger(N - 2)\mathbf{P}(N - 2)\mathbf{x}(N - 2)$$

where

$$\mathbf{P}(N - 2) = [\mathbf{A}(N - 2) + \mathbf{B}(N - 2)\mathbf{E}(N - 2)]^\dagger \mathbf{P}(N - 1)$$
$$\times [\mathbf{A}(N - 2) + \mathbf{B}(N - 2)\mathbf{E}(N - 2)]$$
$$+ \mathbf{E}^\dagger(N - 2)\mathbf{R}(N - 2)\mathbf{E}(N - 2) + \mathbf{Q}(N - 2)$$

The optimal input $\mathbf{u}(N - 3)$ is the solution to a problem of the same form, and so on.

This recursive calculation of the optimal feedback gains for the linear-quadratic regulator is summarized in Table 4-2. Beginning with $\mathbf{P}(N)$, the last feedback gain matrix $\mathbf{E}(N - 1)$ is calculated. Using $\mathbf{E}(N - 1)$, the matrix $\mathbf{P}(N - 1)$ is computed. Then all of the indices are stepped backward one step and, with $\mathbf{P}(N - 1)$, the feedback gain matrix $\mathbf{E}(N - 2)$ is calculated. Using $\mathbf{E}(N - 2)$, $\mathbf{P}(N - 2)$ is calculated. The cycle is continued until $\mathbf{E}(0)$ is found. A formidable amount of algebraic computation is required; the user should therefore have digital computer aid for all but the lowest-order problems.

4.7.3 First-Order Examples

A program in the BASIC language for computing the optimal feedback control gains for a first-order step-invariant plant

$$x(k + 1) = ax(k) + bu(k)$$
$$u(k) = e(k)x(k)$$

with performance measure of the form

$$J = Px^2(N) + \sum_{i=0}^{N-1} [Qx^2(i) + Ru^2(i)]$$

is listed in Table 4-3. The solution for the specific plant

$$x(k + 1) = 0.5x(k) + 3u(k)$$

Table 4-2 Procedure for Backward-in-Time Calculation of Optimal Quadratic Regulator Gains

For the plant

$$\mathbf{x}(k + 1) = \mathbf{A}(k)\mathbf{x}(k) + \mathbf{B}(k)\mathbf{u}(k)$$

with state feedback

$$\mathbf{u}(k) = \mathbf{E}(k)\mathbf{x}(k)$$

and performance measure

$$J = \mathbf{x}^t(N)\mathbf{P}(N)\mathbf{x}(N) + \sum_{i=0}^{N-1} [\mathbf{x}^t(i)\mathbf{Q}(i)\mathbf{x}(i) + \mathbf{u}^t(i)\mathbf{R}(i)\mathbf{u}(i)]$$

begin with $i = 1$ and the known $\mathbf{P}(N)$

1. $\mathbf{E}(N - i) = -[\mathbf{B}^t(N - i)\mathbf{P}(N + 1 - i)\mathbf{B}(N - i) + \mathbf{R}(N - i)]^{-1}$
   $\times \mathbf{B}^t(N - i)\mathbf{P}(N + 1 - i)\mathbf{A}(N - i)$ (4-39)
2. $\mathbf{P}(N - i) = [\mathbf{A}(N - i) + \mathbf{B}(N - i)\mathbf{E}(N - i)]^t\mathbf{P}(N + 1 - i)$
   $\times [\mathbf{A}(N - i) + \mathbf{B}(N - i)\mathbf{E}(N - i)]$
   $+ \mathbf{E}^t(N - i)\mathbf{R}(N - i)\mathbf{E}(N - i) + \mathbf{Q}(N - i)$ (4-40)
3. Increment i and repeat steps 1, 2, and 3 until $\mathbf{E}(0)$ and (if desired) $\mathbf{P}(0)$ have been calculated.

The minimum performance measure is

$$\min_{\mathbf{u}(0), \ldots, \mathbf{u}(N - 1)} \{J\} = \mathbf{x}^t(0)\mathbf{P}(0)\mathbf{x}(0)$$

Table 4-3 Program in Basic to Calculate Optimal Control Gains for a Step-Invariant System

```
100   PRINT "SCALAR OPTIMAL CONTROL CALCULATION"
120   A = 0.5
130   B = 3
140   P = 10
150   Q = 1
160   R = 4
170   N = 10
180   FOR I = 1 to N
190   e = -B*P*A/(B*P*B + R)
200   P = (A + B*e)*P*(A + B*e) + e*R*e + Q
210   PRINT " "
220   PRINT "AT STEP";N - I
230   PRINT "e = ";e
240   PRINT "P = ";P
250   NEXT I
260   END
```

with

$$N = 10 \qquad P = 10 \qquad Q = 1 \qquad R = 4$$

is listed and plotted in Figure 4-17 for $x(0) = 100$. The feedback gain is step-varying. Moving backward in time, from step $k = N$ toward $k = 0$, there is an initial transient, after which the gain is very nearly constant.

For a step-invariant plant, the optimum feedback gains, from $e(N)$ backward, are not changed if the final step N is changed. An example, when N is changed from 10 to 4 is shown in Figure 4-18. When $Q = 0$ so that there is no weighting of the state error except at the last step, as in Figure 4-19, the feedback gains tend to be very small at first. This lets the plant state decay as much as it can without feedback (and without cost). This would not be case, of course, if the plant were unstable. When $R = 0$ so that there is no weighting of the control effort in the performance measure, the feedback gains make the system deadbeat, as shown in Figure 4-20. The plant state is driven to zero right away so that there is no state error cost on succeeding steps.

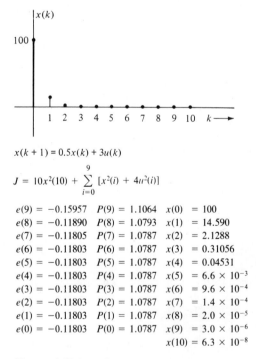

$$x(k + 1) = 0.5x(k) + 3u(k)$$

$$J = 10x^2(10) + \sum_{i=0}^{9} [x^2(i) + 4u^2(i)]$$

$e(9) = -0.15957$	$P(9) = 1.1064$	$x(0) = 100$
$e(8) = -0.11890$	$P(8) = 1.0793$	$x(1) = 14.590$
$e(7) = -0.11805$	$P(7) = 1.0787$	$x(2) = 2.1288$
$e(6) = -0.11803$	$P(6) = 1.0787$	$x(3) = 0.31056$
$e(5) = -0.11803$	$P(5) = 1.0787$	$x(4) = 0.04531$
$e(4) = -0.11803$	$P(4) = 1.0787$	$x(5) = 6.6 \times 10^{-3}$
$e(3) = -0.11803$	$P(3) = 1.0787$	$x(6) = 9.6 \times 10^{-4}$
$e(2) = -0.11803$	$P(2) = 1.0787$	$x(7) = 1.4 \times 10^{-4}$
$e(1) = -0.11803$	$P(1) = 1.0787$	$x(8) = 2.0 \times 10^{-5}$
$e(0) = -0.11803$	$P(0) = 1.0787$	$x(9) = 3.0 \times 10^{-6}$
		$x(10) = 6.3 \times 10^{-8}$

Figure 4-17 A scalar optimal regulation example.

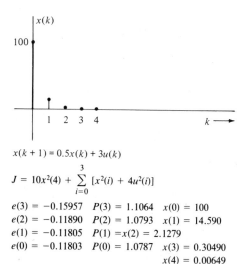

$$x(k + 1) = 0.5x(k) + 3u(k)$$

$$J = 10x^2(4) + \sum_{i=0}^{3} [x^2(i) + 4u^2(i)]$$

$e(3) = -0.15957$	$P(3) = 1.1064$	$x(0) = 100$
$e(2) = -0.11890$	$P(2) = 1.0793$	$x(1) = 14.590$
$e(1) = -0.11805$	$P(1) = x(2) = 2.1279$	
$e(0) = -0.11803$	$P(0) = 1.0787$	$x(3) = 0.30490$
		$x(4) = 0.00649$

Figure 4-18 Scalar optimal regulation example with a different number of steps, $N = 4$.

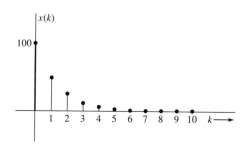

$$x(k + 1) = 0.5x(k) + 3u(k)$$

$$J = 10x^2(10) + \sum_{i=0}^{9} 4u^2(i)$$

$e(9) = -0.15957$	$P(9) = -0.1064$	$x(0) = 100$
$e(8) = -3.2193 \times 10^{-2}$	$P(8) = 2.146 \times 10^{-2}$	$x(1) = 50.000$
$e(7) = -7.6768 \times 10^{-3}$	$P(7) = 5.118 \times 10^{-3}$	$x(2) = 25.000$
$e(6) = -1.8974 \times 10^{-3}$	$P(6) = 1.265 \times 10^{-3}$	$x(3) = 12.500$
$e(5) = -4.72977 \times 10^{-4}$	$P(5) = 3.153 \times 10^{-4}$	$x(4) = 6.249$
$e(4) = -1.18160 \times 10^{-5}$	$P(4) = 7.877 \times 10^{-5}$	$x(5) = 3.122$
$e(3) = -2.95349 \times 10^{-5}$	$P(3) = 1.969 \times 10^{-5}$	$x(6) = 1.557$
$e(2) = -7.38339 \times 10^{-6}$	$P(2) = 4.922 \times 10^{-6}$	$x(7) = 0.769$
$e(1) = -1.84583 \times 10^{-6}$	$P(1) = 1.231 \times 10^{-6}$	$x(8) = 0.367$
$e(0) = -4.61456 \times 10^{-7}$	$P(0) = 3.076 \times 10^{-7}$	$x(9) = 0.148$
		$x(10) = 3.15 \times 10^{-3}$

Figure 4-19 Optimal regulation when there is no state error cost except at the last step.

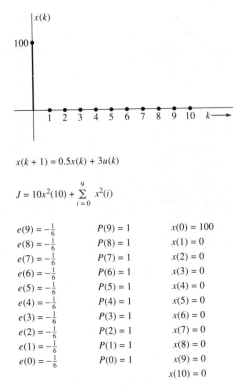

$x(k + 1) = 0.5x(k) + 3u(k)$

$$J = 10x^2(10) + \sum_{i=0}^{9} x^2(i)$$

$e(9) = -\frac{1}{6}$	$P(9) = 1$	$x(0) = 100$
$e(8) = -\frac{1}{6}$	$P(8) = 1$	$x(1) = 0$
$e(7) = -\frac{1}{6}$	$P(7) = 1$	$x(2) = 0$
$e(6) = -\frac{1}{6}$	$P(6) = 1$	$x(3) = 0$
$e(5) = -\frac{1}{6}$	$P(5) = 1$	$x(4) = 0$
$e(4) = -\frac{1}{6}$	$P(4) = 1$	$x(5) = 0$
$e(3) = -\frac{1}{6}$	$P(3) = 1$	$x(6) = 0$
$e(2) = -\frac{1}{6}$	$P(2) = 1$	$x(7) = 0$
$e(1) = -\frac{1}{6}$	$P(1) = 1$	$x(8) = 0$
$e(0) = -\frac{1}{6}$	$P(0) = 1$	$x(9) = 0$
		$x(10) = 0$

Figure 4-20 Optimal regulation when there is no control cost.

These simple examples serve to demonstrate that an optimal quadratic feedback regulator performs reasonably and relatively well. In this approach, the designer selects the performance measure instead of choosing state feedback convergence properties directly.

4.7.4 Roll Attitude Control of a Spacecraft

The formation of an optimal control problem sometimes is a natural consequence of the objectives of a control system design. A simplified state space model for the roll attitude control of a spacecraft is

$$\begin{bmatrix} \dot{x}_1(t) \\ \dot{x}_2(t) \end{bmatrix} = \begin{bmatrix} 0 & 1 \\ 0 & 0 \end{bmatrix} \begin{bmatrix} x_1(t) \\ x_2(t) \end{bmatrix} + \begin{bmatrix} 0 \\ \dfrac{1}{J} \end{bmatrix} u(t) \tag{4-41}$$

$$y(t) = \begin{bmatrix} 1 & 0 \end{bmatrix} \begin{bmatrix} x_1(t) \\ x_2(t) \end{bmatrix}$$

where

x_1 = Roll attitude of the spacecraft in radians

x_2 = Roll rate in radians per second

u = Control torque about the roll axis produced by the thrusters in foot-pounds

and

J = Moment of inertia of the vehicle about the roll axis at the vehicle center of mass in slug-feet squared.

The s-transfer function relating the spacecraft attitude to the torque input is

$$G(s) = \frac{Y(s)}{U(s)} = \frac{1}{Js^2}$$

indicating that the open-loop system has two poles at the origin of the s-plane. When the control $u(t)$ is periodically sampled and held, the discrete-time equivalence of equation (4-41) is

$$\mathbf{x}[(k + 1)T] = \mathbf{A}\mathbf{x}(kT) + \mathbf{b}u(kT)$$
$$y(kT) = \mathbf{c}^t\mathbf{x}(kT)$$

where, using equation (3-20),

$$\mathbf{A} = \mathbf{I} + \mathcal{A}T + \frac{\mathcal{A}^2T^2}{2!} + \cdots$$

$$= \begin{bmatrix} 1 & 0 \\ 0 & 1 \end{bmatrix} + \begin{bmatrix} 0 & 1 \\ 0 & 0 \end{bmatrix} T = \begin{bmatrix} 1 & T \\ 0 & 1 \end{bmatrix}$$

because $\mathcal{A}^2 = \mathbf{0}$, and using equation (3-21a)

$$\mathbf{b} = \left[\mathbf{I}T + \frac{\mathcal{A}T^2}{2!} + \frac{\mathcal{A}^2T^3}{3!} + \cdots + \frac{\mathcal{A}^iT^{i+1}}{(i + 1)!} + \cdots \right] \boldsymbol{\ell}$$

$$= \left\{ \begin{bmatrix} T & 0 \\ 0 & T \end{bmatrix} + \begin{bmatrix} 0 & 1 \\ 0 & 0 \end{bmatrix} \frac{T^2}{2!} \right\} \begin{bmatrix} 0 \\ \frac{1}{J} \end{bmatrix}$$

$$= \begin{bmatrix} \dfrac{T^2}{2J} \\ \dfrac{T}{J} \end{bmatrix}$$

Therefore, the discrete-time model of the spacecraft roll axis is

$$\begin{bmatrix} x_1(kT + T) \\ x_2(kT + T) \end{bmatrix} = \begin{bmatrix} 1 & T \\ 0 & 1 \end{bmatrix} \begin{bmatrix} x_1(kT) \\ x_2(kT) \end{bmatrix} + \begin{bmatrix} \dfrac{T^2}{2J} \\ \dfrac{T}{J} \end{bmatrix} u(kT)$$

$$y(kT) = \begin{bmatrix} 1 & 0 \end{bmatrix} \begin{bmatrix} x_1(kT) \\ x_2(kT) \end{bmatrix}$$

Normalizing the torque input by the moment of inertia,

$$u_n(kT) = \frac{1}{J} u(kT)$$

and for a sampling rate $T = 0.05$ sec, the model becomes

$$\begin{bmatrix} x_1(k + 1) \\ x_2(k + 1) \end{bmatrix} = \begin{bmatrix} 1 & 0.05 \\ 0 & 1 \end{bmatrix} \begin{bmatrix} x_1(k) \\ x_2(k) \end{bmatrix} + \begin{bmatrix} 0.00125 \\ 0.05 \end{bmatrix} u_n(k)$$

$$y(k) = \begin{bmatrix} 1 & 0 \end{bmatrix} \begin{bmatrix} x_1(k) \\ x_2(k) \end{bmatrix} \qquad\qquad \textbf{(4-42)}$$

When it is desired to bring the spacecraft from an initial roll attitude and rate to near zero attitude and rate in a fixed number of steps,
the optimal regulator solution is attractive. The state variables can be
defined so that the zero attitude is any angle desired. The performance
measure

$$J = 100x_1^2(40) + 10x_2^2(40) + \sum_{i=0}^{39} [100x_1^2(i) + 10x_2^2(i) + u_n^2(i)]$$

is for a 41-step interval from step 0 through step 40 with the control
input u_n applied at steps 0 through 39. The final penalties at step 40 for
attitude and rate errors have weights of 100 and 10, respectively. In
terms of matrices

$$J = \mathbf{x}^\dagger(40)\mathbf{P}(40)\mathbf{x}(40) + \sum_{i=0}^{39} [\mathbf{x}^\dagger(i)\mathbf{Q}\mathbf{x}(i) + u^\dagger(i)Ru(i)]$$

where

$$\mathbf{P}(40) = \begin{bmatrix} 100 & 0 \\ 0 & 10 \end{bmatrix} \quad \mathbf{Q} = \begin{bmatrix} 100 & 0 \\ 0 & 10 \end{bmatrix} \quad R = 1$$

The optimal quadratic regulator gains are given by

$$\mathbf{e}^\dagger(39) = -[\mathbf{b}^\dagger\mathbf{P}(40)\mathbf{b} + R]^{-1}\mathbf{b}^\dagger\mathbf{P}(40)\mathbf{A}$$

$$\mathbf{P}(39) = [\mathbf{A} + \mathbf{b}\mathbf{e}^\dagger(39)]^\dagger\mathbf{P}(40)[\mathbf{A} + \mathbf{b}\mathbf{e}^\dagger(39)] + \mathbf{e}(39)R\mathbf{e}^\dagger(39) + \mathbf{Q}$$

$$\mathbf{e}^\dagger(38) = -[\mathbf{b}^\dagger\mathbf{P}(39)\mathbf{b} + R]^{-1}\mathbf{b}^\dagger\mathbf{P}(39)\mathbf{A}$$

$$\mathbf{P}(38) = [\mathbf{A} + \mathbf{b}\mathbf{e}^\dagger(38)]^\dagger\mathbf{P}(39)[\mathbf{A} + \mathbf{b}\mathbf{e}^\dagger(38)] + \mathbf{e}(38)R\mathbf{e}^\dagger(38) + \mathbf{Q}$$

and so on, through

$$\mathbf{e}^\dagger(0) = -[\mathbf{b}^\dagger\mathbf{P}(1)\mathbf{b} + R]^{-1}\mathbf{b}^\dagger\mathbf{P}(1)\mathbf{A}$$

Table 4-4 gives numerical results for the gains $\mathbf{e}^\dagger$. For the initial conditions

$$x_1(0) = 0.1$$

$$x_2(0) = 0$$

the performance of this regulator is shown in Figure 4-21.

Table 4-4 Optimal Feedback Gains for the Spacecraft Roll Attitude Control

Step	e_1	e_2
39	−0.1219	−0.4938
38	−0.4728	−0.9811
37	−1.0185	−1.4593
36	−1.7124	−1.9237
35	−2.5002	−2.3671
$\vdots$		
5	−8.7182	−5.0037
4	−8.7190	−5.0040
3	−8.7196	−5.0042
2	−8.7201	−5.0044
1	−8.7204	−5.0045
0	−8.7207	−5.0046

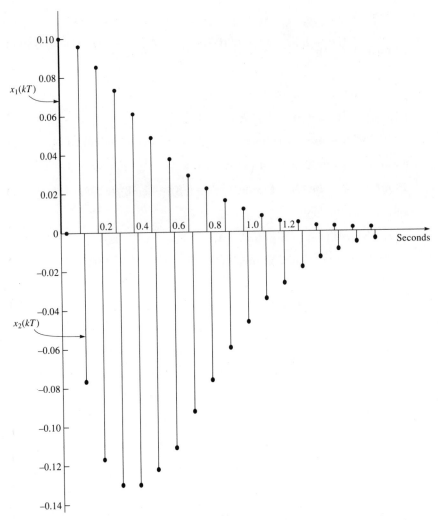

Figure 4-21 Representative performance of the optimal roll attitude control for the spacecraft.

4.8 Closed-Form Solution for Optimal Gain

The procedure summarized in Table 4-2 for calculating the optimal regulator gain results, in general, in a numerical sequence of gain matrices. When the matrices **A, B, Q,** and **R** are constant, it may be possible to generate an analytic expression for the optimal gain from which the numerical value can be calculated for any point of time.

4.8.1 The Approach

Consider an nth-order, completely controllable, step-invariant system described by

$$\mathbf{x}(k + 1) = \mathbf{A}\mathbf{x}(k) + \mathbf{B}\mathbf{u}(k)$$

where $\mathbf{A}$ and $\mathbf{B}$ are constant matrices of appropriate dimensions, and $\mathbf{A}$ is nonsingular. As in Section 4.7, our objective here is to choose the control sequence $\mathbf{u}(0), \mathbf{u}(1), \ldots, \mathbf{u}(N - 1)$, such that the scalar quadratic performance measure

$$J = \mathbf{x}^\dagger(N)\mathbf{P}(N)\mathbf{x}(N) + \sum_{k=0}^{N-1} [\mathbf{x}^\dagger(k)\mathbf{Q}\mathbf{x}(k) + \mathbf{u}^\dagger(k)\mathbf{R}\mathbf{u}(k)] \qquad \text{(4-43)}$$

is minimized where $\mathbf{Q}$ and $\mathbf{R}$ are constant matrices and $\mathbf{P}(N)$ and $\mathbf{R}$ are positive definite.

We start the development from equations (4-39) and (4-40) in Table 4-2 by letting $\mathbf{A}, \mathbf{B}, \mathbf{Q},$ and $\mathbf{R}$ be constant matrices. Then,

$$\mathbf{E}(N - i) = -[\mathbf{B}^\dagger\mathbf{P}(N + 1 - i)\mathbf{B} + \mathbf{R}]^{-1}\mathbf{B}^\dagger\mathbf{P}(N + 1 - i)\mathbf{A} \qquad \text{(4-44)}$$

and

$$\begin{aligned}
\mathbf{P}(N - i) &= [\mathbf{A} + \mathbf{B}\mathbf{E}(N - i)]^\dagger\mathbf{P}(N + 1 - i)[\mathbf{A} + \mathbf{B}\mathbf{E}(N - i)] \\
&\quad + \mathbf{E}^\dagger(N - i)\mathbf{R}\mathbf{E}(N - i) + \mathbf{Q} \\
&= [\mathbf{A}^\dagger + \mathbf{E}^\dagger(N - i)\mathbf{B}^\dagger]\mathbf{P}(N + 1 - i)[\mathbf{A} + \mathbf{B}\mathbf{E}(N - i)] \\
&\quad + \mathbf{E}^\dagger(N - i)\mathbf{R}\mathbf{E}(N - i) + \mathbf{Q} \qquad \text{(4-45)}
\end{aligned}$$

Substituting $\mathbf{E}(N - i)$ in equation (4-44) into equation (4-45) for $\mathbf{P}(N - i)$, we obtain, after some algebra, the equation

$$\begin{aligned}
\mathbf{P}(N - i) &= \{\mathbf{A}^\dagger - \mathbf{A}^\dagger\mathbf{P}(N + 1 - i)\mathbf{B}^\dagger \\
&\quad \times [\mathbf{B}^\dagger\mathbf{P}(N + 1 - i)\mathbf{B} + \mathbf{R}]^{-1}\mathbf{B}^\dagger\} \\
&\quad \times \mathbf{P}(N + 1 - i)\{\mathbf{A} - \mathbf{B}[\mathbf{B}^\dagger\mathbf{P}(N + 1 - i)\mathbf{B} + \mathbf{R}]^{-1} \\
&\quad \times \mathbf{B}^\dagger\mathbf{P}(N + 1 - i)\mathbf{A}\} \\
&\quad + \mathbf{A}^\dagger\mathbf{P}(N + 1 - i)\mathbf{B}[\mathbf{B}^\dagger\mathbf{P}(N + 1 - i)\mathbf{B} + \mathbf{R}]^{-1}\mathbf{R} \\
&\quad \times [\mathbf{B}^\dagger\mathbf{P}(N + 1 - i)\mathbf{B} + \mathbf{R}]^{-1}\mathbf{B}^\dagger\mathbf{P}(N + 1 - i)\mathbf{A} + \mathbf{Q} \\
&= \mathbf{A}^\dagger\{\mathbf{P}(N + 1 - i) - \mathbf{P}(N + 1 - i)\mathbf{B} \\
&\quad \times [\mathbf{B}^\dagger\mathbf{P}(N + 1 - i)\mathbf{B} + \mathbf{R}]^{-1}\mathbf{B}^\dagger\mathbf{P}(N + 1 - i)\}\mathbf{A} + \mathbf{Q}
\end{aligned}$$

$$\text{(4-46)}$$

Defining the expression within braces in equation (4-46) as

$$\Gamma(N + 1 - i) = \mathbf{P}(N + 1 - i) - \mathbf{P}(N + 1 + i)\mathbf{B}$$
$$\times [\mathbf{B}^\dagger\mathbf{P}(N + 1 - i)\mathbf{B} + \mathbf{R}]^{-1}\mathbf{B}^\dagger\mathbf{P}(N + 1 - i) \quad (4\text{-}47)$$

we now show that the inverse of $\Gamma(N + 1 - i)$ in equation (4-47) is given by

$$\Gamma^{-1}(N + 1 - i) = \mathbf{P}^{-1}(N + 1 - i) + \mathbf{B}\mathbf{R}^{-1}\mathbf{B}^\dagger \quad (4\text{-}48)$$

provided that the matrices $\mathbf{P}$ and $\mathbf{R}$ are both invertible.

Because $\mathbf{R}$ is, by assumption, positive definite, then $\mathbf{R}^{-1}$ exists. The proof that $\mathbf{P}^{-1}(N + 1 - i)$ exists is shown in Subsection 4.8.2. Using equations (4-47) and (4-48), we form the product:

$$\Gamma^{-1}(N + 1 - i)\Gamma(N + 1 - i)$$
$$= [\mathbf{P}^{-1}(N + 1 - i) + \mathbf{B}\mathbf{R}^{-1}\mathbf{B}^\dagger]\{\mathbf{P}(N + 1 - i)$$
$$- \mathbf{P}(N + 1 - i)\mathbf{B}[\mathbf{B}^\dagger\mathbf{P}(N + 1 - i)\mathbf{B} + \mathbf{R}]^{-1}\mathbf{B}^\dagger\mathbf{P}(N + 1 - i)\}$$
$$= \mathbf{I} - \mathbf{B}\mathbf{R}^{-1}\{\mathbf{R}[\mathbf{B}^\dagger\mathbf{P}(N + 1 - i)\mathbf{B} + \mathbf{R}]^{-1} - \mathbf{I} + \mathbf{B}^\dagger\mathbf{P}(N + 1 - i)$$
$$\times \mathbf{B}[\mathbf{B}^\dagger\mathbf{P}(N + 1 - i)\mathbf{B} + \mathbf{R}]^{-1}\}\mathbf{B}^\dagger\mathbf{P}(N + 1 - i)$$
$$= \mathbf{I} - \mathbf{B}\mathbf{R}^{-1}\{[\mathbf{B}^\dagger\mathbf{P}(N + 1 - i)\mathbf{B} + \mathbf{R}]$$
$$\times [\mathbf{B}^\dagger\mathbf{P}(N + 1 - i)\mathbf{B} + \mathbf{R}]^{-1} - \mathbf{I}\}\mathbf{B}^\dagger\mathbf{P}(N + 1 - i)$$
$$= \mathbf{I}$$

and therefore the inverse of $\Gamma(N + 1 - i)$ is given by equation (4-48). Hence, according to equation (4-48), the expression within the braces in equation (4-46) can be replaced by

$$\Gamma(N + 1 - i) = [\mathbf{P}^{-1}(N + 1 - i) + \mathbf{B}\mathbf{R}^{-1}\mathbf{B}^\dagger]^{-1}$$

Therefore, equation (4-46) becomes

$$\mathbf{P}(N - i) = \mathbf{A}^\dagger[\mathbf{P}^{-1}(N + 1 - i) + \mathbf{B}\mathbf{R}^{-1}\mathbf{B}^\dagger]^{-1}\mathbf{A} + \mathbf{Q} \quad (4\text{-}49)$$

Equation (4-49) is equivalent to a pair of linear matrix equations of the form

$$\begin{bmatrix} \mathbf{Y}(N - i) \\ \mathbf{Z}(N - i) \end{bmatrix} = \begin{bmatrix} \mathbf{A}^\dagger + \mathbf{Q}\mathbf{A}^{-1}\mathbf{B}\mathbf{R}^{-1}\mathbf{B}^\dagger & \mathbf{Q}\mathbf{A}^{-1} \\ \mathbf{A}^{-1}\mathbf{B}\mathbf{R}^{-1}\mathbf{B}^\dagger & \mathbf{A}^{-1} \end{bmatrix} \begin{bmatrix} \mathbf{Y}(N + 1 - i) \\ \mathbf{Z}(N + 1 - i) \end{bmatrix}$$
$$(4\text{-}50)$$

where

$$\mathbf{P}(N - i) = \mathbf{Y}(N - i)\mathbf{Z}^{-1}(N - i) \quad (4\text{-}51)$$

and

$$\mathbf{Y}(N) = \mathbf{P}(N)$$

and

$$\mathbf{Z}(N) = \mathbf{I}$$

The equivalence of equations (4-49) and (4-50), including the existence of $\mathbf{Z}^{-1}(N - i)$, is shown in Subsection 4.8.2. On first reading of this chapter, one may choose to defer reading Subsection 4.8.2 and proceed directly to Subsection 4.8.3, where a closed-form solution for $\mathbf{E}(N - i)$ is developed.

4.8.2 Equation Equivalence

We now prove the equivalence of the nonlinear equation (4-49) to the pair of linear equations (4-50) using induction. Assuming that $\mathbf{Y}(N + 1 - i)$, $\mathbf{Z}(N + 1 - i)$, and $\mathbf{P}(N + 1 - i)$ are available and that

(1) $\mathbf{Y}(N + 1 - i) = \mathbf{P}(N + 1 - i)\mathbf{Z}(N + 1 - i)$
(2) $\mathbf{Y}^{-1}(N + 1 - i)$ exists
(3) $\mathbf{Z}^{-1}(N + 1 - i)$ exists
(4) $\mathbf{P}^{-1}(N + 1 - i)$ exists

we will then show that $\mathbf{Y}(N - i)$, $\mathbf{Z}(N - i)$, and $\mathbf{P}(N - i)$, which are calculated from equations (4-50) and (4-49), respectively, are such that

(1) $\mathbf{Y}(N - i) = \mathbf{P}(N - i)\mathbf{Z}(N - i)$
(2) $\mathbf{Y}^{-1}(N - i)$ exists
(3) $\mathbf{Z}^{-1}(N - i)$ exists
(4) $\mathbf{P}^{-1}(N - i)$ exists

Because

$$\mathbf{P}(N) = \mathbf{Y}(N)$$

is positive definite and

$$\mathbf{Z}(N) = \mathbf{I}$$

then

(1) $\mathbf{Y}(N) = \mathbf{P}(N)\mathbf{Z}(N)$
(2) $\mathbf{Y}^{-1}(N)$ exists
(3) $\mathbf{Z}^{-1}(N)$ exists
(4) $\mathbf{P}^{-1}(N)$ exists

and the induction proof is complete.

To show the induction step, we now consider the right-hand terms of the two equations in (4-50):

$$\mathbf{Y}(N - i) = (\mathbf{A}^{\dagger} + \mathbf{QA}^{-1}\mathbf{BR}^{-1}\mathbf{B}^{\dagger})\mathbf{Y}(N + 1 - i)$$
$$+ \mathbf{QA}^{-1}\mathbf{Z}(N + 1 - i) \qquad (4\text{-}52)$$

and

$$\mathbf{Z}(N - i) = \mathbf{A}^{-1}\mathbf{BR}^{-1}\mathbf{B}^{\dagger}\mathbf{Y}(N + 1 - i) + \mathbf{A}^{-1}\mathbf{Z}(N + 1 - i) \qquad (4\text{-}53)$$

Premultiplying both sides of equation (4-53) by $\mathbf{P}(N - i)$ gives

$$\mathbf{P}(N - i)\mathbf{Z}(N - i)$$
$$= \mathbf{P}(N - i)[\mathbf{A}^{-1}\mathbf{BR}^{-1}\mathbf{B}^{\dagger}\mathbf{Y}(N + 1 - i) + \mathbf{A}^{-1}\mathbf{Z}(N + 1 - i)] \qquad (4\text{-}54)$$

Letting

$$\mathbf{Y}(N + 1 - i) = \mathbf{P}(N + 1 - i)\mathbf{Z}(N + 1 - i)$$

then equations (4-52) and (4-54) become

$$\mathbf{Y}(N - i) = [(\mathbf{A}^{\dagger} + \mathbf{QA}^{-1}\mathbf{BR}^{-1}\mathbf{B}^{\dagger})\mathbf{P}(N + 1 - i) + \mathbf{QA}^{-1}]$$
$$\times \mathbf{Z}(N + 1 - i) \qquad (4\text{-}55)$$

and

$$\mathbf{P}(N - i)\mathbf{Z}(N - i) = \mathbf{P}(N - i)[\mathbf{A}^{-1}\mathbf{BR}^{-1}\mathbf{B}^{\dagger}\mathbf{P}(N + 1 - i) + \mathbf{A}^{-1}]$$
$$\times \mathbf{Z}(N + 1 - i) \qquad (4\text{-}56)$$

respectively.

We now successively postmultiply both sides of equations (4-55) and (4-56) by

$$\mathbf{Z}^{-1}(N + 1 - i), \quad \mathbf{P}^{-1}(N + 1 - i), \quad [\mathbf{BR}^{-1}\mathbf{B}^{\dagger} + \mathbf{P}^{-1}(N + 1 - i)]^{-1},$$

and $\mathbf{A}$, thus forming the two right-hand terms

$$\mathbf{Q} + \mathbf{A}^{\dagger}[\mathbf{BR}^{-1}\mathbf{B}^{\dagger} + \mathbf{P}^{-1}(N + 1 - i)]^{-1}\mathbf{A}$$

and

$$\mathbf{P}(N - i)$$

which by equation (4-49) are equal. That is,

$$\mathbf{Y}(N - i) = \mathbf{P}(N - i)\mathbf{Z}(N - i)$$

and, therefore, equations (4-49) and (4-50) are equivalent.

To complete the proof, we must show that $\mathbf{Y}^{-1}(N - i)$, $\mathbf{Z}^{-1}(N - i)$, and $\mathbf{P}^{-1}(N - i)$ exist.

Substituting

$$\mathbf{Z}(N + 1 - i) = \mathbf{P}^{-1}(N + 1 - i)\mathbf{Y}(N + 1 - i)$$

into equation (4-52), we get

$$\mathbf{Y}(N - i) = \mathbf{A}^{\dagger}\{[\mathbf{BR}^{-1}\mathbf{B}^{\dagger} + \mathbf{P}^{-1}(N + 1 - i)]^{-1} + (\mathbf{A}^{\dagger})^{-1}\mathbf{QA}^{-1}\}$$
$$\times [\mathbf{BR}^{-1}\mathbf{B}^{\dagger} + \mathbf{P}^{-1}(N + 1 - i)]\mathbf{Y}(N + 1 - i) \qquad \textbf{(4-57)}$$

Taking the inverse of equation (4-57) gives

$$\mathbf{Y}^{-1}(N - i) = \mathbf{Y}^{-1}(N + 1 - i)[\mathbf{BR}^{-1}\mathbf{B}^{\dagger} + \mathbf{P}^{-1}(N + 1 - i)]^{-1}$$
$$\times \{[\mathbf{BR}^{-1}\mathbf{B}^{\dagger} + \mathbf{P}^{-1}(N + 1 - i)]^{-1}$$
$$+ (\mathbf{A}^{\dagger})^{-1}\mathbf{QA}^{-1}\}^{-1}(\mathbf{A}^{\dagger})^{-1} \qquad \textbf{(4-58)}$$

But the existence of all of the inverse terms on the right-hand side of equation (4-58) is guaranteed by our earlier discussion. Therefore, the inverse of $\mathbf{Y}(N - i)$ exists.

Next, substituting

$$\mathbf{Y}(N + 1 - i) = \mathbf{P}(N + 1 - i)\mathbf{Z}(N + 1 - i)$$

into equation (4-53) gives

$$\mathbf{Z}(N - i) = \mathbf{A}^{-1}[\mathbf{BR}^{-1}\mathbf{B}^{\dagger} + \mathbf{P}^{-1}(N + 1 - i)]$$
$$\times \mathbf{P}(N + 1 - i)\mathbf{Z}(N + 1 - i) \qquad \textbf{(4-59)}$$

Taking the inverse in equation (4-59), therefore,

$$\mathbf{Z}^{-1}(N - i) = \mathbf{Z}^{-1}(N + 1 - i)\mathbf{P}^{-1}(N + 1 - i)$$
$$\times [\mathbf{BR}^{-1}\mathbf{B}^{\dagger} + \mathbf{P}^{-1}(N + 1 - i)]^{-1}\mathbf{A} \qquad \textbf{(4-60)}$$

where the existence of all of the inverse terms on the right-hand side is guaranteed by our earlier discussion.

Finally, because

$$\mathbf{Y}(N - i) = \mathbf{P}(N - i)\mathbf{Z}(N - i)$$

then

$$\mathbf{P}^{-1}(N - i) = \mathbf{Z}(N - i)\mathbf{Y}^{-1}(N - i)$$

and therefore the induction proof is complete.

4.8.3 Closed-Form Solution for Optimal Gain

In this section, a closed-form solution for the optimal gain $\mathbf{E}(N - i)$ is developed using equations (4-50). First, we define the matrix

$$\mathbf{H} = \begin{bmatrix} \mathbf{A}^\dagger + \mathbf{Q}\mathbf{A}^{-1}\mathbf{B}\mathbf{R}^{-1}\mathbf{B}^\dagger & \mathbf{Q}\mathbf{A}^{-1} \\ \mathbf{A}^{-1}\mathbf{B}\mathbf{R}^{-1}\mathbf{B}^\dagger & \mathbf{A}^{-1} \end{bmatrix}$$

The inverse of $\mathbf{H}$ exists and is given by

$$\mathbf{H}^{-1} = \begin{bmatrix} (\mathbf{A}^\dagger)^{-1} & -(\mathbf{A}^\dagger)^{-1}\mathbf{Q} \\ -\mathbf{B}\mathbf{R}^{-1}\mathbf{B}^\dagger(\mathbf{A}^\dagger)^{-1} & \mathbf{A} + \mathbf{B}\mathbf{R}^{-1}\mathbf{B}^\dagger(\mathbf{A}^\dagger)^{-1}\mathbf{Q} \end{bmatrix} \quad \text{(4-61)}$$

This is easily shown by forming the matrix product $\mathbf{H}\mathbf{H}^{-1}$, which equals $\mathbf{I}$, the identity matrix.

We claim that the eigenvalues of the matrix $\mathbf{H}$ are such that the reciprocal of every eigenvalue of $\mathbf{H}$ is also an eigenvalue of $\mathbf{H}$. To show this, we first recall from Table A-8, property E in Appendix A, that if λ is an eigenvalue of a nonsingular matrix, then $1/\lambda$ is an eigenvalue of the matrix inverse. Using this result, we proceed to prove our claim by letting λ be an eigenvalue of $\mathbf{H}$ and $[\mathbf{f} \ \mathbf{g}]^\dagger$ be the corresponding eigenvector in partitioned form. Then

$$\begin{bmatrix} \mathbf{A}^\dagger + \mathbf{Q}\mathbf{A}^{-1}\mathbf{B}\mathbf{R}^{-1}\mathbf{B}^\dagger & \mathbf{Q}\mathbf{A}^{-1} \\ \mathbf{A}^{-1}\mathbf{B}\mathbf{R}^{-1}\mathbf{B}^\dagger & \mathbf{A}^{-1} \end{bmatrix} \begin{bmatrix} \mathbf{f} \\ \mathbf{g} \end{bmatrix} = \lambda \begin{bmatrix} \mathbf{f} \\ \mathbf{g} \end{bmatrix}$$

which can be written in terms of the two matrix equations

$$(\mathbf{A}^\dagger + \mathbf{Q}\mathbf{A}^{-1}\mathbf{B}\mathbf{R}^{-1}\mathbf{B}^\dagger)\mathbf{f} + \mathbf{Q}\mathbf{A}^{-1}\mathbf{g} = \lambda\mathbf{f} \quad \text{(4-62)}$$

$$\mathbf{A}^{-1}\mathbf{B}\mathbf{R}^{-1}\mathbf{B}^\dagger\mathbf{f} + \mathbf{A}^{-1}\mathbf{g} = \lambda\mathbf{g} \quad \text{(4-63)}$$

On the other hand, consider the eigenvalue equation

$$(\mathbf{H}^\dagger)^{-1} \begin{bmatrix} \mathbf{g} \\ -\mathbf{f} \end{bmatrix} = \lambda \begin{bmatrix} \mathbf{g} \\ -\mathbf{f} \end{bmatrix} \quad \text{(4-64)}$$

where $\mathbf{H}^{-1}$ is given in equation (4-61). Hence, equation (4-64) can be written in terms of the two equations

$$\mathbf{A}^{-1}\mathbf{g} + \mathbf{A}^{-1}\mathbf{B}\mathbf{R}^{-1}\mathbf{B}^\dagger\mathbf{f} = \lambda\mathbf{g} \quad \text{(4-65)}$$

$$-\mathbf{Q}\mathbf{A}^{-1}\mathbf{g} - (\mathbf{A}^\dagger\mathbf{Q}\mathbf{A}^{-1}\mathbf{B}\mathbf{R}^{-1}\mathbf{B}^\dagger)\mathbf{f} = \lambda\mathbf{f} \quad \text{(4-66)}$$

which are equal to equations (4-63) and (4-64), respectively. Therefore, λ is also an eigenvalue of $(\mathbf{H}^\dagger)^{-1}$ and thus of $\mathbf{H}^{-1}$. But this implies that $1/\lambda$ is an eigenvalue of $\mathbf{H}$, that is, if λ is an eigenvalue of $\mathbf{H}$, then $1/\lambda$ is also an eigenvalue of $\mathbf{H}$.

In the remainder of this development, it is assumed that the eigenvalues of $\mathbf{H}$ are distinct. If they are not distinct, we can still proceed in a similar manner to the following.

The eigenvalues of the $\mathbf{H}$ matrix can be arranged in a diagonal matrix of the form

$$\mathbf{D} = \begin{bmatrix} \Lambda & 0 \\ 0 & \Lambda^{-1} \end{bmatrix} \tag{4-67}$$

where Λ is a diagonal matrix of the eigenvalues that are outside the unit circle and Λ^{-1} is a diagonal matrix of the eigenvalues that are inside the unit circle. We will not consider the case where eigenvalues are on the unit circle.

Returning to equation (4-50), because the $\mathbf{H}$ matrix has distinct eigenvalues, there then exists a nonsingular matrix $\mathbf{W}$ of eigenvectors such that

$$\mathbf{D} = \mathbf{W}^{-1}\mathbf{H}\mathbf{W} \tag{4-68}$$

The similarity transformation

$$\begin{bmatrix} \mathbf{Y}(N - i) \\ \mathbf{Z}(N - i) \end{bmatrix} = \mathbf{W} \begin{bmatrix} \mathbf{U}(N - i) \\ \mathbf{V}(N - i) \end{bmatrix} \tag{4-69}$$

substituted into equation (4-50) gives

$$\mathbf{W} \begin{bmatrix} \mathbf{U}(N - i) \\ \mathbf{V}(N - i) \end{bmatrix} = \mathbf{H}\mathbf{W} \begin{bmatrix} \mathbf{U}(N + 1 - i) \\ \mathbf{V}(N + 1 - i) \end{bmatrix}$$

or

$$\begin{bmatrix} \mathbf{U}(N - i) \\ \mathbf{V}(N - i) \end{bmatrix} = \mathbf{W}^{-1}\mathbf{H}\mathbf{W} \begin{bmatrix} \mathbf{U}(N + 1 - i) \\ \mathbf{V}(N + 1 - i) \end{bmatrix}$$

Then

$$\begin{bmatrix} \mathbf{U}(N - i) \\ \mathbf{V}(N - i) \end{bmatrix} = \begin{bmatrix} \Lambda & 0 \\ 0 & \Lambda^{-1} \end{bmatrix} \begin{bmatrix} \mathbf{U}(N + 1 - i) \\ \mathbf{V}(N + 1 - i) \end{bmatrix}$$

which can be easily solved to give

$$\begin{bmatrix} \mathbf{U}(N - i) \\ \mathbf{V}(N - i) \end{bmatrix} = \begin{bmatrix} \Lambda^i & 0 \\ 0 & \Lambda^{-i} \end{bmatrix} \begin{bmatrix} \mathbf{U}(N) \\ \mathbf{V}(N) \end{bmatrix} \tag{4-70}$$

On the other hand, partitioning the $\mathbf{W}$ matrix in equation (4-69) into four $n \times n$ submatrices gives

$$\begin{bmatrix} \mathbf{Y}(N - i) \\ \mathbf{Z}(N - i) \end{bmatrix} = \begin{bmatrix} \mathbf{W}_{11} & \mathbf{W}_{12} \\ \mathbf{W}_{21} & \mathbf{W}_{22} \end{bmatrix} \begin{bmatrix} \mathbf{U}(N - i) \\ \mathbf{V}(N - i) \end{bmatrix}$$

or

$$Y(N - i) = W_{11}U(N - i) + W_{12}V(N - i) \tag{4-71}$$

$$Z(N - i) = W_{21}U(N - i) + W_{22}V(N - i) \tag{4-72}$$

But because

$$Y(N) = P(N)Z(N)$$

then equations (4-71) and (4-72) can be arranged to give

$$[W_{11} - P(N)W_{21}]U(N) = [P(N)W_{22} - W_{12}]V(N) \tag{4-73}$$

Solving for $U(N)$ in equation (4-73) results in

$$U(N) = [W_{11} - P(N)W_{21}]^{-1}[P(N)W_{22} - W_{12}]V(N)$$

$$= MV(N) \tag{4-74}$$

But from equation (4-70)

$$U(N) = \Lambda^{-i}U(N - i)$$

and

$$V(N) = \Lambda^{i}V(N - i)$$

Substituting these two expressions into equation (4-74) results in

$$\Lambda^{-i}U(N - i) = M\Lambda^{i}V(N - i)$$

or

$$V(N - i) = \Lambda^{-i}M^{-1}\Lambda^{-i}U(N - i)$$

$$= G(i)U(N - i) \tag{4-75}$$

where

$$\lim_{i \to \infty} G(i) = 0$$

because the eigenvalues of the Λ^{-1} matrix are all inside the unit circle. Substituting equation (4-75) into equation (4-71) gives

$$Y(N - i) = W_{11}U(N - i) + W_{12}V(N - i)$$

$$= [W_{11} + W_{12}G(i)]U(N - i) \tag{4-76}$$

and substituting equation (4-75) into equation (4-72) gives

$$Z(N - i) = W_{21}U(N - i) + W_{22}V(N - i)$$

$$= [W_{21} + W_{22}G(i)]U(N - i) \tag{4-77}$$

But

$$\mathbf{P}(N - i) = \mathbf{Y}(N - i)\mathbf{Z}^{-1}(N - i)$$

Therefore, using equations (4-76) and (4-77), we arrive at

$$\mathbf{P}(N - i) = [\mathbf{W}_{11} + \mathbf{W}_{12}\mathbf{G}(i)][\mathbf{W}_{21} + \mathbf{W}_{22}\mathbf{G}(i)]^{-1} \qquad \textbf{(4-78)}$$

the required result.

The procedure for the calculation of the optimal regulator gain for a step-invariant system with constant cost weighting matrices is summarized in Table 4-5.

As an example of this technique, consider again the system presented in Subsection 4.7.3, where

$$x(k + 1) = 0.5x(k) + 3u(k)$$

with

$$N = 10 \qquad P(N) = 10 \qquad Q = 1 \qquad R = 4$$

Using the procedure in Table 4-5, the optimal gain is calculated as follows:

1. Form **H**

$$\mathbf{H} = \begin{bmatrix} [0.5 + (1)(2)(3)(0.25)(3)] & (1)(2) \\ [(2)(3)(0.25)(3)] & 2 \end{bmatrix}$$

$$= \begin{bmatrix} 5 & 2 \\ 4.5 & 2 \end{bmatrix}$$

2. Find the eigenvalues and eigenvectors of **H**

$$\begin{vmatrix} (\lambda - 5) & -2 \\ -4.5 & (\lambda - 2) \end{vmatrix} = \lambda^2 - 7\lambda + 1 = 0$$

and $\lambda_1 = 6.8541$ and $\lambda_2 = 0.1459$. Note that

$$\lambda_2 = \frac{1}{\lambda_1}$$

$$\Lambda = \lambda_1 = 6.8541$$

$$\Lambda^{-1} = \lambda_2 = 0.1459$$

For $\lambda_1 = 6.8541$

$$\begin{bmatrix} (\lambda_1 - 5) & -2 \\ -4.5 & (\lambda_1 - 2) \end{bmatrix} \begin{bmatrix} p_1^1 \\ p_2^1 \end{bmatrix} = \mathbf{0}$$

Table 4-5 Procedure for Calculation of Optimal Regulator Gain

For the nth-order plant

$\mathbf{x}(k + 1) = \mathbf{A}\mathbf{x}(k) + \mathbf{B}\mathbf{u}(k)$

with state feedback

$\mathbf{u}(k) = \mathbf{E}(k)\mathbf{x}(k)$

and performance measure

$$J = \mathbf{x}^\dagger(N)\mathbf{P}(N)\mathbf{x}(N) + \sum_{i=0}^{N-1} [\mathbf{x}^\dagger(i)\mathbf{Q}\mathbf{x}(i) + \mathbf{u}^\dagger(i)\mathbf{R}\mathbf{u}(i)]$$

begin with $i = 1$ and the known $\mathbf{P}(N)$

1. Form the matrix

$$\mathbf{H} = \begin{bmatrix} \mathbf{A}^\dagger + \mathbf{Q}\mathbf{A}^{-1}\mathbf{B}\mathbf{R}^{-1}\mathbf{B}^\dagger & \mathbf{Q}\mathbf{A}^{-1} \\ \mathbf{A}^{-1}\mathbf{B}\mathbf{R}^{-1}\mathbf{B}^\dagger & \mathbf{A}^{-1} \end{bmatrix}$$

2. Find the eigenvalues and the corresponding eigenvectors of $\mathbf{H}$.

3. Generate the matrix $\mathbf{W}$ from eigenvectors such that

$$\mathbf{W}^{-1}\mathbf{H}\mathbf{W} = \mathbf{D} = \begin{bmatrix} \mathbf{\Lambda} & 0 \\ 0 & \mathbf{\Lambda}^{-1} \end{bmatrix}$$

Where $\mathbf{\Lambda}$ is the diagonal matrix of eigenvalues outside the unit circle on the z-plane.

4. Partition $\mathbf{W}$ into four $n \times n$ submatrices as

$$\mathbf{W} = \begin{bmatrix} \mathbf{W}_{11} & \mathbf{W}_{12} \\ \mathbf{W}_{21} & \mathbf{W}_{22} \end{bmatrix}$$

5. Form

$\mathbf{G}(i) = \mathbf{\Lambda}^{-i}[\mathbf{P}(N)\mathbf{W}_{22} - \mathbf{W}_{12}]^{-1}[\mathbf{W}_{11} - \mathbf{P}(N)\mathbf{W}_{21}]\mathbf{\Lambda}^{-i}$

6. Form

$\mathbf{P}(N - i) = [\mathbf{W}_{11} + \mathbf{W}_{12}\mathbf{G}(i)][\mathbf{W}_{21} + \mathbf{W}_{22}\mathbf{G}(i)]^{-1}$

7. Form

$\mathbf{E}(N - i) = -[\mathbf{B}^\dagger\mathbf{P}(N + 1 - i)\mathbf{B} + \mathbf{R}]^{-1}\mathbf{B}^\dagger\mathbf{P}(N + 1 - i)\mathbf{A}$

Where

$\mathbf{P}(N + 1 - i) = [\mathbf{W}_{11} + \mathbf{W}_{12}\mathbf{G}(i - 1)][\mathbf{W}_{21} + \mathbf{W}_{22}\mathbf{G}(i - 1)]^{-1}$

or

$$\mathbf{p}^1 = \begin{bmatrix} 1.0787 \\ 1 \end{bmatrix}$$

For $\lambda_2 = 0.1459$

$$\begin{bmatrix} (\lambda_2 - 5) & -2 \\ -4.5 & (\lambda_2 - 2) \end{bmatrix} \begin{bmatrix} p_1^2 \\ p_2^2 \end{bmatrix} = 0$$

or

$$\mathbf{p}^2 = \begin{bmatrix} -0.412 \\ 1 \end{bmatrix}$$

3. Form $\mathbf{W}$

$$W = \begin{bmatrix} 1.0787 & -0.412 \\ 1 & 1 \end{bmatrix}$$

4. Partition $\mathbf{W}$

$W_{11} = 1.0787 \qquad W_{12} = -0.412$

$W_{21} = 1 \qquad\qquad W_{22} = 1$

5. Form $\mathbf{G}(i)$ beginning with $i = 1$

$$\begin{aligned} G(i) &= (\lambda^{-1})^i [P(N)W_{22} - W_{12}]^{-1}[W_{11} - P(N)W_{21}](\lambda^{-1})^i \\ &= (0.1459)^i [(10)(1) + 0.412]^{-1}[1.0787 - (10)(1)](0.1459)^i \\ &= \left(-\frac{8.9213}{10.412} \right)(0.1459)^{2i} \\ &= (-0.8568)(0.0213)^i \end{aligned}$$

6. Form $\mathbf{P}(N - i)$

$$\begin{aligned} P(N - i) = &[1.0787 + (0.412)(0.8568)(0.0213)^i] \\ &\times [1 - (0.8568)(0.0213)^i]^{-1} \end{aligned}$$

For $i = 1$, and $N = 10$

$P(9) = 1.1064$

For $i = 2$, and $N = 10$

$P(8) = 1.0793$

and so on. The remaining values are identical to those shown in Figure 4-17.

7. Form $\mathbf{E}(N - i)$

$$e(N - i) = -[(3)P(N + 1 - i)(3) + 4]^{-1}(3)P(N + 1 - i)(0.5)$$

where

$$P(N + 1 - i) = [1.0787 - 0.412(-0.8568)(0.0213)^{i-1}]$$
$$\times [1 + (1)(-0.8568)(0.0213)^{i-1}]^{-1}$$

$$e(N - i) = \frac{-1.5[1.0787 + 0.353(0.0213)^{i-1}]}{13.7056 - 0.2502(0.0213)^{i-1}}$$

For $i = 1$, and $N = 10$

$$e(9) = -0.15957$$

For $i = 2$,

$$e(8) = -0.1189$$

and so on. The remaining values are, of course, identical to those shown in Figure 4-17.

4.8.4 Steady State Regulation

For a completely controllable, step-invariant plant and constant cost weighting matrices $\mathbf{Q}$ and $\mathbf{R}$, the optimum feedback gains, from $\mathbf{E}(N)$ backward, are not changed if the final step N is changed. This is apparent by examining equations (4-78) and (4-44) for $\mathbf{P}(N - i)$ and $\mathbf{E}(N - i)$, because, under the stated conditions, the values of $\mathbf{P}(N - i)$ and $\mathbf{E}(N - i)$ depend only on $\mathbf{P}(N)$ and the value of i. This is to say that the sequence of gains starting at N and proceeding backward is always the same independent of the value of N. Figure 4-18 is an example of what happens when N is changed from 10 to 4. Furthermore, the matrix for the optimal quadratic regulator solution approaches a constant feedback gain matrix as we proceed backward from the final step ($k = N$). Figure 4-22 illustrates this property for roll attitude control of a spacecraft. As the final step N goes to infinity, in what is termed the *infinite horizon* problem, these steady state gains become optimal. Steady state optimal gains (with regard to some performance measure) are occasionally used in lieu of placing state feedback eigenvalues for step-invariant regulator design. Of course, the steady state optimal gains result in some certain eigenvalue placement, just as any eigenvalue placement minimizes some quadratic performance measure. The designer's choice is whether to select the system eigenvalues or the weighting matrices $\mathbf{Q}$ and $\mathbf{R}$.

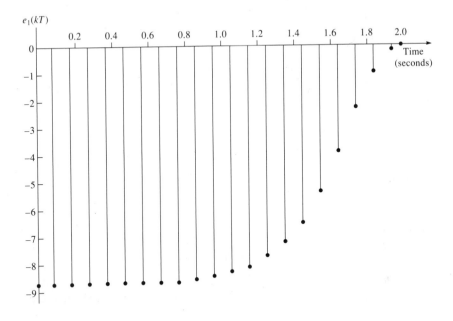

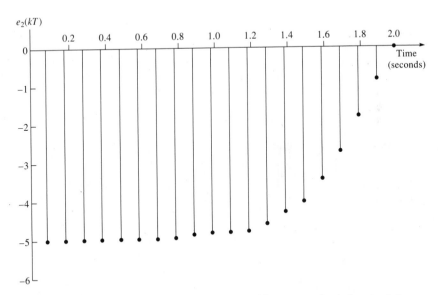

Figure 4-22 Optimal feedback gains approaching a constant steady state, backward from the final time.

The procedure developed in the previous subsection can be easily adapted to the steady state regulator problem. If we let i approach infinity in equation (4-75), then $\mathbf{G}(i)$ goes to zero and

$$\lim_{i \to \infty} \mathbf{P}(N - i) = \mathbf{P} = \mathbf{W}_{11}\mathbf{W}_{21}^{-1} \tag{4-79}$$

$$\lim_{i \to \infty} \mathbf{E}(N - i) = \mathbf{E} = -[\mathbf{B}^\dagger\mathbf{PB} + \mathbf{R}]^{-1}\mathbf{B}^\dagger\mathbf{PA}$$

$$= -[\mathbf{B}^\dagger\mathbf{W}_{11}\mathbf{W}_{21}^{-1}\mathbf{B} + \mathbf{R}]^{-1}\mathbf{B}^\dagger\mathbf{W}_{11}\mathbf{W}_{21}^{-1}\mathbf{A} \tag{4-80}$$

The procedure summarized in Table 4-5 for calculating the optimal regulator gain can be adapted to the calculation of the steady state regulator gain by replacing steps 5, 6, and 7 with the following single step, 5.

5. Form

$$\mathbf{E} = -[\mathbf{B}^\dagger\mathbf{W}_{11}\mathbf{W}_{21}^{-1}\mathbf{B} + \mathbf{R}]^{-1}\mathbf{B}^\dagger\mathbf{W}_{11}\mathbf{W}_{21}^{-1}\mathbf{A}$$

As an example of this technique, consider again the first-order system discussed in Subsections 4.7.3 and 4.8.3 in which now we assume that N approaches infinity. According to equation (4-80), the steady state gain is

$$e = -[(3)(1.0787)(1)(3) + 4]^{-1}(3)(1.0787)(1)(0.5)$$
$$= -0.1180$$

which agrees with all the previous results. For the roll attitude control system in Section 4.7.4, the steady state gains can also be calculated using the previous procedure. These gains are

$$e_1 = -8.7215 \quad \text{and} \quad e_2 = -5.0050$$

and result in the feedback system

$$\begin{bmatrix} x_1(k+1) \\ x_2(k+1) \end{bmatrix} = \begin{bmatrix} 1 & 0.05 \\ 0 & 1 \end{bmatrix}\begin{bmatrix} x_1(k) \\ x_2(k) \end{bmatrix} + \begin{bmatrix} 0.00125 \\ 0.05 \end{bmatrix}[-8.7215 \quad -5.005]\begin{bmatrix} x_1(k) \\ x_2(k) \end{bmatrix}$$
$$= \begin{bmatrix} 0.9891 & 0.0437 \\ -0.4361 & 0.7497 \end{bmatrix}\begin{bmatrix} x_1(k) \\ x_2(k) \end{bmatrix}$$

which has eigenvalues

$$\lambda_1 = 0.8694 + j0.0689$$

and

$$\lambda_2 = 0.8694 - j0.0689$$

Mapping these eigenvalues to the s-domain using

$$z = e^{sT}$$

$$s = \frac{1}{T} \ln z$$

gives

$$\lambda_{1,2} = \frac{1}{T} \ln(0.8694 \pm j0.0689)$$

$$= \frac{1}{T} \ln(0.872 e^{\pm j0.0791})$$

$$= -2.7358 \pm j1.5828$$

indicating that the closed-loop bandwidth is 3.16 radians/sec, and the damping ratio of the closed-loop poles is $\zeta = 0.865$.

4.9 Summary

If an nth-order, linear, step-invariant plant

$$\mathbf{x}(k + 1) = \mathbf{A}\mathbf{x}(k) + \mathbf{B}\mathbf{u}(k)$$
$$\mathbf{y}(k) = \mathbf{C}\mathbf{x}(k) + \mathbf{D}\mathbf{u}(k)$$

is completely observable, its state at any step can be calculated from a finite number of its most recent inputs and outputs. The linear algebraic equations for the state $\mathbf{x}(k)$ at any step k in terms of known plant inputs and outputs from step k onward have the plant's observability matrix,

$$\mathbf{M}_0 = \begin{bmatrix} \mathbf{C} \\ \mathbf{C}\mathbf{A} \\ \mathbf{C}\mathbf{A}^2 \\ \vdots \\ \mathbf{C}\mathbf{A}^{n-1} \end{bmatrix}$$

as the coefficient matrix. Hence the plant is completely observable, and $\mathbf{x}(k)$ can be found from measurements of $\mathbf{y}(k)$, $\mathbf{y}(k + 1)$, . . . , $\mathbf{y}(k + n - 1)$ and $\mathbf{u}(k)$, $\mathbf{u}(k + 1)$, . . . , $\mathbf{u}(k + n - 1)$, if and only if $\mathbf{M}_0$ is of full rank. From $\mathbf{x}(k)$ and the inputs, $\mathbf{x}(k + 1)$, $\mathbf{x}(k + 2)$, . . . can also be computed. The rank test of $\mathbf{M}_0$ for complete observability also applies to continuous-time systems.

Plant input and output measurements can also be used to identify plant parameters. If the parameters to be found are z-transfer function coefficients (which are also the input–output equation coefficients), the solution involves solving linear algebraic equations. When more than the minimal number of measurements is available, the method of least squares estimation (discussed in Chapter 8) can be applied. Once the parameters are identified, the same input and output data can be used to determine or to estimate the plant state.

A step-varying system is observable at step k if $\mathbf{x}(k)$ can be determined from the system outputs $\mathbf{y}(k), \mathbf{y}(k + 1), \ldots, \mathbf{y}(k + N - 1)$ and inputs $\mathbf{u}(k), \mathbf{u}(k + 1), \ldots, \mathbf{u}(k + N - 1)$ for some finite number of steps N. It is uniformly observable if it is observable at every step.

If an nth-order, linear, step-invariant plant is completely controllable, there are plant inputs that bring the plant state from an initial state at any initial step to any desired later state in a finite number of steps. The linear algebraic equations for the required input sequence have the system's controllability matrix

$$\mathbf{M}_c = [\mathbf{B} \mid \mathbf{AB} \mid \mathbf{A}^2\mathbf{B} \mid \cdots \mid \mathbf{A}^{n-1}\mathbf{B}]$$

as the coefficient matrix. There is a solution for an input sequence, and the plant is therefore completely controllable if and only if $\mathbf{M}_c$ is of full rank. The rank test of the controllability matrix also applies to continuous-time plants.

A step-varying system is controllable at step k if there are input sequences $\mathbf{u}(k), \mathbf{u}(k + 1), \ldots, \mathbf{u}(k + N - 1)$ that move the state from any known $\mathbf{x}(k)$ to any desired $\mathbf{x}(N)$ in some finite number of steps N. It is uniformly controllable if it is controllable at every step. More general observability and controllability matrices than those for step-invariant systems can be used to test observability and controllability of step-varying systems.

When the state of a plant is available and is used for feedback

$$\mathbf{u}(k) = \mathbf{Ex}(k) + \boldsymbol{\rho}(k)$$

where $\boldsymbol{\rho}(k)$ is an external input, the state equation for the plant with feedback is

$$\mathbf{x}(k + 1) = (\mathbf{A} + \mathbf{BE})\mathbf{x}(k) + \mathbf{B}\boldsymbol{\rho}(k)$$

If the plant is completely controllable, the eigenvalues of the feedback system, those of $(\mathbf{A} + \mathbf{BE})$, can be placed at any locations selected by the designer by appropriately choosing the feedback gain matrix $\mathbf{E}$. Eigenvalue placement with state feedback for a single-input plant in controllable form is especially simple because, in that form, each element of the feedback gain vector determines one coefficient of the feedback system's characteristic equation. Transformation to and from controllable form can be used for feedback system eigenvalue placement design for any completely controllable single-input plant.

State feedback regulation

$$\mathbf{u}(k) = \mathbf{E}(k)\mathbf{x}(k)$$

so that

$$\mathbf{x}(k + 1) = [\mathbf{A}(k) + \mathbf{B}(k)\mathbf{E}(k)]\mathbf{x}(k) = \mathbf{\Phi}(k)\mathbf{x}(k)$$

and

$$\mathbf{x}(k) = \mathbf{\Phi}(k - 1)\mathbf{\Phi}(k - 2) \ldots \mathbf{\Phi}(0)\mathbf{x}(0)$$

is characterized by backward-in-step gain calculations. The algorithm of Table 4-1 is for step-varying deadbeat state feedback, for which

$$\mathbf{\Phi}(n - 1)\mathbf{\Phi}(n - 2) \ldots \mathbf{\Phi}(0) = \mathbf{0}$$

If an nth-order plant has n linearly independent outputs, its state is just a nonsingular linear transformation of the outputs minus any direct plant input-to-output coupling. The state is then obtainable from the output, and output feedback can be used to place all the eigenvalues of the plant with feedback. When the plant has fewer than n linearly independent outputs, its state cannot be obtained from an output transformation. If the plant has a single input, it is not possible to place all the eigenvalues of the output feedback system arbitrarily. If the plant has multiple inputs, arbitrary placement of output feedback eigenvalues may or may not be possible.

The simplest optimization involves minimizing a quadratic function of the state variables and inputs because the minimum is then described by linear equations. The quadratic optimal regulation algorithm (Table 4-2) is derived by applying the principle of optimality. This approach gives added insight into the solution and into why the result is backward in step. When the matrices $\mathbf{A}$, $\mathbf{B}$, $\mathbf{Q}$, and $\mathbf{R}$ are constant, a closed-form solution for the optimal gain is derived (Table 4-5) from which the numerical value of the gain can be calculated for any point in time. First-order examples were seen to give reasonable and expected results, and an example of optimal roll attitude control of a spacecraft provided a more involved example.

References

Key references on observability and controllability were listed in Chapter 3. Results for eigenvalue placement with state feedback were developed by a number of researchers at about the same time.

W. M. Wonham, "On Pole Assignment in Multi-Input Controllable Linear Systems," *IEEE Trans. Automatic Control*, Vol. AC-12, Dec. 1967, pp. 660–665.

C.-T. Chen, "A Note on Pole Assignment," *IEEE Trans. Automatic Control*, Vol. AC-13, Oct. 1968, p. 597–598;

E. J. Davison, "On Pole Assignment in Multivariable Linear Systems," *IEEE Trans. Automatic Control*, Vol. AC-13, Dec. 1968, pp. 747–748;

M. Heymann, "Comments on Pole Assignment in Multi-Input Controllable Linear Systems," *IEEE Trans. Automatic Control*, Vol. AC-13, Dec. 1968, pp. 748–749;

Both continuous-time and discrete-time optimal control are covered in

M. Athans and P. L. Falb, *Optimal Control: An Introduction to the Theory and Its Applications*. New York: McGraw-Hill, 1966;

D. E. Kirk, *Optimal Control Theory*. Englewood Cliffs, N.J.: Prentice-Hall, 1970;

A. P. Sage and C. C. White, *Optimum Systems Control*, 2nd edition. Englewood Cliffs, NJ: Prentice-Hall, 1977,

and a comprehensive survey of most of the subject is

M. Athans, "The Status of Optimal Control Theory and Applications for Deterministic Systems," *IEEE Trans. Automatic Control*, Vol. AC-11, July 1966, pp. 580–596.

Dynamic programming is discussed in detail in

R. E. Bellman, *Dynamic Programming*. Princeton, NJ: Princeton University Press, 1957,

and

R. E. Bellman and R. E. Kalaba, *Dynamic Programming and Modern Control Theory*. New York: Academic Press, 1965.

The equivalence of eigenvalue placement and steady state optimal control was first given in

R. E. Kalman, "When is a Linear Control System Optimal?" *Trans. ASME J. Basic Eng.*, Ser. D, 86, March 1964, pp. 1–10.

Additional features can be included in the quadratic optimal control problem, to account both for system inputs other than the control and for a penalty on the rate of change of the control. A recent paper that examines these features is

D. A. Pierre, "Properties of a Discrete-Time Enhanced Linear-Quadratic Controller," *Automatica*, Vol. 27, No. 6, 1991, pp. 1029–1034.

Chapter Four Problems

4-1. Use observability matrix rank tests to determine whether the following systems are completely observable:

a.
$$\begin{bmatrix} x_1(k+1) \\ x_2(k+1) \end{bmatrix} = \begin{bmatrix} 5 & 7 \\ -1 & 0 \end{bmatrix} \begin{bmatrix} x_1(k) \\ x_2(k) \end{bmatrix} + \begin{bmatrix} 14 \\ -4 \end{bmatrix} u(k)$$

$$y(k) = \begin{bmatrix} 0 & 1 \end{bmatrix} \begin{bmatrix} x_1(k) \\ x_2(k) \end{bmatrix} + 4u(k)$$

b.
$$\begin{bmatrix} x_1(k+1) \\ x_2(k+1) \\ x_3(k+1) \end{bmatrix} = \begin{bmatrix} -7 & -2 & 1 \\ -21 & 10 & 3 \\ -27 & 6 & 13 \end{bmatrix} \begin{bmatrix} x_1(k) \\ x_2(k) \\ x_3(k) \end{bmatrix} + \begin{bmatrix} -1 & 2 \\ 0 & 0 \\ 3 & 0 \end{bmatrix} \begin{bmatrix} u_1(k) \\ u_2(k) \end{bmatrix}$$

$$\begin{bmatrix} y_1(k) \\ y_2(k) \end{bmatrix} = \begin{bmatrix} 3 & 1 & 1 \\ -9 & 1 & 2 \end{bmatrix} \begin{bmatrix} x_1(k) \\ x_2(k) \\ x_3(k) \end{bmatrix} + \begin{bmatrix} 4 & -1 \\ -1 & \frac{1}{2} \end{bmatrix} \begin{bmatrix} u_1(k) \\ u_2(k) \end{bmatrix}$$

4-2. Use observability matrix rank tests to determine whether the following systems are completely observable:

a.
$$\begin{bmatrix} \dot{x}_1(t) \\ \dot{x}_2(t) \end{bmatrix} = \begin{bmatrix} -1 & 4 \\ 0 & 1 \end{bmatrix} \begin{bmatrix} x_1(t) \\ x_2(t) \end{bmatrix} + \begin{bmatrix} 1 \\ 1 \end{bmatrix} u(t)$$

$$y(t) = \begin{bmatrix} 0 & 1 \end{bmatrix} \begin{bmatrix} x_1(t) \\ x_2(t) \end{bmatrix}$$

b.
$$\begin{bmatrix} \dot{x}_1(t) \\ \dot{x}_2(t) \\ \dot{x}_3(t) \end{bmatrix} = \begin{bmatrix} 1 & 0 & 1 \\ -1 & 1 & 2 \\ 0 & 1 & 0 \end{bmatrix} \begin{bmatrix} x_1(t) \\ x_2(t) \\ x_3(t) \end{bmatrix} + \begin{bmatrix} 0 \\ 0 \\ 1 \end{bmatrix} u(t)$$

$$\begin{bmatrix} y_1(t) \\ y_2(t) \end{bmatrix} = \begin{bmatrix} 0 & 1 & 1 \\ 1 & 0 & 1 \end{bmatrix} \begin{bmatrix} x_1(t) \\ x_2(t) \\ x_3(t) \end{bmatrix}$$

4-3. For the system

$$\begin{bmatrix} x_1(k+1) \\ x_2(k+1) \end{bmatrix} = \begin{bmatrix} 2 & 1 \\ -1 & 1 \end{bmatrix} \begin{bmatrix} x_1(k) \\ x_2(k) \end{bmatrix} + \begin{bmatrix} 2 \\ 3 \end{bmatrix} u(k)$$

$$y(k) = \begin{bmatrix} 0 & 1 \end{bmatrix} \begin{bmatrix} x_1(k) \\ x_2(k) \end{bmatrix} + 2u(k)$$

find $\mathbf{x}(0)$ if $u(0) = 0$, $u(1) = 1$, $y(0) = -1$, and $y(1) = 4$. Then find $\mathbf{x}(1)$.

4-4. For the system of problem 4-3, find $\mathbf{x}(5)$ if $u(4) = -1$, $u(5) = 2$, $y(4) = 0$, and $y(5) = 3$.

4-5. For the system

$$
\begin{bmatrix} x_1(k+1) \\ x_2(k+1) \\ x_3(k+1) \end{bmatrix} = \begin{bmatrix} 1 & 0 & 1 \\ -1 & 1 & 2 \\ 0 & 1 & 3 \end{bmatrix} \begin{bmatrix} x_1(k) \\ x_2(k) \\ x_3(k) \end{bmatrix} + \begin{bmatrix} 3 \\ 1 \\ 0 \end{bmatrix} u(k)
$$

$$
y(k) = \begin{bmatrix} 0 & 1 & 1 \end{bmatrix} \begin{bmatrix} x_1(k) \\ x_2(k) \\ x_3(k) \end{bmatrix} + 2u(k)
$$

find $\mathbf{x}(0)$ if

$y(0) = 1 \qquad y(1) = 2 \qquad y(2) = -1$
$u(0) = 1 \qquad u(1) = -2 \qquad u(2) = 2$

4-6. For the system

$$
\begin{bmatrix} x_1(k+1) \\ x_2(k+1) \end{bmatrix} = \begin{bmatrix} 1 & -1 \\ -2 & 0 \end{bmatrix} \begin{bmatrix} x_1(k) \\ x_2(k) \end{bmatrix} + \begin{bmatrix} 1 \\ -4 \end{bmatrix} u(k)
$$

$$
y(k) = \begin{bmatrix} -1 & 1 \end{bmatrix} \begin{bmatrix} x_1(k) \\ x_2(k) \end{bmatrix} + 2u(k)
$$

find $\mathbf{x}(4)$ if

$u(0) = 2$, $u(1) = 0$, $u(2) = 1$, $u(3) = 2$, $y(1) = 3$, and $y(2) = 0$.

4-7. For the system

$$
\begin{bmatrix} x_1(k+1) \\ x_2(k+1) \\ x_3(k+1) \end{bmatrix} = \begin{bmatrix} 0 & 0 & 1 \\ 1 & -1 & 0 \\ -2 & 0 & -1 \end{bmatrix} \begin{bmatrix} x_1(k) \\ x_2(k) \\ x_3(k) \end{bmatrix} + \begin{bmatrix} 1 \\ -1 \\ 2 \end{bmatrix} u(k)
$$

$$
\begin{bmatrix} y_1(k) \\ y_2(k) \end{bmatrix} = \begin{bmatrix} 1 & 0 & 2 \\ -1 & 1 & 0 \end{bmatrix} \begin{bmatrix} x_1(k) \\ x_2(k) \\ x_3(k) \end{bmatrix}
$$

find $\mathbf{x}(2)$ if $u(0) = 1$, $u(1) = -1$,

$$
\mathbf{y}(0) = \begin{bmatrix} 2 \\ -1 \end{bmatrix} \qquad \mathbf{y}(1) = \begin{bmatrix} 1 \\ 3 \end{bmatrix}
$$

4-8. A causal second-order, single-input/single-output system has inputs and outputs as follows:

$y(0) = 3 \qquad u(0) = 2$

$$y(1) = 0 \qquad u(1) = 1$$
$$y(2) = -1 \qquad u(2) = 6$$
$$y(3) = 4 \qquad u(3) = 2$$
$$y(4) = 0 \qquad u(4) = 0$$
$$y(5) = -1 \qquad u(5) = -1$$
$$y(6) = 1 \qquad u(6) = 0$$

a. Find the system's input–output equation.

b. If the system's state variable equations are in controllable form, find its state at steps 0, 1, and 2.

4-9. A single-input/two-output system has z-transfer functions of the form

$$T_1(z) = \frac{b}{z + a_0}$$

$$T_2(z) = \frac{b_1 z + b_0}{z + a_0}$$

Find the coefficients a_0, b, b_1, and b_0 if the system has the following outputs and inputs:

$$y_1(3) = 2 \qquad y_2(3) = 2 \qquad u(3) = 1$$
$$y_1(4) = -1 \qquad y_2(4) = -1 \qquad u(4) = -1$$
$$y_1(5) = 4 \qquad y_2(5) = 4 \qquad u(5) = 0$$

4-10. A two-input/one-output system has z-transfer functions of the form

$$T_1(z) = \frac{b_1 z + b_0}{z + a_0}$$

$$T_2(z) = \frac{b}{z + a_0}$$

Find the coefficients a_0, b_1, b_0, and b if the system has the following outputs and inputs:

$$y(0) = 0 \qquad u_1(0) = 2 \qquad u_2(0) = -1$$
$$y(1) = -1 \qquad u_1(1) = -3 \qquad u_2(1) = -1$$
$$y(2) = 2 \qquad u_1(2) = 1 \qquad u_2(2) = 4$$
$$y(3) = 1 \qquad u_1(3) = 1 \qquad u_2(3) = 0$$
$$y(4) = 3 \qquad u_1(4) = 2$$

4-11. A system with z-transfer function

$$T(z) = \frac{b_1 z + b_0}{z + a_0} = \frac{3z + 6}{z + 2} = 3$$

has a pole-zero cancellation. Show that if the initial conditions are zero, the parameters a_0 and b_0 cannot be separately identified from subsequent input and output data, but that the data will show that $b_0 = 3a_0$. Then let the initial conditions be $y(-1) = 4$, $u(-1) = 0$ and show that a_0, b_1, and b_0 can be identified from the outputs and inputs at steps 0, 1, and 2.

4-12. If the parameters of a system vary slowly enough with step, they can be identified accurately, as they vary, by simply using the minimum amount of most recent data to solve for them at each step. For a system with the slowly step-varying input–output equation

$$y(k + 2) + a_1(k)y(k + 1) + a_0(k)y(k) = b_0(k)u(k)$$

generate (with the aid of a computer) input and output data for steps $k = 0$ through $k = 99$. Then use the most recent data available to compute $a_1(k)$, $a_0(k)$, and $b_0(k)$ at each step from $k = 4$ on. Use the following parameters and inputs, and compare the actual and the computed parameters.

a. $a_1(k) = 0.5 + 0.1k$ $a_0(k) = 0.1$
   $b_0(k) = 4 - 0.1k$ $u(k) = 0.92^k - 0.8^k$

b. $a_1(k) = 0.5 + 0.2k$ $a_0(k) = 0.2$
   $b_0(k) = 4 - 0.3k$ $u(k) = 0.96^k$

4-13. Describe how the parameters of a *continuous-time* system can be found from samples of its outputs. Apply the method to a second-order example system.

4-14. For the system

$$\begin{bmatrix} x_1(k+1) \\ x_2(k+1) \end{bmatrix} = \begin{bmatrix} (-1)^k & -1 \\ 0 & 1 \end{bmatrix} \begin{bmatrix} x_1(k) \\ x_2(k) \end{bmatrix} + \begin{bmatrix} k \\ 2 \end{bmatrix} u(k)$$

$$y(k) = \begin{bmatrix} \frac{1}{k+1} & 0 \end{bmatrix} \begin{bmatrix} x_1(k) \\ x_2(k) \end{bmatrix} - u(k)$$

find $\mathbf{x}(0)$, $\mathbf{x}(1)$, and $\mathbf{x}(2)$ if

$$y(0) = 0 \qquad u(0) = 1$$

$$y(1) = -\frac{1}{3} \qquad u(1) = -1$$

$$y(2) = -1 \qquad u(2) = 4$$
$$y(3) = 2.1 \qquad u(3) = -2$$

4-15. Use controllability matrix rank tests to determine whether the systems with the following state equations are completely controllable:

a.
$$\begin{bmatrix} x_1(k+1) \\ x_2(k+1) \\ x_3(k+1) \end{bmatrix} = \begin{bmatrix} 0 & 2 & -1 \\ -1 & 2 & 0 \\ 0 & 1 & -3 \end{bmatrix} \begin{bmatrix} x_1(k) \\ x_2(k) \\ x_3(k) \end{bmatrix} + \begin{bmatrix} 1 \\ 0 \\ -2 \end{bmatrix} u(k)$$

b.
$$\begin{bmatrix} x_1(k+1) \\ x_2(k+1) \\ x_3(k+1) \end{bmatrix} = \begin{bmatrix} 0 & 0 & 3 \\ -1 & -1 & 5 \\ -1 & 1 & -1 \end{bmatrix} \begin{bmatrix} x_1(k) \\ x_2(k) \\ x_3(k) \end{bmatrix} + \begin{bmatrix} 0 & 1 \\ 0 & 1 \\ 1 & 0 \end{bmatrix} \begin{bmatrix} u_1(k) \\ u_2(k) \end{bmatrix}$$

4-16. Use controllability matrix rank tests to determine whether the systems with the following state equations are completely controllable:

a.
$$\begin{bmatrix} \dot{x}_1(t) \\ \dot{x}_2(t) \end{bmatrix} = \begin{bmatrix} -2 & -4 \\ 0 & -4 \end{bmatrix} \begin{bmatrix} x_1(t) \\ x_2(t) \end{bmatrix} + \begin{bmatrix} 3 \\ -1 \end{bmatrix} u(t)$$

b.
$$\begin{bmatrix} \dot{x}_1(t) \\ \dot{x}_2(t) \\ \dot{x}_3(t) \end{bmatrix} = \begin{bmatrix} -1 & 1 & 0 \\ 0 & 1 & -2 \\ 1 & 0 & 1 \end{bmatrix} \begin{bmatrix} x_1(t) \\ x_2(t) \\ x_3(t) \end{bmatrix} + \begin{bmatrix} 1 & 0 \\ 1 & -1 \\ 2 & 1 \end{bmatrix} \mathbf{u}(t)$$

4-17. For the system with state equations

$$\begin{bmatrix} x_1(k+1) \\ x_2(k+1) \end{bmatrix} = \begin{bmatrix} 1 & -1 \\ 2 & -1 \end{bmatrix} \begin{bmatrix} x_1(k) \\ x_2(k) \end{bmatrix} + \begin{bmatrix} 1 \\ 1 \end{bmatrix} u(k)$$

find the inputs $u(0)$ and $u(1)$ that will bring the state from

$$\mathbf{x}(0) = \begin{bmatrix} 1 \\ 3 \end{bmatrix} \qquad \text{to} \qquad \mathbf{x}(2) = \begin{bmatrix} -1 \\ 2 \end{bmatrix}$$

4-18. For the system of problem 4-17, find the inputs $u(4)$ and $u(5)$ that will bring the state from

$$\mathbf{x}(4) = \begin{bmatrix} 1 \\ 0 \end{bmatrix} \qquad \text{to} \qquad \mathbf{x}(6) = \begin{bmatrix} 0 \\ 1 \end{bmatrix}$$

4-19. For the system with state equations

$$\begin{bmatrix} x_1(k+1) \\ x_2(k+1) \\ x_3(k+1) \end{bmatrix} = \begin{bmatrix} 1 & 2 & 0 \\ 1 & -1 & 0 \\ 0 & 1 & 3 \end{bmatrix} \begin{bmatrix} x_1(k) \\ x_2(k) \\ x_3(k) \end{bmatrix} + \begin{bmatrix} -1 \\ 1 \\ 0 \end{bmatrix} u(k)$$

find the inputs $u(0)$, $u(1)$, and $u(2)$ that will bring the state from

$$\begin{bmatrix} x_1(0) \\ x_2(0) \\ x_3(0) \end{bmatrix} = \begin{bmatrix} 2 \\ -3 \\ 1 \end{bmatrix} \quad \text{to} \quad \begin{bmatrix} x_1(3) \\ x_2(3) \\ x_3(3) \end{bmatrix} = \begin{bmatrix} 2 \\ 0 \\ -2 \end{bmatrix}$$

4-20. Find another set of inputs in problem 4-19 that will bring the system from the given $\mathbf{x}(0)$ to a state where the *output*

$$\begin{bmatrix} y_1(k) \\ y_2(k) \end{bmatrix} = \begin{bmatrix} 3 & 0 & -2 \\ 1 & 1 & 0 \end{bmatrix} \begin{bmatrix} x_1(k) \\ x_2(k) \\ x_3(k) \end{bmatrix}$$

is

$$\begin{bmatrix} y_1 \\ y_2 \end{bmatrix} = \begin{bmatrix} 3 \\ 4 \end{bmatrix}$$

Can this be done in two steps?

4-21. For the system with state equations

$$\begin{bmatrix} x_1(k+1) \\ x_2(k+1) \end{bmatrix} = \begin{bmatrix} 1 & 0 \\ 2 & -1 \end{bmatrix} \begin{bmatrix} x_1(k) \\ x_2(k) \end{bmatrix} + \begin{bmatrix} 1 & 2 \\ -1 & 0 \end{bmatrix} \begin{bmatrix} u_1(k) \\ u_2(k) \end{bmatrix}$$

find inputs $\mathbf{u}(1)$, $\mathbf{u}(2)$, and $\mathbf{u}(3)$ that will bring the state $\mathbf{x}$ from

$$\mathbf{x}(1) = \begin{bmatrix} 1 \\ -1 \end{bmatrix} \quad \text{to} \quad \mathbf{x}(4) = \begin{bmatrix} 0 \\ 3 \end{bmatrix}$$

4-22. For the system of problem 4-14, find $u(0)$, $u(1)$, and $u(2)$ that will take the system from

$$\mathbf{x}(0) = \begin{bmatrix} 0 \\ -1 \end{bmatrix} \quad \text{to} \quad \mathbf{x}(3) = \begin{bmatrix} 1 \\ 0 \end{bmatrix}$$

4-23. For the discrete-time system with state equations

$$\begin{bmatrix} x_1(k+1) \\ x_2(k+1) \\ x_3(k+1) \end{bmatrix} = \begin{bmatrix} 0 & 1 & 0 \\ 0 & 0 & 1 \\ -1 & 2 & -3 \end{bmatrix} \begin{bmatrix} x_1(k) \\ x_2(k) \\ x_3(k) \end{bmatrix} + \begin{bmatrix} 0 \\ 0 \\ 1 \end{bmatrix} u(k)$$

find the state feedback gains $\mathbf{e}$ in

$$u(k) = \mathbf{e}^\dagger \mathbf{x}(k) + \rho(k)$$

such that the feedback system has eigenvalues at $\lambda = \frac{1}{3}$ and at $\lambda = \frac{1}{3} \pm j\frac{1}{2}$.

4-24. For the discrete-time system with state equations

$$\begin{bmatrix} x_1(k+1) \\ x_2(k+1) \\ x_3(k+1) \end{bmatrix} = \begin{bmatrix} 0 & -1 & 1 \\ 2 & 2 & 0 \\ -1 & 0 & -1 \end{bmatrix} \begin{bmatrix} x_1(k) \\ x_2(k) \\ x_3(k) \end{bmatrix} + \begin{bmatrix} 0 \\ 1 \\ 0 \end{bmatrix} u(k)$$

find the state feedback gains **e** in

$$u(k) = \mathbf{e}^t\mathbf{x}(k) + \rho(k)$$

such that the feedback system has eigenvalues at $\lambda = 0$ and at $\lambda = \pm j\frac{1}{2}$.

4-25. For the systems in each of the following problems, find state feedback gains **e** using Ackermann's formula

a. Problem 4-23

b. Problem 4-24

4-26. For the discrete-time system with state equations

$$\begin{bmatrix} x_1(k+1) \\ x_2(k+1) \\ x_3(k+1) \end{bmatrix} = \begin{bmatrix} \frac{1}{3} & 0 & 0 \\ 0 & \frac{1}{2} & 0 \\ 0 & 0 & -1 \end{bmatrix} \begin{bmatrix} x_1(k) \\ x_2(k) \\ x_3(k) \end{bmatrix} + \begin{bmatrix} 1 & 0 \\ -1 & 1 \\ 0 & 2 \end{bmatrix} \begin{bmatrix} u_1(k) \\ u_2(k) \end{bmatrix}$$

find the state feedback gains **E** in

$$\begin{bmatrix} u_1(k) \\ u_2(k) \end{bmatrix} = \begin{bmatrix} e_{11} & e_{12} & e_{13} \\ e_{21} & e_{22} & e_{23} \end{bmatrix} \begin{bmatrix} x_1(k) \\ x_2(k) \\ x_3(k) \end{bmatrix} + \begin{bmatrix} \rho_1(k) \\ \rho_2(k) \end{bmatrix} = \mathbf{E}\mathbf{x}(k) + \boldsymbol{\rho}(k)$$

such that the feedback system has all eigenvalues at $\lambda = 0$.

4-27. Find a *different* set of feedback gains **E** for problem 4-26.

4-28. Find the state feedback gains **e**(0), **e**(1), and **e**(2) that make the system with state equations

$$\begin{bmatrix} x_1(k+1) \\ x_2(k+1) \\ x_3(k+1) \end{bmatrix} = \begin{bmatrix} \frac{1}{3} & (-1)^k & 0 \\ 0 & 0 & 0 \\ 0 & \frac{1}{3} & -\frac{1}{3} \end{bmatrix} \begin{bmatrix} x_1(k) \\ x_2(k) \\ x_3(k) \end{bmatrix} + \begin{bmatrix} 0 \\ 1 \\ 0 \end{bmatrix} u(k)$$

and input

$$u(k) = \mathbf{e}^t(k)\mathbf{x}(k) + \rho(k)$$

where ρ is an external input, deadbeat.

4-29. Find state feedback gains $\mathbf{E}(0)$ and $\mathbf{E}(1)$ that make the system with state equations

$$\begin{bmatrix} x_1(k+1) \\ x_2(k+1) \\ x_3(k+1) \end{bmatrix} = \begin{bmatrix} \frac{1}{3} & 2k & 0 \\ 0 & -1 & 1 \\ 0 & (-1)^k & \frac{1}{3} \end{bmatrix} \begin{bmatrix} x_1(k) \\ x_2(k) \\ x_3(k) \end{bmatrix} + \begin{bmatrix} 1 & 0 \\ k & -1 \\ 0 & 0 \end{bmatrix} \begin{bmatrix} u_1(k) \\ u_2(k) \end{bmatrix}$$

and input

$$\mathbf{u}(k) = \mathbf{E}(k)\mathbf{x}(k) + \boldsymbol{\rho}(k)$$

where $\boldsymbol{\rho}(k)$ is an external input, deadbeat in two steps.

4-30. For the system

$$\begin{bmatrix} x_1(k+1) \\ x_2(k+1) \\ x_3(k+1) \end{bmatrix} = \begin{bmatrix} 0 & 1 & 0 \\ 0 & 0 & 1 \\ -3 & 2 & -1 \end{bmatrix} \begin{bmatrix} x_1(k) \\ x_2(k) \\ x_3(k) \end{bmatrix} + \begin{bmatrix} 0 \\ 0 \\ 1 \end{bmatrix} u(k)$$

$$\begin{bmatrix} y_1(k) \\ y_2(k) \\ y_3(k) \end{bmatrix} = \begin{bmatrix} 1 & -1 & 0 \\ 3 & 0 & 1 \\ 0 & 1 & -1 \end{bmatrix} \begin{bmatrix} x_1(k) \\ x_2(k) \\ x_3(k) \end{bmatrix} + \begin{bmatrix} 3 \\ -2 \\ 4 \end{bmatrix} u(k)$$

find the output feedback gains $\mathbf{f}$ in

$$u(k) = \mathbf{f}^\dagger[\mathbf{y}(k) - \mathbf{d}u(k)] + \rho(k)$$

such that the feedback system has eigenvalues at $\lambda = 0$ and at $\lambda = \pm j\frac{1}{2}$.

4-31. For the system of Problem 4-30, suppose instead that the outputs

$$\begin{bmatrix} y_1(k) \\ y_2(k) \\ y_3(k) \\ y_4(k) \end{bmatrix} = \begin{bmatrix} 1 & 2 & 3 \\ 2 & 1 & 6 \\ 1 & 0 & -2 \\ -1 & 3 & 7 \end{bmatrix} \begin{bmatrix} x_1(k) \\ x_2(k) \\ x_3(k) \end{bmatrix} + \begin{bmatrix} 6 \\ -3 \\ 0 \\ 4 \end{bmatrix} u(k)$$

are available. Design an output feedback system such that the overall system's characteristic equation is

$$\lambda^3 = 0$$

Verify that the system you have designed has this property.

4-32. Repeat problem 4-31, designing a different feedback system for which the characteristic equation of the overall system is

$$\lambda^3 - \lambda^2 + \lambda + \frac{1}{2} = 0$$

In this design, have *all four* outputs couple to the input $u(k)$ with nonzero coefficients.

4-33. For the system

$$\begin{bmatrix} x_1(k+1) \\ x_2(k+1) \\ x_3(k+1) \end{bmatrix} = \begin{bmatrix} 1 & 1 & 0 \\ 1 & 0 & 1 \\ -2 & 0 & 0 \end{bmatrix} \begin{bmatrix} x_1(k) \\ x_2(k) \\ x_3(k) \end{bmatrix} + \begin{bmatrix} 1 & 0 & -1 \\ -1 & 1 & 2 \\ 2 & 0 & 1 \end{bmatrix} \begin{bmatrix} u_1(k) \\ u_2(k) \\ u_3(k) \end{bmatrix}$$

$$y(k) = \begin{bmatrix} 1 & 0 & 0 \end{bmatrix} \begin{bmatrix} x_1(k) \\ x_2(k) \\ x_3(k) \end{bmatrix}$$

find the output feedback gains **f** in

$$\mathbf{u}(k) = \mathbf{f}y(k)$$

such that the feedback system has eigenvalues at $\lambda = \frac{1}{2}$ and at $\lambda = \pm j\frac{1}{3}$.

4-34. For the system shown below:

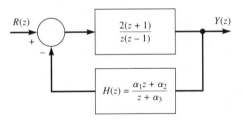

a. Find the coefficients of the first-order compensator $H(z)$ such that the feedback system poles are at $\lambda = 0.4 \pm j0.2$ and $\lambda = \frac{1}{2}$.

b. Is it possible to achieve satisfactory pole locations for the feedback system of part (a) with a zero-order compensator $H(z)$? If so, find it.

c. Repeat parts (a) and (b) above for the following set of feedback system poles:

$$\lambda_{1,2} = 0.1 \pm j0.1 \qquad \lambda_3 = 0.2$$

d. Plot the root locus for changes in the plant gain for part (b).

4-35. For the first-order plant

$$x(k+1) = 0.4x(k) + u(k)$$

calculate the optimal state feedback gains using the principle of optimality for each of the following performance measures:

a. $J = 10x^2(4) + \sum_{i=0}^{3} u^2(i)$

b. $J = 10x^2(4) + \sum_{i=0}^{3} 5x^2(i)$

c. $J = 10x^2(4) + \sum_{i=0}^{3} [5x^2(i) + u^2(i)]$

d. $J = 10x^2(4) + \sum_{i=0}^{3} [x^2(i) + 5u^2(i)]$

4-36. Repeat problem 4-35 for the unstable plant

$$x(k + 1) = 1.2x(k) + u(k)$$

4-37. Repeat problem 4-35 using the method of Section 4.8.

4-38. Repeat problem 4-36 using the method of Section 4.8.

4-39. Find the optimal steady state feedback gain for the first-order plant

$$x(k + 1) = 0.7x(k) + u(k)$$

corresponding to the performance measure

$$J = \sum_{i=0}^{\infty} [x^2(i) + 3u^2(i)]$$

a. Find e to three significant figures by calculating e, with the aid of a computer or calculator, for enough backward-in-step iterations.

b. Find e by letting e and P in (4-39) and (4-40) be constant and solving the resulting quadratic equation.

4-40. Repeat problem 4-39 using the method of Section 4.8.

4-41. For the spacecraft roll attitude equation (4-41), find and plot the optimal state feedback gains for each of the following performance measures:

$$J = \mathbf{x}^\dagger(10)\mathbf{P}(10)\mathbf{X}(10)\mathbf{x} + \sum_{i=0}^{9} [\mathbf{x}^\dagger(i)\mathbf{Q}\mathbf{x}(i) + u^\dagger(i)Ru(i)]$$

a. $P(10) = \begin{bmatrix} 1000 & 0 \\ 0 & 100 \end{bmatrix}$ $Q = \begin{bmatrix} 10 & 0 \\ 0 & 20 \end{bmatrix}$ $R = 3$

b. $P(10) = \begin{bmatrix} 1000 & 0 \\ 0 & 100 \end{bmatrix}$ $Q = \begin{bmatrix} 20 & 0 \\ 0 & 40 \end{bmatrix}$ $R = 3$

c. $P(10) = \begin{bmatrix} 1000 & 0 \\ 0 & 100 \end{bmatrix}$ $Q = \begin{bmatrix} 10 & 0 \\ 0 & 20 \end{bmatrix}$ $R = 5$

d. $P(10) = \begin{bmatrix} 100 & 0 \\ 0 & 10 \end{bmatrix}$ $Q = \begin{bmatrix} 10 & 0 \\ 0 & 20 \end{bmatrix}$ $R = 3$

4-42. Repeat problem 4-41 for the optimal steady state feedback gains, based on the method developed in Section 4.8, and, for each case, find the feedback system eigenvalues.

5

Digital Observers and Regulator Design

5.1 Preview

In 1964, David Luenberger of Stanford University put forth the idea of *observers*, systems that recursively estimate the state of other systems. It was soon realized that observers offer a powerful, unified framework for feedback control system design.

This chapter addresses the first of two concerns of digital tracking system design, that of obtaining acceptable feedback system zero-input response. The systems that result are *regulators*. When the plant state is not entirely accessible, as is usually the case, the state is estimated with an observer, and the estimated state is used in place of the actual state for feedback. Basic observer theory is developed and a simple, complete observer design procedure, yielding observers of minimal order, if desired, is applied. It is then shown that by using the plant state estimate for feedback in place of the measured plant state itself, the designer has complete freedom to place all of the feedback system eigenvalues.

5.2 Full-Order State Observers

When a plant's state is not available for feedback, a suitably formed estimate of the state, from the plant outputs and inputs, can be used in place of the state itself. An *observer* of a plant is another system with inputs that are the plant inputs and the plant outputs. The observer produces an estimate of the plant state or of a transformation of that state.

5.2.1 Basic Theory

For an nth-order plant

$$\mathbf{x}(k + 1) = \mathbf{A}\mathbf{x}(k) + \mathbf{B}\mathbf{u}(k)$$
$$\mathbf{y}(k) = \mathbf{C}\mathbf{x}(k) + \mathbf{D}\mathbf{u}(k) \qquad (5\text{-}1)$$

another nth-order system, driven by the inputs and outputs of the plant as in Figure 5-1,

$$\boldsymbol{\xi}(k + 1) = \mathbf{F}\boldsymbol{\xi}(k) + \mathbf{G}\mathbf{y}(k) + \mathbf{H}\mathbf{u}(k) \qquad (5\text{-}2)$$

is termed *a full-order state observer* of the plant, provided that the error between the plant and observer states

$$
\begin{aligned}
\mathbf{x}(k + 1) - \boldsymbol{\xi}(k + 1) &= \mathbf{A}\mathbf{x}(k) + \mathbf{B}\mathbf{u}(k) - \mathbf{F}\boldsymbol{\xi}(k) - \mathbf{G}\mathbf{y}(k) - \mathbf{H}\mathbf{u}(k) \\
&= \mathbf{A}\mathbf{x}(k) + \mathbf{B}\mathbf{u}(k) - \mathbf{F}\boldsymbol{\xi}(k) - \mathbf{G}\mathbf{C}\mathbf{x}(k) \\
&\quad - \mathbf{G}\mathbf{D}\mathbf{u}(k) - \mathbf{H}\mathbf{u}(k) \\
&= (\mathbf{A} - \mathbf{G}\mathbf{C})\mathbf{x}(k) - \mathbf{F}\boldsymbol{\xi}(k) + (\mathbf{B} - \mathbf{G}\mathbf{D} - \mathbf{H})\mathbf{u}(k)
\end{aligned}
$$

is governed by an autonomous (zero-input) equation. When $\mathbf{F}$ and $\mathbf{H}$ are chosen as

$$\mathbf{F} = \mathbf{A} - \mathbf{G}\mathbf{C}$$
$$\mathbf{H} = \mathbf{B} - \mathbf{G}\mathbf{D}$$

so that the error signal satisfies

$$\mathbf{x}(k + 1) - \boldsymbol{\xi}(k + 1) = (\mathbf{A} - \mathbf{G}\mathbf{C})[\mathbf{x}(k) - \boldsymbol{\xi}(k)]$$

or

$$
\begin{aligned}
\mathbf{x}(k) - \boldsymbol{\xi}(k) &= (\mathbf{A} - \mathbf{G}\mathbf{C})^k[\mathbf{x}(0) - \boldsymbol{\xi}(0)] \\
&= \mathbf{F}^k[\mathbf{x}(0) - \boldsymbol{\xi}(0)]
\end{aligned}
$$

then the system in equation (5-2) is a full-order state observer of the plant in equation (5-1), if the matrix $\mathbf{G}$ can be chosen so that all the eigenvalues of $\mathbf{F} = \mathbf{A} - \mathbf{G}\mathbf{C}$ are inside the unit circle in the complex plane. Then the error approaches zero with step regardless of the initial

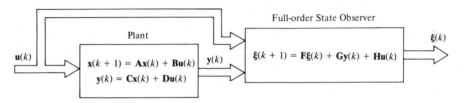

Figure 5-1 A full-order state observer of a plant.

values of $\mathbf{x}(0)$ and $\xi(0)$. Of course, the closer the observer eigenvalues are to the origin of the unit circle, the faster the convergence of the error to zero. *Full-order* means that the observer is of the same order as the plant; *state observer* means that the output (which is here the observer state) approaches the plant state rather than some transformation of the plant state. The full-order state observer relations are summarized in Table 5-1.

Table 5-1 Full-Order State Observer Relations

Plant Model

$\mathbf{x}(k + 1) = \mathbf{Ax}(k) + \mathbf{Bu}(k)$

$\quad \mathbf{y}(k) = \mathbf{Cx}(k) + \mathbf{Du}(k)$

Observer

$\xi(k + 1) = \mathbf{F}\xi(k) + \mathbf{G}\mathbf{y}(k) + \mathbf{H}\mathbf{u}(k)$

where

$\mathbf{F} = \mathbf{A} - \mathbf{GC}$

$\mathbf{H} = \mathbf{B} - \mathbf{GD}$

Observer Error

$\mathbf{x}(k + 1) - \xi(k + 1) = \mathbf{F}[\mathbf{x}(k) - \xi(k)]$

$\mathbf{x}(k) - \xi(k) = \mathbf{F}^k[\mathbf{x}(0) - \xi(0)]$

5.2.2 Observers of Systems in Observable Form

For a single-output plant, as shown in Figure 5-2

$\mathbf{x}(k + 1) = \mathbf{Ax}(k) + \mathbf{Bu}(k)$

$\quad y(k) = \mathbf{c}^\dagger\mathbf{x}(k) + \mathbf{d}^\dagger\mathbf{u}(k)$

the full-order state observer relations are

$\xi(k + 1) = \mathbf{F}\xi(k) + \mathbf{g}y(k) + \mathbf{H}\mathbf{u}(k)$

$\qquad = (\mathbf{A} - \mathbf{gc}^\dagger)\xi(k) + \mathbf{g}y(k) + (\mathbf{B} - \mathbf{gd}^\dagger)\mathbf{u}(k)$

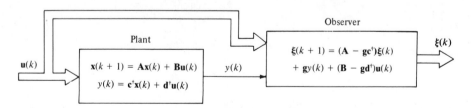

Figure 5-2 Full-order state observer of a single-output plant.

and the error between the plant state and the observer state is governed by

$$\mathbf{x}(k + 1) - \boldsymbol{\xi}(k + 1) = (\mathbf{A} - \mathbf{gc}^\dagger)[\mathbf{x}(k) - \boldsymbol{\xi}(k)]$$

The eigenvalues of $\mathbf{F} = \mathbf{A} - \mathbf{gc}^\dagger$ can be placed arbitrarily by the choice of $\mathbf{g}$, provided that $(\mathbf{A}, \mathbf{c})$ is completely observable. This is similar to the situation with state feedback, in which the eigenvalues of $\mathbf{A} + \mathbf{be}^\dagger$ can be placed arbitrarily by the choice of $\mathbf{e}$, providing that the system $(\mathbf{A}, \mathbf{b})$ is completely controllable.

For example, consider the plant

$$\begin{bmatrix} x_1(k + 1) \\ x_2(k + 1) \\ x_3(k + 1) \end{bmatrix} = \begin{bmatrix} \frac{1}{2} & 1 & 0 \\ -1 & 0 & 1 \\ 0 & 0 & 0 \end{bmatrix} \begin{bmatrix} x_1(k) \\ x_2(k) \\ x_3(k) \end{bmatrix} + \begin{bmatrix} 1 & 4 \\ 0 & 0 \\ -3 & 2 \end{bmatrix} \begin{bmatrix} u_1(k) \\ u_2(k) \end{bmatrix}$$

$$= \mathbf{Ax}(k) + \mathbf{Bu}(k)$$

$$y(k) = \begin{bmatrix} 1 & 0 & 0 \end{bmatrix} \begin{bmatrix} x_1(k) \\ x_2(k) \\ x_3(k) \end{bmatrix} + \begin{bmatrix} 0 & 4 \end{bmatrix} \begin{bmatrix} u_1(k) \\ u_2(k) \end{bmatrix}$$

$$= \mathbf{c}^\dagger \mathbf{x}(k) + \mathbf{d}^\dagger \mathbf{u}(k) \tag{5-3}$$

which is in observable form. A full-order state observer of this plant is another third-order sytem. If the observer output is the observer state, then

$$\boldsymbol{\xi}(k + 1) = \mathbf{F}\boldsymbol{\xi}(k) + \mathbf{g}y(k) + \mathbf{Hu}(k)$$
$$= (\mathbf{A} - \mathbf{gc}^\dagger)\boldsymbol{\xi}(k) + \mathbf{g}y(k) + (\mathbf{B} - \mathbf{gd}^\dagger)\mathbf{u}(k)$$

or

$$\begin{bmatrix} \xi_1(k + 1) \\ \xi_2(k + 1) \\ \xi_3(k + 1) \end{bmatrix} = \left(\begin{bmatrix} \frac{1}{2} & 1 & 0 \\ -1 & 0 & 1 \\ 0 & 0 & 0 \end{bmatrix} - \begin{bmatrix} g_1 \\ g_2 \\ g_3 \end{bmatrix} \begin{bmatrix} 1 & 0 & 0 \end{bmatrix} \right) \begin{bmatrix} \xi_1(k) \\ \xi_2(k) \\ \xi_3(k) \end{bmatrix}$$

$$+ \begin{bmatrix} g_1 \\ g_2 \\ g_3 \end{bmatrix} y(k) + \left(\begin{bmatrix} 1 & 4 \\ 0 & 0 \\ -3 & 2 \end{bmatrix} - \begin{bmatrix} g_1 \\ g_2 \\ g_3 \end{bmatrix} \begin{bmatrix} 0 & 4 \end{bmatrix} \right) \begin{bmatrix} u_1(k) \\ u_2(k) \end{bmatrix}$$

$$= \begin{bmatrix} (\frac{1}{2} - g_1) & 1 & 0 \\ (-1 - g_2) & 0 & 1 \\ -g_3 & 0 & 0 \end{bmatrix} \begin{bmatrix} \xi_1(k) \\ \xi_2(k) \\ \xi_3(k) \end{bmatrix} + \begin{bmatrix} g_1 \\ g_2 \\ g_3 \end{bmatrix} y(k)$$

$$+ \begin{bmatrix} 1 & (4 - 4g_1) \\ 0 & -4g_2 \\ -3 & (2 - 4g_3) \end{bmatrix} \begin{bmatrix} u_1(k) \\ u_2(k) \end{bmatrix}$$

The characteristic equation of the observer is, in terms of the gains g_1, g_2, and g_3,

$$\lambda^3 + \left(-\frac{1}{2} + g_1\right)\lambda^2 + (1 + g_2)\lambda + g_3 = 0$$

If it is desired that the observer have eigenvalues of 0 and $-\frac{1}{2} \pm j\frac{1}{4}$, then the desired characteristic equation is

$$\lambda\left(\lambda + \frac{1}{2} + j\frac{1}{4}\right)\left(\lambda + \frac{1}{2} - j\frac{1}{4}\right) = \lambda^3 + \lambda^2 + \frac{5}{16}\lambda = 0$$

which is achieved for

$$\mathbf{g} = \begin{bmatrix} g_1 \\ g_2 \\ g_3 \end{bmatrix} = \begin{bmatrix} \frac{3}{2} \\ -\frac{11}{16} \\ 0 \end{bmatrix}$$

The observer with these eigenvalues is

$$\begin{bmatrix} \xi_1(k + 1) \\ \xi_2(k + 1) \\ \xi_3(k + 1) \end{bmatrix} = \begin{bmatrix} -1 & 1 & 0 \\ -\frac{5}{16} & 0 & 1 \\ 0 & 0 & 0 \end{bmatrix}\begin{bmatrix} \xi_1(k) \\ \xi_2(k) \\ \xi_3(k) \end{bmatrix} + \begin{bmatrix} \frac{3}{2} \\ -\frac{11}{16} \\ 0 \end{bmatrix}y(k)$$

$$+ \begin{bmatrix} 1 & -2 \\ 0 & \frac{11}{4} \\ -3 & 2 \end{bmatrix}\begin{bmatrix} u_1(k) \\ u_2(k) \end{bmatrix} \tag{5-4}$$

In Figure 5-3, the states of the plant in equation (5-3) and of the observer in equation (5-4) are shown for arbitrarily chosen plant inputs and initial conditions. The error between the state of the plant and the observer state is described by

$$\mathbf{x}(k + 1) - \xi(k + 1) = \mathbf{F}[\mathbf{x}(k) - \xi(k)]$$

or

$$\mathbf{x}(k) - \xi(k) = \mathbf{F}^k[\mathbf{x}(0) - \xi(0)]$$

and decays as $\mathbf{F}^k$. If the eigenvalues of $\mathbf{F}$ are inside the unit circle, then $\mathbf{F}^k$ decays to zero as k approaches infinity.

5.2.3 Deadbeat Observation

If the characteristic equation of an nth-order observer is chosen to be

$$\lambda^n = 0$$

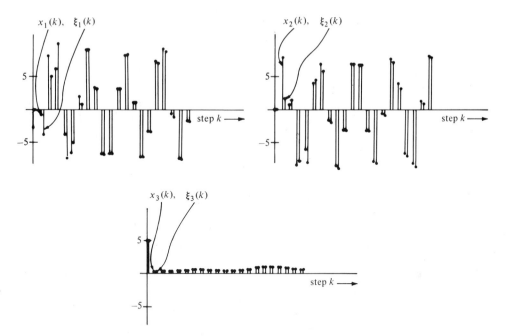

Figure 5-3 Comparison of plant and observer states.

that is, with all eigenvalues at $\lambda = 0$, then applying the Cayley-Hamilton theorem

$$\mathbf{F}^n = \mathbf{0}$$

At the nth step, the error between the plant state and the observer state is given by

$$\mathbf{x}(n) - \xi(n) = \mathbf{F}^n[\mathbf{x}(0) - \xi(0)] = \mathbf{0}$$

so

$$\xi(n) = \mathbf{x}(n)$$

and the observer state equals the plant state. Such an observer is termed *deadbeat*. In subsequent steps, the observer state continues to equal the plant state.

By placing all observer eigenvalues at $\lambda = 0$, the observer state is made to converge to that of the plant in n steps. The same can be done with conventional calculation; the state of a completely observable single-output plant can be determined from its inputs and outputs over the preceding n steps by solving the n simultaneous linear algebraic equations involving the observability matrix, as was done earlier. A deadbeat observer is thus a recursive way to solve a set of linear algebraic equations.

A full-order state observer of the plant in equation (5-3), when designed to be deadbeat, is governed by

$$
\begin{bmatrix} \xi_1(k+1) \\ \xi_2(k+1) \\ \xi_3(k+1) \end{bmatrix} = \begin{bmatrix} 0 & 1 & 0 \\ 0 & 0 & 1 \\ 0 & 0 & 0 \end{bmatrix} \begin{bmatrix} \xi_1(k) \\ \xi_2(k) \\ \xi_3(k) \end{bmatrix} + \begin{bmatrix} \frac{1}{2} \\ -1 \\ 0 \end{bmatrix} y(k) + \begin{bmatrix} 1 & 2 \\ 0 & 4 \\ -3 & 2 \end{bmatrix} \begin{bmatrix} u_1(k) \\ u_2(k) \end{bmatrix}
$$

and has the typical response shown in Figure 5-4. The observer state converges to that of the plant in three steps, regardless of inputs and initial conditions.

The control system designer may or may not wish to design an observer to be deadbeat. A more gradual approach to the plant state by the state of the observer might be preferred. Conventional step-invariant deadbeat observers can exhibit very large errors during the steps before convergence occurs and can be overly susceptible to errors due to noise affecting the plant state and measurements.

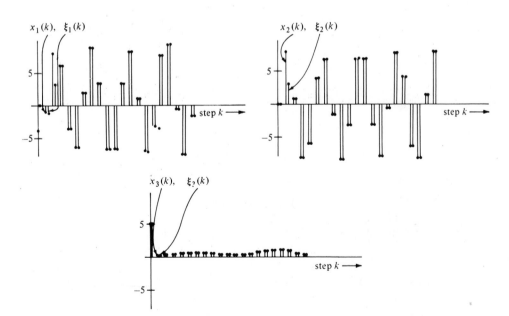

Figure 5-4 Typical response of a deadbeat observer of a third-order plant.

5.2.4 Observers of General Single-Output Systems

One method of designing an observer for a completely observable single-output plant that is not in observable form is to change state variables to the observable form, design the observer in that form, and then convert back to the original system realization. The state variables of a completely observable single-output plant of the form

$$\mathbf{x}(k + 1) = \mathbf{A}\mathbf{x}(k) + \mathbf{B}\mathbf{u}(k)$$
$$y(k) = \mathbf{c}^\dagger\mathbf{x}(k) + \mathbf{d}^\dagger\mathbf{u}(k)$$

are changed via

$$\mathbf{x}(k) = \mathbf{P}\mathbf{x}'(k) = \mathbf{Q}^{-1}\mathbf{x}'(k) \qquad \mathbf{x}'(k) = \mathbf{P}^{-1}\mathbf{x}(k) = \mathbf{Q}\mathbf{x}(k)$$

to place it in observable form

$$\mathbf{x}'(k + 1) = \mathbf{A}'\mathbf{x}'(k) + \mathbf{B}'\mathbf{u}(k)$$
$$y(k) = \mathbf{c}'^\dagger\mathbf{x}'(k) + \mathbf{d}^\dagger\mathbf{u}(k)$$

where

$$\mathbf{A}' = \mathbf{Q}\mathbf{A}\mathbf{Q}^{-1}$$
$$\mathbf{B}' = \mathbf{Q}\mathbf{B}$$
$$\mathbf{c}'^\dagger = \mathbf{c}^\dagger\mathbf{Q}^{-1}$$

A full-order observer of $\mathbf{x}'(k)$, with eigenvalues selected by the designer, is formed

$$\boldsymbol{\xi}'(k + 1) = \mathbf{F}'\boldsymbol{\xi}'(k) + \mathbf{g}'y(k) + \mathbf{H}'\mathbf{u}(k)$$

where

$$\mathbf{F}' = \mathbf{A}' - \mathbf{g}'\mathbf{c}'^\dagger$$
$$\mathbf{H}' = \mathbf{B}' - \mathbf{g}'\mathbf{d}^\dagger$$

Using the observer output equation

$$\mathbf{w}(k) = \mathbf{P}\boldsymbol{\xi}'(k) = \mathbf{Q}^{-1}\boldsymbol{\xi}'(k)$$

transforms the observer state $\boldsymbol{\xi}'(k)$ that observes $\mathbf{x}'(k)$ to $\mathbf{w}(k)$ that observes $\mathbf{x}(k)$, as shown in Figure 5-5(a). Or, transforming the observer state according to

$$\boldsymbol{\xi}(k) = \mathbf{P}\boldsymbol{\xi}'(k) = \mathbf{Q}^{-1}\boldsymbol{\xi}'(k) \qquad \boldsymbol{\xi}'(k) = \mathbf{P}^{-1}\boldsymbol{\xi}(k) = \mathbf{Q}\boldsymbol{\xi}(k)$$

to give

$$\boldsymbol{\xi}(k + 1) = \mathbf{F}\boldsymbol{\xi}(k) + \mathbf{g}y(k) + \mathbf{H}\mathbf{u}(k)$$

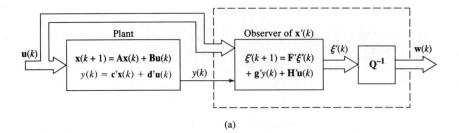

(a)

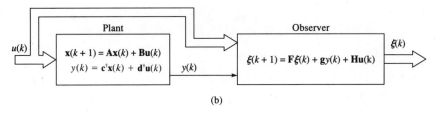

(b)

Figure 5-5 Full-order observers of a single-output system. (a) State observer with output that converges to the plant state. (b) Alternative where the observer state converges to the plant state.

where

$$g = Q^{-1}g'$$
$$F = Q^{-1}F'Q = Q^{-1}A'Q - Q^{-1}g'c'^{\dagger}Q = A - gc^{\dagger}$$
$$H = Q^{-1}H' = Q^{-1}B' - Q^{-1}g'd^{\dagger} = B - gd^{\dagger}$$

results in a full-order state observer of $x(k)$. This arrangement is shown in Figure 5-5(b).

From an implementation viewpoint, the observer arrangement with the lowest number of computations is obviously desirable.

For example, the plant

$$\begin{bmatrix} x_1(k+1) \\ x_2(k+1) \\ x_3(k+1) \end{bmatrix} = \begin{bmatrix} 0 & 3 & 0 \\ 0 & 2 & 1 \\ 1 & 1 & 0 \end{bmatrix} \begin{bmatrix} x_1(k) \\ x_2(k) \\ x_3(k) \end{bmatrix} + \begin{bmatrix} 2 \\ 1 \\ -1 \end{bmatrix} u(k) = Ax(k) + bu(k)$$

$$y(k) = \begin{bmatrix} 0 & 2 & 1 \end{bmatrix} \begin{bmatrix} x_1(k) \\ x_2(k) \\ x_3(k) \end{bmatrix} + u(k) = c^{\dagger}x(k) + du(k) \qquad (5\text{-}5)$$

has the characteristic equation

$$\begin{vmatrix} \lambda & -3 & 0 \\ 0 & (\lambda - 2) & -1 \\ -1 & -1 & \lambda \end{vmatrix} = \lambda^3 - 2\lambda^2 - \lambda - 3$$

$$= \lambda^3 + \alpha_2 \lambda^2 + \alpha_1 \lambda + \alpha_0 = 0$$

The transformation that takes this plant to observable form is given by

$$\mathbf{Q} = \begin{bmatrix} \mathbf{q}_1^\dagger \\ \hline \mathbf{q}_2^\dagger \\ \hline \mathbf{q}_3^\dagger \end{bmatrix}$$

where

$$\mathbf{q}_1^\dagger = \mathbf{c}^\dagger = [0 \quad 2 \quad 1]$$

$$\mathbf{q}_2^\dagger = \mathbf{q}_1^\dagger \mathbf{A} + \alpha_2 \mathbf{c}^\dagger = [0 \quad 2 \quad 1] \begin{bmatrix} 0 & 3 & 0 \\ 0 & 2 & 1 \\ 1 & 1 & 0 \end{bmatrix} + (-2)[0 \quad 2 \quad 1]$$

$$= [1 \quad 1 \quad 0]$$

$$\mathbf{q}_3^\dagger = \mathbf{q}_2^\dagger \mathbf{A} + \alpha_1 \mathbf{c}^\dagger = [1 \quad 1 \quad 0] \begin{bmatrix} 0 & 3 & 0 \\ 0 & 2 & 1 \\ 1 & 1 & 0 \end{bmatrix} + (-1)[0 \quad 2 \quad 1]$$

$$= [0 \quad 3 \quad 0]$$

So

$$\mathbf{Q} = \mathbf{P}^{-1} = \begin{bmatrix} 0 & 2 & 1 \\ 1 & 1 & 0 \\ 0 & 3 & 0 \end{bmatrix} \qquad \mathbf{Q}^{-1} = \mathbf{P} = \begin{bmatrix} 0 & 1 & -\frac{1}{3} \\ 0 & 0 & \frac{1}{3} \\ 1 & 0 & -\frac{2}{3} \end{bmatrix}$$

The transformed plant is

$$\begin{bmatrix} x_1'(k+1) \\ x_2'(k+1) \\ x_3'(k+1) \end{bmatrix} = \begin{bmatrix} 2 & 1 & 0 \\ 1 & 0 & 1 \\ 3 & 0 & 0 \end{bmatrix} \begin{bmatrix} x_1'(k) \\ x_2'(k) \\ x_3'(k) \end{bmatrix} + \begin{bmatrix} 1 \\ 3 \\ 3 \end{bmatrix} u(k) = \mathbf{A}'\mathbf{x}'(k) + \mathbf{b}'u(k)$$

$$y(k) = [1 \quad 0 \quad 0] \begin{bmatrix} x_1'(k) \\ x_2'(k) \\ x_3'(k) \end{bmatrix} + u(k) = \mathbf{c}'^\dagger\mathbf{x}(k) + du(k)$$

and because it is in observable form, a full-order state observer of $\mathbf{x}'(k)$ is easy to design.

Suppose that it is desired that all three observer eigenvalues be at $\lambda = 0$, so that the desired observer is deadbeat, with characteristic equation

$$\lambda^3 = 0$$

Then an observer of the $\mathbf{x}'$ state has state coupling matrix

$$\mathbf{F}' = \mathbf{A}' - \mathbf{g}'\mathbf{c}'^\dagger = \begin{bmatrix} 2 & 1 & 0 \\ 1 & 0 & 1 \\ 3 & 0 & 0 \end{bmatrix} - \begin{bmatrix} g_1' \\ g_2' \\ g_3' \end{bmatrix} [1 \quad 0 \quad 0] = \begin{bmatrix} (2 - g_1') & 1 & 0 \\ (1 - g_2') & 0 & 1 \\ (3 - g_3') & 0 & 0 \end{bmatrix}$$

The observer's characteristic equation, in terms of g_1', g_2', and g_3', is

$$\lambda^3 + (g_1' - 2)\lambda^2 + (g_2' - 1)\lambda + (g_3' - 3) = 0$$

and becomes the desired

$$\lambda^3 = 0$$

for

$$\mathbf{g}' = \begin{bmatrix} g_1' \\ g_2' \\ g_3' \end{bmatrix} = \begin{bmatrix} 2 \\ 1 \\ 3 \end{bmatrix}$$

With this choice of $\mathbf{g}'$, the coupling of the input $u(k)$ into the observer is

$$\mathbf{h}' = \mathbf{b}' - \mathbf{g}'d = \begin{bmatrix} 1 \\ 3 \\ 3 \end{bmatrix} - \begin{bmatrix} 2 \\ 1 \\ 3 \end{bmatrix} (1) = \begin{bmatrix} -1 \\ 2 \\ 0 \end{bmatrix}$$

so that the observer of the $\mathbf{x}'$ state has equations

$$\begin{bmatrix} \xi_1'(k + 1) \\ \xi_2'(k + 1) \\ \xi_3'(k + 1) \end{bmatrix} = \begin{bmatrix} 0 & 1 & 0 \\ 0 & 0 & 1 \\ 0 & 0 & 0 \end{bmatrix} \begin{bmatrix} \xi_1'(k) \\ \xi_2'(k) \\ \xi_3'(k) \end{bmatrix} + \begin{bmatrix} 2 \\ 1 \\ 3 \end{bmatrix} y(k) + \begin{bmatrix} -1 \\ 2 \\ 0 \end{bmatrix} u(k)$$

$$= \mathbf{F}'\boldsymbol{\xi}'(k) + \mathbf{g}'y(k) + \mathbf{h}'u(k)$$

To observe the original state $\mathbf{x}$ rather than $\mathbf{x}'$, this observer's state can be transformed with an output equation

$$\mathbf{w}(k) = \mathbf{P}\boldsymbol{\xi}'(k) = \mathbf{Q}^{-1}\boldsymbol{\xi}'(k)$$

or

$$\begin{bmatrix} w_1(k) \\ w_2(k) \\ w_3(k) \end{bmatrix} = \begin{bmatrix} 0 & 1 & -\frac{1}{3} \\ 0 & 0 & \frac{1}{3} \\ 1 & 0 & -\frac{2}{3} \end{bmatrix} \begin{bmatrix} \xi_1'(k) \\ \xi_2'(k) \\ \xi_3'(k) \end{bmatrix}$$

This observer then has output that converges to the plant state. Or the change of variables

$$\xi = \mathbf{P}\xi' = \mathbf{Q}^{-1}\xi' \qquad \xi' = \mathbf{P}^{-1}\xi = \mathbf{Q}\xi$$

can be made in the observer state

$$\mathbf{Q}\xi(k + 1) = \mathbf{F}'\mathbf{Q}\xi(k) + \mathbf{g}'y(k) + \mathbf{h}'u(k)$$
$$\xi(k + 1) = (\mathbf{Q}^{-1}\mathbf{F}'\mathbf{Q})\xi(k) + (\mathbf{Q}^{-1}\mathbf{g}')y(k) + (\mathbf{Q}^{-1}\mathbf{h}')u(k)$$
$$= \mathbf{F}\xi(k) + \mathbf{g}y(k) + \mathbf{h}u(k)$$

where

$$\mathbf{F} = \mathbf{Q}^{-1}\mathbf{F}'\mathbf{Q} = \begin{bmatrix} 0 & 3 & 0 \\ 0 & 0 & 0 \\ 1 & 1 & 0 \end{bmatrix}$$

$$\mathbf{g} = \mathbf{Q}^{-1}\mathbf{g}' = \begin{bmatrix} 0 & 1 & -\frac{1}{3} \\ 0 & 0 & \frac{1}{3} \\ 1 & 0 & -\frac{2}{3} \end{bmatrix} \begin{bmatrix} 2 \\ 1 \\ 3 \end{bmatrix} = \begin{bmatrix} 0 \\ 1 \\ 0 \end{bmatrix}$$

$$\mathbf{h} = \mathbf{Q}^{-1}\mathbf{h}' = \begin{bmatrix} 0 & 1 & -\frac{1}{3} \\ 0 & 0 & \frac{1}{3} \\ 1 & 0 & -\frac{2}{3} \end{bmatrix} \begin{bmatrix} -1 \\ 2 \\ 0 \end{bmatrix} = \begin{bmatrix} 2 \\ 0 \\ -1 \end{bmatrix}$$

This results in an observer where the observer state approaches the plant state

$$\begin{bmatrix} \xi_1(k + 1) \\ \xi_2(k + 1) \\ \xi_3(k + 1) \end{bmatrix} = \begin{bmatrix} 0 & 3 & 0 \\ 0 & 0 & 0 \\ 1 & 1 & 0 \end{bmatrix} \begin{bmatrix} \xi_1(k) \\ \xi_2(k) \\ \xi_3(k) \end{bmatrix} + \begin{bmatrix} 0 \\ 1 \\ 0 \end{bmatrix} y(k) + \begin{bmatrix} 2 \\ 0 \\ -1 \end{bmatrix} u(k) \qquad \textbf{(5-6)}$$

Because this is a deadbeat observer, it converges to the plant state in three steps.

5.2.5 Ackermann's Formula

Another method of observer design for single-output plants that does not require transforming the system to observable canonical form is to use Ackermann's formula, which was discussed and proved in Chapter

4. Similar to the situation with state feedback, where the eigenvalues of $\mathbf{A} + \mathbf{b}\mathbf{e}^\dagger$ can be placed arbitrarily by choice of $\mathbf{e}$ (see equation 4-11), providing that the system is completely controllable, the eigenvalues of $\mathbf{F} = \mathbf{A} - \mathbf{g}\mathbf{c}^\dagger$ can be placed arbitrarily by choice of $\mathbf{g}$ as given by Ackermann's formula

$$\mathbf{g} = \Delta_o(\mathbf{A})\mathbf{M}_o^{-1}\mathbf{j}_n \tag{5-7}$$

provided that $(\mathbf{A}, \mathbf{c}^\dagger)$ is completely observable. In equation (5-7), $\Delta_o(\mathbf{A})$ is the desired characteristic equation of the observer eigenvalues with the matrix $\mathbf{A}$ substituted for the variable z, $\mathbf{M}_o$ is the observability matrix

$$\mathbf{M}_o = \begin{bmatrix} \mathbf{c}^\dagger \\ \hline \mathbf{c}^\dagger\mathbf{A} \\ \hline \mathbf{c}^\dagger\mathbf{A}^2 \\ \hline \vdots \\ \hline \mathbf{c}^\dagger\mathbf{A}^{n-1} \end{bmatrix}$$

and $\mathbf{j}_n$ is the nth-unit coordinate vector

$$\mathbf{j}_n = \begin{bmatrix} 0 \\ 0 \\ 0 \\ \vdots \\ 1 \end{bmatrix}$$

The proof of equation (5-7) is very similar to the proof of equation (4-11) for the state feedback gain $\mathbf{e}$, and, therefore, it is omitted.

As an example, consider the system (5-5) again, with all observer eigenvalues required to be at $\lambda = 0$. According to Ackermann's formula (5-7), the observer gain matrix $\mathbf{g}$ is

$$
\begin{aligned}
\mathbf{g} &= \mathbf{A}^3\mathbf{M}_o^{-1}\mathbf{j}_3 \\
&= \left(\frac{1}{15}\right) \begin{bmatrix} 3 & 15 & 6 \\ 2 & 15 & 5 \\ 1 & 11 & 5 \end{bmatrix} \begin{bmatrix} -25 & 25 & -5 \\ -5 & -10 & 5 \\ 25 & 20 & -10 \end{bmatrix} \begin{bmatrix} 0 \\ 0 \\ 1 \end{bmatrix} \\
&= \begin{bmatrix} 0 \\ 1 \\ 0 \end{bmatrix}
\end{aligned}
$$

Hence,

$$\mathbf{F} = \mathbf{A} - \mathbf{g}\mathbf{c}^\dagger = \begin{bmatrix} 0 & 3 & 0 \\ 0 & 2 & 1 \\ 1 & 1 & 0 \end{bmatrix} - \begin{bmatrix} 0 \\ 1 \\ 0 \end{bmatrix} [0 \ 2 \ 1]$$

$$= \begin{bmatrix} 0 & 3 & 0 \\ 0 & 0 & 0 \\ 1 & 1 & 0 \end{bmatrix}$$

and

$$\mathbf{h} = \mathbf{b} - \mathbf{g}d = \begin{bmatrix} 2 \\ 1 \\ -1 \end{bmatrix} - \begin{bmatrix} 0 \\ 1 \\ 0 \end{bmatrix} [1] = \begin{bmatrix} 2 \\ 0 \\ -1 \end{bmatrix}$$

so that the observer has equations

$$\begin{bmatrix} \xi_1(k+1) \\ \xi_2(k+1) \\ \xi_3(k+1) \end{bmatrix} = \begin{bmatrix} 0 & 3 & 0 \\ 0 & 0 & 0 \\ 1 & 1 & 0 \end{bmatrix} \begin{bmatrix} \xi_1(k) \\ \xi_2(k) \\ \xi_3(k) \end{bmatrix} + \begin{bmatrix} 0 \\ 1 \\ 0 \end{bmatrix} y(k) + \begin{bmatrix} 2 \\ 0 \\ -1 \end{bmatrix} u(k)$$

which is identical to the observer in equation (5-6) derived earlier.

5.3 More About Observers

We now extend the basic ideas about observers. Using observable form to design a full-order observer requires that the plant have a single output. One way to accommodate multiple-output plants is to arrange them as equivalent single-output plants when possible. However, some of the available design freedom is used in combining multiple outputs before observation, so doing so may not result in the best possible design.

An observer's state and output can be made to converge to a linear transformation of the plant state instead of the state itself. Relations and properties for this more general kind of observer are developed in this section. These are used in the following sections as the basis for simple but powerful general observer design methods.

5.3.1 Observers for Multiple-Output Systems

If a multiple-output plant is completely observable from one of its outputs, then that single output can be used to drive an observer of the plant state. For example, the plant

$$\begin{bmatrix} x_1(k+1) \\ x_2(k+1) \\ x_3(k+1) \end{bmatrix} = \begin{bmatrix} -1 & 1 & 0 \\ 0 & 0 & 1 \\ 1 & 0 & 0 \end{bmatrix} \begin{bmatrix} x_1(k) \\ x_2(k) \\ x_3(k) \end{bmatrix} + \begin{bmatrix} 3 \\ -1 \\ 0 \end{bmatrix} u(k) = \mathbf{A}\mathbf{x}(k) + \mathbf{b}u(k)$$

$$\begin{bmatrix} y_1(k) \\ y_2(k) \end{bmatrix} = \begin{bmatrix} 1 & -1 & 2 \\ 1 & 0 & 0 \end{bmatrix} \begin{bmatrix} x_1(k) \\ x_2(k) \\ x_3(k) \end{bmatrix} + \begin{bmatrix} 2 \\ -2 \end{bmatrix} u(k) = \mathbf{C}\mathbf{x}(k) + \mathbf{d}u(k)$$

is completely observable from the output $y_2(k)$. For convenience, the plant with single output y_2 is in observable form; if it were not originally in that form, the transformation to observable form, as described in the previous section, is straightforward. A full-order state observer of the form

$$\xi(k+1) = \mathbf{F}\xi(k) + \mathbf{g}y_2(k) + \mathbf{h}u(k)$$

can be designed with the usual methods. Suppose that the desired observer characteristic equation is

$$\lambda^3 - \frac{1}{4}\lambda^2 + \frac{1}{4}\lambda + \frac{1}{16} = 0$$

Then

$$\mathbf{F} = \mathbf{A} - \mathbf{g}\mathbf{c}_2^\dagger = \begin{bmatrix} (-1 - g_1) & 1 & 0 \\ -g_2 & 0 & 1 \\ (1 - g_3) & 0 & 0 \end{bmatrix}$$

has the characteristic equation

$$\lambda^3 + (g_1 + 1)\lambda^2 + (g_2)\lambda + (g_3 - 1) = 0$$

so that

$$\mathbf{g} = \begin{bmatrix} g_1 \\ g_2 \\ g_3 \end{bmatrix} = \begin{bmatrix} -\frac{5}{4} \\ \frac{1}{4} \\ \frac{17}{16} \end{bmatrix}$$

and

$$\mathbf{h} = \mathbf{b} - \mathbf{g}d_2 = \begin{bmatrix} 3 \\ -1 \\ 0 \end{bmatrix} - \begin{bmatrix} -\frac{5}{4} \\ \frac{1}{4} \\ \frac{17}{16} \end{bmatrix} (-2) = \begin{bmatrix} \frac{1}{2} \\ -\frac{1}{2} \\ \frac{17}{8} \end{bmatrix}$$

The observed plant is temporarily imagined to have only the output y_2, as illustrated in Figure 5-6, so only the second rows of $\mathbf{C}$ and $\mathbf{d}$ are

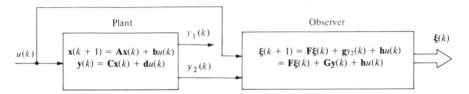

Figure 5-6 Use of a single plant output for observations.

used in these computations. The observer is then described by

$$
\begin{bmatrix} \xi_1(k+1) \\ \xi_2(k+1) \\ \xi_3(k+1) \end{bmatrix} = \begin{bmatrix} \frac{1}{4} & 1 & 0 \\ -\frac{1}{4} & 0 & 1 \\ -\frac{1}{16} & 0 & 0 \end{bmatrix} \begin{bmatrix} \xi_1(k) \\ \xi_2(k) \\ \xi_2(k) \end{bmatrix} + \begin{bmatrix} 0 & -\frac{5}{4} \\ 0 & \frac{1}{4} \\ 0 & \frac{17}{16} \end{bmatrix} \begin{bmatrix} y_1(k) \\ y_2(k) \end{bmatrix} + \begin{bmatrix} \frac{1}{2} \\ -\frac{1}{2} \\ \frac{17}{8} \end{bmatrix} u(k)
$$

$$
= \mathbf{F}\xi(k) + \mathbf{G}y(k) + \mathbf{h}u(k)
$$

where a column of zeros has been included in the matrix $\mathbf{G}$ to represent not using the output y_1.

If the plant to be observed is completely observable but not completely observable from a single output, then a single output that is a linear combination of the individual multiple outputs might be formed from which the plant is completely observable. For example, the plant

$$
\begin{bmatrix} x_1(k+1) \\ x_2(k+1) \\ x_3(k+1) \end{bmatrix} = \begin{bmatrix} \frac{1}{2} & 0 & 0 \\ 0 & -\frac{1}{2} & 0 \\ 0 & 0 & 1 \end{bmatrix} \begin{bmatrix} x_1(k) \\ x_2(k) \\ x_3(k) \end{bmatrix} + \begin{bmatrix} 2 \\ 1 \\ -3 \end{bmatrix} u(k) = \mathbf{A}x(k) + \mathbf{b}u(k)
$$

$$
\begin{bmatrix} y_1(k) \\ y_2(k) \end{bmatrix} = \begin{bmatrix} 1 & 0 & 3 \\ -1 & 2 & 0 \end{bmatrix} \begin{bmatrix} x_1(k) \\ x_2(k) \\ x_3(k) \end{bmatrix} + \begin{bmatrix} -4 \\ -5 \end{bmatrix} u(k) = \mathbf{C}x(k) + \mathbf{d}u(k)
$$

which is in diagonal form, is not completely observable from either of its two outputs. Any linear combination of these two outputs, except one proportional to $y_1 + y_2$, produces a new output from which the plant is completely observable. Choosing

$$
\bar{y}(k) = y_1(k) - y_2(k) = [2 \quad -2 \quad 3] \begin{bmatrix} x_1(k) \\ x_2(k) \\ x_3(k) \end{bmatrix} + u(k)
$$

$$
= \tilde{\mathbf{c}}^\dagger x(k) + \tilde{d}u(k)
$$

as indicated in Figure 5-7, and temporarily taking the plant to have the single output $\bar{y}(k)$, an observer is designed in the usual way.

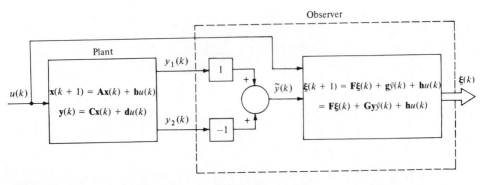

Figure 5-7 State observation using multiple plant outputs.

The transformation

$$\mathbf{x} = \mathbf{Px'} = \mathbf{Q}^{-1}\mathbf{x'} \qquad \mathbf{x'} = \mathbf{P}^{-1}\mathbf{x} = \mathbf{Qx}$$

that takes the plant with output $\bar{y}(k)$ to observable canonical form is given by

$$\mathbf{Q} = \mathbf{P}^{-1} = \begin{bmatrix} 2 & -2 & 3 \\ -1 & 3 & 0 \\ -1 & -1 & -\frac{3}{4} \end{bmatrix}$$

The transformed plant is

$$\begin{bmatrix} x_1'(k+1) \\ x_2'(k+1) \\ x_3'(k+1) \end{bmatrix} = \begin{bmatrix} 1 & 1 & 0 \\ \frac{1}{4} & 0 & 1 \\ -\frac{1}{4} & 0 & 0 \end{bmatrix} \begin{bmatrix} x_1'(k) \\ x_2'(k) \\ x_3'(k) \end{bmatrix} + \begin{bmatrix} -7 \\ 1 \\ -\frac{3}{4} \end{bmatrix} u(k) = \mathbf{A'x'}(k) + \mathbf{b'}u(k)$$

$$\bar{y}(k) = \begin{bmatrix} 1 & 0 & 0 \end{bmatrix} \begin{bmatrix} x_1'(k) \\ x_2'(k) \\ x_3'(k) \end{bmatrix} + u(k) = \tilde{\mathbf{c}}'^{t}\mathbf{x'}(k) + \tilde{d}u(k) \qquad \textbf{(5-8)}$$

and being in observable form makes the observer design simple. Let the desired observer characteristic equation be

$$\lambda^3 - \frac{1}{4}\lambda^2 = 0$$

Then the observer of the state $\mathbf{x'}$ from the output $\bar{y}$ has state coupling matrix

$$\mathbf{F'} = \mathbf{A'} - \mathbf{g'}\tilde{\mathbf{c}}'^{t} = \begin{bmatrix} 1 & 1 & 0 \\ \frac{1}{4} & 0 & 1 \\ -\frac{1}{4} & 0 & 0 \end{bmatrix} - \begin{bmatrix} g_1' \\ g_2' \\ g_3' \end{bmatrix} \begin{bmatrix} 1 & 0 & 0 \end{bmatrix} = \begin{bmatrix} (1 - g_1') & 1 & 0 \\ (\frac{1}{4} - g_2') & 0 & 1 \\ (-\frac{1}{4} - g_3') & 0 & 0 \end{bmatrix}$$

The observer's characteristic equation is

$$\lambda^3 + (g_1' - 1)\lambda^2 + \left(g_2' - \frac{1}{4}\right)\lambda + \left(g_3' + \frac{1}{4}\right) = 0$$

from which

$$\mathbf{g}' = \begin{bmatrix} g_1' \\ g_2' \\ g_3' \end{bmatrix} = \begin{bmatrix} \frac{3}{4} \\ \frac{1}{4} \\ -\frac{1}{4} \end{bmatrix} \qquad \mathbf{h}' = \mathbf{b}' - \mathbf{g}'\tilde{d} = \begin{bmatrix} -7 \\ 1 \\ -\frac{3}{4} \end{bmatrix} - \begin{bmatrix} \frac{3}{4} \\ \frac{1}{4} \\ -\frac{1}{4} \end{bmatrix} = \begin{bmatrix} -\frac{31}{4} \\ \frac{3}{4} \\ -\frac{1}{2} \end{bmatrix}$$

and

$$\begin{bmatrix} \xi_1'(k+1) \\ \xi_2'(k+1) \\ \xi_3'(k+1) \end{bmatrix} = \begin{bmatrix} \frac{1}{4} & 1 & 0 \\ 0 & 0 & 1 \\ 0 & 0 & 0 \end{bmatrix} \begin{bmatrix} \xi_1'(k) \\ \xi_2'(k) \\ \xi_3'(k) \end{bmatrix} + \begin{bmatrix} \frac{3}{4} \\ \frac{1}{4} \\ -\frac{1}{4} \end{bmatrix} \tilde{y}(k) + \begin{bmatrix} -\frac{31}{4} \\ \frac{3}{4} \\ -\frac{1}{2} \end{bmatrix} u(k)$$

If $\tilde{y}(k)$ is expressed in terms of the original system outputs in equation (5-8), the observer is

$$\begin{bmatrix} \xi_1'(k+1) \\ \xi_2'(k+1) \\ \xi_3'(k+1) \end{bmatrix} = \begin{bmatrix} \frac{1}{4} & 1 & 0 \\ 0 & 0 & 1 \\ 0 & 0 & 0 \end{bmatrix} \begin{bmatrix} \xi_1'(k) \\ \xi_2'(k) \\ \xi_3'(k) \end{bmatrix}$$

$$+ \begin{bmatrix} \frac{3}{4} & -\frac{3}{4} \\ \frac{1}{4} & -\frac{1}{4} \\ -\frac{1}{4} & \frac{1}{4} \end{bmatrix} \begin{bmatrix} y_1(k) \\ y_2(k) \end{bmatrix} + \begin{bmatrix} -\frac{31}{4} \\ \frac{3}{4} \\ -\frac{1}{2} \end{bmatrix} u(k)$$

To converge to the original plant's state $\mathbf{x}$ rather than $\mathbf{x}'$, an output equation

$$\mathbf{w}(k) = \mathbf{Q}^{-1}\boldsymbol{\xi}'(k)$$

$$\begin{bmatrix} w_1(k) \\ w_2(k) \\ w_3(k) \end{bmatrix} = \frac{1}{9} \begin{bmatrix} -\frac{9}{4} & -\frac{9}{2} & -9 \\ -\frac{3}{4} & \frac{3}{2} & -3 \\ 4 & 4 & 4 \end{bmatrix} \begin{bmatrix} \xi_1'(k) \\ \xi_2'(k) \\ \xi_3'(k) \end{bmatrix}$$

can be added to the observer, or the change of variables

$$\boldsymbol{\xi}(k) = \mathbf{Q}^{-1}\boldsymbol{\xi}'(k)$$

can be made so that the observer state converges to $\mathbf{x}$ instead of $\mathbf{x}'$.

Observers designed in this way have proportional columns in the $\mathbf{G}$ matrix. An observer of a multiple-output system need not have such a restricted kind of $\mathbf{G}$ matrix; this is simply a convenient design method. Even when a multiple-output system is completely observable from a single output, one might choose to use more than the single output to drive an observer to achieve such benefits as greater reliability and improved performance in the presence of noise.

Not all completely observable plants can be made observable from a single linear combination of their outputs. If, in a block diagonal plant realization, there is more than one Jordan block involving the same eigenvalue, and if two such blocks couple to different outputs (as they must for observability), then any linear combination of the outputs destroys observability of the repeated mode.

5.3.2 Observers as Error Feedback Systems

For a plant

$$\mathbf{x}(k + 1) = \mathbf{A}\mathbf{x}(k) + \mathbf{B}\mathbf{u}(k)$$
$$\mathbf{y}(k) = \mathbf{C}\mathbf{x}(k) + \mathbf{D}\mathbf{u}(k)$$

it is enlightening to express the full-order state observer equations

$$\xi(k + 1) = \mathbf{F}\xi(k) + \mathbf{G}\mathbf{y}(k) + \mathbf{H}\mathbf{u}(k)$$

where

$$\mathbf{F} = \mathbf{A} - \mathbf{G}\mathbf{C}$$
$$\mathbf{H} = \mathbf{B} - \mathbf{G}\mathbf{D}$$

in the form

$$\xi(k + 1) = (\mathbf{A} - \mathbf{G}\mathbf{C})\xi(k) + \mathbf{G}\mathbf{y}(k) + (\mathbf{B} - \mathbf{G}\mathbf{D})\mathbf{u}(k)$$
$$= \mathbf{A}\xi(k) + \mathbf{B}\mathbf{u}(k) + \mathbf{G}[\mathbf{y}(k) - \mathbf{w}(k)]$$

where

$$\mathbf{w}(k) = \mathbf{C}\xi(k) + \mathbf{D}\mathbf{u}(k)$$

As illustrated in Figure 5-8, the observer consists of a model of the plant driven, as the plant is, by the input $\mathbf{u}(k)$ and the error between the plant output $\mathbf{y}(k)$ and the plant output that is estimated by the model $\mathbf{w}(k)$.

5.3.3 Observing Linear State Transformations

When an observer's state

$$\xi(k + 1) = \mathbf{F}\xi(k) + \mathbf{G}\mathbf{y}(k) + \mathbf{H}\mathbf{u}(k)$$

estimates a linear transformation $\mathbf{M}\mathbf{x}(k)$ of the plant state rather than the plant state itself, the error between the observer state and the plant

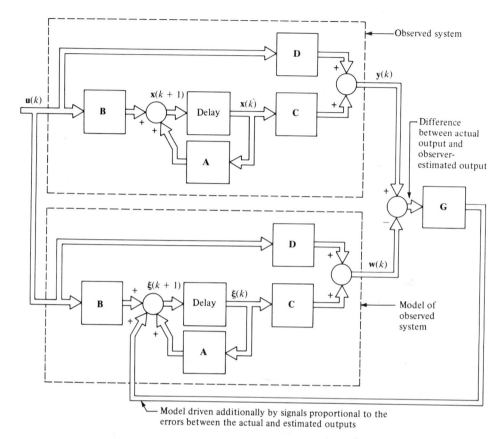

Figure 5-8 Full-order state observer arranged as an error feedback system.

state transformation is given by

$$\mathbf{M}\mathbf{x}(k+1) - \boldsymbol{\xi}(k+1) = \mathbf{M}\mathbf{A}\mathbf{x}(k) + \mathbf{M}\mathbf{B}\mathbf{u}(k) - \mathbf{F}\boldsymbol{\xi}(k) - \mathbf{G}\mathbf{y}(k)$$
$$- \mathbf{H}\mathbf{u}(k)$$
$$= (\mathbf{M}\mathbf{A} - \mathbf{G}\mathbf{C})\mathbf{x}(k) - \mathbf{F}\boldsymbol{\xi}(k) + (\mathbf{M}\mathbf{B} - \mathbf{G}\mathbf{D}$$
$$- \mathbf{H})\mathbf{u}(k)$$

Requiring that the error satisfy an autonomous equation

$$\mathbf{M}\mathbf{x}(k+1) - \boldsymbol{\xi}(k+1) = \mathbf{F}[\mathbf{M}\mathbf{x}(k) - \boldsymbol{\xi}(k)]$$

gives

$$\mathbf{F}\mathbf{M} = \mathbf{M}\mathbf{A} - \mathbf{G}\mathbf{C}$$
$$\mathbf{H} = \mathbf{M}\mathbf{B} - \mathbf{G}\mathbf{D}$$

where the eigenvalues of $\mathbf{F}$ are inside the unit circle on the complex plane.

These are generalizations of the results for full-order state observers, where the *observation matrix* $\mathbf{M}$ is the identity matrix. The relations for observers of linear state transformations are summarized in Table 5-2.

Table 5-2 Relations for Observers of Linear State Transformations

---

Plant Model

$\mathbf{x}(k + 1) = \mathbf{A}\mathbf{x}(k) + \mathbf{B}\mathbf{u}(k)$

$\quad \mathbf{y}(k) = \mathbf{C}\mathbf{x}(k) + \mathbf{D}\mathbf{u}(k)$

Observer

$\boldsymbol{\xi}(k + 1) = \mathbf{F}\boldsymbol{\xi}(k) + \mathbf{G}\mathbf{y}(k) + \mathbf{H}\mathbf{u}(k)$

$\quad \mathbf{w}(k) = \mathbf{L}\boldsymbol{\xi}(k) + \mathbf{N}\mathbf{x}(k)$

where

$\mathbf{FM} = \mathbf{MA} - \mathbf{GC}$

$\;\mathbf{H} = \mathbf{MB} - \mathbf{GD}$

Observer Error

$\mathbf{M}\mathbf{x}(k + 1) - \boldsymbol{\xi}(k + 1) = \mathbf{F}[\mathbf{M}\mathbf{x}(k) - \boldsymbol{\xi}(k)]$

$\qquad \mathbf{M}\mathbf{x}(k) - \boldsymbol{\xi}(k) = \mathbf{F}^k[\mathbf{M}\mathbf{x}(0) - \boldsymbol{\xi}(0)]$

or

$\boldsymbol{\xi}(k) \rightarrow \mathbf{M}\mathbf{x}(k)$

$\mathbf{w}(k) \rightarrow (\mathbf{LM} + \mathbf{N})\mathbf{x}(k) = \mathbf{E}\mathbf{x}(k)$

The notation

$$\boldsymbol{\xi}(k) \rightarrow \mathbf{M}\mathbf{x}(k)$$

is used to indicate that the observer state signals *observe* the state transformation $\mathbf{M}\mathbf{x}(k)$. That is, the error between $\mathbf{M}\mathbf{x}(k)$ and $\boldsymbol{\xi}(k)$ is governed by the observer state coupling matrix $\mathbf{F}$, the eigenvalues of which can be selected by the designer. In general, the observation matrix $\mathbf{M}$ need not be square.

If the observer has an output equation of the form

$$\mathbf{w}(k) = \mathbf{L}\boldsymbol{\xi}(k) + \mathbf{N}\mathbf{x}(k)$$

then the observer output observes

$$\mathbf{w}(k) \rightarrow (\mathbf{LM} + \mathbf{N})\mathbf{x}(k)$$

as indicated in Table 5-2.

5.4 Observer Design

For observers with distinct eigenvalues, a very convenient method of observer design is to consider a collection of first-order observers, each with one of the desired observer eigenvalues. From the individual plant state transformation observed by each first-order observer, the observer gains and a transformation of the observer state is found to achieve the observation desired.

5.4.1 First-Order Observers

For a discrete-time plant

$$\mathbf{x}(k + 1) = \mathbf{Ax}(k) + \mathbf{Bu}(k)$$
$$\mathbf{y}(k) = \mathbf{Cx}(k) + \mathbf{Du}(k)$$

a first-order observer

$$\xi(k + 1) = f\xi(k) + \mathbf{g}^\dagger\mathbf{y}(k) + \mathbf{h}^\dagger\mathbf{u}(k)$$

where $\xi(k)$ is a scalar, observes the linear state transformation

$$\xi \rightarrow \mathbf{m}^\dagger\mathbf{x}$$

given by (applying the results summarized in Table 5-2)

$$\mathbf{m}^\dagger\mathbf{A} - \mathbf{g}^\dagger\mathbf{C} = f\mathbf{m}^\dagger$$
$$\mathbf{m}^\dagger = \mathbf{g}^\dagger\mathbf{C}(\mathbf{A} - f\mathbf{I})^{-1}$$

The observer coupling of the plant input $\mathbf{h}$ is given by

$$\mathbf{h}^\dagger = \mathbf{m}^\dagger\mathbf{B} - \mathbf{g}^\dagger\mathbf{D}$$

These *first-order* observer relations are collected in Table 5-3.

The matrix inverse involved exists whenever the observer eigenvalue does not equal a plant eigenvalue. If an observer and the plant have an eigenvalue in common, the corresponding mode in the observer error occurs even if the observer is not connected to the plant. The part of the observer with the common eigenvalue is thus redundant.

For example, for the plant

$$\begin{bmatrix} x_1(k + 1) \\ x_2(k + 1) \end{bmatrix} = \begin{bmatrix} 1 & 0 \\ 2 & -1 \end{bmatrix}\begin{bmatrix} x_1(k) \\ x_2(k) \end{bmatrix} + \begin{bmatrix} 0 \\ 3 \end{bmatrix} u(k) = \mathbf{Ax}(k) + \mathbf{b}u(k)$$

$$y(k) = \begin{bmatrix} -1 & 2 \end{bmatrix}\begin{bmatrix} x_1(k) \\ x_2(k) \end{bmatrix} + 4u(k) = \mathbf{c}^\dagger\mathbf{x}(k) + du(k)$$

Table 5-3 First-Order Observer Relations

Plant Model

$\mathbf{x}(k + 1) = \mathbf{A}\mathbf{x}(k) + \mathbf{B}\mathbf{u}(k)$

$\quad \mathbf{y}(k) = \mathbf{C}\mathbf{x}(k) + \mathbf{D}\mathbf{u}(k)$

Observer

$\xi(k + 1) = f\xi(k) + \mathbf{g}^\dagger\mathbf{y}(k) + \mathbf{h}^\dagger\mathbf{u}(k)$

where

$\mathbf{m}^\dagger = \mathbf{g}^\dagger\mathbf{C}(\mathbf{A} - f\mathbf{I})^{-1}$

$\mathbf{h}^\dagger = \mathbf{m}^\dagger\mathbf{B} - \mathbf{g}^\dagger\mathbf{D}$

Observer Error

$\mathbf{m}^\dagger\mathbf{x}(k + 1) - \xi(k + 1) = f[\mathbf{m}^\dagger\mathbf{x}(k) - \xi(k)]$

$\quad\quad \mathbf{m}^\dagger\mathbf{x}(k) - \xi(k) = f^k[\mathbf{m}^\dagger\mathbf{x}(0) - \xi(0)]$

or

$\xi(k) \rightarrow \mathbf{m}^\dagger\mathbf{x}(k)$

---

a first-order observer with eigenvalue $\lambda = \frac{1}{4}$, of the form

$$\xi(k + 1) = \frac{1}{4}\xi(k) + gy(k) + hu(k)$$

observes the linear state transformation $\mathbf{m}^\dagger\mathbf{x}$, where the observation matrix is

$$\mathbf{m}^\dagger = g\mathbf{c}^\dagger(\mathbf{A} - f\mathbf{I})^{-1} = g[-1 \quad 2]\begin{bmatrix} \frac{3}{4} & 0 \\ 2 & -\frac{5}{4} \end{bmatrix}^{-1}$$

$$= g[-1 \quad 2]\begin{bmatrix} \frac{4}{3} & 0 \\ \frac{32}{15} & -\frac{4}{5} \end{bmatrix} = g[\frac{44}{15} \quad -\frac{8}{5}]$$

The coupling of the plant input to the observer is

$$h = \mathbf{m}^\dagger\mathbf{b} - gd = g[\frac{44}{15} \quad -\frac{8}{5}]\begin{bmatrix} 0 \\ 3 \end{bmatrix} - 4g = -\frac{44}{5}g$$

This observer observes any linear transformation of the state that is proportional to

$$\left(\frac{1}{g}\right)\mathbf{m}^\dagger = [\frac{44}{15} \quad -\frac{8}{5}]$$

5.4.2 Collections of First-Order Observers

We now form full-order state observers with distinct eigenvalues by designing n first-order observers for an nth-order plant. The collection of these first-order component observers constitutes an nth-order observer in diagonal form. It has the selected eigenvalues, and it observes a transformation of the plant state that is always nonsingular if the plant is completely observable and the observer eigenvalues are different from those of the plant. Forming an observer output equation involving the inverse of this transformation or changing observer state variables produces an observer of the plant state.

As an example, consider the plant

$$\begin{bmatrix} x_1(k + 1) \\ x_2(k + 1) \end{bmatrix} = \begin{bmatrix} 0 & 2 \\ 1 & -1 \end{bmatrix} \begin{bmatrix} x_1(k) \\ x_2(k) \end{bmatrix} + \begin{bmatrix} 0 & 0 \\ 2 & -1 \end{bmatrix} \begin{bmatrix} u_1(k) \\ u_2(k) \end{bmatrix} = \mathbf{A}\mathbf{x}(k) + \mathbf{B}\mathbf{u}(k)$$

$$y(k) = \begin{bmatrix} 0 & -2 \end{bmatrix} \begin{bmatrix} x_1(k) \\ x_2(k) \end{bmatrix} + \begin{bmatrix} 0 & 1 \end{bmatrix} \begin{bmatrix} u_1(k) \\ u_2(k) \end{bmatrix} = \mathbf{c}^\dagger \mathbf{x}(k) + \mathbf{d}^\dagger \mathbf{u}(k)$$

and suppose that it is desired to design a full-order state observer of this plant with eigenvalues $\lambda = \pm\frac{1}{2}$. A first-order observer with eigenvalue $\frac{1}{2}$ has the form

$$\xi_1(k + 1) = \frac{1}{2} \xi_1(k) + g_1 y(k) + \mathbf{h}_1^\dagger \mathbf{u}(k)$$

and observes the scalar linear state transformation proportional to g_1 that is given by

$$\mathbf{m}_1^\dagger = g_1 \mathbf{c}^\dagger \left(\mathbf{A} - \frac{1}{2}\mathbf{I} \right)^{-1} = g_1 \begin{bmatrix} 0 & -2 \end{bmatrix} \begin{bmatrix} -\frac{1}{2} & 2 \\ 1 & -\frac{3}{2} \end{bmatrix}^{-1} = g_1 \begin{bmatrix} -\frac{8}{5} & -\frac{4}{5} \end{bmatrix}$$

The input coupling to this observer is

$$\mathbf{h}_1^\dagger = \mathbf{m}_1^\dagger \mathbf{B} - g_1 \mathbf{d}^\dagger = g_1 \begin{bmatrix} -\frac{8}{5} & -\frac{4}{5} \end{bmatrix}$$

giving

$$\xi_1(k + 1) = \frac{1}{2} \xi_1(k) + g_1 y(k) + g_1 \begin{bmatrix} -\frac{8}{5} & -\frac{4}{5} \end{bmatrix} \begin{bmatrix} u_1(k) \\ u_2(k) \end{bmatrix} \qquad (5\text{-}9)$$

Another first-order observer with eigenvalue $-\frac{1}{2}$ has the form

$$\xi_2(k + 1) = -\frac{1}{2} \xi_2(k) + g_2 y(k) + \mathbf{h}_2^\dagger \mathbf{u}(k)$$

and observes the scalar state transformation given by

$$
\mathbf{m}_2^t = g_2 \mathbf{c}^t \left(\mathbf{A} + \frac{1}{2}\mathbf{I} \right)^{-1} = g_2[0 \quad -2] \begin{bmatrix} \frac{1}{2} & 2 \\ 1 & -\frac{1}{2} \end{bmatrix}^{-1} = g_2[-\frac{8}{9} \quad \frac{4}{9}]
$$

This observer has coupling of the plant input

$$
\mathbf{h}_2^t = \mathbf{m}_2^t \mathbf{B} - g_2 \mathbf{d}^t = g_2[\frac{8}{9} \quad -\frac{13}{9}]
$$

giving

$$
\xi_2(k + 1) = -\frac{1}{2}\xi_2(k) + g_2 y(k) + g_2[\frac{8}{9} \quad -\frac{13}{9}] \begin{bmatrix} u_1(k) \\ u_2(k) \end{bmatrix} \qquad \text{(5-10)}
$$

Because the vectors $[-\frac{8}{5} \quad -\frac{4}{5}]$ and $[-\frac{8}{9} \quad \frac{4}{9}]$ are linearly independent, any plant state transformation can be obtained with a linear combination of $\xi_1(k)$ and $\xi_2(k)$. The observer gains g_1 and g_2 only serve to scale the observer state signals, so they can each be chosen to be unity. Then the observer combining the first-order subsystems (5-9) and (5-10)

$$
\begin{bmatrix} \xi_1(k + 1) \\ \xi_2(k + 1) \end{bmatrix} = \begin{bmatrix} \frac{1}{2} & 0 \\ 0 & -\frac{1}{2} \end{bmatrix} \begin{bmatrix} \xi_1(k) \\ \xi_2(k) \end{bmatrix} + \begin{bmatrix} 1 \\ 1 \end{bmatrix} y(k) + \begin{bmatrix} -\frac{8}{5} & -\frac{1}{5} \\ \frac{8}{9} & -\frac{13}{9} \end{bmatrix} \begin{bmatrix} u_1(k) \\ u_2(k) \end{bmatrix}
$$
$$
= \mathbf{F}\xi(k) + \mathbf{g}y(k) + \mathbf{H}\mathbf{u}(k)
$$

observes

$$
\xi(k) = \begin{bmatrix} \xi_1(k) \\ \xi_2(k) \end{bmatrix} \rightarrow \begin{bmatrix} -\frac{8}{5} & -\frac{4}{5} \\ -\frac{8}{9} & \frac{4}{9} \end{bmatrix} \begin{bmatrix} x_1(k) \\ x_2(k) \end{bmatrix} = \mathbf{M}\mathbf{x}(k)
$$

To observe the state $\mathbf{x}(k)$ rather than $\mathbf{M}\mathbf{x}(k)$, one only need form the observer output signals

$$
\begin{bmatrix} w_1(k) \\ w_2(k) \end{bmatrix} = \begin{bmatrix} -\frac{8}{5} & -\frac{4}{5} \\ -\frac{8}{9} & \frac{4}{9} \end{bmatrix}^{-1} \begin{bmatrix} \xi_1(k) \\ \xi_2(k) \end{bmatrix} = \frac{1}{64} \begin{bmatrix} -20 & -36 \\ -40 & 72 \end{bmatrix} \begin{bmatrix} \xi_1(k) \\ \xi_2(k) \end{bmatrix} = \mathbf{M}^{-1}\xi(k)
$$

This arrangement is indicated in Figure 5-9(a). Or the change of state variables

$$
\begin{bmatrix} \xi_1'(k) \\ \xi_2'(k) \end{bmatrix} = \frac{1}{64} \begin{bmatrix} -20 & -36 \\ -40 & 72 \end{bmatrix} \begin{bmatrix} \xi_1(k) \\ \xi_2(k) \end{bmatrix} = \mathbf{M}^{-1}\xi(k)
$$

can be made so that the observer output is its state $\xi'(k)$ and the

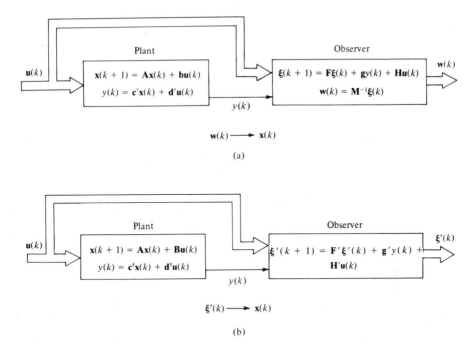

Figure 5-9 Observing a plant state with a collection of first-order observers. (a) Diagonalized observer with an output equation. (b) Change of observer state variables so that the observer state observes the plant state.

observer state observes the state $\mathbf{x}(k)$

$$\mathbf{M}\boldsymbol{\xi}'(k + 1) = \mathbf{F}\mathbf{M}\boldsymbol{\xi}'(k) + \mathbf{g}y(k) + \mathbf{H}u(k)$$
$$\boldsymbol{\xi}'(k + 1) = (\mathbf{M}^{-1}\mathbf{F}\mathbf{M})\boldsymbol{\xi}'(k) + (\mathbf{M}^{-1}\mathbf{g})y(k) + (\mathbf{M}^{-1}\mathbf{H})u(k)$$
$$= \mathbf{F}'\boldsymbol{\xi}'(k) + \mathbf{g}'y(k) + \mathbf{H}'\mathbf{u}(k)$$

as shown in Figure 5-9(b).

When complex conjugate pairs of observer eigenvalues are desired, the observer, being in diagonal form, has matrices with complex elements. Transformation to a block diagonal form or to an observer with state that observes $\mathbf{x}(k)$ results in an observer with real matrices.

5.4.3 Repeated Observer Eigenvalues

When it is desired that an observer have repeated eigenvalues, these methods must be modified slightly. An observer of ith order, with all eigenvalues equal to f, with state coupling matrix in upper block Jordan form, has the structure

$$
\begin{bmatrix} \xi_1(k+1) \\ \xi_2(k+1) \\ \vdots \\ \xi_{i-1}(k+1) \\ \xi_i(k+1) \end{bmatrix} = \begin{bmatrix} f & 1 & 0 & \cdots & 0 & 0 \\ 0 & f & 1 & \cdots & 0 & 0 \\ \vdots & & & & & \\ 0 & 0 & 0 & \cdots & f & 1 \\ 0 & 0 & 0 & \cdots & 0 & f \end{bmatrix} \begin{bmatrix} \xi_1(k) \\ \xi_2(k) \\ \vdots \\ \xi_{i-1}(k) \\ \xi_i(k) \end{bmatrix} + \begin{bmatrix} \mathbf{g}_1^\dagger \\ \hline \mathbf{g}_2^\dagger \\ \hline \vdots \\ \hline \mathbf{g}_{i-1}^\dagger \\ \hline \mathbf{g}_i^\dagger \end{bmatrix} \mathbf{y}(k)
$$

$$
+ \begin{bmatrix} \mathbf{h}_1^\dagger \\ \hline \mathbf{h}_2^\dagger \\ \hline \vdots \\ \hline \mathbf{h}_{i-1}^\dagger \\ \hline \mathbf{h}_i^\dagger \end{bmatrix} \mathbf{u}(k) = \mathbf{F}\boldsymbol{\xi}(k) + \mathbf{G}\mathbf{y}(k) + \mathbf{H}\mathbf{u}(k)
$$

where $\mathbf{G}$ and $\mathbf{H}$ have been partitioned into rows. This observer observes some plant state transformation $\mathbf{M}\mathbf{x}(k)$ where the observation matrix $\mathbf{M}$ satisfies

$$
\mathbf{M}\mathbf{A} - \mathbf{G}\mathbf{C} = \mathbf{F}\mathbf{M}
$$

Partitioning $\mathbf{M}$ into rows

$$
\begin{bmatrix} \mathbf{m}_1^\dagger \\ \hline \mathbf{m}_2^\dagger \\ \hline \vdots \\ \hline \mathbf{m}_{i-1}^\dagger \\ \hline \mathbf{m}_i^\dagger \end{bmatrix} \mathbf{A} - \begin{bmatrix} \mathbf{g}_1^\dagger \\ \hline \mathbf{g}_2^\dagger \\ \hline \vdots \\ \hline \mathbf{g}_{i-1}^\dagger \\ \hline \mathbf{g}_i^\dagger \end{bmatrix} \mathbf{C} = \begin{bmatrix} f & 1 & 0 & \cdots & 0 & 0 \\ 0 & f & 1 & \cdots & 0 & 0 \\ \vdots & & & & & \\ 0 & 0 & 0 & \cdots & f & 1 \\ 0 & 0 & 0 & \cdots & 0 & f \end{bmatrix} \begin{bmatrix} \mathbf{m}_1^\dagger \\ \hline \mathbf{m}_2^\dagger \\ \hline \vdots \\ \hline \mathbf{m}_{i-1}^\dagger \\ \hline \mathbf{m}_i^\dagger \end{bmatrix}
$$

results in the relations

$$
\begin{cases}
\mathbf{m}_i^\dagger = \mathbf{g}_i^\dagger \mathbf{C}(\mathbf{A} - f\mathbf{I})^{-1} \\
\mathbf{m}_{i-1}^\dagger = (\mathbf{g}_{i-1}^\dagger \mathbf{C} + \mathbf{m}_i^\dagger)(\mathbf{A} - f\mathbf{I})^{-1} \\
\vdots \\
\mathbf{m}_2^\dagger = (\mathbf{g}_2^\dagger \mathbf{C} + \mathbf{m}_3^\dagger)(\mathbf{A} - f\mathbf{I})^{-1} \\
\mathbf{m}_1^\dagger = (\mathbf{g}_1^\dagger \mathbf{C} + \mathbf{m}_2^\dagger)(\mathbf{A} - f\mathbf{I})^{-1}
\end{cases}
$$

The first of these is the relation for an ordinary first-order observer; the remaining ones are for repetitions of the observer eigenvalue.

The observer plant input gains are given by

$$\mathbf{H} = \mathbf{MB} - \mathbf{GD}$$

$$
\begin{bmatrix}
\mathbf{h}_1^\dagger \\ \hline
\mathbf{h}_2^\dagger \\ \hline
\vdots \\ \hline
\mathbf{h}_{i-1}^\dagger \\ \hline
\mathbf{h}_i^\dagger
\end{bmatrix}
=
\begin{bmatrix}
\mathbf{m}_1^\dagger \\ \hline
\mathbf{m}_2^\dagger \\ \hline
\vdots \\ \hline
\mathbf{m}_{i-1}^\dagger \\ \hline
\mathbf{m}_i^\dagger
\end{bmatrix}
\mathbf{B} -
\begin{bmatrix}
\mathbf{g}_1^\dagger \\ \hline
\mathbf{g}_2^\dagger \\ \hline
\vdots \\ \hline
\mathbf{g}_{i-1}^\dagger \\ \hline
\mathbf{g}_i^\dagger
\end{bmatrix}
\mathbf{D}
$$

or

$$
\begin{cases}
\mathbf{h}_i^\dagger = \mathbf{m}_i^\dagger \mathbf{B} - \mathbf{g}_i^\dagger \mathbf{D} \\
\mathbf{h}_{i-1}^\dagger = \mathbf{m}_{i-1}^\dagger \mathbf{B} - \mathbf{g}_{i-1}^\dagger \mathbf{D} \\
\vdots \\
\mathbf{h}_2^\dagger = \mathbf{m}_2^\dagger \mathbf{B} - \mathbf{g}_2^\dagger \mathbf{D} \\
\mathbf{h}_1^\dagger = \mathbf{m}_1^\dagger \mathbf{B} - \mathbf{g}_1^\dagger \mathbf{D}
\end{cases}
$$

The equations for observer design with repeated observer eigenvalues are summarized in Table 5-4.

As a numerical example, consider the plant

$$
\begin{bmatrix} x_1(k+1) \\ x_2(k+1) \end{bmatrix} =
\begin{bmatrix} \frac{1}{2} & 0 \\ 1 & -\frac{1}{2} \end{bmatrix}
\begin{bmatrix} x_1(k) \\ x_2(k) \end{bmatrix} +
\begin{bmatrix} 1 \\ -2 \end{bmatrix} u(k) = \mathbf{Ax}(k) + \mathbf{b}u(k)
$$

$$
y(k) = [2 \quad 1] \begin{bmatrix} x_1(k) \\ x_2(k) \end{bmatrix} + 2u(k) = \mathbf{c}^\dagger \mathbf{x}(k) + du(k)
$$

for which it is desired to design a full-order state observer with both eigenvalues at $\lambda = \frac{1}{4}$. An observer of the form

$$
\begin{bmatrix} \xi_1(k+1) \\ \xi_2(k+1) \end{bmatrix} =
\begin{bmatrix} \frac{1}{4} & 1 \\ 0 & \frac{1}{4} \end{bmatrix}
\begin{bmatrix} \xi_1(k) \\ \xi_2(k) \end{bmatrix} +
\begin{bmatrix} g_1 \\ g_2 \end{bmatrix} y(k) +
\begin{bmatrix} h_1 \\ h_2 \end{bmatrix} u(k)
$$

observes the linear plant state transformation

$$
\begin{bmatrix} \xi_1(k) \\ \xi_2(k) \end{bmatrix} \rightarrow
\begin{bmatrix} \mathbf{m}_1^\dagger \\ \mathbf{m}_2^\dagger \end{bmatrix}
\begin{bmatrix} x_1(k) \\ x_2(k) \end{bmatrix} = \mathbf{Mx}(k)
$$

Table 5-4 Relations for an Observer with All Eigenvalues Identical

Plant Model

$$\mathbf{x}(k + 1) = \mathbf{A}\mathbf{x}(k) + \mathbf{B}\mathbf{u}(k)$$
$$\mathbf{y}(k) = \mathbf{C}\mathbf{x}(k) + \mathbf{D}\mathbf{u}(k)$$

Observer

$$\boldsymbol{\xi}(k + 1) = \mathbf{F}\boldsymbol{\xi}(k) + \mathbf{G}\mathbf{y}(k) + \mathbf{H}\mathbf{u}(k)$$

where

$$\mathbf{F} = \begin{bmatrix} f & 1 & 0 & \cdots & 0 & 0 \\ 0 & f & 1 & \cdots & 0 & 0 \\ & \vdots & & & & \\ 0 & 0 & 0 & \cdots & f & 1 \\ 0 & 0 & 0 & \cdots & 0 & f \end{bmatrix} \qquad \mathbf{G} = \begin{bmatrix} \mathbf{g}_1^\dagger \\ \hline \mathbf{g}_2^\dagger \\ \hline \vdots \\ \hline \mathbf{g}_{i-1}^\dagger \\ \hline \mathbf{g}_i^\dagger \end{bmatrix} \qquad \mathbf{H} = \begin{bmatrix} \mathbf{h}_1^\dagger \\ \hline \mathbf{h}_2^\dagger \\ \hline \vdots \\ \hline \mathbf{h}_{i-1}^\dagger \\ \hline \mathbf{h}_i^\dagger \end{bmatrix}$$

Observer Error

$$\mathbf{M}\mathbf{x}(k + 1) - \boldsymbol{\xi}(k + 1) = \mathbf{F}[\mathbf{M}\mathbf{x}(k) - \boldsymbol{\xi}(k)]$$
$$\mathbf{M}\mathbf{x}(k) - \boldsymbol{\xi}(k) = \mathbf{F}^k[\mathbf{M}\mathbf{x}(0) - \boldsymbol{\xi}(0)]$$

or

$$\boldsymbol{\xi}(k) \to \mathbf{M}\mathbf{x}(k)$$

where

$$\mathbf{M} = \begin{bmatrix} \mathbf{m}_1^\dagger \\ \hline \mathbf{m}_2^\dagger \\ \hline \vdots \\ \hline \mathbf{m}_{i-1}^\dagger \\ \hline \mathbf{m}_i^\dagger \end{bmatrix}$$

Recursive Relations for Observer Parameters

$$\begin{cases} \mathbf{m}_i^\dagger = \mathbf{g}_i^\dagger \mathbf{C}(\mathbf{A} - f\mathbf{I})^{-1} \\ \mathbf{m}_{i-1}^\dagger = (\mathbf{g}_{i-1}^\dagger \mathbf{C} + \mathbf{m}_i^\dagger)(\mathbf{A} - f\mathbf{I})^{-1} \\ \quad \vdots \\ \mathbf{m}_2^\dagger = (\mathbf{g}_2^\dagger \mathbf{C} + \mathbf{m}_3^\dagger)(\mathbf{A} - f\mathbf{I})^{-1} \\ \mathbf{m}_1^\dagger = (\mathbf{g}_1^\dagger \mathbf{C} + \mathbf{m}_2^\dagger)(\mathbf{A} - f\mathbf{I})^{-1} \end{cases}$$
$$\mathbf{h}_j^\dagger = \mathbf{m}_j^\dagger \mathbf{B} - \mathbf{g}_j^\dagger \mathbf{D} \quad j = 1, 2, \ldots, i$$

where the observation matrix has rows

$$\mathbf{m}_2^\dagger = g_2 \mathbf{c}^\dagger (\mathbf{A} - f\mathbf{I})^{-1} = g_2[\tfrac{40}{3} \quad -\tfrac{4}{3}]$$

and

$$\mathbf{m}_1^\dagger = (g_1 \mathbf{c}^\dagger + \mathbf{m}_2^\dagger)(\mathbf{A} - f\mathbf{I})^{-1} = g_1[\tfrac{40}{3} \quad -\tfrac{4}{3}] + g_2[\tfrac{416}{9} \quad \tfrac{16}{9}]$$

The corresponding plant input gains to the observer are

$$h_2 = \mathbf{m}_2^\dagger \mathbf{b} - g_2 d = 14 g_2$$

$$h_1 = \mathbf{m}_1^\dagger \mathbf{b} - g_1 d = 14 g_1 + \frac{128}{3} g_2$$

Choosing

$$g_1 = 3 \qquad g_2 = 9$$

for convenience, gives the observer

$$\begin{bmatrix} \xi_1(k+1) \\ \xi_2(k+1) \end{bmatrix} = \begin{bmatrix} \tfrac{1}{4} & 1 \\ 0 & \tfrac{1}{4} \end{bmatrix} \begin{bmatrix} \xi_1(k) \\ \xi_2(k) \end{bmatrix} + \begin{bmatrix} 3 \\ 9 \end{bmatrix} y(k) + \begin{bmatrix} 426 \\ 126 \end{bmatrix} u(k) \qquad \textbf{(5-11)}$$

which observes the linear plant state transformation

$$\begin{bmatrix} \xi_1(k) \\ \xi_2(k) \end{bmatrix} \rightarrow \begin{bmatrix} 456 & 12 \\ 120 & -12 \end{bmatrix} \begin{bmatrix} x_1(k) \\ x_2(k) \end{bmatrix} = \mathbf{M}\mathbf{x}(k)$$

so that the observer with state variables

$$\xi'(k) = \mathbf{M}^{-1}\xi(k)$$

or

$$\begin{bmatrix} \xi_1'(k) \\ \xi_2'(k) \end{bmatrix} = \frac{1}{576} \begin{bmatrix} 1 & 1 \\ 10 & -38 \end{bmatrix} \begin{bmatrix} \xi_1(k) \\ \xi_2(k) \end{bmatrix}$$

has a state that observes $\mathbf{x}(k)$

$$\xi'(k) \rightarrow \mathbf{x}(k)$$

Typical response of this observer for arbitrarily chosen plant initial conditions and input is shown in Figure 5-10. Alternatively, the observer of equation (5-11) with output equation

$$\mathbf{w}(k) = \mathbf{M}^{-1}\xi(k)$$

could be used, and the observer output would observe the plant state

$$\mathbf{w}(k) \rightarrow \mathbf{x}(k)$$

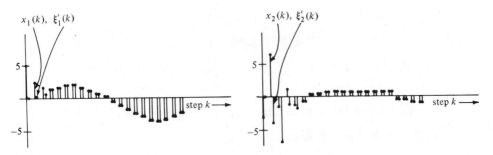

Figure 5-10 Response of an observer with repeated eigenvalues.

For the plant

$$
\begin{bmatrix} x_1(k+1) \\ x_2(k+1) \\ x_3(k+1) \end{bmatrix} = \begin{bmatrix} -1 & 2 & 0 \\ 0 & 1 & 0 \\ -1 & 1 & 1 \end{bmatrix} \begin{bmatrix} x_1(k) \\ x_2(k) \\ x_3(k) \end{bmatrix} + \begin{bmatrix} 2 & 0 \\ -1 & 3 \\ 0 & -2 \end{bmatrix} \begin{bmatrix} u_1(k) \\ u_2(k) \end{bmatrix}
$$

$$
\begin{bmatrix} y_1(k) \\ y_2(k) \end{bmatrix} = \begin{bmatrix} 0 & 1 & -1 \\ 1 & 2 & 0 \end{bmatrix} \begin{bmatrix} x_1(k) \\ x_2(k) \\ x_3(k) \end{bmatrix} + \begin{bmatrix} 1 & 3 \\ 0 & 0 \end{bmatrix} \begin{bmatrix} u_1(k) \\ u_2(k) \end{bmatrix}
$$

suppose that it is desired to design a full-order state observer with eigenvalues 0, 0, and $\frac{1}{2}$. A second-order observer with both eigenvalues at $\lambda = 0$,

$$
\begin{bmatrix} \xi_1(k+1) \\ \xi_2(k+1) \end{bmatrix} = \begin{bmatrix} 0 & 1 \\ 0 & 0 \end{bmatrix} \begin{bmatrix} \xi_1(k) \\ \xi_2(k) \end{bmatrix} + \begin{bmatrix} \mathbf{g}_1^\dagger \\ \mathbf{g}_2^\dagger \end{bmatrix} \begin{bmatrix} y_1(k) \\ y_2(k) \end{bmatrix} + \begin{bmatrix} \mathbf{h}_1^\dagger \\ \mathbf{h}_2^\dagger \end{bmatrix} \begin{bmatrix} u_1(k) \\ u_2(k) \end{bmatrix}
$$

observing the linear plant state transformation

$$
\begin{bmatrix} \xi_1(k) \\ \xi_2(k) \end{bmatrix} \rightarrow \begin{bmatrix} \mathbf{m}_1^\dagger \\ \mathbf{m}_2^\dagger \end{bmatrix} \begin{bmatrix} x_1(k) \\ x_2(k) \\ x_3(k) \end{bmatrix}
$$

is designed, as in the previous example. For this observer subsystem

$$
\mathbf{m}_2^\dagger = \mathbf{g}_2^\dagger \mathbf{C} \mathbf{A}^{-1} = [g_{21} \quad g_{22}] \begin{bmatrix} 0 & 1 & -1 \\ 1 & 2 & 0 \end{bmatrix} \begin{bmatrix} -1 & 2 & 0 \\ 0 & 1 & 0 \\ -1 & 1 & 1 \end{bmatrix}^{-1}
$$

$$
= [(g_{21} - g_{22}) \quad 4g_{22} \quad -g_{21}]
$$

$$\mathbf{m}_1^\dagger = (\mathbf{g}_1^\dagger \mathbf{C} + \mathbf{m}_2^\dagger)\mathbf{A}^{-1}$$

$$= \left([g_{11} \quad g_{12}] \begin{bmatrix} 0 & 1 & -1 \\ 1 & 2 & 0 \end{bmatrix} + [(g_{21} - g_{22}) \quad 4g_{22} \quad -g_{21}] \right) \begin{bmatrix} -1 & 2 & 0 \\ 0 & 1 & 0 \\ -1 & 1 & 1 \end{bmatrix}^{-1}$$

$$= [(g_{11} - g_{12} + g_{22}) \quad (4g_{12} + g_{21} + 2g_{22}) \quad (-g_{11} - g_{21})]$$

The corresponding plant input gains are

$$\mathbf{h}_2^\dagger = \mathbf{m}_2^\dagger \mathbf{B} - \mathbf{g}_2^\dagger \mathbf{D} \qquad \mathbf{h}_1^\dagger = \mathbf{m}_1^\dagger \mathbf{B} - \mathbf{g}_1^\dagger \mathbf{D}$$

Next, a first-order observer subsystem with eigenvalue $\lambda = \frac{1}{2}$,

$$\xi_3(k + 1) = \frac{1}{2} \xi_3(k) + \mathbf{g}_3^\dagger \mathbf{y}(k) + \mathbf{h}_3^\dagger \mathbf{u}(k)$$

observing

$$\xi_3(k) \to \mathbf{m}_3^\dagger \mathbf{x}(k)$$

is designed. For this observer

$$\mathbf{m}_3^\dagger = \mathbf{g}_3^\dagger \mathbf{C}(\mathbf{A} - f\mathbf{I})^{-1} = [g_{31} \quad g_{32}] \begin{bmatrix} 0 & 1 & -1 \\ 1 & 2 & 0 \end{bmatrix} \begin{bmatrix} -\frac{3}{2} & 2 & 0 \\ 0 & \frac{1}{2} & 0 \\ -1 & 1 & \frac{1}{2} \end{bmatrix}^{-1}$$

$$= [(\tfrac{4}{3}g_{31} - \tfrac{2}{3}g_{32}) \quad (\tfrac{2}{3}g_{31} + \tfrac{20}{3}g_{32}) \quad -2g_{31}]$$

and

$$\mathbf{h}_3^\dagger = \mathbf{m}_3^\dagger \mathbf{B} - \mathbf{g}_3 \mathbf{D}$$

The gain vectors $\mathbf{g}_1, \mathbf{g}_2, \mathbf{g}_3$ are next chosen so that the collection of observers observes a nonsingular transformation of the plant state. Arbitrarily choosing

$$g_{11} = 1 \qquad g_{12} = 1$$
$$g_{21} = 1 \qquad g_{22} = 1$$
$$g_{31} = 3 \qquad g_{32} = 0$$

the observer having these gains

$$\begin{bmatrix} \xi_1(k + 1) \\ \xi_2(k + 1) \\ \xi_3(k + 1) \end{bmatrix} = \begin{bmatrix} 0 & 1 & 0 \\ 0 & 0 & 0 \\ 0 & 0 & \frac{1}{2} \end{bmatrix} \begin{bmatrix} \xi_1(k) \\ \xi_2(k) \\ \xi_3(k) \end{bmatrix} + \begin{bmatrix} 1 & 1 \\ 1 & 1 \\ 3 & 0 \end{bmatrix} \begin{bmatrix} y_1(k) \\ y_2(k) \end{bmatrix} + \begin{bmatrix} -6 & 22 \\ -5 & 11 \\ 3 & 9 \end{bmatrix} \begin{bmatrix} u_1(k) \\ u_2(k) \end{bmatrix}$$

$$(5\text{-}12)$$

observes

$$\begin{bmatrix} \xi_1(k) \\ \xi_2(k) \\ \xi_3(k) \end{bmatrix} \rightarrow \begin{bmatrix} \mathbf{m}_1^\dagger \\ \mathbf{m}_2^\dagger \\ \mathbf{m}_3^\dagger \end{bmatrix} \begin{bmatrix} x_1(k) \\ x_2(k) \\ x_3(k) \end{bmatrix} = \begin{bmatrix} 1 & 7 & -2 \\ 0 & 4 & -1 \\ 4 & 2 & -6 \end{bmatrix} \begin{bmatrix} x_1(k) \\ x_2(k) \\ x_3(k) \end{bmatrix} = \mathbf{M}\mathbf{x}(k)$$

Finally, an observer output equation

$$\mathbf{w}(k) = \mathbf{M}^{-1}\boldsymbol{\xi}(k)$$

$$\begin{bmatrix} w_1(k) \\ w_2(k) \\ w_3(k) \end{bmatrix} = \begin{bmatrix} 1.222 & -2.111 & -0.056 \\ 0.222 & -0.111 & -0.056 \\ 0.889 & -1.444 & 0.222 \end{bmatrix} \begin{bmatrix} \xi_1(k) \\ \xi_2(k) \\ \xi_3(k) \end{bmatrix}$$

is added so that

$$\mathbf{w}(k) \rightarrow \mathbf{x}(k)$$

Alternatively, the change of variables

$$\boldsymbol{\xi}'(k) = \mathbf{M}^{-1}\boldsymbol{\xi}(k)$$

can be used on equations (5-12) so that the observer state observes the plant state

$$\boldsymbol{\xi}'(k) \rightarrow \mathbf{x}(k)$$

5.5 Lower-Order Observers

Plant outputs themselves involve linear transformations of the plant state. If these are coupled directly to the observer output, the observer's order can be reduced. In this section, we first consider reduction in the order of state observers. Then we develop methods for further lowering the order of an observer used for placing feedback system eigenvalues by estimating only the linear state transformation to be fed back rather than the entire plant state. There is a simple solution for the minimal-order observer when the plant is single-input so that the needed observer output is a scalar. When there are multiple plant inputs, it is possible to lower the necessary observer order still further.

5.5.1 Reduced-Order State Observers

If a completely observable plant has m linearly independent outputs, a *reduced order* observer, of order $n - m$, having an output that observes the plant state can be constructed. Reduced order means that the state observer order (which without using the plant outputs directly is n) is reduced by the number of linearly independent plant outputs to order $n - m$. The $n - m$ observer eigenvalues can be placed arbitrarily by the designer as long as none equals plant eigenvalues. The order reduction occurs because the observer uses the m linearly independent scalar state transformations involved in the plant outputs. These are coupled to the observer output through direct input-to-output coupling, which is not present in a full-order state observer.

For the plant

$$\mathbf{x}(k + 1) = \mathbf{A}\mathbf{x}(k) + \mathbf{B}\mathbf{u}(k)$$
$$\mathbf{y}(k) = \mathbf{C}\mathbf{x}(k) + \mathbf{D}\mathbf{u}(k)$$

and an $(n - m)$th-order observer with state equation of the form

$$\boldsymbol{\xi}(k + 1) = \mathbf{F}\boldsymbol{\xi}(k) + \mathbf{G}\mathbf{y}(k) + \mathbf{H}\mathbf{u}(k)$$

having designer-selected eigenvalues, the combination of plant outputs and observer state observes

$$\mathbf{w}(k) = \begin{bmatrix} \mathbf{y}(k) - \mathbf{D}\mathbf{u}(k) \\ \hline \boldsymbol{\xi}(k) \end{bmatrix} \rightarrow \mathbf{E}\mathbf{x}(k) = \begin{bmatrix} \mathbf{C} \\ \hline \mathbf{M} \end{bmatrix} \mathbf{x}(k)$$

where the observer state observes

$$\boldsymbol{\xi}(k) \rightarrow \mathbf{M}\mathbf{x}(k)$$

The symbol $\mathbf{E}$ is used for the observation of the n-vector observer output, and $\mathbf{M}$ is the observation matrix of the observer state.

If the plant is completely observable, it is always possible to choose observer gains so that $\mathbf{E}$ is nonsingular. With such a choice

$$\mathbf{w}'(k) = \mathbf{E}^{-1}\mathbf{w}(k) \rightarrow \mathbf{x}(k)$$

A summary of this general method of state observer design is given in Table 5-5.

Table 5-5 General Design Method for State Observers

1. To construct a state observer for the nth-order plant

$$\mathbf{x}(k + 1) = \mathbf{A}\mathbf{x}(k) + \mathbf{B}\mathbf{u}(k)$$

$$\mathbf{y}(k) = \mathbf{C}\mathbf{x}(k) + \mathbf{D}\mathbf{u}(k) = \begin{bmatrix} \mathbf{c}_1^\dagger \\ \hline \vdots \\ \hline \mathbf{c}_m^\dagger \end{bmatrix} \mathbf{x}(k) + \begin{bmatrix} \mathbf{d}_1^\dagger \\ \hline \vdots \\ \hline \mathbf{d}_m^\dagger \end{bmatrix} \mathbf{u}(k)$$

Table 5-5 continued

with m linearly independent outputs, first form a nonsingular observation matrix $\mathbf{E}$ and the associated first-order observer state and observer output equations as follows:

A. If the observer is to be of *reduced order*, the linearly independent rows of $\mathbf{C}$ are included in $\mathbf{E}$. For each, an observer output signal

$$w_i(k) = y_i(k) - \mathbf{d}_i^\dagger \mathbf{u}(k)$$

is formed. If the observer is to be of full order, rows of $\mathbf{C}$ are not included in $\mathbf{E}$.

B. For each distinct observer eigenvalue, a first-order observer of the form

$$\xi_i(k + 1) = f_i \xi_i(k) + \mathbf{g}_i^\dagger \mathbf{y}(k) + \mathbf{h}_i^\dagger \mathbf{u}(k)$$

is designed. The individual observer observation matrix

$$\mathbf{m}_i^\dagger = \mathbf{g}_i^\dagger \mathbf{C}(\mathbf{A} - f_i\mathbf{I})^{-1}$$

which depends on $\mathbf{g}_i$, is included as a row of the overall observation matrix $\mathbf{E}$. The observer state becomes another observer output

$$w_i(k) = \xi_i(k)$$

C. If observer eigenvalue repetitions are desired, the observer design is governed by the recursive relations of Table 5-4. Each of the individual observed state transformations is included in $\mathbf{E}$, and the observer state variables are each added to the collection of observer outputs.

2. For a reduced-order state observer, the observer order is $n - m$. For a full-order state observer, the observer order is n. In either case, $\mathbf{E}$ is $n \times n$. Make convenient but arbitrary choices for all the observer gains $\mathbf{g}_i$ so that $\mathbf{E}$ is nonsingular.

3. The observer output equation is

$$\mathbf{w}'(k) = \mathbf{E}^{-1}\mathbf{w}(k)$$

and

$$\mathbf{w}'(k) \rightarrow \mathbf{x}(k)$$

Or the change of variables

$$\xi'(k) = \mathbf{E}^{-1}\xi(k)$$

can be made so that

$$\xi'(k) \rightarrow \mathbf{x}(k)$$

For the system

$$\begin{bmatrix} x_1(k+1) \\ x_2(k+1) \end{bmatrix} = \begin{bmatrix} 1 & 0 \\ -2 & 3 \end{bmatrix} \begin{bmatrix} x_1(k) \\ x_2(k) \end{bmatrix} + \begin{bmatrix} -1 \\ 2 \end{bmatrix} u(k) = \mathbf{A}\mathbf{x}(k) + \mathbf{b}u(k)$$

$$y(k) = [1 \quad 2] \begin{bmatrix} x_1(k) \\ x_2(k) \end{bmatrix} - 3u(k) = \mathbf{c}^\dagger \mathbf{x}(k) + du(k)$$

a first-order observer with eigenvalue $\lambda = \frac{1}{2}$, of the form

$$\xi(k+1) = \frac{1}{2} \xi(k) + gy(k) + hu(k)$$

observes

$$\xi(k) \to \mathbf{m}^\dagger \mathbf{x}(k)$$

where

$$\mathbf{m}^\dagger = g\mathbf{c}^\dagger \left(\mathbf{A} - \frac{1}{2}\mathbf{I} \right)^{-1} = g[1 \quad 2] \begin{bmatrix} \frac{1}{2} & 0 \\ -2 & \frac{5}{2} \end{bmatrix}^{-1} = g[\frac{26}{5} \quad \frac{4}{5}]$$

As the observer gain g only serves to scale the observed functional, it will be set to unity so that

$$\mathbf{m}^\dagger = [\frac{26}{5} \quad \frac{4}{5}]$$

For this observer

$$h = \mathbf{m}^\dagger \mathbf{b} - d = [\frac{26}{5} \quad \frac{4}{5}] \begin{bmatrix} -1 \\ 2 \end{bmatrix} + 3 = -\frac{3}{5}$$

The two signals observe

$$\begin{bmatrix} [y(k) + 3u(k)] \\ \xi(k) \end{bmatrix} \to \begin{bmatrix} \mathbf{c}^\dagger \\ \hline \mathbf{m}^\dagger \end{bmatrix} \mathbf{x}(k) = \begin{bmatrix} 1 & 2 \\ \frac{26}{5} & \frac{4}{5} \end{bmatrix} \begin{bmatrix} x_1(k) \\ x_2(k) \end{bmatrix} = \mathbf{E}\mathbf{x}(k)$$

hence, the two-output first-order observer

$$\xi(k+1) = \frac{1}{2} \xi(k) + y(k) - \frac{3}{5} u(k)$$

$$\begin{bmatrix} w_1(k) \\ w_2(k) \end{bmatrix} = \begin{bmatrix} 0 \\ 1 \end{bmatrix} \xi(k) + \begin{bmatrix} 1 \\ 0 \end{bmatrix} y(k) + \begin{bmatrix} 3 \\ 0 \end{bmatrix} u(k)$$

has output that observes

$$\mathbf{w}(k) \to \begin{bmatrix} 1 & 2 \\ \frac{26}{5} & \frac{4}{5} \end{bmatrix} \mathbf{x}(k) = \mathbf{E}\mathbf{x}(k)$$

Transforming the output to form

$$\mathbf{w}'(k) = \mathbf{E}^{-1}\mathbf{w}(k) = \frac{1}{48}\begin{bmatrix} -4 & 10 \\ 26 & -5 \end{bmatrix}\mathbf{w}(k)$$

$$\begin{bmatrix} w_1'(k) \\ w_2'(k) \end{bmatrix} = \frac{1}{48}\begin{bmatrix} -4 & 10 \\ 26 & -5 \end{bmatrix}\begin{bmatrix} 0 \\ 1 \end{bmatrix}\xi(k) + \frac{1}{48}\begin{bmatrix} -4 & 10 \\ 26 & -5 \end{bmatrix}\begin{bmatrix} 1 \\ 0 \end{bmatrix}y(k)$$

$$+ \frac{1}{48}\begin{bmatrix} -4 & 10 \\ 26 & -5 \end{bmatrix}\begin{bmatrix} 3 \\ 0 \end{bmatrix}u(k)$$

$$= \begin{bmatrix} \frac{10}{48} \\ -\frac{5}{48} \end{bmatrix}\xi(k) + \begin{bmatrix} -\frac{4}{48} \\ \frac{26}{48} \end{bmatrix}y(k) + \begin{bmatrix} -\frac{12}{48} \\ \frac{78}{48} \end{bmatrix}u(k)$$

results in an observer with output that observes the state $\mathbf{x}(k)$.

For the third-order two-output plant

$$\begin{bmatrix} x_1(k+1) \\ x_2(k+1) \\ x_3(k+1) \end{bmatrix} = \begin{bmatrix} 2 & 0 & 3 \\ 1 & 0 & 1 \\ 0 & -2 & 0 \end{bmatrix}\begin{bmatrix} x_1(k) \\ x_2(k) \\ x_3(k) \end{bmatrix} + \begin{bmatrix} 1 \\ 0 \\ 0 \end{bmatrix}u(k) = \mathbf{A}\mathbf{x}(k) + \mathbf{b}u(k)$$

$$\begin{bmatrix} y_1(k) \\ y_2(k) \end{bmatrix} = \begin{bmatrix} 0 & 2 & 1 \\ 1 & -1 & 2 \end{bmatrix}\begin{bmatrix} x_1(k) \\ x_2(k) \\ x_3(k) \end{bmatrix} + \begin{bmatrix} 3 \\ 0 \end{bmatrix}u(k) = \mathbf{C}\mathbf{x}(k) + \mathbf{d}u(k)$$

a first-order observer with eigenvalue $\lambda = 0$, of the form

$$\xi(k+1) = (0)\xi(k) + \mathbf{g}^t\mathbf{y}(k) + hu(k)$$

observes the scalar state transformation

$$\mathbf{m}^t = \mathbf{g}^t\mathbf{C}(\mathbf{A} - 0 \cdot \mathbf{I})^{-1} = [g_1 \quad g_2]\begin{bmatrix} 0 & 2 & 1 \\ 1 & -1 & 2 \end{bmatrix}\begin{bmatrix} 2 & 0 & 3 \\ 1 & 0 & 1 \\ 0 & -2 & 0 \end{bmatrix}^{-1}$$

$$= g_1[1 \quad -2 \quad -1] + g_2[1 \quad -1 \quad \tfrac{1}{2}]$$

For this observer

$$h = \mathbf{m}^t\mathbf{b} - \mathbf{g}^t\mathbf{d} = -2g_1 + g_2$$

The three signals

$$\begin{bmatrix} w_1(k) \\ w_2(k) \\ w_3(k) \end{bmatrix} = \begin{bmatrix} [y_1(k) - 3u(k)] \\ y_2(k) \\ \xi(k) \end{bmatrix} = \begin{bmatrix} 0 \\ 0 \\ 1 \end{bmatrix}\xi(k)$$

$$+ \begin{bmatrix} 1 & 0 \\ 0 & 1 \\ 0 & 0 \end{bmatrix}\begin{bmatrix} y_1(k) \\ y_2(k) \end{bmatrix} + \begin{bmatrix} -3 \\ 0 \\ 0 \end{bmatrix}u(k)$$

observe

$$
\mathbf{w}(k) \rightarrow \left[\begin{array}{c} \mathbf{C} \\ \hline \mathbf{m}^\dagger \end{array}\right] \mathbf{x}(k) = \left[\begin{array}{ccc} 0 & 2 & 1 \\ 1 & -1 & 2 \\ (g_1 + g_2) & (-2g_1 - g_2) & (-g_1 + \tfrac{1}{2}g_2) \end{array}\right] \left[\begin{array}{c} x_1(k) \\ x_2(k) \\ x_3(k) \end{array}\right]
$$

$$
= \mathbf{E}\mathbf{x}(k)
$$

Arbitrarily choosing

$$g_1 = -1 \qquad g_2 = 0$$

which is one choice of observer gains for which $\mathbf{E}$ is nonsingular

$$
\mathbf{E} = \left[\begin{array}{ccc} 0 & 2 & 1 \\ 1 & -1 & 2 \\ -1 & 2 & 1 \end{array}\right] \qquad \mathbf{E}^{-1} = \left[\begin{array}{ccc} 1 & 0 & -1 \\ \frac{3}{5} & -\frac{1}{5} & -\frac{1}{5} \\ -\frac{1}{5} & \frac{2}{5} & \frac{2}{5} \end{array}\right]
$$

gives the observer

$$
\xi(k + 1) = \begin{bmatrix} -1 & 0 \end{bmatrix} \begin{bmatrix} y_1(k) \\ y_2(k) \end{bmatrix} + 2u(k)
$$

$$
\mathbf{w}'(k) = \mathbf{E}^{-1}\mathbf{w}(k) = \left[\begin{array}{ccc} 1 & 0 & -1 \\ \frac{3}{5} & -\frac{1}{5} & -\frac{1}{5} \\ -\frac{1}{5} & \frac{2}{5} & \frac{2}{5} \end{array}\right] \left[\begin{array}{c} 0 \\ 0 \\ 1 \end{array}\right] \xi(k)
$$

$$
+ \left[\begin{array}{ccc} 1 & 0 & -1 \\ \frac{3}{5} & -\frac{1}{5} & -\frac{1}{5} \\ -\frac{1}{5} & \frac{2}{5} & \frac{2}{5} \end{array}\right] \left[\begin{array}{cc} 1 & 0 \\ 0 & 1 \\ 0 & 0 \end{array}\right] \left[\begin{array}{c} y_1(k) \\ y_2(k) \end{array}\right]
$$

$$
+ \left[\begin{array}{ccc} 1 & 0 & -1 \\ \frac{3}{5} & -\frac{1}{5} & -\frac{1}{5} \\ -\frac{1}{5} & \frac{2}{5} & \frac{2}{5} \end{array}\right] \left[\begin{array}{c} -3 \\ 0 \\ 0 \end{array}\right] u(k)
$$

$$
= \left[\begin{array}{c} -1 \\ -\frac{1}{5} \\ \frac{2}{5} \end{array}\right] \xi(k) + \left[\begin{array}{cc} 1 & 0 \\ \frac{3}{5} & -\frac{1}{5} \\ -\frac{1}{5} & \frac{2}{5} \end{array}\right] \left[\begin{array}{c} y_1(k) \\ y_2(k) \end{array}\right] + \left[\begin{array}{c} -3 \\ -\frac{9}{5} \\ \frac{3}{5} \end{array}\right] u(k)
$$

which observes $\mathbf{x}(k)$.

5.5.2 Minimal-Order Observers of a Scalar State Transformation

Placing feedback eigenvalues for a single-input plant requires only a single scalar linear transformation of the plant state for feedback. We now develop observer design methods similar to the previous ones for designing observers of a given scalar linear transformation of the plant state.

If an nth-order plant has a single output, then obtaining observation of an arbitrary scalar linear plant state transformation generally requires n first-order observers if the plant output is not used directly, if the plant output is not used directly. If the plant output is used directly, $n - 1$ first-order observers are generally needed. It would be a lucky circumstance if the desired scalar state transformation were a linear combination of the individual scalar state transformations observed by a lesser number of component first-order observers. Thus, one may as well construct a state observer, then form the desired scalar transformation from the observed plant state. When the plant has more than one output, however, each first-order observer observes a linear combination of m scalar state transformations, where m is the number of plant outputs. The specific linear combination is determined by the choice of observer gains, so each first-order observer contributes an arbitrary linear combination of m scalar transformations to a single output from a collection of observers.

This procedure for design of a minimum-order observer of a scalar linear plant state transformation is summarized in Table 5-6.

As an example, consider the system

$$\begin{bmatrix} x_1(k+1) \\ x_2(k+1) \\ x_3(k+1) \end{bmatrix} = \begin{bmatrix} 0 & 1 & 0 \\ 0 & 0 & 1 \\ 0 & 0 & 0 \end{bmatrix} \begin{bmatrix} x_1(k) \\ x_2(k) \\ x_3(k) \end{bmatrix} + \begin{bmatrix} 2 \\ 0 \\ -2 \end{bmatrix} u(k) = \mathbf{A}\mathbf{x}(k) + \mathbf{b}u(k)$$

$$\begin{bmatrix} y_1(k) \\ y_2(k) \end{bmatrix} = \begin{bmatrix} 0 & 1 & -1 \\ 1 & 2 & 0 \end{bmatrix} \begin{bmatrix} x_1(k) \\ x_2(k) \\ x_3(k) \end{bmatrix} + \begin{bmatrix} 0 \\ 3 \end{bmatrix} u(k) = \mathbf{C}\mathbf{x}(k) + \mathbf{d}u(k)$$

(5-14)

where two outputs are available to the observer.

A zeroth-order observer (one with no state equation) of the form

$$w(k) = \gamma_1 y_1(k) + \gamma_2 [y_2(k) - 3u(k)]$$

where γ_1 and γ_2 are arbitrary constants, equals (and thus observes) any scalar linear transformation of the plant state $\mathbf{e}^t\mathbf{x}$ of the form

$$\mathbf{e}^t = \gamma_1 [0 \quad 1 \quad -1] + \gamma_2 [1 \quad 2 \quad 0] \qquad \text{(5-15)}$$

If the plant state transformation to be observed, $\mathbf{e}^t$, can be expressed in this form, equation (5-15) is solved for γ_1 and γ_2 and the resulting observer has no dynamics; the state transformation can be derived from the two plant outputs alone.

Table 5-6 Design of Minimal-Order Observers of a Scalar Linear State Transformation

To construct a minimal-order observer of a scalar linear state transformation

$$w(k) \rightarrow \mathbf{e}^\dagger \mathbf{x}(k)$$

for the plant

$$\mathbf{x}(k + 1) = \mathbf{A}\mathbf{x}(k) + \mathbf{B}\mathbf{u}(k)$$

$$\mathbf{y}(k) = \mathbf{C}\mathbf{x}(k) + \mathbf{D}\mathbf{u}(k) = \begin{bmatrix} \mathbf{c}_1^\dagger \\ \vdots \\ \mathbf{c}_m^\dagger \end{bmatrix} \mathbf{x}(k) + \begin{bmatrix} \mathbf{d}_1^\dagger \\ \vdots \\ \mathbf{d}_m^\dagger \end{bmatrix} \mathbf{u}(k)$$

collect terms in the equation

$$\gamma_1 \mathbf{c}_1^\dagger + \gamma_2 \mathbf{c}_2^\dagger + \cdots + \gamma_m \mathbf{c}_m^\dagger + \mathbf{m}_1^\dagger + \mathbf{m}_2^\dagger + \cdots + \mathbf{m}_i^\dagger = \mathbf{e}^\dagger \qquad (5\text{-}13)$$

until a solution exists for the γ's and the observer gains. A solution to equation (5-13) defines a specific linear combination of the plant output state transformations and a choice of observer gains for which the observer output observes the scalar state transformation $\mathbf{e}^\dagger \mathbf{x}(k)$.

   If the plant outputs are to be used by the observer, each row of $\mathbf{C}$ is included in equation (5-13) and, for each output, a term

$$w(k) = \cdots + \gamma_i [y_i(k) - \mathbf{d}_i^\dagger \mathbf{u}(k)] + \cdots$$

is added to the single observer output, where γ_i is a constant to be determined. Otherwise, these terms are not used. For each distinct observer eigenvalue, a first-order observer of the form

$$\xi_i(k + 1) = f_i \xi_i(k) + \mathbf{g}_i^\dagger \mathbf{y}(k) + \mathbf{h}_i^\dagger \mathbf{u}(k)$$

is designed. The observation matrix

$$\mathbf{m}_i^\dagger = \mathbf{g}_i^\dagger \mathbf{C}(\mathbf{A} - f_i \mathbf{I})^{-1}$$

involves a linear combination of the elements of the observer gain vector $\mathbf{g}_i$ to be determined. Each observation matrix $\mathbf{m}_i$ is added to equation (5-13), and the observer state is added to the single observer output

$$w(k) = \cdots + \xi_i(k) + \cdots$$

When enough terms are added to equation (5-13), a solution for the variables $\gamma_1, \gamma_2, \ldots$ and the elements of $\mathbf{g}_1, \mathbf{g}_2, \ldots$ exists and is found.

   If observer eigenvalue repetitions are desired, the observer design is governed by the recursive relations of Table 5-4. Each of the observed scalar state transformations is added to equation (5-13), and each of the observer states is added to the observer output $w(k)$.

If, as is likely, equation (5-15) does not have a solution, a first-order observer is added. If it is desired that the observer eigenvalue be $\lambda = -\frac{1}{4}$, then this observer has the form

$$\xi_1(k + 1) = -\frac{1}{4}\xi_1(k) + [g_{11} \quad g_{12}]\begin{bmatrix} y_1(k) \\ y_2(k) \end{bmatrix} + h_1 u(k)$$

and has the observation matrix

$$\mathbf{m}_1^\dagger = \mathbf{g}_1^\dagger \mathbf{C}\left(\mathbf{A} + \frac{1}{4}\mathbf{I}\right)^{-1} = [g_{11} \quad g_{12}]\begin{bmatrix} 0 & 1 & -1 \\ 1 & 2 & 0 \end{bmatrix}\begin{bmatrix} \frac{1}{4} & 1 & 0 \\ 0 & \frac{1}{4} & 1 \\ 0 & 0 & \frac{1}{4} \end{bmatrix}^{-1}$$

$$= g_{11}[0 \quad 4 \quad -20] + g_{12}[4 \quad -8 \quad 32]$$

The plant input coupling to this observer is

$$h_1 = \mathbf{m}_1^\dagger \mathbf{b} - \mathbf{g}_1^\dagger \mathbf{d} = 40g_{11} - 59g_{12}$$

giving

$$\xi_1(k + 1) = -\frac{1}{4}\xi_1(k) + [g_{11} \quad g_{12}]\begin{bmatrix} y_1(k) \\ y_2(k) \end{bmatrix} + (40g_{11} - 59g_{12})u(k) \tag{5-16}$$

The observer output

$$w(k) = \gamma_1 y_1(k) + \gamma_2[y_2(k) - 3u(k)] + \xi_1(k)$$

observes a linear transformation of the plant state $\mathbf{e}^\dagger\mathbf{x}$ of the form

$$\mathbf{e}^\dagger = \gamma_1[0 \quad 1 \quad -1] + \gamma_2[1 \quad 2 \quad 0]$$
$$+ g_{11}[0 \quad 4 \quad -20] + g_{12}[4 \quad -8 \quad 32] \tag{5-17}$$

which, with appropriate choice of γ_1, γ_2, g_{11}, and g_{12}, can observe any scalar linear plant state transformation. If it is desired that

$$w(k) \to [1 \quad 0 \quad 0]\mathbf{x}(k) = \mathbf{e}^\dagger\mathbf{x}(k)$$

the γ's and g's must satisfy

$$\begin{cases} \gamma_2 \quad\quad + 4g_{12} = 1 \\ \gamma_1 + 2\gamma_2 + 4g_{11} - 8g_{12} = 0 \\ -\gamma_1 \quad\quad - 20g_{11} + 32g_{12} = 0 \end{cases} \tag{5-18}$$

one solution to which is

$$\gamma_1 = -\frac{5}{2} \quad \gamma_2 = 1$$

$$g_{11} = \frac{1}{8} \quad g_{12} = 0$$

This solution defines the observer

$$\xi_1(k + 1) = -\frac{1}{4}\xi(k) + [\frac{1}{8} \quad 0]\mathbf{y}(k) + u(k)$$

$$w(k) = \xi_1(k) + [-\frac{5}{2} \quad 1]\mathbf{y}(k) - 3u(k)$$

There are many other solutions to equations (5-18), of course.

For the plant in equation (5-14), suppose instead that a solution is desired that does not use the plant outputs directly. That is, there is to be no direct coupling of $\mathbf{y}(k)$ to the observer output. We may wish to have this property because it tends to reduce the effects of plant output noise. Beginning with the first-order observer in equation (5-16), the observer output

$$w(k) = \xi_1(k)$$

observes any scalar linear plant state transformation $\mathbf{e}^\dagger\mathbf{x}$ of the form (setting $\gamma_1 = \gamma_2 = 0$ in equation (5-17))

$$\mathbf{e}^\dagger = g_{11}[0 \quad 4 \quad -20] + g_{12}[4 \quad -8 \quad 32]$$

If the state transformation to be observed is of this form, then g_{11} and g_{12} are found. If not, as is likely because $\mathbf{e}$ has three components and there are only two variables, another first-order observer is added to the design.

Another first-order observer, with eigenvalue chosen to be $\lambda = \frac{1}{2}$, has the form

$$\xi_2(k + 1) = \frac{1}{2}\xi_2(k) + [g_{21} \quad g_{22}]\begin{bmatrix} y_1(k) \\ y_2(k) \end{bmatrix} + h_2u(k)$$

and has the observation matrix

$$\mathbf{m}_2^\dagger = \mathbf{g}_2^\dagger\mathbf{C}\left(\mathbf{A} - \frac{1}{2}\mathbf{I}\right)^{-1} = [g_{21} \quad g_{22}]\begin{bmatrix} 0 & 1 & -1 \\ 1 & 2 & 0 \end{bmatrix}\begin{bmatrix} -\frac{1}{2} & 1 & 0 \\ 0 & -\frac{1}{2} & 1 \\ 0 & 0 & -\frac{1}{2} \end{bmatrix}^{-1}$$

$$= g_{21}[0 \quad -2 \quad -2] + g_{22}[-2 \quad -8 \quad -16]$$

The plant input coupling to this observer is

$$h_2 = \mathbf{m}_2^\dagger\mathbf{b} - \mathbf{g}_2^\dagger\mathbf{d} = 4g_{21} + 25g_{22}$$

giving

$$\xi_2(k + 1) = \frac{1}{2}\xi_2(k) + [g_{21} \quad g_{22}]\begin{bmatrix} y_1(k) \\ y_2(k) \end{bmatrix} + (4g_{21} + 25g_{22})u(k)$$

The observer output

$$w(k) = \xi_1(k) + \xi_2(k)$$

observes any scalar linear plant state transformation $e^\dagger x$ of the form

$$e^\dagger = g_{11}[0 \quad 4 \quad -20] + g_{12}[4 \quad -8 \quad 32]$$
$$+ \, g_{21}[0 \quad -2 \quad -2] + g_{22}[-2 \quad -8 \quad -16]$$

For

$$e^\dagger = [1 \quad 0 \quad 0]$$

the g's must satisfy

$$\begin{cases} 4g_{12} \quad\quad - \quad 2g_{22} = 1 \\ 4g_{11} - 8g_{12} - 2g_{21} - 8g_{22} = 0 \\ -20g_{11} + 32g_{12} - 2g_{21} - 16g_{22} = 0 \end{cases}$$

one solution to which is

$$g_{11} = \frac{5}{12} \quad g_{12} = \frac{1}{4} \quad g_{21} = -\frac{1}{6} \quad g_{22} = 0$$

The observer

$$\begin{bmatrix} \xi_1(k+1) \\ \xi_2(k+1) \end{bmatrix} = \begin{bmatrix} -\frac{1}{4} & 0 \\ 0 & \frac{1}{2} \end{bmatrix} \begin{bmatrix} \xi_1(k) \\ \xi_2(k) \end{bmatrix} + \begin{bmatrix} \frac{5}{12} & \frac{1}{4} \\ -\frac{1}{6} & 0 \end{bmatrix} \begin{bmatrix} y_1(k) \\ y_2(k) \end{bmatrix} + \begin{bmatrix} \frac{23}{12} \\ -\frac{1}{6} \end{bmatrix} u(k)$$

$$w(k) = [1 \quad 1] \begin{bmatrix} \xi_1(k) \\ \xi_2(k) \end{bmatrix}$$

then has output $w(k)$ that observes $e^\dagger x$.

For a completely observable plant of order n, with m outputs and no observer eigenvalue equaling a plant eigenvalue, the order j of a minimal-order observer of an arbitrary scalar plant state transformation is given by

$$m + jm \geqslant n \quad j \geqslant \frac{n}{m} - 1 = \frac{n-m}{m}$$

If the observer is to have no direct coupling of plant signals to its output, then

$$jm \geqslant n \quad j \geqslant \frac{n}{m}$$

Table 5-7 lists observer orders for a tenth-order plant with various numbers of outputs, as a representative comparison.

Table 5-7 Observer Orders for Feedback Eigenvalue Placement for a Tenth-Order Plant

Number of Plant Outputs	Order of the Full-Order State Observer	Order of the Reduced-Order State Observer	Order of the Minimal-Order Observer of a Scalar Transformation of the Plant State	
			Without Direct Plant Coupling to Its Output	With Direct Plant Coupling to Its Output
1	10	9	10	9
2	10	8	5	4
3	10	7	4	3
4	10	6	3	2
5	10	5	2	1
6	10	4	2	1
7	10	3	2	1
8	10	2	2	1
9	10	1	2	1
10	10	0	1	0

5.5.3 Observers of Multiple Scalar State Transformations

If an observer is used for placing feedback system eigenvalues for a multiple-input plant, it may be desirable to observe more than one scalar transformation of the plant state, but not n of them, because that would amount to observing the entire plant state. For feedback eigenvalue placement, the additional freedom in the choice of individual state feedback vectors e_i, one for each plant input, can also be exploited in reducing the required observer order. A minimal-order feedback observer solution for eigenvalue placement does exist, but it is fairly involved. We will not pursue the topic of minimal-order observers further here because most often, as we will see, the higher-order solutions give better overall tracking system performance.

5.6 Eigenvalue Placement with Observer Feedback

The eigenvalues of a plant, or of a plant with feedback, determine the character of the system's zero-input response. Normally, the initial conditions of the plant are not entirely known, and because the amplitudes of each zero-input response mode depend on these, the best one can usually do is to select the modes themselves. If the plant state is available for feedback, the designer can place all of the feedback system eigenvalues as desired and thus completely design the form of a controlled system's zero-input response. When, as is usual, the plant state is not accessible for feedback, an observer's estimate of the plant state can be fed back in place of the actual state.

It should be noted that we normally avoid choosing observer eigenvalues equal to any of the original plant eigenvalues. There is no reason to avoid having observer eigenvalues equal to those of the plant with feedback.

5.6.1 The Separation Theorem

We now prove the separation theorem: When observer feedback is used in place of plant state feedback, the eigenvalues of the feedback system are those the plant would have if state feedback were used and those of the observer. The arrangement of the plant and the observer is shown in Figure 5-11. To see the principle involved easily, the result is first shown for feedback by a full-order state observer.

Consider the nth-order plant

$$\mathbf{x}(k + 1) = \mathbf{Ax}(k) + \mathbf{Bu}(k)$$

$$\mathbf{y}(k) = \mathbf{Cx}(k) + \mathbf{Du}(k)$$

with feedback

$$\mathbf{u}(k) = \mathbf{E}\,\boldsymbol{\xi}(k) + \boldsymbol{\rho}(k)$$

from a full-order state observer

$$\boldsymbol{\xi}(k + 1) = \mathbf{F}\boldsymbol{\xi}(k) + \mathbf{Gy}(k) + \mathbf{Hu}(k)$$

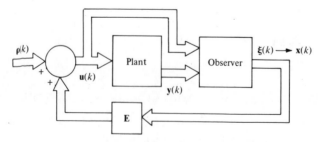

Figure 5-11 An observer estimate of the plant state, fed back in place of the state.

where

$$\mathbf{F} = \mathbf{A} - \mathbf{GC}$$

$$\mathbf{H} = \mathbf{B} - \mathbf{GD}$$

Rather than dealing with the $2n$-state variables $\mathbf{x}$ and $\boldsymbol{\xi}$, we use $\mathbf{x}$ and the n observer error variables

$$\boldsymbol{\varepsilon}(k) = \mathbf{x}(k) - \boldsymbol{\xi}(k)$$

to describe the feedback system. Either set of variables can be used, but using the latter results in simplified equations. One can view this as a choice of new state variables, $\mathbf{x}$ and $\boldsymbol{\varepsilon}$, that are related to the old ones, $\mathbf{x}$ and $\boldsymbol{\xi}$, by a nonsingular transformation. Substituting for $\mathbf{u}(k)$ in the plant state equations in terms of $\boldsymbol{\varepsilon}(k)$ results in

$$\mathbf{x}(k + 1) = \mathbf{Ax}(k) + \mathbf{BE}\boldsymbol{\xi}(k) + \mathbf{B}\rho(k)$$

$$= (\mathbf{A} + \mathbf{BE})\mathbf{x}(k) - \mathbf{BE}\boldsymbol{\varepsilon}(k) + \mathbf{B}\rho(k)$$

The observer error is governed by

$$\boldsymbol{\varepsilon}(k + 1) = \mathbf{F}\boldsymbol{\varepsilon}(k)$$

so that equations for the feedback system are

$$\begin{bmatrix} \mathbf{x}(k + 1) \\ \hline \boldsymbol{\varepsilon}(k + 1) \end{bmatrix} = \begin{bmatrix} \mathbf{A} + \mathbf{BE} & -\mathbf{BE} \\ \hline \mathbf{0} & \mathbf{F} \end{bmatrix} \begin{bmatrix} \mathbf{x}(k) \\ \hline \boldsymbol{\varepsilon}(k) \end{bmatrix} + \begin{bmatrix} \mathbf{B} \\ \hline \mathbf{0} \end{bmatrix} \rho(k) \qquad (5\text{-}19)$$

The characteristic equation for (5-19) is given by

$$\begin{vmatrix} \lambda\mathbf{I} - (\mathbf{A} + \mathbf{BE}) & \mathbf{BE} \\ \hline \mathbf{0} & \lambda\mathbf{I} - \mathbf{F} \end{vmatrix} = |\lambda\mathbf{I} - (\mathbf{A} + \mathbf{BE})||\lambda\mathbf{I} - \mathbf{F}| = 0$$

which demonstrates that the eigenvalues of the plant with full-order state observer feedback are those of the state feedback system

$$|\lambda\mathbf{I} - (\mathbf{A} + \mathbf{BE})| = 0$$

and those of the observer

$$|\lambda\mathbf{I} - \mathbf{F}| = 0$$

We next consider observer feedback when the observer is possibly of different order than that of the plant and involves an output equation.

$$\boldsymbol{\xi}(k + 1) = \mathbf{F}\boldsymbol{\xi}(k) + \mathbf{Gy}(k) + \mathbf{Hu}(k)$$

$$\mathbf{w}(k) = \mathbf{L}\boldsymbol{\xi}(k) + \mathbf{Nx}(k)$$

Here the observer output equation is expressed in a form emphasizing that the feedback state estimate $\mathbf{w}(k)$ is a combination of observer state and plant state feedback. The observer state observes a linear transformation of the plant state

$$\xi(k) \rightarrow \mathbf{Mx}(k)$$

and the error in this observation is governed by

$$\varepsilon(k + 1) \doteq \mathbf{Mx}(k + 1) - \xi(k + 1) = \mathbf{F}[\mathbf{Mx}(k) - \xi(k)] = \mathbf{F}\varepsilon(k)$$

The observer output equation is formed so that

$$\mathbf{w}(k) \rightarrow \mathbf{x}(k)$$

That is, because

$$\xi(k) \rightarrow \mathbf{Mx}(k)$$

then

$$\mathbf{LM} + \mathbf{N} = \mathbf{I}$$

Substituting

$$\mathbf{u}(k) = \mathbf{Ew}(k) + \rho(k)$$

in the plant state equations in terms of $\varepsilon(k)$ results in

$$
\begin{aligned}
\mathbf{x}(k + 1) &= \mathbf{Ax}(k) + \mathbf{BEw}(k) + \mathbf{B}\rho(k) \\
&= \mathbf{Ax}(k) + \mathbf{BEL}\xi(k) + \mathbf{BENx}(k) + \mathbf{B}\rho(k) \\
&= \mathbf{Ax}(k) + \mathbf{BEL}[\mathbf{Mx}(k) - \varepsilon(k)] + \mathbf{BENx}(k) + \mathbf{B}\rho(k) \\
&= [\mathbf{A} + \mathbf{BE}(\mathbf{LM} + \mathbf{N})]\mathbf{x}(k) - \mathbf{BEL}\varepsilon(k) + \mathbf{B}\rho(k) \\
&= (\mathbf{A} + \mathbf{BE})\mathbf{x}(k) - \mathbf{BEL}\varepsilon(k) + \mathbf{B}\rho(k)
\end{aligned}
$$

The feedback system equations are then

$$
\left[\begin{array}{c} \mathbf{x}(k + 1) \\ \hline \varepsilon(k + 1) \end{array}\right] = \left[\begin{array}{c:c} \mathbf{A} + \mathbf{BE} & -\mathbf{BEL} \\ \hdashline \mathbf{0} & \mathbf{F} \end{array}\right]\left[\begin{array}{c} \mathbf{x}(k) \\ \hline \varepsilon(k) \end{array}\right] + \left[\begin{array}{c} \mathbf{B} \\ \hline \mathbf{0} \end{array}\right]\rho(k)
$$

and, as in the case of the full-order state observer, have eigenvalues that are those of the state feedback system and those of the observer.

5.6.2 Transfer Function Matrix of the Feedback Compensator

When a full-order state observer is arranged as an error feedback system, as in Figure 5-8, its estimate of the plant state can be fed back in place of the actual state to locate all of the feedback system eigen-

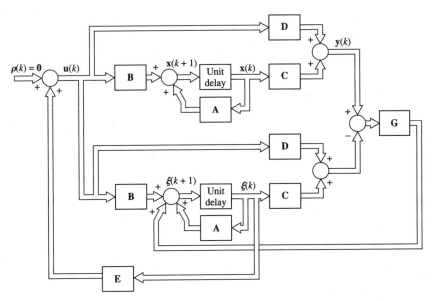

Figure 5-12 Eigenvalue placement with full-order state observer feedback.

values as desired. This arrangement is shown in Figure 5-12. An equivalent frequency domain representation of this arrangement is shown in Figure 5-13 where the transfer function matrix of the feedback compensator is determined as follows:

As given in Section 5.3.2, the full-order state observer equation is

$$\xi(k + 1) = A\xi(k) + Bu(k) + G[y(k) - C\xi(k) - Du(k)] \qquad (5\text{-}20)$$

and the state feedback equation is

$$u(k) = E\xi(k) + \rho(k) \qquad (5\text{-}21)$$

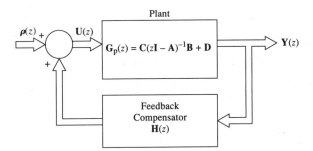

Figure 5-13 Feedback compensation for pole placement.

Substituting equation (5-21) into equation (5-20) and assuming that $\rho(k) = \mathbf{0}$ gives

$$\xi(k + 1) = \mathbf{A}\xi(k) + \mathbf{BE}\xi(k) + \mathbf{G}y(k) - \mathbf{GC}\xi(k) - \mathbf{GDE}\xi(k)$$

or, collecting terms,

$$\xi(k + 1) = (\mathbf{A} + \mathbf{BE} - \mathbf{GC} - \mathbf{GDE})\xi(k) + \mathbf{G}y(k) \tag{5-22}$$

The z-transfer function matrix of the feedback compensator is determined by z-transforming equation (5-22) with zero initial conditions. Hence,

$$z\xi(z) = (\mathbf{A} + \mathbf{BE} - \mathbf{GC} - \mathbf{GDE})\xi(z) + \mathbf{G}Y(z)$$

or, collecting terms,

$$[z\mathbf{I} - \mathbf{A} - \mathbf{BE} + \mathbf{GC} + \mathbf{GDE}]\xi(z) = \mathbf{G}Y(z)$$

then,

$$\xi(z) = [z\mathbf{I} - \mathbf{A} - \mathbf{BE} + \mathbf{GC} + \mathbf{GDE}]^{-1}\mathbf{G}Y(z) \tag{5-23}$$

Premultiplying both sides of equation (5-23) with the feedback gain matrix $\mathbf{E}$ gives

$$U(z) = \mathbf{E}\xi(z) = \mathbf{E}[z\mathbf{I} - \mathbf{A} - \mathbf{BE} + \mathbf{GC} + \mathbf{GDE}]^{-1}\mathbf{G}Y(z)$$

and, therefore, the transfer function matrix of the feedback compensator is

$$\mathbf{H}(z) = \mathbf{E}[z\mathbf{I} - \mathbf{A} - \mathbf{BE} + \mathbf{GC} + \mathbf{GDE}]^{-1}\mathbf{G} \tag{5-24}$$

with

$$U(z) = \mathbf{H}(z)Y(z)$$

The arrangement shown in Figure 5-13 allows one to analyze the behavior of the feedback control system using frequency domain techniques, such as Bode or root locus methods.

5.6.3 Example of Full-Order State Observer Feedback

Even though a lower-order observer might suffice, full-order state observers are often used for feedback in practice because of the close relationship that can be arranged between the plant state variables and the state variables of the observer. If x_1 is a temperature, it can be arranged so that ξ_1 is the observer estimate of that temperature, and so on. Full-order observers also need not have direct input-to-output coupling, so they can smooth noisy signals better.

As an example of full-order state observer feedback, consider the plant

$$\begin{bmatrix} x_1(k+1) \\ x_2(k+1) \end{bmatrix} = \begin{bmatrix} 2 & -1 \\ -1 & 1 \end{bmatrix} \begin{bmatrix} x_1(k) \\ x_2(k) \end{bmatrix} + \begin{bmatrix} 4 \\ 3 \end{bmatrix} u(k) = \mathbf{A}\mathbf{x}(k) + \mathbf{b}u(k)$$

$$y(k) = \begin{bmatrix} 1 & 1 \end{bmatrix} \begin{bmatrix} x_1(k) \\ x_2(k) \end{bmatrix} + 7u(k) = \mathbf{c}^\dagger\mathbf{x}(k) + du(k) \qquad \textbf{(5-25)}$$

which is completely controllable and completely observable. The plant's characteristic equation is

$$|\lambda\mathbf{I} - \mathbf{A}| = \begin{vmatrix} (\lambda - 2) & 1 \\ 1 & (\lambda - 1) \end{vmatrix} = \lambda^2 - 3\lambda + 1 = \lambda^2 + \alpha_1\lambda + \alpha_0 = 0$$

In this example, transformations to controllable and observable form are used to design the plant feedback and the observer. The transformation to controllable form is

$$\mathbf{x} = \mathbf{P}\mathbf{x}' \qquad \mathbf{x}' = \mathbf{P}^{-1}\mathbf{x}$$

$$\mathbf{P} = \begin{bmatrix} -7 & 4 \\ -10 & 3 \end{bmatrix} \qquad \mathbf{P}^{-1} = \frac{1}{19}\begin{bmatrix} 3 & -4 \\ 10 & -7 \end{bmatrix}$$

The plant state equations in controllable form are

$$\mathbf{x}'(k+1) = \mathbf{P}^{-1}\mathbf{A}\mathbf{P}\mathbf{x}'(k) + \mathbf{P}^{-1}\mathbf{b}u(k) = \mathbf{A}'\mathbf{x}'(k) + \mathbf{b}'u(k)$$

or

$$\begin{bmatrix} x_1'(k+1) \\ x_2'(k+1) \end{bmatrix} = \begin{bmatrix} 0 & 1 \\ -1 & 3 \end{bmatrix} \begin{bmatrix} x_1'(k) \\ x_2'(k) \end{bmatrix} + \begin{bmatrix} 0 \\ 1 \end{bmatrix} u(k)$$

Feedback of the primed plant state

$$u(k) = \mathbf{e}'^\dagger\mathbf{x}'(k) + \rho(k)$$

results in the feedback system

$$\mathbf{x}'(k+1) = (\mathbf{A}' + \mathbf{b}'\mathbf{e}'^\dagger)\mathbf{x}'(k) + \mathbf{b}'\rho(k)$$

$$= \begin{bmatrix} 0 & 1 \\ (e_1' - 1) & (e_2' + 3) \end{bmatrix} \mathbf{x}'(k) + \mathbf{b}'\rho(k)$$

which, in terms of the feedback gains e_1' and e_2', has the characteristic equation

$$\lambda^2 + (-3 - e_2')\lambda + (1 - e_1') = 0$$

If the desired eigenvalues are $\pm j\frac{1}{2}$, the desired characteristic equation is

$$\left(\lambda + j\frac{1}{2}\right)\left(\lambda - j\frac{1}{2}\right) = \lambda^2 + \frac{1}{4} = 0$$

and the feedback gains for the $\mathbf{x}'$ state are

$$\begin{bmatrix} e_1' \\ e_2' \end{bmatrix} = \begin{bmatrix} \frac{3}{4} \\ -3 \end{bmatrix}$$

If, instead, the $\mathbf{x}$ state is fed back, the gains required are given by

$$\mathbf{e}'^\dagger\mathbf{x}'(k) = \mathbf{e}'^\dagger\mathbf{P}^{-1}\mathbf{x}(k) = \mathbf{e}^\dagger\mathbf{x}(k)$$

$$\mathbf{e}^\dagger = \mathbf{e}'^\dagger\mathbf{P}^{-1} = \begin{bmatrix} \frac{3}{4} & -3 \end{bmatrix}\begin{bmatrix} 3 & -4 \\ 10 & -7 \end{bmatrix}\left(\frac{1}{19}\right) = \begin{bmatrix} -\frac{111}{76} & \frac{18}{19} \end{bmatrix}$$

To design an observer for the plant in equation (5-25), we transform to the observable form using

$$\mathbf{x}'' = \mathbf{Q}\mathbf{x} \qquad \mathbf{x} = \mathbf{Q}^{-1}\mathbf{x}''$$

where

$$\mathbf{Q} = \begin{bmatrix} 1 & 1 \\ -2 & -3 \end{bmatrix} \qquad \mathbf{Q}^{-1} = \begin{bmatrix} 3 & 1 \\ -2 & -1 \end{bmatrix}$$

The plant equations in observable form are thus

$$\mathbf{x}''(k + 1) = \mathbf{Q}\mathbf{A}\mathbf{Q}^{-1}\mathbf{x}''(k) + \mathbf{Q}\mathbf{b}u(k) = \mathbf{A}''\mathbf{x}''(k) + \mathbf{b}''u(k)$$

$$y(k) = \mathbf{c}^\dagger\mathbf{Q}^{-1}\mathbf{x}''(k) + du(k) = \mathbf{c}''^\dagger\mathbf{x}''(k) + du(k)$$

or

$$\begin{bmatrix} x_1''(k + 1) \\ x_2''(k + 1) \end{bmatrix} = \begin{bmatrix} 3 & 1 \\ -1 & 0 \end{bmatrix}\begin{bmatrix} x_1''(k) \\ x_2''(k) \end{bmatrix} + \begin{bmatrix} 7 \\ -17 \end{bmatrix}u(k) = \mathbf{A}''\mathbf{x}''(k) + \mathbf{b}''u(k)$$

$$y(k) = \begin{bmatrix} 1 & 0 \end{bmatrix}\begin{bmatrix} x_1''(k) \\ x_2''(k) \end{bmatrix} + 7u(k) = \mathbf{c}''^\dagger\mathbf{x}''(k) + du(k)$$

An observer of the $\mathbf{x}''$ plant state has the form

$$\xi''(k + 1) = \mathbf{F}''\xi''(k) + \mathbf{g}''y(k) + \mathbf{h}''u(k)$$

where

$$\mathbf{F}'' = \mathbf{A}'' - \mathbf{g}''\mathbf{c}''^\dagger = \begin{bmatrix} 3 & 1 \\ -1 & 0 \end{bmatrix} - \begin{bmatrix} g_1'' \\ g_2'' \end{bmatrix}\begin{bmatrix} 1 & 0 \end{bmatrix} = \begin{bmatrix} (3 - g_1'') & 1 \\ (-1 - g_2'') & 0 \end{bmatrix}$$

The characteristic equation of the observer is, in terms of g_1'' and g_2''

$$\lambda^2 + (g_1'' - 3)\lambda + (g_2'' + 1) = 0$$

If the desired observer characteristic equation is

$$\left(\lambda - \frac{1}{2}\right)\left(\lambda - \frac{1}{4}\right) = \lambda^2 - \frac{3}{4}\lambda + \frac{1}{8} = 0$$

then the observer gains are

$$\mathbf{g}'' = \begin{bmatrix} g_1'' \\ g_2'' \end{bmatrix} = \begin{bmatrix} \frac{9}{4} \\ -\frac{7}{8} \end{bmatrix}$$

$$\mathbf{h}'' = \mathbf{b}'' - \mathbf{g}''d = \begin{bmatrix} 7 \\ -17 \end{bmatrix} - \begin{bmatrix} \frac{9}{4} \\ -\frac{7}{8} \end{bmatrix}(7) = \begin{bmatrix} -\frac{35}{4} \\ -\frac{87}{8} \end{bmatrix}$$

so that the observer equations are

$$\begin{bmatrix} \xi_1''(k+1) \\ \xi_2''(k+1) \end{bmatrix} = \begin{bmatrix} \frac{3}{4} & 1 \\ -\frac{1}{8} & 0 \end{bmatrix}\begin{bmatrix} \xi_1''(k) \\ \xi_2''(k) \end{bmatrix} + \begin{bmatrix} \frac{9}{4} \\ -\frac{7}{8} \end{bmatrix}y(k) + \begin{bmatrix} -\frac{35}{4} \\ -\frac{87}{8} \end{bmatrix}u(k)$$

$$= \mathbf{F}''\boldsymbol{\xi}''(k) + \mathbf{g}''y(k) + \mathbf{h}''u(k)$$

The variables

$$\boldsymbol{\xi}(k) = \mathbf{Q}^{-1}\boldsymbol{\xi}''(k)$$

observe the **x** plant state, and in terms of them, the observer equations are

$$\mathbf{Q}\boldsymbol{\xi}(k+1) = \mathbf{F}''\mathbf{Q}\boldsymbol{\xi}(k) + \mathbf{g}''y(k) + \mathbf{h}''u(k)$$

$$\boldsymbol{\xi}(k+1) = (\mathbf{Q}^{-1}\mathbf{F}''\mathbf{Q})\boldsymbol{\xi}(k) + (\mathbf{Q}^{-1}\mathbf{g}'')y(k) + (\mathbf{Q}^{-1}\mathbf{h}'')u(k)$$

$$= \mathbf{F}\boldsymbol{\xi}(k) + \mathbf{g}y(k) + \mathbf{h}u(k)$$

where

$$\mathbf{F} = \mathbf{Q}^{-1}\mathbf{F}''\mathbf{Q} = \begin{bmatrix} -\frac{31}{8} & -\frac{55}{8} \\ \frac{21}{8} & \frac{37}{8} \end{bmatrix}$$

$$\mathbf{g} = \mathbf{Q}^{-1}\mathbf{g}'' = \begin{bmatrix} \frac{47}{8} \\ -\frac{29}{8} \end{bmatrix} \qquad \mathbf{h} = \mathbf{Q}^{-1}\mathbf{h}'' = \begin{bmatrix} -\frac{297}{8} \\ \frac{227}{8} \end{bmatrix}$$

The plant, observer, and feedback relationships are summarized in Figure 5-14, where plant state feedback is replaced by feedback of the observer estimate of the plant state.

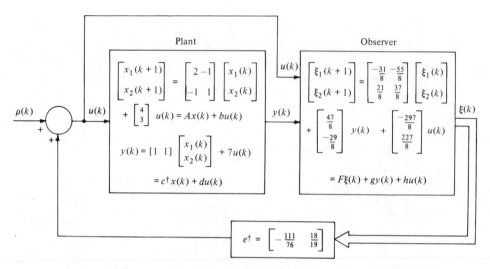

Figure 5-14 A full-order state observer used for feedback.

5.6.4 Example of Reduced-Order Observer Feedback

The separation theorem applies to feedback by any observer, including feedback by reduced-order state observers and observers of linear plant state transformations. As another numerical example of observer feedback, consider the plant

$$\begin{bmatrix} x_1(k+1) \\ x_2(k+1) \\ x_3(k+1) \end{bmatrix} = \begin{bmatrix} 0 & 1 & 0 \\ 0 & 0 & 1 \\ 3 & -1 & 2 \end{bmatrix} \begin{bmatrix} x_1(k) \\ x_2(k) \\ x_3(k) \end{bmatrix} + \begin{bmatrix} 0 \\ 0 \\ 1 \end{bmatrix} u(k) = \mathbf{A}x(k) + \mathbf{b}u(k)$$

$$\begin{bmatrix} y_1(k) \\ y_2(k) \end{bmatrix} = \begin{bmatrix} 2 & -1 & 0 \\ 1 & 0 & 0 \end{bmatrix} \begin{bmatrix} x_1(k) \\ x_2(k) \\ x_3(k) \end{bmatrix} + \begin{bmatrix} -2 \\ 0 \end{bmatrix} u(k) = \mathbf{C}x(k) + \mathbf{d}u(k)$$

which is conveniently in controllable form. State feedback of the form

$$u(k) = \mathbf{e}^t \mathbf{x}(k) + \rho(k)$$

gives

$$\begin{bmatrix} x_1(k+1) \\ x_2(k+1) \\ x_3(k+1) \end{bmatrix} = \begin{bmatrix} 0 & 1 & 0 \\ 0 & 0 & 0 \\ (3 + e_1) & (-1 + e_2) & (2 + e_3) \end{bmatrix} \begin{bmatrix} x_1(k) \\ x_2(k) \\ x_3(k) \end{bmatrix} + \begin{bmatrix} 0 \\ 0 \\ 1 \end{bmatrix} \rho(k)$$

which has the characteristic equation

$$\lambda^3 + (-2 - e_3)\lambda^2 + (1 - e_2)\lambda + (-3 - e_1) = 0$$

If it is desired that all feedback system eigenvalues be at $\lambda = 0$, the desired characteristic equation is

$$\lambda^3 = 0$$

and the state feedback gains needed are

$$\mathbf{e}^\dagger = [-3 \quad 1 \quad -2]$$

A first-order observer with eigenvalue $\lambda = 0$, of the form

$$\xi(k + 1) = 0 \cdot \xi(k) + [g_1 \quad g_2]\begin{bmatrix} y_1(k) \\ y_2(k) \end{bmatrix} + hu(k) = \mathbf{g}^\dagger\mathbf{y}(k) + hu(k)$$

observes the scalar linear state transformation $\mathbf{m}^\dagger\mathbf{x}(k)$, where

$$\mathbf{m}^\dagger = \mathbf{g}^\dagger\mathbf{C}(\mathbf{A} - 0 \cdot \mathbf{I})^{-1} = [g_1 \quad g_2]\begin{bmatrix} 2 & -1 & 0 \\ 1 & 0 & 0 \end{bmatrix}\begin{bmatrix} 0 & 1 & 0 \\ 0 & 0 & 1 \\ 3 & -1 & 2 \end{bmatrix}^{-1}$$

$$= g_1[-\tfrac{1}{3} \quad -\tfrac{4}{3} \quad \tfrac{2}{3}] + g_2[\tfrac{1}{3} \quad -\tfrac{2}{3} \quad \tfrac{1}{3}]$$

For this observer

$$h = \mathbf{m}^\dagger\mathbf{b} - \mathbf{g}^\dagger\mathbf{d} = \frac{8}{3}g_1 + \frac{1}{3}g_2$$

The collection of signals

$$\begin{bmatrix} w_1(k) \\ w_2(k) \\ w_3(k) \end{bmatrix} = \begin{bmatrix} y_1(k) + 2u(k) \\ y_2(k) \\ \xi(k) \end{bmatrix} = \begin{bmatrix} 0 \\ 0 \\ 1 \end{bmatrix}\xi(k) + \begin{bmatrix} 1 & 0 \\ 0 & 1 \\ 0 & 0 \end{bmatrix}\begin{bmatrix} y_1(k) \\ y_2(k) \end{bmatrix} + \begin{bmatrix} 2 \\ 0 \\ 0 \end{bmatrix}u(k)$$

observes

$$\begin{bmatrix} w_1(k) \\ w_2(k) \\ w_3(k) \end{bmatrix} \rightarrow \begin{bmatrix} \mathbf{C} \\ \hline \mathbf{m}^\dagger \end{bmatrix}\mathbf{x}(k)$$

$$= \begin{bmatrix} 2 & -1 & 0 \\ 1 & 0 & 0 \\ (-\tfrac{1}{3}g_1 + \tfrac{1}{3}g_2) & (-\tfrac{4}{3}g_1 - \tfrac{2}{3}g_2) & (\tfrac{2}{3}g_1 + \tfrac{1}{3}g_2) \end{bmatrix}\mathbf{x}(k)$$

$$= \mathbf{M}\mathbf{x}(k)$$

Choosing one set of gains g_1 and g_2 for which $\mathbf{M}$ is nonsingular

$$g_1 = 1 \qquad g_2 = 1$$

gives the observer

$$\xi(k + 1) = [1 \quad 1]\begin{bmatrix} y_1(k) \\ y_2(k) \end{bmatrix} + 3u(k)$$

$$\begin{bmatrix} w_1(k) \\ w_2(k) \\ w_3(k) \end{bmatrix} = \begin{bmatrix} 0 \\ 0 \\ 1 \end{bmatrix} \xi(k) + \begin{bmatrix} 1 & 0 \\ 0 & 1 \\ 0 & 0 \end{bmatrix} \begin{bmatrix} y_1(k) \\ y_2(k) \end{bmatrix} + \begin{bmatrix} 2 \\ 0 \\ 0 \end{bmatrix} u(k)$$

which observes

$$\mathbf{w}(k) \to \mathbf{Mx}(k) = \begin{bmatrix} 2 & -1 & 0 \\ 1 & 0 & 0 \\ 0 & -2 & 1 \end{bmatrix} \begin{bmatrix} x_1(k) \\ x_2(k) \\ x_3(k) \end{bmatrix}$$

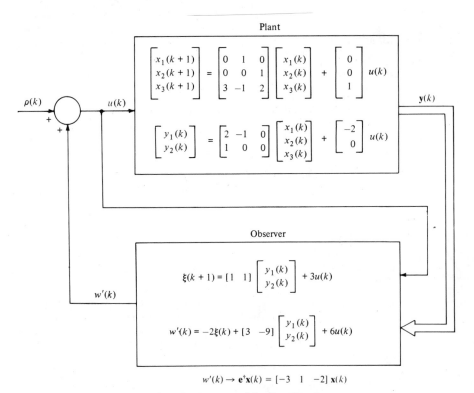

Figure 5-15 A reduced-order observer used for feedback.

An observer output

$$w'(k) = \mathbf{e}^\dagger \mathbf{M}^{-1}\mathbf{w}(k) = \begin{bmatrix} -3 & 1 & -2 \end{bmatrix}\begin{bmatrix} 0 & 1 & 0 \\ -1 & 2 & 0 \\ -2 & 4 & 1 \end{bmatrix}\mathbf{w}(k)$$

observes

$$w'(k) = \mathbf{e}^\dagger \mathbf{x}(k)$$

This output is

$$w'(k) = -2\xi(k) + \begin{bmatrix} 3 & -9 \end{bmatrix}\begin{bmatrix} y_1(k) \\ y_2(k) \end{bmatrix} + 6u(k)$$

and the completed observer design is shown in Figure 5-15. The feedback system, which is of fourth order, has all eigenvalues placed at $\lambda = 0$.

As in the case of eigenvalue placement with state feedback, the feedback system eigenvalues are usually chosen as a compromise between speed of response (that is, rate of decay of the zero-input response component) and susceptibility to errors or "noise."

5.7 Step-Varying Observers

The observer concept generalizes nicely for the linear, step-varying case. However, eigenvalue placement is no longer of direct interest because the observer error is not generally governed by powers of a matrix. We concentrate in this section on deadbeat step-varying observer design, although other observer error responses can be obtained if desired. Methods for observing linear state transformations, paralleling those for step-invariant observers, allow state observer order reduction and direct observation of a state transformation needed for feedback. In the step-invariant case, continuing convergence of the observer is guaranteed. That is, for a pth-order observer with state $\xi(k)$ and a plant with state $\mathbf{x}(k)$, if

$$\mathbf{Mx}(p) - \xi(p) = \mathbf{F}^p[\mathbf{Mx}(0) - \xi(0)]$$

then

$$\mathbf{Mx}(p + 1) - \xi(p + 1) = \mathbf{F}^p[\mathbf{Mx}(1) - \xi(1)]$$

and so on. For step-varying observers, we must specifically design continued convergence; it is not automatic.

Step-varying observation is desirable for step-invariant plants, too, because it can give an error response with a relatively small transient response amplitude.

5.7.1 State Observation

For an nth-order linear, step-varying plant

$$\mathbf{x}(k + 1) = \mathbf{A}(k)\mathbf{x}(k) + \mathbf{B}(k)\mathbf{u}(k)$$

$$\mathbf{y}(k) = \mathbf{C}(k)\mathbf{x}(k) + \mathbf{D}(k)\mathbf{u}(k)$$

a full-order state observer is another nth-order linear, step-varying system of the form

$$\boldsymbol{\xi}(k + 1) = \mathbf{F}(k)\boldsymbol{\xi}(k) + \mathbf{G}(k)\mathbf{y}(k) + \mathbf{H}(k)\mathbf{u}(k) \tag{5-26}$$

as shown in Figure 5-16. The error between the plant state and the observer state is governed by

$$\mathbf{x}(k + 1) - \boldsymbol{\xi}(k + 1) = [\mathbf{A}(k) - \mathbf{G}(k)\mathbf{C}(k)]\mathbf{x}(k) - \mathbf{F}(k)\boldsymbol{\xi}(k)$$
$$+ [\mathbf{B}(k) - \mathbf{H}(k) - \mathbf{G}(k)\mathbf{D}(k)]\mathbf{u}(k)$$

Requiring that the error equation be autonomous, then if

$$\mathbf{F}(k) = \mathbf{A}(k) - \mathbf{G}(k)\mathbf{C}(k)$$

$$\mathbf{H}(k) = \mathbf{B}(k) - \mathbf{G}(k)\mathbf{D}(k) \tag{5-27}$$

where the step-varying matrix $\mathbf{G}(k)$ is arbitrary, the observer error is then governed by

$$\mathbf{x}(k + 1) - \boldsymbol{\xi}(k + 1) = \mathbf{F}(k)[\mathbf{x}(k) - \boldsymbol{\xi}(k)]$$

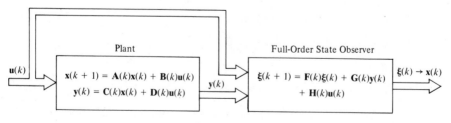

Figure 5-16 Plant and step-varying observer.

or

$$\mathbf{x}(k) - \boldsymbol{\xi}(k) = \mathbf{F}(k - 1)\mathbf{F}(k - 2) \ldots \mathbf{F}(0)[\mathbf{x}(0) - \boldsymbol{\xi}(0)]$$

Rewriting the observer state equation (5-26) using equation (5-27) and the observer estimate of the plant outputs

$$\mathbf{w}(k) = \mathbf{C}(k)\boldsymbol{\xi}(k) + \mathbf{D}(k)\mathbf{u}(k)$$

results in

$$\boldsymbol{\xi}(k + 1) = \mathbf{A}(k)\boldsymbol{\xi}(k) + \mathbf{B}(k)\mathbf{u}(k) + \mathbf{G}(k)[\mathbf{y}(k) - \mathbf{w}(k)]$$

which demonstrates that the full-order state observer can be viewed as a model of the plant, driven by the difference between the actual plant output and the model plant output. This arrangement is illustrated in Figure 5-17.

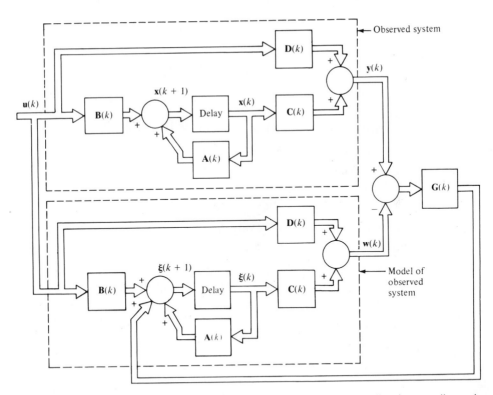

Figure 5-17 State observer as a plant model driven by the output estimate error through a step-varying gain.

5.7.2 Gain Calculations for Single-Output Plants

Although the observer gain sequence $G(k)$ can be chosen to place eigenvalues of the observer state coupling matrices $F(k)$, this is of little direct help in designing the observer so that its state converges to that of the plant unless F does not vary with step. A particularly simple and effective general method for achieving convergence is the following: For a single-output nth-order plant

$$\mathbf{x}(k + 1) = \mathbf{A}(k)\mathbf{x}(k) + \mathbf{B}(k)\mathbf{u}(k)$$

$$y(k) = \mathbf{c}^\dagger(k)\mathbf{x}(k) + \mathbf{d}^\dagger(k)\mathbf{u}(k) \tag{5-28}$$

and full-order state observer

$$\boldsymbol{\xi}(k + 1) = \mathbf{F}(k)\boldsymbol{\xi}(k) + \mathbf{g}(k)y(k) + \mathbf{H}(k)\mathbf{u}(k)$$

the error between the plant state and the observer state at the nth step is

$$\mathbf{x}(n) - \boldsymbol{\xi}(n) = \mathbf{F}(n - 1)\mathbf{F}(n - 2) \dots \mathbf{F}(1)\mathbf{F}(0)[\mathbf{x}(0) - \boldsymbol{\xi}(0)]$$

If $\mathbf{g}(0), \mathbf{g}(1), \dots, \mathbf{g}(n - 1)$ can be chosen so that

$$\mathbf{F}(n - 1)\mathbf{F}(n - 2) \dots \mathbf{F}(1)\mathbf{F}(0) = 0 \tag{5-29}$$

then the observer state equals the plant state at the nth step and beyond.

The desired relation in equation (5-29) is obtained if

$$\begin{cases} \mathbf{F}(n - 1)\mathbf{F}(n - 2) \dots \mathbf{F}(1)\mathbf{F}(0)\mathbf{j}_1 = 0 \\ \mathbf{F}(n - 1)\mathbf{F}(n - 2) \dots \mathbf{F}(1)\mathbf{F}(0)\mathbf{j}_2 = 0 \\ \quad \vdots \\ \mathbf{F}(n - 1)\mathbf{F}(n - 2) \dots \mathbf{F}(1)\mathbf{F}(0)\mathbf{j}_{n-1} = 0 \\ \mathbf{F}(n - 1)\mathbf{F}(n - 2) \dots \mathbf{F}(1)\mathbf{F}(0)\mathbf{j}_n = 0 \end{cases} \tag{5-30}$$

where $\mathbf{j}_1, \mathbf{j}_2, \dots, \mathbf{j}_n$ are any n linearly independent n-vectors. These are called *basis vectors* because they span the n-dimensional space. Relations (5-30) are, in turn, obtained if

$$\begin{cases} \mathbf{F}(0)\mathbf{j}_1 = 0 \\ \mathbf{F}(1)\mathbf{F}(0)\mathbf{j}_2 = 0 \\ \quad \vdots \\ \mathbf{F}(n - 2)\mathbf{F}(n - 3) \dots \mathbf{F}(1)\mathbf{F}(0)\mathbf{j}_{n-1} = 0 \\ \mathbf{F}(n - 1)\mathbf{F}(n - 2) \dots \mathbf{F}(1)\mathbf{F}(0)\mathbf{j}_n = 0 \end{cases} \tag{5-31}$$

At step 0 the observer state coupling matrix is

$$\mathbf{F}(0) = \mathbf{A}(0) - \mathbf{g}(0)\mathbf{c}^\dagger(0)$$

where the observer gain $\mathbf{g}(0)$ is an n-vector. Substituting into the first equation of (5-31) gives

$$\mathbf{A}(0)\mathbf{j}_1 - \mathbf{g}(0)\mathbf{c}^\dagger(0)\mathbf{j}_1 = \mathbf{0}$$

$$\mathbf{g}(0) = \frac{\mathbf{A}(0)\mathbf{j}_1}{\mathbf{c}^\dagger(0)\mathbf{j}_1}$$

where the quantity in the denominator of $\mathbf{g}(0)$ is a scalar. Similarly, given $\mathbf{g}(0)$ and thus $\mathbf{F}(0)$, the second equation of (5-31) is

$$[\mathbf{A}(1) - \mathbf{g}(1)\mathbf{c}^\dagger(1)]\mathbf{F}(0)\mathbf{j}_2 = \mathbf{0}$$

$$\mathbf{g}(1) = \frac{\mathbf{A}(1)\mathbf{F}(0)\mathbf{j}_2}{\mathbf{c}^\dagger(1)\mathbf{F}(0)\mathbf{j}_2}$$

In general

$$\mathbf{g}(k) = \frac{\mathbf{A}(k)\mathbf{F}(k-1) \dots \mathbf{F}(1)\mathbf{F}(0)\mathbf{j}_{k+1}}{\mathbf{c}^\dagger(k)\mathbf{F}(k-1) \dots \mathbf{F}(1)\mathbf{F}(0)\mathbf{j}_{k+1}} \qquad (5\text{-}32)$$

Table 5-8 summarizes this recursive process of determining observer gain vectors $\mathbf{g}(0), \mathbf{g}(1), \dots, \mathbf{g}(n-1)$. The resulting observer is termed *deadbeat*. Like the case of a deadbeat step-invariant system, its error is zero in n steps or less.

Each observer gain vector $\mathbf{g}(i)$ is that needed to drive the observer error, due to a nonzero initial error in the $\mathbf{j}_{i+1}$ direction, to zero. The transients in this step-varying observer's error response tend to be much smaller in amplitude than those typical for step-invariant observers.

After n steps, the observer error ideally has been driven to zero. However, if there are inaccuracies, it is expedient to attempt to drive any remaining error toward zero on succeeding steps. One strategy is to restart the previous observer gain calculation algorithm by computing

$$\mathbf{g}(n+1) = \frac{\mathbf{A}(n+1)\mathbf{j}_1}{\mathbf{c}^\dagger(n+1)\mathbf{j}_1}$$

$$\mathbf{F}(n+1) = \mathbf{A}(n+1) - \mathbf{g}(n+1)\mathbf{c}^\dagger(n+1)$$

$$\mathbf{g}(n+2) = \frac{\mathbf{A}(n+2)\mathbf{F}(n+1)\mathbf{j}_2}{\mathbf{c}^\dagger(n+2)\mathbf{F}(n+1)\mathbf{j}_2}$$

and so on. Another strategy is to require *ongoing* observer convergence, in which, for each successive step after step $n-1$, the next

Table 5-8 Deadbeat Observer Design for Single-Output Plants

Plant Model

$$\mathbf{x}(k + 1) = \mathbf{A}(k)\mathbf{x}(k) + \mathbf{B}(k)\mathbf{u}(k)$$

$$y(k) = \mathbf{c}^\dagger(k)\mathbf{x}(k) + \mathbf{d}^\dagger(k)\mathbf{u}(k)$$

where $\mathbf{x}(k)$ is an n-vector and $y(k)$ is a scalar.

Observer Structure

$$\boldsymbol{\xi}(k + 1) = \mathbf{F}(k)\boldsymbol{\xi}(k) + \mathbf{g}(k)y(k) + \mathbf{H}(k)\mathbf{u}(k)$$

where

$$\mathbf{F}(k) = \mathbf{A}(k) - \mathbf{g}(k)\mathbf{c}^\dagger(k)$$

$$\mathbf{H}(k) = \mathbf{B}(k) - \mathbf{g}(k)\mathbf{d}^\dagger(k)$$

Observer Gain Calculation

Choose any set of linearly independent n-vectors (basis vectors)

$$\mathbf{j}_1, \mathbf{j}_2, \ldots, \mathbf{j}_n$$

Then

$$\begin{cases} \mathbf{g}(0) = \dfrac{\mathbf{A}(0)\mathbf{j}_1}{\mathbf{c}^\dagger(0)\mathbf{j}_1} \\[2mm] \mathbf{F}(0) = \mathbf{A}(0) - \mathbf{g}(0)\mathbf{c}^\dagger(0) \\[1mm] \mathbf{H}(0) = \mathbf{B}(0) - \mathbf{g}(0)\mathbf{d}^\dagger(0) \end{cases}$$

$$\begin{cases} \mathbf{g}(1) = \dfrac{\mathbf{A}(1)\mathbf{F}(0)\mathbf{j}_2}{\mathbf{c}^\dagger(1)\mathbf{F}(0)\mathbf{j}_2} \\[2mm] \mathbf{F}(1) = \mathbf{A}(1) - \mathbf{g}(1)\mathbf{c}^\dagger(1) \\[1mm] \mathbf{H}(1) = \mathbf{B}(1) - \mathbf{g}(1)\mathbf{d}^\dagger(1) \end{cases}$$

$$\vdots$$

$$\begin{cases} \mathbf{g}(k) = \dfrac{\mathbf{A}(k)\mathbf{F}(k-1) \ldots \mathbf{F}(1)\mathbf{F}(0)\mathbf{j}_{k+1}}{\mathbf{c}^\dagger(k)\mathbf{F}(k-1) \ldots \mathbf{F}(1)\mathbf{F}(0)\mathbf{j}_{k+1}} \\[2mm] \mathbf{F}(k) = \mathbf{A}(k) - \mathbf{g}(k)\mathbf{c}^\dagger(k) \\[1mm] \mathbf{H}(k) = \mathbf{B}(k) - \mathbf{g}(k)\mathbf{d}^\dagger(k) \end{cases}$$

$$\vdots$$

$$\mathbf{g}(n-1) = \frac{\mathbf{A}(n-1)\mathbf{F}(n-2) \ldots \mathbf{F}(1)\mathbf{F}(0)\mathbf{j}_n}{\mathbf{c}^\dagger(n-1)\mathbf{F}(n-2) \ldots \mathbf{F}(1)\mathbf{F}(0)\mathbf{j}_n}$$

If, for any gain calculation $\mathbf{g}(k)$, the scalar divisor involved is zero

$$\mathbf{c}^\dagger(k)\mathbf{F}(k-1) \ldots \mathbf{F}(1)\mathbf{F}(0)\mathbf{j}_{k+1} = 0$$

reordering the *remaining* n-vectors $\mathbf{j}_{k+1}, \mathbf{j}_{k+2}, \ldots, \mathbf{j}_n$ gives a solution whenever one exists. If all remaining n-vectors give a zero scalar divisor, the plant output at this step is not linearly independent of previously used outputs and can be discarded.

observer gain is chosen so that

$$\mathbf{F}(n - 1 + i)\mathbf{F}(n - 2 + i) \ldots \mathbf{F}(i) = 0 \qquad i = 1, 2, \ldots$$

The observer gain needed is

$$\mathbf{g}(n - 1 + i) = \frac{\mathbf{A}(n - 1 - i)\mathbf{F}(n - 2 + i) \ldots \mathbf{F}(i)\mathbf{j}_\nu}{\mathbf{c}^\dagger(n - 1 + i)\mathbf{F}(n - 2 + i) \ldots \mathbf{F}(i)\mathbf{j}_\nu}$$

$$i = 1, 2, \ldots$$

where $\mathbf{j}_\nu$ is any basis vector for which the scalar denominator is non-zero. The combination of the initial and ongoing observer gain algorithms is summarized in Table 5-9.

Table 5-9 Initial and Ongoing Deadbeat Observer Design

Choose any set of linearly independent n-vectors (basis vectors) $\mathbf{j}_1, \mathbf{j}_2, \ldots,$ $\mathbf{j}_n$. Then

$$\mathbf{g}(k) = \frac{\mathbf{A}(k)\mathbf{F}(k - 1) \ldots \mathbf{F}(k - n + 2)\mathbf{F}(k - n + 1)\mathbf{j}_i}{\mathbf{c}^\dagger(k)\mathbf{F}(k - 1) \ldots \mathbf{F}(k - n + 2)\mathbf{F}(k - n + 1)\mathbf{j}_i}$$

where $\mathbf{j}_i$ is any vector for which the scalar divisor in the above is nonzero and where

$$\mathbf{F}(i) = \begin{cases} \mathbf{A}(i) - \mathbf{g}(i)\mathbf{c}^\dagger(i) & i = 0, 1, 2, \ldots \\ \mathbf{I} & i = -1, -2, \ldots \end{cases}$$

If the scalar divisor is zero for all of the n-vectors $\mathbf{j}_1, \mathbf{j}_2, \ldots, \mathbf{j}_n$, the plant output at this step is not linearly independent of previously used outputs and can be ignored.

As a numerical example for calculating deadbeat observer gain, consider the single-output step-varying system

$$\begin{bmatrix} x_1(k + 1) \\ x_2(k + 1) \end{bmatrix} = \begin{bmatrix} k & 2 \\ 1 & -1 \end{bmatrix} \begin{bmatrix} x_1(k) \\ x_2(k) \end{bmatrix} + \begin{bmatrix} -1 \\ 1 \\ k + 1 \end{bmatrix} u(k)$$

$$= \mathbf{A}(k)\mathbf{x}(k) + \mathbf{b}(k)u(k)$$

$$y(k) = [(\tfrac{1}{2})^k \quad 0] \begin{bmatrix} x_1(k) \\ x_2(k) \end{bmatrix} + u(k) = \mathbf{c}^\dagger(k)\mathbf{x}(k) + d(k)u(k)$$

For a deadbeat state observer of this plant

$$\xi(k + 1) = \mathbf{F}(k)\,\xi(k) + \mathbf{g}(k)y(k) + \mathbf{h}(k)u(k)$$

choosing the basis vectors $\mathbf{j}_1$ and $\mathbf{j}_2$ to be unit coordinate vectors

$$\mathbf{j}_1^\dagger = [1 \quad 0] \qquad \mathbf{j}_2^\dagger = [0 \quad 1]$$

gives the following two deadbeat observer gains and plant input couplings:

$$\mathbf{g}(0) = \frac{\mathbf{A}(0)\mathbf{j}_1}{\mathbf{c}^\dagger(0)\mathbf{j}_1} = \frac{\begin{bmatrix} 0 & 2 \\ 1 & -1 \end{bmatrix}\begin{bmatrix} 1 \\ 0 \end{bmatrix}}{[1 \quad 0]\begin{bmatrix} 1 \\ 0 \end{bmatrix}} = \begin{bmatrix} 0 \\ 1 \end{bmatrix}$$

$$\mathbf{F}(0) = \mathbf{A}(0) - \mathbf{g}(0)\mathbf{c}^\dagger(0) = \begin{bmatrix} 0 & 2 \\ 1 & -1 \end{bmatrix} - \begin{bmatrix} 0 \\ 1 \end{bmatrix}[1 \quad 0] = \begin{bmatrix} 0 & 2 \\ 0 & -1 \end{bmatrix}$$

$$\mathbf{h}(0) = \mathbf{b}(0) - \mathbf{g}(0)d(0) = \begin{bmatrix} -1 \\ 1 \end{bmatrix} - \begin{bmatrix} 0 \\ 1 \end{bmatrix} = \begin{bmatrix} -1 \\ 0 \end{bmatrix}$$

$$\mathbf{g}(1) = \frac{\mathbf{A}(1)\mathbf{F}(0)\mathbf{j}_2}{\mathbf{c}^\dagger(1)\mathbf{F}(0)\mathbf{j}_2} = \frac{\begin{bmatrix} 1 & 2 \\ 1 & -1 \end{bmatrix}\begin{bmatrix} 0 & 2 \\ 0 & -1 \end{bmatrix}\begin{bmatrix} 0 \\ 1 \end{bmatrix}}{[\frac{1}{2} \quad 0]\begin{bmatrix} 0 & 2 \\ 0 & -1 \end{bmatrix}\begin{bmatrix} 0 \\ 1 \end{bmatrix}} = \begin{bmatrix} 0 \\ 3 \end{bmatrix}$$

$$\mathbf{F}(1) = \mathbf{A}(1) - \mathbf{g}(1)\mathbf{c}^\dagger(1) = \begin{bmatrix} 1 & 2 \\ 1 & -1 \end{bmatrix} - \begin{bmatrix} 0 \\ 3 \end{bmatrix}[\frac{1}{2} \quad 0] = \begin{bmatrix} 1 & 2 \\ -\frac{1}{2} & -1 \end{bmatrix}$$

$$\mathbf{h}(1) = \mathbf{b}(1) - \mathbf{g}(1)d(1) = \begin{bmatrix} -1 \\ \frac{1}{2} \end{bmatrix} - \begin{bmatrix} 0 \\ 3 \end{bmatrix} = \begin{bmatrix} -1 \\ -\frac{5}{2} \end{bmatrix}$$

For ongoing deadbeat observation with the same basis vector ordering, the next observer gain is

$$\mathbf{g}(2) = \frac{\mathbf{A}(2)\mathbf{F}(1)\mathbf{j}_1}{\mathbf{c}^\dagger(2)\mathbf{F}(1)\mathbf{j}_1} = \frac{\begin{bmatrix} 2 & 2 \\ 1 & -1 \end{bmatrix}\begin{bmatrix} 1 & 2 \\ -\frac{1}{2} & -1 \end{bmatrix}\begin{bmatrix} 1 \\ 0 \end{bmatrix}}{[\frac{1}{4} \quad 0]\begin{bmatrix} 1 & 2 \\ -\frac{1}{2} & -1 \end{bmatrix}\begin{bmatrix} 1 \\ 0 \end{bmatrix}} = \begin{bmatrix} 4 \\ 6 \end{bmatrix}$$

and so on.

Other observer error responses besides deadbeat ones can be obtained, if desired. For example, the algorithm for which

$$\begin{cases} \mathbf{F}(0)\mathbf{j}_1 = \alpha\mathbf{j}_1 \\ \mathbf{F}(1)\mathbf{F}(0)\mathbf{j}_2 = \alpha\mathbf{j}_2 \\ \mathbf{F}(2)\mathbf{F}(1)\mathbf{F}(0)\mathbf{j}_3 = \alpha\mathbf{j}_3 \\ \quad \vdots \end{cases}$$

(5-33)

where

$$|\alpha| < 1$$

progressively reduces, but does not drive to zero, the observer error. Observer gains that give equation (5-33) are

$$\begin{cases} \mathbf{g}(0) = \dfrac{\mathbf{A}(0)\mathbf{j}_1 - \alpha\mathbf{j}_1}{\mathbf{c}^\dagger(0)\mathbf{j}_1} \\ \mathbf{F}(0) = \mathbf{A}(0) - \mathbf{g}(0)\mathbf{c}^\dagger(0) \end{cases}$$

$$\begin{cases} \mathbf{g}(1) = \dfrac{\mathbf{A}(1)\mathbf{F}(0)\mathbf{j}_2 - \alpha\mathbf{j}_2}{\mathbf{c}^\dagger(1)\mathbf{F}(0)\mathbf{j}_2} \\ \mathbf{F}(1) = \mathbf{A}(1) - \mathbf{g}(1)\mathbf{c}^\dagger(1) \end{cases}$$

$$\mathbf{g}(2) = \dfrac{\mathbf{A}(2)\mathbf{F}(1)\mathbf{F}(0)\mathbf{j}_3 - \alpha\mathbf{j}_3}{\mathbf{c}^\dagger(2)\mathbf{F}(1)\mathbf{F}(0)\mathbf{j}_3}$$

$$\vdots$$

When the algorithm for deadbeat observer gain in equation (5-32) is used for observation of a plant (5-28) with matrices $\mathbf{A}$ and $\mathbf{c}$ constant, the last observer gain obtained, $\mathbf{g}(n-1)$, is the constant gain for which all the eigenvalues of

$$\mathbf{F} = \mathbf{F}(n-1) = \mathbf{A} - \mathbf{g}(n-1)\mathbf{c}^\dagger$$

are at $\lambda = 0$. That gain can simply be held for steps beyond the $(n-1)$th to obtain ongoing observer convergence. For example, for the step-invariant system

$$\begin{bmatrix} x_1(k+1) \\ x_2(k+1) \end{bmatrix} = \begin{bmatrix} 1 & 2 \\ 0 & -1 \end{bmatrix}\begin{bmatrix} x_1(k) \\ x_2(k) \end{bmatrix} + \begin{bmatrix} 1 \\ 2 \end{bmatrix} u(k)$$

$$y(k) = \begin{bmatrix} 1 & -1 \end{bmatrix}\begin{bmatrix} x_1(k) \\ x_2(k) \end{bmatrix}$$

and unit coordinate basis vectors $\mathbf{j}_1$ and $\mathbf{j}_2$ as before, equation (5-32) gives

$$\mathbf{g}(0) = \dfrac{\begin{bmatrix} 1 & 2 \\ 0 & -1 \end{bmatrix}\begin{bmatrix} 1 \\ 0 \end{bmatrix}}{\begin{bmatrix} 1 & -1 \end{bmatrix}\begin{bmatrix} 1 \\ 0 \end{bmatrix}} = \begin{bmatrix} 1 \\ 0 \end{bmatrix}$$

$$\mathbf{F}(0) = \begin{bmatrix} 1 & 2 \\ 0 & -1 \end{bmatrix} - \begin{bmatrix} 1 \\ 0 \end{bmatrix}\begin{bmatrix} 1 & -1 \end{bmatrix} = \begin{bmatrix} 0 & 3 \\ 0 & -1 \end{bmatrix}$$

$$g(1) = \frac{\begin{bmatrix} 1 & 2 \\ 0 & -1 \end{bmatrix}\begin{bmatrix} 0 & 3 \\ 0 & -1 \end{bmatrix}\begin{bmatrix} 0 \\ 1 \end{bmatrix}}{[1 \quad -1]\begin{bmatrix} 0 & 3 \\ 0 & -1 \end{bmatrix}\begin{bmatrix} 0 \\ 1 \end{bmatrix}} = \begin{bmatrix} \frac{1}{4} \\ \frac{1}{4} \end{bmatrix}$$

$$F(1) = \begin{bmatrix} 1 & 2 \\ 0 & -1 \end{bmatrix} - \begin{bmatrix} \frac{1}{4} \\ \frac{1}{4} \end{bmatrix}[1 \quad -1] = \begin{bmatrix} \frac{3}{4} & \frac{9}{4} \\ -\frac{1}{4} & -\frac{3}{4} \end{bmatrix}$$

The observer state coupling maxtrix $F(1)$ has the characteristic equation

$$\lambda^2 = 0$$

Using the algorithm of equation (5-32) is also a handy way of calculating step-invariant deadbeat observer gains.

5.7.3 Gain Calculations for Multiple-Output Plants

The same general approach to observer design can be used for multiple-output plants. To illustrate the ideas involved, we consider the initial deadbeat gain sequence for a two-output plant. Extensions to plants with more than two outputs and to ongoing observation are straightforward.

For an nth-order two-output plant

$$x(k + 1) = A(k)x(k) + B(k)u(k)$$

$$y(k) = \begin{bmatrix} y_1(k) \\ \hline y_2(k) \end{bmatrix} = \begin{bmatrix} c_1^\dagger(k) \\ \hline c_2^\dagger(k) \end{bmatrix} x(k) + \begin{bmatrix} d_1^\dagger(k) \\ \hline d_2^\dagger(k) \end{bmatrix} u(k)$$

$$= C(k)x(k) + D(k)u(k)$$

a full-order state observer has the form

$$\xi(k + 1) = F(k)\xi(k) + [g_1(k) \mathbin{\vdots} g_2(k)]y(k) + H(k)u(k) \tag{5-34}$$

where

$$F(k) = A(k) - g_1(k)c_1^\dagger(k) - g_2(k)c_2^\dagger(k)$$

$$H(k) = B(k) - g_1(k)d_1^\dagger(k) - g_2(k)d_2^\dagger(k) \tag{5-35}$$

The error between the plant state and the observer state is governed by

$$x(k) - \xi(k) = F(k - 1)F(k - 2) \ldots F(1)F(0)[x(0) - \xi(0)]$$

Choosing n linearly independent basis vectors $\mathbf{j}_1, \mathbf{j}_2, \ldots, \mathbf{j}_n$ and requiring that

$$
\begin{cases}
\mathbf{F}(0)\mathbf{j}_1 = \mathbf{0} \\
\mathbf{F}(0)\mathbf{j}_2 = \mathbf{0} \\
\mathbf{F}(1)\mathbf{F}(0)\mathbf{j}_3 = \mathbf{0} \\
\mathbf{F}(1)\mathbf{F}(0)\mathbf{j}_4 = \mathbf{0} \\
\mathbf{F}(2)\mathbf{F}(1)\mathbf{F}(0)\mathbf{j}_5 = \mathbf{0} \\
\quad \vdots
\end{cases}
$$

gives

$$
\begin{cases}
[\mathbf{A}(0) - \mathbf{g}_1(0)\mathbf{c}_1^\dagger(0) - \mathbf{g}_2(0)\mathbf{c}_2^\dagger(0)]\mathbf{j}_1 = \mathbf{0} \\
[\mathbf{A}(0) - \mathbf{g}_1(0)\mathbf{c}_1^\dagger(0) - \mathbf{g}_2(0)\mathbf{c}_2^\dagger(0)]\mathbf{j}_2 = \mathbf{0}
\end{cases}
$$

$$
\begin{cases}
[\mathbf{A}(1) - \mathbf{g}_1(1)\mathbf{c}_1^\dagger(1) - \mathbf{g}_2(1)\mathbf{c}_2^\dagger(1)]\mathbf{F}(0)\mathbf{j}_3 = \mathbf{0} \\
[\mathbf{A}(1) - \mathbf{g}_1(1)\mathbf{c}_1^\dagger(1) - \mathbf{g}_2(1)\mathbf{c}_2^\dagger(1)]\mathbf{F}(0)\mathbf{j}_4 = \mathbf{0}
\end{cases}
$$

$$
\begin{cases}
[\mathbf{A}(2) - \mathbf{g}_1(2)\mathbf{c}_1^\dagger(2) - \mathbf{g}_2(2)\mathbf{c}_2^\dagger(2)]\mathbf{F}(1)\mathbf{F}(0)\mathbf{j}_5 = \mathbf{0} \\
[\mathbf{A}(2) - \mathbf{g}_1(2)\mathbf{c}_1^\dagger(2) - \mathbf{g}_2(2)\mathbf{c}_2^\dagger(2)]\mathbf{F}(1)\mathbf{F}(0)\mathbf{j}_6 = \mathbf{0}
\end{cases}
$$

$$\vdots$$

and so on.

Each pair of simultaneous vector equations is a set of $2n$ linear algebraic equations in the $2n$ unknown elements of two observer gain vectors of the form

$$
\begin{cases}
\delta_{11}\mathbf{g}_1(k) + \delta_{12}\mathbf{g}_2(k) = \mathbf{A}(k)\mathbf{F}(k-1) \ldots \mathbf{F}(0)\mathbf{j}_a \\
\delta_{21}\mathbf{g}_1(k) + \delta_{22}\mathbf{g}_2(k) = \mathbf{A}(k)\mathbf{F}(k-1) \ldots \mathbf{F}(0)\mathbf{j}_b
\end{cases}
$$

where $\mathbf{j}_a$ and $\mathbf{j}_b$ are the two n-vectors involved. Defining the 2×2 matrix

$$
\Delta = \begin{bmatrix} \delta_{11} & \delta_{12} \\ \delta_{21} & \delta_{22} \end{bmatrix}
$$

then

$$
\begin{bmatrix} \{\mathbf{g}_1(k)\}i\text{th element} \\ \{\mathbf{g}_2(k)\}i\text{th element} \end{bmatrix} = \Delta^{-1} \begin{bmatrix} \{\mathbf{A}(k)\mathbf{F}(k-1) \ldots \mathbf{F}(0)\mathbf{j}_a\}i\text{th element} \\ \{\mathbf{A}(k)\mathbf{F}(k-1) \ldots \mathbf{F}(0)\mathbf{j}_b\}i\text{th element} \end{bmatrix}
$$

or

$$\begin{bmatrix} \mathbf{g}_1^\dagger(k) \\ \hline \mathbf{g}_2^\dagger(k) \end{bmatrix} = \boldsymbol{\Delta}^{-1} \begin{bmatrix} \{\mathbf{A}(k)\mathbf{F}(k-1) \dots \mathbf{F}(0)\mathbf{j}_a\}^\dagger \\ \{\mathbf{A}(k)\mathbf{F}(k-1) \dots \mathbf{F}(0)\mathbf{j}_b\}^\dagger \end{bmatrix}$$

Only a 2×2 matrix need be inverted to find each pair of observer gains.

Consider, for example, the third-order/two-output plant

$$\begin{bmatrix} x_1(k+1) \\ x_2(k+1) \\ x_3(k+1) \end{bmatrix} = \begin{bmatrix} (\frac{1}{2})^k & 0 & 1 \\ -1 & -1 & 0 \\ 2 & 0 & k \end{bmatrix} \begin{bmatrix} x_1(k) \\ x_2(k) \\ x_3(k) \end{bmatrix} + \begin{bmatrix} 1 \\ 1 \\ 1 \end{bmatrix} u(k)$$

$$\begin{bmatrix} y_1(k) \\ y_2(k) \end{bmatrix} = \begin{bmatrix} 0 & (k-2) & 1 \\ -1 & 3 & 0 \end{bmatrix} \begin{bmatrix} x_1(k) \\ x_2(k) \\ x_3(k) \end{bmatrix} + \begin{bmatrix} 0 \\ 4 \end{bmatrix} u(k)$$

and a third-order state observer of the form of equation (5-34) with equation (5-35). Choosing the basis vectors to be unit coordinate vectors

$$\mathbf{j}_1 = \begin{bmatrix} 1 \\ 0 \\ 0 \end{bmatrix} \qquad \mathbf{j}_2 = \begin{bmatrix} 0 \\ 1 \\ 0 \end{bmatrix} \qquad \mathbf{j}_3 = \begin{bmatrix} 0 \\ 0 \\ 1 \end{bmatrix}$$

and requiring that the observer gain vector at step 0 satisfy

$$\left\{ \left[\begin{bmatrix} 1 & 0 & 1 \\ -1 & -1 & 0 \\ 2 & 0 & 0 \end{bmatrix} - \begin{bmatrix} g_{11}(0) \\ g_{21}(0) \\ g_{31}(0) \end{bmatrix} [0 \ -2 \ 1] - \begin{bmatrix} g_{12}(0) \\ g_{22}(0) \\ g_{32}(0) \end{bmatrix} [-1 \ 3 \ 0] \right] \begin{bmatrix} 1 \\ 0 \\ 0 \end{bmatrix} \right\} = \begin{bmatrix} 0 \\ 0 \\ 0 \end{bmatrix}$$

$$\left\{ \left[\begin{bmatrix} 1 & 0 & 1 \\ -1 & -1 & 0 \\ 2 & 0 & 0 \end{bmatrix} - \begin{bmatrix} g_{11}(0) \\ g_{21}(0) \\ g_{31}(0) \end{bmatrix} [0 \ -2 \ 1] - \begin{bmatrix} g_{12}(0) \\ g_{22}(0) \\ g_{32}(0) \end{bmatrix} [-1 \ 3 \ 0] \right] \begin{bmatrix} 0 \\ 1 \\ 0 \end{bmatrix} \right\} = \begin{bmatrix} 0 \\ 0 \\ 0 \end{bmatrix}$$

gives

$$\mathbf{G}(0) = \begin{bmatrix} g_{11}(0) & g_{12}(0) \\ g_{21}(0) & g_{22}(0) \\ g_{31}(0) & g_{32}(0) \end{bmatrix} = \begin{bmatrix} -\frac{3}{2} & -1 \\ 2 & 1 \\ -3 & -2 \end{bmatrix}$$

and

$$\mathbf{F}(0) = \begin{bmatrix} 1 & 0 & 1 \\ -1 & -1 & 0 \\ 2 & 0 & 0 \end{bmatrix} - \begin{bmatrix} -\frac{3}{2} & -1 \\ 2 & 1 \\ -3 & -2 \end{bmatrix} \begin{bmatrix} 0 & -2 & 1 \\ -1 & 3 & 0 \end{bmatrix} = \begin{bmatrix} 0 & 0 & \frac{3}{2} \\ 0 & 0 & -2 \\ 0 & 0 & 3 \end{bmatrix}$$

Requiring that, on step 1,

$$\begin{cases} [\mathbf{A}(1) - \mathbf{g}_1(1)\mathbf{c}_1^\dagger(1) - \mathbf{g}_2(1)\mathbf{c}_2^\dagger(1)]\mathbf{F}(0)\mathbf{j}_3 = \mathbf{0} \\ [\mathbf{A}(1) - \mathbf{g}_1(1)\mathbf{c}_1^\dagger(1) - \mathbf{g}_2(1)\mathbf{c}_2^\dagger(1)]\mathbf{F}(0)\mathbf{j}_1 = \mathbf{0} \end{cases}$$

drives the observer error to zero as well as zeroing again any residual observer error in the $\mathbf{j}_1$ direction.

5.7.4 Observing Linear State Transformations

When the state of a step-varying observer, which is not necessarily of the same order as the plant, observes a linear transformation of the plant state

$$\boldsymbol{\xi}(k) \rightarrow \mathbf{M}(k)\mathbf{x}(k)$$

the observer error is governed by

$$\mathbf{M}(k+1)\mathbf{x}(k+1) - \boldsymbol{\xi}(k+1)$$
$$= [\mathbf{M}(k+1)\mathbf{A}(k) - \mathbf{G}(k)\mathbf{C}(k)]\mathbf{x}(k) - \mathbf{F}(k)\boldsymbol{\xi}(k)$$
$$+ [\mathbf{M}(k+1)\mathbf{B}(k) - \mathbf{G}(k)\mathbf{D}(k) - \mathbf{H}(k)]\mathbf{u}(k)$$

Requiring that the error be autonomous, with

$$\mathbf{M}(k+1)\mathbf{x}(k+1) - \boldsymbol{\xi}(k+1) = \mathbf{F}(k)[\mathbf{M}(k)\mathbf{x}(k) - \boldsymbol{\xi}(k)]$$

gives

$$\mathbf{F}(k)\mathbf{M}(k) = \mathbf{M}(k+1)\mathbf{A}(k) - \mathbf{G}(k)\mathbf{C}(k)$$

$$\mathbf{H}(k) = \mathbf{M}(k+1)\mathbf{B}(k) - \mathbf{G}(k)\mathbf{D}(k) \tag{5-36}$$

The observer error is then governed by

$$\mathbf{M}(k)\mathbf{x}(k) - \boldsymbol{\xi}(k) = \mathbf{F}(k-1)\mathbf{F}(k-2) \ldots \mathbf{F}(0)[\mathbf{M}(0)\mathbf{x}(0) - \boldsymbol{\xi}(0)]$$

These relations are summarized in Table 5-10. As with their counterparts in the step-invariant case, these equations can be used to design state observers of reduced order and observers of linear plant state transformations.

5.7.5 Observer Feedback

Suppose that state estimates from a pth-order deadbeat observer are substituted for the state in a plant feedback arrangement as in Figure 5-18. The character of the observer feedback system's zero-input response is the same as the zero-input response of the state feedback

Table 5-10 Step-Varying Observers of Plant State Transformations

---

Plant Model

$\mathbf{x}(k + 1) = \mathbf{A}(k)\mathbf{x}(k) + \mathbf{B}(k)\mathbf{u}(k)$

$\mathbf{y}(k) = \mathbf{C}(k)\mathbf{x}(k) + \mathbf{D}(k)\mathbf{u}(k)$

Observer Structure

$\xi(k + 1) = \mathbf{F}(k)\xi(k) + \mathbf{G}(k)\mathbf{y}(k) + \mathbf{H}(k)\mathbf{u}(k)$

$\xi(k) \rightarrow \mathbf{M}(k)\mathbf{x}(k)$

where

$\mathbf{F}(k)\mathbf{M}(k) = \mathbf{M}(k + 1)\mathbf{A}(k) - \mathbf{G}(k)\mathbf{C}(k)$

$\mathbf{H}(k) = \mathbf{M}(k + 1)\mathbf{B}(k) - \mathbf{G}(k)\mathbf{D}(k)$

Observer Error

$\mathbf{M}(k + 1)\mathbf{x}(k + 1) - \xi(k + 1) = \mathbf{F}(k)[\mathbf{M}(k)\mathbf{x}(k) - \xi(k)]$

$\mathbf{M}(k)\mathbf{x}(k) - \xi(k) = \mathbf{F}(k - 1)\mathbf{F}(k - 2) \ldots \mathbf{F}(0)[\mathbf{M}(0)\mathbf{x}(0) - \xi(0)]$

---

system after p-steps because, then, the observer state and the plant state are identical. The observer feedback system has a p-step initial transient interval during which the observer is converging to the plant state. Thereafter, the feedback system behaves as if it had state feedback. If the plant is nth-order and the state feedback has also been designed for deadbeat response, the $(n + p)$th-order observer feedback system is deadbeat.

5.8 Control of Flexible Spacecrafts

In this section, we consolidate the material presented and the techniques developed earlier in the chapter by means of a detailed example that involves the control of flexible spacecraft.

Many of the recent applications of space involve spacecraft that have to point *very precisely* at some object on earth, in the ocean, or in space. As shown in Figure 5-19, a satellite of this class typically consists of a *central rigid body*, gimballed telescopes or antennas, and one or more solar arrays that generate electric energy from the sun. These solar arrays are usually large and *flexible*. When thrusters are fired to maneuver the spacecraft from one position to another, the solar arrays, if not controlled, bend and oscillate, and consequently the precision performance of the vehicle may be degraded significantly.

Plant with Observer Feedback

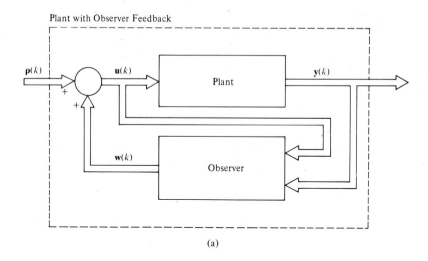

(a)

Plant with Observer Feedback

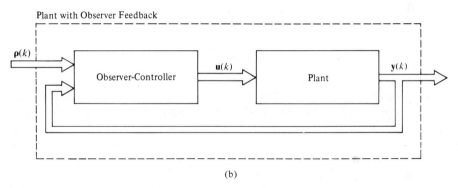

(b)

Figure 5-18 Step-varying plant with observer feedback in place of state feedback. (a) Plant with observer feedback. (b) Observer rearranged.

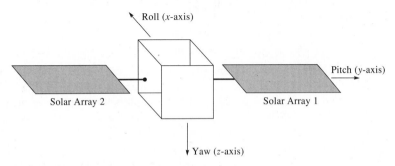

Figure 5-19 A spacecraft with two flexible solar arrays.

The dynamic behavior of a spacecraft is described in terms of its so-called *rigid modes* and *flexible modes*. In the simplest situations, the control system designer ignores all flexible modes and designs the control system as if the vehicle were entirely a rigid body. The designer then hopes that the effect of the flexible modes on the spacecraft behavior is insignificant. For precision pointing applications this approach may not be possible. As the bandwidth of the control system for the rigid body modes of the spacecraft approaches or overlaps the flexible mode frequencies, the control signals produced by the thrusters may excite the flexible modes. It is, therefore, necessary to design a feedback control system that damps out the flexible modes as quickly as possible so that the effects of modal excitation on the attitude error are minimized.

As a simple example of designing a control system for two flexible modes of the spacecraft, consider the state equations

$$
\begin{bmatrix} \dot{x}_1(t) \\ \dot{x}_2(t) \\ \hline \dot{x}_3(t) \\ \dot{x}_4(t) \end{bmatrix} = \left[\begin{array}{cc|cc} 0 & 1 & 0 & 0 \\ -\omega_1^2 & 0 & 0 & 0 \\ \hline 0 & 0 & 0 & 1 \\ 0 & 0 & -\omega_2^2 & 0 \end{array} \right] \begin{bmatrix} x_1(t) \\ x_2(t) \\ \hline x_3(t) \\ x_4(t) \end{bmatrix} + \begin{bmatrix} 0 \\ \phi_1 \\ \hline 0 \\ \phi_2 \end{bmatrix} u(t)
$$

where x_1 and x_2 are the generalized modal position and modal rate of mode 1, while x_3 and x_4 represent the modal position and modal rate, respectively, of mode 2. The natural radian frequencies of mode 1 and mode 2 are ω_1 and ω_2, respectively. The terms ϕ_1 and ϕ_2 are the mode shapes of modes 1 and 2, respectively, at the actuator location. The desired control torque $u(t)$ is produced by an actuator that is properly located on the spacecraft.

The output equations

$$
\begin{bmatrix} y_1(t) \\ y_2(t) \end{bmatrix} = \begin{bmatrix} \phi_1 & 0 & \phi_2 & 0 \\ \phi_3 & 0 & \phi_4 & 0 \end{bmatrix} \begin{bmatrix} x_1(t) \\ x_2(t) \\ x_3(t) \\ x_4(t) \end{bmatrix}
$$

represent two position sensors. One sensor is placed at the same location as the actuator, and the other sensor is conveniently located on the spacecraft to provide additional measurements of the modes.

The system is completely controllable and completely observable and has open-loop eigenvalues at $\lambda_1 = j\omega_1$, $\lambda_2 = -j\omega_1$, $\lambda_3 = j\omega_2$, and $\lambda_4 = -j\omega_2$. If not controlled, these modes will oscillate with amplitudes depending on the initial conditions.

A discrete-time model of the continuous-time system in terms of the sampling rate, the natural frequencies ω_1, ω_2, and the modal shapes

ϕ_1, ϕ_2, ϕ_3, and ϕ_4 is given by equations (3-20) and (3-21) as

$$
\begin{bmatrix} x_1(k+1) \\ x_2(k+1) \\ \hline x_3(k+1) \\ x_4(k+1) \end{bmatrix}
$$

$$
= \begin{bmatrix} \cos \omega_1 T & \left(\dfrac{1}{\omega_1}\right) \sin \omega_1 T & 0 & 0 \\ -\omega_1 \sin \omega_1 T & \cos \omega_1 T & 0 & 0 \\ \hline 0 & 0 & \cos \omega_2 T & \left(\dfrac{1}{\omega_2}\right) \sin \omega_2 T \\ 0 & 0 & -\omega_2 \sin \omega_2 T & \cos \omega_2 T \end{bmatrix} \begin{bmatrix} x_1(k) \\ x_2(k) \\ x_3(k) \\ x_4(k) \end{bmatrix}
$$

$$
+ \begin{bmatrix} \left(\dfrac{\phi_1}{\omega_1^2}\right)(1 - \cos \omega_1 T) \\ \left(\dfrac{\phi_1}{\omega_1}\right) \sin \omega_1 T \\ \hline \left(\dfrac{\phi_2}{\omega_2^2}\right)(1 - \cos \omega_2 T) \\ \left(\dfrac{\phi_2}{\omega_2}\right) \sin \omega_2 T \end{bmatrix} u(kT) = \mathbf{A}x(k) + \mathbf{b}u(k)
$$

$$
\begin{bmatrix} y_1(kT) \\ y_2(kT) \end{bmatrix} = \begin{bmatrix} \phi_1 & 0 & \phi_2 & 0 \\ \phi_3 & 0 & \phi_4 & 0 \end{bmatrix} \begin{bmatrix} x_1(kT) \\ x_2(kT) \\ x_3(kT) \\ x_4(kT) \end{bmatrix} = \begin{bmatrix} \mathbf{c}_1^\dagger \\ \mathbf{c}_2^\dagger \end{bmatrix} \mathbf{x}(k)
$$

One method of controlling the flexible modes is to find a feedback control law of the form

$$u(k) = \mathbf{e}^\dagger \mathbf{x}(k)$$

so that the eigenvalues of the modes are placed at desired locations. The desired locations depend on how quickly one wishes the modes to decay with step.

Suppose that the modal parameters for the two flexible modes are

$\omega_1 = 6.4$ rad/sec

$\omega_2 = 18.7$ rad/sec

$\phi_1 = 0.6 \qquad \phi_2 = 1.0$

$\phi_3 = 0.5 \qquad \phi_4 = 0.4$

If the sampling interval is $T = 0.01$ sec, the state feedback gain vector

$$\mathbf{e}^\dagger = [-4.76 \quad -4.77 \quad -3.85 \quad -1.64]$$

places the feedback eigenvalues at

$$z_{1,2} = 0.98 \pm j0.064 \qquad z_{3,4} = 0.97 \pm j0.18$$

in the z-domain. Mapping these eigenvalues to the s-domain via

$$s = \left(\frac{1}{T}\right) \ln z$$

gives

$$s_{1,2} = -1.44 \pm j6.56 \qquad s_{3,4} = -0.84 \pm j18.73$$

and, therefore, the eigenvalues of the feedback control system have damping ratios of $\zeta_1 = 21\%$ and $\zeta_2 = 4.48\%$, respectively.

For the initial state vector

$$\mathbf{x}(0) = \begin{bmatrix} 0.05 \\ 0 \\ 0.05 \\ 0 \end{bmatrix}$$

the states of the closed-loop system, assuming all the states are available for feedback, are shown in Figure 5-20. The control sequence $u(k)$ and the outputs $y_1(k)$ and $y_2(k)$ are shown in Figures 5-21(a) and 5-21(b), respectively.

Because the state variables are not all available for feedback, the second step of the design process is to construct an observer of minimal order that observes the control input $u(k)$. Based on the discussion in Section 5.5.2, a first-order functional observer is adequate to observe the control input $u(k)$.

Arbitrarily selecting the observer eigenvalue at $z = 0.5$, an observer

$$\xi(k + 1) = 0.5\xi(k) + [g_{11} \quad g_{12}] \begin{bmatrix} y_1(k) \\ y_2(k) \end{bmatrix} + hu(k)$$

with the output

$$w(k) = [\gamma_{11} \quad \gamma_{12}] \begin{bmatrix} y_1(k) \\ y_2(k) \end{bmatrix} + \delta_{11} \xi(k)$$

may be designed to observe the control input $u(k)$, which is a linear combination of the states.

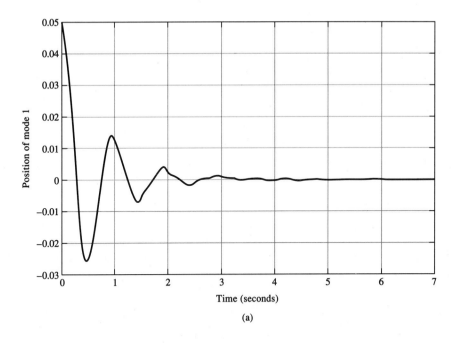

(a)

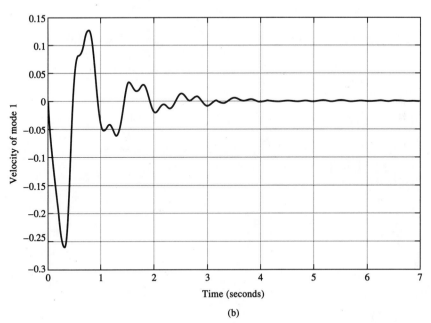

(b)

Figure 5-20 Controlled modal positions and model rates of mode 1 and mode 2.

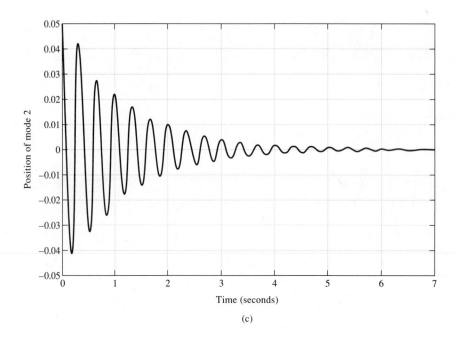

(c)

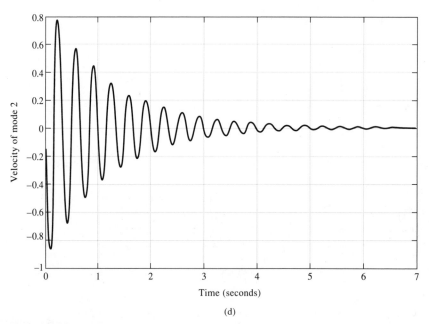

(d)

Figure 5-20 (continued)

The observer output in terms of the state is

$$w(k) = \left\{ [\gamma_{11} \quad \gamma_{12}] \begin{bmatrix} \mathbf{c}_1^\dagger \\ \mathbf{c}_2^\dagger \end{bmatrix} + \delta_{11} \, \mathbf{m}_1^\dagger \right\} \mathbf{x}(k) \tag{5-37}$$

where

$$\xi(k) \to \mathbf{m}^\dagger \mathbf{x}(k)$$

Hence, if the quantity within the braces in equation (5-37) is equated to $\mathbf{e}^\dagger$, then

$$w(k) \to \mathbf{e}^\dagger \mathbf{x}(k)$$

and, therefore, the observer output approaches the control input $u(k)$ asymptomatically.

Substituting

$$\mathbf{m}_1^\dagger = [g_{11} \quad g_{12}] \begin{bmatrix} \mathbf{c}_1^\dagger \\ \mathbf{c}_2^\dagger \end{bmatrix} (\mathbf{A} - 0.5\mathbf{I})^{-1}$$

into equation (5-37) gives

$$\gamma_{11}\mathbf{c}_1^\dagger + \gamma_{12}\mathbf{c}_2^\dagger + \delta_{11} \, \mathbf{m}_1^\dagger = \mathbf{e}^\dagger \tag{5-38}$$

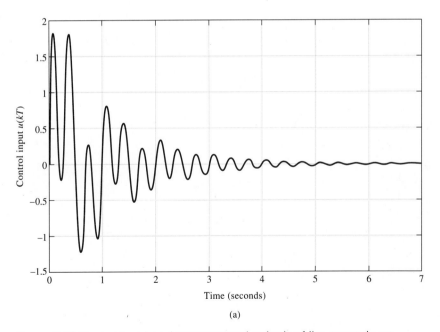

(a)

Figure 5-21 Control input and measurement outputs of the example system.

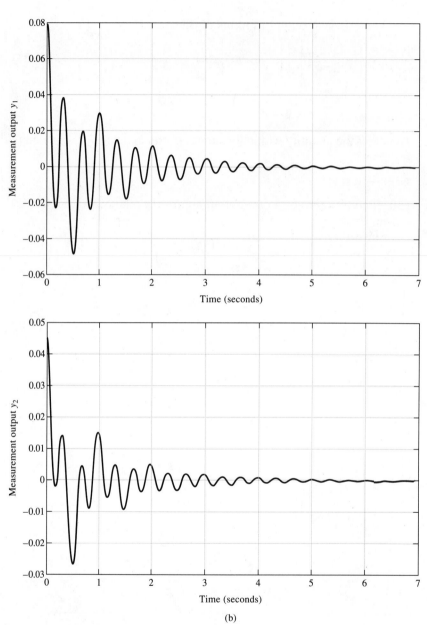

(b)

Figure 5-21 (continued)

Arbitrarily selecting $\delta_{11} = 100$ and solving equation (5-38) gives

$g_{11} = -1.0107$ $g_{12} = 3.6239$
$\gamma_{11} = 213.9883$ $\gamma_{12} = -742.6269$

Hence,

$h = \mathbf{m}_1^\dagger \mathbf{b} = -3.3788 \times 10^{-4}$

and therefore the observer is completely specified as

$$\xi(k + 1) = 0.5\xi(k)$$
$$+ [-1.0107 \quad 3.6239]\begin{bmatrix} y_1(k) \\ y_2(k) \end{bmatrix} - 3.3788 \times 10^{-4}u(k)$$

with the output

$$w(k) = [213.9883 \quad -742.6269]\begin{bmatrix} y_1(k) \\ y_2(k) \end{bmatrix} + 100\xi(k)$$

In Figure 5-22(a), the observer state is shown for the initial observer state given by

$$\xi(0) \rightarrow \mathbf{m}^\dagger \mathbf{x}(0)$$

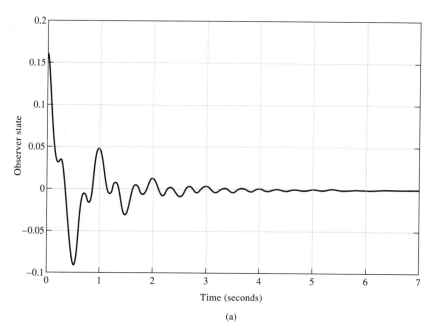

(a)

Figure 5-22 Observer state, estimate of the control input, and control estimate error.

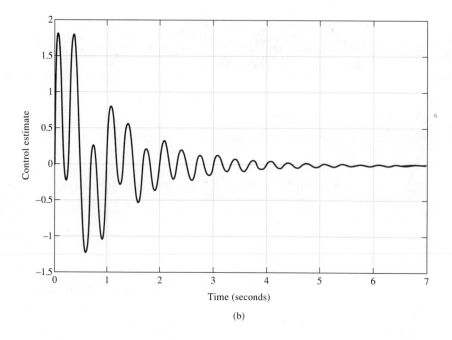

(b)

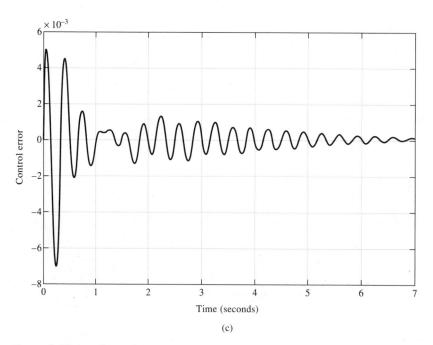

(c)

Figure 5-22 (continued)

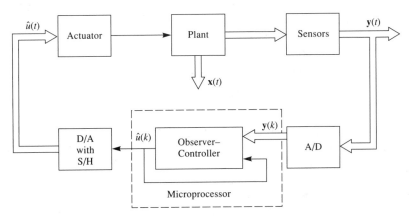

Figure 5-23 A block diagram for the spacecraft digital control system.

The estimate of the control input as given by the observer output is shown in Figure 5-22(b), and the error in the estimate of the control input is shown in Figure 5-22(c). The control estimation error is the difference between the observer output $w(k)$ and the control input $u(k)$ based on state feedback. A block diagram for the observer–controller and the controlled plant is shown in Figure 5-23. One can also plot the state variables of the system controlled with the observer. The plots for the state variables are almost identical (within state estimation error) to those shown in Figure 5-20, and therefore we elected not to show them.

In this simple example, only the flexible portion of a spacecraft with only two modes is considered. In practice, the number of flexible modes that must be retained to describe the dynamic behavior of a flexible spacecraft is probably higher. Model-order reduction, minimization of number of sensors and number of actuators, control spillover, and observation spillover are some of the problems encountered during control system design for flexible spacecrafts.

5.9 Summary

Digital tracking system design for linear systems can be divided into two parts:

1. Obtaining acceptable zero-input response of the tracking outputs
2. Obtaining acceptable zero-state response of the tracking outputs

The first of these two parts is addressed for step-invariant plants and controllers in this chapter. Then the methods are extended to step-varying plants and controllers. When the plant initial conditions are unknown, the best one can do is to shape the zero-input response by selecting the eigenvalues of the combination of the plant and controller. The zero-input response is thereby made to decay to zero in an acceptable manner.

In the previous chapter, it was shown that for a completely controllable plant, state feedback places the feedback system eigenvalues at any locations desired. When, as is usual, the state is not available for feedback, an observer's estimate of the state can be fed back in place of the state. For a completely observable, completely controllable plant, this feedback system has the observer eigenvalues and the eigenvalues that would have resulted from the corresponding plant feedback, a property known as the separation theorem. Eigenvalue placement with observer feedback can thus be accomplished in two parts:

1. Plant eigenvalue placement with state feedback
2. Replacement of the state feedback with an observer's estimate of the state feedback

For an nth-order plant

$$\mathbf{x}(k + 1) = \mathbf{A}\mathbf{x}(k) + \mathbf{B}\mathbf{u}(k)$$
$$\mathbf{y}(k) = \mathbf{C}\mathbf{x}(k) + \mathbf{D}\mathbf{u}(k)$$

a full-order state observer is another nth-order system, driven by the plant inputs and outputs

$$\boldsymbol{\xi}(k + 1) = \mathbf{F}\boldsymbol{\xi}(k) + \mathbf{G}\mathbf{y}(k) + \mathbf{H}\mathbf{u}(k)$$

for which the error between the observer state and the plant state is autonomous. For arbitrary $\mathbf{G}$ and

$$\mathbf{F} = \mathbf{A} - \mathbf{GC}$$
$$\mathbf{H} = \mathbf{B} - \mathbf{GD}$$

the error is autonomous, being governed by

$$\mathbf{x}(k + 1) - \boldsymbol{\xi}(k + 1) = \mathbf{F}[\mathbf{x}(k) - \boldsymbol{\xi}(k)]$$

The observer consists of a model of the plant, driven by the error between the plant output and the observer estimate of that output

$$\boldsymbol{\xi}(k + 1) = \mathbf{A}\boldsymbol{\xi}(k) + \mathbf{B}\mathbf{u}(k) + \mathbf{G}[\mathbf{y}(k) - \mathbf{w}(k)]$$

where

$$\mathbf{w}(k) = \mathbf{C}\boldsymbol{\xi}(k) + \mathbf{D}\mathbf{u}(k)$$

If the plant is completely observable, the observer gain matrix $\mathbf{G}$ can always be chosen to place the observer eigenvalues, the eigenvalues of $\mathbf{F}$, at any locations selected by the designer. When all observer eigenvalues are placed at $\lambda = 0$, the observer is deadbeat

$$\mathbf{F}^n = \mathbf{0}$$

and the observer state equals the plant state after n steps. If desired, a full-order state observer can have an output equation

$$\mathbf{w}(k) = \mathbf{P}\boldsymbol{\xi}(k)$$

and it can be arranged so that its n-vector output, rather than its state, converges to the plant state.

When the plant has a single output and its state variable equations are in observable form, it is simple to design a full-order state observer because, in that form, each observer gain vector element determines one coefficient of the observer's characteristic equation. Transformation to and from observable form can be used to design a full-order state observer of any completely observable single-output plant. Ackermann's formula can also be used to design full-order observers for any completely observable single-output plants. Either the observer state or a nonsingular transformation of that state, given by an observer output equation, can be made to observe the plant state.

Observers of linear transformations of the plant state

$$\boldsymbol{\xi}(k + 1) = \mathbf{F}\boldsymbol{\xi}(k) + \mathbf{G}\mathbf{y}(k) + \mathbf{H}\mathbf{u}(k)$$

where the observer is not necessarily of the same order as the plant, have

$$\mathbf{FM} = \mathbf{MA} - \mathbf{GC}$$
$$\mathbf{H} = \mathbf{MB} - \mathbf{GD}$$

so that the error between the observer state and the transformation $\mathbf{Mx}(k)$ of the plant state is governed by

$$\mathbf{Mx}(k + 1) - \boldsymbol{\xi}(k + 1) = \mathbf{F}[\mathbf{Mx}(k) - \boldsymbol{\xi}(k)]$$

Observation of a transformation of the plant state is denoted by

$$\boldsymbol{\xi}(k) \rightarrow \mathbf{Mx}(k)$$

A first-order observer with eigenvalue f, distinct from the plant eigenvalues,

$$\boldsymbol{\xi}(k + 1) = f\boldsymbol{\xi}(k) + \mathbf{g}^\dagger\mathbf{y}(k) + \mathbf{h}^\dagger\mathbf{u}(k)$$

observes $\mathbf{m}^\dagger\mathbf{x}(k)$, where

$$\mathbf{m}^\dagger = \mathbf{g}^\dagger\mathbf{C}(\mathbf{A} - f\mathbf{I})^{-1}$$

providing that

$$\mathbf{h}^\dagger = \mathbf{m}^\dagger\mathbf{B} - \mathbf{g}^\dagger\mathbf{D}$$

A powerful general method of observer design is to form a collection of first-order observers, each having one desired eigenvalue. For repeated observer eigenvalues, the relations of Table 5-4 are applied. The collection of observer states observes

$$\xi(k) \rightarrow \mathbf{M}\mathbf{x}(k)$$

where the rows of $\mathbf{M}$ are the individual first-order observers' observation matrices. With n such first-order observers having eigenvalues different from those of the plant, the gains $\mathbf{g}$ can always be chosen so that $\mathbf{M}$ is nonsingular. Then the observer output

$$\mathbf{w}(k) = \mathbf{M}^{-1}\xi(k) \rightarrow \mathbf{x}(k)$$

observes the plant state. If the state transformation associated with the m-vector of linearly independent plant outputs

$$\mathbf{y}(k) - \mathbf{D}\mathbf{u}(k) = \mathbf{C}\mathbf{x}(k)$$

is included in $\mathbf{M}$ and is supplied to the observer output, a reduced-order state observer, of order $n - m$, can be constructed.

To observe a scalar transformation of the plant state, a sum of first-order observer states is formed, and the observer gains are selected to form the desired observed transformation. A linear combination of the scalar plant state transformations derived from the plant outputs can also be included, if desired. When the first-order component observers are included one at a time until the desired observation is achieved, the resulting design is of minimal order.

Step-varying observers recursively estimate the state of a plant, whether it is step-varying or not. A full-order state observer of the form

$$\xi(k + 1) = \mathbf{F}(k)\xi(k) + \mathbf{G}(k)\mathbf{y}(k) + \mathbf{H}(k)\mathbf{u}(k)$$

is governed by

$$\mathbf{F}(k) = \mathbf{A}(k) - \mathbf{G}(k)\mathbf{C}(k)$$
$$\mathbf{H}(k) = \mathbf{B}(k) - \mathbf{G}(k)\mathbf{D}(k)$$

where $\mathbf{G}(k)$ is arbitrary. It has error given by

$$\mathbf{x}(k) - \xi(k) = \mathbf{F}(k - 1)\mathbf{F}(k - 2) \ldots \mathbf{F}(0)[\mathbf{x}(0) - \xi(0)]$$

The algorithm of Table 5-8 gives gains for a deadbeat observer. For an nth-order plant

$$\mathbf{F}(n - 1)\mathbf{F}(n - 2) \ldots \mathbf{F}(0) = \mathbf{0}$$

so that

$$\xi(n) = \mathbf{x}(n)$$

For ongoing deadbeat convergence

$$\mathbf{F}(n + i)\mathbf{F}(n + i - 1) \ldots \mathbf{F}(i) = \mathbf{0} \quad i = 0, 1, 2, \ldots$$

the observer gain calculation summarized in Table 5-9 can be used. Reduced-order observers and observers of state transformations (Table 5-10) also have their step-varying counterparts.

The chapter concluded with a detailed example of eigenvalue placement with observer feedback control of a flexible spacecraft.

References

The theory of observers is summarized in

D. G. Luenberger, "Observers for Multivariable Systems," *IEEE Trans. Automatic Control*, Vol. AC-11, April 1966, pp. 190–197;

D. G. Luenberger, "An Introduction to Observers," *IEEE Trans. Automatic Control*, Vol. AC-16, Dec. 1971, pp. 596–602;

J. O'Reilly, *Observers for Linear Systems*. New York: Academic Press, 1983.

The application of observers to placing feedback system eigenvalues is discussed in

J. D. Ferguson and Z. V. Rekasius, "Optimal Linear Control Systems with Incomplete State Measurements," *IEEE Trans. Automatic Control*, Vol. AC-14, April 1969, pp. 135–140;

E. J. Davison, "On Pole Assignment in Linear Systems with Incomplete State Feedback," *IEEE Trans. Automatic Control*, Vol. AC-15, June 1970, pp. 348–351;

B. Gopinath, "On the Control of Linear Multiple Input-Output Systems," *Bell System Tech. J.*, March 1971, pp. 1101–1113;

T. E. Fortmann and D. Williamson, "Design of Low-Order Observers for Linear Feedback Control Laws," *IEEE Trans. Automatic Control*, Vol. AC-17, April 1972, pp. 255–256;

B. D. O. Anderson and J. B. Moore, *Linear Optimal Control*. Englewood Cliffs, NJ: Prentice-Hall, 1971, pp. 149–169; 190–224;

and in many later texts.

An early observer design procedure used a special canonical form for multiple-output systems which is described in

D. G. Luenberger, "Canonical Forms for Linear Multivariable Systems," *IEEE Trans. Automatic Control*, Vol. AC-12, June 1967, pp. 290–293.

The design method used here was first described in

G. H. Hostetter and R. T. Stefani, "Observer and Controller Design Methods and Examples," *ASEE CoEd Trans.*, Vol. 8, no. 5, May 1976, pp. 53–60;

M. S. Santina and G. H. Hostetter, "New Algorithms for Minimal-Order Observer-Controller Design," Proceedings of the 1986 IEEE conference on *Decision and Control*, Dec. 1986.

An outstanding textbook on spacecraft dynamics is

P. C. Hughes, *Spacecraft Attitude Dynamics*. New York: Wiley, 1986. This text is quite useful and should be read thoroughly.

Chapter Five Problems

5-1. Design a full-order state observer of the plant

$$\begin{bmatrix} x_1(k+1) \\ x_2(k+1) \end{bmatrix} = \begin{bmatrix} \frac{1}{2} & 1 \\ -\frac{1}{2} & 0 \end{bmatrix} \begin{bmatrix} x_1(k) \\ x_2(k) \end{bmatrix} + \begin{bmatrix} 1 \\ -2 \end{bmatrix} u(k)$$

$$y(k) = [1 \quad 0] \begin{bmatrix} x_1(k) \\ x_2(k) \end{bmatrix} + u(k)$$

Choose the observer eigenvalues to be $\lambda = 0.1 \pm j\, 0.1$.

5-2. Design a full-order state observer of the plant

$$\begin{bmatrix} x_1(k+1) \\ x_2(k+1) \\ x_3(k+1) \end{bmatrix} = \begin{bmatrix} 3 & 1 & 0 \\ -3 & 0 & 1 \\ 1 & 0 & 0 \end{bmatrix} \begin{bmatrix} x_1(k) \\ x_2(k) \\ x_3(k) \end{bmatrix} + \begin{bmatrix} 2 \\ 1 \\ 3 \end{bmatrix} u(k)$$

$$y(k) = [1 \quad 0 \quad 1] \begin{bmatrix} x_1(k) \\ x_2(k) \\ x_3(k) \end{bmatrix} - 2u(k)$$

Choose the observer eigenvalues all to be $\lambda = 0.5$.

5-3. Design a full-order deadbeat state observer of the plant

$$\begin{bmatrix} x_1(k+1) \\ x_2(k+1) \end{bmatrix} = \begin{bmatrix} -1 & 3 \\ -3 & -1 \end{bmatrix} \begin{bmatrix} x_1(k) \\ x_2(k) \end{bmatrix} + \begin{bmatrix} 1 \\ 3 \end{bmatrix} u(k)$$

$$y(k) = [1 \quad 1] \begin{bmatrix} x_1(k) \\ x_2(k) \end{bmatrix} - u(k)$$

5-4. Using Ackermann's formula, design a full-order state observer of the plant

$$
\begin{bmatrix} x_1(k+1) \\ x_2(k+1) \\ x_3(k+1) \end{bmatrix} = \begin{bmatrix} 6 & 1 & 0 \\ -11 & 0 & 1 \\ 6 & 0 & 0 \end{bmatrix} \begin{bmatrix} x_1(k) \\ x_2(k) \\ x_3(k) \end{bmatrix} + \begin{bmatrix} 1 & -1 \\ 0 & 1 \\ 3 & -2 \end{bmatrix} \begin{bmatrix} u_1(k) \\ u_2(k) \end{bmatrix}
$$

$$
y(k) = \begin{bmatrix} 1 & 0 & 0 \end{bmatrix} \begin{bmatrix} x_1(k) \\ x_2(k) \\ x_3(k) \end{bmatrix} + \begin{bmatrix} 2 & 0 \end{bmatrix} \begin{bmatrix} u_1(k) \\ u_2(k) \end{bmatrix}
$$

Choose the observer eigenvalues all to be $\lambda = -\frac{1}{4}$.

5-5. For the matrices

$$
\mathbf{B} = \begin{bmatrix} 3 & -2 \\ 6 & -1 \end{bmatrix} \qquad \mathbf{c} = \begin{bmatrix} 0 \\ -3 \end{bmatrix}
$$

find a vector $\mathbf{d}$ such that the eigenvalues of $\mathbf{A}$ are all at $\lambda = -\frac{1}{2}$, where

a. $\mathbf{A} = \mathbf{B} + \mathbf{cd}^\dagger$

b. $\mathbf{A} = \mathbf{B} + \mathbf{dc}^\dagger$

5-6. A finite impulse response (FIR) system has the property that a unit pulse at any input produces a response at every output that is nonzero only for a finite number of steps. Using z-transfer functions, show that linear, step-invariant, *deadbeat* systems are FIR.

5-7. Design a full-order state observer of the plant

$$
\begin{bmatrix} x_1(k+1) \\ x_2(k+1) \\ x_3(k+1) \end{bmatrix} = \begin{bmatrix} 0 & 3 & 3 \\ -2 & 0 & 1 \\ 6 & 0 & 2 \end{bmatrix} \begin{bmatrix} x_1(k) \\ x_2(k) \\ x_3(k) \end{bmatrix} + \begin{bmatrix} 1 \\ 0 \\ 4 \end{bmatrix} u(k)
$$

$$
\begin{bmatrix} y_1(k) \\ y_2(k) \end{bmatrix} = \begin{bmatrix} 1 & -2 & 0 \\ 0 & -1 & 1 \end{bmatrix} \begin{bmatrix} x_1(k) \\ x_2(k) \\ x_3(k) \end{bmatrix} + \begin{bmatrix} 1 \\ 3 \end{bmatrix} u(k)
$$

with all eigenvalues at $\lambda = 0$.

5-8. Find another, *different* solution to problem 5-7 that uses both outputs to drive the observer with nonzero gains.

5-9. Design a full-order state observer of the plant

$$\begin{bmatrix} x_1(k+1) \\ x_2(k+1) \\ x_3(k+1) \end{bmatrix} = \begin{bmatrix} -1 & 0 & 2 \\ 0 & 0 & -1 \\ 3 & 1 & 0 \end{bmatrix} \begin{bmatrix} x_1(k) \\ x_2(k) \\ x_3(k) \end{bmatrix} + \begin{bmatrix} 1 & -1 \\ 2 & 2 \\ 0 & 0 \end{bmatrix} \begin{bmatrix} u_1(k) \\ u_2(k) \end{bmatrix}$$

$$\begin{bmatrix} y_1(k) \\ y_2(k) \end{bmatrix} = \begin{bmatrix} -1 & 0 & 1 \\ 3 & -1 & 2 \end{bmatrix} \begin{bmatrix} x_1(k) \\ x_2(k) \\ x_3(k) \end{bmatrix} + \begin{bmatrix} 0 & 1 \\ 1 & -1 \end{bmatrix} \begin{bmatrix} u_1(k) \\ u_2(k) \end{bmatrix}$$

with eigenvalues at $\lambda = \frac{1}{2}, \pm j\frac{1}{2}$.

5-10. An observer of the plant

$$\begin{bmatrix} x_1(k+1) \\ x_2(k+1) \\ x_3(k+1) \end{bmatrix} = \begin{bmatrix} 1 & 2 & 3 \\ 0 & -1 & 0 \\ 3 & 0 & -1 \end{bmatrix} \begin{bmatrix} x_1(k) \\ x_2(k) \\ x_3(k) \end{bmatrix} + \begin{bmatrix} 1 \\ -2 \\ 0 \end{bmatrix} u(k)$$

$$y(k) = \begin{bmatrix} 1 & 2 & -2 \end{bmatrix} \begin{bmatrix} x_1(k) \\ x_2(k) \\ x_3(k) \end{bmatrix} - u(k)$$

has the form

$$\begin{bmatrix} \xi_1(k+1) \\ \xi_2(k+1) \end{bmatrix} = \begin{bmatrix} \frac{1}{2} & 0 \\ 0 & \frac{1}{4} \end{bmatrix} \begin{bmatrix} \xi_1(k) \\ \xi_2(k) \end{bmatrix} + \mathbf{g}y(k) + \mathbf{h}u(k)$$

Find the observation matrix $\mathbf{M}$ in the linear state information

$$\xi(k) \rightarrow \mathbf{M}x(k)$$

for this observer, and find the required input coupling matrix $\mathbf{h}$ in terms of $\mathbf{g}$.

5-11. For the plant

$$\begin{bmatrix} x_1(k+1) \\ x_2(k+1) \\ x_3(k+1) \end{bmatrix} = \begin{bmatrix} 1 & 0 & -2 \\ 1 & 3 & 1 \\ 0 & 0 & -1 \end{bmatrix} \begin{bmatrix} x_1(k) \\ x_2(k) \\ x_3(k) \end{bmatrix} + \begin{bmatrix} 2 & 0 \\ -1 & 0 \\ 1 & 1 \end{bmatrix} \begin{bmatrix} u_1(k) \\ u_2(k) \end{bmatrix}$$

$$y(k) = \begin{bmatrix} 1 & -1 & 0 \end{bmatrix} \begin{bmatrix} x_1(k) \\ x_2(k) \\ x_3(k) \end{bmatrix} + \begin{bmatrix} 3 & -2 \end{bmatrix} \begin{bmatrix} u_1(k) \\ u_2(k) \end{bmatrix}$$

find the observation matrix $\mathbf{m}$ in

$$\xi(k) \rightarrow \mathbf{m}^t x(k)$$

where $\xi(k)$ is the state of a first-order observer of the form

$$\xi(k + 1) = \frac{1}{4}\,\xi(k) + y(k) + \mathbf{h}^t\mathbf{u}(k)$$

Also find the required observer input coupling $\mathbf{h}$.

5-12. Use a collection of first-order observers to design a full-order state observer of the plant

$$\begin{bmatrix} x_1(k + 1) \\ x_2(k + 1) \end{bmatrix} = \begin{bmatrix} 0 & 2 \\ -1 & 1 \end{bmatrix}\begin{bmatrix} x_1(k) \\ x_2(k) \end{bmatrix} + \begin{bmatrix} 1 \\ 0 \end{bmatrix}u(k)$$

$$y(k) = \begin{bmatrix} 1 & -1 \end{bmatrix}\begin{bmatrix} x_1(k) \\ x_2(k) \end{bmatrix} + u(k)$$

Choose the observer eigenvalues to be $\lambda = \pm j\frac{1}{4}$.

5-13. Use a collection of first-order observers to design a full-order state observer of the plant

$$\begin{bmatrix} x_1(k + 1) \\ x_2(k + 1) \\ x_3(k + 1) \end{bmatrix} = \begin{bmatrix} 2 & 1 & 0 \\ 1 & 0 & 1 \\ -2 & 0 & 0 \end{bmatrix}\begin{bmatrix} x_1(k) \\ x_2(k) \\ x_3(k) \end{bmatrix} + \begin{bmatrix} 1 \\ 3 \\ 0 \end{bmatrix}u(k)$$

$$\begin{bmatrix} y_1(k) \\ y_2(k) \end{bmatrix} = \begin{bmatrix} 2 & -1 & 0 \\ 0 & 0 & 1 \end{bmatrix}\begin{bmatrix} x_1(k) \\ x_2(k) \\ x_3(k) \end{bmatrix} + \begin{bmatrix} 1 \\ -2 \end{bmatrix}u(k)$$

Choose the observer eigenvalues to be $\lambda = 0, \pm\frac{1}{2}$.

5-14. Design a full-order deadbeat state observer of the plant

$$\begin{bmatrix} x_1(k + 1) \\ x_2(k + 1) \end{bmatrix} = \begin{bmatrix} \frac{1}{2} & 0 \\ \frac{1}{3} & 1 \end{bmatrix}\begin{bmatrix} x_1(k) \\ x_2(k) \end{bmatrix} + \begin{bmatrix} 0 & -1 \\ 3 & 1 \end{bmatrix}\begin{bmatrix} u_1(k) \\ u_2(k) \end{bmatrix}$$

$$y(k) = \begin{bmatrix} 1 & 1 \end{bmatrix}\begin{bmatrix} x_1(k) \\ x_2(k) \end{bmatrix} + \begin{bmatrix} 0 & 2 \end{bmatrix}\begin{bmatrix} u_1(k) \\ u_2(k) \end{bmatrix}$$

using the methods of this chapter.

5-15. Design a full-order state observer of the plant

$$\begin{bmatrix} x_1(k + 1) \\ x_2(k + 1) \\ x_3(k + 1) \end{bmatrix} = \begin{bmatrix} 2 & -1 & 1 \\ 0 & 1 & 2 \\ 0 & -2 & 0 \end{bmatrix}\begin{bmatrix} x_1(k) \\ x_2(k) \\ x_3(k) \end{bmatrix} + \begin{bmatrix} 2 \\ 1 \\ 0 \end{bmatrix}u(k)$$

$$y(k) = \begin{bmatrix} 1 & 0 & 1 \end{bmatrix}\begin{bmatrix} x_1(k) \\ x_2(k) \\ x_3(k) \end{bmatrix} + u(k)$$

using the methods of this chapter. Choose the observer eigenvalues to be $\lambda = 0, 0, -\frac{1}{2}$.

5-16. Design a reduced-order state observer of the plant

$$\begin{bmatrix} x_1(k+1) \\ x_2(k+1) \end{bmatrix} = \begin{bmatrix} 2 & 2 \\ 0 & -1 \end{bmatrix} \begin{bmatrix} x_1(k) \\ x_2(k) \end{bmatrix} + \begin{bmatrix} 4 \\ -1 \end{bmatrix} u(k)$$

$$y(k) = \begin{bmatrix} 2 & 1 \end{bmatrix} \begin{bmatrix} x_1(k) \\ x_2(k) \end{bmatrix}$$

Let the observer have the eigenvalue $\lambda = \frac{1}{2}$.

5-17. Design a reduced-order state observer of the plant

$$\begin{bmatrix} x_1(k+1) \\ x_2(k+1) \\ x_3(k+1) \end{bmatrix} = \begin{bmatrix} 3 & 0 & 0 \\ 2 & -1 & 0 \\ 0 & 1 & 1 \end{bmatrix} \begin{bmatrix} x_1(k) \\ x_2(k) \\ x_3(k) \end{bmatrix} + \begin{bmatrix} 3 & -1 \\ 0 & 1 \\ -1 & -2 \end{bmatrix} \begin{bmatrix} u_1(k) \\ u_2(k) \end{bmatrix}$$

$$y(k) = \begin{bmatrix} 0 & 1 & 1 \end{bmatrix} \begin{bmatrix} x_1(k) \\ x_2(k) \\ x_3(k) \end{bmatrix} + \begin{bmatrix} 2 & 0 \end{bmatrix} \begin{bmatrix} u_1(k) \\ u_2(k) \end{bmatrix}$$

Let the observer eigenvalues be $\pm j\frac{1}{4}$.

5-18. Design a minimal-order state observer of the plant

$$\begin{bmatrix} x_1(k+1) \\ x_2(k+1) \\ x_3(k+1) \end{bmatrix} = \begin{bmatrix} 1 & 1 & 0 \\ 4 & 0 & 1 \\ -4 & 0 & 0 \end{bmatrix} \begin{bmatrix} x_1(k) \\ x_2(k) \\ x_3(k) \end{bmatrix} + \begin{bmatrix} 1 \\ 0 \\ -1 \end{bmatrix} u(k)$$

$$\begin{bmatrix} y_1(k) \\ y_2(k) \end{bmatrix} = \begin{bmatrix} 1 & 0 & 0 \\ 0 & 2 & 1 \end{bmatrix} \begin{bmatrix} x_1(k) \\ x_2(k) \\ x_3(k) \end{bmatrix} + \begin{bmatrix} 2 \\ 3 \end{bmatrix} u(k)$$

Let the observer eigenvalue(s) be from the set $\{0, \pm\frac{1}{2}, \pm\frac{1}{4}\}$.

5-19. For the plant of problem 5-18, and using eigenvalues from those listed for that problem, design a minimal-order observer of the state transformation $e^\dagger x(k)$ where

$$e^\dagger = \begin{bmatrix} -1 & 0 & 2 \end{bmatrix}$$

5-20. For the plant

$$\begin{bmatrix} x_1(k+1) \\ x_2(k+1) \end{bmatrix} = \begin{bmatrix} -1 & 0 \\ 1 & 1 \end{bmatrix} \begin{bmatrix} x_1(k) \\ x_2(k) \end{bmatrix} + \begin{bmatrix} 2 \\ 1 \end{bmatrix} u(k)$$

$$y(k) = \begin{bmatrix} 1 & 2 \end{bmatrix} \begin{bmatrix} x_1(k) \\ x_2(k) \end{bmatrix} - 3u(k)$$

design a full-order state feedback observer arrangement so that the feedback system has the eigenvalues $\lambda = 0, 0, \frac{1}{2}, \frac{1}{4}$.

5-21. For the system of Subsection 5.6.3, find, using equation (5-24), the transfer function of the feedback compensator relating the plant input $\mathbf{U}(z)$ and output $\mathbf{Y}(z)$. Then draw the root locus of the system by introducing a gain K and finding the roots for $K = 1$.

5-22. For the system of problem 5-20, find, using equation (5-24), the transfer function of the feedback compensator relating the plant input $\mathbf{U}(z)$ and output $\mathbf{Y}(z)$.

5-23. Design a reduced-order observer for the plant of problem 5-20 so that the feedback system has the eigenvalues $\lambda = 0, 0, \frac{1}{2}$.

5-24. For the plant

$$\begin{bmatrix} x_1(k+1) \\ x_2(k+1) \\ x_3(k+1) \end{bmatrix} = \begin{bmatrix} 1 & 0 & 2 \\ 0 & 1 & 0 \\ 1 & 0 & 2 \end{bmatrix} \begin{bmatrix} x_1(k) \\ x_2(k) \\ x_3(k) \end{bmatrix} + \begin{bmatrix} 1 \\ 0 \\ 1 \end{bmatrix} u(k)$$

$$y(k) = [3 \quad 1 \quad 1] \begin{bmatrix} x_1(k) \\ x_2(k) \\ x_3(k) \end{bmatrix} + 4u(k)$$

design a full-order state observer feedback arrangement so that the composite system has the eigenvalues $\lambda = 0, 0, \frac{1}{2}, \frac{1}{2}, \pm j\frac{1}{2}$.

5-25. Design a second-order observer for the plant of problem 5-24 so that the feedback system has the eigenvalues $\lambda = 0, 0, \frac{1}{2}, \pm j\frac{1}{2}$.

5-26. For the plant

$$\begin{bmatrix} x_1(k+1) \\ x_2(k+1) \\ x_3(k+1) \end{bmatrix} = \begin{bmatrix} 2 & -3 & 1 \\ k & -1 & 1 \\ -1 & 0 & 0 \end{bmatrix} \begin{bmatrix} x_1(k) \\ x_2(k) \\ x_3(k) \end{bmatrix} + \begin{bmatrix} \frac{1}{k+2} \\ 0 \\ -1 \end{bmatrix} u(k)$$

$$y(k) = [2 \quad (-1)^k \quad 0] \begin{bmatrix} x_1(k) \\ x_2(k) \\ x_3(k) \end{bmatrix} - u(k)$$

design a deadbeat state observer and find the observer gains $\mathbf{g}(0)$, $\mathbf{g}(1)$, and $\mathbf{g}(2)$. The solution obtained depends on your choice of basis vectors.

5-27. For the plant and observer of problem 5-26, find the next three observer gains, $\mathbf{g}(3)$, $\mathbf{g}(4)$, and $\mathbf{g}(5)$ by

    a. restarting the initial observer gain algorithm;

    b. using the ongoing algorithm;

    c. using the ongoing algorithm with a different ordering of the basis vectors.

5-28. Repeat problem 5-26 for the step-invariant plant

$$\begin{bmatrix} x_1(k+1) \\ x_2(k+1) \\ x_3(k+1) \end{bmatrix} = \begin{bmatrix} 1 & -1 & 1 \\ 0 & 1 & 1 \\ 3 & 0 & -1 \end{bmatrix} \begin{bmatrix} x_1(k) \\ x_2(k) \\ x_3(k) \end{bmatrix} + \begin{bmatrix} 1 \\ 2 \\ 0 \end{bmatrix} u(k)$$

$$y(k) = \begin{bmatrix} -1 & 0 & 1 \end{bmatrix} \begin{bmatrix} x_1(k) \\ x_2(k) \\ x_3(k) \end{bmatrix}$$

Verify that the gain $\mathbf{g}(2)$ places all the eigenvalues of

$$\mathbf{F} = \mathbf{A} - \mathbf{g}(2)\mathbf{c}^{\dagger}$$

at $\lambda = 0$.

5-29. For the plant with the state equations in problem 5-26 but outputs

$$\begin{bmatrix} y_1(k) \\ y_2(k) \end{bmatrix} = \begin{bmatrix} 2 & (-1)^k & 0 \\ 0 & 1 & k \end{bmatrix} \begin{bmatrix} x_1(k) \\ x_2(k) \\ x_3(k) \end{bmatrix} + \begin{bmatrix} 0 \\ 1 \end{bmatrix} u(k)$$

design a deadbeat state observer that converges in two steps. Find also an ongoing observer gain for the next step.

5-30. For the plant of problem 5-26, specify the various matrices defining a reduced-order deadbeat state observer for steps 0 through 4.

5-31. Design observer feedback so that the plant

$$\begin{bmatrix} x_1(k+1) \\ x_2(k+1) \end{bmatrix} = \begin{bmatrix} \dfrac{1}{k+1} & 1 \\ 0 & \dfrac{1}{4} \end{bmatrix} \begin{bmatrix} x_1(k) \\ x_2(k) \end{bmatrix} + \begin{bmatrix} 1 \\ (k+1) \end{bmatrix} u(k)$$

$$y(k) = \begin{bmatrix} 2 & -3 \end{bmatrix} \begin{bmatrix} x_1(k) \\ x_2(k) \end{bmatrix} + \frac{1}{3} u(k)$$

with the feedback is deadbeat. Specify the various matrices involved for steps 0 through 4. Find also the state equations that describe the composite feedback system.

5-32. For the spacecraft example in Section 5.8, repeat the simulation so that the observer has zero initial conditions. Compare the results with those in Figure 5-22.

5-33. For the spacecraft example in Section 5.8, design a digital controller for the flexible modes that uses a full-order state observer instead. Let the observer eigenvalues be from the set $\{\pm 0.3, \pm 0.4, \pm 0.5, \pm 0.6\}$. Compare the performances of the two controllers.

5-34. For the spacecraft example in Section 5.8, design a digital controller for the flexible modes that uses a reduced-order state observer instead. Let the observer eigenvalues be from the set $\{\pm 0.3, \pm 0.4, \pm 0.5, \pm 0.6\}$. Which observer would you select?

5-35. A single-axis dynamic model of a flexible spacecraft is shown in the figure below where the subscripts r and f denote rigid and flexible quantities, respectively. The symbols ϕ_{si} and ϕ_{ai} are the modal coefficients of mode i at the sensor and actuator locations, respectively. Two colocated sensors are available, one that measures the spacecraft attitude $\theta_{s/c}$ and the other that measures the spacecraft rate $\dot{\theta}_{s/c}$.

    a. Write state variable equations for the spacecraft using state variables θ_r, $\dot{\theta}_r$, η_1, $\dot{\eta}_1$, η_2, $\dot{\eta}_2$, . . . , η_n, $\dot{\eta}_n$.

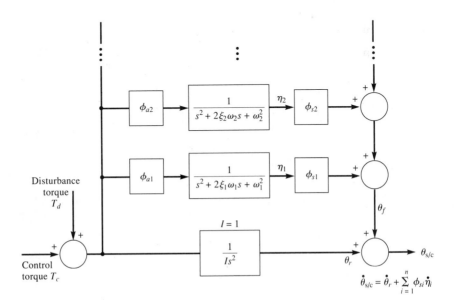

b. Ignoring all flexible modes, design a digital controller so that the bandwidth of the rigid body control system is 2.5 rad/sec and the closed-loop damping ratio is 0.707. Choose the sampling interval $T = 0.04$ sec. Assume reasonable initial conditions.

c. Through simulation, investigate the controller performance in the presence of one mode with parameters

$$\zeta_1 = 0.0025 \qquad \omega_1 = 6.28 \text{ rad/sec} \qquad \phi_{si} = \phi_{ai} = 0.03$$

and the same initial conditions as in part **b**.

d. Suppose that the flexible mode in part **c** has the following parameters instead:

$$\zeta_1 = 0.0025 \qquad \omega_1 = 1.26 \text{ rad/sec} \qquad \phi_{si} = \phi_{ai} = 0.6$$

Using the same control requirements as in part **b**, investigate the performance of the control system in the presence of the flexible mode. Modify the control system if necessary, so that the flexible mode has a damping ratio greater than or equal to 0.4.

Digital Tracking System Design

6.1 Preview

In this chapter, it is assumed that the first concern of tracking system design, satisfactory zero-input response by feedback system eigenvalue placement, has been achieved. The second concern, the zero-state tracking response, is the subject of this chapter. Three basic tracking system design methods are explored in detail:

1. Ideal tracking system design
2. Response model design
3. Reference model design

When a solution exists, ideal tracking system design achieves exact zero-state tracking of any reference input. It involves constructing an inverse filter for the plant. This may require an unstable or noncausal solution. An ideal tracking solution can also have other undesirable properties, such as unreasonably large gains, highly oscillatory plant control inputs, and the necessity of canceling plant poles and zeros when the plant model is not known accurately.

In response model design, the entire tracking system is made to have the same z-transfer functions as an acceptable model system. The added design freedom under this approach over ideal tracking system design allows solutions when the ideal system is unstable, noncausal, or otherwise undesirable. Higher-order solutions can improve performance. Given the plant and the model, design feasibility is governed by simultaneous linear algebraic equations. The difficulty with this method is in choosing suitable model systems.

Reference model design concentrates on the class of representative reference input signals that are to be tracked exactly by the plant's zero-state response. A state variable model is made for the class of inputs of interest, then the plant system is itself considered to be an observer of the reference input signal model. Thus, the powerful methods of observer theory are used twice in the design process. First, observer feedback is used to place the feedback system eigenvalues at locations chosen by the designer. The character of the feedback system's zero-input response is thereby selected. Then the feedback system, with the original plant inputs and additional inputs to each of the observer states, is made an observer of the input signal model by selecting plant and observer input gains. If the order of the reference model is high, the order of the controller is raised until the reference model can be observed as desired. All design options are expressed as solutions of linear algebraic equations.

Plant disturbances are undesired, inaccessible plant input signals that the plant should *not* track. When they are modeled, disturbance effects on the plant state can be observed and reduced. A phase-locked loop system design provides an interesting disturbance model application example.

A great deal of creativity and engineering judgment is involved in modern control system design, perhaps now more than ever. These sections on tracking system design are intended to convey not only the methods, but some of the flavor of the design process.

6.2 Ideal Tracking System Design

A *tracking system* is one in which the plant outputs are controlled so that they become and remain nearly equal to externally applied *reference* signals $r(k)$ as shown in Figure 6-1(a). These outputs are said to "track" or "follow" the reference inputs. As a plant signal that is to track an external input is not necessarily one of the accessible plant outputs, we distinguish between the *measurement outputs* $y(k)$ that are available for processing and feedback and the *tracking outputs* $\bar{y}(k)$ that are to perform the tracking. It is generally a good idea to measure the tracking outputs and to feed them back, but it is not necessary to do so.

A typical tracking control system design problem is to determine and specify a controller that results in a feedback tracking system with prescribed performance requirements.

Tracking system design has two basic concerns:

1. Obtaining acceptable zero-input system response
2. Obtaining acceptable zero-state system response to reference inputs

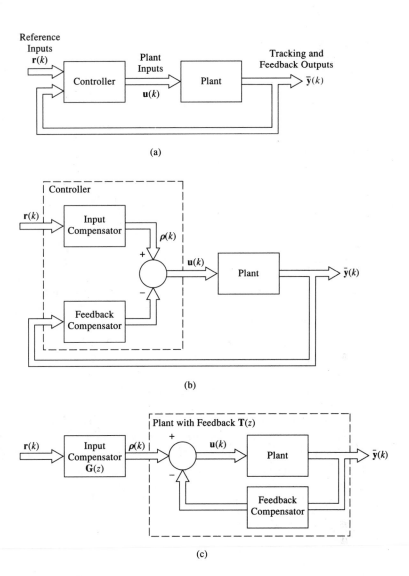

(a)

(b)

(c)

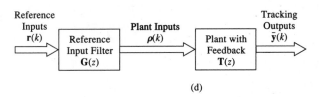

(d)

Figure 6-1 Controlling a multiple-input/multiple-output plant. The output $y(k)$ is to track the reference input $r(k)$. (a) A tracking system using the reference inputs and plant outputs. (b) Representing a controller with a feedback compensator and an input compensator. (c) Feedback compensator combined with plant to produce a plant-with-feedback transfer function matrix $T(z)$. (d) Using a reference input filter for tracking.

407

The character of a system's zero-input response is determined by its eigenvalue locations, so the first concern of tracking system design is choosing a feedback portion (state feedback or a feedback compensator) that results in an acceptable transfer function. The zero-input system response was discussed in the previous chapter. The zero-state tracking system response is the subject of this chapter.

In the feedback design of the previous chapter, it was important to distinguish between the plant inputs **u** and the feedback system inputs ρ. When observer feedback was employed, the plant state vector was **x** and the observer state was ξ. In tracking system design, it is important to distinguish between the plant inputs **u** and the feedback system inputs ρ. The relationship between **u** and ρ is illustrated in Figure 6-1. The original plant input is **u**, and the input to the plant with feedback is ρ. Now we assume that any feedback has already been designed. The "plant" will now mean the original plant and its feedback, if used. Hence, the plant inputs are ρ.

As indicated in Figure 6-1(b), a linear, time-invariant feedback controller of a multiple-input/multiple-output plant is described by two transfer function matrices: one relating the reference inputs to the plant inputs, and the other relating the output feedback vector to the plant inputs. The feedback compensator is used for system pole placement, as it was in the previous chapters. The input compensator, on the other hand, is designed to achieve good tracking of the reference inputs by the system outputs.

The output of any linear system can always be decomposed into two parts, the zero-input component, due to the initial conditions alone, and the zero-state component, due to the input alone. That is,

$$\overline{\mathbf{y}}(k) = \overline{\mathbf{y}}_{\text{zero-input}}(k) + \overline{\mathbf{y}}_{\text{zero-state}}(k)$$

Usually, the plant initial conditions are unknown, and there is little one can do about the zero-input response term beyond selecting its modes through placing the feedback system eigenvalues in the regulator portion of the design. *Ideal tracking* is obtained if we can arrange things so that

$$\overline{\mathbf{y}}_{\text{zero-state}}(k) = \mathbf{r}(k)$$

The tracking outputs $\overline{\mathbf{y}}(k)$ have an initial transient error due to any nonzero plant initial conditions, after which they are equal to the reference inputs $\mathbf{r}(k)$, no matter what those inputs are.

Suppose that a plant with feedback has the z-transfer function matrix $\mathbf{T}(z)$ relating the tracking outputs to the plant inputs as shown in Figure 6-1(c). Then

$$\overline{\mathbf{Y}}(z) = \mathbf{T}(z)\boldsymbol{\rho}(z)$$

An input compensator or a *reference input filter* as shown in Figure 6-1(d) with z-transfer function matrix $\mathbf{G}(z)$, for which

$$\boldsymbol{\rho}(z) = \mathbf{G}(z)\mathbf{R}(z)$$

gives

$$\overline{\mathbf{Y}}(z) = \mathbf{T}(z)\mathbf{G}(z)\mathbf{R}(z)$$

The reference input filter does not change the plant eigenvalues which are assumed to have been previously placed with output or observer feedback. Ideal tracking is achieved if

$$\mathbf{T}(z)\mathbf{G}(z) = \mathbf{I}$$

where $\mathbf{I}$ is the identity matrix of dimension equal to the number of reference inputs and tracking outputs. That is, ideal tracking is obtained if the reference input filter is an *inverse filter* for the plant.

In this section, we first consider the case of a single-input plant with a single tracking output. Then multiple-input plants with a single tracking output are shown to offer additional design freedom. In subsequent sections, it will be shown that observers used for plant feedback provide additional plant inputs and how arrangements other than the reference input filter can be used.

6.2.1 Ideal Single-Input/Single-Output Tracking

For a single-input plant with a single tracking output, if the z-transfer function that relates the tracking output to the plant input is

$$T(z) = \frac{\overline{Y}(z)}{\rho(z)}\bigg|_{\substack{\text{zero initial} \\ \text{conditions}}}$$

then

$$\overline{Y}_{\text{zero-state}}(z) = T(z)\rho(z)$$

For ideal tracking, the inverse filter has the z-transfer function

$$G(z) = \frac{\rho(z)}{R(z)}\bigg|_{\substack{\text{zero initial} \\ \text{conditions}}} = \frac{1}{T(z)}$$

so that

$$\overline{Y}_{\text{zero-state}}(z) = T(z)G(z)R(z) = R(z)$$

For example, consider the second-order/single-input plant

$$\begin{bmatrix} x_1(k+1) \\ x_2(k+1) \end{bmatrix} = \begin{bmatrix} 0 & 1 \\ \frac{1}{4} & 0 \end{bmatrix} \begin{bmatrix} x_1(k) \\ x_2(k) \end{bmatrix} + \begin{bmatrix} 0 \\ 1 \end{bmatrix} \rho(k)$$

$$\overline{y}(k) = \begin{bmatrix} \frac{7}{18} & -\frac{1}{3} \end{bmatrix} \begin{bmatrix} x_1(k) \\ x_2(k) \end{bmatrix} + 2\rho(k)$$

where it is desired that the single tracking output $\overline{y}(k)$ track a reference input signal $r(k)$. For this example, the system description is in controllable form, which is convenient, but the use of special forms is not necessary for this design process. The z-transfer function of the plant is

$$T(z) = \frac{-\frac{1}{3}z + 7/18}{z^2 - \frac{1}{4}} + 2 = \frac{2(z - \frac{1}{3})(z + \frac{1}{6})}{(z + \frac{1}{2})(z - \frac{1}{2})}$$

The inverse filter has the z-transfer function

$$G(z) = \frac{1}{T(z)} = \frac{\frac{1}{2}(z + \frac{1}{2})(z - \frac{1}{2})}{(z - \frac{1}{3})(z + \frac{1}{6})} = \frac{1}{2} + \frac{(1/12)z - (7/72)}{z^2 - \frac{1}{6}z - (1/18)}$$

It is realized by the filter (also chosen to be in controllable form, for convenience)

$$\begin{bmatrix} \mu_1(k+1) \\ \mu_2(k+1) \end{bmatrix} = \begin{bmatrix} 0 & 1 \\ \frac{1}{18} & \frac{1}{6} \end{bmatrix} \begin{bmatrix} \mu_1(k) \\ \mu_2(k) \end{bmatrix} + \begin{bmatrix} 0 \\ 1 \end{bmatrix} r(k)$$

$$\rho(k) = \begin{bmatrix} -\frac{7}{72} & \frac{1}{12} \end{bmatrix} \begin{bmatrix} \mu_1(k) \\ \mu_2(k) \end{bmatrix} + \frac{1}{2}r(k)$$

The inverse filter gives ideal tracking

$$\overline{y}_{\text{zero-state}}(k) = r(k)$$

as is demonstrated in the typical response plot of Figure 6-2. For this plot and others later, the inverse filter initial conditions $\mu(0)$ were chosen to be zero, as is easily arranged in practice. The plant initial conditions were chosen arbitrarily and are not zero.

If there are any plant z-transfer function zeros outside the unit circle on the complex plane, the inverse reference input filter will be unstable because the filter has poles where the plant transfer function has zeros. For example, the plant with the transfer function

$$T(z) = \frac{(z - 2)(z + \frac{1}{3})}{(z + \frac{1}{2})(z - \frac{1}{2})}$$

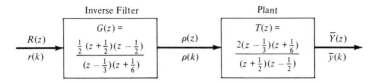

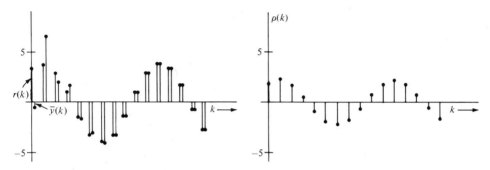

Figure 6-2 Response of a tracking system that uses an inverse filter.

does not have all of its zeros inside the unit circle. As a consequence, the inverse filter, which has the z-transfer function

$$G(z) = \frac{1}{T(z)} = \frac{(z + \frac{1}{2})(z - \frac{1}{2})}{(z - 2)(z + \frac{1}{3})}$$

is unstable. The plant input signal $\rho(k)$ generated by this filter grows without bound, as in the typical response shown in Figure 6-3. Although the entire system may work for a while, eventually the plant inputs will be too large for the plant to accommodate.

If the plant z-transfer function has a lower-order polynomial in the numerator than in the denominator, the resulting inverse reference input filter has a lower-order polynomial in the denominator than in the numerator, making the filter noncausal. The reference input must then be available to the filter one or more steps in advance. In some applications, the necessity of having advance samples of the reference input $r(k)$ available is a disadvantage; in others, it is not.

For example, consider the plant with the z-transfer function

$$T(z) = \frac{z - \frac{1}{3}}{z^2 - \frac{1}{4}} = \frac{z - \frac{1}{3}}{(z + \frac{1}{2})(z - \frac{1}{2})}$$

Fortunately, the plant zero is inside the unit circle on the complex plane; but the inverse of $T(z)$ has a numerator polynomial of higher

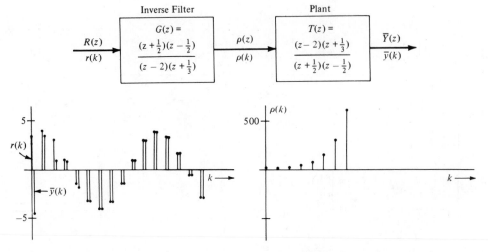

Figure 6-3 Response of a tracking system where the inverse filter is unstable.

degree than its denominator polynomial

$$G(z) = \frac{1}{T(z)} = \frac{(z + \frac{1}{2})(z - \frac{1}{2})}{z - \frac{1}{3}}$$

Rather than dealing with a noncausal reference input filter, the causal filter

$$\tilde{G}(z) = \frac{1}{zT(z)} = \frac{(z + \frac{1}{2})(z - \frac{1}{2})}{z(z - \frac{1}{3})}$$

is used, but then $r(k + 1)$ is needed as the filter input. In terms of z-transforms

$$\rho(z) = zR(z)\tilde{G}(z) = \frac{1}{T(z)} R(z)$$

The typical response of this system is shown in Figure 6-4.

Suppose an nth-order plant is deadbeat (all of its eigenvalues at $\lambda = 0$) and has an inverse reference input filter that is started with zero initial conditions. Then because the filter's zero-state response is zero and the plant's zero-state response dies to zero after n steps, exact tracking is achieved after n steps.

As an example of a tracking system with a deadbeat plant with

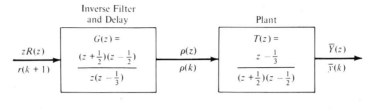

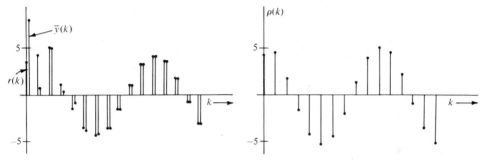

Figure 6-4 Response of a tracking system where the tracking input $r(k)$ must be supplied one step in advance if the inverse filter is to be causal.

feedback, consider the plant and accessible outputs

$$\begin{bmatrix} x_1(k+1) \\ x_2(k+1) \end{bmatrix} = \begin{bmatrix} 0 & 1 \\ \frac{1}{3} & \frac{2}{3} \end{bmatrix} \begin{bmatrix} x_1(k) \\ x_2(k) \end{bmatrix} + \begin{bmatrix} 0 \\ 1 \end{bmatrix} u(k) = \mathbf{A}\mathbf{x}(k) + \mathbf{b}u(k)$$

$$\bar{y}(k) = \begin{bmatrix} \frac{1}{2} & \frac{3}{2} \end{bmatrix} \begin{bmatrix} x_1(k) \\ x_2(k) \end{bmatrix} + u(k) = \bar{\mathbf{c}}^\dagger \mathbf{x}(k) + \mathbf{e}^\dagger \mathbf{x}(k) + \rho(k)$$

which are in controllable form, for convenience. The state feedback

$$u(k) = -\frac{1}{3}x_1(k) - \frac{2}{3}x_2(k) + \rho(k) = \mathbf{e}^\dagger \mathbf{x}(k) + \rho(k)$$

where $\rho(k)$ is an external input, places both plant eigenvalues at $\lambda = 0$ which can be easily verified by substitution:

$$\begin{bmatrix} x_1(k+1) \\ x_2(k+1) \end{bmatrix} = \left\{ \begin{bmatrix} 0 & 1 \\ \frac{1}{3} & \frac{2}{3} \end{bmatrix} + \begin{bmatrix} 0 \\ 1 \end{bmatrix} \begin{bmatrix} -\frac{1}{3} & -\frac{2}{3} \end{bmatrix} \right\} \begin{bmatrix} x_1(k) \\ x_2(k) \end{bmatrix} + \begin{bmatrix} 0 \\ 1 \end{bmatrix} \rho(k)$$

With feedback, then

$$\begin{bmatrix} x_1(k+1) \\ x_2(k+1) \end{bmatrix} = \begin{bmatrix} 0 & 1 \\ 0 & 0 \end{bmatrix} \begin{bmatrix} x_1(k) \\ x_2(k) \end{bmatrix} + \begin{bmatrix} 0 \\ 1 \end{bmatrix} \rho(k)$$

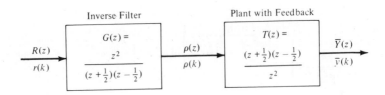

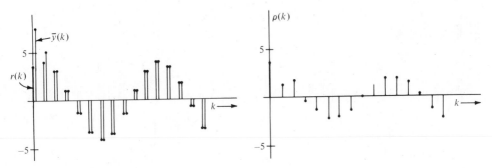

Figure 6-5 Response of a tracking system consisting of an inverse filter and a deadbeat plant.

The output that is to track the input $r(k)$ in a tracking system may or may not be accessible for feedback. For this example, suppose that it is the signal $\bar{y}(k)$ that is to track $r(k)$. The z-transfer function relating $\bar{y}(k)$ and $\rho(k)$ is

$$T(z) = \frac{z^2 + \frac{5}{6}z + \frac{1}{6}}{z^2} = \frac{(z + \frac{1}{2})(z + \frac{1}{3})}{z^2}$$

so that the inverse reference input filter has the z-transfer function

$$G(z) = \frac{1}{T(z)} = \frac{z^2}{(z + \frac{1}{2})(z + \frac{1}{3})}$$

which is both stable and causal. The typical response of this system is shown in Figure 6-5. Because the reference input filter is begun with zero initial conditions and the second-order plant is deadbeat, after two steps the tracking output $\bar{y}(k)$ equals the reference input $r(k)$.

6.2.2 Ideal Tracking with Multiple Plant Inputs

When more than one plant input can be used to achieve tracking of a single reference input, there is freedom in the choice of an inverse filter, and it may be possible to obtain an acceptable design when the use of a single plant input alone would not yield a useable solution. For example, consider the following two-input plant with feedback, for which it is desired that the tracking output $\bar{y}(k)$ track a reference input $r(k)$

$$\begin{bmatrix} x_1(k+1) \\ x_2(k+1) \end{bmatrix} = \begin{bmatrix} 0 & 1 \\ \frac{1}{4} & 0 \end{bmatrix}\begin{bmatrix} x_1(k) \\ x_2(k) \end{bmatrix} + \begin{bmatrix} 1 & 2 \\ -1 & 0 \end{bmatrix}\begin{bmatrix} \rho_1(k) \\ \rho_2(k) \end{bmatrix} = \mathbf{A}\mathbf{x}(k) + \mathbf{B}\boldsymbol{\rho}(k)$$

$$\bar{y}(k) = \begin{bmatrix} 1 & -1 \end{bmatrix}\begin{bmatrix} x_1(k) \\ x_2(k) \end{bmatrix} + \begin{bmatrix} 0 & 1 \end{bmatrix}\begin{bmatrix} \rho_1(k) \\ \rho_2(k) \end{bmatrix} = \bar{\mathbf{c}}^\dagger\mathbf{x}(k) + \bar{\mathbf{d}}^\dagger\boldsymbol{\rho}(k)$$

The plant z-transfer functions are

$$\mathbf{T}(z) = \bar{\mathbf{c}}^\dagger(z\mathbf{I} - \mathbf{A})^{-1}\mathbf{B} + \bar{\mathbf{d}}^\dagger = \begin{bmatrix} \dfrac{2z - \frac{5}{4}}{z^2 - \frac{1}{4}} & \dfrac{z^2 + 2z - \frac{3}{4}}{z^2 - \frac{1}{4}} \end{bmatrix} = \begin{bmatrix} T_1(z) & T_2(z) \end{bmatrix}$$

The first z-transfer function $T_1(z)$ has a numerator polynomial of an order one less than that of the denominator polynomial order. If the plant input $\rho_1(k)$ was used alone to obtain tracking, the reference input $r(k)$ would have to be provided to the filter one step in advance of the step at which the plant was to track it. The second z-transfer function $T_2(z)$ has a zero outside the unit circle on the complex plane. If the plant input ρ_2 alone was used for tracking, the inverse filter would be unstable.

If both inputs to this plant were used to obtain tracking, the single-input/two-output inverse filter has the z-transfer functions $G_1(z)$ and $G_2(z)$, as in Figure 6-6, that must satisfy

$$\bar{Y}(z) = [T_1(z)G_1(z) + T_2(z)G_2(z)]R(z) = R(z)$$

or

$$T_1(z)G_1(z) + T_2(z)G_2(z) = 1$$

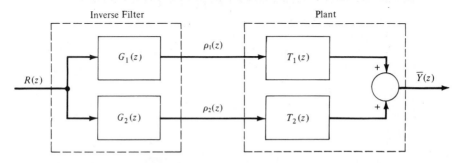

Figure 6-6 Inverse filter and plant z-transfer functions for a two-input plant with one tracking output.

or

$$G_1(z)\left(\frac{2z - \frac{5}{4}}{z^2 - \frac{1}{4}}\right) + G_2(z)\left(\frac{z^2 + 2z - \frac{3}{4}}{z^2 - \frac{1}{4}}\right) = 1$$

In terms of the numerator and denominator polynomials of the inverse filter z-transfer functions

$$G_1(z) = \frac{n_1(z)}{d(z)} \qquad G_2(z) = \frac{n_2(z)}{d(z)}$$

where each shares the denominator polynomial $d(z)$, ideal tracking requires that

$$n_1(z)\left(2z - \frac{5}{4}\right) + n_2(z)\left(z^2 + 2z - \frac{3}{4}\right) = d(z)\left(z^2 - \frac{1}{4}\right) \qquad \textbf{(6-1)}$$

To design an inverse filter of the lowest possible order, we begin by attempting to satisfy equation (6-1) with a zero-order filter with "polynomials" that are constants

$$\begin{cases} n_1(z) = \alpha_1 \\ n_2(z) = \alpha_2 \\ d(z) = 1 \end{cases}$$

In terms of these, equation (6-1) becomes

$$\alpha_1 \left(2z - \frac{5}{4}\right) + \alpha_2 \left(z^2 + 2z - \frac{3}{4}\right) = z^2 - \frac{1}{4}$$

Equating coefficients of like powers of z, the following simultaneous linear algebraic equations result:

$$\begin{cases} \alpha_2 = 1 \\ 2\,\alpha_1 + 2\,\alpha_2 = 0 \\ -\frac{5}{4}\,\alpha_1 - \frac{3}{4}\,\alpha_2 = -\frac{1}{4} \end{cases}$$

These equations do not have a solution, so we try a first-order filter with

$$\begin{cases} n_1(z) = \alpha_1 z + \alpha_2 \\ n_2(z) = \alpha_3 z + \alpha_4 \\ d(z) = z + \alpha_5 \end{cases}$$

instead. The requirement of equation (6-1) is now

$$(\alpha_1 z + \alpha_2)\left(2z - \frac{5}{4}\right) + (\alpha_3 z + \alpha_4)\left(z^2 + 2z - \frac{3}{4}\right) = (z + \alpha_5)\left(z^2 - \frac{1}{4}\right)$$

and equating coefficients yields

$$
\begin{cases}
\alpha_3 = 1 \\
2\,\alpha_1 + 2\,\alpha_3 + \alpha_4 = \alpha_5 \\
-\dfrac{5}{4}\,\alpha_1 + 2\,\alpha_2 - \dfrac{3}{4}\,\alpha_3 + 2\,\alpha_4 = -\dfrac{1}{4} \\
\dfrac{5}{4}\,\alpha_2 + \dfrac{3}{4}\,\alpha_4 = \dfrac{1}{4}\,\alpha_5
\end{cases}
$$

These are four simultaneous linear algebraic equations in five unknowns.

For the choice $\alpha_5 = 0$ so that

$$d(z) = z$$

the equations have the solution

$$
\alpha_1 = -\frac{14}{19} \qquad \alpha_2 = \frac{6}{19} \qquad \alpha_3 = 1
$$

$$
\alpha_4 = -\frac{10}{19} \qquad \alpha_5 = 0
$$

which specifies the inverse filter transfer functions

$$
G_1(z) = \frac{\alpha_1 z + \alpha_2}{z + \alpha_5} = \frac{-(14/19)z + (6/19)}{z} = -\frac{14}{19} + \frac{(6/19)}{z}
$$

$$
G_2(z) = \frac{\alpha_3 z + \alpha_4}{z + \alpha_5} = \frac{z - (10/19)}{z} = 1 + \frac{-(10/19)}{z}
$$

These can be realized by the state variable equations

$$\mu(k + 1) = r(k)$$

$$
\begin{bmatrix} \rho_1(k) \\ \rho_2(k) \end{bmatrix} = \begin{bmatrix} \frac{6}{19} \\ -\frac{10}{19} \end{bmatrix} \mu(k) + \begin{bmatrix} -\frac{14}{19} \\ 1 \end{bmatrix} r(k)
$$

The typical response of the plant with this inverse reference input filter is shown in Figure 6-7. By making use of the added design freedom of the additional input, it was possible in this case to give tracking with a stable inverse filter that does not require knowing the reference input $r(k)$ in advance.

6.2.3 Ideal Tracking of Multiple Reference Signals

It is sometimes possible to achieve simultaneous ideal tracking of two or more different reference inputs by an equal number of different plant tracking outputs. Figure 6-8(a) shows the interconnection of a two-

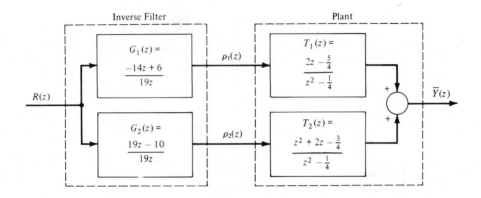

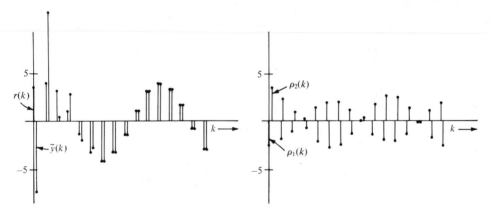

Figure 6-7 Response of an ideal tracking system for a two-input plant with a single tracking output.

input/two-tracking-output plant and a two-input/two-output filter. To achieve simultaneous tracking of two reference inputs with this arrangement, the inverse reference input filter z-transfer functions G_{11}, G_{21}, G_{12}, and G_{22} must be chosen so that

$$
\begin{cases}
\begin{aligned}
\overline{Y}_1(z) &= T_{11}(z)\rho_1(z) + T_{12}(z)\rho_2(z) \\
&= T_{11}(G_{11}R_1 + G_{12}R_2) + T_{12}(G_{21}R_1 + G_{22}R_2) \\
&= (T_{11}G_{11} + T_{12}G_{21})R_1 + (T_{11}G_{12} + T_{12}G_{22})R_2 = R_1 \\
\overline{Y}_2(z) &= T_{21}(z)\rho_1(z) + T_{22}(z)\rho_2(z) \\
&= T_{21}(G_{11}R_1 + G_{12}R_2) + T_{22}(G_{21}R_1 + G_{22}R_2) \\
&= (T_{21}G_{11} + T_{22}G_{21})R_1 + (T_{21}G_{12} + T_{22}G_{22})R_2 = R_2
\end{aligned}
\end{cases}
$$

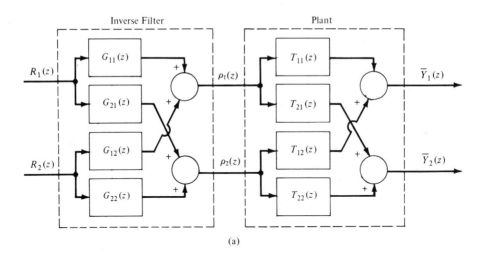

(a)

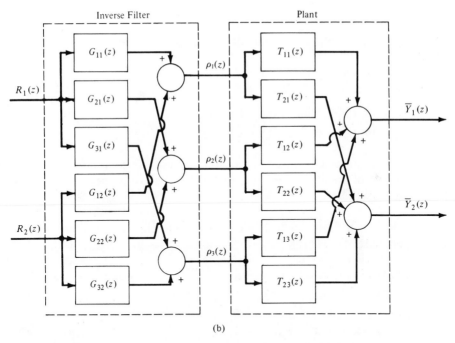

(b)

Figure 6-8 Simultaneous tracking of two reference signals. (a) System with two plant inputs. (b) System with three plant inputs.

or

$$\begin{cases} T_{11}(z)G_{11}(z) + T_{12}(z)G_{21}(z) = 1 \\ T_{11}(z)G_{12}(z) + T_{12}(z)G_{22}(z) = 0 \\ T_{21}(z)G_{11}(z) + T_{22}(z)G_{21}(z) = 0 \\ T_{21}(z)G_{12}(z) + T_{22}(z)G_{22}(z) = 1 \end{cases} \qquad \textbf{(6-2)}$$

Should a noncausal filter result, the actual filter used would incorporate additional $z = 0$ poles as necessary and a corresponding advance of the reference input $r(k)$. Expressing each of these transfer functions as ratios of polynomials

$$T_{11}(z) = \frac{p_{11}(z)}{q(z)} \qquad T_{21}(z) = \frac{p_{21}(z)}{q(z)} \qquad T_{12}(z) = \frac{p_{12}(z)}{q(z)} \qquad T_{22}(z) = \frac{p_{22}(z)}{q(z)}$$

$$G_{11}(z) = \frac{g_{11}(z)}{d(z)} \qquad G_{21}(z) = \frac{g_{21}(z)}{d(z)} \qquad G_{12}(z) = \frac{g_{12}(z)}{d(z)} \qquad G_{22}(z) = \frac{g_{22}(z)}{d(z)}$$

indicated that the requirements of equation (6-2) are

$$\begin{cases} g_{11}(z)p_{11}(z) + g_{21}(z)p_{12}(z) = d(z)q(z) \\ g_{12}(z)p_{11}(z) + g_{22}(z)p_{12}(z) = 0 \\ g_{11}(z)p_{21}(z) + g_{21}(z)p_{22}(z) = 0 \\ g_{12}(z)p_{21}(z) + g_{22}(z)p_{22}(z) = d(z)q(z) \end{cases}$$

These requirements often do not have acceptable solutions in practice because of the likelihood that any solution will result in an unstable or noncausal filter.

When two reference signals are to be tracked with a three-input plant, as in Figure 6-8(b), the inverse reference input filter requirements are

$$\begin{cases} T_{11}G_{11} + T_{12}G_{21} + T_{13}G_{31} = 1 \\ T_{11}G_{12} + T_{12}G_{22} + T_{13}G_{32} = 0 \\ T_{21}G_{11} + T_{22}G_{21} + T_{23}G_{31} = 0 \\ T_{21}G_{12} + T_{22}G_{22} + T_{23}G_{32} = 1 \end{cases} \qquad \textbf{(6-3)}$$

In terms of the numerator and denominator polynomials of each z-transfer function

$$T_{ij}(z) = \frac{p_{ij}(z)}{q(z)} \qquad G_{ij}(z) = \frac{g_{ij}(z)}{d(z)}$$

the requirements of equation (6-3) are

$$\begin{cases} g_{11}p_{11} + g_{21}p_{12} + g_{31}p_{13} = dq \\ g_{12}p_{11} + g_{22}p_{12} + g_{32}p_{13} = 0 \\ g_{11}p_{21} + g_{21}p_{22} + g_{31}p_{23} = 0 \\ g_{12}p_{21} + g_{22}p_{22} + g_{32}p_{23} = dq \end{cases}$$

These are more likely to have an acceptable solution.

One should certainly have digital computer aid for the computations needed for design of all but the most simple tracking systems, especially those that are to track two or more reference signals. Design is straightforward, though, being governed by linear algebraic equations.

6.3 Response Model Tracking System Design

Inverse filters are useful in some situations, but unfortunately a solution can be unstable and noncausal. Even when a stable, causal inverse reference input filter exists, the design might not be satisfactory if the filter is required to have poles within but near the unit circle on the complex plane, resulting in the filter having lightly damped zero-input response. A stable, causal filter design might also be unsatisfactory because a plant z-transfer function is not known accurately so that important pole-zero cancellations cannot be assured.

When ideal tracking is not possible or desirable, the designer can elect to design *response model tracking*, for which

$$\overline{\mathbf{Y}}_{\text{zero-state}}(z) = \mathbf{\Omega}(z)\mathbf{R}(z)$$

where the response model z-transfer function matrix $\mathbf{\Omega}(z)$ characterizes an acceptable relation between the zero-state components of the plant's tracking outputs and the reference inputs. The combination of a reference input filter and the plant does not have the ideal identity z-transfer function matrix that the inverse filter has. Instead, the filter and plant together behave as the response model. The reference input filter is designed so that it is stable and (if needed) causal and so that the response model, of which it is part, has acceptable response to important reference inputs, such as powers of time. In this respect, response model design is a generalization of the classical design technique of imposing requirements for a controller's steady state response to power-of-time inputs.

In this section, we first design reference input filters for response model tracking system design. We then consider the use of additional control inputs to a feedback observer's states.

6.3.1 Reference Input Filters for Model Response

For a single-input plant with a single tracking input, response model tracking system design involves relaxing the reference input filter requirement from that of the inverse filter

$$T(z)G(z) = 1$$

to

$$T(z)G(z) = \Omega(z)$$

where $\Omega(z)$ is an acceptable *response model* z-transfer function relating the tracking output to the reference input

$$\overline{Y}(z) = \Omega(z)R(z)$$

For example, consider again the single-input plant with single tracking output and the z-transfer function

$$T(z) = \frac{(z - 2)(z + \frac{1}{3})}{(z + \frac{1}{2})(z - \frac{1}{2})}$$

Requiring that a reference input filter $G(z)$ result in an acceptable response model allows the designer to avoid an unstable filter due to the zero of $T(z)$ that is outside the unit circle on the complex plane. Choosing $G(z)$ to have the factors

$$G(z) = \frac{a(z + \frac{1}{2})(z - \frac{1}{2})}{(z - b)(z + \frac{1}{3})}$$

where a and b are constants to be chosen and where the $(z - b)$ factor in the denominator has been added to make $G(z)$ causal, gives a response model z-transfer function of the form

$$T(z)G(z) = \Omega(z) = \frac{a(z - 2)}{z - b}$$

This is the z-transfer function that relates the plant tracking output to the reference input.

There are many other possibilities for $G(z)$, including not canceling one or both of the $T(z)$ poles, not canceling the $T(z)$ zero at $z = \frac{1}{2}$, and for which $G(z)$ has additional poles and zeros. This choice, however, results in the lowest-order response model for which $G(z)$ can be both stable and causal. The "best" choice of the constants a and b depends on the design objective for the system. Certainly a choice with

$$|b| < 1$$

is made so that the reference input filter is stable. Suppose the choice $b = 0$ is made, giving a pole at $z = 0$. Then $G(z)$ is of the form

$$G(z) = \frac{a(z + \frac{1}{2})(z - \frac{1}{2})}{z(z + \frac{1}{3})}$$

and $\Omega(z)$ has the form

$$\Omega(z) = \frac{a(z - 2)}{z}$$

If it is desired that any constant input sequence

$$r(k) = \beta u(k) \qquad R(z) = \frac{\beta z}{z - 1}$$

where β is a constant, produce the same steady state system output, then using

$$\overline{Y}(z) = \Omega(z)R(z) = \left[\frac{a(z - 2)}{z}\right]\left(\frac{\beta z}{z - 1}\right)$$

$$\lim_{k \to \infty} \bar{y}(k) = \lim_{z \to 1} \left[\left(\frac{z - 1}{z}\right) \overline{Y}(z)\right] = -a\beta = \beta$$

requires that

$$a = -1$$

With these considerations, the response model chosen is

$$\Omega(z) = \frac{-(z - 2)}{z}$$

and the corresponding reference input filter has the z-transfer function

$$G(z) = \frac{-(z + \frac{1}{2})(z - \frac{1}{2})}{z(z + \frac{1}{3})}$$

Figure 6-9(a) shows the step response of this tracking system. In Figure 6-9(b), the response to an arbitrary tracking input is shown. The input used is the same one used in the examples in the previous section, including one with this same plant, in Figure 6-3. Tracking of the input is not good because the input varies so rapidly. In Figure 6-9(c), the response of this system to a more slowly varying reference input is shown. Clearly, the price one pays for the added design freedom of a response model can be poorer tracking performance. Performance can be improved, however, by increasing the order of the reference input filter.

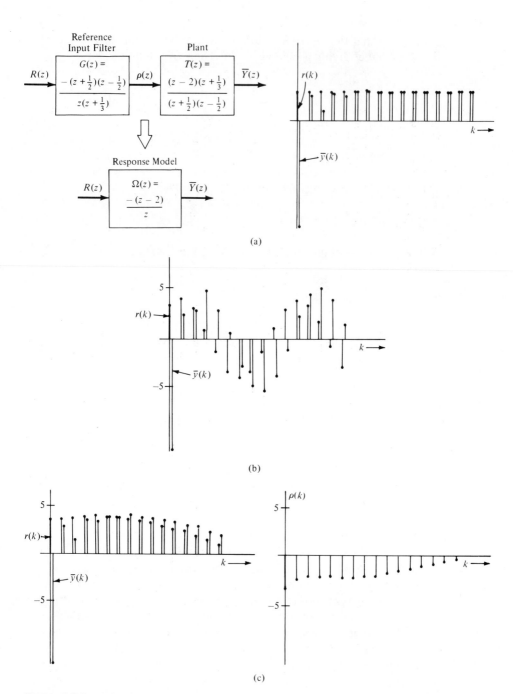

Figure 6-9 Response model tracking system in which the inverse filter for the plant is unstable. (a) Step response. (b) Response to any arbitrary tracking input. (c) Response to a more slowly varying tracking input.

For a plant in which an inverse filter is noncausal, as it is for the plant with the z-transfer function

$$T(z) = \frac{z - \frac{1}{3}}{(z + \frac{1}{2})(z - \frac{1}{2})}$$

considered earlier, a response model design

$$T(z)G(z) = \Omega(z)$$

can result in a causal reference input filter. Choosing

$$G(z) = \frac{\frac{3}{2}(z - \frac{1}{2})}{z - \frac{1}{3}}$$

so that

$$T(z)G(z) = \Omega(z) = \frac{\frac{3}{2}}{z + \frac{1}{2}}$$

gives a causal first-order plant input filter and a tracking system with zero steady state error to a constant reference input. For

$$R(z) = \frac{\beta z}{z - 1}$$

$$R(z) - \overline{Y}(z) = [1 - \Omega(z)]R(z) = \left(\frac{z - 1}{z + \frac{1}{2}}\right)\left(\frac{\beta z}{z - 1}\right)$$

$$\lim_{k \to \infty} [r(k) - \bar{y}(k)] = \lim_{z \to 1} \left[\left(\frac{z - 1}{z}\right)\left(\frac{z - 1}{z + \frac{1}{2}}\right)\left(\frac{\beta z}{z - 1}\right)\right] = 0$$

Figure 6-10 shows a typical response for this design.

One can add one or more additional pairs of poles and zeros to the response model $\Omega(z)$ to improve performance. For the previous example, choosing instead

$$G(z) = \frac{(\frac{5}{2})(z - \frac{2}{5})(z - \frac{1}{2})}{z(z - \frac{1}{3})}$$

so that

$$T(z)G(z) = \Omega(z) = \frac{\frac{5}{2}(z - \frac{2}{5})}{z(z + \frac{1}{2})}$$

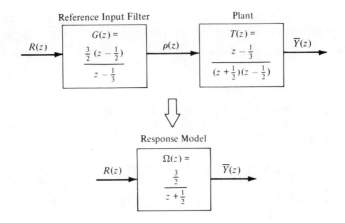

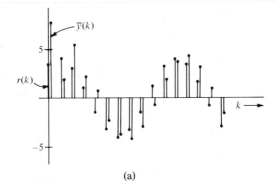

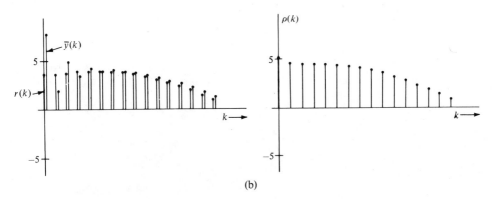

Figure 6-10 Another response model tracking system. The inverse filter for the plant is noncausal. (a) Response to any arbitrary input. (b) Response to a more slowly varying tracking input.

results in a tracking system with zero steady state error to any constant-plus-ramp reference sequence. For

$$R(z) = \frac{\beta_1 z^2 + \beta_2 z}{(z - 1)^2}$$

$$R(z) - \overline{Y}(z) = [1 - \Omega(z)]R(z) = \left[\frac{z^2 - 2z + 1}{z(z + \frac{1}{2})}\right]\left[\frac{\beta_1 z^2 + \beta_2 z}{(z - 1)^2}\right]$$

$$\lim_{k \to \infty} [r(k) - \overline{y}(k)] = \lim_{z \to 1} \left\{ \left(\frac{z - 1}{z}\right)\left[\frac{\beta_1 z^2 + \beta_2 z}{z(z + \frac{1}{2})}\right] \right\} = 0$$

Typical response of this improved tracking system is shown in Figure 6-11.

These methods are easily generalized to systems with more than a single plant input and where simultaneous tracking of two or more reference signals is desired. For the case of a two-input plant that is to track a single reference input as in Figure 6-12(a)

$$T_1(z)G_1(z) + T_2(z)G_2(z) = \Omega(z)$$

where $\Omega(z)$ is the response model. For a two-input/two-output plant, as in Figure 6-12(b), the reference input filter z-transfer functions satisfy

$$\begin{cases} T_{11}(z)G_{11}(z) + T_{12}(z)G_{21}(z) = \Omega_{11}(z) \\ T_{11}(z)G_{12}(z) + T_{12}(z)G_{22}(z) = \Omega_{12}(z) \\ T_{21}(z)G_{11}(z) + T_{22}(z)G_{21}(z) = \Omega_{21}(z) \\ T_{21}(z)G_{12}(z) + T_{22}(z)G_{22}(z) = \Omega_{22}(z) \end{cases}$$

when using a response model, rather than the inverse filter conditions where $\Omega_{11} = \Omega_{22} = 1$ and $\Omega_{12} = \Omega_{21} = 0$. The response model relates the two reference inputs to the two plant outputs as

$$\overline{Y}_1(z) = \Omega_{11}(z)R_1(z) + \Omega_{12}(z)R_2(z)$$
$$\overline{Y}_2(z) = \Omega_{21}(z)R_1(z) + \Omega_{22}(z)R_2(z)$$

For a three-input/two-output plant, the reference input filter z-transfer functions satisfy

$$\begin{cases} T_{11}G_{11} + T_{12}G_{21} + T_{13}G_{31} = \Omega_{11} \\ T_{11}G_{12} + T_{12}G_{22} + T_{13}G_{32} = \Omega_{12} \\ T_{21}G_{11} + T_{22}G_{21} + T_{23}G_{31} = \Omega_{21} \\ T_{21}G_{12} + T_{22}G_{22} + T_{23}G_{32} = \Omega_{22} \end{cases}$$

Equating coefficients of like powers of z in the transfer function results in design options that are governed by linear algebraic equations. For the more complicated response model designs, however, it may be

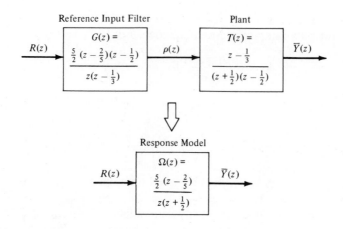

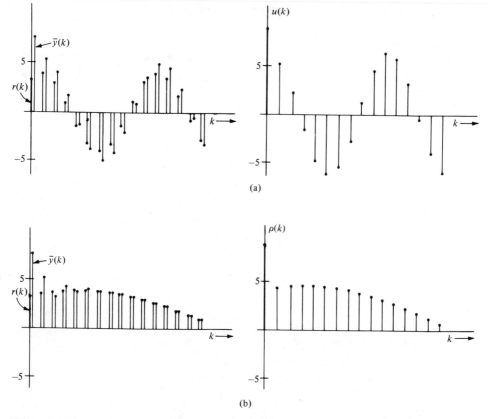

Figure 6-11 Response model tracking system for the previous plant with a higher-order reference input filter. (a) Response to an arbitrary tracking input. (b) Response to a more slowly varying tracking input.

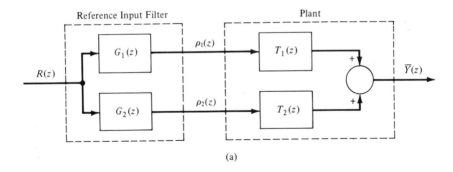

(a)

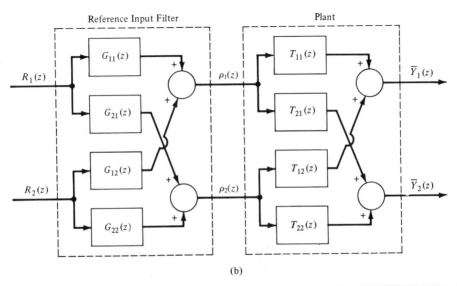

(b)

Figure 6-12 Use of a response model for more complicated design problems.
(a) Two-input/one-output plant. (b) Two-input/two-output plant.

difficult to select the best solution when the model is incompletely
specified.

There is, too, the problem of determining acceptable models. For
the case of a single reference input, filter pole location and steady state
performance requirements can be imposed, as in the examples here. Or
some other criteria can be used. When more than one reference input
signal is to be tracked simultaneously, the response model z-transfer
functions to be selected include not only those relating plant tracking
outputs and the reference inputs they are to track, but also those relating unwanted couplings between each tracking output and the other
reference inputs.

6.3.2 Using Feedback Observer Inputs

When there are multiple plant inputs available for a response model tracking system design, there is more design freedom for the filter, just as there is in the case of ideal reference input filter design. If an observer is used for feedback for plant eigenvalue placement, an additional independent external input can be added to each individual observer state equation, increasing the number of inputs available to be driven by a reference input filter.

As a simple numerical example of using both plant and feedback observer inputs in reference input filter design, consider the plant with identical measurement and tracking outputs

$$\begin{bmatrix} x_1(k+1) \\ x_2(k+1) \end{bmatrix} = \begin{bmatrix} 0 & 1 \\ -\frac{1}{4} & -1 \end{bmatrix} \begin{bmatrix} x_1(k) \\ x_2(k) \end{bmatrix} + \begin{bmatrix} 0 \\ 1 \end{bmatrix} u(k) = \mathbf{A}\mathbf{x}(k) + \mathbf{b}u(k)$$

$$\bar{y}(k) = y(k) = \begin{bmatrix} 3 & -1 \end{bmatrix} \begin{bmatrix} x_1(k) \\ x_2(k) \end{bmatrix} + u(k) = \mathbf{c}^\dagger \mathbf{x}(k) + du(k) \quad \textbf{(6-4)}$$

Suppose that it is desired that all system eigenvalues be at $\lambda = \frac{1}{4}$. If the plant state were available for feedback, the state feedback

$$u(k) = \begin{bmatrix} \frac{3}{16} & \frac{3}{2} \end{bmatrix} \begin{bmatrix} x_1(k) \\ x_2(k) \end{bmatrix}$$

would place both plant eigenvalues at $\lambda = \frac{1}{4}$

$$\begin{bmatrix} x_1(k+1) \\ x_2(k+1) \end{bmatrix} = \left\{ \begin{bmatrix} 0 & 1 \\ -\frac{1}{4} & -1 \end{bmatrix} + \begin{bmatrix} 0 \\ 1 \end{bmatrix} \begin{bmatrix} \frac{3}{16} & \frac{3}{2} \end{bmatrix} \right\} \begin{bmatrix} x_1(k) \\ x_2(k) \end{bmatrix}$$

A first-order observer with eigenvalue $\lambda = \frac{1}{4}$ and unit gain $g = 1$

$$\xi(k+1) = \frac{1}{4} \xi(k) + y(k) + hu(k)$$

observes the single linear plant state transformation

$$\xi(k) \rightarrow \mathbf{m}^\dagger \mathbf{x}(k)$$

given by

$$\mathbf{m}^\dagger = \mathbf{c}^\dagger \left(\mathbf{A} - \frac{1}{4} \mathbf{I} \right)^{-1} = \begin{bmatrix} -\frac{64}{9} & -\frac{44}{9} \end{bmatrix}$$

where

$$h = \mathbf{m}^\dagger \mathbf{b} - gd = -\frac{53}{9}$$

A linear combination of this observer state and the signal $y(k) - u(k)$ observes the desired plant state feedback transformation

$$-0.215\xi(k) - 0.448[y(k) - u(k)] \rightarrow [\tfrac{3}{16} \quad \tfrac{3}{2}]\mathbf{x}(k)$$

Adding external inputs $\rho_1(k)$ to the plant input and $\rho_2(k)$ to the observer, the plant with observer feedback is described by the plant equations (6-4) and

$$\xi(k + 1) = \frac{1}{4} \xi(k) + y(k) - \frac{53}{9} u(k) + \rho_2(k)$$

$$u(k) = -0.215\xi(k) - 0.448[y(k) - u(k)] + \rho_1(k) \qquad (6\text{-}5)$$

as shown in Figure 6-13(a). Eliminating $u(k)$ on the right side of both equations (6-5) allows the controller to be described by

$$\xi(k + 1) = 2.543\xi(k) + 5.779y(k) - 10.668\rho_1(k) + \rho_2(k)$$
$$u(k) = -0.3895\xi(k) - 0.8116y(k) + 1.8116\rho_1(k)$$

as shown in Figure 6-13(b).

The z-transfer functions of this system, consisting of the plant with observer feedback, are

$$\begin{cases} T_1(z) = \dfrac{\overline{Y}(z)}{\rho_1(z)} \begin{vmatrix} \text{Zero Initial Conditions} \\ \text{and } \rho_2 = 0 \end{vmatrix} = \dfrac{z^3 - 0.25z^2 + 3.25z - 0.816}{z^3 - 0.75z^2 + 0.188z - 0.0156} \\[4mm] T_2(z) = \dfrac{\overline{Y}(z)}{\rho_2(z)} \begin{vmatrix} \text{Zero Initial Conditions} \\ \text{and } \rho_1 = 0 \end{vmatrix} = \dfrac{-0.215z^2 - 0.700}{z^3 - 0.75z^2 + 0.188z - 0.0156} \end{cases}$$

For a reference input filter with the structure of Figure 6-13(c), ideal tracking requires that

$$T_1(z)G_1(z) + T_2(z)G_2(z) = 1$$

or, in terms of the individual z-transfer function numerator and denominator polynomials

$$T_1(z) = \frac{p_1(z)}{q(z)} \qquad T_2(z) = \frac{p_2(z)}{q(z)}$$

$$G_1(z) = \frac{g_1(z)}{d(z)} \qquad G_2(z) = \frac{g_2(z)}{d(z)}$$

then

$$g_1(z)p_1(z) + g_2(z)p_2(z) = d(z)q(z) \qquad (6\text{-}6)$$

A zero-order reference input filter will not satisfy equation (6-6). A first-order filter will satisfy equation (6-6) only if it has a pole at $z = 2.6$, which is unstable.

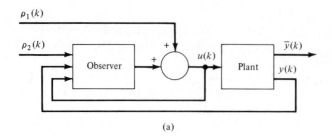

(a)

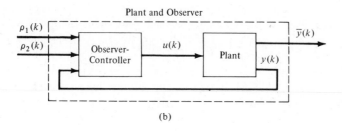

(b)

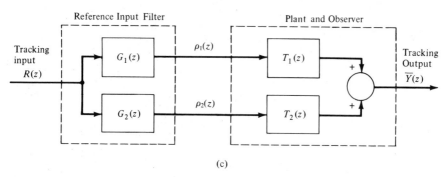

(c)

Figure 6-13 Use of both observer and plant inputs for tracking. (a) Plant with observer feedback. (b) The observer rearranged. (c) Reference input filter for the plant with observer feedback.

Choosing a second-order filter with both poles at $z = 0$, of the form

$$G_1(z) = \frac{\alpha_1 z^2 + \alpha_2 z + \alpha_3}{z^2} \qquad G_2(z) = \frac{\alpha_4 z^2 + \alpha_5 z + \alpha_6}{z^2}$$

results in a requirement equation (6-6) for ideal tracking that becomes

$$(\alpha_1 z^2 + \alpha_2 z + \alpha_3)(z^3 - 0.25z^2 + 3.25z - 0.816)$$
$$+ (\alpha_4 z^2 + \alpha_5 z + \alpha_6)(-0.215z^2 - 0.7)$$
$$= z^2(z^3 - 0.75z^2 + 0.188z - 0.0156)$$

Equating coefficients and solving

$$\alpha_1 = 1 \qquad \alpha_2 = -445.8 \qquad \alpha_3 = -1557.8$$
$$\alpha_4 = -2071 \qquad \alpha_5 = -6713 \qquad \alpha_6 = 1816$$

results in filter z-transfer functions

$$G_1(z) = \frac{z^2 - 445.8z - 1557.8}{z^2} \qquad G_2(z) = \frac{-2071z^2 - 6713z + 1816}{z^2}$$

This solution is probably not acceptable because of the large gain coefficients involved. The unit pulse response of the filter has huge output samples and large changes from sample to sample. Not only might these signals overload the plant inputs for typical reference inputs, but the precision required of the plant model and the filter is likely to be unacceptably high because differences of large numbers must be involved in the plant to produce relatively small outputs from such large inputs.

We now attempt to find a lower-order reference input filter for an acceptable response model design rather than the previous ideal tracking system design. Requiring that

$$T_1(z)G_1(z) + T_2(z)G_2(z) = \Omega(z) = \frac{\gamma(z)}{\delta(z)}$$

gives the polynomial equation

$$[g_1(z)p_1(z) + g_2(z)p_2(z)]\delta(z) = d(z)q(z)\gamma(z)$$

Choosing a zero-order filter and letting

$$\delta(z) = q(z) = \left(z - \frac{1}{4}\right)^3 = z^3 - 0.75z^2 + 0.188z - 0.0156$$

the polynomial equation becomes

$$\alpha_1 p_1(z) + \alpha_2 p_2(z) = \gamma(z) \tag{6-7}$$

where $\gamma(z)$ is the numerator polynomial of the response model z-transfer function

$$\Omega(z) = \frac{\gamma(z)}{(z - \frac{1}{4})^3} = \frac{z^3 + \alpha_3 z^2 + \alpha_4 z + \alpha_5}{(z - \frac{1}{4})^3}$$

For the response model to have zero steady state error to a constant input,

$$\lim_{z \to 1} \left\{ \left(\frac{z-1}{z}\right)\left(\frac{z}{z-1}\right)[1 - \Omega(z)] \right\} = 0$$

or

$$\alpha_3 + \alpha_4 + \alpha_5 = -0.578 \qquad\qquad (6\text{-}8)$$

In terms of the α's, the design equation (6-7) is

$$\alpha_1(z^3 - 0.25z^2 + 3.25z - 0.816) + \alpha_2(-0.215z^2 - 0.700)$$
$$= z^3 + \alpha_3 z^2 + \alpha_4 z + \alpha_5$$

Equating coefficients and appending the steady state error requirement from equation (6-8) results in these simultaneous linear algebraic equations:

$$\begin{cases}
\alpha_1 & & & = 1 \\
-0.25\alpha_1 & -0.215\alpha_2 & -\alpha_3 & & = 0 \\
3.25\alpha_1 & & & -\alpha_4 & = 0 \\
-0.816\alpha_1 & -0.700\alpha_2 & & & -\alpha_5 = 0 \\
& & \alpha_3 & +\alpha_4 & +\alpha_5 = -0.578
\end{cases}$$

The solution is

$$\alpha_1 = 1 \qquad \alpha_2 = 3.02 \qquad \alpha_3 = -0.899$$
$$\alpha_4 = 3.25 \qquad \alpha_5 = -2.93$$

and corresponds to the reference input filter with

$$G_1(z) = \alpha_1 = 1 \qquad G_2(z) = \alpha_2 = 3.02$$

6.3.3 Tracking Error Feedback

If the tracking output of a plant is available for feedback, its steady state error to a constant reference input can be made to be zero, even when the plant parameters are not known accurately. The tracking system of Figure 6-14(a), for example, has the overall z-transfer function

$$T(z) = \frac{\left(\dfrac{\alpha}{z-1}\right)\left[\dfrac{p(z)}{q(z)}\right]}{1 + \left(\dfrac{\alpha}{z-1}\right)\left[\dfrac{p(z)}{q(z)}\right]} = \frac{\alpha p(z)}{(z-1)q(z) + \alpha p(z)}$$

where the plant z-transfer function is the ratio of polynomials

$$G_p(z) = \frac{p(z)}{q(z)}$$

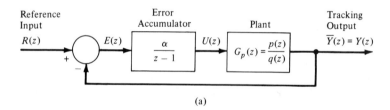

(a)

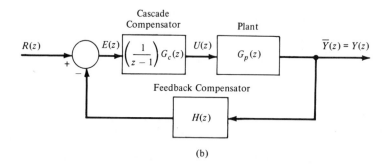

(b)

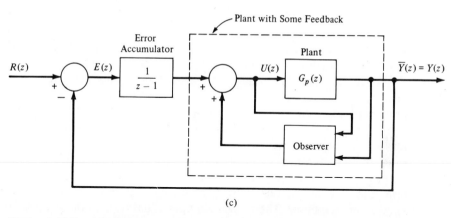

(c)

Figure 6-14 Tracking error feedback. (a) Driving a plant with accumulated tracking error. (b) Error accumulation and compensation. (c) Pole (eigenvalue) placement and error accumulation.

The tracking error of this system is given by

$$E(z) = \overline{Y}(z) - R(z) = [1 - T(z)]R(z) = \frac{(z-1)q(z)}{(z-1)q(z) + \alpha p(z)} R(z)$$

For a constant reference input

$$R(z) = \frac{z}{z - 1}$$

the steady state tracking error is

$$\lim_{k \to \infty} e(k) = \lim_{z \to 1} \left(\frac{z-1}{z} \right) \frac{(z-1)q(z)}{(z-1)q(z) + \alpha p(z)} \left(\frac{z}{z-1} \right) = 0$$

provided that the feedback system is stable. The character of the feedback system's zero-input response is determined by its pole locations (eigenvalues), so the designer is usually interested in achieving more than just stability.

In classical tracking system design, one attempts to adjust the gain α to give acceptable feedback system pole locations. More generally, the orders and parameters of the compensators $G_c(z)$ and $H(z)$ in Figure 6-14(b), where

$$\lim_{z \to 1} H(z) = 1$$

can be chosen to give desirable feedback system poles. If the plant parameters differ from those used for design, the overall poles are different, but there is a range of plant parameters for which the feedback system has acceptable pole locations. Regardless of the plant parameters, so long as the feedback system is stable, there is zero steady state error to any constant reference input. Such a design is said to be *robust* with regard to the plant parameters.

An equivalent way of achieving acceptable feedback system pole locations is shown in Figure 6-14(c). Feedback is placed around the original plant in such a way that the additional tracking error feedback results in desired overall pole locations. Again, different plant parameters result in different overall feedback system poles, but there is a range of plant parameters for which the feedback system pole locations are acceptable. For any plant parameters for which the feedback system poles are acceptable, there is zero steady state error to any constant reference input. The tracking output is fed back to the plant both through the observer in the feedback subsystem and in the error signal difference with $r(k)$. The reference input $r(k)$, however, only occurs in the error signal, never alone.

To relate this kind of robustness to the observer feedback and plant input filter design of Figure 6-15(a), let the overall single-input/single-output tracking system have z-transfer function $T(z)$, and let the tracking error z-transfer function be of the form

$$\left. \frac{R(z) - \bar{Y}(z)}{R(z)} \right|_{\substack{\text{Zero Initial} \\ \text{Conditions}}} = 1 - T(z) = \frac{(z-1)n(z)}{d(z)}$$

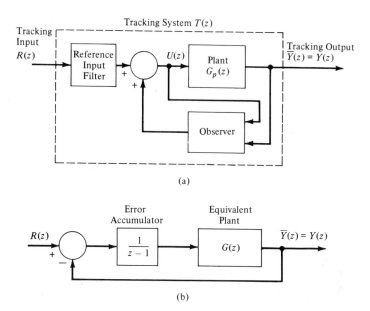

(a)

(b)

Figure 6-15 Arranging a tracking system to have tracking error feedback and accumulation. (a) Tracking system in which the tracking output is available for feedback. (b) Rearranged system.

The zero at $z = 1$ gives zero steady state error to any constant reference input, and $n(z)$ and $d(z)$ are polynomials. Then

$$T(z) = \frac{\overline{Y}(z)}{R(z)}\bigg|_{\substack{\text{Zero Initial} \\ \text{Conditions}}} = \frac{d(z) - (z - 1)n(z)}{d(z)}$$

Expressing $T(z)$ as a unity feedback system with an accumulator in the forward loop as in Figure 6-15(b), then

$$G(z) = \frac{d(z) - (z - 1)n(z)}{n(z)}$$

Such a tracking system can thus be rearranged into the error feedback form, if desired.

In view of the inaccuracies and drift common with analog elements, for continuous-time control it can be important to form a tracking error signal and process it as a whole, rather than processing the reference input and the tracking output separately. Accuracy and freedom from drift are routinely obtained with digital control, however.

6.4 Reference Model Tracking System Design

The awkwardness of practical response model design arises because of the difficulty in relating performance criteria to the z-transfer functions of response models. An alternative design method models the reference input signals $r(k)$ instead of the system response. This method allows the designer to specify the class of representative reference inputs that are to be tracked perfectly, rather than having to specify acceptable response models for all possible inputs. Linear algebraic design equations result, so determining the feasibility of design options is straightforward.

6.4.1 Reference Signal Models

A scalar reference input signal can be modeled as the response of an autonomous (zero-input) system of the form

$$\sigma(k + 1) = \psi\sigma(k)$$
$$r(k) = \theta^\dagger\sigma(k) \tag{6-9}$$

The model response is of the form

$$r(k) = \theta^\dagger\psi^k\sigma(0)$$

which, for distinct eigenvalues $\lambda_1, \lambda_2, \ldots, \lambda_i$, is of the form

$$r(k) = \beta_1\lambda_1^k + \beta_2\lambda_2^k + \cdots + \beta_i\lambda_i^k$$

where the β's are arbitrary constants dependent on the initial conditions $\sigma(0)$. When eigenvalues of the model are complex, a damped sinusoidal form for the corresponding response term is usually preferred. For repeated eigenvalues, response terms are as listed in Table 6-1. For example, a tracking input signal model with response of the form

$$r(k) = \beta_1 + \beta_2 k + \beta_3 \left(\frac{1}{2}\right)^k + \beta_4 k \left(\frac{1}{2}\right)^k$$

where $\beta_1, \beta_2, \beta_3,$ and β_4 are any constants, is an autonomous system as in equation (6-9) with characteristic equation

$$(\lambda - 1)^2 \left(\lambda - \frac{1}{2}\right)^2 = \lambda^4 - 3\lambda^3 + \frac{13}{4}\lambda^2 - \frac{3}{2}\lambda + \frac{1}{4} = 0$$

Table 6-1 Relations Between Eigenvalues and Zero-Input System Response

Eigenvalues	Characteristic Equation Factor(s)	Form of the Response
1	$z - 1$	β_1
1, 1	$(z - 1)^2$	$\beta_1 + \beta_2 k$
1, 1, 1	$(z - 1)^3$	$\beta_1 + \beta_2 k + \beta_3 k^2$
c	$(z - c)$	$\beta_1 c^k$
c, c	$(z - c)^2$	$\beta_1 c^k + \beta_2 k c^k$
c, c, c	$(z - c)^3$	$\beta_1 c^k + \beta_2 k c^k + \beta_3 k^2 c^k$
$e^{j\omega}, e^{-j\omega}$	$(z - \cos \omega + j \sin \omega)$ $(z - \cos \omega - j \sin \omega)$ $= (z^2 - 2z \cos \omega + 1)$	$\beta_1 \cos \omega k + \beta_2 \sin \omega k$ $= \beta_1' \cos(\omega k + \beta_2')$
$ce^{j\omega}, ce^{-j\omega}$	$[z^2 - (2c \cos \omega)z + c^2]$	$c^k(\beta_1 \cos \omega k + \beta_2 \sin \omega k)$
$e^{j\omega}, e^{j\omega},$ $e^{-j\omega}, e^{-j\omega}$	$(z^2 - 2z \cos \omega + 1)^2$	$\beta_1 \cos \omega k + \beta_2 \sin \omega k$ $+ \beta_3 k \cos \omega k + \beta_4 k \sin \omega k$
$ce^{j\omega}, ce^{j\omega},$ $ce^{-j\omega}, ce^{-j\omega}$	$[z^2 - (2c \cos \omega)z + c^2]^2$	$c^k(\beta_1 \cos \omega k + \beta_2 \sin \omega k$ $+ \beta_3 k \cos \omega k + \beta_4 k \sin \omega k)$

A convenient state variable model is the observable form

$$\begin{bmatrix} \sigma_1(k + 1) \\ \sigma_2(k + 1) \\ \sigma_3(k + 1) \\ \sigma_4(k + 1) \end{bmatrix} = \begin{bmatrix} 3 & 1 & 0 & 0 \\ -\frac{13}{4} & 0 & 1 & 0 \\ \frac{3}{2} & 0 & 0 & 1 \\ -\frac{1}{4} & 0 & 0 & 0 \end{bmatrix} \begin{bmatrix} \sigma_1(k) \\ \sigma_2(k) \\ \sigma_3(k) \\ \sigma_4(k) \end{bmatrix}$$

$$r(k) = \begin{bmatrix} 1 & 0 & 0 & 0 \end{bmatrix} \begin{bmatrix} \sigma_1(k) \\ \sigma_2(k) \\ \sigma_3(k) \\ \sigma_4(k) \end{bmatrix}$$

6.4.2 Reference Input Model Following

In reference model tracking system design, an autonomous model for the reference input is selected by the designer. If the tracking system is to track constant signals well, a model with response of the form

$$r(k) = \beta$$

is chosen. If the tracking system is to track arbitrary constant and arbitrary ramp signals well, then a model with response of the form

$$r(k) = \beta_1 + \beta_2 k$$

is used, and so on. Once the plant's zero-input response has been improved as necessary by measurement or observer feedback, the feedback system external input gains are selected so that the plant is an observer of the reference signal model, as indicated in Figure 6-16. After its zero-input response has died out, the plant then tracks any signal in the class of modeled signals exactly.

The autonomous reference input model has no physical existence; the actual reference input $r(k)$ likely deviates somewhat from the predictions of the model. The designer deals with representative reference inputs, such as constants and ramps, and, by designing for exact tracking of these, obtains acceptable tracking performance for other reference inputs.

As an example, consider the plant and tracking output

$$\begin{bmatrix} x_1(k+1) \\ x_2(k+1) \\ x_3(k+1) \end{bmatrix} = \begin{bmatrix} \frac{5}{6} & 1 & 0 \\ -\frac{1}{6} & 0 & 1 \\ 0 & 0 & 0 \end{bmatrix} \begin{bmatrix} x_1(k) \\ x_2(k) \\ x_3(k) \end{bmatrix} + \begin{bmatrix} 1 & -1 \\ 1 & 0 \\ 0 & 1 \end{bmatrix} \begin{bmatrix} \rho_1(k) \\ \rho_2(k) \end{bmatrix}$$

$$\bar{y}(k) = \begin{bmatrix} 1 & -2 & 0 \end{bmatrix} \begin{bmatrix} x_1(k) \\ x_2(k) \\ x_3(k) \end{bmatrix} + \begin{bmatrix} 2 & 0 \end{bmatrix} \begin{bmatrix} \rho_1(k) \\ \rho_2(k) \end{bmatrix}$$

where it is assumed that any feedback, with or without an observer, is included in this model. When the reference input $r(k)$ is supplied to the plant through gains α_1 and α_2, to be determined

$$\rho_1(k) = \alpha_1 r(k) \qquad \rho_2(k) = \alpha_2 r(k)$$

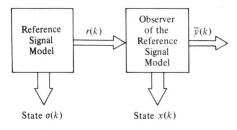

State $o(k)$ State $x(k)$

Figure 6-16 Observing a reference signal model.

the state and tracking output equations are, in terms of $r(k)$,

$$
\begin{bmatrix} x_1(k+1) \\ x_2(k+1) \\ x_3(k+1) \end{bmatrix} = \begin{bmatrix} \tfrac{5}{6} & 1 & 0 \\ -\tfrac{1}{6} & 0 & 1 \\ 0 & 0 & 0 \end{bmatrix} \begin{bmatrix} x_1(k) \\ x_2(k) \\ x_3(k) \end{bmatrix} + \begin{bmatrix} (\alpha_1 - \alpha_2) \\ \alpha_1 \\ \alpha_2 \end{bmatrix} r(k)
$$

$$
= \mathbf{A}\mathbf{x}(k) + \mathbf{b}r(k)
$$

$$
\bar{y}(k) = \begin{bmatrix} 1 & -2 & 0 \end{bmatrix} \begin{bmatrix} x_1(k) \\ x_2(k) \\ x_3(k) \end{bmatrix} + 2\alpha_1 r(k) = \bar{\mathbf{c}}^{\dagger}\mathbf{x}(k) + \bar{d}r(k) \quad \textbf{(6-10)}
$$

It is now required that this system be an observer of a suitable reference signal model. The first-order model

$$
\sigma(k+1) = \sigma(k) = \psi\sigma(k)
$$
$$
r(k) = \sigma(k) = \theta\sigma(k)
$$

is of an arbitrary constant reference input. When driven by $r(k)$, the plant state observes

$$
\mathbf{x}(k) \rightarrow \mathbf{m}\sigma(k) = \mathbf{m}r(k)
$$

where

$$
\mathbf{m}\psi - \mathbf{A}\mathbf{m} = \mathbf{b}\theta
$$

or

$$
\begin{bmatrix} m_1 \\ m_2 \\ m_3 \end{bmatrix} - \begin{bmatrix} \tfrac{5}{6} & 1 & 0 \\ -\tfrac{1}{6} & 0 & 1 \\ 0 & 0 & 0 \end{bmatrix} \begin{bmatrix} m_1 \\ m_2 \\ m_3 \end{bmatrix} = \begin{bmatrix} (\alpha_1 - \alpha_2) \\ \alpha_1 \\ \alpha_2 \end{bmatrix} \quad \textbf{(6-11)}
$$

The plant tracking output observes

$$
\bar{y}(k) = \bar{\mathbf{c}}^{\dagger}\mathbf{x}(k) + \bar{d}r(k) \rightarrow (\bar{\mathbf{c}}^{\dagger}\mathbf{m} + \bar{d})r(k)
$$

so it is necessary that

$$
\bar{\mathbf{c}}^{\dagger}\mathbf{m} + \bar{d} = 1
$$

or

$$
\begin{bmatrix} 1 & -2 & 0 \end{bmatrix} \begin{bmatrix} m_1 \\ m_2 \\ m_3 \end{bmatrix} + 2\alpha_1 = 1 \quad \textbf{(6-12)}
$$

for

$$
\bar{y}(k) \rightarrow r(k)
$$

Collecting the design equations from (6-11) and (6-12) results in

$$
\begin{cases}
\dfrac{1}{6} m_1 - m_2 \quad\quad - \alpha_1 + \alpha_2 = 0 \\[2mm]
\dfrac{1}{6} m_1 + m_2 - m_3 - \alpha_1 \quad\quad = 0 \\[2mm]
\quad\quad\quad m_3 \quad\quad - \alpha_2 = 0 \\[2mm]
m_1 - 2m_2 \quad\quad + 2\alpha_1 \quad\quad = 1
\end{cases}
\tag{6-13}
$$

one solution to which is

$$
m_1 = \frac{3}{4} \quad\quad m_2 = 0 \quad\quad m_3 = 0
$$

$$
\alpha_1 = \frac{1}{8} \quad\quad \alpha_2 = 0
$$

Figure 6-17 shows this controller and its typical response. It responds to an arbitrary constant signal as expected. Its response to other reference inputs is poor if the input varies too quickly.

Because the design equations in (6-13) were underdetermined, we now attempt to improve the design by using a higher-order reference signal model. The second-order model

$$
\begin{bmatrix} \sigma_1(k+1) \\ \sigma_2(k+1) \end{bmatrix} = \begin{bmatrix} 2 & 1 \\ -1 & 0 \end{bmatrix} \begin{bmatrix} \sigma_1(k) \\ \sigma_2(k) \end{bmatrix} = \boldsymbol{\psi}\boldsymbol{\sigma}(k)
$$

$$
r(k) = [1 \quad 0] \begin{bmatrix} \sigma_1(k) \\ \sigma_2(k) \end{bmatrix} = \boldsymbol{\theta}^\dagger \boldsymbol{\sigma}(k)
$$

is of an arbitrary constant plus an arbitrary ramp tracking input signal. When driven by $r(k)$, the plant state equation (6-10) observes

$$
\mathbf{x}(k) \rightarrow \mathbf{M}\boldsymbol{\sigma}(k)
$$

where $\mathbf{M}$ is 3×2 and where

$$
\mathbf{M}\boldsymbol{\psi} - \mathbf{A}\mathbf{M} = \mathbf{b}\boldsymbol{\theta}^\dagger
$$

or

$$
\begin{bmatrix} m_{11} & m_{12} \\ m_{21} & m_{22} \\ m_{31} & m_{32} \end{bmatrix} \begin{bmatrix} 2 & 1 \\ -1 & 0 \end{bmatrix} - \begin{bmatrix} \frac{5}{6} & 1 & 0 \\ -\frac{1}{6} & 0 & 1 \\ 0 & 0 & 0 \end{bmatrix} \begin{bmatrix} m_{11} & m_{12} \\ m_{21} & m_{22} \\ m_{31} & m_{32} \end{bmatrix} = \begin{bmatrix} (\alpha_1 - \alpha_2) \\ \alpha_1 \\ \alpha_2 \end{bmatrix} [1 \quad 0]
\tag{6-14}
$$

The plant tracking output observes

$$
\bar{y}(k) = \bar{\mathbf{c}}^\dagger \mathbf{x}(k) + \bar{d}r(k) \rightarrow \bar{\mathbf{c}}^\dagger \mathbf{M}\boldsymbol{\sigma}(k) + \bar{d}r(k)
$$

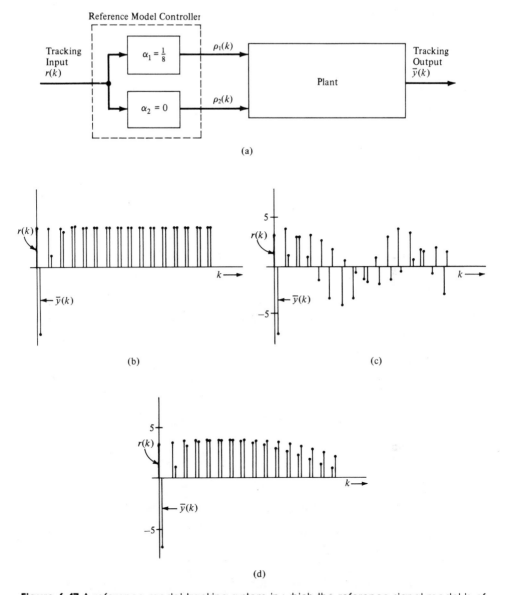

Figure 6-17 A reference model tracking system in which the reference signal model is of an arbitrary constant. (a) Block diagram. (b) Step response. (c) Response to an arbitrary reference input. (d) Response to a more slowly varying reference input.

and for

$$\bar{y}(k) \rightarrow r(k)$$

it is necessary that

$$\bar{c}^{\dagger}\mathbf{M}\sigma(k) + \bar{d}r(k) = r(k)$$

Because

$$\sigma_1(k) = r(k)$$

we know that

$$[1 \quad -2 \quad 0] \begin{bmatrix} m_{11} & m_{12} \\ m_{21} & m_{22} \\ m_{31} & m_{32} \end{bmatrix} \begin{bmatrix} r(k) \\ \sigma_2(k) \end{bmatrix} + 2\alpha_1 r(k) = r(k)$$

or

$$\begin{cases} m_{11} - 2m_{21} + 2\alpha_1 = 1 \\ m_{12} - 2m_{22} \quad\quad\quad = 0 \end{cases} \tag{6-15}$$

Collecting the design equations (6-14) and (6-15) results in

$$\begin{cases} \dfrac{7}{6}m_{11} - m_{12} - m_{21} & - \alpha_1 + \alpha_2 = 0 \\[2mm] \dfrac{1}{6}m_{11} + 2m_{21} - m_{22} - m_{31} & - \alpha_1 \quad\quad = 0 \\[2mm] 2m_{31} - m_{32} & - \alpha_2 = 0 \\[2mm] m_{11} - \dfrac{5}{6}m_{12} - m_{22} & = 0 \\[2mm] \dfrac{1}{6}m_{12} + m_{21} - m_{32} & = 0 \\[2mm] m_{31} & = 0 \\[2mm] m_{11} - 2m_{21} + 2\alpha_1 & = 1 \\[2mm] m_{12} - 2m_{22} & = 0 \end{cases}$$

which has the unique solution

$$m_{11} = 1.20 \quad\quad m_{12} = 0.90 \quad\quad m_{21} = 0.15$$
$$m_{22} = 0.45 \quad\quad m_{31} = 0 \quad\quad m_{32} = 0.3$$
$$\alpha_1 = 0.05 \quad\quad \alpha_2 = -0.3$$

Figure 6-18 shows this controller and its typical response, which is clearly improved over that of the previous design. A still higher-order

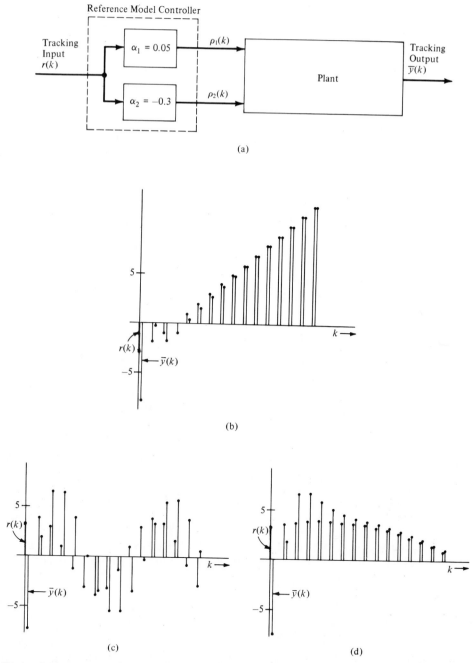

Figure 6-18 A reference model tracking system for the previous plant, in which the signal model is of a constant plus a ramp. (a) Block diagram. (b) Response to an arbitrary constant plus an arbitrary ramp reference input. (c) Response to an arbitrary reference input. (d) Response to a more slowly varying reference input.

reference signal model cannot be used (without modifying the plant) because it would result in overdetermined design equations.

In reference model tracking system design, the concept of an observer is used in a new way; it is the *plant* or the plant with feedback that is the observer now. If the available plant inputs are sufficient, the plant becomes an observer of a fictitious reference input model system.

The basic process for designing a reference model tracking system consists of these two steps:

1. Design plant measurement or observer feedback to place the feedback system eigenvalues as desired for the system zero-input response.
2. Choose observer and plant external input gains so that the feedback system observes a suitable reference input signal model. The model represents the class of reference input signals that will be tracked exactly, after the plant's zero-input response dies out.

These two steps are a state space multiple-input/multiple-output generalization of the classical approach (usually tractable only for low-order single-input/single-output plants) of seeking a controller that simultaneously meets requirements for feedback system eigenvalue (or pole) location and zero-state (or steady state) performance. The design methods here result in simultaneous linear algebraic equations so that it is straightforward to determine design options.

6.4.3 Higher-Order Controllers

When, with a given plant and feedback, a reference model of sufficiently high order is not possible, the order of the observer used for feedback can be raised, which generally raises the highest feasible order of the reference signal model. As a simple example, consider the first-order plant

$$x(k + 1) = x(k) + u(k)$$
$$y(k) = x(k)$$
$$\bar{y}(k) = x(k) + 3u(k) \tag{6-16}$$

and feedback of the accessible output according to

$$u(k) = -\frac{4}{5} y(k) + \alpha r(k)$$

where the gain α of the external input is to be determined. This gives a feedback system described by

$$x(k + 1) = \frac{1}{5} x(k) + \alpha r(k) = ax(k) + br(k)$$

$$\bar{y}(k) = -\frac{7}{5} x(k) + 3\alpha r(k) = \bar{c}x(k) + \bar{d}r(k)$$

A first-order reference model

$$\sigma(k + 1) = \sigma(k) = \psi\sigma(k)$$
$$r(k) = \sigma(k) = \theta\sigma(k)$$

models an arbitrary constant reference input $r(k)$. The feedback system observes the model state according to

$$x(k) \rightarrow m\sigma(k)$$

where

$$m\psi - am = b\theta$$

or

$$\left(1 - \frac{1}{5}\right) m = \alpha \tag{6-17}$$

The tracking output observes

$$\bar{y}(k) = -\frac{7}{5} x(k) + 3\alpha r(k) \rightarrow -\frac{7}{5} mr(k) + 3\alpha r(k)$$

and

$$\bar{y}(k) \rightarrow r(k)$$

requires that

$$-\frac{7}{5} m + 3\alpha = 1 \tag{6-18}$$

Solving the two simultaneous linear algebraic design equations (6-17) and (6-18) gives the unique solution

$$m = 1 \qquad \alpha = \frac{4}{5}$$

Figure 6-19 shows this tracking system and its typical response.

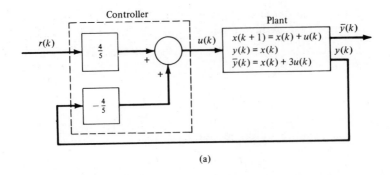

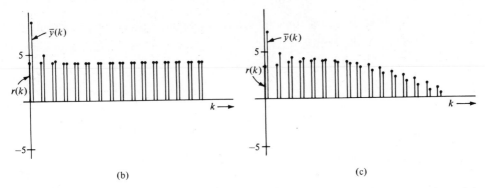

Figure 6-19 A reference model tracking system using a first-order reference input model. (a) Block diagram. (b) Step response. (c) Response to a slowly varying reference input.

The feedback system cannot observe a reference input model of an order greater than 1. However, replacing the output feedback by a combination of output and observer feedback raises the order of the resulting feedback system to 2 and allows observation of a higher-order model.

A first-order observer of the plant state (6-16) with eigenvalue $\lambda = 0$ is

$$\xi(k + 1) = y(k) + u(k) + \alpha_1 r(k)$$

where α_1 is an external input gain to be determined. For this observer,

$$\xi(k) \rightarrow x(k)$$

The plant (6-16) with the observer feedback

$$u(k) = -\frac{4}{5}\xi(k) + \alpha_2 r(k)$$

where α_2 is an external input gain, places the eigenvalues of the second-order system consisting of the first-order plant and the first-order observer at $\lambda = \frac{1}{5}$ and $\lambda = 0$. This observer feedback system is described by

$$\begin{bmatrix} x(k+1) \\ \xi(k+1) \end{bmatrix} = \begin{bmatrix} 1 & -\frac{4}{5} \\ 1 & -\frac{4}{5} \end{bmatrix} \begin{bmatrix} x(k) \\ \xi(k) \end{bmatrix} + \begin{bmatrix} \alpha_2 \\ (\alpha_1 + \alpha_2) \end{bmatrix} r(k)$$

$$= \mathbf{Ax}(k) + \mathbf{b}r(k)$$

$$\bar{y}(k) = [1 \quad -\tfrac{12}{5}] \begin{bmatrix} x(k) \\ \xi(k) \end{bmatrix} + 3\alpha_2 r(k) = \bar{\mathbf{c}}^\dagger \mathbf{x}(k) + \bar{d}r(k) \qquad \textbf{(6-19)}$$

The state vector $\mathbf{x}(k)$ now includes the original scalar plant state $x(k)$ as one of its two components.

The second-order reference input model

$$\begin{bmatrix} \sigma_1(k+1) \\ \sigma_2(k+1) \end{bmatrix} = \begin{bmatrix} 2 & 1 \\ -1 & 0 \end{bmatrix} \begin{bmatrix} \sigma_1(k) \\ \sigma_2(k) \end{bmatrix} = \boldsymbol{\psi}\boldsymbol{\sigma}(k)$$

$$r(k) = [1 \quad 0] \begin{bmatrix} \sigma_1(k) \\ \sigma_2(k) \end{bmatrix} = \boldsymbol{\theta}^\dagger \boldsymbol{\sigma}(k)$$

is of an arbitrary constant plus an arbitrary ramp input. When driven by $r(k)$, the observer feedback system (6-19) observes

$$\mathbf{x}(k) \to \mathbf{M}\boldsymbol{\sigma}(k)$$

where the 2×2 matrix $\mathbf{M}$ satisfies

$$\mathbf{M}\boldsymbol{\psi} - \mathbf{A}\mathbf{M} = \mathbf{b}\boldsymbol{\theta}^\dagger$$

or

$$\begin{bmatrix} m_{11} & m_{12} \\ m_{21} & m_{22} \end{bmatrix} \begin{bmatrix} 2 & 1 \\ -1 & 0 \end{bmatrix} - \begin{bmatrix} 1 & -\frac{4}{5} \\ 1 & -\frac{4}{5} \end{bmatrix} \begin{bmatrix} m_{11} & m_{12} \\ m_{21} & m_{22} \end{bmatrix} = \begin{bmatrix} \alpha_2 \\ (\alpha_1 + \alpha_2) \end{bmatrix} [1 \quad 0]$$

$$\textbf{(6-20)}$$

The plant tracking output $\bar{y}(k)$ observes

$$\bar{y}(k) = \bar{\mathbf{c}}^\dagger \mathbf{x}(k) + \bar{d}r(k) \to \bar{\mathbf{c}}^\dagger \mathbf{M}\boldsymbol{\sigma}(k) + \bar{d}r(k)$$

and for

$$\bar{y}(k) \to r(k)$$

it is necessary that

$$[1 \quad -\tfrac{12}{5}] \begin{bmatrix} m_{11} & m_{12} \\ m_{21} & m_{22} \end{bmatrix} \begin{bmatrix} r(k) \\ \sigma_2(k) \end{bmatrix} + 3\alpha_2 r(k) = r(k) \qquad \textbf{(6-21)}$$

Collecting the design equations (6-20) and (6-21) results in

$$
\begin{cases}
m_{11} - m_{12} + \dfrac{4}{5} m_{21} & - \alpha_2 = 0 \\[2mm]
-m_{11} & + \dfrac{14}{5} m_{21} - & m_{22} - \alpha_1 - \alpha_2 = 0 \\[2mm]
m_{11} - m_{12} & + \dfrac{4}{5} m_{22} & = 0 \\[2mm]
- m_{12} + & m_{21} + \dfrac{4}{5} m_{22} & = 0 \\[2mm]
m_{11} & - \dfrac{12}{5} m_{21} & + 3\alpha_2 = 1 \\[2mm]
m_{12} & - \dfrac{12}{5} m_{22} & = 0
\end{cases}
$$

which has the unique solution

$$m_{11} = -2 \qquad m_{12} = -3 \qquad m_{21} = -2$$
$$m_{22} = -1.25 \qquad \alpha_1 = -1.75 \qquad \alpha_2 = -0.6$$

Figure 6-20 shows typical response of this higher-order observer–controller, which is improved over that of the previous lower-order design.

6.4.4 Tracking Multiple Inputs

The simultaneous tracking of multiple reference input signals can also be accommodated within the framework of reference model tracking system design. In practice, situations occur in which two or more reference signals are related to one another, such as the position and velocity of a single mass. Then one simply models both the reference inputs and their relation to one another with state equations of the form

$$\boldsymbol{\sigma}(k + 1) = \boldsymbol{\psi}\boldsymbol{\sigma}(k)$$
$$\mathbf{r}(k) = \boldsymbol{\Theta}\boldsymbol{\sigma}(k) \tag{6-22}$$

where $\mathbf{r}(k)$ is now a vector. When two reference inputs are not related, they can be modeled separately as

$$\sigma_1(k + 1) = \psi_1 \sigma_1(k)$$
$$r_1(k) = \theta_1 \sigma_1(k)$$

and

$$\sigma_2(k + 1) = \psi_2 \sigma_2(k)$$
$$r_2(k) = \theta_2 \sigma_2(k)$$

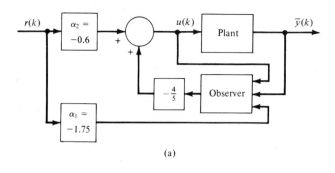

(a)

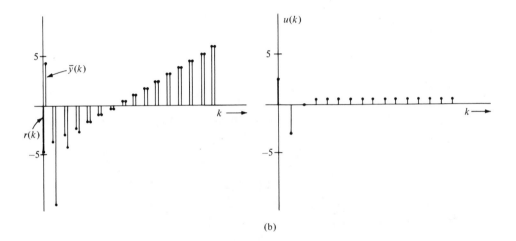

(b)

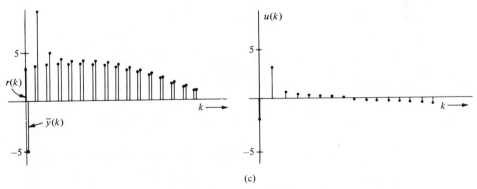

(c)

Figure 6-20 A reference model tracking system using a second-order signal model. The plant is the same as for the previous system. (a) Block diagram. (b) Response to a reference input with a constant plus ramp. (c) Response to a slowly varying reference input.

Then the uncoupled models can be combined in the form given in equation (6-22).

For example, consider a two-input/third-order plant with feedback with two tracking outputs

$$\begin{bmatrix} x_1(k+1) \\ x_2(k+1) \\ x_3(k+1) \end{bmatrix} = \begin{bmatrix} \frac{1}{4} & 1 & 0 \\ 0 & 0 & 1 \\ 0 & 0 & 0 \end{bmatrix} \begin{bmatrix} x_1(k) \\ x_2(k) \\ x_3(k) \end{bmatrix} + \begin{bmatrix} 1 & -1 \\ 0 & 1 \\ 2 & 0 \end{bmatrix} \begin{bmatrix} p_1(k) \\ p_2(k) \end{bmatrix}$$

$$\begin{bmatrix} \bar{y}_1(k) \\ \bar{y}_2(k) \end{bmatrix} = \begin{bmatrix} 1 & 0 & 0 \\ -2 & -1 & 1 \end{bmatrix} \begin{bmatrix} x_1(k) \\ x_2(k) \\ x_3(k) \end{bmatrix}$$

If two reference inputs, $r_1(k)$ and $r_2(k)$, are coupled to this plant according to

$$p_1(k) = \alpha_1 r_1(k) + \alpha_2 r_2(k)$$
$$p_2(k) = \alpha_3 r_1(k) + \alpha_4 r_2(k)$$ **(6-23)**

the plant with feedback and the controller given in equation (6-23) are described by

$$\begin{bmatrix} x_1(k+1) \\ x_2(k+1) \\ x_3(k+1) \end{bmatrix} = \begin{bmatrix} \frac{1}{4} & 1 & 0 \\ 0 & 0 & 1 \\ 0 & 0 & 0 \end{bmatrix} \begin{bmatrix} x_1(k) \\ x_2(k) \\ x_3(k) \end{bmatrix} + \begin{bmatrix} (\alpha_1 - \alpha_3) & (\alpha_2 - \alpha_4) \\ \alpha_3 & \alpha_4 \\ 2\alpha_1 & 2\alpha_2 \end{bmatrix} \begin{bmatrix} r_1(k) \\ r_2(k) \end{bmatrix}$$

$$= \mathbf{A}\mathbf{x}(k) + \mathbf{B}\mathbf{r}(k)$$

$$\begin{bmatrix} \bar{y}_1(k) \\ \bar{y}_2(k) \end{bmatrix} = \begin{bmatrix} 1 & 0 & 0 \\ -2 & -1 & 1 \end{bmatrix} \begin{bmatrix} x_1(k) \\ x_2(k) \\ x_3(k) \end{bmatrix} = \mathbf{\bar{C}}\mathbf{x}(k)$$

Let the reference model be

$$\begin{bmatrix} \sigma_1(k+1) \\ \sigma_2(k+1) \end{bmatrix} = \begin{bmatrix} 1 & 0 \\ 0 & 1 \end{bmatrix} \begin{bmatrix} \sigma_1(k) \\ \sigma_2(k) \end{bmatrix} = \boldsymbol{\psi}\boldsymbol{\sigma}(k)$$

$$\begin{bmatrix} r_1(k) \\ r_2(k) \end{bmatrix} = \begin{bmatrix} 1 & 0 \\ 0 & 1 \end{bmatrix} \begin{bmatrix} \sigma_1(k) \\ \sigma_2(k) \end{bmatrix} = \boldsymbol{\Theta}\boldsymbol{\sigma}(k)$$

which models each of the two reference inputs as an independent arbitrary constant.

The plant with controller observes

$$\mathbf{x}(k) \rightarrow \mathbf{M}\mathbf{r}(k)$$

where the 3 × 2 matrix $\mathbf{M}$ satisfies

$$\mathbf{M}\boldsymbol{\psi} - \mathbf{A}\mathbf{M} = \mathbf{B}\boldsymbol{\Theta}$$

or

$$\begin{bmatrix} m_{11} & m_{12} \\ m_{21} & m_{22} \\ m_{31} & m_{32} \end{bmatrix} \begin{bmatrix} 1 & 0 \\ 0 & 1 \end{bmatrix} - \begin{bmatrix} \frac{1}{4} & 1 & 0 \\ 0 & 0 & 1 \\ 0 & 0 & 0 \end{bmatrix} \begin{bmatrix} m_{11} & m_{12} \\ m_{21} & m_{22} \\ m_{31} & m_{32} \end{bmatrix}$$

$$= \begin{bmatrix} (\alpha_1 - \alpha_3) & (\alpha_2 - \alpha_4) \\ \alpha_3 & \alpha_4 \\ 2\alpha_1 & 2\alpha_2 \end{bmatrix} \begin{bmatrix} 1 & 0 \\ 0 & 1 \end{bmatrix} \qquad (6\text{-}24)$$

The plant tracking outputs observe

$$\overline{\mathbf{C}}\mathbf{x}(k) \rightarrow \overline{\mathbf{C}}\mathbf{M}\boldsymbol{\sigma}(k) = \overline{\mathbf{C}}\mathbf{M}\mathbf{r}(k)$$

so

$$\overline{\mathbf{y}}(k) \rightarrow \mathbf{r}(k)$$

requires that

$$\overline{\mathbf{C}}\mathbf{M} = \mathbf{I}$$

or

$$\begin{bmatrix} 1 & 0 & 0 \\ -2 & -1 & 1 \end{bmatrix} \begin{bmatrix} m_{11} & m_{12} \\ m_{21} & m_{22} \\ m_{31} & m_{32} \end{bmatrix} = \begin{bmatrix} 1 & 0 \\ 0 & 1 \end{bmatrix} \qquad (6\text{-}25)$$

Collecting the design equations (6-24) and (6-25) results in the following 10 linear algebraic equations in 10 unknowns:

$$\begin{cases}
\frac{3}{4}m_{11} & - m_{21} & & - \alpha_1 & + \alpha_3 & = 0 \\
& m_{21} & - m_{31} & & - \alpha_3 & = 0 \\
& & m_{31} & - 2\alpha_1 & & = 0 \\
\frac{3}{4}m_{12} & - m_{22} & & - \alpha_2 & + \alpha_4 = 0 \\
& m_{22} & - m_{32} & & -\alpha_4 = 0 \\
& & m_{32} & - 2\alpha_2 & = 0 \\
m_{11} & & & & = 1 \\
-2m_{11} & - m_{21} & + m_{31} & & = 0 \\
m_{12} & & & & = 0 \\
- 2m_{12} & - m_{22} & + m_{32} & & = 1
\end{cases}$$

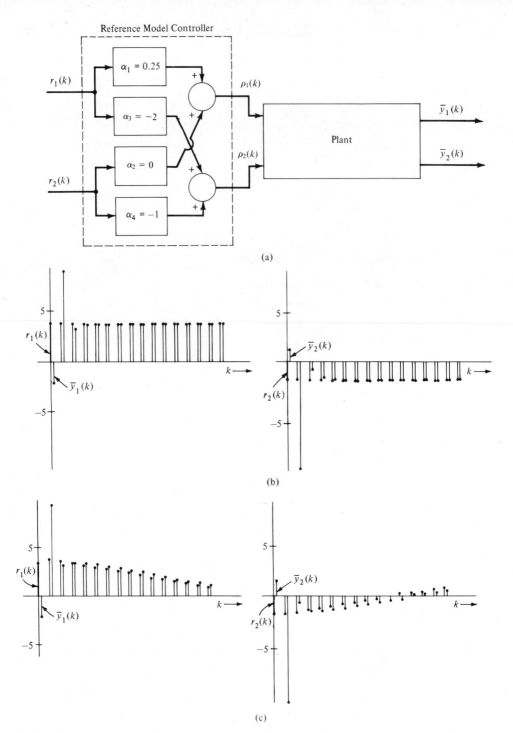

Figure 6-21 A reference model controller for the simultaneous tracking of two reference inputs. (a) Block diagram. (b) Response to two constant reference inputs. (c) Response to two slowly varying reference inputs.

The unique solution to these equations is

$$m_{11} = 1 \qquad m_{12} = 0 \qquad m_{21} = -\frac{3}{2}$$

$$m_{22} = -1 \qquad m_{31} = \frac{1}{2} \qquad m_{32} = 0$$

$$\alpha_1 = 0.25 \qquad \alpha_2 = 0$$
$$\alpha_3 = -2 \qquad \alpha_4 = -1$$

and specifies the reference model controller of Figure 6-21.

6.5 Disturbance Rejection

Disturbances are inputs to the plant that are not accessible for control. Examples of disturbances are wind gusts buffeting a positioning system for a microwave antenna, "noise" in electronic amplifiers, unknown offset calibration errors in sensors, and digital computation inaccuracies, such as numerical roundoff and truncation. Disturbances are inputs that we do *not* want the plant to track. Like initial conditions, the specific disturbance signals are normally unknown, although something is probably known about their character, their statistics, or both.

There are four basic ways to deal with plant disturbances:

1. *Ignore them* in the hope that their effects on system performance will be slight. Simulation studies can be used to evaluate the wisdom of this choice.
2. *Structure the control system* so that the effects of disturbances on those plant signals of concern are smaller. For example, negative feedback around a disturbance often reduces its effects. Unfortunately, little can be said in general about selecting control structures because practical design problems typically have many specific structural constraints. One usually must rely on common sense and experience to determine when it is possible and useful to vary the control structure. This is most often done with simple subsystems, before overall plant control design is begun.
3. *Model the disturbances as stochastic processes* and choose controller parameters (including the feedback system eigenvalues) to minimize disturbance effects in a statistical sense. This is very effective for disturbances, such as amplifier noise, that are adequately described by known short-term means and variances. It is not very effective for disturbances like wind gusts and crosstalk that have unknown and widely varying short-term statistics.

4. *Model the disturbances as plant modes with unknown amplitudes.*
The possibilities include constant, ramp, exponential, and sinusoidal sequences and sums of these. The disturbance model then becomes part of the plant model. This is effective for disturbances
that are highly correlated from step to step, and not as effective
otherwise.

A combination of these approaches may be needed. Some disturbance
sources are ignored, some are reduced by local feedback, some are
smoothed by the choice of controller parameters, and some are partially or completely canceled through modeling them as additional
plant modes.

The theory and applications associated with item number 3, stochastic disturbance modeling, is the subject of Chapter 8. One of the
results of this method is of interest in our present study. For zero-mean
white noise sequence disturbances ("white" means that the signal has
no sample-to-sample correlation) with constant variances, the linear
step-invariant estimator that gives minimum state estimate error variance is a full-order state observer with eigenvalue locations that depend on the disturbance variances. That is, for disturbances of this
type, the "best" observer to use (best in the sense of minimum state
estimate variance) is a full-order one with certain specific eigenvalues.
If the state estimator is allowed to vary with step, the estimates can be
further improved. Step-varying plants and disturbance variances that
change with step can then be accommodated, too. More about this
topic will appear in Chapter 8, in connection with recursive least
squares estimation.

The remainder of this section concerns item number 4, deterministic disturbance modeling.

6.5.1 State Disturbance Models

Deterministic disturbance modeling assumes that disturbances $\mathbf{v}(k)$ can
be modeled as the response of a jth-order autonomous system of the
form

$$\boldsymbol{\varepsilon}(k + 1) = \mathbf{T}\boldsymbol{\varepsilon}(k)$$

$$\mathbf{v}(k) = \boldsymbol{\Pi}\boldsymbol{\varepsilon}(k) \tag{6-26}$$

The response of this model is, for distinct eigenvalues $\lambda_1, \lambda_2, \ldots, \lambda_j$,

$$\mathbf{v}(k) = \boldsymbol{\beta}_1\lambda_1^k + \boldsymbol{\beta}_1\lambda_2^k + \cdots + \boldsymbol{\beta}_j\lambda_j^k$$

where the $\boldsymbol{\beta}$'s are arbitrary vector constants dependent on the model
initial conditions $\boldsymbol{\varepsilon}(0)$. For repeated eigenvalues, the corresponding

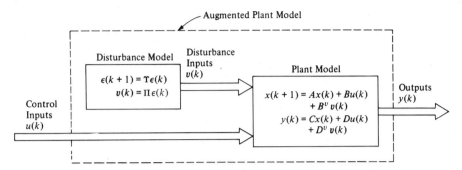

Figure 6-22 Modeling disturbances as the response of an autonomous disturbance model.

response terms are as listed in Table 6-1. This model, shown in Figure 6-22, is similar to that used for designing reference model disturbances to account for their effects on the state of a plant and so that their effects on plant tracking outputs can be wholly or partially canceled.

For a plant

$$\mathbf{x}(k + 1) = \mathbf{A}\mathbf{x}(k) + \mathbf{B}\mathbf{u}(k) + \mathbf{B}^v\mathbf{v}(k)$$

$$\mathbf{y}(k) = \mathbf{C}\mathbf{x}(k) + \mathbf{D}\mathbf{u}(k) + \mathbf{D}^v\mathbf{v}(k) \qquad (6\text{-}27)$$

with disturbance inputs $\mathbf{v}(k)$ modeled in this way, one can combine equations (6-26) and (6-27) into an *augmented* plant model

$$\begin{bmatrix} \mathbf{x}(k + 1) \\ \hline \boldsymbol{\varepsilon}(k + 1) \end{bmatrix} = \begin{bmatrix} \mathbf{A} & \mathbf{B}^v\boldsymbol{\Pi} \\ \hline \mathbf{0} & \mathbf{T} \end{bmatrix} \begin{bmatrix} \mathbf{x}(k) \\ \hline \boldsymbol{\varepsilon}(k) \end{bmatrix} + \begin{bmatrix} \mathbf{B} \\ \hline \mathbf{0} \end{bmatrix} \mathbf{u}(k)$$

$$\mathbf{y}(k) = [\mathbf{C} \mid \mathbf{D}^v\boldsymbol{\Pi}] \begin{bmatrix} \mathbf{x}(k) \\ \hline \boldsymbol{\varepsilon}(k) \end{bmatrix} + \mathbf{D}\mathbf{u}(k) \qquad (6\text{-}28)$$

that can be handled in much the same way as the original plant model. If the disturbances are modeled accurately by equation (6-26), then they are fully accounted for in the augmented plant model. The disturbance modes are not controllable, however. If the disturbances are given only approximately by their model, then using the augmented plant model only partially accounts for them. Better approximations give better results.

6.5.2 Disturbance Rejection with Observer Feedback

Improved state estimates are obtained by modeling disturbances according to equation (6-26) and observing the augmented plant model consisting of the combination of the original plant model and the distur-

bance model. More accurate state estimates generally result in feedback systems that are less susceptible to the modeled disturbances.
Consider the plant

$$\begin{bmatrix} x_1(k+1) \\ x_2(k+1) \end{bmatrix} = \begin{bmatrix} 1 & 1 \\ -\frac{1}{4} & 0 \end{bmatrix} \begin{bmatrix} x_1(k) \\ x_2(k) \end{bmatrix} + \begin{bmatrix} 1 \\ 0 \end{bmatrix} u(k) + \begin{bmatrix} 2 \\ -1 \end{bmatrix} v(k)$$

$$y(k) = \begin{bmatrix} 1 & 0 \end{bmatrix} \begin{bmatrix} x_1(k) \\ x_2(k) \end{bmatrix} + v(k) \tag{6-29}$$

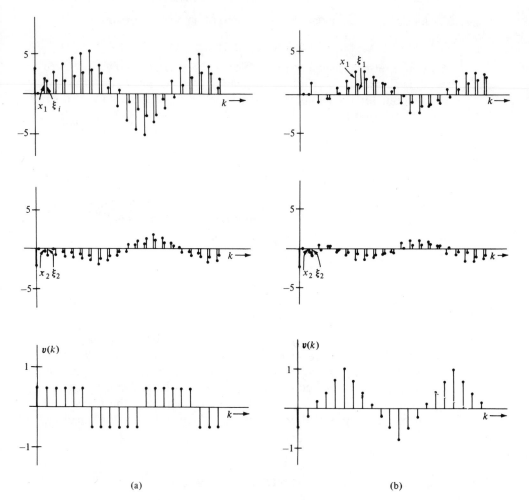

(a)   (b)

Figure 6-23 Response of a plant and observer when the plant has a disturbance input. The control input is zero. (a) Squarewave disturbance. (b) Triangular wave disturbance.

where $v(k)$ is a disturbance input. The full-order state observer of equation (6-29), with both eigenvalues $\frac{1}{4}$, that ignores the inaccessible disturbance input $v(k)$ is

$$\begin{bmatrix} \xi_1(k+1) \\ \xi_2(k+1) \end{bmatrix} = \begin{bmatrix} \frac{1}{2} & 1 \\ -\frac{1}{16} & 0 \end{bmatrix} \begin{bmatrix} \xi_1(k) \\ \xi_2(k) \end{bmatrix} + \begin{bmatrix} \frac{1}{2} \\ -\frac{3}{16} \end{bmatrix} y(k) + \begin{bmatrix} 1 \\ 0 \end{bmatrix} u(k)$$

Response of this observer for squarewave and triangular wave disturbances is shown in Figure 6-23. The plant control input $u(k)$ is temporarily set to zero so that the disturbance effects can be seen clearly. The designer hopes that the amplitude of any actual disturbance will be small so that errors in the observer estimate of the plant state will be small.

When the plant in equation (6-29) has the state feedback

$$u(k) = [-1 \quad -1] \begin{bmatrix} x_1(k) \\ x_2(k) \end{bmatrix}$$

the feedback plant eigenvalues are both at $\lambda = 0$. So when the observer's state estimate is fed back in place of the state (which is not accessible)

$$u(k) = [-1 \quad -1] \begin{bmatrix} \xi_1(k) \\ \xi_2(k) \end{bmatrix}$$

the feedback system eigenvalues are 0, 0, $\frac{1}{4}$, and $\frac{1}{4}$. Response of this observer feedback system with the squarewave disturbance $v(k)$ is shown in Figure 6-24.

If, instead, the disturbance is modeled as an unknown constant,

$$\varepsilon(k+1) = \varepsilon(k)$$

$$v(k) = \varepsilon(k)$$

the augmented plant model is

$$\begin{bmatrix} x_1(k+1) \\ x_2(k+1) \\ \varepsilon(k+1) \end{bmatrix} = \begin{bmatrix} 1 & 1 & 2 \\ -\frac{1}{4} & 0 & -1 \\ 0 & 0 & 1 \end{bmatrix} \begin{bmatrix} x_1(k) \\ x_2(k) \\ \varepsilon(k) \end{bmatrix} + \begin{bmatrix} 1 \\ 0 \\ 0 \end{bmatrix} u(k)$$

$$= \mathbf{A}\mathbf{x}'(k) + \mathbf{b}u(k)$$

$$y(k) = [1 \quad 0 \quad 1] \begin{bmatrix} x_1(k) \\ x_2(k) \\ \varepsilon(k) \end{bmatrix} = \mathbf{c}^\dagger \mathbf{x}'(k)$$

The disturbance state $\varepsilon(k)$ is uncontrollable, as one would expect, but it is observable.

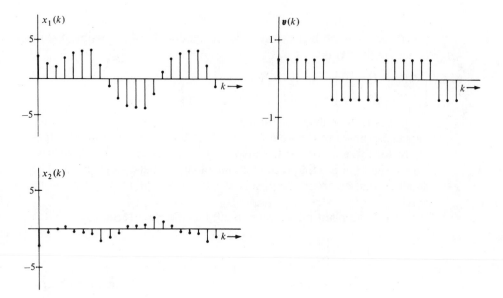

Figure 6-24 Response of the observer feedback system to a squarewave when the disturbance is not modeled.

A full-order state observer of this augmented plant, with all three eigenvalues $\frac{1}{4}$, is

$$
\begin{bmatrix} \xi_1(k+1) \\ \xi_2(k+1) \\ \xi_3(k+1) \end{bmatrix} = \begin{bmatrix} \frac{7}{80} & 1 & \frac{87}{80} \\ \frac{19}{80} & 0 & -\frac{41}{80} \\ -\frac{27}{80} & 0 & \frac{53}{80} \end{bmatrix} \begin{bmatrix} \xi_1(k) \\ \xi_2(k) \\ \xi_3(k) \end{bmatrix} + \begin{bmatrix} \frac{73}{80} \\ -\frac{39}{80} \\ \frac{27}{80} \end{bmatrix} y(k) + \begin{bmatrix} 1 \\ 0 \\ 0 \end{bmatrix} u(k)
$$

(6-30)

If the disturbance v is indeed a constant, the observer estimate converges to the plant state and to the constant disturbance, as shown in Figure 6-25(a). If the disturbance is not a constant, the estimate exhibits errors, but these can be much reduced from the error that results when the disturbance is simply ignored. Figure 6-25(b) shows the response of this observer with the same plant initial conditions and squarewave disturbance.

If state feedback were available for the augmented plant, a control input of the form

$$
u(k) = \begin{bmatrix} e_1 & e_2 & e_3 \end{bmatrix} \begin{bmatrix} x_1(k) \\ x_2(k) \\ \varepsilon(k) \end{bmatrix} = \mathbf{e}^\dagger \mathbf{x}'(k)
$$

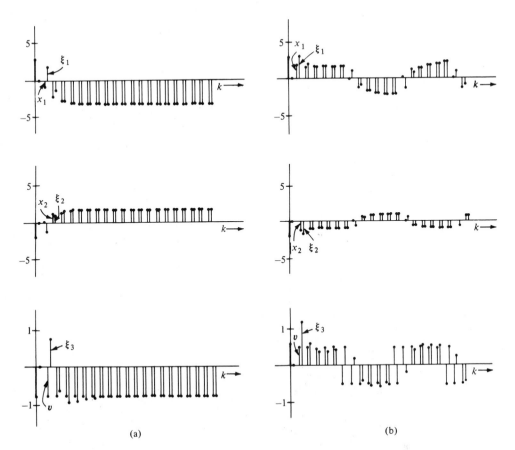

Figure 6-25 Response of the plant and an observer augmented by a disturbance model for an unknown constant disturbance. (a) Constant disturbance. (b) Squarewave disturbance.

would only place the two eigenvalues associated with the plant in equation (6-29), which is the controllable part of the augmented plant

$$|\lambda \mathbf{I} - (\mathbf{A} + \mathbf{b}\mathbf{e}^\dagger)| = \begin{vmatrix} (\lambda - 1 - e_1) & -(1 + e_2) & -(2 + e_3) \\ \frac{1}{4} & \lambda & 1 \\ 0 & 0 & (\lambda - 1) \end{vmatrix}$$

$$= (\lambda - 1)[\lambda^2 - (1 + e_1)\lambda + (\tfrac{1}{4} + \tfrac{1}{4}e_2)]$$

Any feedback of the form

$$u(k) = \begin{bmatrix} -1 & -1 & e_3 \end{bmatrix} \begin{bmatrix} x_1(k) \\ x_2(k) \\ \varepsilon(k) \end{bmatrix}$$

places both plant eigenvalues at $\lambda = 0$, as before; hence, any feedback from the observer in equation (6-30) of the form

$$u(k) = [-1 \quad -1 \quad e_3] \begin{bmatrix} \xi_1(k) \\ \xi_2(k) \\ \xi_3(k) \end{bmatrix} \tag{6-31}$$

results in an observer feedback system having three $\lambda = \frac{1}{4}$ observer eigenvalues and two $\lambda = 0$ plant eigenvalues.

Figure 6-26 shows the squarewave response of the plant and this observer for

$$u(k) = [-1 \quad -1 \quad 0] \begin{bmatrix} \xi_1(k) \\ \xi_2(k) \\ \xi_3(k) \end{bmatrix}$$

If, instead, the disturbance is modeled as an arbitrary constant plus an arbitrary ramp sequence

$$\begin{bmatrix} \varepsilon_1(k+1) \\ \varepsilon_2(k+1) \end{bmatrix} = \begin{bmatrix} 2 & 1 \\ -1 & 0 \end{bmatrix} \begin{bmatrix} \varepsilon_1(k) \\ \varepsilon_2(k) \end{bmatrix}$$

$$v(k) = [1 \quad 0] \begin{bmatrix} \varepsilon_1(k) \\ \varepsilon_2(k) \end{bmatrix}$$

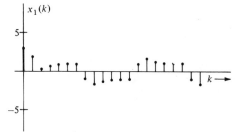
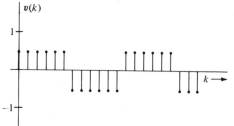

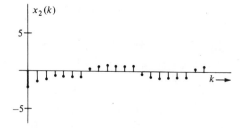

Figure 6-26 Response of the observer feedback system to a squarewave disturbance when the disturbance is modeled as an unknown constant.

the augmented plant model is

$$
\begin{bmatrix} x_1(k+1) \\ x_2(k+1) \\ \varepsilon_1(k+1) \\ \varepsilon_2(k+1) \end{bmatrix} = \begin{bmatrix} 1 & 1 & 2 & 0 \\ -\frac{1}{4} & 0 & -1 & 0 \\ 0 & 0 & 2 & 1 \\ 0 & 0 & -1 & 0 \end{bmatrix} \begin{bmatrix} x_1(k) \\ x_2(k) \\ \varepsilon_1(k) \\ \varepsilon_2(k) \end{bmatrix} + \begin{bmatrix} 1 \\ 0 \\ 0 \\ 0 \end{bmatrix} u(k)
$$

$$
y(k) = \begin{bmatrix} 1 & 0 & 1 & 0 \end{bmatrix} \begin{bmatrix} x_1(k) \\ x_2(k) \\ \varepsilon_1(k) \\ \varepsilon_2(k) \end{bmatrix}
$$

and an observer of this model gives still better disturbance rejection performance when used for feedback.

6.5.3 Disturbance Cancellation

To the extent that a disturbance is correctly modeled as in equation (6-26), its effect on the plant state is correctly calculated. Although the disturbance model modes cannot be changed by feedback, because these modes are uncontrollable, feedback of observer estimates of disturbances can be used to reduce the effects of the disturbances on plant tracking outputs. Disturbance estimates can be used much as reference inputs, but only where the objective is cancellation of their effects on the tracking outputs.

When disturbance estimates are fed back, the eigenvalue placement obtained through feedback of plant state estimates is not changed. For the general augmented plant in equation (6-28), plant and disturbance model state feedback

$$
\mathbf{u}(k) = \mathbf{E}_x \mathbf{x}(k) + \mathbf{E}_\varepsilon \boldsymbol{\varepsilon}(k) + \boldsymbol{\rho}(k)
$$

gives the feedback system

$$
\begin{bmatrix} \mathbf{x}(k+1) \\ \boldsymbol{\varepsilon}(k+1) \end{bmatrix} = \begin{bmatrix} \mathbf{A} + \mathbf{BE}_x & \mathbf{B}^v \boldsymbol{\Pi} + \mathbf{BE}_\varepsilon \\ \mathbf{0} & \mathbf{T} \end{bmatrix} \begin{bmatrix} \mathbf{x}(k) \\ \boldsymbol{\varepsilon}(k) \end{bmatrix} + \begin{bmatrix} \mathbf{B} \\ \mathbf{0} \end{bmatrix} \boldsymbol{\rho}(k)
$$

which has eigenvalues that are those of the plant with state feedback $\mathbf{A} + \mathbf{BE}_x$ (independent of the disturbance feedback) and of the disturbance model $\mathbf{T}$.

Consider again the plant in equation (6-29) augmented by a constant disturbance model and the full-order state observer in equation (6-30) of the augmented plant. The observer variable $\xi_3(k)$ estimates any unknown constant disturbance input v. By observing v as well as the plant state, the effect of any arbitrary constant disturbance v on the plant state is accounted for, and the observer generates correct state estimates for feedback. The observer estimates of v can be fed back, too, through the gain e_3 in equation (6-31), without affecting the feedback system eigenvalues. This estimate $\xi_3(k)$ can be used to cancel a constant disturbance's effect on some plant signal of interest, as the observer estimate converges. For example, for the choice

$$e_3 = -2$$

the steady state effect of any constant disturbance v on the state variable $x_1(k)$ is zero. Response of the feedback system consisting of the plant, the observer of the augmented plant, and observer feedback

$$u(k) = [-1 \quad -1 \quad -2] \begin{bmatrix} \xi_1(k) \\ \xi_2(k) \\ \xi_3(k) \end{bmatrix}$$

to a squarewave disturbance $v(k)$ is shown in Figure 6-27. The effect of this disturbance is much reduced from that with the same plant initial conditions and disturbance in Figure 6-26.

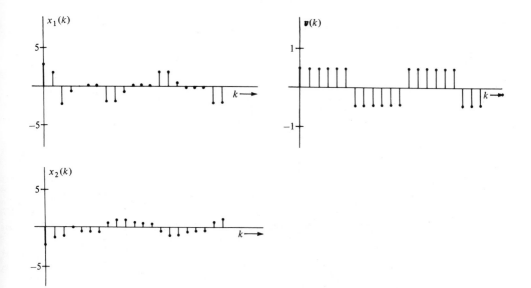

Figure 6-27 Response of the plant with observer feedback when the observer estimate of the disturbance is used for cancellation in the $x_1(k)$ signal.

6.6 A Digital Phase-Locked Loop

A digital phase-locked loop (PLL) produces a time-varying binary output signal in response to a nearly periodic time-varying binary input signal. The intervals between the switching times of the input waveform are to change relatively slowly, but there may be small but rapid fluctuations from the ideal. The PLL produces an output waveform with switching times that are approximately at the same times as the input switchings but with any fluctuations in the switching times smoothed. Phase-locked loops are used to demodulate the stereo subcarrier in FM broadcast receivers, to detect the color information in TV receivers, to aid in synchronizing alternating current power distribution, to recover the master clock in digital data transmission and recording, and in many other applications.

Figure 6-28(a) is a block diagram of a digital phase-locked loop. The PLL input waveform switches at nearly constant intervals that are given by the input sequence $v(k)$. The running sum (accumulation) of

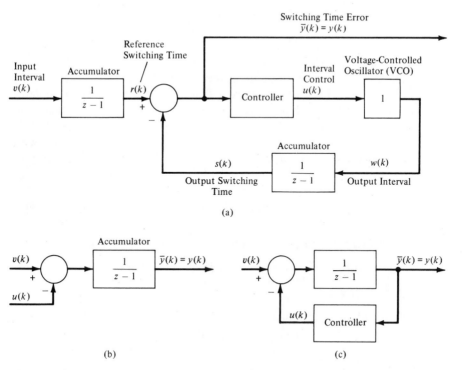

Figure 6-28 Phase-locked loop block diagrams. (a) A digital phase-locked loop. (b) Equivalent block diagram. (c) The plant and controller.

the input switching intervals is the sequence of input switching times, which is nearly a ramp. Similarly, the accumulation of output intervals is the sequence of output switching times. The differences between the input and output switching times is the switching time error, $\bar{y}(k) - y(k)$. The switching time error sequence is both accessible for feedback and is the system's output to be regulated. The controller controls the period of a voltage-controlled oscillator (VCO) that produces the output waveform.

The switching time error $\bar{y}(k)$ should decay to zero whenever $v(k)$ is constant. When the input interval $v(k)$ changes, perhaps erratically, the corresponding changes in the output interval $w(k)$ should be smoothed. One does not want simply to let $w(k)$ be a copy of $v(k)$ because, then, there is no smoothing action by the PLL. Every little fluctuation in $v(k)$ would also appear in $w(k)$. Instead, the controller should allow only relatively slow changes in the VCO output interval.

In Figure 6-28(b), the plant portion of the digital phase-locked loop has been redrawn with the two separate accumulations combined. State variable equations for the plant are

$$x(k + 1) = x(k) + v(k) - u(k)$$

$$\bar{y}(k) = y(k) = x(k) \tag{6-32}$$

The plant equations are unusual in that the step index k is the oscillation interval number, not a time step. When the interval lengths differ, each step k represents a different length step in time. Figure 6-28(c) shows this plant with a controller that is to be designed. Although any of the other tracking system design methods could be applied to this problem, we choose to consider the design to be that of a regulator with output $\bar{y}(k)$ and disturbance input $v(k)$. The regulator output $\bar{y}(k)$ should decay to zero for any constant disturbance $v(k)$. The modes of the overall system should be such that for rapid fluctuations in the input interval $v(k)$, the output interval $w(k)$ changes relatively slowly.

6.6.1 A First-Order Observer-Controller

If the disturbance $v(k)$ is modeled as an unknown constant

$$v(k + 1) = v(k)$$

the augmented plant model is

$$\begin{bmatrix} x(k + 1) \\ v(k + 1) \end{bmatrix} = \begin{bmatrix} 1 & 1 \\ 0 & 1 \end{bmatrix} \begin{bmatrix} x(k) \\ v(k) \end{bmatrix} + \begin{bmatrix} -1 \\ 0 \end{bmatrix} u(k) = \mathbf{A}\mathbf{x}(k) + \mathbf{b}u(k)$$

$$\bar{y}(k) = y(k) = \begin{bmatrix} 1 & 0 \end{bmatrix} \begin{bmatrix} x(k) \\ v(k) \end{bmatrix} = \mathbf{c}^\dagger\mathbf{x}(k)$$

This augmented plant is not completely controllable, as expected, because the disturbance input model is autonomous. It is completely observable, however.

If the augmented plant's state is available for feedback, state feedback of the form

$$u(k) = [e_1 \quad e_2] \begin{bmatrix} x(k) \\ v(k) \end{bmatrix} = \mathbf{e}^\dagger \mathbf{x}(k)$$

results in the feedback system state coupling matrix

$$\mathbf{A} + \mathbf{be}^\dagger = \begin{bmatrix} (1 - e_1) & (1 - e_2) \\ 0 & 1 \end{bmatrix}$$

which has the characteristic equation

$$(\lambda - 1)(\lambda - 1 + e_1) = 0$$

The $\lambda = 1$ eigenvalue of the uncontrollable disturbance model part of the augmented plant is not affected by any feedback. The original plant eigenvalue can be placed by feedback of the original plant state. It is not affected by feedback of the disturbance model state. Any state feedback with $e_1 = \frac{1}{2}$ results in the augmented plant eigenvalues $\lambda = 1$ and $\lambda = \frac{1}{2}$.

For $e_1 = \frac{1}{2}$, the augmented plant with feedback is

$$\begin{bmatrix} x(k + 1) \\ v(k + 1) \end{bmatrix} = \begin{bmatrix} \frac{1}{2} & (1 - e_2) \\ 0 & 1 \end{bmatrix} \begin{bmatrix} x(k) \\ v(k) \end{bmatrix}$$

$$\bar{y}(k) = [1 \quad 0] \begin{bmatrix} x(k) \\ v(k) \end{bmatrix}$$

and its tracking output is related to the initial conditions by

$$\bar{y}(k) = \left(\frac{1}{2}\right)^k x(0) + 2(1 - e_2)\left[1 - \left(\frac{1}{2}\right)^k\right] v(0)$$

Choosing $e_2 = 1$ gives zero steady state error to any constant disturbance v.

The augmented plant state is not available for feedback, though, so we will design an observer of the state and feed back the observer's state estimate in place of the actual state. A first-order observer with eigenvalue $\lambda = \frac{1}{2}$ and gain $g = 1$ is of the form

$$\xi(k + 1) = \frac{1}{2}\xi(k) + y(k) + hu(k)$$

It observes

$$\xi(k) \to \mathbf{m}^\dagger \mathbf{x}(k)$$

where

$$\mathbf{m}^\dagger = \mathbf{c}^\dagger \left(\mathbf{A} - \frac{1}{2}\mathbf{I} \right)^{-1} = [2 \quad -4]$$

and

$$h = \mathbf{m}^\dagger\mathbf{b} = -2$$

The collection of signals

$$\begin{bmatrix} w_1(k) \\ w_2(k) \end{bmatrix} = \begin{bmatrix} 0 \\ -\frac{1}{4} \end{bmatrix} \xi(k) + \begin{bmatrix} 1 \\ \frac{1}{2} \end{bmatrix} y(k)$$

observes the augmented plant state

$$\mathbf{w}(k) \rightarrow \mathbf{x}(k)$$

The observer feedback

$$u(k) = \mathbf{e}^\dagger\mathbf{w}(k) = [\tfrac{1}{2} \quad 1]\mathbf{w}(k) = -\frac{1}{4}\xi(k) + y(k)$$

places the plant eigenvalue at $\lambda = \frac{1}{2}$ and gives a tracking output signal $\bar{y}(k)$ that decays to zero for any constant disturbance input $v(k)$.

The observer-controller is described by

$$\xi(k + 1) = \frac{1}{2}\xi(k) + y(k) - 2u(k)$$

$$u(k) = -\frac{1}{4}\xi(k) + y(k)$$

Eliminating $u(k)$ from the first of these equations results in the controller state variable equations

$$\xi(k + 1) = \xi(k) - y(k)$$

$$u(k) = -\frac{1}{4}\xi(k) + y(k)$$

Using these and the plant equations in (6-32), the zero-state tracking output and the disturbance input are related by

$$\bar{Y}(z) = \left[\frac{z - 1}{(z - \frac{1}{2})^2} \right] V(z) = T(z)V(z)$$

For any constant disturbance

$$V(z) = \frac{\beta z}{z - 1}$$

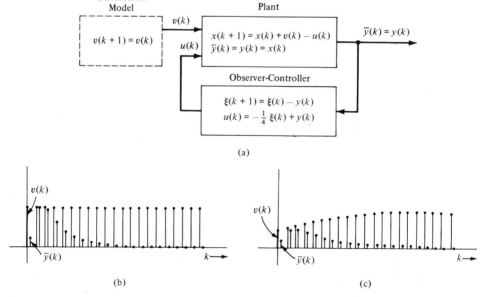

Figure 6-29 The digital phase-locked loop with a controller based on a first-order disturbance signal model. Ideally, $y(k)$ is zero. (a) Block diagram. (b) Response when the input interval is constant. (c) Response when the input interval varies slowly.

then

$$\lim_{k \to \infty} \bar{y}(k) = \lim_{z \to 1} \left[\frac{z-1}{z} V(z)T(z) \right] = 0$$

as expected.

Figure 6-29(a) shows the plant with this controller. Response of the system to a constant input is shown in Figure 6-29(b). As expected, the switching time error, which is the system's tracking output, decays to zero. Figure 6-29(c) shows the response of the system to a slowly varying input.

6.6.2 A Higher-Order Controller

If, instead, the order of the disturbance signal model is raised so that $v(k)$ is modeled as an unknown constant plus an unknown ramp sequence, a disturbance model is

$$\begin{bmatrix} \sigma_1(k+1) \\ \sigma_2(k+1) \end{bmatrix} = \begin{bmatrix} 0 & 1 \\ -1 & 2 \end{bmatrix} \begin{bmatrix} \sigma_1(k) \\ \sigma_2(k) \end{bmatrix}$$

$$v(k) = \sigma_1(k)$$

The augmented plant model is

$$
\begin{bmatrix} x(k+1) \\ \sigma_1(k+1) \\ \sigma_2(k+1) \end{bmatrix} = \begin{bmatrix} 1 & 1 & 0 \\ 0 & 0 & 1 \\ 0 & -1 & 2 \end{bmatrix} \begin{bmatrix} x(k) \\ \sigma_1(k) \\ \sigma_2(k) \end{bmatrix} + \begin{bmatrix} -1 \\ 0 \\ 0 \end{bmatrix} u(k) = \mathbf{A}\mathbf{x}(k) + \mathbf{b}u(k)
$$

$$
y(k) = \begin{bmatrix} 1 & 0 & 0 \end{bmatrix} \begin{bmatrix} x(k) \\ \sigma_1(k) \\ \sigma_2(k) \end{bmatrix} = \mathbf{c}^t\mathbf{x}(k)
$$

State feedback of the form

$$
u(k) = \begin{bmatrix} e_1 & e_2 & e_3 \end{bmatrix} \begin{bmatrix} x(k) \\ \sigma_1(k) \\ \sigma_2(k) \end{bmatrix} = \mathbf{e}^t\mathbf{x}(k)
$$

gives an augmented plant with feedback having the characteristic equation

$$
(\lambda - 1)^2(\lambda - 1 + e_1) = 0
$$

so that the eigenvalue of the original plant, which is the controllable part of the augmented plant, can be placed by selecting the plant state gain e_1. The disturbance state gains e_2 and e_3 do not affect this eigenvalue placement.

For the choice

$$
\mathbf{e}^t = \begin{bmatrix} \frac{3}{4} & 1 & 0 \end{bmatrix}
$$

the eigenvalue of the controllable part of the augmented plant is moved to $\lambda = \frac{1}{4}$, and the tracking output is related to the augmented plant initial conditions by

$$
\bar{y}(k) = \mathbf{c}^t(\mathbf{A} + \mathbf{b}\mathbf{e}^t)^k\mathbf{x}(0) = \left(\frac{1}{4}\right)^k x(0) + 0\sigma_1(0) + 0\sigma_2(0)
$$

There is then no steady state tracking error for any disturbance that is a constant plus a ramp.

All of the augmented plant state is not available for feedback, so we will design an observer to estimate the state. A full-order state observer of the augmented plant, with all three eigenvalues $\lambda = \frac{1}{4}$, is

$$
\begin{bmatrix} \xi_1(k+1) \\ \xi_2(k+1) \\ \xi_3(k+1) \end{bmatrix} = \begin{bmatrix} -2.75 & 1 & 0 \\ -4.69 & 0 & 1 \\ -6.46 & -1 & 2 \end{bmatrix} \begin{bmatrix} \xi_1(k) \\ \xi_2(k) \\ \xi_3(k) \end{bmatrix}
$$

$$
+ \begin{bmatrix} 3.75 \\ 4.69 \\ 6.46 \end{bmatrix} y(k) + \begin{bmatrix} -1 \\ 0 \\ 0 \end{bmatrix} u(k)
$$

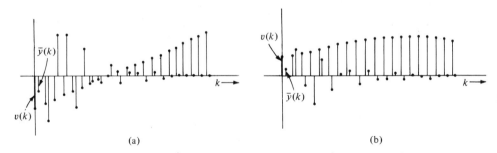

(a)   (b)

Figure 6-30 Response of the digital phase-locked loop with a controller based on a second-order disturbance signal model. (a) Response when the input interval is a constant plus a ramp sequence. (b) Response when the input interval varies slowly.

For

$$u(k) = \frac{3}{4} \xi_1(k) + \xi_2(k)$$

the original plant with observer feedback has all four eigenvalues at $\lambda = \frac{1}{4}$ and has zero steady state error to any constant-plus-ramp input sequence. Typical response is shown in Figure 6-30 and is seen to be improved over that of the lower-order observer–controller.

In certain applications, for example, data clock recovery from rotating disk drives, the input signal $v(k)$ has a small but significant sinusoidal component. For a disk, the frequency of the sinusoidal component is the drive rotation frequency and results mainly from eccentricities in the circular recorded tracks. An appropriate model for $v(k)$ is then an arbitrary constant plus a sinusoidal sequence of arbitrary amplitude and phase.

6.6.3 Error Quantization and Measurement Interruption

It is common in a system such as this one for the switching time error measurement $y(k)$ to be made by a counter driven by a high-speed clock. The switching time error is made by counting the clock pulses between the input and output switchings. This error measurement, consisting of a count of clock cycles, could then be highly quantized. The output switching times might also be quantized, synchronized to a master clock. It is then appropriate to consider the measurements $y(k)$ to consist of the accurate switching time error plus a random "noise" component representing the quantization error. Figure 6-31 shows the effect of quantization on the performance of the system with the second-order disturbance input model. Response of the PLL to slowly

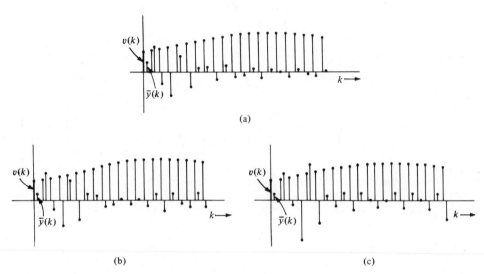

Figure 6-31 Effects on performance of using quantized switching-time error measurements. (a) Eight bits. (b) Six bits. (c) Four bits.

varying input intervals is shown when the switching time error measurements are truncated to 8, 6, and 4 binary digits (bits) of accuracy. Under these circumstances, performance would likely be improved by placing the observer and plant eigenvalues still further from the origin of the complex plane so that the reference signal estimate involves greater smoothing of the measurements. The disadvantage of doing this is that the speed of response to changes in $v(k)$ is slowed.

It is also common for there to be interruptions in the PLL measurements $y(k)$. A PLL used to recover the clock signal from transmitted or recorded data, for instance, can make switching time comparisons only when the data changes from a binary 1 to a 0 or from a 0 to a 1. If the observer–controller is arranged in the form of the augmented plant model driven by measurement error (as in Figure 5-8), new measurement data need be incorporated only when it is available. When it is not available, the plant model predicts the behavior of the plant, including the input model, on the basis of the previous measurements.

6.7 Tracking System Design for Step-Varying Systems

Most step-invariant tracking system design principles are also applicable to step-varying tracking system design. Each of the three basic tracking system design methods

1. Ideal tracking
2. Response model
3. Reference input model

can be generalized to step-varying systems. Existence of an ideal tracking filter depends on whether the plant with feedback has a suitable step-varying inverse filter. Design using a response model requires, as before, selecting a response model that is both suitable and feasible, a process that, in general, is even more difficult than in the step-invariant case. In reference input model design, the plant to be controlled is made to be an observer of a model of the reference signals of particular interest. The system then tracks ideally all reference inputs that are modeled.

6.7.1 Observing Tracking Signal Models

Having designed an acceptable zero-input response, the zero-state response of a tracking system can be designed by requiring that the tracking outputs observe signals from a suitable reference signal model, as in Figure 6-32. In the step-varying case, observation of a signal model

$$\boldsymbol{\sigma}(k + 1) = \boldsymbol{\Psi}(k)\boldsymbol{\sigma}(k)$$

$$\mathbf{r}(k) = \boldsymbol{\Theta}(k)\boldsymbol{\sigma}(k) \tag{6-33}$$

by the plant

$$\mathbf{x}(k + 1) = \mathbf{A}(k)\mathbf{x}(k) + \mathbf{B}(k)\mathbf{r}(k)$$

$$\bar{\mathbf{y}}(k) = \bar{\mathbf{C}}(k)\mathbf{x}(k) + \bar{\mathbf{D}}(k)\mathbf{r}(k)$$

where

$$\mathbf{x}(k) \rightarrow \mathbf{M}(k)\boldsymbol{\sigma}(k)$$

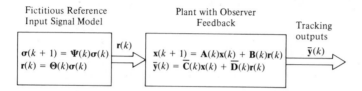

Figure 6-32 Observation of a step-varying, reference input signal model.

requires that

$\mathbf{A}(k)\mathbf{M}(k) = \mathbf{M}(k + 1)\mathbf{\Psi}(k) - \mathbf{B}(k)\mathbf{\Theta}(k)$

For the plant tracking outputs to observe the model reference signal

$\bar{\mathbf{y}}(k) \rightarrow \mathbf{r}(k)$

constrains the observation transformation matrices $\mathbf{M}(k)$ according to

$\bar{\mathbf{y}}(k) = \bar{\mathbf{C}}(k)\mathbf{x}(k) + \bar{\mathbf{D}}(k)\mathbf{r}(k) \rightarrow \bar{\mathbf{C}}(k)\mathbf{M}(k)\boldsymbol{\sigma}(k) + \bar{\mathbf{D}}(k)\mathbf{r}(k) = \mathbf{r}(k)$

Substituting from equation (6-33) results in

$\mathbf{C}(k)\mathbf{M}(k)\boldsymbol{\sigma}(k) = [\mathbf{I} - \bar{\mathbf{D}}(k))]\mathbf{\Theta}(k)\boldsymbol{\sigma}(k)$

or

$\mathbf{C}(k)\mathbf{M}(k) = [\mathbf{I} - \bar{\mathbf{D}}(k)]\mathbf{\Theta}(k)$

   As an example, consider the plant with feedback

$$\begin{bmatrix} x_1(k + 1) \\ x_2(k + 1) \end{bmatrix} = \begin{bmatrix} \frac{1}{2} & 1 \\ 0 & \frac{1}{k + 1} \end{bmatrix} \begin{bmatrix} x_1(k) \\ x_2(k) \end{bmatrix} + \begin{bmatrix} \alpha(k) \\ 2\alpha(k) \end{bmatrix} r(k)$$

$$= \mathbf{A}(k)\mathbf{x}(k) + \mathbf{b}(k)r(k)$$

$$\bar{y}(k) = \begin{bmatrix} 1 & k \end{bmatrix}\begin{bmatrix} x_1(k) \\ x_2(k) \end{bmatrix} = \bar{\mathbf{c}}^\dagger(k)\mathbf{x}(k)$$

which includes output or observer feedback, if used, and has a scalar control input gain sequence $\alpha(k)$ to be selected. Suppose it is desired that the zero-state component of the plant's tracking output $\bar{y}(k)$ tracks an alternating reference input of the form

$r(k) = \beta(-1)^k$

where β is any constant. A model for this class of reference input signals is the first-order system

$\sigma(k + 1) = \sigma(k) = \mathbf{\Psi}\sigma(k)$

$\quad r(k) = (-1)^k\sigma(k) = \theta(k)\sigma(k)$

   Denoting the state transformation of $\sigma(k)$ observed by $\mathbf{x}(k)$ as

$\mathbf{x}(k) \rightarrow \mathbf{m}(k)\sigma(k)$

for the plant to be an observer of the reference signal model requires that

$\mathbf{A}(k)\mathbf{m}(k) = \mathbf{m}(k + 1)\mathbf{\Psi} - \mathbf{b}(k)\theta(k)$

or

$$A \begin{bmatrix} m_1(k) \\ m_2(k) \end{bmatrix} = \begin{bmatrix} m_1(k+1) \\ m_2(k+1) \end{bmatrix} - \begin{bmatrix} \alpha(k) \\ 2\alpha(k) \end{bmatrix} (-1)^k \qquad \text{(6-34)}$$

If

$$\bar{y}(k) \rightarrow r(k)$$

it is also required that

$$\bar{y}(k) = \bar{c}^\dagger(k)x(k) \rightarrow \bar{c}^\dagger(k)m(k)\sigma(k) = r(k)$$

or

$$c^\dagger(k)m(k) = \theta(k)$$

or

$$[1 \quad k] \begin{bmatrix} m_1(k) \\ m_2(k) \end{bmatrix} = m_1(k) + km_2(k) = (-1)^k \qquad \text{(6-35)}$$

At each step k, equation (6-35) constrains the two components of the observation matrix m such that

$$m_1(k) = (-1)^k - km_2(k) \qquad \text{(6-36)}$$

Combining this constraint at step $k + 1$ with the other two linear algebraic equations in (6-34) gives

$$\begin{cases} m_1(k+1) + (k+1)m_2(k+1) = (-1)^{k+1} \\ m_1(k+1) - (-1)^k\alpha(k) = \dfrac{1}{2} m_1(k) + m_2(k) \\ m_2(k+1) - 2(-1)^k\alpha(k) = \dfrac{1}{k+1} m_2(k) \end{cases} \qquad \text{(6-37)}$$

which is a recursive arrangement for computing $m(k + 1)$ and $\alpha(k)$ from any initial $m(0)$ satisfying equation (6-36). If

$$m(0) = \begin{bmatrix} 1 \\ 0 \end{bmatrix}$$

for instance, then equation (6-37) with $k = 0$ gives

$$\begin{cases} m_1(1) + m_2(1) = -1 \\ m_1(1) - \alpha(0) = \dfrac{1}{2} \\ m_2(1) - 2\alpha(0) = 0 \end{cases}$$

which has the solution

$$\mathbf{m}(1) = \begin{bmatrix} 0 \\ -1 \end{bmatrix} \qquad \alpha(0) = -\frac{1}{2}$$

Then equation (6-37) with $k = 1$ gives

$$\begin{cases} m_1(2) + 2m_2(2) = 1 \\ m_1(2) + \alpha(1) = \dfrac{1}{2} m_1(1) + m_2(1) = -1 \\ m_2(2) + 2\alpha(1) = \dfrac{1}{2} m_2(1) = -\dfrac{1}{2} \end{cases}$$

which has the solution

$$\mathbf{m}(2) = \begin{bmatrix} -0.4 \\ 0.7 \end{bmatrix} \qquad \alpha(1) = -0.6$$

and so on. The gains $\alpha(k)$ make the plant's zero-state response track any alternating sequence perfectly.

6.7.2 Disturbance Models

The approaches to tracking system disturbance rejection

1. Ignoring the disturbances
2. Structuring the plant and controller (when possible) to reduce the undesirable effects of disturbances
3. Modeling disturbances as stochastic processes
4. Modeling disturbances as the output of an autonomous system

apply also to step-varying tracking system design. As always, simulation studies are helpful for evaluating choice 1. Choice 2 is difficult to discuss in general because system configurations have highly individual costs and constraints. We touch on the third choice in Chapter 8.

The fourth choice, which is especially suitable for disturbances that are highly correlated from step to step, is pictured in Figure 6-33. It is actually easier to design step-varying state models of disturbances than step-invariant ones. For example, any scalar disturbance of the form

$$v(k) = \beta_1 f_1(k) + \beta_2 f_2(k) + \cdots + \beta_i f_i(k)$$

with constants $\beta_1, \beta_2, \ldots, \beta_i$ arbitrary and $f_1(k), f_2(k), \ldots, f_i(k)$ any functions, is easily modeled with an ith-order linear, step-varying system

$$\varepsilon(k + 1) = \mathbf{T}(k)\varepsilon(k)$$
$$v(k) = \boldsymbol{\pi}^{\dagger}(k)\varepsilon(k)$$

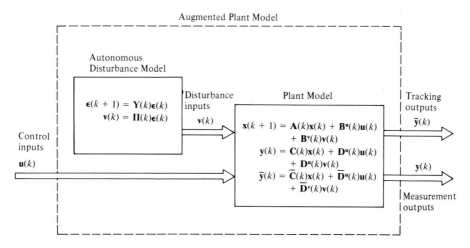

Figure 6-33 Incorporating a disturbance model.

by choosing

$$T(k) = I$$

and

$$\pi^\dagger(k) = [f_1(k) \quad f_2(k) \cdots f_i(k)]$$

For these choices, the model state $\varepsilon(k) = \varepsilon(0)$ is constant and equal to the initial conditions

$$v(k) = \varepsilon_1(0)f_1(k) + \varepsilon_2(0)f_2(k) + \cdots + \varepsilon_i(0)f_i(k)$$

For example, a disturbance of the form

$$v(k) = \beta_1 e^{-0.1k^2} + \beta_2 \left(\frac{1}{k!}\right)$$

can only be modeled approximately with a step-invariant system, even one of very high order. It is modeled exactly by the linear, step-varying model

$$\begin{bmatrix} \varepsilon_1(k + 1) \\ \varepsilon_2(k + 1) \end{bmatrix} = \begin{bmatrix} \varepsilon_1(k) \\ \varepsilon_2(k) \end{bmatrix}$$

$$v(k) = \begin{bmatrix} e^{-0.1k^2} & \dfrac{1}{k!} \end{bmatrix} \begin{bmatrix} \varepsilon_1(k) \\ \varepsilon_2(k) \end{bmatrix}$$

As with step-invariant systems, observer estimates of disturbances, obtained by observing a plant augmented with a state disturbance model, can be used to cancel the effect of a disturbance on a tracking output, to the extent that the model is accurate.

6.8 Summary

For a plant

$$\mathbf{x}(k + 1) = \mathbf{A}\mathbf{x}(k) + \mathbf{B}\mathbf{u}(k)$$

we distinguish between the measurement outputs

$$\mathbf{y}(k) = \mathbf{C}\mathbf{x}(k) + \mathbf{D}\mathbf{u}(k)$$

that are available for feedback control and the tracking outputs

$$\bar{\mathbf{y}}(k) = \bar{\mathbf{C}}\mathbf{x}(k) + \bar{\mathbf{D}}\mathbf{u}(k)$$

which are to become and remain nearly equal to externally supplied reference inputs $\mathbf{r}(k)$. The zero-input component of $\bar{\mathbf{y}}(k)$ depends on the initial conditions which are usually unknown. Having shaped the plant's zero-input response by placing the feedback system eigenvalues using the methods of the previous chapter, the zero-state response of the tracking outputs is now designed. If, instead, the initial conditions are known, the zero-input components of the tracking outputs can be calculated and the system designed to track

$$\bar{\mathbf{r}}(k) = \mathbf{r}(k) - \bar{\mathbf{y}}_{\text{zero-input}}(k)$$

When it can be arranged so that

$$\bar{\mathbf{y}}_{\text{zero-state}}(k) = \mathbf{r}(k)$$

ideal tracking is obtained. The tracking outputs have decaying zero-input response shaped by the selection of feedback system eigenvalues, after which they equal the reference inputs. In terms of transfer function matrices, for a plant with the z-transfer function matrix $\mathbf{T}(z)$, ideal tracking requires a reference input filter with the z-transfer function matrix $\mathbf{G}(z)$ such that

$$\mathbf{T}(z)\mathbf{G}(z) = \mathbf{I}$$

and $\mathbf{G}(z)$ is an inverse filter for the plant.

Any reference input filter does not change the feedback system eigenvalues, so the feedback eigenvalue placement part of the design is not affected. The ideal tracking design equations are linear but solutions can, unfortunately, be unstable and noncausal.

Response model tracking is obtained when the reference input filter gives

$$\bar{\mathbf{Y}}_{\text{zero-state}}(z) = \mathbf{\Omega}(z)\mathbf{R}(z)$$

where $\mathbf{\Omega}(z)$ characterizes an acceptable z-transfer function matrix relation between the tracking outputs and the reference inputs that is less stringent than the

$$\Omega(z) = \mathbf{I}$$

for ideal tracking. As with ideal tracking, the response model tracking design equations are linear. Unstable, noncausal, and noise-magnifying filter solutions can be avoided but can result in less than perfect tracking.

The more plant inputs that are available, the more design options there are for a reference input filter. When an observer is used for placing feedback eigenvalues, external inputs can be added to each scalar observer state equation to provide more inputs for the filter.

In reference model tracking, the feedback plant input gains (including gains to feedback observer states) are selected so that the plant is an observer of a designer-selected autonomous (with no input) reference input model. Typical models are of step and step-plus-ramp reference inputs. The zero-state components of the plant tracking outputs track perfectly any reference inputs from the class of modeled inputs. This approach allows the designer to specify those types of inputs the tracking system should track well. If the order of the reference signal model is so high as to preclude a solution with the number of available inputs to the plant with feedback, the order of the observer used for plant feedback can be raised to provide the needed additional inputs.

Disturbances are inaccessible plant inputs that the designer does not want to have much influence on the tracking outputs. Of the several ways of handling disturbances, we concentrated on the design technique of modeling disturbances as autonomous signal models.* To the extent that the disturbance is accurately modeled, its effects on the plant state can be accounted for and its effects on the tracking outputs can be canceled. Because the disturbance models are autonomous, they are not controllable, so feedback of a disturbance estimate does not affect previously placed feedback system eigenvalues. So far as the model is accurate, its disturbance estimate can be used as a reference input to cancel the actual disturbance's effects on the tracking outputs.

A digital phase-locked loop (PLL) provided an example of tracking system design. In a somewhat unconventional approach, the PLL phase error was modeled as a disturbance. The first controller design incorporated a first-order, arbitrary constant, disturbance model. Then a higher-order controller was designed with an arbitrary constant-plus-ramp disturbance model.

---

*The "whitening filters" used in Kalman filtering to accommodate noise that is correlated from step to step are precisely deterministic disturbance models driven, in that case, by white noise.

Except in very simple situations, one cannot be very specific about most quantization effects except perhaps in a statistical sense. The quantization (e.g., roundoff error) associated with ordinary digital processing is likely to be of diminishing concern in the future because of the wide availability of inexpensive floating-point hardware, particularly in signal processing chips. In the PLL example, however, as in similar situations with other systems, the quantization is inherent in the measuring process. Simulation is an important means of study.

There are many other practical systems for which some or all of the measurements are not available at every step. The PLL example demonstrated that a state observer can easily be used to predict the plant state at future steps, providing feedback of the estimated plant state when measurement outputs are not available at those steps.

Step-varying disturbance modeling and tracking system design parallel that for the step-invariant case. Even for a step-invariant plant, the additional design freedom in step-varying observation and feedback can result in better performance.

References

Ideal and response model tracking system design has been used for a long time, sometimes explicitly, sometimes not. State variable viewpoints of inverse filters are the subject of the following papers:

L. M. Silverman, "Inversion of Multivariable Systems," *IEEE Trans. Automatic Control,* Vol. AC-14, June 1969, pp. 270–276;

M. K. Sain and J. L. Massey, "Invertibility of Linear, Time-Invariant Dynamic Systems," *IEEE Trans. Automatic Conrol,* Vol. AC-14, April 1969, pp. 141–149;

L. M. Silverman, "Properties and Application of Inverse Systems," *IEEE Trans. Automatic Control,* Vol. AC-13, August 1968, pp. 436–437.

The text

C. T. Chen, *Linear System Theory and Design.* New York: Holt, Rinehart and Winston, 1984

has an extended discussion of these tracking system methods.

The article

G. H. Hostetter and J. S. Meditch, "Generalized Inverse Filtering," *Proc. 10th Asilomar Conf. on Circuits, Systems and Computers,* Nov. 1976, pp. 234–239.

addresses basic issues and methods in tracking control system design, with emphasis on the logical nature of the design process.

Reference model tracking system design, the observation of a reference signal model, was first discussed in

G. H. Hostetter, "Controller Design for Exact Signal Model Tracking," *Proc. 18th Asilomar Conf. on Circuits, Systems and Computers,* Nov. 1984, pp. 449–453;

C. T. Chen and G. H. Hostetter, "Design of Two-Input, One-Output Compensators to Achieve Asymptotic Tracking," *Int. J. Control,* 1987, Vol. 46, No. 6, pp. 1883–1887;

G. H. Hostetter and M. S. Santina, "Rational Linear Algebraic Tracking Control System Design," *IEEE Control System Magazine,* August 1988, pp. 34–42

Deterministic disturbance models have been used for many years. A modern viewpoint of the estimation of unknown constant disturbances is discussed in

A. E. Bryson and D. G. Luenberger, "The Synthesis of Regulator Logic Using State-Variable Concepts," *Proc. IEEE,* Vol. 58, Nov. 1970, pp. 1803–1811.

More general disturbance models, especially polynomial ones, and their observation are discussed in

G. H. Hostetter and J. S. Meditch, "On the Generalization of Observers to Systems with Unmeasurable, Unknown Inputs," *Automatica,* Vol. 9, July 1973, pp. 721–724;

J. S. Meditch and G. H. Hostetter, "Observers for Systems with Unknown and Inaccessible Inputs," *Int. J. Control,* Vol. 19, March 1974, pp. 473–480.

The text

B. Friedland, *Control System Design.* New York: McGraw-Hill, 1986

also considers deterministic disturbances and reference signal models, using the term "exogenous variables."

The digital phase-locked loop example is based in part on work by Alan H. Ross and Gene H. Hostetter for Western Digital Corporation. More information about phase-locked loops can be found in

F. M. Gardner, *Phaselock Techniques,* 2nd edition. New York: Wiley, 1979.

Chapter Six Problems

6-1. Design inverse reference input filters for single-input/single-output plants with the following z-transfer functions. Specify the number of steps in advance, if any, that the reference input must be provided to each filter. Determine whether or not each filter is stable.

a. $T(z) = \dfrac{z^2 - \frac{1}{2}z + \frac{1}{2}}{z^3 - z^2 - z}$

b. $T(z) = \dfrac{2z^2 + 3z + 1}{4z^3 - 2z^2 + 2z - 1}$

c. $T(z) = \dfrac{z^2 - 3z + 1}{z^3}$

d. $T(z) = \dfrac{z^4 + 3z^2 + 2}{z^4 - 1}$

6-2. Design inverse reference input filters for two-input/single-output plants with the following z-transfer functions. In order of importance, the design objectives are

1. Use a stable filter
2. Use a causal filter
3. Use a filter with all poles at $z = 0$

a. $T_1(z) = \dfrac{2z - 1}{z^2 - \frac{1}{4}}$

  $T_2(z) = \dfrac{3z + 1}{z^2 - \frac{1}{4}}$

b. $T_1(z) = \dfrac{z^2 + 2z - 1}{(2z - 1)(3z + 1)}$

  $T_2(z) = \dfrac{2z - 1}{(2z - 1)(3z + 1)}$

c. $T_1(z) = \dfrac{2z^3 - 3z + 1}{z^3 - \frac{5}{6}z^2 + \frac{1}{6}z}$

  $T_2(z) = \dfrac{2z + 1}{z^2 - \frac{1}{2}z}$

6-3. For the plant

$$\begin{bmatrix} x_1(k+1) \\ x_2(k+1) \end{bmatrix} = \begin{bmatrix} 0 & 1 \\ \frac{1}{2} & \frac{1}{2} \end{bmatrix} \begin{bmatrix} x_1(k) \\ x_2(k) \end{bmatrix} + \begin{bmatrix} 0 \\ 1 \end{bmatrix} u(k)$$

$$y(k) = \begin{bmatrix} 2 & 1 \end{bmatrix} \begin{bmatrix} x_1(k) \\ x_2(k) \end{bmatrix}$$

$$\bar{y}(k) = \begin{bmatrix} 2 & -3 \end{bmatrix} \begin{bmatrix} x_1(k) \\ x_2(k) \end{bmatrix}$$

design feedback with a first-order observer so that all eigenvalues of the feedback system are at $\lambda = 0$. Then, using both the plant and an observer input, design an inverse reference input filter. Specify state variable equations for the controller which is the combination of the observer and the filter.

6-4. Design response model reference input filters for single-input/single-tracking-output plants with the following z-transfer functions. The filters must be stable, causal, and must result in systems with zero steady state error to a constant reference input.

a. $T(z) = \dfrac{(z - \frac{1}{2})}{(z + \frac{1}{3})(z - \frac{1}{3})}$

b. $T(z) = \dfrac{2z - 1}{z^2 - \frac{1}{2}z + 1}$

c. $T(z) = \dfrac{(z - 2)(z^2 - z + \frac{1}{3})}{z^3 - \frac{1}{3}z}$

d. $T(z) = \dfrac{(z - 2)(z + \frac{1}{2})}{z^3}$

6-5. Design response model reference input filters for two-input/single-tracking-output plants with the following z-transfer functions. The filters must be causal and are to have all poles inside $|z| = \frac{1}{2}$ on the complex plane. Each overall system must have zero steady state error to a constant reference input.

a. $T_1(z) = \dfrac{z - 1}{z + \frac{1}{3}}$

$T_2(z) = \dfrac{z}{z^2 - \frac{1}{6}z - \frac{1}{6}}$

b. $T_1(z) = \dfrac{2z^2 - z + 3}{z^2 - \frac{2}{3}z + \frac{1}{9}}$

$T_2(z) = \dfrac{z}{z^2 - \frac{2}{3}z + \frac{1}{9}}$

c. $T_1(z) = \dfrac{z^2 - 4}{z^3}$

$T_2(z) = \dfrac{-z + 1}{z^3}$

6-6. Design observer feedback that will place all eigenvalues of the feedback system with plant

$$\begin{bmatrix} x_1(k+1) \\ x_2(k+1) \end{bmatrix} = \begin{bmatrix} 0 & 1 \\ -4 & 4 \end{bmatrix} \begin{bmatrix} x_1(k) \\ x_2(k) \end{bmatrix} + \begin{bmatrix} 0 \\ 1 \end{bmatrix} u(k)$$

$$y(k) = \begin{bmatrix} 1 & 1 \end{bmatrix} \begin{bmatrix} x_1(k) \\ x_2(k) \end{bmatrix}$$

$$\bar{y}(k) = \begin{bmatrix} 0 & 1 \end{bmatrix} \begin{bmatrix} x_1(k) \\ x_2(k) \end{bmatrix} + 2u(k)$$

at $\lambda = 0$. Then design an ideal reference input filter for the resulting two-input feedback plant. Specify state equations for the filter.

6-7. For the observer feedback system of problem 6-6, design instead a stable, causal response model reference input filter. Specify state equations for the filter.

6-8. Design a response model tracking system for a plant with the z-transfer function

$$T(z) = \frac{z - \frac{5}{4}}{(z + \frac{1}{6})(z + \frac{1}{3})}$$

using a reference input filter with a z-transfer function of the form

$$G(z) = \frac{\alpha_1 z^2 + \alpha_2 z + \alpha_3}{(z + \frac{1}{2})^2}$$

The model's zero-input response should decay at least as rapidly as $(\frac{3}{4})^k$, and it should have zero steady state error to a step input.

6-9. There is a duality between plant and reference input filter transfer functions whereby the two can sometimes be interchanged. For example, consider a single-input plant with a single tracking output, having the z-transfer function $T(z)$. If an acceptable response model results with a filter z-transfer function $G(z)$, then the same model is produced if the plant z-transfer function is $G(z)$ and the filter is chosen to have the plant z-transfer function. Generalize this result to a filter for a two-input/two-tracking-output plant. Carefully explain and interpret the interchange that is involved.

6-10. Design signal models in state variable form having outputs of the following forms, where the β's are arbitrary constants

a. $\beta_1 + \beta_2 k + \beta_3 k^2 + \beta_4 k^3$

b. $\beta_1 + \beta_2 \cos\left(\dfrac{k}{2}\right) + \beta_3 \sin\left(\dfrac{k}{2}\right)$

c. $\beta_1 + \beta_2 e^{-(k/10)} + \beta_3 k e^{-(k/10)}$

d. $\beta_1 e^{-(k/3)} \cos\left[\left(\dfrac{k}{4}\right) + \beta_2\right]$

6-11. Find the form of the output of the signal model

$$\begin{bmatrix} \sigma_1(k+1) \\ \sigma_2(k+1) \\ \sigma_3(k+1) \end{bmatrix} = \begin{bmatrix} \frac{1}{4} & 0 & 0 \\ 0 & \frac{1}{4} & 1 \\ 0 & 1 & \frac{1}{4} \end{bmatrix} \begin{bmatrix} \sigma_1(k) \\ \sigma_2(k) \\ \sigma_3(k) \end{bmatrix}$$

$$r(k) = \begin{bmatrix} 1 & 0 & 1 \end{bmatrix} \begin{bmatrix} \sigma_1(k) \\ \sigma_2(k) \\ \sigma_3(k) \end{bmatrix}$$

6-12. For each of the following plants, choose the input gains, if possible, so that the plant tracking output $\bar{y}(k)$ observes a reference signal of the form $r(k)$, where the β's are arbitrary constants

a.
$$\begin{bmatrix} x_1(k+1) \\ x_2(k+1) \end{bmatrix} = \begin{bmatrix} 0 & 1 \\ \frac{1}{2} & \frac{2}{3} \end{bmatrix} \begin{bmatrix} x_1(k) \\ x_2(k) \end{bmatrix} + \begin{bmatrix} 1 \\ -2 \end{bmatrix} u(k)$$

$$\bar{y}(k) = \begin{bmatrix} 2 & 1 \end{bmatrix} \begin{bmatrix} x_1(k) \\ x_2(k) \end{bmatrix}$$

$$r(k) = \beta_1$$

b.
$$\begin{bmatrix} x_1(k+1) \\ x_2(k+1) \end{bmatrix} = \begin{bmatrix} \frac{5}{6} & 1 \\ \frac{1}{6} & 0 \end{bmatrix} \begin{bmatrix} x_1(k) \\ x_2(k) \end{bmatrix} + \begin{bmatrix} 1 \\ -1 \end{bmatrix} u(k)$$

$$\bar{y}(k) = \begin{bmatrix} 2 & 1 \end{bmatrix} \begin{bmatrix} x_1(k) \\ x_2(k) \end{bmatrix} + u(k)$$

$$r(k) = \beta_1$$

c.
$$\begin{bmatrix} x_1(k+1) \\ x_2(k+1) \\ x_3(k+1) \end{bmatrix} = \begin{bmatrix} 1 & 1 & 0 \\ \frac{1}{4} & 0 & 1 \\ -1 & 0 & 0 \end{bmatrix} \begin{bmatrix} x_1(k) \\ x_2(k) \\ x_3(k) \end{bmatrix} + \begin{bmatrix} 0 & 1 \\ -1 & -2 \\ 1 & 0 \end{bmatrix} \begin{bmatrix} u_1(k) \\ u_2(k) \end{bmatrix}$$

$$\bar{y}(k) = \begin{bmatrix} 0 & 1 & -1 \end{bmatrix} \begin{bmatrix} x_1(k) \\ x_2(k) \\ x_3(k) \end{bmatrix}$$

$$r(k) = \beta_1 + \beta_2 k$$

6-13. For each of the following plants, design output or observer feedback to place the composite system eigenvalues all at $\lambda = 0$. Then choose the input gains (including those to any observer) so that the plant tracking output signal $\bar{y}(k)$ observes the highest possible order polynomial signal model.

a.
$$\begin{bmatrix} x_1(k+1) \\ x_2(k+1) \end{bmatrix} = \begin{bmatrix} 0 & 1 \\ 0 & -\frac{1}{9} \end{bmatrix} \begin{bmatrix} x_1(k) \\ x_2(k) \end{bmatrix} + \begin{bmatrix} 0 \\ 1 \end{bmatrix} u(k)$$

$$y(k) = \begin{bmatrix} 1 & -1 \end{bmatrix} \begin{bmatrix} x_1(k) \\ x_2(k) \end{bmatrix}$$

$$\bar{y}(k) = x_1(k)$$

b.
$$\begin{bmatrix} x_1(k+1) \\ x_2(k+1) \end{bmatrix} = \begin{bmatrix} 0 & 1 \\ 1 & -\frac{1}{2} \end{bmatrix} \begin{bmatrix} x_1(k) \\ x_2(k) \end{bmatrix} + \begin{bmatrix} 0 \\ 1 \end{bmatrix} u(k)$$

$$\begin{bmatrix} y_1(k) \\ y_2(k) \end{bmatrix} = \begin{bmatrix} 1 & 2 \\ -2 & 0 \end{bmatrix} \begin{bmatrix} x_1(k) \\ x_2(k) \end{bmatrix}$$

$$\bar{y}(k) = x_2(k) + u(k)$$

c.
$$\begin{bmatrix} x_1(k+1) \\ x_2(k+1) \end{bmatrix} = \begin{bmatrix} 0 & 1 \\ \frac{1}{2} & -\frac{1}{2} \end{bmatrix} \begin{bmatrix} x_1(k) \\ x_2(k) \end{bmatrix} + \begin{bmatrix} 0 & 2 \\ 1 & 3 \end{bmatrix} \begin{bmatrix} u_1(k) \\ u_2(k) \end{bmatrix}$$

$$y(k) = \begin{bmatrix} 1 & 0 \end{bmatrix} \begin{bmatrix} x_1(k) \\ x_2(k) \end{bmatrix} + \begin{bmatrix} 0 & 1 \end{bmatrix} \begin{bmatrix} u_1(k) \\ u_2(k) \end{bmatrix}$$

$$\bar{y}(k) = 2x_1(k) - x_2(k)$$

6-14. Design an observer–controller for the plant

$$\begin{bmatrix} x_1(k+1) \\ x_2(k+1) \\ x_3(k+1) \end{bmatrix} = \begin{bmatrix} 0 & 1 & 0 \\ 0 & 0 & 1 \\ -\frac{1}{12} & -\frac{1}{4} & -\frac{1}{3} \end{bmatrix} \begin{bmatrix} x_1(k) \\ x_2(k) \\ x_3(k) \end{bmatrix} + \begin{bmatrix} 0 & 0 \\ 0 & 1 \\ 2 & -1 \end{bmatrix} \begin{bmatrix} u_1(k) \\ u_2(k) \end{bmatrix}$$

$$\begin{bmatrix} y_1(k) \\ y_2(k) \end{bmatrix} = \begin{bmatrix} 1 & 1 & -2 \\ 2 & 3 & 0 \end{bmatrix} \begin{bmatrix} x_1(k) \\ x_2(k) \\ x_3(k) \end{bmatrix} + \begin{bmatrix} 0 & 2 \\ -2 & 1 \end{bmatrix} \begin{bmatrix} u_1(k) \\ u_2(k) \end{bmatrix}$$

such that the tracking output

$$\bar{y}(k) = x_1(k) - x_2(k)$$

tracks an arbitrary ramp signal and has a zero-input component that decays with step at least as quickly as $\kappa(\frac{1}{2})^k$, where the constant κ depends on the initial conditions.

6-15. Design an observer–controller for the plant of problem 6-14 such that $\bar{y}(k)$ exactly tracks any sinusoidal sequence of the form

$$r(k) = \beta_1 \cos\left(\frac{k}{5}\right) + \beta_2 \sin\left(\frac{k}{5}\right)$$

after six or fewer steps.

6-16. Design an observer–controller for the plant of problem 6-14 with tracking outputs

$$\begin{bmatrix} \bar{y}_1(k) \\ \bar{y}_2(k) \end{bmatrix} = \begin{bmatrix} 0 & 1 & 0 \\ -1 & 2 & 1 \end{bmatrix} \begin{bmatrix} x_1(k) \\ x_2(k) \\ x_3(k) \end{bmatrix} + \begin{bmatrix} 0 & 0 \\ 3 & 0 \end{bmatrix} \begin{bmatrix} u_1(k) \\ u_2(k) \end{bmatrix}$$

such that $\bar{y}_1(k)$ tracks $r_1(k)$ and $\bar{y}_2(k)$ tracks $r_2(k)$, where $r_1(k)$ and $r_2(k)$ are each modeled as arbitrary constants.

6-17. Show that when a single controller gain is to be chosen, the choice that gives zero steady state error to a constant input is the same as the choice for which the tracking output observes an arbitrary constant tracking input.

6-18. The term *model algorithmic control* denotes a response model tracking system for which the model plant has all of its eigenvalues at $\lambda = 0$. The model plant is thus a deadbeat system. Show that any input model tracking system for which all of the eigenvalues of the plant with observer feedback have been placed at $\lambda = 0$ is a model algorithmic controller.

6-19. Find state equations for disturbance models with responses having the following forms

a. $\beta_1 + \beta_2 e^{-(k/2)} + \beta_3 k e^{-(k/2)}$

b. $\beta_1 + \beta_2 \left(\frac{1}{4}\right)^k \cos\left(\frac{k}{2}\right)$

6-20. For each of the following plants, find an autonomous state variable model for disturbances $v(k)$ of the form given, then design a full-order state observer of the augmented plant

a. $\begin{bmatrix} x_1(k + 1) \\ x_2(k + 1) \end{bmatrix} = \begin{bmatrix} \frac{1}{3} & 1 \\ 0 & 1 \end{bmatrix} \begin{bmatrix} x_1(k) \\ x_2(k) \end{bmatrix} + \begin{bmatrix} 0 \\ 1 \end{bmatrix} u(k) + \begin{bmatrix} 1 \\ -1 \end{bmatrix} v(k)$

$$y(k) = \begin{bmatrix} 2 & 1 \end{bmatrix} \begin{bmatrix} x_1(k) \\ x_2(k) \end{bmatrix}$$

$$v(k) = \beta_1$$

b.
$$\begin{bmatrix} x_1(k+1) \\ x_2(k+1) \end{bmatrix} = \begin{bmatrix} 1 & 1 \\ -\frac{1}{2} & 0 \end{bmatrix}\begin{bmatrix} x_1(k) \\ x_2(k) \end{bmatrix} + \begin{bmatrix} 3 & 0 \\ 0 & 2 \end{bmatrix}\begin{bmatrix} u_1(k) \\ u_2(k) \end{bmatrix} + \begin{bmatrix} 4 \\ 0 \end{bmatrix}v(k)$$

$$y(k) = \begin{bmatrix} 2 & 0 \end{bmatrix}\begin{bmatrix} x_1(k) \\ x_2(k) \end{bmatrix} + \begin{bmatrix} 0 & 3 \end{bmatrix}\begin{bmatrix} u_1(k) \\ u_2(k) \end{bmatrix}$$

$$v(k) = \beta_1 + \beta_2 k$$

c.
$$\begin{bmatrix} x_1(k+1) \\ x_2(k+1) \end{bmatrix} = \begin{bmatrix} 0 & 1 \\ \frac{1}{2} & -\frac{1}{4} \end{bmatrix}\begin{bmatrix} x_1(k) \\ x_2(k) \end{bmatrix} + \begin{bmatrix} 0 \\ 1 \end{bmatrix}u(k)$$

$$+ \begin{bmatrix} 2 & 0 \\ -3 & 2 \end{bmatrix}\begin{bmatrix} v_1(k) \\ v_2(k) \end{bmatrix}$$

$$y(k) = \begin{bmatrix} 2 & 3 \end{bmatrix}\begin{bmatrix} x_1(k) \\ x_2(k) \end{bmatrix}$$

$$v_1(k) = \beta_1$$

$$v_2(k) = \beta_2 + \beta_3 k$$

6-21. For the plant and disturbance model of problem 6-20(a), design observer feedback to place the feedback system eigenvalues all at $\lambda = 0$ and to produce zero steady state error to any constant disturbance v in the tracking output

$$\bar{y}(k) = x_2(k) + u(k)$$

6-22. Design a controller for the digital phase-locked loop that models the input interval sequence as an arbitrary constant plus an arbitrary ramp and uses a reduced-order observer.

6-23. Arrange the higher-order digital phase-locked loop controller of Section 6.6.2 in the form of a plant model driven by observer error feedback. Then carefully explain how the controller should operate if switching-time error measurements $y(k)$ are occasionally not available.

6-24. The initial "lock" of the digital phase-locked loop (called its *acquisition*, or *capture*, performance) can be improved by making the observer–controller deadbeat at first, then changing the observer gains to obtain better smoothing of the VCO period control signal $u(k)$. Design such a controller using the first-order observer structure of Section 6.6.1.

6-25. For the digital phase-locked loop, design the controller as a tracking system where $s(k)$ in Figure 6-28(a) is to track $r(k)$.

6-26. Design observer feedback so that the plant

$$\begin{bmatrix} x_1(k+1) \\ x_2(k+1) \end{bmatrix} = \begin{bmatrix} \dfrac{1}{k+2} & -1 \\ 0 & \dfrac{1}{3} \end{bmatrix} \begin{bmatrix} x_1(k) \\ x_2(k) \end{bmatrix} + \begin{bmatrix} 1 \\ k+1 \end{bmatrix} u(k)$$

$$y(k) = \begin{bmatrix} -1 & 2 \end{bmatrix} \begin{bmatrix} x_1(k) \\ x_2(k) \end{bmatrix} + \frac{1}{3} u(k)$$

with the feedback is deadbeat. Specify the various matrices involved for steps 0 through 4. Find also the state equations that describe the composite feedback system.

6-27. Design a linear, step-varying model of the vector of disturbance signals

$$\begin{bmatrix} v_1(k) \\ v_2(k) \end{bmatrix} = \begin{bmatrix} \beta_1 \sin(k^2) + \beta_2(-\frac{1}{2})^k \\ 2\beta_2 \sin k + \beta_3 \ln(k+1) \end{bmatrix}$$

where β_1, β_2, and β_3 are arbitrary constants.

6-28. For the plant

$$\begin{bmatrix} x_1(k+1) \\ x_2(k+1) \end{bmatrix} = \begin{bmatrix} 4 & -1 \\ 0 & \dfrac{1}{k+1} \end{bmatrix} \begin{bmatrix} x_1(k) \\ x_2(k) \end{bmatrix} + \begin{bmatrix} \alpha(k) \\ 1 + \alpha(k) \end{bmatrix} r(k)$$

$$\bar{y}(k) = \begin{bmatrix} 4(\frac{1}{2})^k & 1 \end{bmatrix} \begin{bmatrix} x_1(k) \\ x_2(k) \end{bmatrix}$$

find a control input gain sequence $\alpha(k)$ for which $\bar{y}(k)$ observes any constant reference input r. Find $\alpha(k)$ for steps 0 through 4.

6-29. Repeat problem 6-28, but find a control input gain sequence $\alpha(k)$ for which $\bar{y}(k)$ observes any alternating reference input sequence of the form

$$r(k) = \beta(-1)^k$$

where β is a constant.

7

Digital Control of Continuous-Time Systems

7.1 Preview

The first of the two design problems for a tracking system, that of obtaining acceptable digital zero-input response, was solved in Chapter 5. If the designer is willing to provide the needed controller complexity, observer feedback can be used to place the feedback system eigenvalues at any desired locations. Sometimes output feedback will suffice, but in general one is likely to need an observer. The second design problem, obtaining acceptable zero-state tracking performance, was the subject of Chapter 6.

When a digital controller is to control a continuous-time plant, there is a third design problem, that of achieving good between-sample response of the continuous time plant. That concern is addressed in this chapter. A good discrete-time design will ensure that samples of the plant response are well-behaved, but it is also necessary to have satisfactory response between the discrete-time steps. For a fixed plant, this third design concern is addressed by shaping the plant input waveforms and raising the sample rate as necessary. Significant increases in effective sample rate can be achieved at low cost and without greatly increasing the required controller throughput by updating the plant control signals at a higher rate than that for the bulk of the controller computations.

Digital controllers for continuous-time plants can be designed by beginning with an analog controller design, then deriving a digital controller that closely approximates the analog controller's response. This approach is especially useful when a portion of an existing control system is to be replaced, but usually, even for relatively high sample rates, the digital approximation performs less well than the analog controller from which it was derived.

In this chapter, we first develop several methods for digitizing analog controllers. Digital computer simulation of continuous-time plant response is discussed in Section 7.3. In Section 7.4, we examine the response of continuous-time plants to sampled-and-held inputs. The between-sample response of a continuous-time plant can be improved, if necessary, by changing the plant (perhaps with continuous-time feedback), by increasing the sampling rate, and by changing the shapes of the sampled-and-held plant inputs with a plant input filter. This latter approach is examined in some detail.

Sections 7.5 and 7.6 discuss basic hardware and software considerations in digital controller design. A wide variety of hardware is now available, including general purpose computers and microprocessors, dedicated signal processing chips, and custom logic. These are assembled in a suitable architecture for the needed processing. Data acquisition and distribution are of key importance to hardware design, so some of the practical aspects of these are discussed at length. Software design might be with a high-level language, or it may be necessary to develop a program in an assembly language or the machine code of a particular processor. In any case, given the hardware architecture, there are usually many options for program, subroutine, and interrupt structures. The final hardware and software design usually must also include extensive alarm, test, and external selection and override provisions.

Finally, the design of two practical discrete-time controllers of continuous-time plants is discussed in Section 7.7. The first of these is an advanced elevator positioning controller with a limited speed of travel. The second example is of a highly underdamped torsional positioning mechanism. The otherwise unacceptable between-sample response of the torsional positioner is improved with a continuous-time plant input filter.

7.2 Digitizing Analog Controllers

A simple method of designing digital controllers for continuous-time plants is first to design a continuous-time (or *analog*) controller for the plant, then approximate the behavior of the analog controller with a digital one. As indicated in Figure 7-1, a digital controller that might approximate the behavior of a single-input/single-output analog controller consists of an A/D converter driving a discrete-time system with the z-transform function $H(z)$, followed by D/A conversion with sample

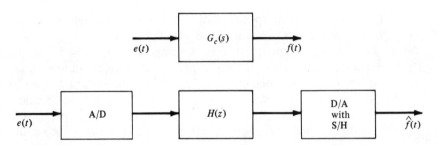

Figure 7-1 Digitizing an analog controller.

and hold. This configuration is also called a *digital filter*. If the sample rate is sufficiently high and the approximation sufficiently good, the behavior of the digital controller will be nearly indistinguishable from that of the analog controller. The digital controller will have such advantages as high reliability, low drift with temperature, power supply and age, the ability to make changes in software, and so on.

We now discuss several methods for discretizing analog controllers. In the material to follow, a "hat" over a symbol denotes an approximation of the quantity.

7.2.1 Numerical Approximation of Differential Equations

One way to approximate an analog controller with a digital one is to convert the analog controller's transfer function to a differential equation and then obtain a numerical approximation of the solution of the differential equation. There are two basic methods of numerical approximation of the solution of differential equations: (1) numerical integration and (2) numerical differentiation. We first discuss numerical integration and then numerical differentiation.

Consider the controller transfer function

$$G_c(s) = \frac{F(s)}{E(s)} = \frac{1}{s} \tag{7-1}$$

which has a corresponding differential equation

$$\frac{df}{dt} = e(t) \tag{7-2}$$

Integrating both sides of equation (7-2) from t_0 to t gives

$$f(t) = f(t_0) + \int_{t_0}^{t} e(t) \, dt \quad t \geq t_0$$

For evenly spaced sample times $t = kT$, $k = 0, 1, 2, \ldots$ and during one sampling interval $t_0 = kT$ to $t = (k + 1)T$, the solution is

$$f(kT + T) = f(kT) + \int_{kT}^{(k+1)T} e(t)\, dt \tag{7-3}$$

We now seek several approximation methods to the integral in equation (7-3).

Euler's Forward Method (One Sample)

The simplest approximation of the integral in equation (7-3) is to approximate the integrand by a constant equal to the value of the integrand at the *left* endpoint of each T subinterval times the sampling interval T as shown in Figure 7-2(a). The controller signal samples are thus related by

$$\hat{f}(kT + T) = \hat{f}(kT) + Te(kT) \tag{7-4}$$

Taking the z-transform of both sides of equation (7-4) gives

$$z\hat{F}(z) - \hat{F}(z) = TE(z)$$

and, therefore,

$$H(z) = \frac{\hat{F}(z)}{E(z)} = \frac{T}{z - 1} \tag{7-5}$$

Comparing equation (7-5) with the analog controller transfer function given by equation (7-1) implies that a discrete-time equivalence of an analog controller can be derived, using Euler's forward rectangular method, by simply replacing each s in the analog controller transfer function with $(z - 1)/T$. That is,

$$H(z) = G_c(s) \Big|_{s = \frac{z-1}{T}} \tag{7-6}$$

For example, the analog controller

$$G_c(s) = \frac{2s + 1}{s + 3}$$

has the equivalent discrete-time controller

$$H(z) = \frac{2(z - 1)/T + 1}{(z - 1)/T + 3} = \frac{2z - 2 + T}{z - 1 + 3T}$$

according to Euler's forward method.

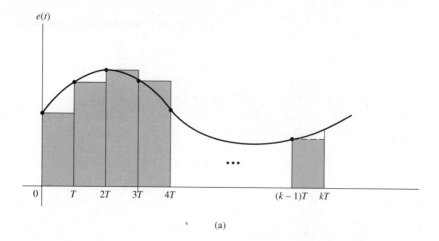

(a)

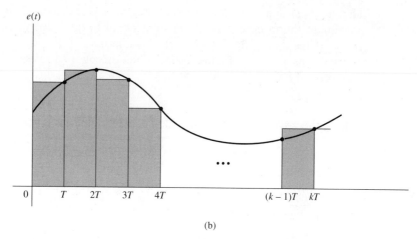

(b)

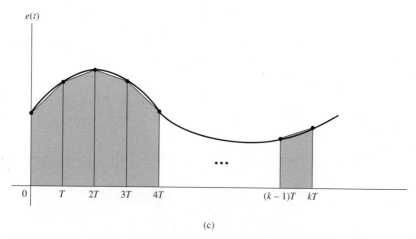

(c)

Figure 7-2 Comparing Euler's and trapezoidal integration approximations. (a) Approximation using Euler's forward method. (b) Approximation using Euler's backward method. (c) Approximation using the trapezoidal method.

Euler's Backward Method (One Sample)

Rather than approximating the integrand in equation (7-3) by a constant at the left endpoint, Euler's *backward* rectangular method approximates the value of the integrand by a constant equal to the value of the integrand at the *right* endpoint of each T subinterval times the sampling interval T, as shown in Figure 7-2(b). Equation (7-3) becomes

$$\hat{f}(kT + T) = \hat{f}(kT) + Te(kT + T)$$

or, using the z-transform, equation (7-3) becomes

$$H(z) = \frac{\hat{F}(z)}{E(z)} = \frac{Tz}{z - 1} \qquad (7\text{-}7)$$

Comparing equation (7-7) with the analog controller transfer function in equation (7-1) implies that the discrete-time controller transfer function can be determined by replacing each s of the analog controller transfer function with $(z - 1)/Tz$. That is,

$$H(z) = G_c(s) \Big|_{s = \frac{z-1}{Tz}} \qquad (7\text{-}8)$$

For the analog controller

$$G_c(s) = \frac{a}{s + a}$$

for example, applying Euler's backward method results in the discrete-time controller

$$H(z) = \frac{a}{(z - 1)/Tz + a} = \frac{aTz}{(1 + aT)z - 1}$$

Trapezoidal Method (Two Samples)

Euler's forward and backward approximation methods are called first order because they use one sample during each sampling interval. When more than one sample is used to update the approximation of the analog controller transfer function during a sampling interval, the performance of the digital controller can be improved over that of the simpler approximation given by either Euler's forward or backward methods. As shown in Figure 7-2(c), the trapezoidal approximation to the integral in equation (7-3) gives

$$\hat{f}(kT + T) = \hat{f}(kT) + \frac{T}{2} \{e(kT) + e(kT + T)\}$$

which has a corresponding z-transform

$$(z - 1)\hat{F}(z) = \frac{T}{2}(z + 1)E(z)$$

or

$$H(z) = \frac{\hat{F}(z)}{E(z)} = \frac{T}{2}\left(\frac{z + 1}{z - 1}\right) \tag{7-9}$$

Comparing equation (7-9) with equation (7-1) implies that a digital controller transfer function can be derived by replacing each s in the analog controller transfer function with $\frac{2}{T}\frac{z-1}{z+1}$. That is,

$$H(z) = G_c(s)\Big|_{s=\frac{2}{T}\frac{z-1}{z+1}}$$

The trapezoidal method is also called the bilinear transformation, or *Tustin's method*.

For example, consider the system shown in Figure 7-3(a), in which a continuous-time plant with the transfer function

$$G_p(s) = \frac{25}{s^2 + 9s + 40}$$

is controlled by an analog controller with the transfer function

$$G_c(s) = \frac{(s + 10)}{s(s + 4)}$$

This feedback system has the overall transfer function

$$T(s) = \frac{G_c(s)G_p(s)}{1 + G_c(s)G_p(s)} = \frac{25s + 250}{s^4 + 13s^3 + 76s^2 + 185s + 250}$$

which has all poles to the left of about $s = -1.5$ on the complex plane and zero steady state error to a step input.

If Euler's forward method is used, the digital controller has the transfer function

$$H(z) = G_c(s)\Big|_{s=\frac{z-1}{T}} = \frac{(z - 1)/T + 10}{[(z - 1)/T][(z - 1)/T + 4]}$$

$$= \frac{T(z - 1 + 10T)}{(z - 1)(z - 1 + 4T)}$$

as shown in Figure 7-3(b). The plant and the digital controller are shown in Figure 7-3(c).

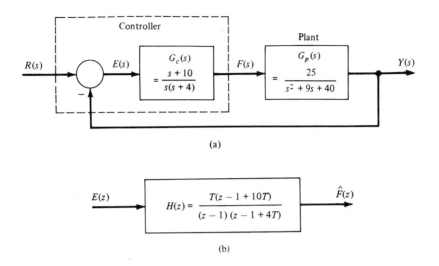

(a)

(b)

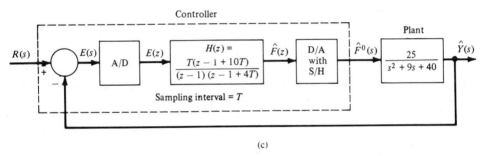

(c)

Figure 7-3 Obtaining the z-transfer function associated with Euler's forward approximation of an analog controller. (a) Analog control system. (b) Digital controller transfer
function using Euler's forward method. (c) Plant with digital controller.

Figure 7-4 shows the step response of this Euler's forward digital
approximation to the original analog controller. Because the analog
controller has a pole at the origin of the complex plane representing
pure integration, its step response has a ramp component. For a relatively large sampling interval T, the controller approximation deviates
significantly from the analog controller. When T is sufficiently small so
that there are several steps during each time constant of the analog
controller's fastest mode e^{-4t}, the analog and the digital controller step
responses are nearly the same. Step responses of the feedback system
for various sampling intervals T are compared in Figure 7-5.

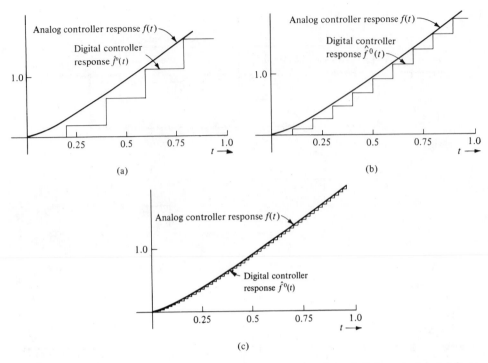

Figure 7-4 Step response of the Euler's forward approximation digital controller. (a) Sampling interval $T = 0.2$. (b) $T = 0.1$. (c) $T = 0.02$.

Suppose we use, instead, the trapezoidal method for each integral. Substituting $1/s$ with

$$\frac{T(z + 1)}{2(z - 1)}$$

into the analog controller transfer function gives

$$H(z) = \frac{T(z + 1)[(5T + 1)z + (5T - 1)]}{(z - 1)[(4T + 2)z + (4T - 2)]}$$

The character of the response of this controller is similar to that of the first-order controller response shown in Figure 7-4. For the same sampling interval, the controller using Tustin's method tends to track the analog controller output more accurately at the sample times because the approximations to the analog integrations are better. Because Euler's and Tustin's approximations result in controllers of the same order, the designer usually opts for Tustin's approximation.

Euler's forward rule has the potential of mapping poles in the left half of the s-plane to poles outside the unit circle on the z-plane, as

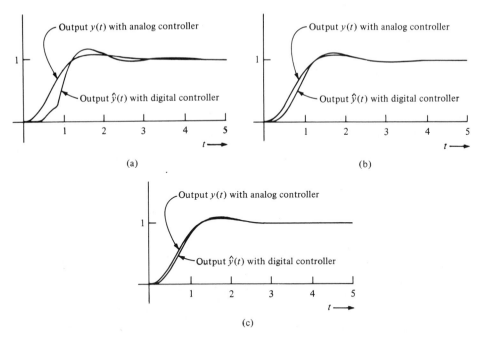

Figure 7-5 Step response of the feedback system with a first-order approximation digital controller. (a) Sampling interval $T = 0.4$. (b) $T = 0.2$. (c) $T = 0.1$.

shown in Figure 7-6(b). Then some stable analog controllers may produce unstable digital controllers, which is not desirable. Euler's backward rule maps the left half of the s-plane to a region inside the unit circle, as shown in Figure 7-6(c). On the other hand, Tustin's rule maps the imaginary axis on the s-plane to the unit circle on the z-plane, the left half of the s-plane to the interior of the unit circle in the z-plane and the right half of the s-plane to the exterior of the unit circle on the z-plane, as shown in Figure 7-6(d). Hence, the trapezoidal method is very popular.

Table 7-1 summarizes some common approximations for integrals, together with the corresponding z-transfer function of each integral approximation. Higher-order approximations result in digital controllers of progressively higher order. The higher the order of the approximation, the better the approximations to the analog integrations and the more accurately the digital controller output tends to track samples of the analog controller output for any input. The digital controller, however, probably has a sampled-and-held output between samples, so accurate tracking of samples is often of less concern to the designer than the sample rate.

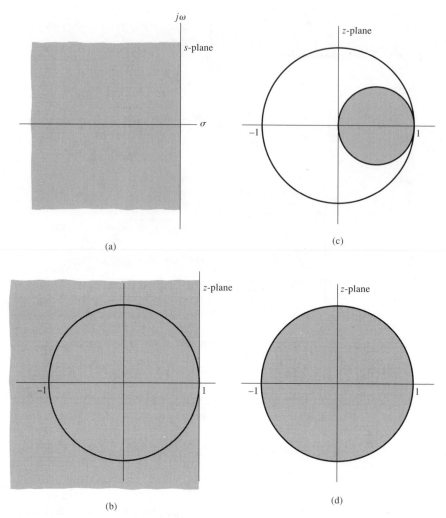

Figure 7-6 Mappings between the s-plane and the z-plane using Euler's and Tustin's methods. (a) Stability region in the s-plane. (b) Corresponding stability region in the z-plane using Euler's forward method. (c) Corresponding region in the z-plane using Euler's backward method. (d) Corresponding region in the z-plane using Tustin's method.

Table 7-1 Some Integral Approximations Using Present and Past Integrand Samples

Approximation to the Integral Over One Step	Difference Equation for the Approximate Integral	Z-Transmittance of the Approximate Integrator
One-Sample		
$\int_{kT}^{kT+T} e(t)\,dt \cong Te(kT)$	$\hat{f}[(k+1)T] = \hat{f}(kT) + Te(kT)$	$\dfrac{T}{z-1}$
$\int_{kT}^{kT+T} e(t)\,dt \cong Te(kT+T)$	$\hat{f}[(k+1)T] = \hat{f}(kT) + Te(kT+T)$	$\dfrac{Tz}{z-1}$
Two-Sample (Tustin approximation)		
$\int_{kT}^{kT+T} e(t)\,dt \cong T\left\{\dfrac{1}{2}\,e[(k+1)T] + \dfrac{1}{2}\,e(kT)\right\}$	$\hat{f}[(k+1)T] = \hat{f}(kT) + \dfrac{T}{2}\,e[(k+1)T]$ $+\dfrac{T}{2}\,e(kT)$	$\dfrac{T(z+1)}{2(z-1)}$
Three-Sample		
$\int_{kT}^{kT+T} e(t)\,dt \cong T\left\{\dfrac{5}{12}\,e[(k+1)T] + \dfrac{8}{12}\,e(kT)\right.$ $\left. -\dfrac{1}{12}\,e[(k-1)T]\right\}$	$\hat{f}[(k+1)T] = \hat{f}(kT) + \dfrac{5T}{12}\,e[(k+1)T]$ $+\dfrac{8T}{12}\,e(kT) + \dfrac{T}{12}\,e[(k-1)T]$	$\dfrac{T[(5/12)z^2 + (8/12)z + (1/12)]}{z(z-1)}$

Bilinear Transformation with Frequency Warping

In many control system applications, a digital filter $G(z)$ is desired that closely approximates the frequency response of a continuous-time transfer function described by $G(z)$ within the bandlimited range

$$G(z = e^{j\omega T}) \cong G(s = j\omega) \quad 0 \le \omega < \omega_0 = \frac{\pi}{T}$$

The bilinear method applies but with slight modifications.

Although the bilinear transformation and the z-transformation both map the left half of the s-plane to the interior of the unit circle on the z-plane, the right half of the s-plane to the exterior of the unit circle in the z-plane, and the s-plane imaginary axis to the z-plane unit circle boundary, there is a nonlinear relationship between the continuous frequency ω_c and the corresponding digital frequency ω_d of the discretized filter obtained by the bilinear transformation. This nonlinear relationship is based on the fact that the *entire jω-axis* of the s-plane is mapped into one complete revolution of the unit circle in the z-plane.

Comparing the frequency response of the continuous filter described by $G(s)$ with the frequency response of the discrete filter $G(z)$, then

$$G(z) = G\left(s = \frac{2}{T}\frac{z-1}{z+1}\right)$$

$$G(z = e^{j\omega_d T}) = G\left(j\omega_c = \frac{2}{T}\frac{e^{j\omega_d T} - 1}{e^{j\omega_d T} + 1}\right)$$

$$= G\left[j\omega_c = \frac{2}{T}\left(\frac{e^{(j\omega_d T/2)} - e^{(-j\omega_d T/2)}}{e^{(j\omega_d T/2)} + e^{(-j\omega_d T/2)}}\right)\right]$$

$$= G\left[j\omega_c = j\frac{2}{T}\frac{\sin \omega_d T/2}{\cos \omega_d T/2}\right]$$

$$= G\left[j\omega_c = j\frac{2}{T}\tan \omega_d T\right]$$

and, therefore,

$$\omega_c = \frac{2}{T}\tan\frac{\omega_d T}{2} \tag{7-10}$$

For relatively small values of ω_d (compared with the folding frequency π/T), then

$$j\omega_c \cong j\frac{2}{T}\frac{\omega_d T}{2} = j\omega_d$$

and the behavior of the discrete-time filter closely approximates the frequency response of the corresponding continuous-time filter.

When the digital frequency ω_d approaches the folding frequency $\omega_d = \pi/T$,

$$j\omega_c = j\frac{2}{T}\tan\frac{\omega_d T}{2}$$

$$= j\infty$$

then the continuous frequency approaches infinity and distortion becomes evident.

If the bilinear transformation is applied in conjunction with equation (7-10) near the frequencies of interest, however, the frequency distortion is reduced considerably. The general design procedure of making a continuous-time filter discrete using the bilinear transformation with frequency *prewarping* is summarized as follows:

For a continuous-time filter $G(s)$

1. Obtain a new continuous-time transfer function $G'(s)$ whose poles and zeros with critical frequencies $(s + \alpha')$ are related to those of the original $G(s)$ by

$$(s + \alpha) \rightarrow (s + \alpha')\big|_{\alpha' = 2/T \tan \alpha\, T/2}$$

in the case of real roots and by

$$s^2 + 2\zeta\omega_n s + \omega_n^2 \rightarrow s^2 + 2\zeta\omega_n' s + \omega_n'^2\big|_{\omega_n' = 2/T \tan \omega_n T/2}$$

in the case of complex roots.

2. Apply the bilinear transformation to $G'(s)$ by simply replacing each s in $G'(s)$ with

$$s = \frac{2}{T}\frac{z-1}{z+1}$$

3. If necessary (and it almost always is), the multiplying constant of the resulting digital filter $G'(z)$ should be scaled to match the multiplying constant of the continuous-time filter at a specific frequency.

If when using this design procedure, the corner frequencies of the continuous-time filter are adjusted to new locations, then the bilinear transformation brings them back to the desired corner frequencies in the discrete domain. For example, for the controller

$$G(s) = \frac{100}{s + 100}$$

a digital controller using the bilinear transformation with frequency prewarping is desired. For a sampling interval of $T = 0.01$ sec, then

$$s + 100 \rightarrow s + \alpha'|\alpha' = (2/0.01) \tan (100)(0.01)/2 = 109$$

and

$$G'(s) = \frac{100}{s + 109}$$

The corresponding digital controller using the bilinear transformation is

$$G'(z) = K \frac{100}{200(z - 1/z + 1) + 109} = \frac{100K(z + 1)}{309z - 91}$$

For unity D.C. gain of the continuous-time filter

$$G'(z = e^{j0} = 1) = \frac{200K}{218} = 1$$

then

$$K = 1.09$$

and, therefore, the digital filter

$$G'(z) = \frac{0.35(z + 1)}{z - 0.294}$$

has a rolloff at the radian frequency $\omega = 100$.

As another example, consider the controller transfer function

$$G(s) = \frac{(200\pi)^2}{s^2 + (200\pi)s + (200\pi)^2}$$

which is low-pass, with unit D.C. gain, undamped natural frequency $f = 100$ Hz, and damping ratio $\zeta = 0.5$. The sampling interval

$$T = \frac{1}{500}$$

is chosen, which gives a folding frequency of $f_0 = 250$ Hz, well above the 100-Hz cutoff of the filter. At $\omega = 200\pi$

$$\omega_n' = \frac{2}{T} \tan \frac{\omega_n T}{2} = 1000 \tan \frac{200\pi}{1000}$$

$$= 726 \text{ radians/sec}$$

and the warped transfer function is

$$G'(s) = \frac{K}{s^2 + 726s + (726)^2}$$

Hence,

$$G'(z) = \frac{K}{[1000(z - 1/z + 1)]^2 + 726{,}000(z - 1/z + 1) + (726)^2}$$

$$= \frac{K(z + 1)^2}{1{,}198{,}924z^2 - 945{,}848z + 801{,}076}$$

For unity D.C. gain of the continuous-time filter

$$G'(1) = \frac{4K}{1{,}054{,}152} = 1$$

or

$$K = 263{,}538$$

and, hence,

$$G'(z) = \frac{263{,}538(z + 1)^2}{1{,}198{,}924z^2 - 945{,}848z + 801{,}076}$$

$$= \frac{0.22(z + 1)^2}{z^2 - 0.789z + 0.668}$$

which is the required filter.

Numerical Differentiation

The other main approach for approximating a numerical differential equation solution is *numerical differentiation*. Here, the numerical solution of the differential equation is obtained by replacing the derivative terms in the differential equation with approximate numerical values. Consequently the resulting difference equation is solved. Three methods of first-order and second-order approximations are given in Table 7-2 and illustrated in Figure 7-7. These are called *finite-difference* approximations of derivatives.

Using the forward-difference approximation of the first-order derivative in equation (7-2) gives

$$\frac{df}{dt} = e(t) = \frac{f[(k + 1)T] - f(kT)}{T}$$

for t in the vicinity of $t = kT$, $k = 0, 1, 2, \cdots$. Then

$$\hat{f}[(k + 1)T] = \hat{f}(kT) + Te(kT)$$

which generates evenly spaced samples of an approximation of $f(t)$.

On the other hand, the backward-difference approximation of the first-order derivative in equation (7-2) gives

$$\frac{df}{dt} = e(t) = \frac{f(kT) - f[(k - 1)T]}{T}$$

Table 7-2 Finite-Difference Approximations of Derivatives

	Derivative Approximation	Z-Transmittance of the Approximate Differentiator
First-Order Derivative		
Forward difference	$\dfrac{f[(k+1)T] - f(kT)}{T}$	$\dfrac{z-1}{T}$
Backward difference	$\dfrac{f(kT) - f[(k-1)T]}{T}$	$\dfrac{z-1}{Tz}$
Central difference	$\dfrac{f[(k+1)T] - f[(k-1)T]}{2T}$	$\dfrac{z^2-1}{2Tz}$
Second-Order Derivative		
Forward difference	$\dfrac{f[(k+2)T] - 2f[(k+1)T] + f(kT)}{T^2}$	$\dfrac{z^2 - 2z + 1}{T^2}$
Backward difference	$\dfrac{f(kT) - 2f[(k-1)T] + f[(k-2)T]}{T^2}$	$\dfrac{z^2 - 2z + 1}{T^2 z^2}$
Central difference	$\dfrac{f[(k+1)T] - 2f(kT) + f[(k-1)T]}{T^2}$	$\dfrac{z^2 - 2z + 1}{T^2 z}$

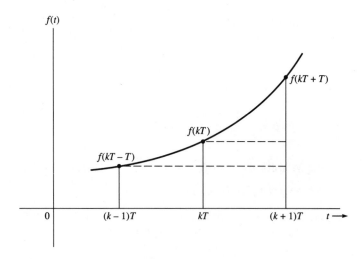

Figure 7-7 Finite-difference approximations of derivatives.

Then

$$\hat{f}(kT) = \hat{f}[(k - 1)T] + Te(kT)$$

which is an approximation of the function $f(t)$ at the evenly spaced samples.

As shown in Figure 7-7, the central-difference approximation of the derivative in equation (7-2) yields

$$\frac{df}{dt} = e(t) = \frac{f[(k + 1)T] - f[(k - 1)T]}{2T}$$

Then

$$\hat{f}[(k + 1)T] = \hat{f}[(k - 1)T] + 2Te(kT)$$

is an approximation of $f(t)$ at evenly spaced samples.

The corresponding z-transform of each of these finite-difference approximations is also shown in Table 7-2.

For the analog controller

$$G_c(s) = \frac{K(s + a)}{(s + b)}$$

for example, the forward-difference approximation gives the digital controller

$$H(z) = \frac{K[(z - 1)/T + a]}{(z - 1)/T + b} = \frac{K(z - 1 + aT)}{z - 1 + bT}$$

and the backward-difference approximation yields the digital controller

$$H(z) = \frac{K[(z - 1)/Tz + a]}{(z - 1)/Tz + b} = \frac{K(1 + aT)}{1 + bT} \left[\frac{z - \dfrac{1}{1 + aT}}{z - \dfrac{1}{1 + bT}} \right]$$

As another example, consider the continuous-time controller

$$G_c(s) = \frac{K}{s^2 + \alpha s + \beta}$$

The central-difference approximation gives

$$H(z) = \frac{K}{(z^2 - 2z + 1)/T^2 z + \alpha(z^2 - 1)/2Tz + \beta}$$

$$= \frac{2KT^2 z}{z^2(2 + \alpha T) + 2z(-2 + \beta T^2) + 2 - \alpha T}$$

7.2.2 Matching Step and Other Responses

Another method of approximating an analog controller by a digital one is to require that, at the sample times, the digital controller step response be samples of the analog controller step response. Figure 7-8(a) shows the unit step response $f_{step}(t)$ of an analog controller with transmittance $G_c(s)$. If, as in Figure 7-8(b), the discrete-time portion $H(z)$ of the digital controller is chosen so that its response to a unit step sequence consists of samples of $f_{step}(t)$, then the digital controller of Figure 7-8(c) has a step response that equals the analog controller step response at the sample times. This is termed a *step-invariant approximation* of an analog system by a digital system.

For example, a continuous-time controller with the transfer function

$$G_c(s) = \frac{4s^2 + 17s + 12}{s^2 + 5s + 6}$$

has the unit step response

$$F_{step}(s) = \left(\frac{1}{s}\right) G_c(s) = \frac{4s^2 + 17s + 12}{s(s + 2)(s + 3)} = \frac{2}{s} + \frac{3}{s + 2} + \frac{-1}{s + 3}$$

For a sampling rate $T = 0.2$, the samples of $f_{step}(t)$ have the z-transform

$$F_{step}(z) = \frac{2z}{z - 1} + \frac{3z}{z - e^{-0.4}} - \frac{z}{z - e^{-0.6}}$$

Taking these samples to be the output of a discrete-time system $H(z)$ driven by a unit step sequence

$$F_{step}(z) = \left(\frac{z}{z - 1}\right) H(z) = \frac{2z}{z - 1} + \frac{3z}{z - 0.67} - \frac{z}{z - 0.54}$$

the step-invariant approximation is

$$H(z) = 2 + \frac{3(z - 1)}{z - 0.67} - \frac{(z - 1)}{z - 0.54} = \frac{4z^2 - 5.37z + 1.67}{(z - 0.67)(z - 0.54)}$$

It is occasionally desirable to design digital controllers so that their response to some input other than a step consists of a sampled-and-held version of an analog controller's response to that input. A *ramp-invariant approximation*, for example, has a discrete-time response to a unit ramp sequence that consists of samples of the unit ramp response of the continuous-time system.

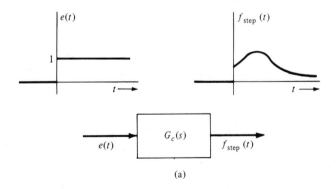

(a)

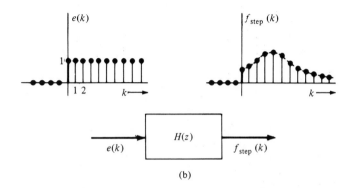

(b)

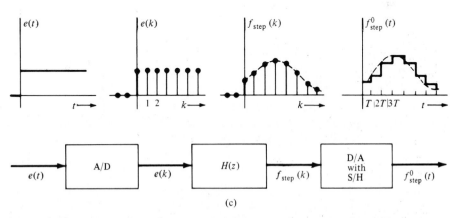

(c)

Figure 7-8 Finding a step-invariant approximation of a continuous-time controller. (a) Analog controller step response. (b) Discrete step response consisting of samples of the analog controller step response. (c) Step-invariant digital controller.

7.2.3 Pole-Zero Matching

For the pole-zero matching method, both the poles and the zeros of the analog controller transfer function $G_c(s)$ are mapped to those of the corresponding digital controller $H(z)$ as follows:

$$(s + a) \rightarrow z - e^{-aT}$$

for real roots and

$$(s + a)^2 + b^2 \rightarrow z^2 - 2(e^{-aT} \cos bT)z + e^{-2aT}$$

for complex conjugate pairs.

For example, for the analog controller

$$G_c(s) = \frac{5(s + 2)}{(s + 10)}$$

if the sampling interval $T = 0.1$ sec, the pole-zero matching method gives

$$H(z) = \frac{K(z - e^{-2/10})}{(z - e^{-10/10})} = \frac{K(z - 0.819)}{(z - 0.368)}$$

At

$$\omega_c = 0$$
$$G_c(s = j0) = 1$$

Requiring the same D.C. gain for the digital controller gives

$$H(z = e^{j0} = 1) = \frac{K(1 - 0.819)}{(1 - 0.368)} = 1$$

then $K = 3.49$, and, hence,

$$H(z) = \frac{3.49(z - 0.819)}{(z - 0.368)}$$

When the analog controller has more finite poles than zeros, its high-frequency response tends to zero as ω_c approaches infinity. The highest possible frequency on the $j\omega$-axis is at $\omega_c = \pi/T$, hence,

$$z = e^{sT} = e^{j(\pi/T)T} = -1$$

Therefore, infinite zeros of the analog controller map into finite zeros of the digital equivalence located at $z = -1$. The resulting digital controller has the number of poles equal to the number of zeros in its transfer function.

For example, the analog controller

$$G_c(s) = \frac{6s + 10}{s^2 + 2s + 5} = \frac{6(s + \frac{5}{3})}{(s + 1)^2 + 2^2}$$

has two finite poles and one finite zero. For a sampling interval $T = 0.1$ sec,

$$H(z) = \frac{K(z + 1)(z - e^{-(5/30)})}{z^2 - 2(e^{-0.1} \cos 2T)z + e^{-0.2}}$$

$$= \frac{K(z + 1)(z - 0.85)}{z^2 - 1.773z + 0.818}$$

The D.C. gain of the analog controller is

$$G_c(s = j0) = 2$$

For identical D.C. gain of the digital controller

$$H(z = 1) = \frac{K(2)(0.15)}{1 - 1.773 + 0.818} = 2$$

then $K = 0.3$, and the digital controller is completely specified as

$$H(z) = \frac{0.3(z + 1)(z - 0.85)}{z^2 - 1.773z + 0.818}$$

In general, the multiplying constant of the digital controller is selected to match the gain of the analog controller with that of the corresponding digital controller at a specific frequency.

At low frequency

$$H(z)\big|_{z = 1} = G_c(s)\big|_{s = 0}$$

and at high frequency

$$H(z)\big|_{z = -1} = G_c(s)\big|_{s = \infty}$$

Almost every digital approximation to an analog controller, one-sample integration approximation, two-sample integration approximation, step-invariant approximation, and so on, tends to perform well for a sufficiently short sampling interval T. For longer sampling intervals, which are often dictated by design and cost constraints, one approximation or another might perform the best in a given situation.

Digitizing an analog controller is not a very good general *design* technique, although it is very useful occasionally, as when the designer is *replacing* an existing analog controller or part of a controller with a digital one. The technique requires beginning with a good analog design, which is probably as difficult as directly creating a good digital design. And it usually performs less well than the analog counterpart from which it was derived. The step-invariant and other approximations are not easily extended to systems with multiple inputs and outputs. When the resulting feedback system performance is inadequate, the designer may have few options besides raising the sampling rate.

7.3 Sampled Continuous-Time Systems

When the plant is sampled, it is important to control its between-sample response as well as its discrete-time response. An important tool in the design process is the computer simulation of discrete-time control of continuous-time plants.

7.3.1 Between-Sample Response

When a continuous-time system in state variable form

$$\dot{\mathbf{x}}(t) = \mathscr{A}\mathbf{x}(t) + \mathscr{B}\mathbf{u}(t)$$
$$\mathbf{y}(t) = \mathbf{C}\mathbf{x}(t) + \mathbf{D}\mathbf{u}(t) \tag{7-11}$$

is sampled such that the plant inputs are driven by sample-and-hold (S/H) devices, the inputs are constant during each sampling interval. The discrete-time model of equation (7-11) is then

$$\mathbf{x}[(k + 1)T] = \mathbf{A}\mathbf{x}(kT) + \mathbf{B}\mathbf{u}(kT)$$
$$\mathbf{y}(kT) = \mathbf{C}\mathbf{x}(kT) + \mathbf{D}\mathbf{u}(kT)$$

or

$$\mathbf{x}(k + 1) = \mathbf{A}\mathbf{x}(k) + \mathbf{B}\mathbf{u}(k)$$
$$\mathbf{y}(k) = \mathbf{C}\mathbf{x}(k) + \mathbf{D}\mathbf{u}(k)$$

where

$$\mathbf{A} = e^{\mathscr{A}T} = \mathbf{I} + \frac{\mathscr{A}T}{1!} + \frac{\mathscr{A}^2 T^2}{2!} + \cdots + \frac{\mathscr{A}^i T^i}{i!} + \cdots$$

$$\mathbf{B} = \left[\mathbf{I}T + \frac{\mathscr{A}T^2}{2!} + \frac{\mathscr{A}^2 T^3}{3!} + \cdots + \frac{\mathscr{A}^i T^{i+1}}{(i + 1)!} + \cdots\right]\mathscr{B} \tag{7-12}$$

and where

$$\mathbf{B} = \mathscr{A}^{-1}[\exp(\mathscr{A}T) - \mathbf{I}]\mathscr{B} = [\exp(\mathscr{A}T) - \mathbf{I}]\mathscr{A}^{-1}\mathscr{B} \tag{7-13}$$

when $\mathscr{A}$ is nonsingular.

Other methods for calculating equations (7-12) and (7-13) are found in Appendix A and can be used here instead.

The plant response during each sampling interval is a step response of the plant *without* the digital controller's feedback. If continuous-time feedback to improve the open-loop plant step response is a design option, it should be considered. When used, the plant to be digitally controlled is then the original plant with the continuous-time feedback. Often, though, continuous-time feedback is not an acceptable option. In this section, it is assumed that any continuous-time feedback around the original plant has been incorporated into the plant model that is now to be digitally controlled.

It is possible for signals in a continuous-time plant to fluctuate wildly, even though discrete-time samples of those signals are very well behaved. The basic problem is illustrated in Figure 7-9 with the zero-input continuous-time system

$$\begin{bmatrix} \dot{x}_1(t) \\ \dot{x}_2(t) \end{bmatrix} = \begin{bmatrix} -0.2 & 1 \\ -1.01 & 0 \end{bmatrix} \begin{bmatrix} x_1(t) \\ x_2(t) \end{bmatrix} = \mathbf{A}\mathbf{x}(t)$$

$$y(t) = \begin{bmatrix} 1 & 0 \end{bmatrix} \begin{bmatrix} x_1(t) \\ x_2(t) \end{bmatrix} = \mathbf{c}^\dagger\mathbf{x}(t)$$

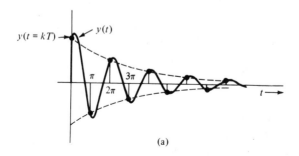

(a)

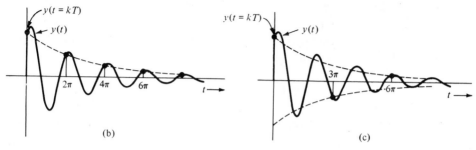

(b)

(c)

Figure 7-9 Hidden oscillations in a sampled continuous-time signal. (a) $T = \pi$. (b) $T = 2\pi$. (c) $T = 3\pi$.

This system has the characteristic equation

$$s^2 + 0.2s + 1.01 = (s + 0.1 + j)(s + 0.1 - j) = 0$$

and thus a response of the form

$$y(t) = Me^{-0.1t} \cos(t + \theta)$$

where the arbitrary constants M and θ depend on the initial conditions $\mathbf{x}(0)$. When the output of this system is sampled with sampling interval $T = \pi$, the output samples are

$$y(k) = y(t = k\pi) = Me^{-0.1k\pi} \cos(k\pi + \theta)$$
$$= M(e^{-0.1\pi})^k(-1)^k \cos\theta = M \cos\theta(-0.73)^k$$

as shown in Figure 7-9(a). These samples are the response of a *first-order* discrete-time system having a single geometric series model. The wide fluctuations of $y(t)$ between sampling times, termed *hidden oscillations*, cannot be determined from the samples $y(k)$.

As one might expect, the discrete-time model of this continuous-time system

$$\mathbf{x}(k + 1) = [\exp(\mathcal{A}T)]\mathbf{x}(k)$$
$$y(k) = \mathbf{c}^\dagger\mathbf{x}(k)$$

or

$$\begin{bmatrix} x_1(k + 1) \\ x_2(k + 1) \end{bmatrix} = \begin{bmatrix} -0.73 & 0 \\ 0 & -0.73 \end{bmatrix} \begin{bmatrix} x_1(k) \\ x_2(k) \end{bmatrix} = \mathbf{A}\mathbf{x}(k)$$

$$y(k) = \begin{bmatrix} 1 & 0 \end{bmatrix} \begin{bmatrix} x_1(k) \\ x_2(k) \end{bmatrix} = \mathbf{c}^\dagger\mathbf{x}(k)$$

is not completely observable in this circumstance because

$$\mathbf{M}_0 = \begin{bmatrix} \mathbf{c}^\dagger \\ \hline \mathbf{c}^\dagger\mathbf{A} \end{bmatrix} = \begin{bmatrix} 1 & 0 \\ -0.73 & 0 \end{bmatrix}$$

This phenomenon is called *loss of observability due to sampling*. The discrete-time system would normally have two modes, those given by its characteristic equation

$$\begin{vmatrix} (z + 0.73) & 0 \\ 0 & (z + 0.73) \end{vmatrix} = (z + 0.73)^2 = 0$$

which are $(-0.73)^k$ and $k(-0.73)^k$. Only the $(-0.73)^k$ mode appears in the output, however.

Hidden oscillations occur at any other integer multiple of the sampling period $T = \pi$. Figure 7-9(b) shows sampling with $T = 2\pi$, for which only a $[(-0.73)^2]^k = (0.533)^k$ mode is observable from $y(k)$. In Figure 7-9(c), with $T = 3\pi$, only a $[(-0.73)^3]^k = (-0.39)^k$ mode is observable from $y(k)$. Although this illustration is with an autonomous (zero-input) continuous-time system, the situation is the same when the plant has inputs, except that steps in the inputs at the sampling times (from D/A conversion of controller outputs) tend to keep the hidden oscillations excited.

For a slightly different sampling interval, for example $T = 3$, where

$$\begin{bmatrix} x_1(k + 1) \\ x_2(k + 1) \end{bmatrix} = \begin{bmatrix} -0.74 & 0.10 \\ -0.11 & -0.73 \end{bmatrix} \begin{bmatrix} x_1(k) \\ x_2(k) \end{bmatrix}$$

$$y(k) = \begin{bmatrix} 1 & 0 \end{bmatrix} \begin{bmatrix} x_1(k) \\ x_2(k) \end{bmatrix}$$

there are no hidden oscillations, and the discrete-time model is completely observable. Hidden oscillations in a continuous-time system and the accompanying loss of observability of the discrete-time model occur only when the continuous-time system has oscillatory modes and then only when the sampling interval is half the period of oscillation of an oscillatory mode or an integer multiple of that period. Although it is very unlikely that the sampling interval chosen for a plant control system would be precisely one resulting in hidden oscillations, intervals close to these result in modes in discrete-time models that are "almost unobservable." With limited numerical precision in measurements and in the controller computations, these modes can be difficult to detect and control.

7.3.2 Hybrid System Simulation

One of the designer's most important tools is the simulation of control systems and the plants they are to control on a digital computer. It is usually through simulation that difficulties with between-sample plant response are uncovered and solved. It is also good practice to investigate carefully the behavior of the controlled system when the arithmetic precision of the controller is reduced, when disturbance signals are injected into the system at likely points, and when the plant model is changed in ways that might occur in practice.

When a continuous-time plant is simulated on a digital computer, its response is computed at closely spaced discrete times. It is plotted by joining the closely spaced calculated response values with straight line segments in approximation of a continuous curve. Thus, a digital computer simulation of discrete-time control of a continuous-time system involves at least two sets of discrete-time calculations. One runs at high rate for simulation of the continuous-time plant. The other runs at a lower rate (say once every 10 or 50 of the former calculations) to generate new control signals at each discrete control step.

The structure of a simple simulation program is shown in Figure 7-10, where it is assumed that the user wishes to have 50 steps of continuous-time response calculation during each discrete-time step. The controller begins with zero initial conditions and obtains its first output from the first reference input and the plant measurement output. That controller output (which is to be applied to the plant input via S/H devices) is the constant plant input during the first sampling interval. The loop to the left is then traversed 50 times, and 50 calculations of the plant state are made and plotted, spaced from the beginning to the end of the discrete-time step. The discrete-time step is then incremented, a new controller state and output are calculated, and this controller output becomes the plant input for the next 50 closely spaced time steps during the second sampling interval. The process continues until program execution is halted.

As an example of a plant and controller simulation, consider the continuous-time plant

$$\begin{bmatrix} \dot{x}_1(t) \\ \dot{x}_2(t) \end{bmatrix} = \begin{bmatrix} -\frac{3}{2} & 1 \\ -\frac{1}{2} & 0 \end{bmatrix} \begin{bmatrix} x_1(t) \\ x_2(t) \end{bmatrix} + \begin{bmatrix} -1 \\ 1 \end{bmatrix} u(t) = \mathscr{A}\mathbf{x}(t) + \mathscr{b}u(t)$$

$$\bar{y}(t) = [1 \quad 0] \begin{bmatrix} x_1(t) \\ x_2(t) \end{bmatrix} = \bar{\mathbf{c}}^t\mathbf{x}(t) \qquad\qquad \textbf{(7-14)}$$

The discrete-time controller operates with a sampling interval $T = 0.1$, and a discrete-time model of this plant with that sampling interval and a sampled-and-held input is of the form

$$\mathbf{x}(k + 1) = [\exp(\mathscr{A}T)]\mathbf{x}(k) + [\exp(\mathscr{A}T) - \mathbf{I}]\mathscr{A}^{-1}\mathscr{b}u(k)$$
$$\bar{y}(k) = \bar{\mathbf{c}}^t\mathbf{x}(k)$$

or

$$\begin{bmatrix} x_1(k + 1) \\ x_2(k + 1) \end{bmatrix} = \begin{bmatrix} 0.858 & 0.0928 \\ -0.0464 & 0.998 \end{bmatrix} \begin{bmatrix} x_1(k) \\ x_2(k) \end{bmatrix} + \begin{bmatrix} -0.087 \\ 0.1 \end{bmatrix} u(k)$$

$$\bar{y}(k) = [1 \quad 0] \begin{bmatrix} x_1(k) \\ x_2(k) \end{bmatrix}$$

Let

$$u(k) = \alpha r(k) = 0.47r(k) \qquad\qquad \textbf{(7-15)}$$

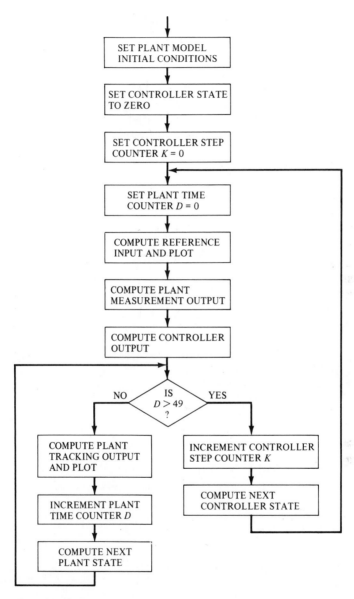

Figure 7-10 Computer program structure for simulation of discrete-time control of a continuous-time plant.

where $r(k)$ is the reference input. The controller gain α is chosen so that the tracking output $\bar{y}(k)$ observes an arbitrary constant reference input.

The simulation shown in Figure 7-11 consists of a plot of the continuous-time tracking output $\bar{y}(t)$ and the discrete-time reference input $r(k)$. The continuous-time input $u(t)$ is also plotted. For the continuous-time plot, 50 values of $\mathbf{x}(t)$ are plotted and connected during each sampling interval $T = 0.1$. The discrete-time model used to plot the continuous-time plant response has the sampling interval

$$\Delta t = \frac{T}{50} = 0.002$$

so that its state variable description is

$$\mathbf{x}'(k' + 1) = [\exp(\mathcal{A} \, \Delta t)]\mathbf{x}'(k') + [\exp(\mathcal{A} \, \Delta t) - \mathbf{I}]\mathcal{A}^{-1}\mathcal{b}u(k')$$
$$\bar{y}(k') = \bar{\mathbf{c}}^t\mathbf{x}'(k')$$

or

$$\begin{bmatrix} x_1'(k' + 1) \\ x_2'(k' + 1) \end{bmatrix} = \begin{bmatrix} 0.997 & (1.997 \times 10^{-3}) \\ (-9.985 \times 10^{-4}) & 1 \end{bmatrix} \begin{bmatrix} x_1'(k') \\ x_2'(k') \end{bmatrix}$$
$$+ \begin{bmatrix} (1.99 \times 10^{-3}) \\ (-2 \times 10^{-3}) \end{bmatrix} u(k')$$
$$\bar{y}(k') = \begin{bmatrix} 1 & 0 \end{bmatrix} \begin{bmatrix} x_1'(k') \\ x_2'(k') \end{bmatrix}$$

Table 7-3 is a digital computer program in the BASIC language for the simulation of the continuous-time plant in equation (7-14) and the

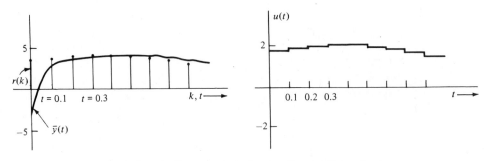

Figure 7-11 Simulating discrete-time control of a continuous-time plant.

Table 7-3 Simulation Program Example in Basic

```
100    DIM X(2), X1(2)
110    REM SET PLANT MODEL INITIAL CONDITIONS
120    X(1) = -3
130    X(2) = 1
140    REM SET CONTROLLER STEP COUNTER TO ZERO
150    K = 0
160    REM SET PLANT TIME COUNTER TO ZERO
170    T = 0
180    REM COMPUTE REFERENCE INPUT AND PLOT
190    R = 4 * COS(9 * K - 30)
200    PRINT 0.1 * K, R
210    REM COMPUTE CONTROLLER OUTPUT
220    U = 0.47 * R
230    REM CHECK WHETHER ITS TIME FOR A NEW
231    REM DISCRETE-TIME STEP
240    IF T > 49 THEN GOTO 400
250    REM COMPUTE PLANT TRACKING OUTPUT
251    REM AND PLOT
260    Y = X(1)
270    PRINT T,Y
280    REM INCREMENT PLANT TIME COUNTER
290    T = T + 0.002
300    REM COMPUTE NEXT PLANT STATE
310    X1(1) = 0.997 * X(1) + 1.997E - 3 * X(2) + 1.99E - 3 * U
320    X1(2) = -9.985E - 4 * X(1) + X(2) - 2E - 3 * U
330    X(1) = X1(1)
340    X(2) = X1(2)
350    REM LOOP BACK TO CHECK TIME
360    GOTO 240

400    REM INCREMENT CONTROLLER STEP COUNTER
410    K = K + 1
420    REM LOOP BACK TO RESET TIME COUNTER
430    GOTO 170

999    END
```

discrete-time controller in equation (7-15). Because plotting commands are peculiar to the computer used, this example program simply prints the coordinates to be plotted. This program follows the structure given in Figure 7-10, but without the need for any controller state calculations.

7.4 Designing Between-Sample Response

The character of the between-sample response of a plant driven by sampled-and-held inputs is the step response of the plant without digital feedback. If between-sample plant response is not acceptable, the designer's options are to

1. *Increase the controller sample rate* so that each step change in the plant input is of smaller amplitude. Often this is undesirable because it requires a major increase in hardware cost.
2. *Change the plant*, perhaps by adding continuous-time feedback. This too is often undesirable because of its susceptibility to noise and drift and the expense of routing analog signals.
3. *Change the shapes of the plant input signals* from having step changes at the controller sample rate to a shape that gives improved plant response.

It is straightforward to raise the sampling rate and redesign a controller for the higher rate, if necessary. It is also straightforward to incorporate analog feedback to improve the plant's step response by placing the continuous-time plant eigenvalues to achieve acceptable zero-input response. The continuous-time plant with feedback, rather than the original plant, then becomes the plant to be controlled digitally. The third option, that of changing the shape of the plant input signals is now examined. First, input signal shaping with analog filters is considered, then shaping with high-speed dedicated digital filters is discussed.

7.4.1 Analog Plant Input Filtering

Analog filters between the D/A converters and the plant inputs, particularly if they are relatively simple and composed of reliable and stable components, are usually acceptable in a controller design. As indicated in Figure 7-12, the idea is to use a filter or filters to smooth or shape the

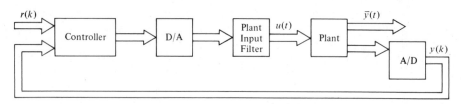

Figure 7-12 Use of an analog plant input filter to improve between-sample response.

plant inputs so that the undesirable modes of the plant's open-loop response are not excited as much as they would be with abrupt changes in the plant inputs at each step.

Figure 7-13(a) shows simulation results for the continuous-time plant

$$\begin{bmatrix} \dot{x}_1(t) \\ \dot{x}_2(t) \end{bmatrix} = \begin{bmatrix} -0.6 & 1 \\ -9 & 0 \end{bmatrix} \begin{bmatrix} x_1(t) \\ x_2(t) \end{bmatrix} + \begin{bmatrix} -1 \\ 1 \end{bmatrix} u(t) = \mathscr{A}\mathbf{x}(t) + \mathbf{b}u(t)$$

$$\bar{y}(t) = \begin{bmatrix} 1 & 0 \end{bmatrix} \begin{bmatrix} x_1(t) \\ x_2(t) \end{bmatrix} = \bar{\mathbf{c}}^\dagger \mathbf{x}(t) \tag{7-16}$$

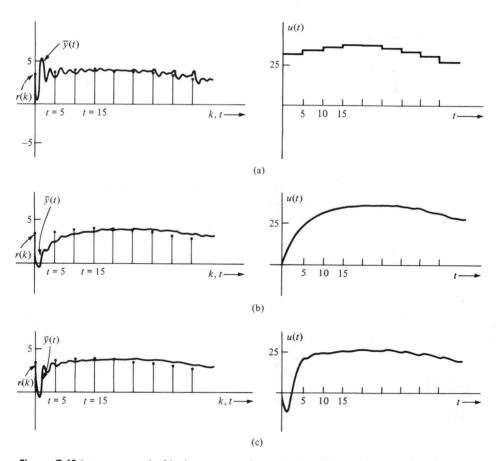

(a)

(b)

(c)

Figure 7-13 Improvement of between-sample response with analog plant input filtering. (a) Response of a system without a plant input filter. (b) Response with an added first-order analog plant input filter. (c) Response with a better analog plant input filter.

driven by a discrete-time control system

$$u(k) = -9r(k)$$

where $r(k)$ is the reference input and where the sampling interval is $T = 5$. The plant characteristic equation is

$$s^2 + 0.6s + 9 = (s + 0.3 + j2.98)(s + 0.3 - j2.98) = 0$$

which means that its zero-input response is of the form

$$\bar{y}_{\text{zero-input}}(t) = Me^{-0.3t} \cos(2.98t + \theta)$$

where M and θ are arbitrary constants. Although for the sampling interval $T = 5$ there are no hidden oscillations, the highly undamped zero-input plant response results in large fluctuations ("ringing") of the tracking output each time there is a step change in the plant input by the controller.

The plant response is improved considerably by the insertion of a plant input filter with the transfer function

$$G(s) = \frac{\frac{1}{3}}{s + \frac{1}{3}}$$

as is shown in Figure 7-13(b). This first-order filter was designed heuristically; its 3-sec time constant was chosen to smooth the plant input waveform during each 5-sec sampling interval, resulting in much less ringing of the tracking output. This filter is realized by the state variable equations

$$\dot{\varepsilon}(t) = -\frac{1}{3}\,\varepsilon(t) + \frac{1}{3}\,w(t)$$

$$u(t) = \varepsilon(t)$$

and now it is the composite third-order system

$$\begin{bmatrix} \dot{x}_1(t) \\ \dot{x}_2(t) \\ \dot{\varepsilon}(t) \end{bmatrix} = \begin{bmatrix} -0.6 & 1 & -1 \\ -9 & 0 & 1 \\ 0 & 0 & -\frac{1}{3} \end{bmatrix} \begin{bmatrix} x_1(t) \\ x_2(t) \\ \varepsilon(t) \end{bmatrix} + \begin{bmatrix} 0 \\ 0 \\ \frac{1}{3} \end{bmatrix} w(t)$$

$$\bar{y}(t) = \begin{bmatrix} 1 & 0 & 0 \end{bmatrix} \begin{bmatrix} x_1(t) \\ x_2(t) \\ \varepsilon(t) \end{bmatrix}$$

that the controller must control. Here

$$w(k) = -9r(k)$$

In general, insertion of a plant input filter with the transfer function matrix $\mathbf{G}(s)$ before a plant with the transfer function matrix $\mathbf{T}(s)$ results in a composite continuous-time *model plant* with the transfer function matrix

$$\mathbf{M}(s) = \mathbf{G}(s)\mathbf{T}(s)$$

Designing analog plant input filters is quite similar to designing discrete-time reference input filters, although the former are continuous-time. The objective of the design is to improve the between-sample plant response, and this is done by obtaining a model plant with acceptable step response. With a plant input filter, the order of the continuous-time plant is raised, which might result in a different controller design than that for the plant without the input filter.

As a simple example of analog plant input filter design, consider again the continuous-time plant (7-16) with the transfer function

$$T(s) = \bar{\mathbf{c}}^\dagger(s\mathbf{I} - \mathcal{A})^{-1}\mathbf{b} = \frac{-s + 1}{s^2 + 0.6s + 9}$$

A plant input filter with the transfer function

$$G(s) = \frac{s^2 + 0.6s + 9}{(s + 1)^2} = 1 + \frac{-1.4s + 8}{s^2 + 2s + 1}$$

cancels the plant poles and results in a model plant transfer function

$$M(s) = G(s)T(s) = \frac{-s + 1}{(s + 1)^2}$$

If the plant model is imperfectly known, exact pole cancellation is not possible, but approximate cancellation will still reduce the effects of those poles on the plant's step response. This plant input filter is realized by the state equations (in observable form for convenience)

$$\begin{bmatrix} \dot{\varepsilon}_1(t) \\ \dot{\varepsilon}_2(t) \end{bmatrix} = \begin{bmatrix} -2 & 1 \\ -1 & 0 \end{bmatrix} \begin{bmatrix} \varepsilon_1(t) \\ \varepsilon_2(t) \end{bmatrix} + \begin{bmatrix} -1.4 \\ 8 \end{bmatrix} w(t)$$

$$u(t) = \begin{bmatrix} 1 & 0 \end{bmatrix} \begin{bmatrix} \varepsilon_1(t) \\ \varepsilon_2(t) \end{bmatrix} + w(t)$$

so that the combination of plant and filter is described by

$$\begin{bmatrix} \dot{x}_1(t) \\ \dot{x}_2(t) \\ \dot{\varepsilon}_1(t) \\ \dot{\varepsilon}_2(t) \end{bmatrix} = \begin{bmatrix} -0.6 & 1 & -1 & 0 \\ -9 & 0 & 1 & 0 \\ 0 & 0 & -2 & 1 \\ 0 & 0 & -1 & 0 \end{bmatrix} \begin{bmatrix} x_1(t) \\ x_2(t) \\ \varepsilon_1(t) \\ \varepsilon_2(t) \end{bmatrix} + \begin{bmatrix} -1 \\ 1 \\ -1.4 \\ 8 \end{bmatrix} w(t)$$

$$\bar{y}(t) = [1 \quad 0 \quad 0 \quad 0] \begin{bmatrix} x_1(t) \\ x_2(t) \\ \varepsilon_1(t) \\ \varepsilon_2(t) \end{bmatrix}$$

A discrete-time controller

$$w(k) = -r(k)$$

like the previous ones, makes the plant an observer of any constant reference input. The resulting response is shown in Figure 7-13(c). The original plant's zero-input response, excited by nonzero plant initial conditions, is apparent at first but eventually decays to zero.

7.4.2 Higher-Rate Digital Filtering and Feedback

Another way of performing plant input filtering is to use a digital filter that operates at many times the rate of the controller, as indicated in Figure 7-14. Analog-to-digital conversion is usually accomplished with repeated approximations, one for each bit. With a given technology, one can perform several D/A conversions as fast as one A/D conversion. The cost of a digital plant input filter is thus relatively low because it requires higher D/A, not A/D, speed.

As an example, consider the continuous-time plant in equation (7-16) again. The discrete-time controller for this plant operates with a

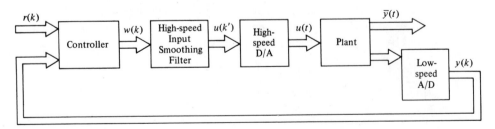

Figure 7-14 Use of high-speed digital plant input filtering to improve between-sample response.

sampling interval $T = 5$. An analog filter with transfer function

$$G(s) = \frac{\frac{1}{3}}{s + \frac{1}{3}}$$

was earlier found to improve the plant's between-sample behavior. The step response of the plant alone and the response of the plant with this input filter are shown in Figure 7-15(a).

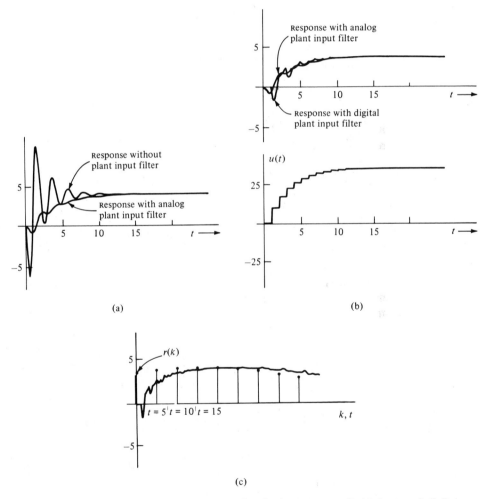

(a)

(b)

(c)

Figure 7-15 Improvement of between-sample plant response with high-speed digital input filtering. (a) Step response of the example system, with and without an analog input filter. (b) Step response with an analog input filter and with a digital input filter. (c) Tracking response of the system with a digital input filter.

A state variable realization of the analog filter is

$$\dot{\varepsilon}(t) = -\frac{1}{3}\varepsilon(t) + \frac{1}{3}w(t)$$

$$u(t) = \varepsilon(t)$$

A discrete-time model of this filter with sampling interval $\Delta t = 1$, one-fifth the controller interval, is

$$\varepsilon(k' + 1) = 0.717\varepsilon(k') + 0.283w(k')$$
$$u(k') = \varepsilon(k')$$

where k' is the index for steps of size Δt. The step response of the plant with the digital filter that approximates the continuous-time filter is shown in Figure 7-15(b). Figure 7-15(c) is a tracking response plot for the combination of plant, high-speed digital plant input filter, and lower-speed digital controller.

A discrete-time model for the plant with sampling interval $\Delta t = 1$ is

$$\mathbf{x}(k' + 1) = [\exp(\mathcal{A}\,\Delta t)]\mathbf{x}(k') + [\exp(\mathcal{A}\,\Delta t) - \mathbf{I}]\mathcal{A}^{-1}\boldsymbol{b}u(k')$$
$$y(k') = \bar{\mathbf{c}}^\dagger\mathbf{x}(k')$$

or

$$\begin{bmatrix} x_1(k' + 1) \\ x_2(k' + 1) \end{bmatrix} = \begin{bmatrix} -0.743 & 0.0387 \\ -0.348 & -0.720 \end{bmatrix}\begin{bmatrix} x_1(k') \\ x_2(k') \end{bmatrix} + \begin{bmatrix} 0.152 \\ 1.87 \end{bmatrix}u(k')$$

The composite system consisting of this model and the digital filter is

$$\begin{bmatrix} x_1(k' + 1) \\ x_2(k' + 1) \\ \varepsilon(k' + 1) \end{bmatrix} = \begin{bmatrix} -0.743 & 0.0387 & 0.152 \\ -0.348 & -0.720 & 1.87 \\ 0 & 0 & 0.717 \end{bmatrix}\begin{bmatrix} x_1(k') \\ x_2(k') \\ \varepsilon(k') \end{bmatrix} + \begin{bmatrix} 0 \\ 0 \\ 0.283 \end{bmatrix}w(k')$$

$$= \mathbf{A}\mathbf{x}(k') + \mathbf{b}w(k')$$

$$\bar{y}(k') = [1 \quad 0 \quad 0]\begin{bmatrix} x_1(k') \\ x_2(k') \\ \varepsilon(k') \end{bmatrix} = \bar{\mathbf{c}}^\dagger\mathbf{x}(k')$$

The discrete-time model of the plant and filter with steps k and the sampling interval $T = 5\Delta t$ is given by

$$\mathbf{x}(k' + 5) = \mathbf{A}^5\mathbf{x}(k') + \mathbf{A}^4\mathbf{b}w(k') + \mathbf{A}^3\mathbf{b}w(k' + 1)$$
$$+ \mathbf{A}^2\mathbf{b}w(k' + 2) + \mathbf{A}\mathbf{b}w(k' + 3) + \mathbf{b}w(k' + 4)$$
$$\bar{y}(k' + 5) = \bar{\mathbf{c}}^\dagger\mathbf{x}(k')$$

but because the controller output $w(k')$ changes only every fifth k' step,

$$\mathbf{x}(k' + 5) = \mathbf{x}(k + 1) = \mathbf{A}^5\mathbf{x}(k) + [\mathbf{A}^4 + \mathbf{A}^3 + \mathbf{A}^2 + \mathbf{A} + \mathbf{I}]\mathbf{b}w(k)$$
$$\bar{y}(k) = \bar{\mathbf{c}}^{\dagger}\mathbf{x}(k)$$

This is now the discrete-time plant model to be used for the controller design.

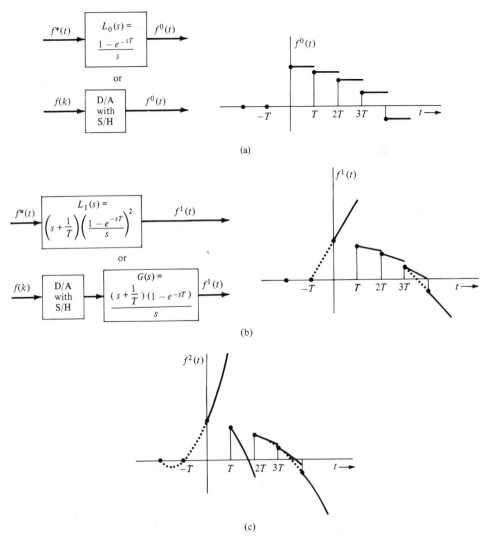

Figure 7-16 Typical response of holds. (a) Zero-order hold. (b) First-order hold. (c) Second-order hold.

Higher-rate digital plant feedback can also be used to improve a plant's between-sample response, but this requires high-rate A/D as well as high-rate D/A conversion. Another possibility is to sample the plant measurement outputs at the lower rate, but estimate the plant state at a high rate, feeding the state estimate back at high rate.

7.4.3 Higher-Order Holds

A traditional approach to improving reconstruction is to employ higher-order holds than the zero-order one. An nth-order hold produces a piecewise nth-degree polynomial output that passes through the most recent $n + 1$ input samples. It can be shown that, as the order of the hold is increased, a well-behaved signal is reconstructed with increasing accuracy. Several holds and their typical responses are shown in Figure 7-16.

Although higher-order holds do have a smoothing effect on the plant inputs, the resulting improvement of plant between-sample response is generally poor compared with that possible with a conventional filter of comparable complexity. Hardware for holds for higher than zero order (which is the sample-and-hold operation) is not routinely available. One approach is to employ high-speed digital devices and D/A conversion, as in the technique of Figure 7-14, but where the high-speed input-smoothing filter performs the interpolation calculations for a hold.

7.5 Digital Hardware for Control

The specifics of the hardware used for realization of a digital control system are highly dependent on the application involved and the state of technology. Nonetheless, it is helpful to discuss some general hardware considerations that are unlikely to be obsolete in the near future. On the horizon, but not yet sufficiently developed for enduringly relevant discussion here, are efficient use of the enormous processing capacity of very large scale integrated (VLSI) circuits, the blinding speed promised by massively parallel architectures, and the seemingly limitless storage capacity of laser disks and similar devices. All of these may greatly influence future directions in digital control.

7.5.1 Processing with General Purpose Computers

The hardware available for the computations involved in digital control includes the following:

large mainframe computers

minicomputers

desktop (personal) computers

microprocessors

special purpose devices, such as signal processing, and multiplier-accumulator chips

In addition, some manufacturers offer computers that are especially designed for control applications.

Figure 7-17 shows a general purpose computer architecture similar to that of many IBM mainframe computers and to that of Intel microprocessors in the 8080A and the 8086 through 80286, 80386, and 80486 series. The *central processor unit* (CPU) accesses memory and peripheral devices through separate channels. *Read-only memory* (ROM), which cannot be erased or changed electronically, is used for the *bootstrap* instructions that begin processing and perhaps for other instructions. *Random access memory* (RAM) is used for temporary high-speed storage. Its contents are probably destroyed whenever the computer power is turned off.

In small microprocessor-based control systems, the main program is usually stored in ROM so that it cannot be inadvertently changed or erased. For larger control systems, the main program is commonly stored on disk memory and is transferred to the faster-operating RAM by bootstrap instructions just after the computer system is turned on. In the system shown, the disk memory can access the RAM directly, without having to pass through (and tie up) the CPU. This capability is called *direct memory access* (DMA). It allows blocks of data to be transferred very rapidly between disk and RAM, usually interleaved between RAM memory accesses by the CPU. The combination of the memory circuits and any DMA circuitry is called the computer's *memory controller*. In microprocessor systems, the CPU, memory controller, perhaps even with some DMA capability, might be combined on the same chip.

Input and output from this computer is via *peripheral interface* circuits. These receive data from external devices, store it, and send it to the CPU when it is needed. They also receive data from the CPU for transmission to external devices. In most computers, the peripheral

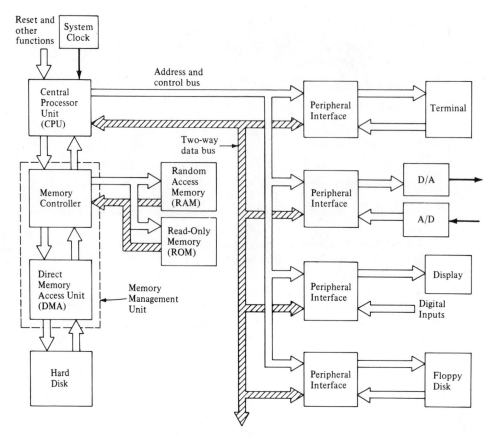

Figure 7-17 A general purpose computer architecture in which the CPU controls memory and peripheral devices separately.

interfaces share common sets of wires, called *busses*, for communicating with the CPU. One bus is an address and control bus from the CPU to the peripheral interfaces, used for designating an interface and transmitting an instruction to it. The other is a two-way data bus for transferring data to and from the CPU.

In the general purpose computer architecture described in Figure 7-18, memory and peripheral devices share the same address and data busses, an arrangement termed *memory-mapped input–output*. The peripheral interfaces contain registers for instructions and to transfer data to and from the CPU. These registers are addressed as if they were memory locations. This architecture is similar to that of many DEC computers and Motorola microprocessors in the 6800 and 68000 series.

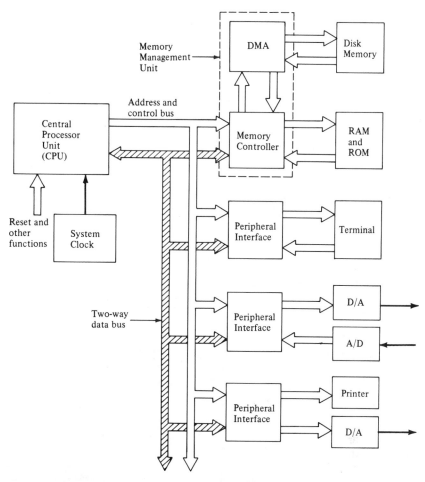

Figure 7-18 A general purpose computer architecture in which memory and peripheral devices share the same address space.

Multiprocessor architectures are common, particularly for microprocessor-based digital control systems. Figure 7-19 describes a system with separate processors for program control and for computation. The program control processor executes the main program for control, but passes the required lengthy observer–controller state and output calculations to a second processor. The second processor is, in turn, aided by a *coprocessor*, which is a device that performs precision arithmetic computations at high speed. The data acquisition and distribution system might involve a third processor. The advantages of multiprocessing include increased effective processing speed and the

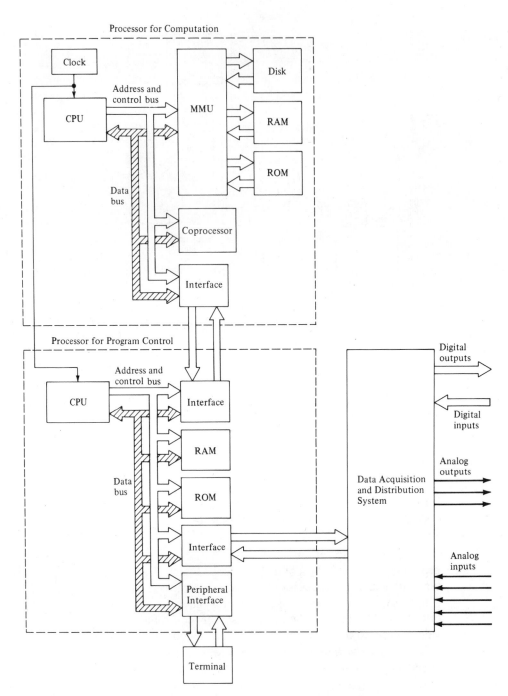

Figure 7-19 A multiprocessor architecture for a centralized controller.

ability to gain and build long-term experience with subsystems for standardized functions.

Another common multiprocessor arrangement is one in which individual processors are placed in several different locations, each near a cluster of plant sensors and actuators. Figure 7-20(a) shows a diagram of such an arrangement. Often the local computers each constitute smaller control systems. These subsystems are then joined and controlled overall by the central computer. Complicated plants, such as those for manufacturing, can thus be divided into smaller, more tractable subsystems to be controlled. Once locally controlled, control of the collection of subsystems is generally much simplified. Changes in subsystem hardware and software are usually easier to make than in a wholly central system because it is easier to isolate each subsystem and its interface.

In a *distributed control system*, all of the control is performed by a collection of local computers that share information; there is no central controller. Figure 7-20(b) describes such a system. In this example, additional storage and computational ability are available to the local computers from a shared resource. It may be possible to design distributed systems that would withstand component failures and have a flexibility and adaptability similar to that exhibited by some organisms and social systems. Each local computer would operate according to local principles and objectives, the collection of which would give good overall system control. A distributed control system's parts can be linked by a *local area network* (LAN).

7.5.2 Data Acquisition and Distribution

Digital control systems require that signal samples be collected from sensors of plant outputs and that samples be delivered to actuators at plant inputs. Most often in practice, the sensors and actuators are analog devices so that analog-to-digital (A/D) and digital-to-analog (D/A) conversion is needed. Data acquisition is the A/D conversion of a set of analog sensor signals and the transmission of their samples to a digital processor. Usually, samples from several sensors are collected and sent one at a time, reducing the number of needed wires and connections to the processor. One type of data acquisition system is shown in Figure 7-21(a). To obtain signal samples all at the same time, the sensor signals are sampled by individual sample-and-hold (S/H) devices and converted to binary digital representations by individual A/D converters. The processor provides a sequence of digital codes that cause a digital multiplexer to transmit the sensor samples, one after the other, to the processor.

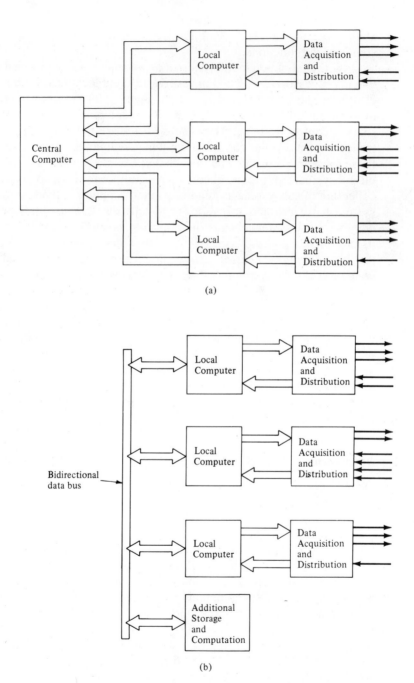

(a)

(b)

Figure 7-20 Multiprocessor arrangements. (a) System with central computer. (b) Distributed system.

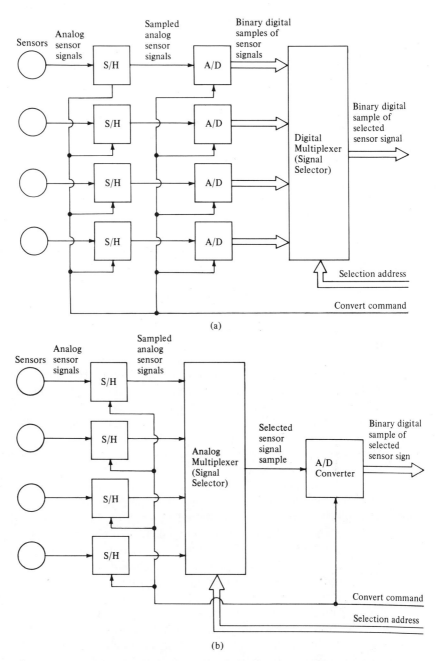

Figure 7-21 Some multiple-input data acquisition arrangements. (a) Simultaneous input sampling using multiple A/D converters. (b) Simultaneous input sampling using a single A/D converter. (c) Nonsimultaneous sampling.

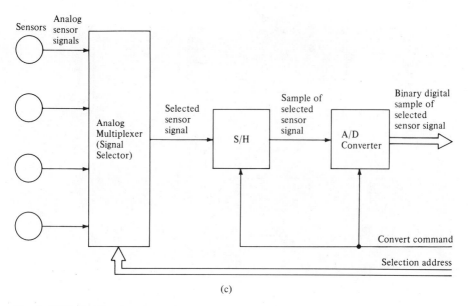

(c)

Figure 7-21 Cont.

A single A/D converter is shared by each of the sensor signal conversions in the configuration of Figure 7-21(b). As before, separate S/H devices are used so that all samples are taken at the same time. An analog multiplexer selects one then the next signal sample for conversion to binary code. To convert *m* sensor signals sequentially during each sampling interval, the A/D converter in this configuration must operate *m* times as rapidly as each A/D converter must in the configuration of Figure 7-21(a).

The arrangement of Figure 7-21(c) is more common than either of the other two because its hardware is simpler. A single S/H device and a single A/D converter are shared by each of the signals to be converted. The resulting samples are taken at different instants during each sampling interval. These *time offsets* complicate the control system design only to the extent of needing a modified discrete-time model of the continuous-time system to be controlled. With time offsets, the discrete-time model can be of higher order because parts of two successive sampled-and-held inputs can affect the plant state at each step.

Several multiple-output data distribution systems are shown in Figure 7-22. In the first, Figure 7-22(a), binary digital samples are sent from the processor, one set at a time, and stored in digital registers. The contents of the registers are converted by individual D/A converters, then individual S/H devices are used so that each of the analog

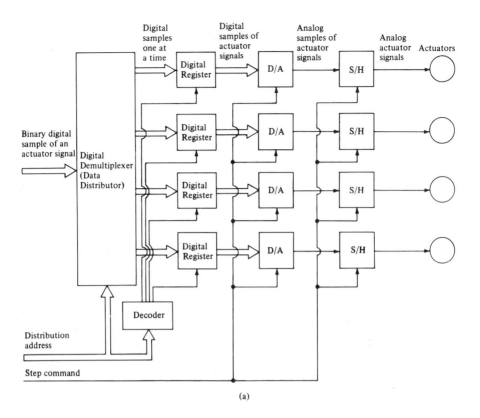

(a)

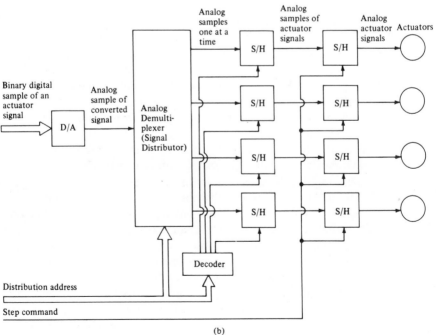

(b)

Figure 7-22 Some multiple-output data distribution arrangements. (a) Simultaneous output changes using multiple D/A converters. (b) Simultaneous output changes using a single D/A converter. (c) Nonsimultaneous output changes.

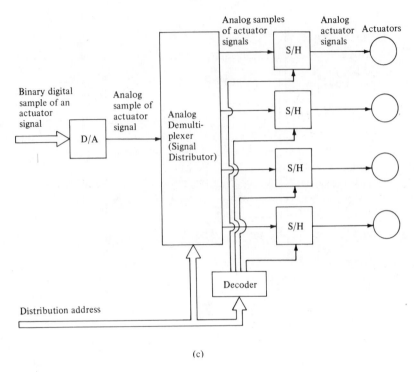

(c)

Figure 7-22 Cont.

output signals changes only at the sampling times. A single D/A converter is shared by each of the actuator signals in the configuration of Figure 7-22(b), where an analog demultiplexer is used to distribute the sequentially converted signals for storage in individual S/H circuits. For *m* signals converted during each sampling interval, the single shared D/A converter must operate *m* times as rapidly as individual D/A converters. So that the analog output signals each change only at the sampling times, a second set of S/H circuits is used.

The most common data distribution arrangement is that of Figure 7-22(c), which uses a single D/A converter and only one S/H device for each output. The output signals change one at a time, as each conversion is completed, not all at the same time. It is straightforward to account for this nonsimultaneous reconstruction when deriving the discrete-time model of the continuous-time plant.

Data acquisition and distribution systems also commonly contain analog *prefilters* and *postfilters*. The prefilters attenuate unwanted high frequencies from incoming continuous-time signals before they are sampled and A/D converted; usually those frequencies above half the sampling rate are removed so that the converted signal is uniquely described by its samples. Postfiltering is analog plant input filtering to smooth changes in the plant inputs after D/A conversion of the plant input samples.

The time used to process incoming and outgoing samples cannot exceed the sampling interval. Similarly, a cycle of data transmissions, A/D, and D/A conversions must be completed during each sampling interval. Even so, especially in high-rate systems, the total time needed to perform both the processing and conversions can be more than one sampling interval, in which event the *processing delay* must be accounted for in the system model.

7.5.3 Processing with Special Purpose Hardware

General purpose computer hardware has the advantages of low cost for the functions provided, a base of application experience, and the availability of sophisticated programming tools. Special purpose hardware is also available for the designer who needs higher-speed operation and for high-volume applications in which the hardware development and manufacturing costs are justified. Among the special purpose devices available are microprocessors designed especially for certain types of signal processing and chips for specific signal processing functions, such as multiplication.

A multiplier–accumulator (M/A) is a device for rapidly calculating sums of products of numbers, a function that is central to the implementation of difference equations. It typically has two sets of input connections, called input *ports*, one for each binary representation of the two numbers to be multiplied. The product of the two incoming numbers is added to any previous result and stored in an accumulator register connected to an output port. Additional connections control the chip's timing and allow the user to reset the accumulator contents to zero.

Figure 7-23 is a diagram of the hardware for a second-order controller

$$\begin{bmatrix} \xi_1(k+1) \\ \xi_2(k+1) \end{bmatrix} = \begin{bmatrix} 0 & 1 \\ -\frac{5}{6} & -\frac{1}{6} \end{bmatrix} \begin{bmatrix} \xi_1(k) \\ \xi_2(k) \end{bmatrix} + \begin{bmatrix} 0 \\ 1 \end{bmatrix} r(k)$$

$$u(k) = \begin{bmatrix} \frac{1}{2} & -\frac{3}{5} \end{bmatrix} \begin{bmatrix} \xi_1(k) \\ \xi_2(k) \end{bmatrix} + r(k) \tag{7-17}$$

using an M/A chip. The clock and counter generate a sequence of ROM addresses, the contents of which control which of several digital

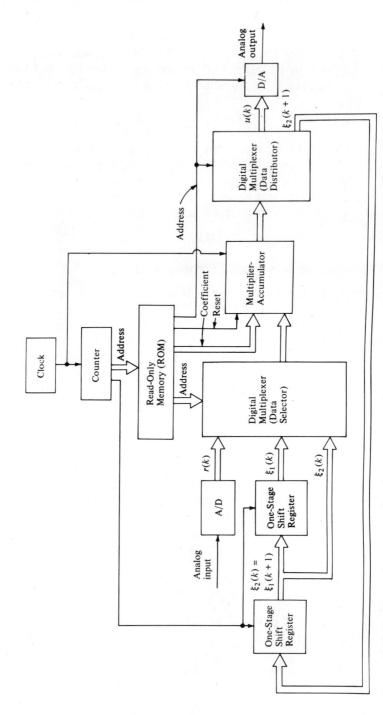

Figure 7-23 Hardware for a second-order, single-input/single-output digital controller using a multiplier–accumulator controlled by a sequencer.

signals is applied to one port of the M/A, the coefficient that is applied to the other M/A port, and where the M/A output is to be sent. The ROM additionally provides a signal that can reset the accumulator to zero and a signal that can activate the shift registers.

At the first count, the accumulator is reset and the binary signal representing r is multiplied by unity and added to the accumulator. At the second count, ξ_1 is selected, multiplied by $\frac{1}{2}$, and added to the accumulator. At count 3, ξ_2 is selected, multiplied by $-\frac{3}{5}$, and added to the accumulator. The M/A output now is the new value of the output u, which is transferred to the D/A converter.

At count 4, the accumulator is reset to zero, r is selected, multiplied by unity, and added to the accumulator. At count 5, ξ_1 is selected, multiplied by $-\frac{5}{6}$, and added to the accumulator. The binary signal representing ξ_2 is selected at count 6, multiplied by $-\frac{1}{6}$, and added to the accumulator so that the M/A output now holds the new value of ξ_2. The shift registers are clocked, and this new value of ξ_2 is stored in the register to the left while the new ξ_1, which is the old ξ_2, is moved to the rightmost shift register. The output generation and state updating cycle then repeats.

This hardware arrangement constitutes a special purpose, stored program computer that executes the same cycle of instructions over and over. The counter specifies the address of the current instruction, and the ROM output is the instruction code.

Another dedicated hardware option for the controller in equation (7-17) is shown in Figure 7-24. Here, individual multipliers and adders are interconnected to generate the needed signals. As in the previous example, these are represented by *fixed-point* binary numbers. The adders and multipliers are arranged to accommodate both negative and positive numbers. The controller state variables ξ_1 and ξ_2 are stored in shift registers as before. At each clock cycle, the old ξ_2 is shifted into the right register to become the new ξ_1, and the new ξ_2 is shifted into the left register. The new ξ_2 is formed by multiplying the old ξ_1 by $-\frac{5}{6}$, multiplying the old ξ_2 by $-\frac{1}{6}$, and adding to r, as shown. To form the output, ξ_2 is multiplied by $-\frac{3}{5}$, added to r, then added to $\frac{1}{2}\xi_1$. Because this design does not share a multiplier–adder, it has the potential of higher-speed operation than the previous hardware.

If the number of bits representing each signal is not to grow as the computations proceed, the results of the multiplications must be rounded or truncated, because the product of two m-bit binary numbers is a $2m$-bit number. Similarly, the sum of two m-bit binary numbers can require $m + 1$ bits, so there is the possibility of an *overflow* condition at each adder. The design in Figure 7-24 shows the roundings or truncations and the possible overflows in the design.

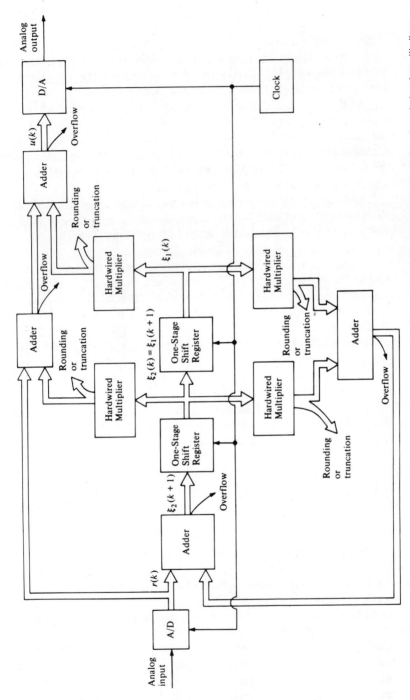

Figure 7-24 Hardware realization of a second-order, single-input/single-output digital controller using hardwired multipliers. No software is involved.

Controller hardware can alternatively be configured for *floating-point* representations and arithmetic. Inaccuracies due to numerical truncations or roundings still occur, but a wider range of numbers can be accommodated with the same number of bits than when the position of the binary point is fixed. Floating-point multiplication involves multiplication of mantissas, addition of exponents, and shifting of the result as necessary to keep the binary point of the mantissa of the product at a fixed position. Floating-point addition requires mantissa addition and possibly shifting. Multiplier, adder, and multiplier–accumulator chips that perform floating-point computations are available. They are generally more complicated (and thus more expensive) and somewhat slower in operation than their fixed-point counterparts.

7.6 Computer Software for Control

Like digital hardware requirements, software needs for computer control are likely to continue to evolve rapidly in the foreseeable future. In this section, some general software considerations are discussed that complement the discussion of digital control system hardware in the previous section. Standard networking interfaces and protocols, virtual memory, expert systems, artificial intelligence, and increasingly capable and sophisticated dedicated signal processing chips promise much for the future.

7.6.1 Main Program Structures

As the art of computer programming has matured, the sophistication of programming languages and other available aid has increased. At the most primitive level, that of *binary machine code*, each string of ones and zeros representing an instruction is stored in memory. It is unusual for a designer to deal directly with binary codes because it is a simple matter to use another computer for aid in translating a more human-friendly alphanumeric *mnemonic machine code* to the binary code equivalents. When the programmer can use symbols to represent addresses and other quantities in a machine code program, the program is said to be written in an *assembly language*, which requires the use of a more sophisticated assembler program to generate the corresponding binary machine code.

High-level languages deal with more abstract instructions and divorce the programmer from the underlying machine code. The needed machine code is obtained with a program that *compiles* the higher-level statements into machine language instructions. The same BASIC, Pascal, or other high-level language program can run on computers with different sets of machine instructions if each has a compiler that converts high-level instructions to the machine instructions peculiar to each computer.

High-level languages are usually less efficient than lower-level ones; that is, they tend to produce machine language programs that take longer to execute than when the program is written directly in machine language. When the required hardware and speed of operation are adequate, however, use of a higher-level language is greatly preferred because the entire program is far easier and faster to develop, debug, and modify. There is the possibility, too, that a high-level language program will be *portable*; that is, it will also run on other computers with different machine language instruction sets.

Programs for digital control are said to operate in *real time*, meaning that the computer processes data as it arrives, not as a "batch." The structure of a simple digital controller program is shown in Figure 7-25. The sampling rate is maintained by a periodic signal called the *step clock* (perhaps generated externally) that informs the processor each time a new set of measurement and reference samples is to be processed and a new set of plant input samples is to be produced. The program begins with initialization. Each controller state variable is set to zero, and other variables used in the program are given initial values. Possibly, digital signals are sent to turn on external equipment and to test whether it is functioning properly.

When the initialization is complete, the program waits for the step clock to indicate that a new set of measurement and reference samples is to be processed. Then those samples are transferred, under program control, to specific processor memory locations. The new controller states and outputs are computed, and the outputs are transferred for data distribution to the plant inputs. This program then checks to see if the system is to be shut down. If so, it executes a series of instructions to turn off the system in an orderly manner. These might involve, for example, a gradual tapering to zero of some or all of the plant inputs. If the system is not to be shut down, the program again waits for the step clock to indicate that new measurement and reference samples are to be processed.

Computers for control are often used to collect data about system performance and to carry out routine testing and other functions. These "housekeeping" tasks are performed when there is otherwise unused processing time available, as when the program is waiting to begin the next discrete-time step. These tasks can be *stacked* so that when the processor is available for them it resumes processing where it last left off, continuing until it must return to its control function. When

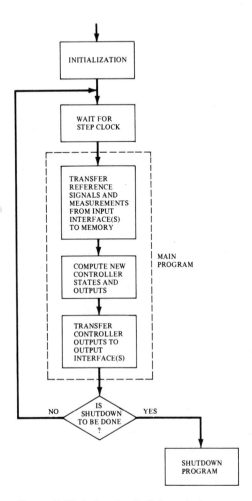

Figure 7-25 A simple digital control program structure.

it must return, the contents of all CPU registers needed, including the one containing the address of the next "housekeeping" instruction, are stored in memory, to be recalled later.

Instead of controlling the sampling times with a step clock, the input, output, and computation operations can be interleaved and the CPU clock used to control the rate and timing of sampling. For a given CPU clock frequency, the time spacing of data transfers and the rate at which the entire loop of the program is traversed depends on the number of instructions executed and the number of CPU clock cycles needed for each instruction. This kind of program structure is simple, but it is difficult to modify because any modification of the program can change the program's timing. It is usually necessary to write the program in assembly language rather than a higher-level language.

7.6.2 Subroutines and Interrupt Structures

Instead of stringing all the instructions in the order of their execution in a complicated program, an experienced programmer likely decomposes the program into coherent parts and codes each of the parts as a subroutine. The main program then consists mostly of subroutine calls, as in the example program structure shown in Figure 7-26. When a

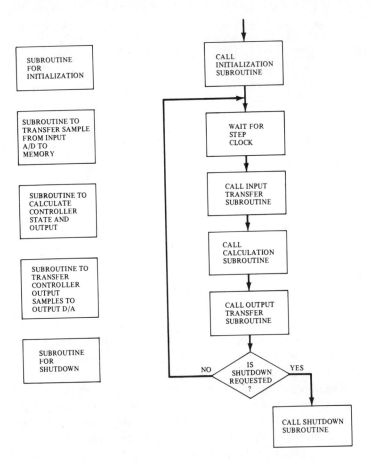

Figure 7-26 A simple subroutine arrangement for a single-input/single-output digital controller.

subroutine is called, the CPU saves the address of the next instruction in the main program and any other CPU register contents being used before it executes the series of instructions in the subroutine. When the subroutine is completed, the CPU returns to execution of the main program. It is a simple matter to arrange things so that subroutines can call other subroutines. One advantage of using subroutines is that the individual parts of a program can be developed and modified separately. Another advantage is that repetitive portions of the code need be included only once. Also, the main program, consisting as it does mostly of subroutine calls, is easier to follow.

An *interrupt* is a method for an external device easily and quickly to signal its need for attention to the CPU. When the CPU receives an interrupt, it completes execution of the current instruction, stores the address of the next instruction in memory, then jumps to the execution of an interrupt service routine. The interrupt service routine saves the contents of any CPU registers that should later be restored, determines which device sent the interrupt, and does the necessary processing required by that interrupt. When the interrupt processing has been completed, the CPU returns to processing the original stream of instructions.

A very useful interrupt device is the *timer* which consists of a counter driven by the system clock. The CPU is able to load the counter with a number specifying the number of clock cycles until a new interrupt signal is to be sent. A timer would likely be used for the step clock in the program structures shown in Figures 7-25 and 7-26.

A *nonmaskable interrupt* is one that will not be ignored by the CPU. On its receipt, without fail, a jump is made the interrupt service routine after execution of the current instruction. A *maskable interrupt* is one that can be turned off (or masked) by the program until an appropriate time for its service. A computer power failure in progress and emergency shutdown commands would likely be arranged as unmaskable interrupts. The signal that a new sample is available from a data acquisition system might be masked until it is time for the sample to be incorporated into the calculations.

Figure 7-27 shows hardware connections for interrupts. In the arrangement of Figure 7-27(a), the interrupts for each of several devices are separately wired to the CPU. When the interrupt is being serviced, the CPU sends an interrupt acknowledge signal back to the device. On receipt of the acknowledge signal, the device transfers data via a common data bus. For more than a very few devices, this arrangement uses many wires, so the configuration of Figure 7-27(b) is often preferable. Here single interrupt request and interrupt acknowledge lines are shared by all the devices. When an interrupt request is received by the

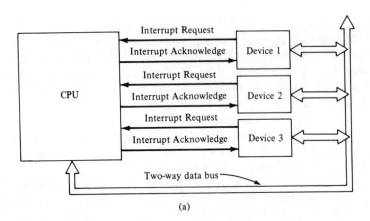

(a)

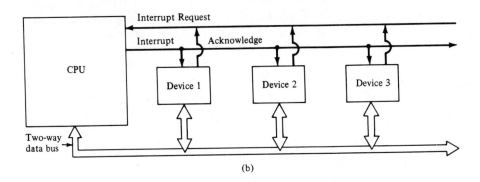

(b)

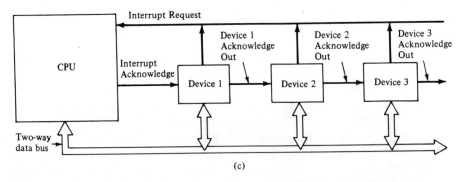

(c)

Figure 7-27 Some interrupt hardware arrangements. (a) Separate request and acknowledge lines for each device. (b) Common request and acknowledge lines. After reception of a request, the processor polls the devices to determine the highest-priority device that has requested an interrupt. (c) Interrupt daisy chain.

CPU, it acknowledges the request, which temporarily prevents the sending of new requests. The CPU then addresses the devices, in order of their priority, via the data bus to determine the highest-priority device that has requested an interrupt, a procedure called *polling*. An interrupting device can identify itself by sending a second interrupt request signal when its address appears on the data bus.

A third hardware connection for interrupts, the daisy chain, is described in Figure 7-27(c). Devices send their interrupt requests on a common line, but the interrupt acknowledge signal is passed on to the next device only if the previous device has not requested an interrupt. Eventually, the acknowledge signal reaches an interrupting device that does not pass it on, and the interrupting device identifies itself to the CPU via the data bus. The order of priority of the interrupts is their order in the chain from the CPU.

An example of the structure of a program for a single-input/single-output digital controller that uses subroutines and interrupts is given in Figure 7-28. Emergency shutdown and power loss conditions are serviced immediately. Each discrete-time step is timed by an external clock that also drives the A/D and D/A converters. When this program is ready to process an input sample, it temporarily unmasks the A/D interrupt and waits until the sample is ready. Similarly, when the program is ready to send a new controller output sample, it temporarily unmasks the D/A interrupt and waits until the sample can be received.

7.6.3 Software Design, Testing, and Modification

Every experienced programmer knows the importance of *designing* software, using many of the same steps that are used in good hardware design. It is particularly important to structure the program carefully at several levels, from the initial "large blocks," to the interrupt structure and subroutines, to the detailed sequences of instructions. Each level of design is modified as conflicts are resolved and as improvements are developed. As there is a high probability that the designer and others will need to understand and perhaps modify the program at a later time, contrary as it is to human nature, clear and complete documentation is essential.

Control system hardware and software are usually tested by simulation in several ways before they are connected to the plant they are to control. At first, the basic controller equations are tested in much the same way as the typical response figures were produced for the design examples in this chapter. The idea is to verify expected properties of the design (such as speed of response and steady state error) and to

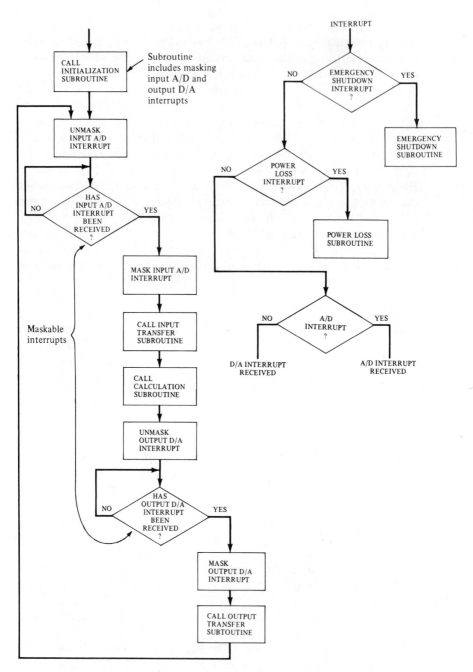

Figure 7-28 A subroutine and interrupt arrangement for a single-input/single-output digital controller.

uncover shortcomings (such as the need for an unreasonably large control signal or high susceptibility to small errors) before proceeding further. Later, most of the complete program is tested using simulated measurement samples. Then the controller hardware–software combination can be tested with simulated analog sensor signals.

A particularly intriguing problem is that of designing hardware and software for control that are automatically able to adapt to failures of some sensors and actuators, still maintaining good control of the plant. Sensor failures usually result in unreasonable (such as zero) or inconsistent measurements. Actuator failures might be inferred from measurements by nearby sensors or from the errors between actual and calculated plant performance. If these failures are identified by the control program, then the controller can be changed to ignore measurements from defective sensors and not to depend on defective actuators.

The design of a program that adapts to failures in a large system such as a power plant is formidable. If there can be 100 different failures and, say, up to 4 failures can be present at once, then the program might be required to have as many as

$$N = 1 + 100 + \frac{1}{2}(100)(99) + \frac{1}{6}(100)(99)(98) + \frac{1}{24}(100)(99)(98)(97)$$

different possibilities for the controller. The program might have only 100 different branches, 1 for each individual failure, but there could be on the order of 4 million different combinations possible. Testing all of these is probably quite impractical, and the alternative is using an untested system for control.

The results of simulation and operation almost invariably lead to hardware and software modifications; this observation underscores the need for well-documented, easily read programs.

7.7 Design Examples

We now consider discrete-time control of two example continuous-time systems. Although these systems are of relatively low order, there is much to learn from each. The first system, a sophisticated elevator position controller, includes a saturation nonlinearity. This kind of model and the control strategy used are quite common in practice. The second example system, a simplified version of an undersea robotic torsional positioning mechanism, has a highly oscillatory continuous-time step response. This response is indicative of poor between-sample tracking, especially at the expected low sample rate that is to be used. A continuous-time plant input filter is used to improve the between-sample response, and because the filter zeros are intended to cancel plant poles, the overall model of the plant and filter has uncontrollable modes.

The discussion of each example system is carried to the point at which discrete-time controllers are to be designed, using the methods of previous chapters. For the elevator controller, the effect of the nonlinearity is taken into account in an observer's state estimate. It is decided that, although the observer might be deadbeat, initially converging quickly to the plant state, the other eigenvalues of the discrete-time feedback system should give a more gradual zero-input response decay. For the torsional positioning mechanism, having designed a continuous-time plant input filter, the designer might elect to ignore the uncontrollable plant modes, which, ideally, decay to zero and stay there. Or recognizing that with errors in the plant model these modes continue to be excited, the modes can be estimated and the estimates used to improve tracking performance.

Some of the problems associated with this section involve simulation, but none concerns hardware or software design. You may wish to extend the problems' scope to include hardware and software design that is compatible with your specific facilities.

7.7.1 An Elevator Controller

A simplified block diagram for elevator positioning is shown in Figure 7-29(a). The voltage for the elevator drive motor armature is supplied by a power amplifier with the saturation characteristic shown in Figure 7-29(b). The armature voltage v is not allowed to exceed ± 200 V which, in steady state, results in an elevator speed of

$$x_1 = \left(\frac{0.001}{s + 0.2}\right)\bigg|_{s=0} (200) = 1 \text{ m/s}$$

The elevator drive motor and cable drum dynamics are modeled as a first-order subsystem with a $1/0.2 = 5$-sec time constant. The elevator speed and position are each sensed separately. The speed sensor, a tachometer on the drive motor shaft with an electrical filter for noise reduction, has a 0.5-sec time constant. The position sensor has a 0.2-sec time constant.

Between floors, the power amplifier is in saturation, supplying either $+200$ V or -200 V to the motor, depending on whether the elevator is moving up or down. As the elevator approaches the floor where it is to stop, the amplifier output voltage comes out of saturation, into the linear region where a controller, using the sensed speed and position, is

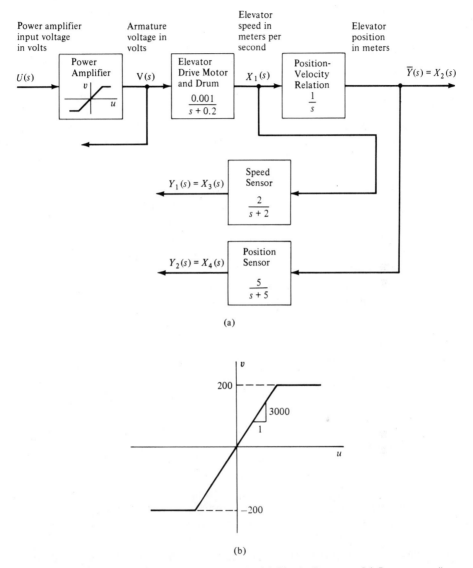

Figure 7-29 Elevator positioning components. (a) Block diagram. (b) Power amplifier input–output characteristic.

to control $u(t)$ in such a way as to bring the elevator smoothly and accurately to rest at the floor. The voltage for the drive motor armature, after the saturation nonlinearity, is easy to measure and can thus be available to the controller.

At first thought, it seems tempting to design the control system so as to track a step input the magnitude of which corresponds to the desired floor. In this case, the steady state error indicates that the elevator car is not aligned properly with the desired floor and may pose danger to the passengers. The disadvantage of such a system, however, is that the input step should be a precise voltage that does not change with time. Any degradation over the years would cause the elevator not to align itself with the desired floor. A better method is to treat the controller as a regulator with nonzero initial conditions. These initial conditions are the distance from the desired floor and the elevator velocity.

Recently, elevator manufacturers have used "fuzzy logic controllers" that respond very quickly to changing demands and thus reduce waiting time. When the elevator is to move from one floor to another, the position sensor is switched so that it measures position to the floor at which the elevator is to stop. The controller is thus to act as a regulator for each movement between floors. The position sensor hardware consists of a subsystem that uses coarse elevator position information derived from electrical switch closures, medium-accuracy position information from an electrooptic range sensor, and short-range fine position information from a differential magnetic sensor on the elevator that is activated by soft iron target strips at each floor.

Figure 7-30(a) shows a diagram of the plant to be controlled, together with a digital controller. A 0.2-sec sampling interval has been chosen for the initial design, based on the fact that the shortest plant time constant associated with the position sensor is 0.2 sec. The data acquisition subsystem is then required to sample three signals, each at this five-sample-per-second rate. The rate is well within the capability of inexpensive dual-slope integrating A/D converters, such as those used in most digital voltmeters. With this sampling rate, the between-sample response of the elevator is expected to be adequate, as can be judged from the elevator velocity unit step response in Figure 7-30(b). The position step response is the integral of this curve.

A control strategy is illustrated in Figure 7-31. The state of the elevator drive and sensor is observed, and a linear transformation of this observed elevator drive and state drives the power amplifier. When this observer–controller attempts to supply the elevator drive motor with more than 200 V, the power amplifier limiting comes into play, but the observer still calculates the proper state because it uses the actual input to the elevator drive. This kind of plant control is common. In a fighter plane's climb control, for example, the aircraft should be allowed to climb no faster than a certain safe rate. When commanded to change altitude quickly, the control system has the aircraft climb at the maximum rate until it nears the new desired altitude.

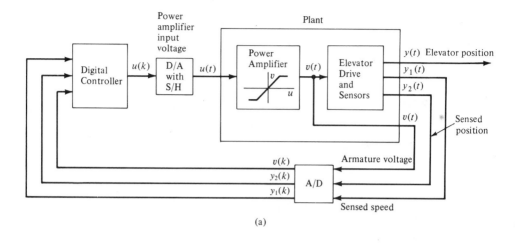

(a)

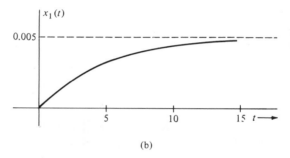

(b)

Figure 7-30 Digital control of elevator positioning. (a) Digital control system. (b) Elevator velocity unit step response.

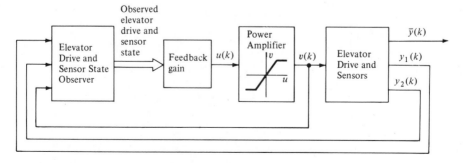

Figure 7-31 Controller arrangement.

Continuous-time state equations for the plant consisting of elevator drive and sensors are

$$
\begin{bmatrix} \dot{x}_1(t) \\ \dot{x}_2(t) \\ \dot{x}_3(t) \\ \dot{x}_4(t) \end{bmatrix} = \begin{bmatrix} -0.2 & 0 & 0 & 0 \\ 1 & 0 & 0 & 0 \\ 2 & 0 & -2 & 0 \\ 0 & 5 & 0 & -5 \end{bmatrix} \begin{bmatrix} x_1(t) \\ x_2(t) \\ x_3(t) \\ x_4(t) \end{bmatrix} + \begin{bmatrix} 10^{-3} \\ 0 \\ 0 \\ 0 \end{bmatrix} v(t)
$$

$$
= \mathscr{A}\mathbf{x}(t) + \mathscr{b}v(t)
$$

$$
\begin{bmatrix} y_1(t) \\ y_2(t) \end{bmatrix} = \begin{bmatrix} 0 & 0 & 1 & 0 \\ 0 & 0 & 0 & 1 \end{bmatrix} \begin{bmatrix} x_1(t) \\ x_2(t) \\ x_3(t) \\ x_4(t) \end{bmatrix} = \mathbf{Cx}(t)
$$

$$
\bar{y}(t) = \begin{bmatrix} 0 & 1 & 0 & 0 \end{bmatrix} \begin{bmatrix} x_1(t) \\ x_2(t) \\ x_3(t) \\ x_4(t) \end{bmatrix} = \bar{\mathbf{c}}^t \mathbf{x}(t)
$$

When the signals of this plant are sampled with sampling interval $T = 0.2$ sec, a discrete-time plant model is

$$
\begin{bmatrix} x_1(k+1) \\ x_2(k+1) \\ x_3(k+1) \\ x_4(k+1) \end{bmatrix} = \begin{bmatrix} 0.961 & 0 & 0 & 0 \\ 0.196 & 1 & 0 & 0 \\ 0.323 & 0 & 0.670 & 0 \\ 0.618 & 0 & 0 & 0.368 \end{bmatrix} \begin{bmatrix} x_1(k) \\ x_2(k) \\ x_3(k) \\ x_4(k) \end{bmatrix}
$$

$$
+ \begin{bmatrix} (0.196 \times 10^{-3}) \\ (0.02 \times 10^{-3}) \\ (0.035 \times 10^{-3}) \\ (0.072 \times 10^{-3}) \end{bmatrix} v(k) = \mathbf{Ax}(k) + \mathbf{b}v(k)
$$

$$
\begin{bmatrix} y_1(k) \\ y_2(k) \end{bmatrix} = \begin{bmatrix} 0 & 0 & 1 & 0 \\ 0 & 0 & 0 & 1 \end{bmatrix} \begin{bmatrix} x_1(k) \\ x_2(k) \\ x_3(k) \\ x_4(k) \end{bmatrix} = \mathbf{Cx}(k)
$$

$$
\bar{y}(k) = \begin{bmatrix} 0 & 1 & 0 & 0 \end{bmatrix} \begin{bmatrix} x_1(k) \\ x_2(k) \\ x_3(k) \\ x_4(k) \end{bmatrix} = \bar{\mathbf{c}}^t \mathbf{x}(k)
$$

Numerical values for the elements of **A** and **b** are found using truncated series, as in Section 7.3. Twenty terms of each series give results that are accurate to at least 10 significant figures.

State feedback can be designed to place the plant's eigenvalues at any desired locations. The locations desired depend on how quickly one wishes the zero-input component of the plant response to decay with step. For example, decay by a factor of one-half in each 1-sec interval might be desirable, requiring eigenvalue locations with radius c from the origin of the complex plane such that

$$c^5 = \frac{1}{2} \qquad c = 0.87$$

because there are five discrete-time steps each second. One possibility is to choose all four feedback system eigenvalues to be real and equal to 0.87.

For this plant with two measurement outputs, a second-order observer could be used to observe the plant state. Only a first-order observer is necessary, however, for observation of a single scalar state transformation $\mathbf{e}^\dagger\mathbf{x}$. The choice of observer eigenvalue(s) depends on the accuracy of the individual measurements. The choice of eigenvalues that give a slow decay of the observer error produces estimates that depend on past measurements as well as the most recent ones; this can have a smoothing effect on the estimates that reduces the influence of measurement "noise." If individual measurement samples are sufficiently accurate, deadbeat design, with all eigenvalues at $\lambda = 0$, is workable. The deadbeat observer uses only the most recent measurements to produce its estimates.

7.7.2 Positioning Mechanism Tracking

A torsional positioning mechanism that is part of an undersea robotic manipulator is described by an input–output differential equation of the form

$$J\frac{d^2\Theta}{dt^2} + D\frac{d\Theta}{dt} + K\Theta = \tau(t) = Li(t)$$

J is the manipulator's moment of inertia, D is its damping constant, K is its spring constant, and $\tau(t)$ is the applied torque, supplied by an electromagnetic actuator. This torque is, in turn, proportional to the actuator drive current $i(t)$. For

$$J = \frac{1}{100} \; \text{N} \cdot \text{m sec}^2/\text{rad}$$

$$D = \frac{2}{100} \text{ N} \cdot \text{m sec/rad}$$

$$K = 1 \text{ N} \cdot \text{m/rad}$$

$$L = 1 \text{ N} \cdot \text{m/a}$$

the input–output equation is

$$\frac{d^2\Theta}{dt^2} + \frac{D}{J}\frac{d\Theta}{dt} + \frac{K}{J}\Theta(t) = \frac{L}{J}i(t)$$

or

$$\frac{d^2\Theta}{dt^2} + 2\frac{d\Theta}{dt} + 100\Theta(t) = 100i(t)$$

Letting the state variables $x_1(t)$ and $x_2(t)$ be, respectively, rotational position and rotational velocity

$$x_1(t) = \Theta(t) \qquad x_2(t) = \frac{d\Theta}{dt}$$

and letting the input be the drive current

$$u(t) = i(t)$$

state equations for the mechanism are

$$\begin{bmatrix} \dot{x}_1(t) \\ \dot{x}_2(t) \end{bmatrix} = \begin{bmatrix} 0 & 1 \\ -100 & -2 \end{bmatrix}\begin{bmatrix} x_1(t) \\ x_2(t) \end{bmatrix} + \begin{bmatrix} 0 \\ 100 \end{bmatrix}u(t) \qquad \textbf{(7-18)}$$

$$= \mathcal{A}x(t) + \mathcal{b}u(t)$$

Rotational position is sensed with an electrical strain gauge, and rotational velocity is sensed with a magnetic motion detector, so the measurement outputs are

$$\begin{bmatrix} y_1(t) \\ y_2(t) \end{bmatrix} = \begin{bmatrix} 1 & 0 \\ 0 & 1 \end{bmatrix}\begin{bmatrix} x_1(t) \\ x_2(t) \end{bmatrix} = \mathbf{C}x(t) \qquad \textbf{(7-19)}$$

The position is to be controlled in this tracking system, so the tracking output is

$$\bar{y}(t) = y_1(t) = [1 \quad 0]\begin{bmatrix} x_1(t) \\ x_2(t) \end{bmatrix} = \bar{\mathbf{c}}^{\dagger}\mathbf{x}(t) \qquad \textbf{(7-20)}$$

The transfer function relating the tracking output $\bar{y}(t)$ to the control input $u(t)$ is

$$T(s) = \bar{\mathbf{c}}^{\dagger}(s\mathbf{I} - \mathcal{A})^{-1}\mathcal{b} = \frac{100}{s^2 + 2s + 100}$$

as indicated in Figure 7-32(a), where this plant's step response is shown. The plant eigenvalues (or poles) are at $-1 \pm j\sqrt{99}$. Mechanical losses of the manipulator are deliberately kept low, with the result that its zero-input response (and thus its step response) are highly oscillatory. If the input $u(t)$ is a sampled-and-held waveform from a digital controller, the between-sample response of $\bar{y}(t)$ exhibits this character. Because a sampling interval no shorter than $T = 0.5$ sec is contemplated, between-sample oscillations in $\bar{y}(t)$ are likely to be substantial.

Figure 7-32(b) shows a plant input filter to improve the between-sample tracking response, with the transfer function

$$G(s) = \frac{s^2 + 2s + 100}{s^2 + 20s + 100}$$

The combination of plant and filter has the model transfer function

$$M(s) = T(s)G(s) = \frac{100}{s^2 + 20s + 100}$$

The filter's zeros cancel the plant poles, and the filter's two poles are repeated, resulting in a critically damped step response. A portion of a simulation study to evaluate the effects of inaccuracies in the plant model is shown in Figure 7-33. Precise pole-zero cancellation is not needed for the filter to be effective; when the filter's zeros are sufficiently close to the unwanted plant poles, the amplitude of the original plant's zero-input response is low. A state variable model of the filter,

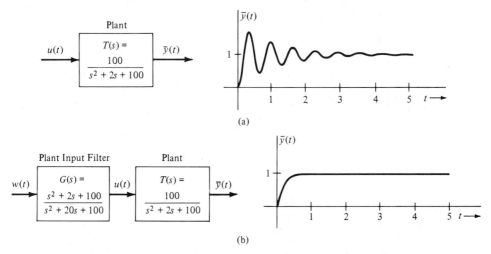

Figure 7-32 Improving between-sample response with a plant input filter. (a) Step response of the plant tracking output. (b) Step response with a plant input filter.

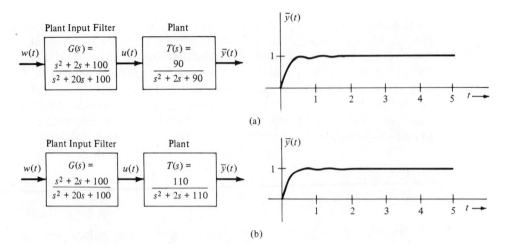

Figure 7-33 Simulation of step response with a plant parameter change. (a) One change. (b) Another change.

in upper Jordan form, is

$$\begin{bmatrix} \dot{x}_3(t) \\ \dot{x}_4(t) \end{bmatrix} = \begin{bmatrix} -10 & 1 \\ 0 & -10 \end{bmatrix} \begin{bmatrix} x_3(t) \\ x_4(t) \end{bmatrix} + \begin{bmatrix} -18 \\ 180 \end{bmatrix} w(t)$$

$$u(t) = \begin{bmatrix} 1 & 0 \end{bmatrix} \begin{bmatrix} x_3(t) \\ x_4(t) \end{bmatrix} + w(t) \tag{7-21}$$

where $w(t)$ is the filter input.

The combined state variable equations of the original plant in equations (7-18) to (7-20) and the plant input filter in equation (7-21) are

$$\begin{bmatrix} \dot{x}_1(t) \\ \dot{x}_2(t) \\ \dot{x}_3(t) \\ \dot{x}_4(t) \end{bmatrix} = \begin{bmatrix} 0 & 1 & 0 & 0 \\ -100 & -2 & 100 & 0 \\ 0 & 0 & -10 & 1 \\ 0 & 0 & 0 & -10 \end{bmatrix} \begin{bmatrix} x_1(t) \\ x_2(t) \\ x_3(t) \\ x_4(t) \end{bmatrix} + \begin{bmatrix} 0 \\ 100 \\ -18 \\ 180 \end{bmatrix} w(t)$$

$$\begin{bmatrix} y_1(t) \\ y_2(t) \end{bmatrix} = \begin{bmatrix} 1 & 0 & 0 & 0 \\ 0 & 1 & 0 & 0 \end{bmatrix} \begin{bmatrix} x_1(t) \\ x_2(t) \\ x_3(t) \\ x_4(t) \end{bmatrix}$$

$$\bar{y}(t) = \begin{bmatrix} 1 & 0 & 0 & 0 \end{bmatrix} \begin{bmatrix} x_1(t) \\ x_2(t) \\ x_3(t) \\ x_4(t) \end{bmatrix} \tag{7-22}$$

The original plant modes with eigenvalues $-1 \pm j\sqrt{99}$ are not controllable because of the pole-zero cancellations of the filter. All four modes are observable from each of the outputs.

If the cancellation of plant poles by the plant input filter is sufficiently precise, the completely controllable and observable part of equation (7-22) can be used as the plant and input filter model for digital controller design. If the initial conditions are zero, the amplitude of the uncontrollable plant modes in $\bar{y}(t)$ is zero. For nonzero initial conditions, the uncontrollable modes contribute a transient of the form

$$\bar{y}_{\text{zero-input}}(t) = Ke^{-t}\cos(\sqrt{99}t + \phi)$$

where K and ϕ are constants. This term decays with a 1-sec time constant, never to be seen again. The continuous-time plant is then described by the state variable equations

$$\begin{bmatrix} \dot{x}_1(t) \\ \dot{x}_2(t) \end{bmatrix} = \begin{bmatrix} 0 & 1 \\ -100 & -20 \end{bmatrix} \begin{bmatrix} x_1(t) \\ x_2(t) \end{bmatrix} + \begin{bmatrix} 0 \\ 100 \end{bmatrix} w(t)$$

$$\begin{bmatrix} y_1(t) \\ y_2(t) \end{bmatrix} = \begin{bmatrix} 1 & 0 \\ 0 & 1 \end{bmatrix} \begin{bmatrix} x_1(t) \\ x_2(t) \end{bmatrix}$$

$$\bar{y}(t) = \begin{bmatrix} 1 & 0 \end{bmatrix} \begin{bmatrix} x_1(t) \\ x_2(t) \end{bmatrix} \tag{7-23}$$

Digital controller design is relatively easy because the state is available for feedback. If the series methods of Section 7.3 are used, a discrete-time model of equation (7-23) with sampling interval $T = \frac{1}{2}$ is

$$\begin{bmatrix} x_1(k + 1) \\ x_2(k + 1) \end{bmatrix} = \begin{bmatrix} 0.040 & (3.37 \times 10^{-3}) \\ -0.337 & (-2.70 \times 10^{-2}) \end{bmatrix} \begin{bmatrix} x_1(k) \\ x_2(k) \end{bmatrix} + \begin{bmatrix} 2.05 \\ -62.3 \end{bmatrix} w(k)$$

$$\begin{bmatrix} y_1(k) \\ y_2(k) \end{bmatrix} = \begin{bmatrix} 1 & 0 \\ 0 & 1 \end{bmatrix} \begin{bmatrix} x_1(k) \\ x_2(k) \end{bmatrix}$$

$$\bar{y}(k) = \begin{bmatrix} 1 & 0 \end{bmatrix} \begin{bmatrix} x_1(k) \\ x_2(k) \end{bmatrix} \tag{7-24}$$

If the plant pole cancellation by the filter is not so accurate, the full model of the plant and filter can be used to advantage. The effects of the modes that would ideally be uncontrollable can be estimated and used to improve the plant state estimate. Effects of the uncontrollable modes on the tracking output might even be canceled. A discrete-time model of the plant and filter in equation (7-22) with sampling interval

$T = \frac{1}{2}$, computed using series as in Section 7.3, is

$$
\begin{bmatrix}
x_1(k+1) \\
x_2(k+1) \\
x_3(k+1) \\
x_4(k+1)
\end{bmatrix}
$$

$$
= \begin{bmatrix}
0.0986 & -0.0589 & -0.378 & (-3.23 \times 10^{-3}) \\
5.89 & 0.216 & -2.11 & -0.346 \\
0 & 0 & (6.74 \times 10^{-3}) & (3.37 \times 10^{-3}) \\
0 & 0 & 0 & (6.74 \times 10^{-3})
\end{bmatrix}
\begin{bmatrix}
x_1(k) \\
x_2(k) \\
x_3(k) \\
x_4(k)
\end{bmatrix}
$$

$$
+ \begin{bmatrix}
1.28 \\
-45.4 \\
11.2 \\
-75.2
\end{bmatrix} w(k)
$$

$$
\begin{bmatrix}
y_1(k) \\
y_2(k)
\end{bmatrix} = \begin{bmatrix}
1 & 0 & 0 & 0 \\
0 & 1 & 0 & 0
\end{bmatrix}
\begin{bmatrix}
x_1(k) \\
x_2(k) \\
x_3(k) \\
x_4(k)
\end{bmatrix}
$$

$$
\bar{y}(k) = \begin{bmatrix} 1 & 0 & 0 & 0 \end{bmatrix}
\begin{bmatrix}
x_1(k) \\
x_2(k) \\
x_3(k) \\
x_4(k)
\end{bmatrix} \tag{7-25}
$$

The eigenvalues of the continuous-time system of equation (7-22)

$$-1 \pm j\sqrt{99} \qquad -10 \qquad -10$$

translate to

$$\exp[(-1 \pm j\sqrt{99})T] \qquad \exp(-10T) \qquad \exp(-10T)$$

or

$$0.157 \pm j0.59 \qquad 6.74 \times 10^{-3} \qquad 6.74 \times 10^{-3}$$

in the discrete-time model.

The continuous-time model assumes plant pole cancellation by the filter, so the original plant modes $-1 \pm j\sqrt{99}$ are uncontrollable. They affect the system (7-24) as if they were produced by an autonomous disturbance model, as in Section 6.5. Even though these modes are actually excited by small unmodeled couplings from the input $w(k)$, they can be treated as if they were produced by initial conditions.

7.8 Summary

Digital controllers for continuous-time plants can be designed by beginning with an analog controller design and approximating it with a digital filter. The filter design can approximate the integrations with discrete-time operations or it can be made to have step (or other) response samples that are equal to samples of the analog controller's step (or other) response. Usually, even for relatively high sample rates, the digital approximation performs less well than the analog controller from which it was derived.

The relationship between a linear, time-invariant, continuous-time plant driven by sampled-and-held inputs

$$\dot{\mathbf{x}}(t) = \mathscr{A}\mathbf{x}(t) + \mathscr{B}\mathbf{u}(t)$$
$$\mathbf{y}(t) = \mathbf{C}\mathbf{x}(t) + \mathbf{D}\mathbf{u}(t)$$

and the linear, step-invariant, discrete-time model of its evenly spaced samples at intervals T

$$\mathbf{x}(k + 1) = \mathbf{A}\mathbf{x}(k) + \mathbf{B}\mathbf{u}(k)$$
$$\mathbf{y}(k) = \mathbf{C}\mathbf{x}(k) + \mathbf{D}\mathbf{u}(k)$$

is

$$\mathbf{A} = e^{\mathscr{A}\mathbf{T}} = \left[\mathbf{I} + \frac{\mathscr{A}T}{1!} + \frac{\mathscr{A}^2 T^2}{2!} + \cdots + \frac{\mathscr{A}^i T^i}{i!} + \cdots \right]$$

and

$$\mathbf{B} = \left[\mathbf{I}T + \frac{\mathscr{A}T^2}{2!} + \frac{\mathscr{A}^2 T^3}{3!} + \cdots + \frac{\mathscr{A}^i T^{i+1}}{(i + 1)!} + \cdots \right] \mathscr{B}$$

Truncated power series are a convenient computational method. Computer simulation of the plant and controller is an important design tool.

When a continuous-time plant is driven by sampled-and-held input signals from a digital controller, the character of the plant's continuous-time response between samples is the plant step response without feedback. It sometimes happens that, although tracking is quite good at the sample times, the plant response is highly oscillatory or in some other way undesirable between samples. If it is feasible to change the plant with analog feedback, this option should be explored.

Raising the sample rate tends to decrease the amplitude of each step input change and thus reduces the amplitude of the undesirable between-sample response, but it is usually expensive to increase the sample rate greatly. If the plant model is sufficiently well known, a continuous-time filter between the controller's D/A conversion and the plant inputs can be designed to improve the plant between-sample responses.

Designing filters at the analog plant inputs to improve plant be-tween-sample response is a problem for continuous-time plant input filter design similar to the reference input filters for discrete-time track-ing discussed in Chapter 6. For a continuous-time plant with the trans-fer function matrix $\mathbf{T}(s)$ relating the tracking outputs to the plant in-puts, an analog plant input filter with the transfer function matrix $\mathbf{G}(s)$ is designed such that the model plant

$$\mathbf{M}(s) = \mathbf{T}(s)\mathbf{G}(s)$$

responds in an acceptable way to sampled-and-held inputs. Usually, but not always, $\mathbf{G}(s)$ is required to be stable and proper. If $\mathbf{G}(s)$ is an inverse filter, so that

$$\mathbf{T}(s)\mathbf{G}(s) = \mathbf{I}$$

then the forced tracking outputs, like the filter inputs, are piecewise constant between sample times. Other choices of $\mathbf{G}(s)$ give other, per-haps more desirable, responses.

When an analog plant input filter is used, it becomes part of the plant to be controlled by the discrete-time controller. Analog plant input filtering should thus be designed first, obtaining acceptable open-loop plant step response, before the discrete-control design.

If adequate analog plant input filtering is undesirable, the designer might elect instead to use high-speed digital filters at the plant inputs. The higher filter speed is generally feasible and attractive because it requires an increased D/A but not an increased A/D rate. A higher-rate digital plant input filter can be incorporated as part of a lower-rate discrete-time model by including the filter state as well as the original plant state in the model.

Some ideas relating to hardware and software for digital control were then discussed. The hardware–software options range from dedi-cated hardwired processing, involving no software, to more general signal processing hardware programmed by ROM "firmware," to highly flexible general purpose hardware with software derived from a high-level language. The timing of digital control is accomplished in some simple programmed systems by arranging instructions in a loop that executes once each sample period. In more sophisticated systems, timing is controlled by interrupts from an external clock or from timers driven by the CPU clock.

Practical data acquisition and distribution subsystems often share single A/D and D/A converters and a data transmission bus. They commonly convert several different signals sequentially rather than all at the same time. It is logical to account for the resulting time offsets in the discrete-time model, but doing so generally results in an increased model order because two successive input samples, not just one, can affect the plant state at each step.

The chapter concluded with two open-ended examples of design of discrete-time controllers of continuous-time systems, an elevator positioning controller and a torsional position mechanism. Like many other practical systems, the elevator's speed should not exceed a certain maximum amount, so the control voltage to its drive motor will be limited. This limiting constitutes an important kind of nonlinearity. The elevator moves at the maximum rate until the motor drive voltage falls below its maximum value as the elevator nears the floor where it stops. The observer's state estimate remains correct during the time of limiting by sensing and using the limited signal to drive the observer. In the torsional positioner example, a continuous-time plant input filter was used to improve between-sample tracking response. The combined system consisting of filter and plant had uncontrollable modes because of the cancellation of plant poles by the filter. It was the designer's option whether to ignore the uncontrollable modes or to observe them and use their estimates to improve tracking performance.

References

Little has been written in the control literature about improving continuous-time system between-sample response with a fixed sample rate. Discussion of interpolation methods and higher-order holds can be found in

S. A. Tretter, *Introduction to Discrete-Time Signal Processing.* New York: Wiley, 1976

and in an occasional paper such as

G. H. Hostetter and A. R. Stubberud, "High-Order Holds Using Polynomial Spectral Observation," *Computers and Electrical Engineering,* Vol. 9, December 1982, pp. 11–18.

A very good discussion of the issues associated with the implementation of digital control systems is found in the paper

H. Hanselmann, "Implementation of Digital Controllers–A Survey," *Automatica,* Vol. 21, No. 1, January 1987, pp. 7–32,

which also includes a wealth of helpful references.

More material on some of the hardware and software considerations in digital control system design can be found in

P. Katz, *Digital Control Using Microprocessors.* Englewood Cliffs, NJ: Prentice-Hall, 1981;

K. J. Åström and B. Wittenmark, *Computer Controlled Systems.* Englewood Cliffs, NJ: Prentice-Hall, 1984;

C. H. Houpis and G. B. Lamont, *Digital Control Systems: Theory, Hardware, Software*, 2nd ed. New York: McGraw-Hill, 1985;

G. Franklin, J. Powell, and M. Workman, *Digital Control of Dynamic Systems*, 2nd ed. Reading, MA: Addison-Wesley, 1990;

C. Phillips and H. Nagle, *Digital Control System Analysis and Design*, 2nd ed. Englewood Cliffs, NJ: Prentice-Hall, 1990.

B. C. Kuo, *Digital Control Systems*, 2nd ed. Philadelphia, PA: Saunders College Publishing, 1992;

The two example systems of Section 7.7 are based on research work for Battelle Institute′and for the Office of Naval Research.

Chapter Seven Problems

7-1. Draw integration diagrams for continuous-time systems with the following transfer functions. Approximate the integrations by Euler's forward method with the indicated sampling interval T and find the corresponding z-transfer function of the approximation.

a. $G_1(s) = \dfrac{\frac{1}{2}s}{s + \frac{1}{2}}$ $T = 0.5$

b. $G_2(s) = \dfrac{\frac{1}{2}s}{s + \frac{1}{2}}$ $T = 0.2$

c. $G_3(s) = \dfrac{2s - 1}{s(s + 2)}$ $T = 0.1$

7-2. Repeat problem 7-1 using Euler's backward method.

7-3. Repeat problem 7-1 using the trapezoidal method.

7-4. Find the z-transfer function for a three-sample digital approximation of a controller transfer function

$$G_c(s) = \frac{3s - 4}{s^2 + 5s + 6}$$

Use the sampling interval $T = 0.3$.

7-5. The discrete approximations to integrals in Table 7-1 are obtained by passing polynomial curves through the integrand samples, then approximating the integral between the most recent two samples by the area under the polynomial. For example, the three-sample approximation is the area between $t = kT$ and $t = (k + 1)T$ under a quadratic curve that passes through the three points $f[(k - 1)T]$, $f(kT)$, and $f[(k + 1)T]$.

Derive this result, then find the corresponding result for the four-sample approximation. For this case, also find the difference equation for the approximate integral and the z-transfer function of the approximate integrator.

7-6. For each of the following continuous-time controllers, design a digital controller using the bilinear transformation with frequency prewarping. Specify the sampling interval.

a. $G_1(s) = \dfrac{10}{s + 10}$

b. $G_2(s) = \dfrac{s}{s + 100}$

c. $G_3(s) = \dfrac{100}{s^2 + 10s + 10}$

d. $G_4(s) = \dfrac{s}{s^2 + 10s + 10}$

7-7. For the controllers of problem 7-6, sketch the frequency response of the analog controller, the desired analog controller with prewarping, and the resulting digital controller.

7-8. Repeat problem 7-1 using the derivative approximations in Table 7-2.

7-9. Repeat problem 7-1 using the central-difference derivative approximation in Table 7-2.

7-10. Find step-invariant digital approximations to continuous-time controllers with the following transfer functions using the indicated sampling intervals T. Then compare the first three samples of the continuous response with the first three digital samples.

a. $G_1(s) = \dfrac{2s}{s^2 + 3s + 2}$ $\qquad T = 0.2$

b. $G_2(s) = \dfrac{4s - 6}{s^2 + \frac{3}{2}s + \frac{1}{2}}$ $\qquad T = 0.2$

c. $G_3(s) = \dfrac{4s - 6}{s^2 + \frac{3}{2}s + \frac{1}{2}}$ $\qquad T = 0.4$

7-11. For a continuous-time controller with the transfer function

$$G_c(s) = \frac{s + 4}{s^2 + 3s + 2}$$

and a sampling interval $T = 0.2$, find the ramp-invariant digital approximation. Then compare the first three samples of the continuous response with the first three samples of the digital response.

7-12. For each of the following analog controllers, find the transfer function of the corresponding digital controller using the pole-zero matching method.

a. $G_1(s) = \dfrac{s + 1}{s + 2}$ $T = 0.1$

b. $G_2(s) = \dfrac{s + 1}{s^2 + 5s + 6}$ $T = 0.01$

c. $G_3(s) = \dfrac{s}{s^2 + 10s + 10}$ $T = 0.1$

d. $G_4(s) = \dfrac{10}{s^2 + 10s + 16}$ $T = 0.2$

7-13. Consider the digital controller transfer function

$$G_c(z) = \frac{Y(z)}{U(z)} = \frac{b_0 + b_1 z^{-1} + \cdots + b_n z^{-n}}{1 + a_1 z^{-1} + \cdots + a_n z^{-n}}$$

where a_i and b_i are real numbers. One way to realize this transfer function is to use the structure shown here, which requires $2n$ delay elements.

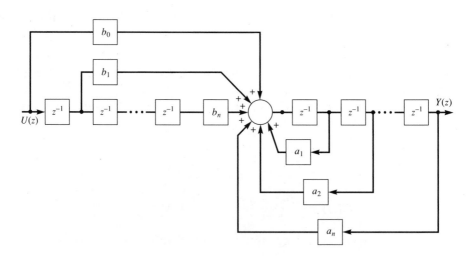

Using this structure, realize each of the following transfer functions:

a. $G_c(z) = \dfrac{z}{z + \frac{1}{2}}$

b. $G_c(z) = \dfrac{6z^2 - 3z + 5}{2z^2 + z - \frac{1}{8}}$

c. $G_c(z) = \dfrac{z}{z^2 - \frac{1}{4}}$

This structure is seldom used in practice because it requires an unnecessarily large number of delay elements.

7-14. The controller transfer function of problem 7-13 may be realized with the structure shown here, which uses the minimum number of unit delay elements. (The minimum number of delay elements is n).

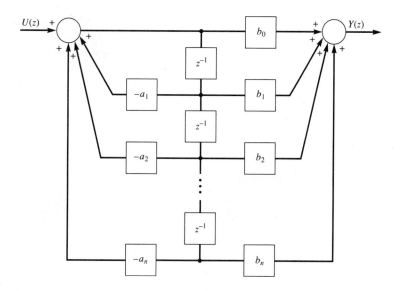

Using this structure, realize each of the following transfer functions:

a. $G_c(z) = \dfrac{-2z + 4}{z - \frac{1}{2}}$

b. $G_c(z) = \dfrac{3z^2 - 4z + 1}{z^2 + \frac{1}{6}z - \frac{1}{6}}$

c. $G_c(z) = \dfrac{3z^2 + 2z - 4}{z - \frac{1}{2}}$

d. $G_c(z) = \dfrac{2z^2 - \frac{2}{9}}{z^2 - \frac{1}{6}z - \frac{1}{6}}$

Note that for part c, one-step-ahead samples of the input are required, or alternatively, the output may be delayed one step.

7-15. The realizations shown in problems 7-13 and 7-14 are sensitive to coefficient variations for large controller order. To avoid the coefficient sensitivity problem, higher-order controller transfer functions are decomposed into *cascaded* first- or second-order transfer functions as shown below.

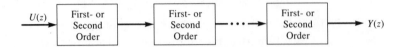

When all poles and zeros of the controller transfer function are real numbers, the cascaded transfer functions may all be of first order. Complex conjugate pairs of roots should be grouped into second-order subsystems to avoid complex number arithmetic operations. Obviously, many different arrangements are generally possible, depending on how poles and zeros are grouped and how the multiplying constant is distributed among the cascaded transfer functions. Realize each of the following transfer functions using the cascaded form:

a. $G_c(z) = \dfrac{(z - \frac{1}{2})(z + \frac{1}{4})}{(z - \frac{1}{3})(z + \frac{1}{2})}$

b. $G_c(z) = \dfrac{(z - \frac{1}{5})(z + \frac{1}{3})}{z(z + \frac{1}{2})}$

c. $G_c(z) = \dfrac{z(z - \frac{1}{4})(z + \frac{1}{2})}{(z - \frac{1}{3})(z^2 + z + \frac{1}{2})}$

d. $G_c(z) = 2\dfrac{(z - \frac{1}{2})(z + \frac{1}{4})(z - \frac{1}{3})}{(z + \frac{1}{3})(z + \frac{1}{5})(z + 1)}$

7-16. Another method to avoid the coefficient sensitivity problems is to use the *parallel* form. The parallel form is obtained by decomposing the controller transfer function using a partial fraction expansion into first- and second-order subsystems as shown on p. 571.

Complex terms should be combined into second-order terms so that complex number operations are avoided. A rich variety of parallel and cascade sections is also generally possible. Using the parallel form, realize the following transfer functions:

a. $G_c(z) = \dfrac{\frac{9}{2}z^2 - (19/6)z + \frac{1}{2}}{z(z - \frac{1}{2})(z - \frac{1}{3})}$

b. $G_c(z) = \dfrac{10(z + \frac{1}{2})^2}{(z - \frac{1}{2})^3}$

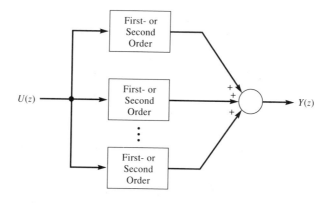

c. $G_c(z) = \dfrac{(29/6)z + \frac{1}{2}}{(z + \frac{1}{3})(z^2 + z + \frac{1}{2})}$

7-17. Find discrete-time models for each of the following continuous-time systems. The inputs are all piecewise-constant during each sampling interval T.

a. $\dot{x}(t) = -4x(t) + 2u(t)$

$y(t) = 3x(t) - u(t)$

$T = 0.2$

b. $\begin{bmatrix} \dot{x}_1(t) \\ \dot{x}_2(t) \end{bmatrix} = \begin{bmatrix} -3 & 4 \\ 1 & -2 \end{bmatrix} \begin{bmatrix} x_1(t) \\ x_2(t) \end{bmatrix} + \begin{bmatrix} 2 \\ -1 \end{bmatrix} u(t)$

$\begin{bmatrix} y_1(t) \\ y_2(t) \end{bmatrix} = \begin{bmatrix} 2 & 0 \\ 1 & -3 \end{bmatrix} \begin{bmatrix} x_1(t) \\ x_2(t) \end{bmatrix} + \begin{bmatrix} 0 \\ -4 \end{bmatrix} u(t)$

$T = 0.1$

c. $\begin{bmatrix} \dot{x}_1(t) \\ \dot{x}_2(t) \\ \dot{x}_3(t) \end{bmatrix} = \begin{bmatrix} 0 & 1 & 0 \\ 0 & 0 & 1 \\ -5 & -2 & -1 \end{bmatrix} \begin{bmatrix} x_1(t) \\ x_2(t) \\ x_3(t) \end{bmatrix} + \begin{bmatrix} 0 & 0 \\ 2 & -1 \\ 3 & 2 \end{bmatrix} \begin{bmatrix} u_1(t) \\ u_2(t) \end{bmatrix}$

$y(t) = [3 \quad 0 \quad -1] \begin{bmatrix} x_1(t) \\ x_2(t) \\ x_3(t) \end{bmatrix} + [0 \quad 1] \begin{bmatrix} u_1(t) \\ u_2(t) \end{bmatrix}$

$T = 0.01$

7-18. Modify the program in Table 7-3 so that the controller output is truncated to give only three decimal digits of precision. Repeat so that the controller output is rounded and has only one significant decimal digit.

7-19. Find *continuous-time* state equation models for plants with the following discrete-time state models with the indicated sampling intervals T:

a. $\begin{bmatrix} x_1(k+1) \\ x_2(k+1) \end{bmatrix} = \begin{bmatrix} -\frac{1}{2} & 1 \\ -\frac{1}{6} & 0 \end{bmatrix} \begin{bmatrix} x_1(k) \\ x_2(k) \end{bmatrix} + \begin{bmatrix} 3 \\ -1 \end{bmatrix} u(k)$

$T = 0.3$

b. $\begin{bmatrix} x_1(k+1) \\ x_2(k+1) \end{bmatrix} = \begin{bmatrix} 0.1 & 1.3 \\ -0.1 & 0.2 \end{bmatrix} \begin{bmatrix} x_1(k) \\ x_2(k) \end{bmatrix} + \begin{bmatrix} 0 & -1 \\ 3 & 2 \end{bmatrix} \begin{bmatrix} u_1(k) \\ u_2(k) \end{bmatrix}$

$T = 1$

c. $\begin{bmatrix} x_1(k+1) \\ x_2(k+1) \\ x_3(k+1) \end{bmatrix} = \begin{bmatrix} 0 & 1 & 0 \\ 0 & 0 & 0 \\ \frac{1}{2} & 0 & 0 \end{bmatrix} \begin{bmatrix} x_1(k) \\ x_2(k) \\ x_3(k) \end{bmatrix} + \begin{bmatrix} 2 \\ -1 \\ 4 \end{bmatrix} u(k)$

$T = 0.1$

7-20. Design analog plant input filters for plants with the following transfer functions, specifying state variable equations for each filter realization. The filter time constants should be no larger than about half of the sampling interval T so that the controlled system can easily respond to significant changes in the tracking input from sample to sample.

a. $T(s) = \dfrac{s^2 + 2}{(s+1)^3(s^2 + 0.2s + 2)}$ $T = 0.1$

b. $T(s) = \dfrac{10^6(s+1)}{s^4 + 20s^3 + 10^4 s^2}$ $T = 10^{-3}$

c. A two-input plant with a single tracking output, with

$T_1(s) = \dfrac{s + 20}{(s^2 + 4s + 20)(s^2 + 2s + 20)}$

$T_2(s) = \dfrac{s^3 + 5s^2 - 30s + 1000}{(s^2 + 2s + 10)(s + 20)}$

and $T = 0.1$

7-21. Design a high-speed discrete-time plant input filter for the system

$\begin{bmatrix} \dot{x}_1(t) \\ \dot{x}_2(t) \end{bmatrix} = \begin{bmatrix} -2 & 1 \\ -3 & 0 \end{bmatrix} \begin{bmatrix} x_1(t) \\ x_2(t) \end{bmatrix} + \begin{bmatrix} -1 \\ 4 \end{bmatrix} u(t)$

$y(t) = \begin{bmatrix} 2 & 0 \end{bmatrix} \begin{bmatrix} x_1(t) \\ x_2(t) \end{bmatrix}$

with a sampling interval $\Delta t = 0.1$, that improves the plant step response. Then find a discrete-time model of the system with sampling interval $T = 5\Delta t = 0.5$, where the discrete-time filter input changes at the lower rate.

7-22. Design high-speed discrete-time measurement of the output feedback for the system

$$\begin{bmatrix} \dot{x}_1(t) \\ \dot{x}_2(t) \end{bmatrix} = \begin{bmatrix} 0 & 1 \\ -4 & -2 \end{bmatrix} \begin{bmatrix} x_1(t) \\ x_2(t) \end{bmatrix} + \begin{bmatrix} 0 \\ 1 \end{bmatrix} u(t)$$

$$y(t) = [1 \quad -1] \begin{bmatrix} x_1(t) \\ x_2(t) \end{bmatrix}$$

with a sampling interval $\Delta t = 0.1$, that improves the plant step response over that without the feedback. Then find a discrete-time model of the system with sampling interval $T = 5\Delta t = 0.5$, in which the plant consists of the sum of the high-speed output feedback and a sampled-and-held external input that changes at the lower rate.

7-23. At the level of Figure 7-19, specify a multiprocessor computer architecture in which separate processors are used for

   1. data acquisition
   2. program control and computation
   3. high-speed output signal generation

7-24. At the level of Figure 7-20, specify a multiprocessor computer architecture for a decentralized distributed system with three local computers. Discuss in some detail how the local computers are to communicate with one another, especially in those situations in which more than one processor wants to communicate with the others at the same time.

7-25. At the level of Figure 7-21, specify a data acquisition arrangement in which there is simultaneous input sampling of a pressure sensor, a voltage sensor, and a temperature sensor, and in which there are two additional temperature sensors, either one of which can be substituted for the temperature sensor currently being used. The substitution should be controllable with an external command. There is time for only three measurements to be sent from the data acquisition system at each step.

7-26. At the level of Figure 7-22, specify a data distribution arrangement for three actuators. Two of the actuator samples are sent to this system at each step, and one of the actuator samples is sent every other step. On in-between steps, the latter actuator is to be driven by the average of the previous two samples received.

7-27. At the level of Figure 7-23 or 7-24, specify a hardware design for a single-input/single-output controller with input $r(k)$ and output $u(k)$, for which

$$
\begin{bmatrix} \xi_1(k + 1) \\ \xi_2(k + 1) \\ \xi_3(k + 1) \end{bmatrix} = \begin{bmatrix} 0 & 2 & 0 \\ -1 & \frac{1}{3} & \frac{4}{3} \\ 2 & 3 & \frac{1}{4} \end{bmatrix} \begin{bmatrix} \xi_1(k) \\ \xi_2(k) \\ \xi_3(k) \end{bmatrix} + \begin{bmatrix} 0 \\ 2 \\ -1 \end{bmatrix} r(k)
$$

$$
u(k) = [1 \quad 0 \quad \tfrac{1}{3}] \begin{bmatrix} \xi_1(k) \\ \xi_2(k) \\ \xi_3(k) \end{bmatrix} + 3r(k)
$$

7-28. Digital controllers are sometimes designed so that they accumulate the error between a desired output and its measured samples and feed this sum of errors back to a plant input. The first-order plant

$$
x(k + 1) = -\frac{1}{2} x(k) + u(k)
$$

$$
\bar{y}(k) = y(k) = x(k)
$$

with the first-order controller

$$
\xi(k + 1) = \xi(k) + [r(k) - y(k)]
$$

$$
u(k) = \frac{1}{16} \xi(k)
$$

is a simple example of such a system. Should the measurement sensor fail, the accumulated error signal ξ can become very large. Once the problem is corrected, it may therefore take a very long time for the accumulated error to return to realistic values. Digital control systems using error accumulation are thus usually designed with provision for resetting accumulators to zero. This provision is called *antiwindup reset*.

Design hardware for the controller discussed above in the form of Figure 7-24 and include an antiwindup reset switch.

7-29. Write a controller program in the BASIC or some other language that realizes

$$\begin{bmatrix} \xi_1(k+1) \\ \xi_2(k+1) \\ \xi_3(k+1) \end{bmatrix} = \begin{bmatrix} \frac{1}{4} & 1 & 0 \\ -\frac{1}{2} & 0 & 1 \\ \frac{1}{3} & 0 & 0 \end{bmatrix} \begin{bmatrix} \xi_1(k) \\ \xi_2(k) \\ \xi_3(k) \end{bmatrix} + \begin{bmatrix} -3 \\ 2 \\ 3 \end{bmatrix} r(k)$$

$$u(k) = \begin{bmatrix} 1 & 0 & 0 \end{bmatrix} \begin{bmatrix} \xi_1(k) \\ \xi_2(k) \\ \xi_3(k) \end{bmatrix} + \frac{1}{2} r(k)$$

Assume that an INPUT R instruction causes the program to wait until a new input sample is available from the input A/D converter, then assigns that sample to the variable R. Assume, too, that a PRINT U instruction causes transfer of the value of the variable U to the output D/A converter. This is essentially as if the samples were received from and delivered to a terminal.

7-30. For the two-input/two-output controller

$$\begin{bmatrix} \xi_1(k+1) \\ \xi_2(k+1) \\ \xi_3(k+1) \end{bmatrix} = \begin{bmatrix} 0.25 & 0.1 & -2.0 \\ 1.25 & -1.2 & 0 \\ 0 & 0.4 & 3.0 \end{bmatrix} \begin{bmatrix} \xi_1(k) \\ \xi_2(k) \\ \xi_3(k) \end{bmatrix} + \begin{bmatrix} 7.31 \\ -0.4 \\ -0.9 \end{bmatrix} \begin{bmatrix} r(k) \\ y(k) \end{bmatrix}$$

$$\begin{bmatrix} u_1(k) \\ u_2(k) \end{bmatrix} = \begin{bmatrix} 1.0 & 2.0 & 0 \\ 0 & -1.5 & 5.5 \end{bmatrix} \begin{bmatrix} \xi_1(k) \\ \xi_2(k) \\ \xi_3(k) \end{bmatrix} + \begin{bmatrix} 0 & 0.3 \\ -0.5 & 0.1 \end{bmatrix} \begin{bmatrix} r(k) \\ y(k) \end{bmatrix}$$

write software in the BASIC or FORTRAN language. The variables R and Y, representing the signal samples $r(k)$ and $y(k)$, respectively, are the contents of certain memory locations that are automatically changed by the data acquisition system whenever new samples are available. When the new samples of each have been entered, the data acquisition system sets the variable P to unity. The program should test to see if P equals one. When it does, the program should set P to zero, then process the new input samples.

Similarly, the data distribution system uses as samples $u_1(k)$ and $u_2(k)$, the contents of the memory locations where the variables U1 and U2 are stored. The program signals to the data distribution system that new output samples are available by setting the variable Q to unity. When the data distribution system has transferred the samples to its D/A converters, it sets Q to zero.

7-31. Design a detailed program flow chart for realizing the controller

$$\begin{bmatrix} \xi_1(k+1) \\ \xi_2(k+1) \\ \xi_3(k+1) \end{bmatrix} = \begin{bmatrix} 0 & 1 & 0 \\ 0 & 0 & 1 \\ -3 & -2 & \frac{1}{4} \end{bmatrix} \begin{bmatrix} \xi_1(k) \\ \xi_2(k) \\ \xi_3(k) \end{bmatrix} + \begin{bmatrix} 2 \\ -4 \\ 3 \end{bmatrix} r(k)$$

$$\begin{bmatrix} u_1(k) \\ u_2(k) \end{bmatrix} = \begin{bmatrix} 2 & 1 & 0 \\ -1 & 4 & 0 \end{bmatrix} \begin{bmatrix} \xi_1(k) \\ \xi_2(k) \\ \xi_3(k) \end{bmatrix} + \begin{bmatrix} 0 \\ 2 \end{bmatrix} r(k)$$

Assume that there is a step clock to indicate when new samples are to be processed and that a loop of the program executes in less than one clock period. Do not use subroutines or interrupts.

7-32. Repeat problem 7-31 but use a subroutine structure similar to that shown in Figure 7-26. Do not use interrupts.

7-33. Write a controller program in the BASIC or some other language that realizes

$$\begin{bmatrix} \xi_1(k+1) \\ \xi_2(k+1) \end{bmatrix} = \begin{bmatrix} 0 & 1 \\ -\frac{1}{4} & -\frac{1}{2} \end{bmatrix} \begin{bmatrix} \xi_1(k) \\ \xi_2(k) \end{bmatrix} + \begin{bmatrix} 4 & 3 \\ -2 & -1 \end{bmatrix} \begin{bmatrix} y_1(k) \\ y_2(k) \end{bmatrix}$$

$$u(k) = \begin{bmatrix} 3 & 1 \end{bmatrix} \begin{bmatrix} \xi_1(k) \\ \xi_2(k) \end{bmatrix}$$

when both the y_1 and y_2 sensors are working properly,

$$\xi(k+1) = \frac{1}{2}\xi(k) + 3y_2(k)$$

$$u(k) = \frac{1}{4}\xi(k)$$

when the y_1 sensor is not working, and

$$\xi(k+1) = \frac{1}{6}\xi(k) + 4y_1(k)$$

$$u(k) = -3\xi(k) + \frac{2}{10}y_1(k)$$

when the y_2 sensor is inoperative. The INPUT A, B, C command (where A, B, C are any variables) causes the program to wait until new samples of y_1 and y_2 are available, then assigns the y_1 sample to A and the y_2 sample to B. When a sensor is not working, input samples are provided, but they are meaningless. The third INPUT variable (C here) indicates which sensors are working properly, as follows:

$0 \leqslant C < 0.5$ y_1 working, y_2 not working

$0.5 \leqslant C < 1.5$ both y_1 and y_2 working

$1.5 \leqslant C < 2.0$ y_1 not working, y_2 working

Output samples are transferred to the data distribution hardware using a PRINT command.

7-34. Write, debug, and test controller and plant simulation programs for the tracking systems in the following figures:

a. Figure 6-2

b. Figure 6-3

c. Figure 6-9

d. Figure 6-10

e. Figure 6-11

7-35. Write, debug, and test controller and plant simulation programs for the tracking systems in the following figures:

a. Figure 7-13(a)

b. Figure 7-13(b)

c. Figure 7-13(c)

7-36. Write, debug, and test a controller and plant simulation program for the tracking system with high-speed plant input filter in Figure 7-15(c).

7-37. Write a controller program in the BASIC language for the two-input/two-output controller

$$\begin{bmatrix} \xi_1(k+1) \\ \xi_2(k+1) \end{bmatrix} = \begin{bmatrix} 0.31 & 1 \\ 0.21 & 0 \end{bmatrix} \begin{bmatrix} \xi_1(k) \\ \xi_2(k) \end{bmatrix} + \begin{bmatrix} 1.1 & 2 \\ -0.3 & -1 \end{bmatrix} \begin{bmatrix} y(k) \\ r(k) \end{bmatrix}$$

$$\begin{bmatrix} u_1(k) \\ u_2(k) \end{bmatrix} = \begin{bmatrix} 1 & 0 \\ 0.1 & -2 \end{bmatrix} \begin{bmatrix} \xi_1(k) \\ \xi_2(k) \end{bmatrix} + \begin{bmatrix} 0 & 3 \\ -0.2 & 0 \end{bmatrix} \begin{bmatrix} y(k) \\ r(k) \end{bmatrix}$$

Assume that at each step, new samples of the plant output $y(k)$ and the reference input $r(k)$ are automatically stored, as the variables Y and R, respectively, by the computer hardware. Similarly, the stored value of U is automatically transferred to the plant input D/A at each step. There is also hardware that causes execution of the program to wait until new input samples are available at each step.

7-38. Provision is usually made in a digital control system for manual control. (There may also be an alternate, simpler backup controller standing by.) When control is switched between manual and the computer, it is undesirable to have the system undergo large transients, behaving as if it were just being started up. Because the controller-generated plant inputs and those produced manually may differ substantially, there can be large step changes in the plant inputs in moving between manual and computer control. When provision has been made to avoid this problem, the control system is said to be capable of *bumpless transfer*. Write a controller program in the BASIC or FORTRAN language for a two-input/single-output controller

$$\begin{bmatrix} \xi_1(k+1) \\ \xi_2(k+1) \end{bmatrix} = \begin{bmatrix} 0.1 & 0.33 \\ 2.1 & -0.02 \end{bmatrix} \begin{bmatrix} \xi_1(k) \\ \xi_2(k) \end{bmatrix} + \begin{bmatrix} 0 & -0.1 \\ 1.0 & 2.2 \end{bmatrix} \begin{bmatrix} y(k) \\ r(k) \end{bmatrix}$$

$$w(k) = \begin{bmatrix} 1 & 3 \end{bmatrix} \begin{bmatrix} \xi_1(k) \\ \xi_2(k) \end{bmatrix} + \begin{bmatrix} 2.0 & -0.7 \end{bmatrix} \begin{bmatrix} y(k) \\ r(k) \end{bmatrix}$$

and arrange for bumpless transfer between manual and computer control. Assume that at each step, new samples of the plant output $y(k)$, the reference input $r(k)$, and the manual plant input $m(k)$ are automatically stored, as the variables Y, R, and M, respectively, by the computer hardware. Similarly, the stored value of U is automatically transferred to the plant input D/A at each step. Just before the input signals are to be automatically updated, there is to be an INPUT L statement, through which the program determines whether the plant input is to be from the controller $w(k)$ or from the manual signal $m(k)$. If L = 0, plant control is to be by the controller; otherwise, plant control is to be manual. Rather than immediately switching the plant input between $w(k)$ and $m(k)$, make the transitions between the two a ramp of appropriate slope.

7-40. Design a second-order deadbeat observer of the four state variables describing the discrete-time model of the elevator drive and sensors in Subsection 7.7.1. Simulate and compare the plant and observer signals for various power amplifier inputs $u(k)$.

7-41. Design state feedback gains **e** for the digital controller arrangement of Figure 7-31 so that the plant with state feedback has the characteristic equation $(\lambda - 0.87)^4 = 0$. For these gains, design a first-order deadbeat observer of $\mathbf{e}^\dagger\mathbf{x}$ for feedback. Simulate this discrete-time control of the continuous-time elevator position for various reasonable initial elevator positions and velocities. Be sure to include the power amplifier nonlinearity.

7-42. Design a discrete-time deadbeat regulator of the positioning mechanism, using the models of equations (7-23) and (7-24). Simulate $\bar{y}(t)$ with this discrete-time control of the plant and plant input filter for various plant initial conditions. Then vary the plant (but not the filter) to give each of the two situations of Figure 7-33.

7-43. Using the models in equations (7-23) and (7-24), design a discrete-time tracking controller for the positioning mechanism. Choose the discrete-time feedback eigenvalues to be other than at the origin and have the tracking output $\bar{y}(k)$ observe any constant-plus-ramp reference input $r(k)$.

7-44. Using the models in equations (7-23) and (7-24), design a deadbeat discrete-time tracking controller for the positioning mechanism. Have the tracking output observe any constant reference input $r(k)$. Simulate this discrete-time control of the continuous-time plant and filter with various plant initial conditions and reference inputs, including sinusoidal ones. Through simulation, investigate the controller's performance with changed plant parameters as in Figure 7-33.

7-45. Using the models in equations (7-22) and (7-25), design a deadbeat tracking controller for the positioning mechanism. Take the uncontrollable modes into account, and have the tracking output observe any constant-plus-ramp reference input $r(k)$.

Stochastic Systems and Recursive Estimation

8.1 Preview

In the previous chapters, we dealt with systems that were entirely deterministic. That is, everything about the plant including its inputs and its outputs were exactly known. Unfortunately, however, there are many systems in practice that have *random* (*stochastic*) disturbances, random initial conditions, or random outputs. In such an event, an explicit representation of the system inputs and the system outputs cannot be obtained.

Basically there are two approaches for analyzing linear systems that are driven by random disturbances: (1) the probabilistic approach and (2) the statistical approach. The probabilistic approach is often not very useful because one is confronted with the problem of determining the probability density functions and higher-order joint density functions of the system outputs based on the probabilistic description of the system inputs. Except for a limited class of problems, it is difficult to obtain these joint densities of the system outputs.

In the statistical approach, however, we are content with knowledge of the first- and second-order moments, such as mean, mean square, autocorrelation, and power spectral density of the random inputs. Consequently, a statistical description of the system outputs can be easily obtained. In this chapter, we consider only the statistical approach.

In Section 8.2, general solution methods are developed for linear, step-invariant, single-input/single-output systems that are driven by random inputs. The statistical properties of the output are derived in terms of the statistical properties of the input and the system transfer function. Analogous results for continuous-time systems are also presented in the section.

Discrete-time state variable models for systems driven by random inputs are introduced in Section 8.3. Recursive algorithms for the propagation of the mean and the covariance are also derived and demonstrated.

The fundamental ideas of least squares estimation are developed in Section 8.4. Least squares methods have become increasingly important in many applications, including navigation, target tracking, signal and image processing, communication, and control systems. The least squares solution involves a linear transformation of the measurements to obtain the optimal estimate. Then a recursive formulation of the least squares solution is derived in which the measurements are processed sequentially. The section concludes with a discussion of probabilistic interpretations of least squares and an indication of how recursive least squares methods can be generalized.

In 1960, Rudolph E. Kalman published his first paper on linear minimum mean square (LMMS) estimation. His approach departed fundamentally from that of Karl Friedrich Gauss in that he began with a stochastic formulation rather than giving a stochastic interpretation to an already developed procedure. The result, now known as the Kalman filter, is an elegant generalization of recursive least squares that nicely unifies and extends many earlier results. It is especially convenient for digital computer implementation. The recursive Kalman filter equations for linear, discrete-time systems are derived in Section 8.5. The Kalman filter produces an optimal estimate of the system state based on all previous noisy measurements through the latest one. With each filter iteration, the estimate is updated and improved by incorporating new measurements. A computer program for demonstrating first-order Kalman filters is provided in Section 8.6. A computer algorithm for programming multivariable filters is also presented.

Section 8.7 introduces extensions of the basic Kalman filter to situations involving noise-coupling matrices, deterministic inputs to the model, nonzero mean values, known initial conditions, correlated noises, and bias estimation. The extended Kalman filter for nonlinear systems is also discussed and applied.

In the final section of the chapter, the solution of the stochastic optimal control problem is formulated. In this case, the Kalman filter is viewed as a state observer. When the plant model is stochastic and the state is not accessible for feedback, the separation principle applies to the solution. The control sequence then consists of the LMMS estimate of the plant state produced by Kalman filter fed back through the optimal gains for the deterministic linear regulator with state feedback.

8.2 Response of Linear Systems to Random Inputs

8.2.1 Discrete-Time Stochastic Processes

As discussed in Appendix B*, a discrete-time stochastic process $x(k)$ is called *wide-sense stationary* if its *mean* is constant and its *autocorrelation* sequence depends on the time index difference n. That is

1. $E[x(k)] = m_x$ a constant
2. $E[x(k + n)x(k)] = R_{xx}(n)$

where E denotes the expected value.

In the z-domain, the autocorrelation sequence $R_{xx}(n)$ of a wide-sense stationary process $x(k)$ and its corresponding *power spectral density* (PSD) $S_{xx}(z)$ are related by

$$S_{xx}(z) = \sum_{n=-\infty}^{\infty} R_{xx}(n)z^{-n}$$

and

$$R_{xx}(n) = \frac{1}{2\pi} \oint_C S_{xx}(z)z^{n-1}\, dz$$

where the contour of integration C is chosen to be the unit circle $|z| = 1$. Consequently, the *mean square value* (or the average power) of a stochastic process $x(k)$ is expressed as

$$E[x^2(k)] = R_{xx}(0) = \frac{1}{2\pi} \oint_C S_{xx}(z)z^{-1}\, dz$$

For example, the autocorrelation sequence

$$R(k) = c^{-|k|} \quad |c| > 1 \tag{8-1}$$

has a PSD given by

$$S(z) = \sum_{k=-\infty}^{\infty} R(k)z^{-k}$$

---

*Appendix B summarizes needed concepts and results of stochastic processes. It should be reviewed before reading Chapter 8.

$$= \sum_{k=-\infty}^{-1} c^k z^{-k} + \sum_{k=0}^{\infty} c^{-k} z^{-k}$$

$$= \sum_{k=1}^{\infty} c^{-k} z^k + \sum_{k=0}^{\infty} c^{-k} z^{-k}$$

$$= \sum_{k=0}^{\infty} (c^{-1} z)^k - 1 + \sum_{k=0}^{\infty} (cz)^{-k}$$

Thus,

$$S(z) = \frac{1}{1 - c^{-1} z} - 1 + \frac{z}{z - c^{-1}}$$

$$= \frac{z}{z - c^{-1}} + \frac{-z}{z - c}$$

$$= \frac{z(c^{-1} - c)}{z^2 - z(c + c^{-1}) + 1} = \frac{z(c^{-1} - c)}{(z - c)(z - c^{-1})} \tag{8-2}$$

The PSD of a real stochastic process is an even function of ω, where $z = e^{j\omega T}$. To show this, one may expand the series

$$S(z) = \sum_{k=-\infty}^{\infty} R(k) z^{-k}$$

$$= \cdots + R(-2) z^2 + R(-1) z^1 + R(0) + R(1) z^{-1}$$
$$+ R(2) z^{-2} + \cdots$$

Because the autocorrelation sequence is an even function, then

$$R(-n) = R(n)$$

and, therefore, collecting terms gives

$$S(z) = R(0) + R(1)[z + z^{-1}] + R(2)[z^2 + z^{-2}] + \cdots$$

But

$$z + z^{-1} = e^{j\omega T} + e^{-j\omega T} = 2\cos \omega T$$

hence,

$$S(e^{j\omega T}) = R(0) + R(1)[2\cos \omega T] + R(2)[2\cos 2\omega T] + \cdots$$

which is a function of cosine terms only, indicating that the PSD is an even function in ω.

For the previous example, dividing the numerator and denominator by z gives

$$S(z)|_{z=e^{j\omega T}} = \frac{c^{-1} - c}{z - (c + c^{-1}) + z^{-1}}\bigg|_{z=e^{j\omega T}} = \frac{c^{-1} - c}{2\cos \omega T - (c + c^{-1})}$$

If a PSD corresponding to a real stochastic process is a rational function of the complex variable z, it can be expressed as a ratio of two polynomials that have real coefficients. Hence, both the poles and zeros of the PSD are either real or, if complex, must appear in complex conjugate pairs. Furthermore, the PSD can be written in the form

$$S(z) = \frac{N(z)N(z^{-1})}{D(z)D(z^{-1})} \tag{8-3}$$

where all the zeros and poles of the minimum phase transfer function $N(z)/D(z)$ are inside the unit circle on the complex plane. The zeros and poles of $N(z^{-1})/D(z^{-1})$ are reciprocals of the zeros and poles of $N(z)/D(z)$ and thus are outside the unit circle. Poles on the unit circle are not allowed.

As an example, the PSD

$$S(z) = \frac{(z^2 - z + 0.5)(z^2 - 2z + 2)}{(z^2 + 0.64)[z^2 + (25/16)]}$$

has the pole-zero plot of Figure 8-1. The poles and zeros are symmetric with respect to the unit circle. For the pole at $z = j0.8$, the symmetric

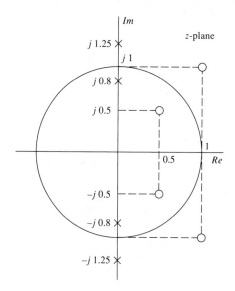

Figure 8-1 Poles and zeros of a power spectral density.

pole is at $z = -j1.25$, and for the zero at $z = 0.5 + j0.5$, the symmetric zero is at $z = 1 - j$, and so on.

8.2.2 Response of Linear, Time-Invariant, Single-Input/Single-Output Systems

Consider a linear, step-invariant, discrete-time, single-input/single-output system driven by a wide-sense stationary random input $w(k)$ as shown in Figure 8-2. Our objective is to determine a statistical model of the system output based on a statistical model of the system input $w(k)$, when the system $G(z)$ is stable with all its poles inside the unit circle on the complex plane.

If a stochastic input $w(k)$ to a linear, step-invariant system is wide-sense stationary, then so is the output $y(k)$. For the system shown in Figure 8-2, the output is given by the convolution solution

$$y(k) = \sum_{n=-\infty}^{k} g(k - n)w(n)$$

where $g(k)$ is the pulse response of the system and $w(n)$ is a sample sequence of the stationary random process. If $g(k)$ is zero prior to $k = 0$, the system is termed *causal* because it does not respond to the pulse before the pulse arrives. Otherwise, the system is termed *non-causal*. On the other hand, because the random input is stationary, the sequence $w(n)$ starts from $n = -\infty$.

If

$$i = k - n$$

then

$$y(k) = \sum_{i=0}^{\infty} g(i)w(k - i)$$

Figure 8-2 A single-input/single-output system driven by a random input.

Mean of the Output

For the system shown in Figure 8-2, the mean of the output is

$$E[y(k)] = E\left[\sum_{i=0}^{\infty} g(i)w(k-i)\right]$$

$$= \sum_{i=0}^{\infty} g(i)E[w(k-i)]$$

But, due to the stationarity of the input, the mean of the input is a constant, therefore,

$$E[y(k)] = E[w(k)] \sum_{i=0}^{\infty} g(i)$$

Moreover,

$$\sum_{i=0}^{\infty} g(i) = \sum_{i=0}^{\infty} g(i)z^{-i}\big|_{z=1} = G(z)\big|_{z=1}$$

Hence,

$$E[y(k)] = E[w(k)]G(1) \tag{8-4}$$

That is, the mean of the output is a constant that equals the mean of the input times the system transfer function evaluated at $z = 1$.

Autocorrelation and Power Spectral Density of the Output

On the other hand, the autocorrelation sequence of the plant output of the system shown in Figure 8-2 is evaluated as

$$R_{yy}(n) = E[y(i)y(i+n)]$$

where

$$y(i) = \sum_{l=0}^{\infty} g(l)w(i-l)$$

and

$$y(i+n) = \sum_{p=0}^{\infty} g(p)w(i+n-p)$$

Hence,

$$E[y(i)y(i + n)] = E\left[\sum_{l=0}^{\infty} g(l)w(i - l) \sum_{p=0}^{\infty} g(p)w(i + n - p)\right]$$

$$= \sum_{l=0}^{\infty} \sum_{p=0}^{\infty} g(l)g(p)E[w(i - l)w(i + n - p)]$$

and, therefore,

$$R_{yy}(n) = \sum_{l=0}^{\infty} \sum_{p=0}^{\infty} g(l)g(p)R_{ww}(n + l - p)$$

which is an expression for the autocorrelation of the plant output in terms of the autocorrelation of the plant input and of the system pulse response.

Because the output is stationary, its PSD is determined by taking the two-sided z-transform of the autocorrelation sequence as follows:

$$S_{yy}(z) = \sum_{n=-\infty}^{\infty} \left[\sum_{l=0}^{\infty} \sum_{p=0}^{\infty} g(l)g(p)R_{ww}(n + l - p)\right]z^{-n}$$

Rearranging variables gives

$$S_{yy}(z) = \sum_{p=0}^{\infty} g(p)z^{-p} \sum_{l=0}^{\infty} g(l)z^{l} \sum_{n=-\infty}^{\infty} R_{ww}(n + l - p)z^{-(n+l-p)}$$

and, therefore,

$$S_{yy}(z) = G(z)G(z^{-1})S_{ww}(z) \tag{8-5}$$

which is an expression for the PSD of the output in terms of the PSD of the input and the plant transfer function.

For the feedback system of Figure 8-3(a), for example, suppose the random disturbance input $w(k)$ has the following statistical properties:

$$E[w(k)] = 1$$

$$E[w(i)w(n)] = \begin{cases} 3 & i = n \\ 0 & \text{otherwise} \end{cases}$$

It is desirable to examine the statistical properties of the system output. Using the concept of superposition, the effect of the deterministic input $u(k)$ on the system output can be determined separately and so the input $u(k)$ is ignored. As shown in Figure 8-3(b), the transfer func-

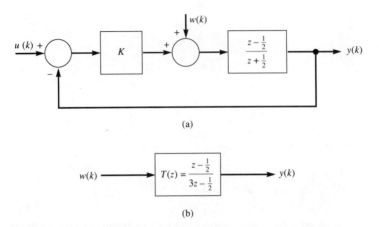

(a)

(b)

Figure 8-3 An example of evaluating the statistical properties of the output in terms of the statistical properties of the random input. (a) An example of a closed-loop system with a random disturbance $w(k)$. (b) A transfer function relating the disturbance $w(k)$ to the output $y(k)$ for $K = 2$.

tion relating the random disturbance $w(k)$ to the output for $K = 2$ is

$$G(z) = \frac{(z - 1/2)/(z + 1/2)}{1 + 2(z - 1/2)/(z + 1/2)} = \frac{z - \frac{1}{2}}{3z - \frac{1}{2}}$$

If equation (8-4) is used, the mean of the output is

$$E[y(k)] = E[w(k)]G(1)$$

$$= \left.\frac{z - \frac{1}{2}}{3z - \frac{1}{2}}\right|_{z=1} \cdot 1 = 0.2$$

and the PSD of the output is determined from equation (8-5) as follows

$$S_{yy}(z) = G(z)G(z^{-1})S_{ww}(z)$$

$$= \left(\frac{z - \frac{1}{2}}{3z - \frac{1}{2}}\right)\left(\frac{z^{-1} - \frac{1}{2}}{3z^{-1} - \frac{1}{2}}\right)3$$

$$= \frac{(z - \frac{1}{2})(z - 2)}{(z - \frac{1}{6})(z - 6)}$$

$$= \frac{z^2 - 2.5z + 1}{z^2 - (37/6)z + 1} = \frac{z - 2.5 + z^{-1}}{z - (37/6) + z^{-1}}$$

Then the PSD of the output is

$$S_{yy}(e^{j\omega T}) = \frac{2\cos \omega T - 2.5}{2\cos \omega T - (37/6)} = \frac{\cos \omega T - 1.25}{\cos \omega T - (37/12)}$$

The autocorrelation sequence of the output may be determined from $S_{yy}(z)$ using a partial fraction expansion. Rather than expanding a z-transform directly in partial fractions, the function $S_{yy}(z)/z$ is expanded so that terms with a z in the numerator result. For the previous example, then

$$\frac{S_{yy}(z)}{z} = \frac{k_1}{z} + \frac{k_2}{z - \frac{1}{6}} + \frac{k_3}{z - 6}$$

$$= \frac{1}{z} + \frac{-(22/35)}{z - \frac{1}{6}} + \frac{(22/35)}{z - 6}$$

Thus,

$$S_{yy}(z) = 1 + \frac{-(22/35)z}{z - \frac{1}{6}} + \frac{(22/35)z}{z - 6}$$

The inverse z-transform of this last equation is determined using equation (8-1):

$$\mathcal{L}^{-1}\left[\frac{z}{z - c^{-1}} - \frac{z}{z - c}\right] = c^{-|k|}$$

Hence

$$R_{yy}(k) = \delta(k) - \frac{22}{35}(6)^{-|k|} \tag{8-6}$$

Mean Square of the Output

For the linear, step-invariant, discrete-time system shown in Figure 8-2, the mean square value of the output is defined as

$$E[y^2(n)] = R_{yy}(0) = \frac{1}{2\pi j}\oint_C S_{yy}(z)z^{-1}\,dz$$

$$= \frac{1}{2\pi j}\oint_C G(z)G(z^{-1})S_{ww}(z)z^{-1}\,dz \tag{8-7}$$

where the contour of integration C is chosen to be the unit circle $|z| = 1$.

If the autocorrelation sequence of the output is known, then the mean square value is simply the value of the autocorrelation sequence at $k = 0$. In the previous example, the autocorrelation sequence is given by equation (8-6); hence,

$$E[y^2(k)] = R_{yy}(0) = \delta(0) - \frac{22}{35} = \frac{13}{35}$$

Apparently, then, the mean square value of the output may be determined by expanding

$$S_{yy}(z)z^{-1} = G(z)G(z^{-1})S_{ww}(z)z^{-1}$$

into partial fractions and evaluating the resulting autocorrelation sequence at $k = 0$.

The method of partial fraction expansion is not easy for complicated PSD functions, however. An efficient numerical method for determining the mean square value of the output as given by equation (8-7) was published by Åström, et al.** It will not be discussed further in this chapter.

As another example, consider the system shown in Figure 8-3 again, with a zero mean unit variance Gaussian white noise input $w(k)$ as shown in Figure 8-4(a). In terms of the gain K, the system transfer function is

$$T(z) = \frac{z - \frac{1}{2}}{(1 + K)z + \frac{1}{2} - \frac{1}{2}K}$$

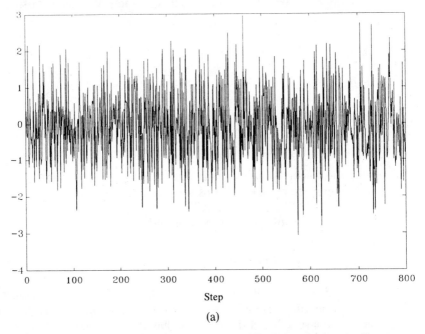

$w(k)$

Step

(a)

Figure 8-4 Response of a system driven by white noise. (a) Input with zero-mean and unit variance white noise. (b) Response of system to random input for $K = 2$. (c) Response of system to random input for $K = 10$.

**K. J. Åström, E. I. Jury, and R. G. Angiel, "A Numerical Method for the Evaluation of Complex Integrals," *IEEE Trans. Automatic Control,* Vol. AC-15, August 1970, pp. 468–471.

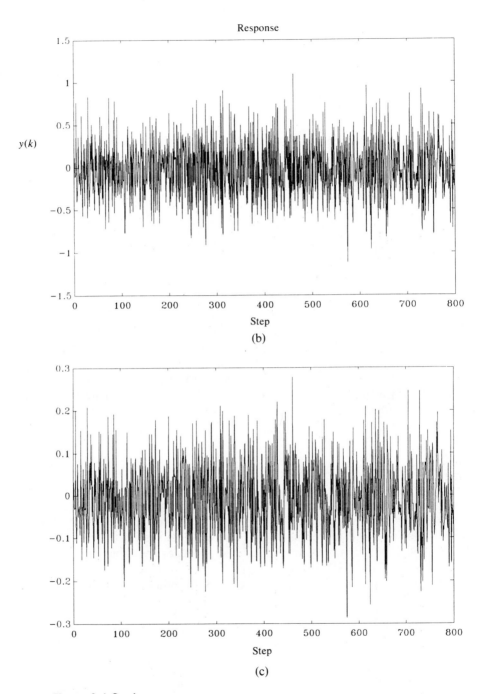

(b)

(c)

Figure 8-4 Cont.

A simple computer program in MATLAB for performing the output calculations of the mean and standard deviation for $K = 2$ follows.

```
% Input Transfer Function Numerator and Denominator
num1 = [1   -0.5];
den1 = [3   -0.5];
% Generate zero mean, unit variance Gaussian White Noise
rand ('normal');
u = rand (800, 1);
m = mean (u)
s = std (u)
y₁ = dlsim (num1, den1, u);
outmean = mean (y₁)
outstd = std (y₁)
```

The program gives the response shown in Figure 8-4(b) for 800 steps. The output has a mean of zero and a standard deviation equal to 0.35, which can be easily verified analytically.

If, instead, the gain K is increased to 10, the disturbance rejection performance of this system is improved, as shown in Figure 8-4(c). For $K = 10$, the program gives a mean of zero and standard deviation equal to 0.0934.

Spectral Factorization

When analyzing control systems that are driven by stochastic disturbance inputs, one is confronted with the problem of generating a noise signal that has a given desired PSD. One approach to this problem is to derive a linear, step-invariant, discrete-time system that when driven by a discrete-time white noise input produces an output signal that has the desired PSD.

Consider a discrete-time, white noise process $w'(k)$ for which

$$S_{w'w'}(e^{j\omega T}) = 1 \quad \text{for all } \omega$$

or, equivalently,

$$R_{w'w'}(n) = \delta(n)$$

as discussed in Appendix B. Then passing the white noise through a system with the transfer function $G(z)$ gives

$$S_{ww}(z) = G(z)G(z^{-1}) \cdot 1 \qquad\qquad (8\text{-}8)$$

Therefore, all that is needed is to derive a minimum phase *shaping filter* $G(z)$ whose poles and zeros are all inside the unit circle, such that when multiplied by $G(z^{-1})$ it produces the desired PSD. Arranging the poles and zeros in the form given by equation (8-3) is called *spectral factorization*.

For example, for the system of Figure 8-5(a) with a *colored* random disturbance input $w(k)$ characterized by the PSD

$$S_{ww}(z) = \frac{-10(z^2 + 2.5z + 1)}{z^2 - 5.2z + 1}$$

one can model the disturbance as the response of a shaping filter driven by white noise $w'(k)$. The PSD can be written in factored form as

$$S_{ww}(z) = \frac{-10(z + 0.5)(z + 2)}{(z - 0.2)(z - 5)}$$

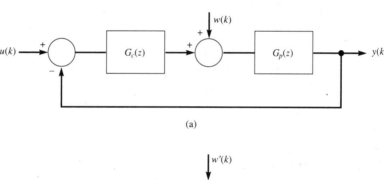

(a)

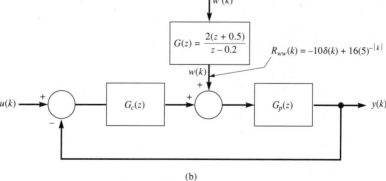

(b)

Figure 8-5 An example of designing a shaping filter. (a) An example of a system with colored disturbance noise input $w(k)$. (b) An equivalent system driven by a zero mean and unit variance white noise input.

Dividing the factors corresponding to the poles and zeros that are outside the unit circle by z and simplifying gives

$$S_{ww}(z) = \frac{4(z + 0.5)(z^{-1} + 0.5)}{(z - 0.2)(z^{-1} - 0.2)}$$

which is in the spectral factorization form. As shown in Figure 8-5 (b), the required shaping filter has the transfer function

$$G(z) = \frac{2(z + 0.5)}{z - 0.2}$$

with poles and zeros that are inside the unit circle.

8.2.3 Response of Continuous-Time Systems to Random Inputs

When a linear, time-invariant, continuous-time, single input/single output system is driven with a random input, as shown in Figure 8-6, the statistical properties of the output are determined much the same way as for the discrete-time systems discussed earlier. Assuming that the system described by $G(s)$ is stable with all its poles to the left of the $j\omega$-axis on the s-plane, and that the random input $w(t)$ is wide-sense stationary, then the mean, autocorrelation, PSD, and mean square of the output are as listed in Table 8-1 on page 595.

As an example, consider the continuous-time system shown in Figure 8-7 with a random input $w(t)$ whose autocorrelation function is described as

$$R_{ww}(\tau) = \sigma_0^2 e^{-\beta|\tau|}$$

The closed-loop transfer function of the system relating the random input to the output $y(t)$ is

$$G(s) = \frac{(1/s)}{1 + (3/s)} = \frac{1}{s + 3}$$

Suppose that the mean of the random input is 2, then, according to equation (8-9) in Table 8-1, the mean of the output is

$$E[y(t)] = E[w(t)]G(0) = \tfrac{2}{3}$$

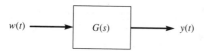

Figure 8-6 A single-input/single-output, continuous-time system driven by a random input.

Table 8-1 Statistical Properties of Linear, Time-Invariant, Continuous-Time Systems Driven by Random Inputs

---

Definition

For a single-input/single-output system that is stable, linear, time-invariant, continuous-time, and driven with a wide-sense stationary random input $w(t)$, the output is

$$y(t) = \int_{-\infty}^{t} g(t - \alpha)w(\alpha)\, d\alpha = \int_{0}^{\infty} g(u)w(t - u)\, du \qquad \alpha = t - u$$

where g is the impulse response of the system.

Mean of the Output

$$E[y(t)] = E\left[\int_{0}^{\infty} g(u)w(t - u)\, du\right] = E[w(t)]G(s)\big|_{s=0} \tag{8-9}$$

Autocorrelation of the Output

$$R_{yy}(\tau) = E[y(t + \tau)y(t)] = E\left[\int_{0}^{\infty}\int_{0}^{\infty} g(u)g(v)w(t + \tau - u)w(t - v)\, du\, dv\right]$$

$$= \int_{0}^{\infty}\int_{0}^{\infty} g(u)g(v)R_{ww}(\tau - u + v)\, du\, dv$$

PSD of the Output

$$S_{yy}(\omega) = \mathcal{F}[R_{yy}(\tau)] = G(j\omega)\, G(-j\omega)\, S_{ww}(\omega) \tag{8-10}$$

$$S_{yy}(s) = G(s)G(-s)S_{ww}(s)$$

Mean Square Value of the Output

$$E[y^2(t)] = R_{yy}(0) = \frac{1}{2\pi j}\int_{-j\infty}^{j\infty} S_{yy}(s)\, ds$$

$$= \frac{1}{2\pi j}\int_{-j\infty}^{j\infty} \frac{N(s)N(-s)}{D(s)D(-s)}\, ds \tag{8-11}$$

where $N(s)$ and $D(s)$ have roots that are in the left half of the s-plane.

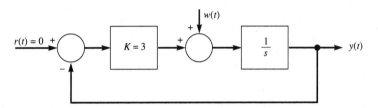

Figure 8-7 An example of a continuous-time system.

On the other hand, the PSD of the input is the Fourier transform of the associated autocorrelation function

$$\mathscr{F}[R_{ww}(\tau)] = \int_{-\infty}^{0} \sigma_0^2 e^{\beta\tau} e^{-j\omega\tau} \, d\tau + \int_{0}^{\infty} \sigma_0^2 e^{-\beta\tau} e^{-j\omega\tau} \, d\tau$$

$$= \frac{\sigma_0^2}{-j\omega + \beta} + \frac{\sigma_0^2}{j\omega + \beta} = \frac{2\beta\sigma_0^2}{\omega^2 + \beta^2} \tag{8-12}$$

Continuing with the example shown in Figure 8-7, suppose that $\sigma_0 = 1$ and $\beta = 1$; then the PSD of the output is

$$S_{yy}(\omega) = \frac{2}{\omega^2 + 1} \cdot \frac{1}{j\omega + 3} \cdot \frac{1}{-j\omega + 3}$$

Using a partial fraction expansion gives

$$S_{yy}(\omega) = \frac{k_1}{j\omega + 1} + \frac{k_2}{-j\omega + 1} + \frac{k_3}{j\omega + 3} + \frac{k_4}{-j\omega + 3}$$

$$= \frac{\frac{1}{8}}{j\omega + 1} + \frac{\frac{1}{8}}{-j\omega + 1} + \frac{-(1/24)}{j\omega + 3} + \frac{-(1/24)}{-j\omega + 3}$$

$$= \frac{\frac{1}{4}}{\omega^2 + 1} + \frac{-\frac{1}{4}}{\omega^2 + 9}$$

Using the inverse Fourier transform in equation (8-12) gives

$$R_{yy}(\tau) = \frac{1}{8} e^{-|\tau|} - \frac{1}{24} e^{-3|\tau|}$$

and, therefore, the mean square value of the output is

$$E[y^2(t)] = R_{yy}(0) = \frac{1}{8} - \frac{1}{24} = \frac{1}{12}$$

8.3 State Variable Representation of Linear Systems Driven by Random Inputs

In this section, discrete-time, state variable, stochastic models are introduced, and the behavior of discrete-time systems with random inputs is analyzed and discussed.

8.3.1 Discrete-Time, State Variable Stochastic Models

In general, state variable equations for a discrete-time stochastic system are of the form

$$\mathbf{x}(k + 1) = \mathbf{F}(k)\mathbf{x}(k) + \mathbf{L}(k)\mathbf{w}(k) \tag{8-13}$$

where the state vector $\mathbf{x}(k)$ is an n-vector, the known state coupling matrix $\mathbf{F}(k)$ is $n \times n$, the input noise vector $\mathbf{w}(k)$ is an r-vector, and the known noise coupling matrix $\mathbf{L}(k)$ is $n \times r$.

A block diagram that shows how the random inputs and state vectors are related in general is given in Figure 8-8. The effects of the known deterministic inputs $\mathbf{u}(k)$ can be accounted for separately, so they are omitted from the model.

It is assumed that the input noise vector $\mathbf{w}(k)$ is a white noise sequence with the known mean

$$E[\mathbf{w}(k)] = \bar{\mathbf{w}}(k)$$

where the bar over the symbol denotes the mean of the symbol, and the known step-by-step covariance matrix

$$E[(\mathbf{w}(j) - \bar{\mathbf{w}}(j))(\mathbf{w}(k) - \bar{\mathbf{w}}(k))^\dagger] = E[\tilde{\mathbf{w}}(j)\tilde{\mathbf{w}}^\dagger(k)]$$

$$= \mathbf{Q}(k)\,\delta_{jk} = \begin{cases} \mathbf{Q}(k) & j = k \\ \mathbf{0} & \text{otherwise} \end{cases}$$

where $\mathbf{Q}(k)$ is symmetric and positive semidefinite, which follows from

$$E[\tilde{\mathbf{w}}(j)\tilde{\mathbf{w}}^\dagger(k)]$$

$$= E\left\{ \begin{bmatrix} \tilde{w}_1(j) \\ \tilde{w}_2(j) \\ \vdots \\ \tilde{w}_r(j) \end{bmatrix} [\tilde{w}_1(k) \quad \tilde{w}_2(k) \quad \cdots \quad \tilde{w}_r(k)] \right\}$$

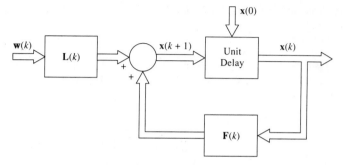

Figure 8-8 Block diagram of a system driven by random inputs.

$$= \begin{bmatrix} E[\tilde{w}_1(j)\tilde{w}_1(k)] & E[\tilde{w}_1(j)\tilde{w}_2(k)] & \cdots & E[\tilde{w}_1(j)\tilde{w}_r(k)] \\ E[\tilde{w}_2(j)\tilde{w}_1(k)] & E[\tilde{w}_2(j)\tilde{w}_2(k)] & \cdots & E[\tilde{w}_2(j)\tilde{w}_r(k)] \\ \vdots & & & \\ E[\tilde{w}_r(j)\tilde{w}_1(k)] & E[\tilde{w}_r(j)\tilde{w}_2(k)] & \cdots & E[\tilde{w}_r(j)\tilde{w}_r(k)] \end{bmatrix}$$

It is also assumed that the initial state vector $\mathbf{x}(0)$ is probabilistic with the known mean

$$E[\mathbf{x}(0)] = \bar{\mathbf{x}}(0)$$

and known covariance matrix

$$E[(\mathbf{x}(0) - \bar{\mathbf{x}}(0))(\mathbf{x}(0) - \bar{\mathbf{x}}(0))^\dagger] = E[\tilde{\mathbf{x}}(0)\tilde{\mathbf{x}}^\dagger(0)] = \mathbf{P}(0)$$

where $\mathbf{P}(0)$ is symmetric and positive semidefinite, which follows from

$$E[\tilde{\mathbf{x}}(0)\tilde{\mathbf{x}}^\dagger(0)]$$

$$= E\left\{ \begin{bmatrix} \tilde{x}_1(0) \\ \tilde{x}_2(0) \\ \vdots \\ \tilde{x}_n(0) \end{bmatrix} [\tilde{x}_1(0) \quad \tilde{x}_2(0) \quad \cdots \quad \tilde{x}_n(0)] \right\}$$

$$= \begin{bmatrix} E[\tilde{x}_1(0)\tilde{x}_1(0)] & E[\tilde{x}_1(0)\tilde{x}_2(0)] & \cdots & E[\tilde{x}_1(0)\tilde{x}_n(0)] \\ E[\tilde{x}_2(0)\tilde{x}_1(0)] & E[\tilde{x}_2(0)\tilde{x}_2(0)] & \cdots & E[\tilde{x}_2(0)\tilde{x}_n(0)] \\ \vdots & & & \\ E[\tilde{x}_n(0)\tilde{x}_1(0)] & E[\tilde{x}_n(0)\tilde{x}_2(0)] & \cdots & E[\tilde{x}_n(0)\tilde{x}_n(0)] \end{bmatrix}$$

Furthermore, it is assumed that the input noise $\mathbf{w}(k)$ and the initial state $\mathbf{x}(0)$ are uncorrelated for all k. That is,

$$E[\tilde{\mathbf{w}}(k)\tilde{\mathbf{x}}^\dagger(0)] = \mathbf{0}$$

Our objective is to determine an expression for the mean $\bar{\mathbf{x}}(k)$ of the state vector and an expression for the covariance

$$\mathbf{P}(k) = E\{[\mathbf{x}(k) - \bar{\mathbf{x}}(k)][\mathbf{x}(k) - \bar{\mathbf{x}}(k)]^\dagger\}$$

The mean provides information about the propagation of the state vector, and the covariance indicates the variations of the state vector about its mean.

8.3.2 Propagation of the Mean

The mean $\bar{x}(k)$ of the state of the system described by equation (8-13) may be determined by taking the expected values as follows:

$$E[x(k + 1)] = E[F(k)x(k) + L(k)w(k)]$$

Thus,

$$\bar{x}(k + 1) = F(k)\bar{x}(k) + L(k)\bar{w}(k) \tag{8-14}$$

Therefore, the mean $\bar{x}(k)$ may be calculated recursively, starting with the known initial mean $\bar{x}(0)$ and the known mean of the input noise $\bar{w}(k)$.

In terms of the initial state $\bar{x}(0)$ and the random inputs $w(k)$ at step 0 and beyond, the solution for the state in equation (8-14) after step 0 is the discrete convolution

$$x(k) = \Phi(k, 0)x(0) + \sum_{i=0}^{k-1} \Phi(k, i + 1)L(i)w(i) \tag{8-15}$$

where the state transition matrices Φ are the $n \times n$ products of state coupling matrices

$$\Phi(k, j) = F(k - 1)F(k - 2) \cdots F(j - 1)F(j) \qquad k > j \tag{8-16}$$

$$\Phi(i, i) = I \tag{8-17}$$

where I is the $n \times n$ identity matrix. Taking the expected values of both sides of equation (8-15) gives

$$\bar{x}(k) = \Phi(k, 0)\bar{x}(0) + \sum_{i=0}^{k-1} \Phi(k, i + 1)L(i)\bar{w}(i) \tag{8-18}$$

which is the desired result. Comparing equations (8-13) and (8-15) with equations (8-14) and (8-18) shows that the solution for the mean $\bar{x}(k)$ is identical in structure to the stochastic solution $x(k)$, except that all random quantities are replaced by their expected values in the expression for $\bar{x}(k)$.

For a linear, step-invariant, discrete-time system, the matrices F and L do not vary with step, and the state model is written as

$$x(k + 1) = Fx(k) + Lw(k)$$

The solution for the state is

$$x(k) = F^k x(0) + \sum_{i=0}^{k-1} F^{k-i-1}Lw(i) \tag{8-19}$$

because the state transition matrices are simply powers of the state coupling matrix $\mathbf{F}$.

Evaluation of the mean of equation (8-19) gives

$$E[\mathbf{x}(k)] = E[\mathbf{F}^k \mathbf{x}(0)] + E\left[\sum_{i=0}^{k-1} \mathbf{F}^{k-i-1}\mathbf{L}\mathbf{w}(i)\right]$$

$$\bar{\mathbf{x}}(k) = \mathbf{F}^k\bar{\mathbf{x}}(0) + \sum_{i=0}^{k-1} \mathbf{F}^{k-i-1}\mathbf{L}\bar{\mathbf{w}}(i)$$

For example, consider the system

$$x(k + 1) = 0.2x(k) + w(k)$$

where $w(k)$ is a zero mean white noise with unity variance. Suppose the initial state $x(0)$ has the mean $\bar{x}(0) = 10$.

The mean of $x(k)$ may be evaluated using equation (8-14), which shows that the mean is governed by

$$\bar{x}(k + 1) = 0.2\bar{x}(k)$$

$$\bar{x}(0) = 10$$

which may be solved using z-transform methods. Then

$$\bar{x}(k) = 10(0.2)^k$$

and is plotted in Figure 8-9(a).

8.3.3. Propagation of the Covariance

In a similar fashion, a recursive algorithm for the covariance matrix $\mathbf{P}(k)$ of the state of the system described by equation (8-13) may be determined.

Knowing that the covariance matrix $\mathbf{P}(k)$ is, at step $k + 1$,

$$\mathbf{P}(k + 1) = E\{[\mathbf{x}(k + 1) - \bar{\mathbf{x}}(k + 1)][\mathbf{x}(k + 1) - \bar{\mathbf{x}}(k + 1)]^\dagger\}$$

$$= E[\tilde{\mathbf{x}}(k + 1)\tilde{\mathbf{x}}^\dagger(k + 1)] \qquad \text{(8-20)}$$

then subtracting equation (8-14) from equation (8-13) and substituting in equation (8-20) gives

$$\mathbf{P}(k + 1) = E\{[\mathbf{F}(k)\tilde{\mathbf{x}}(k) + \mathbf{L}(k)\tilde{\mathbf{w}}(k)][\mathbf{F}(k)\tilde{\mathbf{x}}(k) + \mathbf{L}(k)\tilde{\mathbf{w}}(k)]^\dagger\}$$

$$= \mathbf{F}(k)E[\tilde{\mathbf{x}}(k)\tilde{\mathbf{x}}^\dagger(k)]\mathbf{F}^\dagger(k) + \mathbf{F}(k)E[\tilde{\mathbf{x}}(k)\tilde{\mathbf{w}}^\dagger(k)]\mathbf{L}^\dagger(k)$$

$$+ \mathbf{L}(k)E[\tilde{\mathbf{w}}(k)\tilde{\mathbf{x}}^\dagger(k)]\mathbf{F}^\dagger(k) + \mathbf{L}(k)E[\tilde{\mathbf{w}}(k)\tilde{\mathbf{w}}^\dagger(k)]\mathbf{L}^\dagger(k) \qquad \text{(8-21)}$$

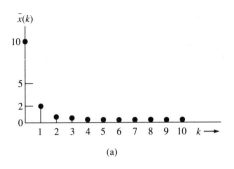

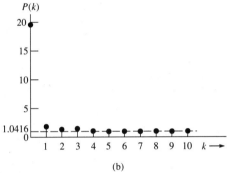

Figure 8-9 Mean and autocovariance of a stochastic system example. (a) Mean of the state x. (b) Autocovariance of the state x.

Because the vector $\bar{\mathbf{x}}(k)$ at the present step depends on previous values of the noise $\bar{\mathbf{w}}(k)$ and because the noise at the present step is uncorrelated with the noise at any other step (white noise assumption), the two middle terms in equation (8-21) vanish.

Therefore, equation (8-21) reduces to

$$\mathbf{P}(k+1) = \mathbf{F}(k)\mathbf{P}(k)\mathbf{F}^{\dagger}(k) + \mathbf{L}(k)\mathbf{Q}(k)\mathbf{L}^{\dagger}(k) \qquad \textbf{(8-22)}$$

If we continue with the previous example and assume that the initial state $x(0)$ is a random variable, independent of $w(k)$ for all k, with covariance

$$E[(x(0) - \bar{x}(0))^2] = 20$$

then equation (8-22) gives

$$P(k+1) = 0.04P(k) + 1$$

which is easily solved for $P(k)$ as follows:

$$P(1) = 0.04P(0) + 1 = 1.8$$

$P(2) = 0.04P(1) + 1 = 1.072$

$P(3) = 0.04P(2) + 1 = 1.04288$

$$\vdots$$

$P(\infty) = 1.041666$

as shown in Figure 8-9(b).

    The solution of $\mathbf{P}(k)$ can be calculated recursively, starting with an initial state covariance and repeatedly using equation (8-22). From $\mathbf{P}(0)$ and $\mathbf{Q}(0)$, $\mathbf{P}(1)$ can be calculated

$$\mathbf{P}(1) = \mathbf{F}(0)\mathbf{P}(0)\mathbf{F}^\dagger(0) + \mathbf{L}(0)\mathbf{Q}(0)\mathbf{L}^\dagger(0)$$

Then using $\mathbf{P}(1)$ and $\mathbf{Q}(1)$ gives

$$\mathbf{P}(2) = \mathbf{F}(1)\mathbf{P}(1)\mathbf{F}^\dagger(1) + \mathbf{L}(1)\mathbf{Q}(1)\mathbf{L}^\dagger(1)$$

$$= \mathbf{F}(1)\mathbf{F}(0)\mathbf{P}(0)\mathbf{F}^\dagger(0)\mathbf{F}^\dagger(1) + \mathbf{F}(1)\mathbf{L}(0)\mathbf{Q}(0)\mathbf{L}^\dagger(0)\mathbf{F}^\dagger(1)$$

$$+ \mathbf{L}(1)\mathbf{Q}(1)\mathbf{L}^\dagger(1)$$

From $\mathbf{P}(2)$ and $\mathbf{Q}(2)$ we get

$$\mathbf{P}(3) = \mathbf{F}(2)\mathbf{P}(2)\mathbf{F}^\dagger(2) + \mathbf{L}(2)\mathbf{Q}(2)\mathbf{L}^\dagger(2)$$

$$= \mathbf{F}(2)\mathbf{F}(1)\mathbf{F}(0)\mathbf{P}(0)\mathbf{F}^\dagger(0)\mathbf{F}^\dagger(1)\mathbf{F}^\dagger(2)$$

$$+ \mathbf{F}(2)\mathbf{F}(1)\mathbf{L}(0)\mathbf{Q}(0)\mathbf{L}^\dagger(0)\mathbf{F}^\dagger(1)\mathbf{F}^\dagger(2)$$

$$+ \mathbf{F}(2)\mathbf{L}(1)\mathbf{Q}(1)\mathbf{L}^\dagger(1)\mathbf{F}^\dagger(2) + \mathbf{L}(2)\mathbf{Q}(2)\mathbf{L}^\dagger(2)$$

and, in general,

$$\mathbf{P}(k) = \Phi(k,\, 0)\mathbf{P}(0)\Phi^\dagger(k,\, 0)$$

$$+ \sum_{i=0}^{k-1} \Phi(k,\, i+1)\mathbf{L}(i)\mathbf{Q}(i)\mathbf{L}^\dagger(i)\Phi^\dagger(k,\, i+1) \qquad \textbf{(8-23)}$$

where the state transition matrices at various steps are given by equations (8-16) and (8-17).

    For a linear, step-invariant, discrete-time system driven with stationary white noise, the matrices $\mathbf{F}$, $\mathbf{L}$, and $\mathbf{Q}$ do not vary with step, hence, equation (8-23) becomes

$$\mathbf{P}(k) = \mathbf{F}^k \mathbf{P}(0)\mathbf{F}^{\dagger k} + \sum_{i=0}^{k-1} \mathbf{F}^{k-i-1}\mathbf{L}\mathbf{Q}\mathbf{L}^\dagger(\mathbf{F}^\dagger)^{k-i-1} \qquad \textbf{(8-24)}$$

8.3.4 Correlated Noise and Shaping Filters

It may happen that the noise sequence $\mathbf{w}(k)$ is not white. That is,

$$E[\mathbf{w}(j)\mathbf{w}^\dagger(k)] \neq \mathbf{0} \qquad j \neq k$$

for some or all j and k. In this event, the basic results for uncorrelated white noise sequences can be applied, provided that a filter can be found that turns an uncorrelated white noise sequence $\mathbf{w}'(k)$ into a sequence that accurately models the correlation of $\mathbf{w}(k)$. State variable equations for the filter have the form

$$\Theta(k + 1) = \mathbf{M}(k)\Theta(k) + \mathbf{B}(k)\mathbf{w}'(k)$$

$$\mathbf{w}(k) = \mathbf{C}(k)\Theta(k) \tag{8-25}$$

Figure 8-10 is a block diagram of such a *shaping filter*.

The system consisting of the original nth-order stochastic model plus an rth-order shaping filter is then described by the $(n + r)$th-order model

$$\begin{bmatrix} \mathbf{x}(k + 1) \\ \hline \Theta(k + 1) \end{bmatrix} = \begin{bmatrix} \mathbf{F}(k) & \vdots & \mathbf{L}(k)\mathbf{C}(k) \\ \hline \mathbf{0} & \vdots & \mathbf{M}(k) \end{bmatrix} \begin{bmatrix} \mathbf{x}(k) \\ \hline \Theta(k) \end{bmatrix} + \begin{bmatrix} \mathbf{0} \\ \hline \mathbf{B}(k) \end{bmatrix} \mathbf{w}'(k)$$

$$= \mathbf{F}'(k)\mathbf{x}'(k) + \mathbf{L}'(k)\mathbf{w}'(k)$$

which fits the basic form given by equation (8-13) with white noise inputs. The fundamental problem is then the determination of a shaping filter that models the actual correlations sufficiently accurately.

In general, if the correlated noise $\mathbf{w}(k)$ is characterized by a rational PSD, the transfer function of the corresponding shaping filter can be

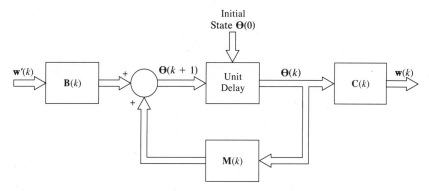

Figure 8-10 Block diagram of a shaping filter.

derived using the spectral factorization method discussed earlier. So it is a simple matter to find state and output equations, in controllable or observable forms, for the filter with a given z-transfer function.

As an example of correlated noise, consider the system

$$\begin{bmatrix} x_1(k+1) \\ x_2(k+1) \end{bmatrix} = \begin{bmatrix} 2 & -1 \\ 3 & 1 \end{bmatrix} \begin{bmatrix} x_1(k) \\ x_2(k) \end{bmatrix} + \begin{bmatrix} 0 \\ 1 \end{bmatrix} w(k)$$

where $w(k)$ is a Gaussian colored noise characterized by the PSD

$$S_{ww}(z) = \frac{-8z}{(z - \frac{1}{2})(z - 2)} = \left(\frac{2}{z - \frac{1}{2}}\right)\left(\frac{2}{z^{-1} - \frac{1}{2}}\right)$$

If a scalar, zero mean, Gaussian white noise $w'(k)$ with unit variance is passed through the shaping filter transfer function

$$G(z) = \frac{2}{z - \frac{1}{2}}$$

the resulting filtered noise $w(k)$ is Gaussian, because Gaussian noise passed through a linear, step-invariant system remains Gaussian. The shaping filter has a delay diagram given in Figure 8-11. Its state and output equations are

$$\Theta(k+1) = \frac{1}{2} \Theta(k) + 2w'(k)$$

$$w(k) = \Theta(k)$$

Combining the shaping filter and original system model into a third-order model with a white noise input yields

$$\begin{bmatrix} x_1(k+1) \\ x_2(k+1) \\ \hline \Theta(k+1) \end{bmatrix} = \begin{bmatrix} 2 & -1 & 0 \\ 3 & 1 & 1 \\ \hline 0 & 0 & \frac{1}{2} \end{bmatrix} \begin{bmatrix} x_1(k) \\ x_2(k) \\ \hline \Theta(k) \end{bmatrix} + \begin{bmatrix} 0 \\ 0 \\ \hline 2 \end{bmatrix} w'(k)$$

In a similar fashion, the results for discrete-time systems with random inputs using state variable stochastic models carry over to continuous time, and, therefore, will not be repeated.

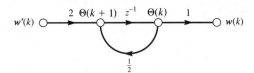

Figure 8-11 Delay diagram of the first-order shaping filter.

8.4 Least Squares Estimation

In the previous section, we examined the behavior of a linear dynamic system in the presence of random inputs. There we assumed that the plant outputs or the state variables were measured exactly. In reality, however, the outputs of physical sensors are usually corrupted by noise.

The method of least squares estimation is a powerful procedure for determining an *optimal* estimate of the state vector based on a set of noisy but known measurements that are linearly related to the state vector. This method is the subject of this section. A recursive formulation of least squares state estimation is then derived in which each additional measurement is incorporated sequentially, as it is available, modifying the previous estimate to produce a new estimate based on the previous measurements and the one additional new measurement.

8.4.1 Least Squares Solution

Karl Friedrich Gauss (1777–1855) invented the method of least squares estimation in 1795 and applied it to the calculation of planetary and comet orbits from telescopic measurement data. Six precise measurements would suffice to determine the six parameters of each orbit, but the individual measurements available were likely to be quite inaccurate. More than the minimum number of measurements were used, and the "best fit" to an orbit was found by minimizing the sum of squares of the parameter measurement errors. The approach adopted by Gauss was to develop the method first, then argue eloquently that it yielded the "most accurate" estimate. Adrien Marie Legendre (1752–1833) independently developed least squares estimation and was the first to publish the method, in 1806.

The basic least squares problem involves the estimation of an n-vector quantity $\mathbf{x}$ from an m-vector of linearly related known measurements

$$\mathbf{z} = \mathbf{Hx} + \mathbf{v} \tag{8-26}$$

where the matrix $\mathbf{H}$ is $m \times n$ and $m \geq n$, and where $\mathbf{v}$ is an m-vector of unknown measurement errors. We want to find, using the measurements $\mathbf{z}$, an estimate of $\mathbf{x}$, denoted by $\hat{\mathbf{x}}$, such that the sum of the squares of the errors between the actual measurements $\mathbf{z}$ and the estimated measurements $\mathbf{H}\hat{\mathbf{x}}$

$$J(\hat{\mathbf{x}}) = (\mathbf{z} - \mathbf{H}\hat{\mathbf{x}})^{\dagger}(\mathbf{z} - \mathbf{H}\hat{\mathbf{x}}) = \mathbf{v}^{\dagger}\mathbf{v} = v_1^2 + v_2^2 + \cdots + v_m^2 \tag{8-27}$$

is minimized. The notation used here, which is common, is that the estimate of a quantity such as $\mathbf{x}$ is denoted by a "hat" over the symbol for the quantity (i.e., $\hat{\mathbf{x}}$). To find the minimum, the partial derivatives of J with respect to each of the elements of $\hat{\mathbf{x}}$ are equated to zero

$$\left(\frac{\partial J}{\partial \hat{\mathbf{x}}}\right)^{\dagger} = \begin{bmatrix} \dfrac{\partial J}{\partial \hat{x}_1} \\ \dfrac{\partial J}{\partial \hat{x}_2} \\ \vdots \\ \dfrac{\partial J}{\partial \hat{x}_n} \end{bmatrix} = -\mathbf{H}^{\dagger}(\mathbf{z} - \mathbf{H}\hat{\mathbf{x}}) = 0$$

Then

$$\mathbf{H}^{\dagger}\mathbf{H}\hat{\mathbf{x}} = \mathbf{H}^{\dagger}\mathbf{z}$$

and, therefore,

$$\hat{\mathbf{x}} = (\mathbf{H}^{\dagger}\mathbf{H})^{-1}\mathbf{H}^{\dagger}\mathbf{z} \tag{8-28}$$

If $\mathbf{H}$ is of full rank, then $\mathbf{H}^{\dagger}\mathbf{H}$, which is $n \times n$, is of full rank, that is, nonsingular. Thus, $(\mathbf{H}^{\dagger}\mathbf{H})^{-1}$ exists, and the least squares estimate $\hat{\mathbf{x}}$, given by equation (8-28), is unique and a minimum. It is a minimum because the matrix of second derivatives of J

$$\frac{\partial^2 J}{\partial \hat{\mathbf{x}}^2} = \frac{\partial}{\partial \hat{\mathbf{x}}}\left(\frac{\partial J}{\partial \hat{\mathbf{x}}}\right) = \mathbf{H}^{\dagger}\mathbf{H}$$

which is symmetric, is positive definite if $\mathbf{H}$ is of full rank. Equation (8-28) shows that the least squares estimate $\hat{\mathbf{x}}$ is linearly related to the measurements $\mathbf{z}$. This is not surprising because derivatives of quadratic functions are linear functions.

As a first-order example of a least squares estimate, consider the three measurements $\mathbf{z}$ and a scalar quantity x to be estimated are related by

$$\begin{cases} z_1 = 4.2 = 3x + v_1 \\ z_2 = -3.3 = -2x + v_2 \\ z_3 = 1.4 = x + v_3 \end{cases} \tag{8-29}$$

which are of the form

$$\mathbf{z} = \begin{bmatrix} 4.2 \\ -3.3 \\ 1.4 \end{bmatrix} = \begin{bmatrix} 3 \\ -2 \\ 1 \end{bmatrix} x + \begin{bmatrix} v_1 \\ v_2 \\ v_3 \end{bmatrix}$$

or

$$\mathbf{z} = \mathbf{h}x + \mathbf{v}$$

These might represent three different measurements of the altitude of an aircraft, each possibly slightly in error. The least squares estimate of x is

$$\hat{x} = (\mathbf{h}^\dagger \mathbf{h})^{-1} \mathbf{h}^\dagger \mathbf{z} = \left(\begin{bmatrix} 3 & -2 & 1 \end{bmatrix} \begin{bmatrix} 3 \\ -2 \\ 1 \end{bmatrix} \right)^{-1} \begin{bmatrix} 3 & -2 & 1 \end{bmatrix} \begin{bmatrix} 4.2 \\ -3.3 \\ 1.4 \end{bmatrix}$$

$$= (14)^{-1}(20.6) = 1.47$$

As an example of vector least squares, suppose it is desired to estimate the quantity

$$\mathbf{x} = \begin{bmatrix} x_1 \\ x_2 \end{bmatrix}$$

from the measurements

$$\begin{cases} z_1 = -4 = 3x_1 - x_2 + v_1 \\ z_2 = 1 = 2x_1 + x_2 + v_2 \\ z_3 = -5 = x_1 - 2x_2 + v_3 \\ z_4 = 1 = 2x_1 + 2x_2 + v_4 \end{cases} \qquad \text{(8-30)}$$

which are of the form

$$\mathbf{z} = \begin{bmatrix} -4 \\ 1 \\ -5 \\ 1 \end{bmatrix} = \begin{bmatrix} 3 & -1 \\ 2 & 1 \\ 1 & -2 \\ 2 & 2 \end{bmatrix} \begin{bmatrix} x_1 \\ x_2 \end{bmatrix} + \begin{bmatrix} v_1 \\ v_2 \\ v_3 \\ v_4 \end{bmatrix} = \mathbf{Hx} + \mathbf{v}$$

The least squares estimate $\hat{\mathbf{x}}$ of the vector $\mathbf{x}$ is

$$\hat{\mathbf{x}} = (\mathbf{H}^\dagger \mathbf{H})^{-1} \mathbf{H}^\dagger \mathbf{z}$$

$$= \left(\begin{bmatrix} 3 & 2 & 1 & 2 \\ -1 & 1 & -2 & 2 \end{bmatrix} \begin{bmatrix} 3 & -1 \\ 2 & 1 \\ 1 & -2 \\ 2 & 2 \end{bmatrix} \right)^{-1} \begin{bmatrix} 3 & 2 & 1 & 2 \\ -1 & 1 & -2 & 2 \end{bmatrix} \begin{bmatrix} -4 \\ 1 \\ -5 \\ 1 \end{bmatrix}$$

$$= \begin{bmatrix} 18 & 1 \\ 1 & 10 \end{bmatrix}^{-1} \begin{bmatrix} -13 \\ 17 \end{bmatrix} = \frac{1}{179} \begin{bmatrix} 10 & -1 \\ -1 & 18 \end{bmatrix} \begin{bmatrix} -13 \\ 17 \end{bmatrix}$$

$$= \begin{bmatrix} -\frac{147}{179} \\ \frac{319}{179} \end{bmatrix} = \begin{bmatrix} -0.821 \\ 1.782 \end{bmatrix}$$

The estimate $\hat{\mathbf{x}}$ might be interpreted as the "most likely" value of the vector $\mathbf{x}$ to have produced a set of measurements $\mathbf{z}$.

8.4.2 State Estimation

In practice, one expects that measured plant outputs will contain slight errors, so that estimates of the plant state can likely be improved by using more than the minimum number of output measurements. If there are errors, however slight, in the outputs, the resulting overdetermined set of linear equations for the initial state will be inconsistent and will not have a solution. We can, however, seek the least squares estimate of the plant state.

Suppose that for the system

$$\begin{bmatrix} x_1(k+1) \\ x_2(k+1) \\ x_3(k+1) \end{bmatrix} = \begin{bmatrix} 0 & 0 & 2 \\ 1 & -1 & 0 \\ -2 & 0 & 1 \end{bmatrix} \begin{bmatrix} x_1(k) \\ x_2(k) \\ x_3(k) \end{bmatrix} = \mathbf{Ax}(k)$$

$$\begin{bmatrix} y_1(k) \\ y_2(k) \end{bmatrix} = \begin{bmatrix} 1 & 0 & 2 \\ -1 & 3 & 0 \end{bmatrix} \begin{bmatrix} x_1(k) \\ x_2(k) \\ x_3(k) \end{bmatrix} = \mathbf{Cx}(k)$$

the system outputs, which should be

$$\mathbf{y}(0) = \begin{bmatrix} 4 \\ 5 \end{bmatrix} \qquad \mathbf{y}(1) = \begin{bmatrix} -7 \\ -1.5 \end{bmatrix}$$

are measured with error as

$$\mathbf{y}(0) = \begin{bmatrix} 4.5 \\ 4.8 \end{bmatrix} \qquad \mathbf{y}(1) = \begin{bmatrix} -7.5 \\ -2 \end{bmatrix}$$

The equations giving the system initial state from the measured outputs are

$$\begin{bmatrix} \mathbf{C} \\ \mathbf{CA} \end{bmatrix} \mathbf{x}(0) = \begin{bmatrix} 1 & 0 & 2 \\ -1 & 3 & 0 \\ -4 & 0 & 4 \\ 3 & -3 & -2 \end{bmatrix} \begin{bmatrix} x_1(0) \\ x_2(0) \\ x_3(0) \end{bmatrix} = \mathbf{Hx}(0) = \begin{bmatrix} 4.5 \\ 4.8 \\ -7.5 \\ -2 \end{bmatrix} = \mathbf{z}$$

These are inconsistent and have no solution $\mathbf{x}(0)$. Their least squares estimate is

$$\hat{\mathbf{x}}(0) = (\mathbf{H}^\dagger \mathbf{H})^{-1} \mathbf{H}^\dagger \mathbf{z}$$

$$= \left(\begin{bmatrix} 1 & -1 & -4 & 3 \\ 0 & 3 & 0 & -3 \\ 2 & 0 & 4 & -2 \end{bmatrix} \begin{bmatrix} 1 & 0 & 2 \\ -1 & 3 & 0 \\ -4 & 0 & 4 \\ 3 & -3 & -2 \end{bmatrix} \right)^{-1}$$

$$
\times \begin{bmatrix} 1 & -1 & -4 & 3 \\ 0 & 3 & 0 & -3 \\ 2 & 0 & 4 & -2 \end{bmatrix} \begin{bmatrix} 4.5 \\ 4.8 \\ -7.5 \\ -2 \end{bmatrix}
$$

$$
= \begin{bmatrix} 27 & -12 & -20 \\ -12 & 18 & 6 \\ -20 & 6 & 24 \end{bmatrix}^{-1} \begin{bmatrix} 23.7 \\ 20.4 \\ -17.0 \end{bmatrix}
$$

$$
= \begin{bmatrix} 0.136 & 0.058 & 0.099 \\ 0.058 & 0.085 & 0.027 \\ 0.099 & 0.027 & 0.117 \end{bmatrix} \begin{bmatrix} 23.7 \\ 20.4 \\ -17.0 \end{bmatrix} = \begin{bmatrix} 2.714 \\ 2.645 \\ 0.892 \end{bmatrix}
$$

8.9.3 Recursive Least Squares

Recursive least squares is an arrangement of the least squares solution in which each new measurement is used to update the previous least squares estimate that was based on previous measurements. Instead of processing all of the measurement data at once, the measurements are processed one at a time, with each new measurement causing a modification in the current estimate. Least squares estimates are linear transformations of the measurements. The least squares estimate based on the first $k + 1$ measurements can, therefore, be expressed as a linear transformation of the least squares estimate based on the first k measurements plus a linear correction term based on the $(k + 1)$th measurements alone.

Denoting the number of measurements used by arguments, the least squares estimate based on k measurements is

$$\hat{\mathbf{x}}(k) = [\mathbf{H}^\dagger(k)\mathbf{H}(k)]^{-1}\mathbf{H}^\dagger(k)\mathbf{z}(k)$$

The least squares estimate based on $k + 1$ measurements is

$$\hat{\mathbf{x}}(k + 1) = [\mathbf{H}^\dagger(k + 1)\mathbf{H}(k + 1)]^{-1}\mathbf{H}^\dagger(k + 1)\mathbf{z}(k + 1)$$

where $\mathbf{H}(k + 1)$ is $\mathbf{H}(k)$ with an additional row $\mathbf{h}^\dagger(k + 1)$

$$\mathbf{H}(k + 1) = \begin{bmatrix} \mathbf{H}(k) \\ \hline \mathbf{h}^\dagger(k + 1) \end{bmatrix}$$

and the vector of measurements $\mathbf{z}(k + 1)$ is the measurement vector $\mathbf{z}(k)$ with one additional scalar measurement z_{k+1}

$$\mathbf{z}(k + 1) = \begin{bmatrix} \mathbf{z}(k) \\ \hline z_{k+1} \end{bmatrix}$$

Then

$$\mathbf{H}^\dagger(k + 1)\mathbf{H}(k + 1) = [\mathbf{H}^\dagger(k) \mid \mathbf{h}(k + 1)]\begin{bmatrix} \mathbf{H}(k) \\ \hline \mathbf{h}^\dagger(k + 1) \end{bmatrix}$$

$$= \mathbf{H}^\dagger(k)\mathbf{H}(k) + \mathbf{h}(k + 1)\mathbf{h}^\dagger(k + 1)$$

Defining

$$\mathbf{P}(k) = [\mathbf{H}^\dagger(k)\mathbf{H}(k)]^{-1}$$

gives

$$\begin{aligned} \mathbf{P}(k + 1) &= [\mathbf{H}^\dagger(k + 1)\mathbf{H}(k + 1)]^{-1} \\ &= [\mathbf{H}^\dagger(k)\mathbf{H}(k) + \mathbf{h}(k + 1)\mathbf{h}^\dagger(k + 1)]^{-1} \\ &= [\mathbf{P}^{-1}(k) + \mathbf{h}(k + 1)\mathbf{h}^\dagger(k + 1)]^{-1} \end{aligned} \qquad \text{(8-31)}$$

Relation (8-31) is in a form for which the matrix inversion lemma

$$(\Gamma + \mathbf{uv}^\dagger)^{-1} = \Gamma^{-1} - \frac{\Gamma^{-1}\mathbf{uv}^\dagger\Gamma^{-1}}{1 + \mathbf{v}^\dagger\Gamma^{-1}\mathbf{u}} \qquad \text{(8-32)}$$

applies. The lemma is proved by multiplying both sides of equation (8-32) by $\Gamma + \mathbf{uv}^\dagger$ to obtain the identity matrix

$$\begin{aligned} (\Gamma + \mathbf{uv}^\dagger)^{-1}(\Gamma + \mathbf{uv}^\dagger) &= \Gamma^{-1}(\Gamma + \mathbf{uv}^\dagger) - \frac{\Gamma^{-1}\mathbf{uv}^\dagger\Gamma^{-1}(\Gamma + \mathbf{uv}^\dagger)}{1 + \mathbf{v}^\dagger\Gamma^{-1}\mathbf{u}} \\ &= \mathbf{I} + \Gamma^{-1}\mathbf{uv}^\dagger - \frac{\Gamma^{-1}\mathbf{u}(\mathbf{v}^\dagger + \mathbf{v}^\dagger\Gamma^{-1}\mathbf{uv}^\dagger)}{1 + \mathbf{v}^\dagger\Gamma^{-1}\mathbf{u}} \\ &= \mathbf{I} + \Gamma^{-1}\mathbf{uv}^\dagger - \frac{\Gamma^{-1}\mathbf{u}(1 + \mathbf{v}^\dagger\Gamma^{-1}\mathbf{u})\mathbf{v}^\dagger}{1 + \mathbf{v}^\dagger\Gamma^{-1}\mathbf{u}} \\ &= \mathbf{I} \end{aligned}$$

Using the matrix inversion lemma on equation (8-31) gives

$$\mathbf{P}(k + 1) = \mathbf{P}(k) - \frac{\mathbf{P}(k)\mathbf{h}(k + 1)\mathbf{h}^\dagger(k + 1)\mathbf{P}(k)}{1 + \mathbf{h}^\dagger(k + 1)\mathbf{P}(k)\mathbf{h}(k + 1)}$$

which is an update equation for $\mathbf{P}(k + 1)$ in terms of $\mathbf{P}(k)$ and the next measurement equation coefficients, $\mathbf{h}(k + 1)$. Defining

$$\delta(k + 1) = 1 + \mathbf{h}^\dagger(k+ 1)\mathbf{P}(k)\mathbf{h}(k + 1)$$

$$\kappa(k + 1) = \mathbf{P}(k)\mathbf{h}(k + 1)\,\delta^{-1}(k + 1)$$

gives an update equation of

$$\begin{aligned} \mathbf{P}(k + 1) &= \mathbf{P}(k) - \mathbf{P}(k)\mathbf{h}(k + 1)\,\delta^{-1}(k + 1)\mathbf{h}^\dagger(k + 1)\mathbf{P}(k) \\ &= [\mathbf{I} - \kappa(k + 1)\mathbf{h}^\dagger(k + 1)]\mathbf{P}(k) \end{aligned} \qquad \text{(8-33)}$$

The least squares estimate at step $k + 1$ is

$$\hat{\mathbf{x}}(k + 1) = \mathbf{P}(k + 1)\mathbf{H}^\dagger(k + 1)\mathbf{z}(k + 1)$$

$$= \mathbf{P}(k + 1)[\mathbf{H}^\dagger(k) \mid \mathbf{h}(k + 1)]\begin{bmatrix} \mathbf{z}(k) \\ \hline z_{k+1} \end{bmatrix}$$

$$= \mathbf{P}(k + 1)[\mathbf{H}^\dagger(k)\mathbf{z}(k) + \mathbf{h}(k + 1)z_{k+1}]$$

$$= [\mathbf{P}(k) - \mathbf{P}(k)\mathbf{h}(k + 1)\ \delta^{-1}(k + 1)\mathbf{h}^\dagger(k + 1)\mathbf{P}(k)]$$

$$\times\ [\mathbf{H}^\dagger(k)\mathbf{z}(k) + \mathbf{h}(k + 1)z_{k+1}]$$

$$= \mathbf{P}(k)\mathbf{H}^\dagger(k)\mathbf{z}(k) + \mathbf{P}(k)\mathbf{h}(k + 1)z_{k+1}$$

$$-\ \mathbf{P}(k)\mathbf{h}(k + 1)\ \delta^{-1}(k + 1)\mathbf{h}^\dagger(k + 1)\mathbf{P}(k)\mathbf{H}^\dagger(k)\mathbf{z}(k)$$

$$-\ \mathbf{P}(k)\mathbf{h}(k + 1)\ \delta^{-1}(k + 1)\mathbf{h}^\dagger(k + 1)\mathbf{P}(k)\mathbf{h}(k + 1)z_{k+1}$$

$$= \hat{\mathbf{x}}(k) + \mathbf{P}(k)\mathbf{h}(k + 1)\ \delta^{-1}(k + 1)[z_{k+1} - \mathbf{h}^\dagger(k + 1)\hat{\mathbf{x}}(k)]$$

$$= \hat{\mathbf{x}}(k) + \boldsymbol{\kappa}(k + 1)[z_{k+1} - \mathbf{h}^\dagger(k + 1)\hat{\mathbf{x}}(k)] \qquad \textbf{(8-34)}$$

The least squares estimate $\hat{\mathbf{x}}(k + 1)$ based on $k + 1$ measurements is the estimate $\hat{\mathbf{x}}(k)$ based on k measurements plus a gain (which does not depend on the measurements)

$$\boldsymbol{\kappa}(k + 1) = \mathbf{P}(k)\mathbf{h}(k + 1)\ \delta^{-1}(k + 1)$$

times the difference between the new measurement and the predicted measurement

$$\mathbf{h}^\dagger(k + 1)\hat{\mathbf{x}}(k)$$

based on the previous estimate.

These equations for recursive least squares estimation are collected in Table 8-2. A block diagram of the recursive least squares estimator is given in Figure 8-12. The filter shown processes one scalar measurement at a time and produces the least squares estimate based on that and all preceding measurements. There are many variations on this basic result, including those in which vector, rather than scalar, measurements are processed at each step.

To estimate an n-vector $\mathbf{x}$ with recursive least squares, one first obtains an initial estimate based on the first n linearly independent measurements

$$\hat{\mathbf{x}}(n) = [\mathbf{H}^\dagger(n)\mathbf{H}(n)]^{-1}\mathbf{H}^\dagger(n)\mathbf{z}(n)$$

and the initial $\mathbf{P}$ matrix

$$\mathbf{P}(n) = [\mathbf{H}^\dagger(n)\mathbf{H}(n)]^{-1}$$

Table 8-2 Recursive Least Squares Computation

Measurement Model and Problem Statement

Consider the scalar measurements

$$z_i = \mathbf{h}^\dagger(i)\mathbf{x} \qquad i = 1, 2, 3, \ldots$$

where the unknown quantity $\mathbf{x}$ is an n-vector and where the scalar measurements z have errors. The least squares solution $\hat{\mathbf{x}}(k)$ based on k measurements satisfies

$$z_i = \mathbf{h}^\dagger(i)\hat{\mathbf{x}}(k) + v_i \qquad i = 1, 2, \ldots, k$$

where the sum of squares of errors

$$J(k) = v_1^2 + v_2^2 + \cdots + v_k^2 = \sum_{i=1}^{k} v_i^2$$

is minimum.

Initialization

$$\mathbf{P}(n) = [\mathbf{H}^\dagger(n)\mathbf{H}(n)]^{-1}$$

$$\hat{\mathbf{x}}(n) = \mathbf{P}(n)\mathbf{H}^\dagger(n)\mathbf{z}(n)$$

where

$$\mathbf{H}(n) = \begin{bmatrix} \mathbf{h}^\dagger(1) \\ \hline \vdots \\ \hline \mathbf{h}^\dagger(n) \end{bmatrix} \qquad \mathbf{z}(n) = \begin{bmatrix} z_1 \\ \vdots \\ z_n \end{bmatrix}$$

Corrector Gain

$$\delta(k + 1) = \mathbf{h}^\dagger(k + 1)\mathbf{P}(k)\mathbf{h}(k + 1) + 1$$

$$\kappa(k + 1) = \mathbf{P}(k)\mathbf{h}(k + 1)\,\delta^{-1}(k + 1)$$

$$\mathbf{P}(k + 1) = [\mathbf{I} - \kappa(k + 1)\mathbf{h}^\dagger(k + 1)]\mathbf{P}(k) \tag{8-33}$$

Predictor–Corrector

$$\hat{\mathbf{x}}(k + 1) = \hat{\mathbf{x}}(k) + \kappa(k + 1)[z_{k+1} - \mathbf{h}^\dagger(k + 1)\hat{\mathbf{x}}(k)] \tag{8-34}$$

Thereafter, the least squares estimate is updated with each new scalar measurement. Except in the first-order case, one cannot begin this recursive least squares algorithm with the first measurement, using

$$\hat{\mathbf{x}}(1) = [\mathbf{H}^\dagger(1)\mathbf{H}(1)]^{-1}\mathbf{H}^\dagger(1)\mathbf{z}(1) = [\mathbf{h}(1)\mathbf{h}^\dagger(1)]^{-1}\mathbf{h}(1)\mathbf{z}(1)$$

$$\mathbf{P}(1) = [\mathbf{H}^\dagger(1)\mathbf{H}(1)]^{-1} = [\mathbf{h}(1)\mathbf{h}^\dagger(1)]^{-1}$$

because $\mathbf{h}(1)\mathbf{h}^\dagger(1)$ is singular. The initial estimate can be obtained recur-

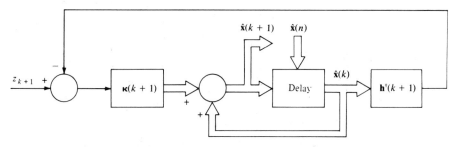

Figure 8-12 Block diagram of the recursive least squares estimator.

sively (for example, with a deadbeat observer), but not with equations (8-33) and (8-34).

As an example of scalar recursive least squares, consider the equations and measurements in equation (8-29) discussed previously. A recursive solution is

1. *Initialization*

$$P(1) = [h^\dagger(1)h(1)]^{-1} = \frac{1}{9}$$

$$\hat{x}(1) = P(1)h^\dagger(1)z_1 = \left(\frac{1}{9}\right)(3)(4.2) = 1.4$$

2. *Update with measurement 2*

$$\delta(2) = h^\dagger(2)P(1)h(2) + 1$$

$$= (-2)\left(\frac{1}{9}\right)(-2) + 1 = \frac{13}{9}$$

$$\kappa(2) = P(1)h(2)\,\delta^{-1}(2) = \frac{1}{9}(-2)\frac{9}{13} = -\frac{2}{13}$$

$$P(2) = [1 - \kappa(2)h^\dagger(2)]P(1)$$

$$= \left[1 + \left(\frac{2}{13}\right)(-2)\right]\frac{1}{9} = \frac{1}{13}$$

$$\hat{x}(2) = \hat{x}(1) + \kappa(2)[z_2 - h(2)\hat{x}(1)]$$

$$= 1.4 - \frac{2}{13}[-3.3 - (-2)(1.4)] = 1.4769$$

3. *Update with measurement 3*

$$\delta(3) = h^\dagger(3)P(2)h(3) + 1$$

$$= (1)\left(\frac{1}{13}\right)(1) + 1 = \frac{14}{13}$$

$$\kappa(3) = P(2)h(3)\,\delta^{-1}(3) = \frac{1}{13}\,(1)\,\frac{13}{14} = \frac{1}{14}$$

$$P(3) = [1 - \kappa(3)h^{\dagger}(3)]P(2)$$

$$= \left[1 - \left(\frac{1}{14}\right)(1)\right]\frac{1}{13}$$

$$= \frac{1}{14}$$

$$\hat{x}(3) = \hat{x}(2) + \kappa(3)[z_3 - h(3)\hat{x}(2)]$$

$$= 1.4769 + \frac{1}{14}[1.4 - (1)(1.4769)]$$

$$= 1.47$$

as was found earlier. The process could be continued if there were more measurements.

As an example of vector recursive least squares, consider the estimation of a two-vector

$$\mathbf{x} = \begin{bmatrix} x_1 \\ x_2 \end{bmatrix}$$

from the following four measurements:

$$\begin{cases} z_1 = \;\;\; 3 = 2x_1 \qquad\quad = \mathbf{h}^{\dagger}(1)\mathbf{x} \\ z_2 = \;\;\; 5 = \;\;x_1 - \;\;\; x_2 = \mathbf{h}^{\dagger}(2)\mathbf{x} \\ z_3 = \;\;\; 4 = 3x_1 + \;\;\; x_2 = \mathbf{h}^{\dagger}(3)\mathbf{x} \\ z_4 = -4 = \;\;x_1 + 2x_2 = \mathbf{h}^{\dagger}(4)\mathbf{x} \end{cases}$$

The initialization step uses the first two of these measurements to obtain an initial estimate of $\mathbf{x}$. Using

$$P(2) = [\mathbf{H}^{\dagger}(2)\mathbf{H}(2)]^{-1} = \left\{\begin{bmatrix} 2 & 1 \\ 0 & -1 \end{bmatrix}\begin{bmatrix} 2 & 0 \\ 1 & -1 \end{bmatrix}\right\}^{-1} = \begin{bmatrix} \frac{1}{4} & \frac{1}{4} \\ \frac{1}{4} & \frac{5}{4} \end{bmatrix}$$

gives

$$\hat{\mathbf{x}}(2) = P(2)\mathbf{H}^{\dagger}(2)\mathbf{z}(2) = \begin{bmatrix} \frac{1}{4} & \frac{1}{4} \\ \frac{1}{4} & \frac{5}{4} \end{bmatrix}\begin{bmatrix} 2 & 1 \\ 0 & -1 \end{bmatrix}\begin{bmatrix} 3 \\ 5 \end{bmatrix} = \begin{bmatrix} \frac{3}{2} \\ -\frac{7}{2} \end{bmatrix}$$

Hereafter, the recursive relations are used. For step 3

$$\delta(3) = 1 + \mathbf{h}^\dagger(3)\mathbf{P}(2)\mathbf{h}(3) = 1 + [3\quad 1]\begin{bmatrix} \frac{1}{4} & \frac{1}{4} \\ \frac{1}{4} & \frac{5}{4} \end{bmatrix}\begin{bmatrix} 3 \\ 1 \end{bmatrix} = 6$$

$$\boldsymbol{\kappa}(3) = \mathbf{P}(2)\mathbf{h}(3)\,\delta^{-1}(3) = \begin{bmatrix} \frac{1}{4} & \frac{1}{4} \\ \frac{1}{4} & \frac{5}{4} \end{bmatrix}\begin{bmatrix} 3 \\ 1 \end{bmatrix}\frac{1}{6} = \begin{bmatrix} \frac{1}{6} \\ \frac{1}{3} \end{bmatrix}$$

$$\mathbf{P}(3) = [\mathbf{I} - \boldsymbol{\kappa}(3)\mathbf{h}^\dagger(3)]\mathbf{P}(2)$$

$$= \left\{\begin{bmatrix} 1 & 0 \\ 0 & 1 \end{bmatrix} - \begin{bmatrix} \frac{1}{6} \\ \frac{1}{3} \end{bmatrix}[3\quad 1]\right\}\begin{bmatrix} \frac{1}{4} & \frac{1}{4} \\ \frac{1}{4} & \frac{5}{4} \end{bmatrix} = \begin{bmatrix} \frac{1}{12} & -\frac{1}{12} \\ -\frac{1}{12} & \frac{7}{12} \end{bmatrix}$$

$$\hat{\mathbf{x}}(3) = \hat{\mathbf{x}}(2) + \boldsymbol{\kappa}(3)[z_3 - \mathbf{h}^\dagger(3)\hat{\mathbf{x}}(2)] = \begin{bmatrix} 2 \\ -\frac{5}{2} \end{bmatrix}$$

For step 4

$$\delta(4) = 1 + \mathbf{h}^\dagger(4)\mathbf{P}(3)\mathbf{h}(4) = 1 + [1\quad 2]\begin{bmatrix} 1 & -1 \\ -1 & 7 \end{bmatrix}\frac{1}{12}\begin{bmatrix} 1 \\ 2 \end{bmatrix} = \frac{37}{12}$$

$$\boldsymbol{\kappa}(4) = \mathbf{P}(3)\mathbf{h}(4)\,\delta^{-1}(4) = \frac{1}{12}\begin{bmatrix} 1 & -1 \\ -1 & 7 \end{bmatrix}\begin{bmatrix} 1 \\ 2 \end{bmatrix}\frac{12}{37} = \frac{1}{37}\begin{bmatrix} -1 \\ 13 \end{bmatrix}$$

$$\mathbf{P}(4) = [\mathbf{I} - \boldsymbol{\kappa}(4)\mathbf{h}^\dagger(4)]\mathbf{P}(3)$$

$$= \left\{\begin{bmatrix} 1 & 0 \\ 0 & 1 \end{bmatrix} - \frac{1}{37}\begin{bmatrix} -1 \\ 13 \end{bmatrix}[1\quad 2]\right\}\begin{bmatrix} 1 & -1 \\ -1 & 7 \end{bmatrix}\frac{1}{12}$$

$$= \frac{1}{444}\begin{bmatrix} 36 & -24 \\ -24 & 90 \end{bmatrix}$$

$$\hat{\mathbf{x}}(4) = \hat{\mathbf{x}}(3) + \boldsymbol{\kappa}(4)[z_4 - \mathbf{h}^\dagger(4)\hat{\mathbf{x}}(3)] = \begin{bmatrix} \frac{75}{37} \\ -\frac{211}{74} \end{bmatrix}$$

and so on. If there were additional measurements, they could be incorporated into the estimate in a similar way.

8.4.4 Probabilistic Interpretation of Least Squares

When a least squares estimate is interpreted as yielding the "best" or "most likely" value of the estimated quantity, probabilistic assumptions are being made about the measurement errors $v_1, v_2, \ldots,$ as Gauss knew and discussed. In the basic least squares problem, equal weightings of the squares of the measurement errors in the performance measure J imply that each measurement has equal likelihood of error and that the errors are independent of each other.

A more general least squares problem minimizes

$$J(\hat{x}) = (z - H\hat{x})^\dagger W(z - H\hat{x}) = v^\dagger W v$$

where W is a symmetric, positive definite weighting matrix. J is then a quadratic form in the measurement errors v. When one is more confident in the accuracy of some of the measurements than of others, the elements of W can be chosen to weigh them more heavily than others.

For the measurements

$$z = Hx + v$$

the estimate $\hat{x}$ that results in minimum weighted sum of squares of measurement error

$$J(\hat{x}) = (z - H\hat{x})^\dagger W(z - H\hat{x}) = v^\dagger W v$$

is

$$\hat{x} = (H^\dagger W H)^{-1} H^\dagger W z \tag{8-35}$$

This result is derived by expressing the positive definite symmetric weighting matrix W as

$$W = \Psi^\dagger \Psi$$

Then

$$J(\hat{x}) = [(\Psi z) - (\Psi H)\hat{x}]^\dagger [(\Psi z) - (\Psi H)\hat{x}]$$

which is in the form of equation (8-27) with z replaced by Ψz and H replaced by ΨH. Making these substitutions into the basic least squares result in equation (8-28)

$$\hat{x} = [(H^\dagger \Psi^\dagger)(\Psi H)]^{-1}(H^\dagger \Psi^\dagger)(\Psi z)$$

gives equation (8-35).

For the measurements

$$z = \begin{bmatrix} -4 \\ 1 \\ -5 \\ 1 \end{bmatrix} = \begin{bmatrix} 3 & -1 \\ 2 & 1 \\ 1 & -2 \\ 2 & 2 \end{bmatrix} \begin{bmatrix} x_1 \\ x_2 \end{bmatrix} + \begin{bmatrix} v_1 \\ v_2 \\ v_3 \\ v_4 \end{bmatrix} = Hx + v$$

for example, suppose that it is desired to weight with

$$W = \begin{bmatrix} 4 & 0 & 0 & 0 \\ 0 & 3 & 0 & 0 \\ 0 & 0 & 2 & 0 \\ 0 & 0 & 0 & 1 \end{bmatrix}$$

In other words, suppose we have the most confidence in the first equation and decreasing confidence in later equations. In general, $\mathbf{W}$ need not be diagonal, but in this simple example, it is particularly easy to interpret the meaning of $\mathbf{W}$ when it is diagonal: we expect the square error in the first equation to be three-quarters as large as the square error in the second equation, and so on. The weighted least squares estimate is

$$
\hat{\mathbf{x}} = \left(\begin{bmatrix} 3 & 2 & 1 & 2 \\ -1 & 1 & -2 & 2 \end{bmatrix} \begin{bmatrix} 4 & 0 & 0 & 0 \\ 0 & 3 & 0 & 0 \\ 0 & 0 & 2 & 0 \\ 0 & 0 & 0 & 1 \end{bmatrix} \begin{bmatrix} 3 & -1 \\ 2 & 1 \\ 1 & -2 \\ 2 & 2 \end{bmatrix} \right)^{-1}
$$

$$
\times \begin{bmatrix} 3 & 2 & 1 & 2 \\ -1 & 1 & -2 & 2 \end{bmatrix} \begin{bmatrix} 4 & 0 & 0 & 0 \\ 0 & 3 & 0 & 0 \\ 0 & 0 & 2 & 0 \\ 0 & 0 & 0 & 1 \end{bmatrix} \begin{bmatrix} -4 \\ 1 \\ -5 \\ 1 \end{bmatrix}
$$

$$
= \begin{bmatrix} 54 & -6 \\ -6 & 19 \end{bmatrix}^{-1} \begin{bmatrix} -50 \\ 41 \end{bmatrix} = \frac{1}{990} \begin{bmatrix} 19 & 6 \\ 6 & 54 \end{bmatrix} \begin{bmatrix} -50 \\ 41 \end{bmatrix}
$$

$$
= \begin{bmatrix} -\frac{704}{990} \\ \frac{1914}{990} \end{bmatrix} = \begin{bmatrix} -0.71 \\ 1.93 \end{bmatrix}
$$

If the errors $\mathbf{v}$ are zero mean

$$E[\mathbf{v}] = \mathbf{0}$$

with known positive definite covariance matrix $\mathbf{R}$

$$E[\mathbf{v}\mathbf{v}^{\dagger}] = \mathbf{R}$$

it is natural to choose

$$\mathbf{W} = \mathbf{R}^{-1}$$

Indeed, for a constant vector $\mathbf{x}$ to be estimated, this gives the same estimate $\hat{\mathbf{x}}$ as the more general result to follow. Recursive least square methods can also be designed to incorporate vector rather than scalar measurements at each step and to estimate a quantity $\mathbf{x}(n)$ that itself changes with the step number n in a known way.

8.5 Linear Minimum Mean Square Estimation

Rather than adopting a least squares criterion and arguing that the resulting estimate is most probable, Kalman began with a stochastic formulation including a probabilistic performance index to be minimized. As R. A. Fisher, A. N. Kolmogorov, N. Wiener, and others did before him, Kalman sought estimates that were linearly related to the measurements, such that the expected sum of squares of the errors between the actual and estimated states

$$J = E[(\mathbf{x} - \hat{\mathbf{x}})^\dagger(\mathbf{x} - \hat{\mathbf{x}})] \tag{8-36}$$

was minimized. The state to be estimated could vary in a known way with the step, the state as well as the measurement could be influenced by noise, and the initial state could be described stochastically. The result is called a *Kalman filter*.

Solving the problem of LMMS estimation is straightforward but involves many details. Consequently, we first develop the solution to the basic problem, then indicate how to extend those results to more complicated situations.

8.5.1 Stochastic System Models

In the following development, we consider an nth-order, linear, discrete-time, stochastic system that has state and output equations of the form

$$\mathbf{x}(k + 1) = \mathbf{F}(k)\mathbf{x}(k) + \mathbf{w}(k) \tag{8-37}$$

$$\mathbf{z}(k + 1) = \mathbf{H}(k + 1)\mathbf{x}(k + 1) + \mathbf{v}(k + 1) \tag{8-38}$$

The state equations of the model are

$$\begin{bmatrix} x_1(k + 1) \\ x_2(k + 1) \\ \vdots \\ x_n(k + 1) \end{bmatrix} = \begin{bmatrix} f_{11}(k) & f_{12}(k) & \cdots & f_{1n}(k) \\ f_{21}(k) & f_{22}(k) & \cdots & f_{2n}(k) \\ \vdots & & & \\ f_{n1}(k) & f_{n2}(k) & \cdots & f_{nn}(k) \end{bmatrix} \begin{bmatrix} x_1(k) \\ x_2(k) \\ \vdots \\ x_n(k) \end{bmatrix} + \begin{bmatrix} w_1(k) \\ w_2(k) \\ \vdots \\ w_n(k) \end{bmatrix} \tag{8-39}$$

where $\mathbf{w}(k)$ is a white noise sequence with zero mean and known step-by-step covariance matrix $\mathbf{Q}(k)$. That is,

$$E[\mathbf{w}(k)] = \mathbf{0}$$

and

$$E[\mathbf{w}(i)\mathbf{w}^\dagger(k)] = \begin{cases} \mathbf{0} & i \neq k \\ \mathbf{Q}(k) & i = k \end{cases}$$

The process covariance matrix $\mathbf{Q}(k)$ is a symmetric, positive semidefinite matrix. By "whiteness" in the sequence $\mathbf{w}(k)$ is meant that it is uncorrelated with itself at any other step.

The measurement equations of the stochastic model are

$$
\begin{bmatrix} z_1(k+1) \\ z_2(k+1) \\ \vdots \\ z_m(k+1) \end{bmatrix} = \begin{bmatrix} h_{11}(k+1) & h_{12}(k+1) & \cdots & h_{1n}(k+1) \\ h_{21}(k+1) & h_{22}(k+1) & \cdots & h_{2n}(k+1) \\ \vdots & & & \\ h_{m1}(k+1) & h_{m2}(k+1) & \cdots & h_{mn}(k+1) \end{bmatrix} \begin{bmatrix} x_1(k+1) \\ x_2(k+1) \\ \vdots \\ x_n(k+1) \end{bmatrix}
$$

$$
+ \begin{bmatrix} v_1(k+1) \\ v_2(k+1) \\ \vdots \\ v_m(k+1) \end{bmatrix} \tag{8-40}
$$

where $\mathbf{v}(k+1)$ is a zero mean white noise sequence with known co-variance matrix $\mathbf{R}(k)$ and is uncorrelated with $\mathbf{w}(k)$. That is,

$$E[\mathbf{v}(k)] = \mathbf{0}$$

$$E[\mathbf{v}(i)\mathbf{v}^\dagger(k)] = \begin{cases} \mathbf{0} & i \neq k \\ \mathbf{R}(k) & i = k \end{cases}$$

and the measurement covariance matrix $\mathbf{R}(k)$ is symmetric and positive semidefinite.

Furthermore, it is assumed that the initial condition vector $\mathbf{x}(0)$ is probabilistic with zero mean, known positive semidefinite covariance matrix $\mathbf{P}(0)$, and uncorrelated with the noise sequences $\mathbf{w}(k)$ and $\mathbf{v}(k)$.

That is,

$$E[\mathbf{x}(0)] = \mathbf{0}$$
$$E[\mathbf{x}(0)\mathbf{x}^\dagger(0)] = \mathbf{P}_0$$
$$E[\mathbf{x}(0)\mathbf{w}^\dagger(k)] = \mathbf{0}$$
$$E[\mathbf{x}(0)\mathbf{v}^\dagger(k)] = \mathbf{0}$$

for all k.

A block diagram showing how the vectors $\mathbf{x}$, $\mathbf{w}$, $\mathbf{z}$, and $\mathbf{v}$ are related is given in Figure 8-13. In the figure, the wide arrows represent vectors of signals. As an example, consider the first-order, stochastic, discrete-time system

$$x(k+1) = 0.5x(k) + w(k)$$
$$z(k+1) = x(k+1) + v(k+1)$$

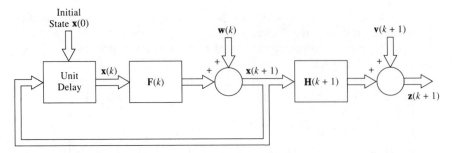

Figure 8-13 Block diagram of a discrete-time stochastic model.

With a probabilistic initial condition $x(0)$, a typical response of the system is plotted in Figure 8-14 for the zero-mean, unit variance, white noise sequences $w(k + 1)$ and $v(k + 1)$ shown.

8.5.2 Statement of the Problem

As noisy measurements are received in time sequence, it is desirable to estimate the state of the system using all measurements currently available. As the system state **x** changes at each time step, a new measurement **z** becomes available.

A formal statement of the problem is as follows:

> **For the stochastic system described by equations (8-37) and (8-38), it is desirable to find a linear minimum mean square (LMMS) estimate of the state x(k + 1), denoted by x̂(k + 1 | k + 1), which is *unbiased*[†], that is**
>
> $$E[\hat{x}(k + 1 \mid k + 1)] = E[x(k + 1)]$$
>
> **and such that the mean square error**
>
> $$J = E\{[x(k + 1) - \hat{x}(k + 1 \mid k + 1)]^{\dagger}[x(k + 1) - \hat{x}(k + 1 \mid k + 1)]\} \quad (8\text{-}41)$$
>
> **is minimized.**

If the estimate x̂(k + 1 | k + 1) is unbiased, the expected value of the estimation error is zero

$$E[x(k + 1) - \hat{x}(k + 1 \mid k + 1)] = 0$$

[†] If on the average (in the probabilistic sense) the state **x** and its estimate **x̂** are equal, then **x̂** is called an *unbiased* estimator of **x**. If $E[\hat{x}] \neq E[x]$, **x̂** is called a *biased* estimator.

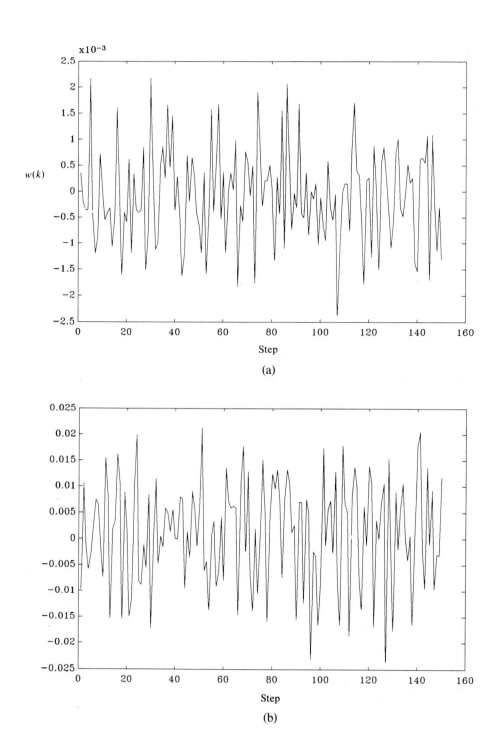

Figure 8-14 Noise input and response of an example of a stochastic discrete-time system. (a) Input noise. (b) Output noise. (c) System state. (d) System output.

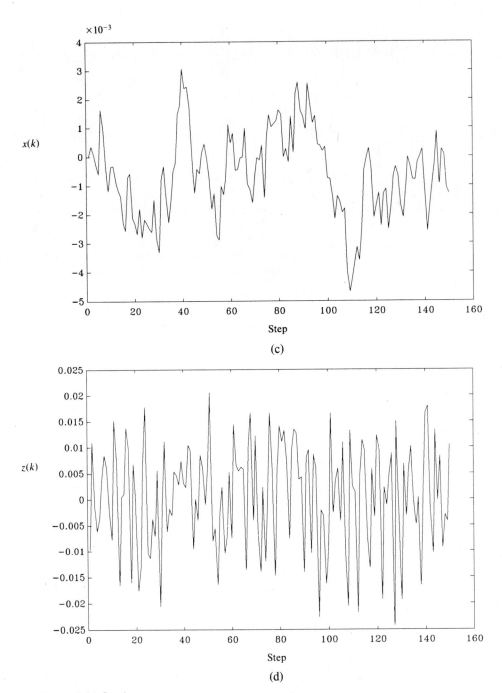

(c)

(d)

Figure 8-14 Cont.

and the mean square error is

$$E\{[\mathbf{x}(k+1) - \hat{\mathbf{x}}(k+1 \mid k+1)]^{\dagger}[\mathbf{x}(k+1) - \hat{\mathbf{x}}(k+1 \mid k+1)]\}$$

$$= E[\mathbf{a}^{\dagger}\mathbf{a}] = E\left\{[a_1 \quad a_2 \quad \cdots \quad a_n]\begin{bmatrix} a_1 \\ a_2 \\ \vdots \\ a_n \end{bmatrix}\right\} = E[a_1^2 + a_2^2 \cdots a_n^2]$$

which is the sum of the variances of the individual components in the estimator error. Thus, we refer to such an estimate as a *minimum variance unbiased estimate*. That is, if $\hat{\mathbf{x}}$ is unbiased, the LMMS estimate equals the minimum variance estimate.

Kalman solved the LMMS estimation problem during a boring train trip in late 1958, and published his results in 1960. As stated earlier, the filter estimate minimizes the expectation of the sum of squares of errors as given by equation (8-41). The Kalman filter for the nth-order, linear, discrete-time, stochastic system described by equations (8-37) and (8-38) is another nth-order, linear, discrete-time system. The filter input is the system measurement, and the filter output is the optimally estimated system state, as indicated in Figure 8-15. Our objective then is to derive Kalman filter equations that achieve the optimal system state estimate. This process is the subject of the next two sections.

8.5.3 Basic Results for Linear Minimum Mean Square Estimation

There are four results for LMMS estimation that are key to the development of the Kalman filter. Because of their importance we elected to include their derivations in Appendix C.

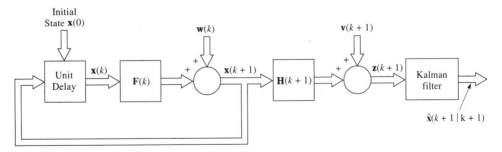

Figure 8-15 Kalman filter driven by the system output to estimate the system state optimally.

1. Minimum Mean Square Estimate

For the random vectors $\mathbf{x}$ and $\mathbf{z}$, the LMMS estimate of $\mathbf{x}$ given $\mathbf{z}$ is

$$\hat{\mathbf{x}} = E[(\mathbf{x} - \mathbf{m_x})(\mathbf{z} - \mathbf{m_z})^\dagger]\{E[(\mathbf{z} - \mathbf{m_z})(\mathbf{z} - \mathbf{m_z})^\dagger]\}^{-1}(\mathbf{z} - \mathbf{m_z}) + \mathbf{m_x}$$

(8-42)

where $\mathbf{m_x}$ is the mean of $\mathbf{x}$ and $\mathbf{m_z}$ is the mean of $\mathbf{z}$. Specifically, if $\mathbf{x}$ and $\mathbf{z}$ are zero-mean random vectors, then the LMMS estimate of $\mathbf{x}$ based on $\mathbf{z}$ is

$$\hat{\mathbf{x}} = E[\mathbf{xz}^\dagger]\{E[\mathbf{zz}^\dagger]\}^{-1}\mathbf{z}$$

2. Orthogonality of the Measurements and Estimation Error

This result, known as the *orthogonality principle,* states that if $\mathbf{x}$, $\mathbf{z}$, and $\hat{\mathbf{x}}$ satisfy equation (8-42), then the measurement vector $\mathbf{z} - \mathbf{m_z}$ is *orthogonal* to the estimation error $\mathbf{x} - \hat{\mathbf{x}}$, that is,

$$E[(\mathbf{z} - \mathbf{m_z})(\mathbf{x} - \hat{\mathbf{x}})^\dagger] = 0$$

(8-43)

3. Estimation of a Linear Composition

For the random vectors $\mathbf{x}$, $\mathbf{y}$, $\mathbf{w}$, and $\mathbf{z}$, if

$$\mathbf{x} = \mathbf{Ay} + \mathbf{Bw}$$

then the LMMS estimate of $\mathbf{x}$ based on $\mathbf{z}$ is given by

$$\hat{\mathbf{x}} = \mathbf{A\hat{y}} + \mathbf{B\hat{w}}$$

(8-44)

where $\hat{\mathbf{y}}$ is the LMMS estimate of $\mathbf{y}$ based on $\mathbf{z}$ and $\hat{\mathbf{w}}$ is the LMMS estimate of $\mathbf{w}$ based on $\mathbf{z}$.

4. Incorporation of Orthogonal Data

For the random vectors $\mathbf{x}$, $\mathbf{z}_1$, and $\mathbf{z}_2$, if $\mathbf{z}_1 - \mathbf{m}_{z1}$ and $\mathbf{z}_2 - \mathbf{m}_{z2}$ are orthogonal,

$$E[(\mathbf{z}_1 - \mathbf{m}_{z1})(\mathbf{z}_2 - \mathbf{m}_{z2})^\dagger] = 0$$

the LMMS estimate of $\mathbf{x}$ based on $\mathbf{z}_1$ and $\mathbf{z}_2$ is

$$\hat{\mathbf{x}} = \hat{\mathbf{x}}_1 + \hat{\mathbf{x}}_2 + \mathbf{m_x}$$

(8-45)

where $\hat{\mathbf{x}}_1$ is the estimate of $\mathbf{x}$ based on $\mathbf{z}_1 - \mathbf{m}_{z1}$ and $\hat{\mathbf{x}}_2$ is the estimate of $\mathbf{x}$ based on $\mathbf{z}_2 - \mathbf{m}_{z2}$.

8.6 The Discrete-Time Kalman Filter

The derivation of the Kalman filter is now presented. Before we proceed, however, we need to explain some of the notation and terminology to be used.

As in common practice, the estimate of a quantity such as $\mathbf{x}$ is denoted by a "hat" over the symbol for the quantity (i.e., $\hat{\mathbf{x}}$). The arguments for all estimates are written in the form $(i \mid k)$, where the first index i is the step of the quantity estimated, and the second index k is the index of the most recent measurement used in making the estimate. Unless otherwise specified, we assume that the first measurement is $\mathbf{z}(1)$. Some examples follow:

$\hat{\mathbf{x}}(k + 1 \mid k)$ is the LMMS estimate of $\mathbf{x}(k + 1)$ based on $\mathbf{z}(1), \mathbf{z}(2),$ $\ldots, \mathbf{z}(k)$

$\hat{\mathbf{x}}(k + 1 \mid k + 1)$ is the LMMS estimate of $\mathbf{x}(k + 1)$ based on $\mathbf{z}(1),$ $\mathbf{z}(2), \ldots, \mathbf{z}(k + 1)$

$\hat{\mathbf{z}}(k + 1 \mid k)$ is the LMMS estimate of $\mathbf{z}(k + 1)$ based on $\mathbf{z}(1), \mathbf{z}(2),$ $\ldots, \mathbf{z}(k)$

Similar definitions are used for estimates $\hat{\mathbf{w}}(k \mid k)$, $\hat{\mathbf{v}}(k + 1 \mid k)$, and so forth. Some other useful definitions follow:

$\boldsymbol{\Delta}\mathbf{x}(k + 1 \mid k) = \mathbf{x}(k + 1) - \hat{\mathbf{x}}(k + 1 \mid k)$ is the state prediction error.

$\boldsymbol{\Delta}\mathbf{x}(k + 1 \mid k + 1) = \mathbf{x}(k + 1) - \hat{\mathbf{x}}(k + 1 \mid k + 1)$ is the state estimation error.

$\boldsymbol{\Delta}\mathbf{z}(k + 1 \mid k) = \mathbf{z}(k + 1) - \hat{\mathbf{z}}(k + 1 \mid k)$ is the measurement prediction error.

$\mathbf{P}(k + 1 \mid k) = E[\boldsymbol{\Delta}\mathbf{x}(k + 1 \mid k)\,\boldsymbol{\Delta}\mathbf{x}^\dagger(k + 1 \mid k)]$ is the state prediction error covariance.

$\mathbf{P}(k + 1 \mid k + 1) = E[\boldsymbol{\Delta}\mathbf{x}(k + 1 \mid k + 1)\,\boldsymbol{\Delta}\mathbf{x}^\dagger(k + 1 \mid k + 1)]$ is the state estimation error covariance.

8.6.1 Prediction and Correction

Prediction

For the system in equation (8-37)

$$\mathbf{x}(k + 1) = \mathbf{F}(k)\mathbf{x}(k) + \mathbf{w}(k)$$

using the linear composition result shown in equation (8-44) results in the optimal estimate of $\mathbf{x}(k + 1)$ given data through the kth step

$$\hat{\mathbf{x}}(k + 1 \mid k) = \mathbf{F}(k)\hat{\mathbf{x}}(k \mid k) + \hat{\mathbf{w}}(k \mid k) \tag{8-46}$$

If equation (8-42) is used, the estimate $\hat{\mathbf{w}}(k \mid k)$ is

$$\hat{\mathbf{w}}(k \mid k) = E[\mathbf{w}(k)\mathbf{z}^\dagger(k)]\{E[\mathbf{z}(k)\mathbf{z}^\dagger(k)]\}^{-1}\mathbf{z}(k) \tag{8-47}$$

Because $\mathbf{w}(k)$ and $\mathbf{z}(i)$, $i = 1, 2, \ldots, k$ are uncorrelated, then

$$E[\mathbf{w}(k)\mathbf{z}^\dagger(k)] = \mathbf{0}$$

Therefore, equation (8-47) gives

$$\hat{\mathbf{w}}(k \mid k) = \mathbf{0}$$

and equation (8-46) reduces to

$$\hat{\mathbf{x}}(k + 1 \mid k) = \mathbf{F}(k)\hat{\mathbf{x}}(k \mid k) \tag{8-48}$$

where $\hat{\mathbf{x}}(0 \mid 0) = E[\mathbf{x}(0)] = \mathbf{0}$. This is to say that the best prediction of the state at the next step is to pass the estimate from the previous step through the system state coupling matrix $\mathbf{F}$.

In a similar fashion, applying the linear composition result in equation (8-44) to the stochastic system output

$$\mathbf{z}(k + 1) = \mathbf{H}(k + 1)\mathbf{x}(k + 1) + \mathbf{v}(k + 1)$$

gives

$$\hat{\mathbf{z}}(k + 1 \mid k) = \mathbf{H}(k + 1)\hat{\mathbf{x}}(k + 1 \mid k) + \hat{\mathbf{v}}(k + 1 \mid k)$$

Because $\mathbf{v}(k)$ and $\mathbf{z}(i)$ are uncorrelated for $k \neq i$,

$$\hat{\mathbf{v}}(k + 1 \mid k) = \mathbf{0}$$

and, therefore,

$$\hat{\mathbf{z}}(k + 1 \mid k) = \mathbf{H}(k + 1)\hat{\mathbf{x}}(k + 1 \mid k) \tag{8-49}$$

indicating that the best prediction of the next measurement is to pass the predicted state through the measurement coupling matrix $\mathbf{H}$.

Correction

We now proceed to derive the corrector equations. The measurement prediction errors

$$\Delta\mathbf{z}(k + 1 \mid k) = \mathbf{z}(k + 1) - \hat{\mathbf{z}}(k + 1 \mid k) \tag{8-50}$$

are also termed the *measurement residuals*, or innovations. Rather than using the original measurements $\mathbf{z}(1), \mathbf{z}(2), \ldots, \mathbf{z}(k), \ldots$, it is expedient to use the measurement residuals $\Delta\mathbf{z}(1 \mid 0), \Delta\mathbf{z}(2 \mid 1), \ldots,$ $\Delta\mathbf{z}(k \mid k - 1), \ldots$ as the measurements. The two are equivalent because either may be found deterministically from the other. Collecting the residuals through step k into a single vector of measurements

$$\Delta\mathbf{z}_k = \begin{bmatrix} \Delta\mathbf{z}(1 \mid 0) \\ \Delta\mathbf{z}(2 \mid 1) \\ \vdots \\ \Delta\mathbf{z}(k \mid k - 1) \end{bmatrix}$$

the quantity $\hat{\mathbf{x}}(k + 1 \mid k)$ then denotes the LMMS estimate of $\mathbf{x}(k + 1)$ based on $\Delta\mathbf{z}_k$.

Using the orthogonality principle in equation (8-43), the measurement residuals and the estimation error are orthogonal that is

$$E\{\Delta\mathbf{z}_k[\mathbf{x}(k + 1) - \hat{\mathbf{x}}(k + 1 \mid k)]^\dagger\} = \mathbf{0} \tag{8-51}$$

Postmultiplying both sides of equation (8-51) by $\mathbf{H}^\dagger(k + 1)$ gives

$$E\{\Delta\mathbf{z}_k[\mathbf{x}(k + 1) - \hat{\mathbf{x}}(k + 1 \mid k)]^\dagger\}\mathbf{H}^\dagger(k + 1) = \mathbf{0}$$

Because $\Delta\mathbf{z}_k$ and $\mathbf{v}(k + 1)$ are uncorrelated

$$E\{\Delta\mathbf{z}_k[\mathbf{z}(k + 1) - \hat{\mathbf{z}}(k + 1 \mid k)]^\dagger\} = E[\Delta\mathbf{z}_k \, \Delta\mathbf{z}^\dagger(k + 1 \mid k)] = \mathbf{0}$$

Because the collection of measurements $\Delta\mathbf{z}_k$ through step k and the measurements $\Delta\mathbf{z}(k + 1 \mid k)$ at step $k + 1$ are orthogonal, any LMMS estimates based on $\Delta\mathbf{z}_k$ and $\Delta\mathbf{z}(k + 1 \mid k)$ are, according to equation (8-45), the sum of the two individual estimates

$$\hat{\mathbf{x}}(k + 1 \mid k + 1) = \hat{\mathbf{x}}(k + 1 \mid k) + E[\mathbf{x}(k + 1) \mid \Delta\mathbf{z}(k + 1 \mid k)] \tag{8-52}$$

which is an expression of result 4, the incorporation of orthogonal data where we define $E[\mathbf{x}(k + 1) \mid \Delta\mathbf{z}(k + 1) \mid k)]$ as the best estimate of $\mathbf{x}(k + 1)$ based on $\Delta\mathbf{z}(k + 1 \mid k)$. The incorporation of new data in the form of the residuals only involves making additive corrections to the previous predictions, not complete recalculations.

Using result 1, for the minimum mean square estimate, then

$$E[\mathbf{x}(k + 1) \mid \Delta\mathbf{z}(k + 1 \mid k)] = E[\mathbf{x}(k + 1) \, \Delta\mathbf{z}^\dagger(k + 1 \mid k)]$$
$$\times \{E[\Delta\mathbf{z}(k + 1 \mid k) \, \Delta\mathbf{z}^\dagger(k + 1 \mid k)]\}^{-1} \, \Delta\mathbf{z}(k + 1 \mid k)$$

If we define the *Kalman gain* $\mathbf{K}(k + 1)$ as

$$\mathbf{K}(k + 1)$$
$$= E[\mathbf{x}(k + 1) \, \Delta\mathbf{z}^\dagger(k + 1 \mid k)]\{E[\Delta\mathbf{z}(k + 1 \mid k) \, \Delta\mathbf{z}^\dagger(k + 1 \mid k)]\}^{-1}$$

then equation (8-52) becomes

$$\hat{\mathbf{x}}(k + 1 \mid k + 1) = \hat{\mathbf{x}}(k + 1 \mid k) + \mathbf{K}(k + 1) \, \Delta\mathbf{z}(k + 1 \mid k) \tag{8-53}$$

8.6.2 Kalman Gain and Error Covariances

Kalman Gain

Finding an expression for the recursive calculations of the Kalman gain sequence $\mathbf{K}(1)$, $\mathbf{K}(2)$, $\mathbf{K}(3)$, . . . , is the most involved part of Kalman filtering. As we shall soon discover, the solution consists of a set of three recursive equations with coupled matrices, from which the Kalman gains can be computed.

Substituting equation (8-38) into the measurement residual in equation (8-50) gives

$$
\begin{aligned}
\mathbf{\Delta z}(k + 1 \mid k) &= \mathbf{z}(k + 1) - \hat{\mathbf{z}}(k + 1 \mid k) \\
&= \mathbf{H}(k + 1)\mathbf{x}(k + 1) + \mathbf{v}(k + 1) \\
&\quad - \mathbf{H}(k + 1)\hat{\mathbf{x}}(k + 1 \mid k) \\
&= \mathbf{H}(k + 1)\,\mathbf{\Delta x}(k + 1 \mid k) + \mathbf{v}(k + 1) \qquad \text{(8-54)}
\end{aligned}
$$

From equation (8-54), we get

$$
\begin{aligned}
E[\mathbf{\Delta z}(k &+ 1 \mid k)\,\mathbf{\Delta z}^\dagger(k + 1 \mid k)] \\
&= \mathbf{H}(k + 1)E[\mathbf{\Delta x}(k + 1 \mid k)\,\mathbf{\Delta x}^\dagger(k + 1 \mid k)] \\
&\quad \times \mathbf{H}^\dagger(k + 1) + \mathbf{H}(k + 1)E[\mathbf{\Delta x}(k + 1 \mid k)\mathbf{v}^\dagger(k + 1)] \\
&\quad + E[\mathbf{v}(k + 1)\,\mathbf{\Delta x}^\dagger(k + 1 \mid k)]\mathbf{H}^\dagger(k + 1) \\
&\quad + E[\mathbf{v}(k + 1)\mathbf{v}^\dagger(k + 1)] \qquad \text{(8-55)}
\end{aligned}
$$

Because $\mathbf{v}(k + 1)$ and $\mathbf{\Delta x}(k + 1 \mid k)$ are uncorrelated

$$
E[\mathbf{\Delta x}(k + 1 \mid k)\mathbf{v}^\dagger(k + 1)] = E[\mathbf{v}(k + 1)\,\mathbf{\Delta x}^\dagger(k + 1 \mid k)]^\dagger = \mathbf{0}
$$

Using the definition of the state prediction error covariance gives

$$
\mathbf{P}(k + 1 \mid k) = E[\mathbf{\Delta x}(k + 1 \mid k)\,\mathbf{\Delta x}^\dagger(k + 1 \mid k)]
$$

and equation (8-55) becomes

$$
\begin{aligned}
E[\mathbf{\Delta z}(k + 1 \mid k)\,&\mathbf{\Delta z}^\dagger(k + 1 \mid k)] \\
&= \mathbf{H}(k + 1)\mathbf{P}(k + 1 \mid k)\mathbf{H}^\dagger(k + 1) + \mathbf{R}(k + 1)
\end{aligned}
$$

Similarly, using the definition of the state prediction error results in

$$
\begin{aligned}
E[\mathbf{x}(k + 1)\,&\mathbf{\Delta z}^\dagger(k + 1 \mid k)] \\
&= E\{[\mathbf{\Delta x}(k + 1 \mid k) + \hat{\mathbf{x}}(k + 1 \mid k)][\mathbf{\Delta z}^\dagger(k + 1 \mid k)]\} \\
&= E[\mathbf{\Delta x}(k + 1 \mid k)\,\mathbf{\Delta z}^\dagger(k + 1 \mid k)] + E[\hat{\mathbf{x}}(k + 1 \mid k)\,\mathbf{\Delta z}^\dagger(k + 1 \mid k)]
\end{aligned}
$$

Because $\mathbf{v}(k + 1)$ is uncorrelated with $\hat{\mathbf{x}}(k + 1 \mid k)$ and because the estimate $\hat{\mathbf{x}}(k + 1 \mid k)$ and the estimation error $\mathbf{\Delta x}(k + 1 \mid k)$ are orthogonal,

$$
E[\hat{\mathbf{x}}(k + 1 \mid k)\,\mathbf{\Delta z}^\dagger(k + 1 \mid k)] = \mathbf{0}
$$

Therefore,

$$E[\mathbf{x}(k + 1) \, \Delta\mathbf{z}^\dagger(k + 1 \mid k)] = E[\Delta\mathbf{x}(k + 1 \mid k) \, \Delta\mathbf{z}^\dagger(k + 1 \mid k)]$$

Using equation (8-54) results in

$$\Delta\mathbf{z}(k + 1 \mid k) = \mathbf{H}(k + 1) \, \Delta\mathbf{x}(k + 1 \mid k) + \mathbf{v}(k + 1)$$

and

$$\begin{aligned}
E[\mathbf{x}(k + 1) \, \Delta\mathbf{z}^\dagger(k + 1 \mid k)] \\
&= E\{\Delta\mathbf{x}(k + 1 \mid k)[\mathbf{H}(k + 1) \, \Delta\mathbf{x}(k + 1 \mid k) + \mathbf{v}(k + 1)]^\dagger\} \\
&= E\{[\Delta\mathbf{x}(k + 1 \mid k) \, \Delta\mathbf{x}^\dagger(k + 1 \mid k)\mathbf{H}^\dagger(k + 1)] \\
&\quad + [\Delta\mathbf{x}(k + 1 \mid k)\mathbf{v}^\dagger(k + 1)]\}
\end{aligned}$$

But $\mathbf{v}(k + 1)$ and $\Delta\mathbf{x}(k + 1 \mid k)$ are uncorrelated, and, therefore,

$$E[\Delta\mathbf{x}(k + 1 \mid k)\mathbf{v}^\dagger(k + 1)] = \mathbf{0}$$

Thus,

$$\begin{aligned}
E[\mathbf{x}(k + 1) \, \Delta\mathbf{z}^\dagger(k + 1 \mid k)] \\
&= E[\Delta\mathbf{x}(k + 1 \mid k) \, \Delta\mathbf{x}^\dagger(k + 1 \mid k)]\mathbf{H}^\dagger(k + 1) \\
&= \mathbf{P}(k + 1 \mid k)\mathbf{H}^\dagger(k + 1)
\end{aligned}$$

and, therefore, the Kalman gain is

$$\begin{aligned}
\mathbf{K}(k + 1) &= E[\mathbf{x}(k + 1) \, \Delta\mathbf{z}^\dagger(k + 1)]\{E[\Delta\mathbf{z}(k + 1)\Delta\mathbf{z}^\dagger(k + 1)]\}^{-1} \\
&= \mathbf{P}(k + 1 \mid k)\mathbf{H}^\dagger(k + 1)[\mathbf{H}(k + 1)\mathbf{P}(k + 1 \mid k)\mathbf{H}^\dagger(k + 1) \\
&\quad + \mathbf{R}(k + 1)]^{-1} \tag{8-56}
\end{aligned}$$

Error Covariances

If we use

$$\mathbf{x}(k + 1) = \mathbf{F}(k)\mathbf{x}(k) + \mathbf{w}(k)$$

and

$$\hat{\mathbf{x}}(k + 1 \mid k) = \mathbf{F}(k)\hat{\mathbf{x}}(k \mid k)$$

then the state prediction error is

$$\begin{aligned}
\Delta\mathbf{x}(k + 1 \mid k) &= \mathbf{x}(k + 1) - \hat{\mathbf{x}}(k + 1 \mid k) \\
&= \mathbf{F}(k) \, \Delta\mathbf{x}(k \mid k) + \mathbf{w}(k)
\end{aligned}$$

and the state prediction error covariance is

$$\begin{aligned}
\mathbf{P}(k + 1 \mid k) &= E[\Delta\mathbf{x}(k + 1 \mid k) \, \Delta\mathbf{x}^\dagger(k + 1 \mid k)] \\
&= E\{[\mathbf{F}(k) \, \Delta\mathbf{x}(k \mid k) + \mathbf{w}(k)][\mathbf{F}(k) \, \Delta\mathbf{x}(k \mid k) + \mathbf{w}(k)]^\dagger\}
\end{aligned}$$

$$= \mathbf{F}(k)E[\Delta\mathbf{x}(k \mid k) \, \Delta\mathbf{x}^\dagger(k \mid k)]\mathbf{F}^\dagger(k)$$
$$+ \mathbf{F}(k)E[\Delta\mathbf{x}(k \mid k)\mathbf{w}^\dagger(k)]$$
$$+ E[\mathbf{w}(k) \, \Delta\mathbf{x}^\dagger(k \mid k)]\mathbf{F}^\dagger(k) + E[\mathbf{w}(k)\mathbf{w}^\dagger(k)]$$

Because $\Delta\mathbf{x}(k \mid k)$ and $\mathbf{w}(k)$ are uncorrelated,

$$E[\Delta\mathbf{x}(k \mid k)\mathbf{w}^\dagger(k)] = E[\mathbf{w}(k) \, \Delta\mathbf{x}^\dagger(k \mid k)] = \mathbf{0}$$

giving

$$\mathbf{P}(k + 1 \mid k) = \mathbf{F}(k)\mathbf{P}(k \mid k)\mathbf{F}^\dagger(k) + \mathbf{Q}(k) \quad \mathbf{P}(0 \mid 0) = \mathbf{P}(0) \qquad \textbf{(8-57)}$$

where $\mathbf{P}(k \mid k)$ is the estimation error covariance.

Finally, for the equations to be recursive, we need an equation for the covariance of the state estimate error $\mathbf{P}(k + 1 \mid k + 1)$.

If we use equation (8-53)

$$\hat{\mathbf{x}}(k + 1 \mid k + 1) = \hat{\mathbf{x}}(k + 1 \mid k) + \mathbf{K}(k + 1) \, \Delta\mathbf{z}(k + 1 \mid k)$$

the state estimate error becomes

$$\Delta\mathbf{x}(k + 1 \mid k + 1) = \mathbf{x}(k + 1) - \hat{\mathbf{x}}(k + 1 \mid k + 1)$$
$$= \Delta\mathbf{x}(k + 1 \mid k) - \mathbf{K}(k + 1) \, \Delta\mathbf{z}(k + 1 \mid k) \qquad \textbf{(8-58)}$$

Substituting the measurement prediction error in equation (8-54)

$$\Delta\mathbf{z}(k + 1 \mid k) = \mathbf{H}(k + 1) \, \Delta\mathbf{x}(k + 1 \mid k) + \mathbf{v}(k + 1)$$

into equation (8-58) gives

$$\Delta\mathbf{x}(k + 1 \mid k + 1) = \Delta\mathbf{x}(k + 1 \mid k) - \mathbf{K}(k + 1)\mathbf{H}(k + 1) \, \Delta\mathbf{x}(k + 1 \mid k)$$
$$- \mathbf{K}(k + 1)\mathbf{v}(k + 1)$$
$$= [\mathbf{I} - \mathbf{K}(k + 1)\mathbf{H}(k + 1)] \, \Delta\mathbf{x}(k + 1 \mid k)$$
$$- \mathbf{K}(k + 1)\mathbf{v}(k + 1)$$

Hence,

$$\mathbf{P}(k + 1 \mid k + 1)$$
$$= E[\Delta\mathbf{x}(k + 1 \mid k + 1) \, \Delta\mathbf{x}^\dagger(k + 1 \mid k + 1)]$$
$$= [\mathbf{I} - \mathbf{K}(k + 1)\mathbf{H}(k + 1)]E[\Delta\mathbf{x}(k + 1 \mid k) \, \Delta\mathbf{x}^\dagger(k + 1 \mid k)]$$
$$\times [\mathbf{I} - \mathbf{K}(k + 1)\mathbf{H}(k + 1)]^\dagger - [\mathbf{I} - \mathbf{K}(k + 1)\mathbf{H}(k + 1)]$$
$$\times E[\Delta\mathbf{x}(k + 1 \mid k)\mathbf{v}^\dagger(k + 1)]\mathbf{K}^\dagger(k + 1) - \mathbf{K}(k + 1)$$
$$\times E[\mathbf{v}(k + 1) \, \Delta\mathbf{x}^\dagger(k + 1 \mid k)][\mathbf{I} - \mathbf{K}(k + 1)\mathbf{H}(k + 1)]^\dagger$$
$$+ \mathbf{K}(k + 1)E[\mathbf{v}(k + 1)\mathbf{v}^\dagger(k + 1)]\mathbf{K}^\dagger(k + 1)$$

Because $\Delta\mathbf{x}(k + 1 \mid k)$ and $\mathbf{v}(k + 1)$ are uncorrelated

$$E[\mathbf{v}(k + 1) \, \Delta\mathbf{x}^\dagger(k + 1 \mid k)] = E[\Delta\mathbf{x}(k + 1 \mid k)\mathbf{v}^\dagger(k + 1)]^\dagger = \mathbf{0}$$

giving

$$\mathbf{P}(k + 1 \mid k + 1) = [\mathbf{I} - \mathbf{K}(k + 1)\mathbf{H}(k + 1)]\mathbf{P}(k + 1 \mid k)$$
$$\times [\mathbf{I} - \mathbf{K}(k + 1)\mathbf{H}(k + 1)]^\dagger + \mathbf{K}(k + 1)\mathbf{R}(k + 1)\mathbf{K}^\dagger(k + 1) \quad \textbf{(8-59)}$$

Equation (8-59) may be put into a simpler form as follows:

$$\mathbf{P}(k + 1 \mid k + 1) = [\mathbf{I} - \mathbf{K}(k + 1)\mathbf{H}(k + 1)]\mathbf{P}(k + 1 \mid k)$$
$$- \mathbf{P}(k + 1 \mid k)\mathbf{H}^\dagger(k + 1)$$
$$\times \mathbf{K}^\dagger(k + 1) + \mathbf{K}(k + 1)[\mathbf{H}(k + 1)$$
$$\times \mathbf{P}(k + 1 \mid k)\mathbf{H}^\dagger(k + 1) + \mathbf{R}(k + 1)]\mathbf{K}^\dagger(k + 1)$$

But from equation (8-56) we get

$$\mathbf{K}(k + 1)[\mathbf{H}(k + 1)\mathbf{P}(k + 1 \mid k)\mathbf{H}^\dagger(k + 1) + \mathbf{R}(k + 1)]$$
$$= \mathbf{P}(k + 1 \mid k)\mathbf{H}^\dagger(k + 1)$$

Therefore, equation (8-59) simplifies to

$$\mathbf{P}(k + 1 \mid k + 1) = [\mathbf{I} - \mathbf{K}(k + 1)\mathbf{H}(k + 1)]\mathbf{P}(k + 1 \mid k) \quad \textbf{(8-60)}$$

and the Kalman filter is completely derived.

8.6.3 The Basic Kalman Filter

The Kalman filter equations derived in the previous section are collected in Table 8–3. The filter consists of a model of the system in equations (8-37) and (8-38) with zero noise inputs replacing the actual unknown system noise inputs and initial conditions, as shown in Figure 8-16. The filter shown processes the measurements and produces the LMMS estimate of the system state based on the measurements from step 1 through the present step $k + 1$.

Figure 8-17 is a block diagram showing how the computations of the Kalman gain and covariance matrices are performed beginning with covariance of the initial state estimation error $\mathbf{P}(0 \mid 0)$.

When the Kalman filter for a linear time-invariant system is processed in real time, the Kalman gains, $\mathbf{K}(1)$, $\mathbf{K}(2)$, . . . and the covariance matrices may be calculated off-line and stored in advance. The off-line calculations involve repetitive cycles of three computations:

1. The state estimation error covariance $\mathbf{P}(k \mid k)$, equation (8-60), beginning with the given initial covariance $\mathbf{P}(0 \mid 0) = \mathbf{P}(0)$
2. The state prediction error covariance $\mathbf{P}(k + 1 \mid k)$, equation (8-57)
3. The Kalman gain, equation (8-56)

Table 8-3 Kalman Filter Equations

Plant Model

$\mathbf{x}(k + 1) = \mathbf{F}(k)\mathbf{x}(k) + \mathbf{w}(k)$

$\mathbf{z}(k + 1) = \mathbf{H}(k + 1)\mathbf{x}(k + 1) + \mathbf{v}(k + 1)$

Prediction

$\hat{\mathbf{x}}(k + 1 \mid k) = \mathbf{F}(k)\hat{\mathbf{x}}(k \mid k) \quad \hat{\mathbf{x}}(0 \mid 0) = \mathbf{0}$

$\hat{\mathbf{z}}(k + 1 \mid k) = \mathbf{H}(k + 1)\hat{\mathbf{x}}(k + 1 \mid k)$

Correction

$\hat{\mathbf{x}}(k + 1 \mid k + 1) = \hat{\mathbf{x}}(k + 1 \mid k) + \mathbf{K}(k + 1) \, \Delta\mathbf{z}(k + 1 \mid k)$

$\Delta\mathbf{z}(k + 1 \mid k) = \mathbf{z}(k + 1) - \hat{\mathbf{z}}(k + 1 \mid k)$

Kalman Filter Gain

$\mathbf{K}(k + 1) = \mathbf{P}(k + 1 \mid k)\mathbf{H}^{\dagger}(k + 1)[\mathbf{H}(k + 1)\mathbf{P}(k + 1 \mid k)\mathbf{H}^{\dagger}(k + 1)$
$$+ \mathbf{R}(k + 1)]^{-1}$$

Covariances

$\mathbf{P}(k + 1 \mid k) = \mathbf{F}(k)\mathbf{P}(k \mid k)\mathbf{F}^{\dagger}(k) + \mathbf{Q}(k) \quad \mathbf{P}(0 \mid 0) = \mathbf{P}(0)$

$\mathbf{P}(k + 1 \mid k + 1) = [\mathbf{I} - \mathbf{K}(k + 1)\mathbf{H}(k + 1)]\mathbf{P}(k + 1 \mid k)$

These calculations are processed recursively as follows:

First Cycle $\begin{cases} \mathbf{P}(0 \mid 0) = \mathbf{P}(0) \\ \mathbf{P}(1 \mid 0) = \mathbf{F}(0)\mathbf{P}(0 \mid 0)\mathbf{F}^{\dagger}(0) + \mathbf{Q}(0) \\ \mathbf{K}(1) \quad = \mathbf{P}(1 \mid 0)\mathbf{H}^{\dagger}(1) \, [\mathbf{H}(1)\mathbf{P}(1 \mid 0)\mathbf{H}^{\dagger}(1) + \\ \qquad\qquad \mathbf{R}(1)]^{-1} \end{cases}$

$$(8\text{-}61)$$

Second Cycle $\begin{cases} \mathbf{P}(1 \mid 1) = [\mathbf{I} - \mathbf{K}(1)\mathbf{H}(1)]\mathbf{P}(1 \mid 0) \\ \mathbf{P}(2 \mid 1) = \mathbf{F}(1)\mathbf{P}(1 \mid 1)\mathbf{F}^{\dagger}(1) + \mathbf{Q}(1) \\ \mathbf{K}(2) \quad = \mathbf{P}(2 \mid 1)\mathbf{H}^{\dagger}(2) \, [\mathbf{H}(2)\mathbf{P}(2 \mid 1)\mathbf{H}^{\dagger}(2) + \\ \qquad\qquad \mathbf{R}(2)]^{-1} \end{cases}$

$$(8\text{-}62)$$

Third Cycle $\begin{cases} \mathbf{P}(2 \mid 2) = [\mathbf{I} - \mathbf{K}(2)\mathbf{H}(2)]\mathbf{P}(2 \mid 1) \\ \mathbf{P}(3 \mid 2) = \mathbf{F}(2)\mathbf{P}(2 \mid 2)\mathbf{F}^{\dagger}(2) + \mathbf{Q}(2) \\ \mathbf{K}(3) \quad = \mathbf{P}(3 \mid 2)\mathbf{H}^{\dagger}(3) \, [\mathbf{H}(3)\mathbf{P}(3 \mid 2)\mathbf{H}^{\dagger}(3) + \\ \qquad\qquad \mathbf{R}(3)]^{-1} \end{cases}$

$$(8\text{-}63)$$

and so on.

Note that even if the linear, discrete-time, stochastic system described by equations (8-37) and (8-38) varies with step so that the state

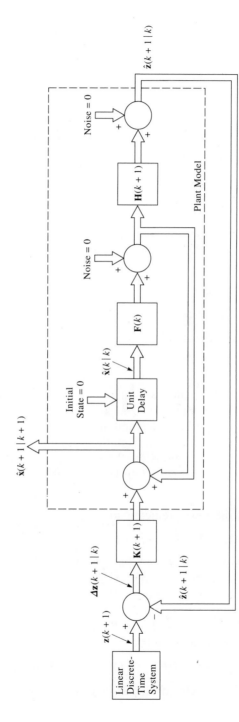

Figure 8-16 Kalman filter block diagram.

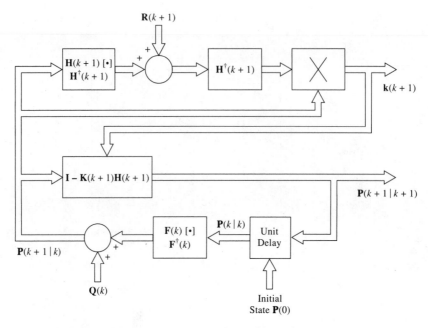

Figure 8-17 Computation of Kalman gain and covariances.

coupling matrix $\mathbf{F}(k)$ and the output coupling matrix $\mathbf{H}(k + 1)$ are not constants, or if the statistics $\mathbf{Q}(k)$ and $\mathbf{R}(k + 1)$ vary with step, then these quantities can be calculated and stored in advance rather than being calculated at each step from the formulas.

The on-line calculations for the basic Kalman filter consist of the predictor and corrector equations listed in Table 8-3. The equations are processed as follows:

1. Compute the next predicted state

$$\hat{\mathbf{x}}(k + 1 \mid k) = \mathbf{F}(k)\hat{\mathbf{x}}(k \mid k)$$

   beginning with $\hat{\mathbf{x}}(0 \mid 0) = E[\mathbf{x}(0)] = \mathbf{0}$
2. Read the next measurement data $\mathbf{z}(k + 1)$
3. Compute the measurement residuals

$$\Delta\mathbf{z}(k + 1 \mid k) = \mathbf{z}(k + 1) - \mathbf{H}(k + 1)\hat{\mathbf{x}}(k + 1 \mid k)$$

4. Compute the state estimate

$$\hat{\mathbf{x}}(k + 1 \mid k + 1) = \hat{\mathbf{x}}(k + 1 \mid k) + \mathbf{K}(k + 1)\,\Delta\mathbf{z}(k + 1 \mid k)$$

   The first few cycles of the on-line calculations are as follows:

$$\text{First Cycle}\begin{cases} \hat{\mathbf{x}}(1 \mid 0) = \mathbf{F}(0)\hat{\mathbf{x}}(0 \mid 0) = \mathbf{0} \\ \text{Read } \mathbf{z}(1) \\ \mathbf{\Delta z}(1 \mid 0) = \mathbf{z}(1) - \mathbf{H}(1)\hat{\mathbf{x}}(1 \mid 0) = \mathbf{z}(1) \\ \hat{\mathbf{x}}(1 \mid 1) = \hat{\mathbf{x}}(1 \mid 0) + \mathbf{K}(1)\,\mathbf{\Delta z}(1 \mid 0) = \\ \qquad\qquad \mathbf{K}(1)\,\mathbf{\Delta z}(1 \mid 0) \end{cases} \quad \text{(8-64)}$$

$$\text{Second Cycle}\begin{cases} \hat{\mathbf{x}}(2 \mid 1) \quad = \mathbf{F}(1)\hat{\mathbf{x}}(1 \mid 1) \\ \text{Read } \mathbf{z}(2) \\ \mathbf{\Delta z}(2 \mid 1) \quad = \mathbf{z}(2) - \mathbf{H}(2)\hat{\mathbf{x}}(2 \mid 1) \\ \hat{\mathbf{x}}(2 \mid 2) \quad = \hat{\mathbf{x}}(2 \mid 1) + \mathbf{K}(2)\,\mathbf{\Delta z}(2 \mid 1) \end{cases} \quad \text{(8-65)}$$

$$\text{Third Cycle}\begin{cases} \hat{\mathbf{x}}(3 \mid 2) \quad = \mathbf{F}(2)\hat{\mathbf{x}}(2 \mid 2) \\ \text{Read } \mathbf{z}(3) \\ \mathbf{\Delta z}(3 \mid 2) \quad = \mathbf{z}(3) - \mathbf{H}(3)\hat{\mathbf{x}}(3 \mid 2) \\ \hat{\mathbf{x}}(3 \mid 3) \quad = \hat{\mathbf{x}}(3 \mid 2) + \mathbf{K}(3)\,\mathbf{\Delta z}(3 \mid 2) \end{cases} \quad \text{(8-66)}$$

and so on.

As a numerical example, consider the first-order system

$$x(k + 1) = 0.9x(k) + w(k)$$

$$z(k + 1) = 4x(k + 1) + v(k + 1)$$

with

$$E[x^2(0)] = P(0) = 9 \times 10^{-4}$$

$$E[w^2(k)] = Q = 10^{-4} \quad \text{for all } k$$

$$E[v^2(k)] = R = 9 \times 10^{-4} \quad \text{for all } k$$

A block diagram of the system is shown in Figure 8-18(a). The sequence of Kalman gains are determined as follows:

$$\text{First Cycle}\begin{cases} P(0 \mid 0) = P(0) = 9 \times 10^{-4} \\ P(1 \mid 0) = 0.81P(0 \mid 0) + 10^{-4} = 8.29 \times 10^{-4} \\ K(1) \quad = \dfrac{4P(1 \mid 0)}{[16P(1 \mid 0) + 9 \times 10^{-4}]} = 0.234114 \end{cases} \quad \text{(8-67)}$$

$$\text{Second Cycle}\begin{cases} P(1 \mid 1) = [1 - 4K(1)]P(1 \mid 0) = 5.2678 \times 10^{-5} \\ P(2 \mid 1) = 0.81P(1 \mid 1) + 10^{-4} = 1.42669 \times 10^{-4} \\ K(2) \quad = \dfrac{4P(2 \mid 1)}{[16P(2 \mid 1) + 9 \times 10^{-4}]} = 0.179305 \end{cases} \quad \text{(8-68)}$$

and so on.

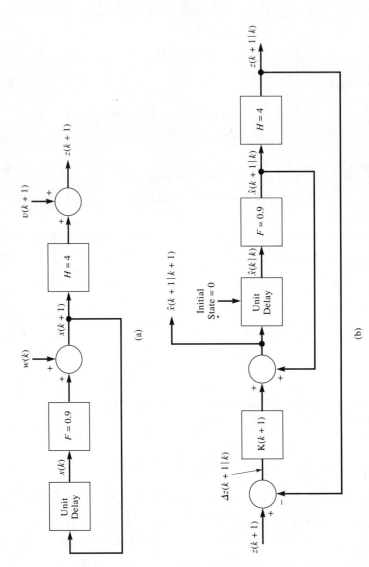

Figure 8-18 A first-order stochastic system and the corresponding Kalman filter. (a) Stochastic system model for the example system. (b) Kalman filter block diagram for the example system.

Computer-generated calculations of the Kalman gain and covariances are given in Table 8-4. The Kalman filter predictor equations for the system are

$$\hat{x}(k + 1 \mid k) = 0.9\hat{x}(k \mid k) \quad \hat{x}(0 \mid 0) = E[x(0)] = 0$$

$$\hat{z}(k + 1 \mid k) = 4\hat{x}(k + 1 \mid k)$$

and the corrector equations are

$$\hat{x}(k + 1 \mid k + 1) = \hat{x}(k + 1 \mid k) + K(k + 1)\,\Delta z(k + 1)$$

$$\Delta z(k + 1) = z(k + 1) - \hat{z}(k + 1 \mid k)$$

A block diagram of the filter is shown in Figure 8-18(b). For sequences $w(k)$ and $v(k + 1)$ with zero mean, unit variance, and Gaussian white noise, the corresponding actual system state $x(k)$, and the Kalman filter estimates of the system state $\hat{x}(k \mid k)$ are plotted in Figure 8-19.

Table 8-4* Covariance and Kalman Gain for the Example System

Step k	$P(k \mid k)$ $(\times\ 10^{-5})$	$P(k + 1 \mid k)$ $(\times\ 10^{-4})$	$K(k + 1)$
0	90	8.29	0.234114656
1	5.267579779	1.426673962	0.179304825
2	4.034358577	1.326783044	0.175567002
3	3.950257558	1.319970862	0.175297648
4	3.944197091	1.319479964	0.175278163
5	3.943758669	1.319444452	0.175276753
6	3.943726944	1.319441882	0.175276651
7	3.943724648	1.319441696	0.175276643
8	3.943724482	1.319441683	0.175276643
9	3.943724470	1.319441682	0.175276643
10			
.			
.			
.			

*In the table, the values of the covariances and the Kalman gain do not change beyond step 9.

8.6.4 Kalman Filter Calculation Programs

A computer program in BASIC, called KALMAN, is listed in Table 8-5. The program generates first-order stochastic system states and measurements, then Kalman filters the measurements to estimate the

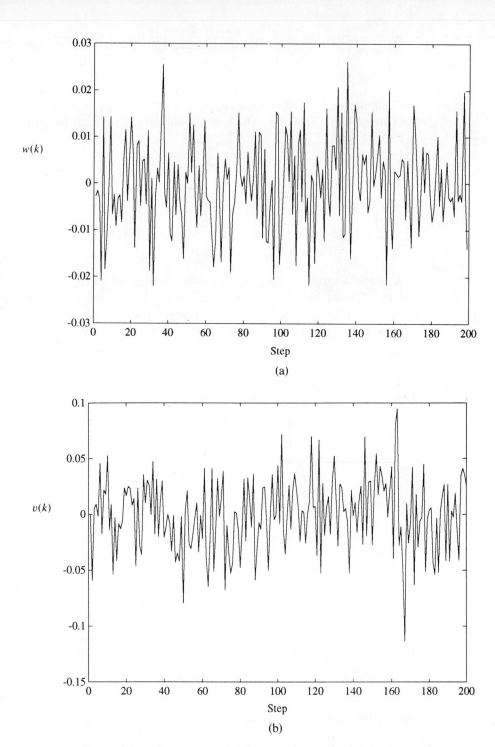

Figure 8-19 System and Kalman filter responses of the example system. (a) White noise input $w(k)$. (b) Measurement noise $v(k)$. (c) System state. (d) State estimate produced by the Kalman filter $x_e(k)$. (e) State error $e(k) = x(k) - x_p(k)$. (f) Measurement residuals $z - z_p$.

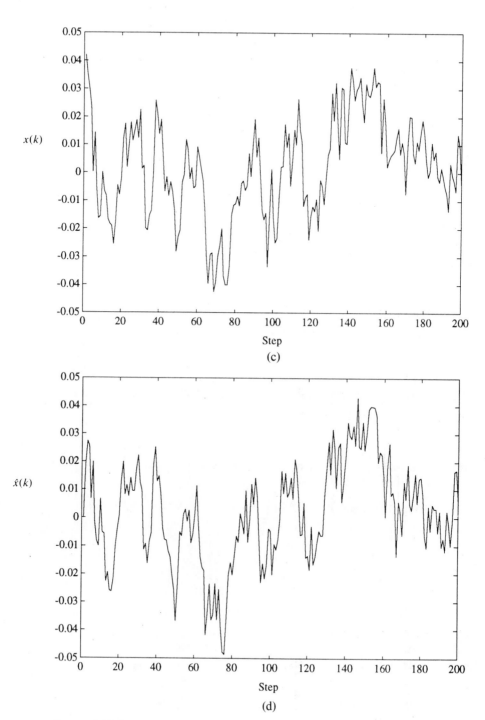

(c)

(d)

Figure 8-19 Cont.

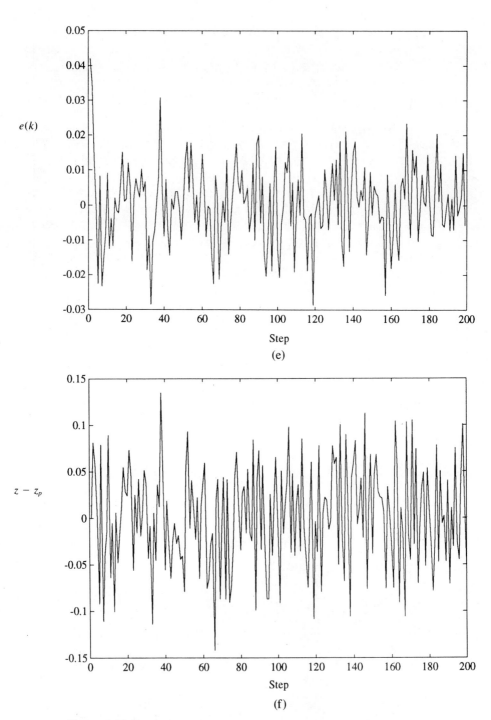

Step

(e)

Step

(f)

Figure 8-19 Cont.

states. Data for the example of the previous section was obtained with this program.

The program uses the RND function which supplies random numbers in the range (0, 1). Twelve such random numbers are added and their mean of 6 subtracted to give a nearly Gaussian probability distribution with unit covariance.‡

Programming higher-order Kalman filters is probably easiest if we organize the computations in matrix form. An algorithm that can be used to perform Kalman filtering according to the equations given in Table 8-3 at the beginning of this section is listed in Table 8-6. The complexity of the operations in these steps is relatively high if matrix operation commands are not available. Some time saving is also possible through clever combination of some of the operations and through exploiting the symmetries of $\mathbf{Q}(k)$, $\mathbf{R}(k)$, $\mathbf{P}(k + 1 \mid k)$, and $\mathbf{P}(k \mid k)$. Nevertheless, it is evident that the computations are not trivial. Clearly, a careful and systematic approach is needed.

Table 8-5 Computer Program in BASIC for First-Order Kalman Filters (KALMAN)

```
100   REM ********************************************************
110   REM *    FIRST-ORDER KALMAN FILTER PROGRAM    *
120   REM ********************************************************
130   REM ENTER SYSTEM PARAMETERS
140     f = .9
150     H = 4
160     P = .0009
170     Q = .0001
180     R = .0009
190   REM Xe IS STATE ESTIMATE X(K/K)
200     K = 0
210     Xe = 0
220   REM ENTER INITIAL SYSTEM STATE
230     C = 0
240     FOR I = 1 TO 12
250     C = C + RND
260     NEXT I
270     X = SQR(P) * (C - 6)
280     PRINT "STEP K", "P(K/K)", "P(K + 1/K)", "K(K + 1)"
290   REM COMPUTE SYSTEM STATE X
300     C = 0
310     FOR I = 1 TO 12
```

---

‡R. W. Hamming, *Introduction to Applied Numerical Analysis*. New York: McGraw-Hill, 1971, Chapter 14.

Table 8-5 Cont.

```
320    C = C + RND
330    NEXT I
340    W = SQR(Q) * (C − 6)
350    X = f * X + W
360    REM COMPUTE SYSTEM MEASUREMENT Z
370    C = 0
380    FOR I = 1 TO 12
390    C = C + RND
400    NEXT I
410    V = SQR(R) * (C − 6)
420    Z = H * X + V
430    REM COMPUTE STATE PREDICTION COVARIANCE Pc
440    Pc = f * P * f + Q
450    REM COMPUTE KALMAN GAIN G
460    G = (Pc * H) / (H * Pc * H + R)
470    REM COMPUTE STATE PREDICTION Xp
480    Xp = f * Xe
490    REM COMPUTE MEASUREMENT PREDICTION Zp
500    Zp = H * Xp
510    REM COMPUTE STATE ESTIMATE Xe
520    Xe = G * (Z − Zp) + Xp
530    PRINT K, P, Pc, G
540    REM UPDATE STATE ESTIMATE COVARIANCE P
550    P = (1 − G * H) * Pc
560    K = K + 1
570    IF K = 100 GOTO 590
580    GOTO 300
590    STOP
600    END
```

8.7 Extensions

In this section, the results of the previous section are extended to systems involving noise-coupling matrices, deterministic inputs, non-zero mean values, known initial conditions, correlated noises, and bias estimation. In each case, very little additional effort is required to modify the basic Kalman filter.

8.7.1 Noise-Coupling Matrices

When a linear, discrete-time, stochastic system model has a matrix $\mathbf{L}(k)$ coupling the white noise input $\mathbf{w}(k)$ to the state equations

$$\mathbf{x}(k + 1) = \mathbf{F}(k)\mathbf{x}(k) + \mathbf{L}(k)\mathbf{w}(k)$$

$$\mathbf{x}(k + 1) = \mathbf{F}(k)\mathbf{x}(k) + \mathbf{w}'(k)$$

Table 8-6 An Algorithm for Multivariable Kalman Filters

1. Initialize the step index $k = 0$
2. Initialize the state estimate vector $\hat{\mathbf{x}}(0 \mid 0)$
3. Initialize the covariance matrix $\mathbf{P}(0 \mid 0) = \mathbf{P}(0)$ of the state estimate error
4. Compute the next covariance of the state prediction error

$$\mathbf{P}(k + 1 \mid k) = \mathbf{F}(k)\mathbf{P}(k \mid k)\mathbf{F}^{\dagger}(k) + \mathbf{Q}(k)$$

5. Compute the matrix

$$\mathbf{S} = \mathbf{H}(k + 1)\mathbf{P}(k + 1 \mid k)\mathbf{H}^{\dagger}(k + 1) + \mathbf{R}(k + 1)$$

6. Compute the matrix inverse $\mathbf{S}^{-1}$
7. Compute the next Kalman gain

$$\mathbf{K}(k + 1) = \mathbf{P}(k + 1 \mid k)\mathbf{H}^{\dagger}(k + 1)\mathbf{S}^{-1}$$

8. Compute the next covariance of state estimator error

$$\mathbf{P}(k + 1 \mid k + 1) = [\mathbf{I} - \mathbf{K}(k + 1)\mathbf{H}(k + 1)]\mathbf{P}(k + 1 \mid k)$$

9. Compute the next predicted state

$$\hat{\mathbf{x}}(k + 1 \mid k) = \mathbf{F}(k)\hat{\mathbf{x}}(k \mid k)$$

10. Read the next measurement $\mathbf{z}(k + 1)$
11. Compute the measurement residual

$$\boldsymbol{\Delta}\mathbf{z}(k + 1 \mid k) = \mathbf{z}(k + 1) - \mathbf{H}(k + 1)\hat{\mathbf{x}}(k + 1 \mid k)$$

12. Compute the state estimate

$$\hat{\mathbf{x}}(k + 1 \mid k + 1) = \hat{\mathbf{x}}(k + 1 \mid k) + \mathbf{K}(k + 1)\boldsymbol{\Delta}\mathbf{z}(k + 1 \mid k)$$

13. Increment k by 1
14. Output the state estimate
15. Return to step 4

the basic Kalman filter equations presented in Table 8-3 apply, except that the state noise covariance matrix $\mathbf{Q}(k)$ is replaced by

$$\mathbf{Q}'(k) = E[\mathbf{w}'(k)\mathbf{w}'^{\dagger}(k)] = E[\mathbf{L}(k)\mathbf{w}(k)\mathbf{w}^{\dagger}(k)\mathbf{L}^{\dagger}(k)]$$

$$= \mathbf{L}(k)E[\mathbf{w}(k)\mathbf{w}^{\dagger}(k)]\mathbf{L}^{\dagger}(k)$$

$$= \mathbf{L}(k)\mathbf{Q}(k)\mathbf{L}^{\dagger}(k)$$

in the covariance equation for state prediction error.

Similarly, if the system output equation has a matrix $\mathbf{M}(k + 1)$ coupling the white noise $\mathbf{v}(k + 1)$ to the measurements

$$\mathbf{z}(k + 1) = \mathbf{H}(k + 1)\mathbf{x}(k + 1) + \mathbf{M}(k + 1)\mathbf{v}(k + 1)$$

$$= \mathbf{H}(k + 1)\mathbf{x}(k + 1) + \mathbf{v}'(k + 1)$$

the matrix $\mathbf{R}(k + 1)$ noise covariance for measurement is replaced in the Kalman filter gain equation by

$$
\begin{aligned}
\mathbf{R}'(k + 1) &= E[\mathbf{v}'(k + 1)\mathbf{v}'^{\dagger}(k + 1)] \\
&= E[\mathbf{M}(k + 1)\mathbf{v}(k + 1)\mathbf{v}^{\dagger}(k + 1)\mathbf{M}^{\dagger}(k + 1)] \\
&= \mathbf{M}(k + 1)\mathbf{R}(k + 1)\mathbf{M}^{\dagger}(k + 1)
\end{aligned}
$$

8.7.2 Systems with Deterministic Inputs

If the discrete-time, stochastic system has additional deterministic inputs $\mathbf{u}(k)$

$$\mathbf{x}(k + 1) = \mathbf{F}(k)\mathbf{x}(k) + \mathbf{w}(k) + \mathbf{B}(k)\mathbf{u}(k)$$

they are incorporated into the system model portion of the Kalman filter, as shown in Figure 8-20. While the filter is adding the known effects of the deterministic inputs to the estimates, it is removing them from the measurement residuals.

Additional known deterministic inputs affect only the Kalman filter prediction equation

$$\hat{\mathbf{x}}(k + 1 \mid k) = \mathbf{F}(k)\hat{\mathbf{x}}(k \mid k) + \mathbf{B}(k)\mathbf{u}(k)$$

When the state of a stochastic system is inaccessible, the optimal feedback system, in the expected mean square sense, for the stochastic plant consists of the Kalman filter estimate of the plant state fed back through the optimal feedback gains $\mathbf{E}(k)$ for the deterministic linear regulator with the state feedback

$$\mathbf{u}(k) = \mathbf{E}(k)\hat{\mathbf{x}}(k \mid k)$$

This problem is discussed in more detail in Section 8.8.

8.7.3 Nonzero Noise and Initial Condition Means

For the stochastic system

$$\mathbf{x}(k + 1) = \mathbf{F}(k)\mathbf{x}(k) + \mathbf{w}(k)$$

if the white noise input $\mathbf{w}(k)$ has a known nonzero mean

$$E[\mathbf{w}(k)] = \mathbf{m}_w(k) \tag{8-69}$$

a new zero-mean white noise sequence

$$\mathbf{w}'(k) = \mathbf{w}(k) - \mathbf{m}_w(k)$$

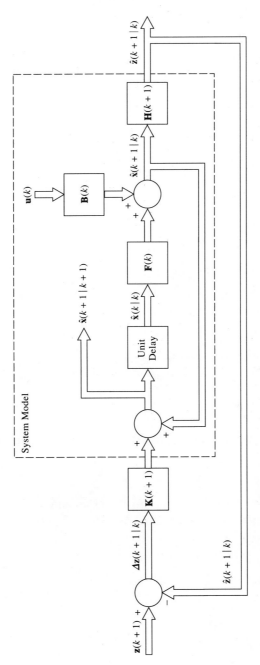

Figure 8-20 Incorporating deterministic inputs into the Kalman filter.

is formed. The system is now viewed as having a zero-mean white noise input $\mathbf{w}'(k)$ plus an additional deterministic input $\mathbf{m}_w(k)$. The noise covariance of $\mathbf{w}'(k)$ is

$$\mathbf{Q}'(k) = E\{[\mathbf{w}(k) - \mathbf{m}_w(k)][\mathbf{w}(k) - \mathbf{m}_w(k)]^\dagger\}$$
$$= \mathbf{Q}(k) - E[\mathbf{w}(k)]\mathbf{m}_w^\dagger(k) - \mathbf{m}_w(k)E[\mathbf{w}^\dagger(k)] + \mathbf{m}_w(k)\mathbf{m}_w^\dagger(k)$$

and using equation (8-69) gives

$$\mathbf{Q}'(k) = \mathbf{Q}(k) - \mathbf{m}_w(k)\mathbf{m}_w^\dagger(k)$$

Therefore, $\mathbf{Q}'(k)$ should replace $\mathbf{Q}(k)$ in the basic Kalman filter equations. Similarly, if the white noise measurement sequence $\mathbf{v}(k + 1)$ has a known nonzero mean

$$E[\mathbf{v}(k + 1)] = \mathbf{m}_v(k + 1)$$

a new measurement sequence with a zero-mean and white noise is formed

$$\mathbf{v}'(k + 1) = \mathbf{v}(k + 1) - \mathbf{m}_v(k + 1)$$

where $\mathbf{m}_v(k + 1)$ is considered as a deterministic input to the measurements. Then the covariance matrix

$$\mathbf{R}'(k + 1) = \mathbf{R}(k + 1) - \mathbf{m}_v(k + 1)\mathbf{m}_v^\dagger(k + 1)$$

should replace $\mathbf{R}(k + 1)$ in the basic Kalman filter equations.

If the initial state vector of a stochastic system has the nonzero mean

$$E[\mathbf{x}(0)] = \mathbf{c}$$

the on-line calculations of the Kalman filter are begun with the initial state vector

$$\hat{\mathbf{x}}(0 \mid 0) = \mathbf{c}$$

Nonzero noise and nonzero initial condition means are incorporated into the Kalman filter as shown in Figure 8-21. Consider the stochastic system

$$\mathbf{x}(k + 1) = \mathbf{F}(k)\mathbf{x}(k) + \mathbf{w}(k)$$
$$\mathbf{z}(k + 1) = \mathbf{H}(k + 1)\mathbf{x}(k + 1) + \mathbf{v}(k + 1)$$

where

$$E[\mathbf{x}(0)] = \mathbf{c}$$
$$E[\mathbf{w}(k)] = \mathbf{m}_w(k)$$
$$E[\mathbf{v}(k + 1)] = \mathbf{m}_v(k + 1)$$

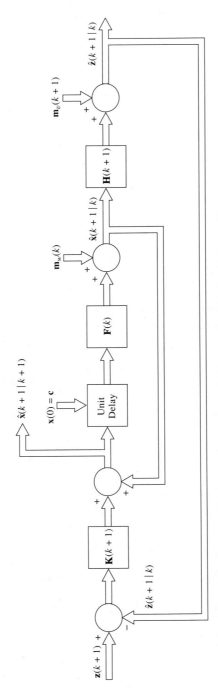

Figure 8-21 Incorporating known noise and initial condition means into the Kalman filter.

and

$$E\{[\mathbf{x}(0) - \mathbf{c}][\mathbf{x}(0) - \mathbf{c}]^\dagger\} = \mathbf{P}'(0)$$

$$E\{[\mathbf{x}(0) - \mathbf{c}][\mathbf{w}(k) - \mathbf{m}_w(k)]^\dagger\} = \mathbf{0} \quad \text{for all } k$$

$$E\{[\mathbf{x}(0) - \mathbf{c}][\mathbf{v}(k+1) - \mathbf{m}_v(k+1)]^\dagger\} = \mathbf{0} \quad \text{for all } k$$

$$E\{[\mathbf{w}(l) - \mathbf{m}_w(l)][\mathbf{w}(k) - \mathbf{m}_w(k)]^\dagger\} = \begin{cases} \mathbf{0} & l \neq k \\ \mathbf{Q}'(k) \end{cases}$$

$$E\{[\mathbf{v}(l) - \mathbf{m}_v(l)][\mathbf{v}(k+1) - \mathbf{m}_v(k+1)]^\dagger\} = \begin{cases} \mathbf{0} & l \neq k+1 \\ \mathbf{R}'(k+1) & l = k+1 \end{cases}$$

$$E\{[\mathbf{w}(l) - \mathbf{m}_w(l)][\mathbf{v}(k+1) - \mathbf{m}_v(k+1)]^\dagger\} = \mathbf{0} \quad \text{for all } l \text{ and } k$$

The Kalman filter prediction equations are then

$$\hat{\mathbf{x}}(k+1 \mid k) = \mathbf{F}(k)\hat{\mathbf{x}}(k \mid k) + \mathbf{m}_w(k) \quad \hat{\mathbf{x}}(0 \mid 0) = \mathbf{c}$$

$$\hat{\mathbf{z}}(k+1 \mid k) = \mathbf{H}(k+1)\hat{\mathbf{x}}(k+1 \mid k) + \mathbf{m}_v(k+1)$$

and the corrector equations are unchanged. The predictor error covariances, estimation error covariances, and Kalman gains are calculated as in the basic filter, with $\mathbf{Q}'(k)$ replacing $\mathbf{Q}(k)$ and $\mathbf{R}'(k+1)$ replacing $\mathbf{R}(k+1)$.

8.7.4 Stochastic Systems with Known Initial Conditions

If the stochastic system has known deterministic initial conditions

$$\mathbf{x}(0) = \mathbf{c}$$

then

$$E[\mathbf{x}(0)] = E[\mathbf{c}] = \mathbf{c}$$

and

$$E\{[\mathbf{x}(0)] - \mathbf{c}][\mathbf{x}(0) - \mathbf{c}]^\dagger\} = \mathbf{P}'(0) = \mathbf{0}$$

The Kalman filter begins with the known mean

$$\hat{\mathbf{x}}(0 \mid 0) = E[\mathbf{x}(0)] = \mathbf{c}$$

and because the variance from the mean is zero the Kalman gain calculations begin with

$$\mathbf{P}(0 \mid 0) = \mathbf{P}(0) = \mathbf{0}$$

8.7.5 Correlated Noises

If a stochastic system has noise sources $\mathbf{w}(k)$ and $\mathbf{v}(k + 1)$ that are not white and are possibly correlated with each other, it may be possible to design an rth-order shaping filter that turns uncorrelated white noise sequences $\mathbf{w}'(k)$ and $\mathbf{v}'(k)$ into sequences that model the correlations of $\mathbf{w}(k)$ and $\mathbf{v}(k + 1)$.

$$\Theta(k + 1) = \mathbf{M}(k)\Theta(k) + \Gamma(k)\mathbf{w}'(k)$$

$$\mathbf{w}(k) = \mathbf{C}(k)\Theta(k)$$

$$\mathbf{v}(k + 1) = \mathbf{D}(k + 1)\Theta(k + 1) + \mathbf{v}'(k + 1) \tag{8-70}$$

The augmented system in equations (8-37), (8-38), and (8-62)

$$\begin{bmatrix} \mathbf{x}(k + 1) \\ \Theta(k + 1) \end{bmatrix} = \begin{bmatrix} \mathbf{F}(k) & \mathbf{C}(k) \\ 0 & \mathbf{M}(k) \end{bmatrix} \begin{bmatrix} \mathbf{x}(k) \\ \Theta(k) \end{bmatrix} + \begin{bmatrix} 0 \\ \Gamma(k) \end{bmatrix} \mathbf{w}'(k)$$

$$= \mathbf{F}'(k)\Psi(k) + \Gamma'(k)\mathbf{w}'(k)$$

$$\mathbf{z}(k + 1) = [\mathbf{H}(k + 1) \; \vdots \; \mathbf{D}(k + 1)] \begin{bmatrix} \mathbf{x}(k + 1) \\ \Theta(k + 1) \end{bmatrix} + \mathbf{v}'(k + 1)$$

$$= \mathbf{H}'(k + 1)\Psi(k + 1) + \mathbf{v}'(k + 1)$$

is of the form for which the basic Kalman filter applies. The $(n + r)$th-order Kalman filter then generates optimal estimates of $\Psi(k)$, that is, of both $\mathbf{x}(k)$ and $\Theta(k)$.

8.7.6 Bias Estimation

In many applications, it is desirable to estimate unknown constant signals, termed *biases*, in a system. In attitude determination systems for spacecraft, for example, there are constant or slowly varying gyroscope bias errors that can be modeled as the response of a system of the form

$$x_i(k + 1) = x_i(k)$$

Each bias term can be modeled by an equation of this form.
For example, consider the stochastic system described by

$$\begin{bmatrix} x_1(k + 1) \\ x_2(k + 1) \end{bmatrix} = \begin{bmatrix} 0.1 & 1 \\ 0.3 & -0.5 \end{bmatrix} \begin{bmatrix} x_1(k) \\ x_2(k) \end{bmatrix} + \begin{bmatrix} w_1(k) \\ w_2(k) \end{bmatrix}$$

$$z(k + 1) = [1 \quad 0] \begin{bmatrix} x_1(k + 1) \\ x_2(k + 1) \end{bmatrix} + v(k + 1) + b$$

where b is an unknown constant bias of the measurement. If b is modeled as

$$b = x_3(k + 1) = x_3(k) + w_3(k)$$

where $w_3(k)$ is added to track slow changes in the bias, the following augmented system gives a model of the standard form

$$\begin{bmatrix} x_1(k + 1) \\ x_2(k + 1) \\ x_3(k + 1) \end{bmatrix} = \begin{bmatrix} 0.1 & 1 & 0 \\ 0.3 & -0.5 & 0 \\ 0 & 0 & 1 \end{bmatrix} \begin{bmatrix} x_1(k) \\ x_2(k) \\ x_3(k) \end{bmatrix} + \begin{bmatrix} w_1(k) \\ w_2(k) \\ w_3(k) \end{bmatrix}$$

$$z(k + 1) = [1 \quad 0 \quad 1] \begin{bmatrix} x_1(k + 1) \\ x_2(k + 1) \\ x_3(k + 1) \end{bmatrix} + v(k + 1)$$

The basic Kalman filter equations then apply with the measurement bias becoming an additional state variable to be estimated.

One must be careful when simultaneously modeling several biases because biases on top of biases may result in an unobservable system. For example, if a system's outputs depend only on the sum of two cumulative unknown constants, say, two bias errors from two different gyroscopes, it is not possible to estimate the unknowns separately; only their sum is observable.

8.7.7 Extended Kalman Filtering

Many of the most successful applications of Kalman filtering have been in areas such as ship, aircraft, missile, and spacecraft navigation; target tracking; and trajectory determination, where the system state and output equations are nonlinear and typically of the form

$$\mathbf{x}(k + 1) = \mathbf{f}[\mathbf{x}(k), k] + \mathbf{L}[\mathbf{x}(k), k]\mathbf{w}(k)$$

$$\mathbf{z}(k + 1) = \mathbf{h}[\mathbf{x}(k + 1), k + 1] + \mathbf{v}(k + 1) \tag{8-71}$$

where $\mathbf{f}$, $\mathbf{L}$, and $\mathbf{h}$ are known deterministic nonlinear functions depending on the system state.

One way of linearizing the system equations is simply to linearize about some nominal state that does not depend on the measurements. The filter is then termed a *linearized Kalman filter*. A more popular way, however, is to linearize the nonlinear equations about the current estimated state trajectory $\hat{\mathbf{x}}(k \mid k)$. In this case, the filter is called an *extended Kalman filter*.

The equations connected with the extended Kalman filter are collected in Table 8-7. The state equations of the system for the next step

Table 8-7 The Extended Kalman Filter Equations

System Model

$$\mathbf{x}(k + 1) = \mathbf{f}[\mathbf{x}(k), k] + \mathbf{L}[\mathbf{x}(k), k]\mathbf{w}(k)$$

$$\mathbf{z}(k + 1) = \mathbf{h}[\mathbf{x}(k + 1), k + 1] + \mathbf{v}(k + 1)$$

Initialization

Given

$$\hat{\mathbf{x}}(0 \mid 0) = E[\mathbf{x}(0)]$$

$$\mathbf{P}(0 \mid 0) = \mathbf{P}(0)$$

Prediction Equations

$$\hat{\mathbf{x}}(k + 1 \mid k) = \mathbf{f}[\hat{\mathbf{x}}(k \mid k), k]$$

$$\mathbf{P}(k + 1 \mid k) = \mathbf{A}(k)\mathbf{P}(k \mid k)\mathbf{A}^\dagger(k) + \mathbf{L}(k)\mathbf{Q}(k)\mathbf{L}^\dagger(k)$$

where

$$\mathbf{Q}(k) = E[\mathbf{w}(k)\mathbf{w}^\dagger(k)]$$

$$\mathbf{A}(k) = \frac{\partial \mathbf{f}[\mathbf{x}(k)]}{\partial \mathbf{x}(k)} \bigg|_{\mathbf{x}(k) = \hat{\mathbf{x}}(k \mid k)}$$

$$\mathbf{L}(k) = \mathbf{L}[\hat{\mathbf{x}}(k \mid k), k]$$

Update Equations

$$\mathbf{H}(k + 1) = \frac{\partial \mathbf{h}[(\mathbf{x}(k + 1), (k + 1)]}{\partial \mathbf{x}(k + 1)} \bigg|_{\mathbf{x}(k) = \hat{\mathbf{x}}(k + 1 \mid k)}$$

Kalman Filter Gain

$$\mathbf{K}(k + 1) = \mathbf{P}(k + 1 \mid k)\mathbf{H}^\dagger(k + 1)[\mathbf{H}(k + 1)\mathbf{P}(k + 1 \mid k)\mathbf{H}^\dagger(k + 1) + \mathbf{R}(k + 1)]^{-1}$$

where

$$\mathbf{R}(k + 1) = E[\mathbf{v}(k + 1)\mathbf{v}^\dagger(k + 1)]$$

Updated Covariance

$$\mathbf{P}(k + 1 \mid k + 1) = [\mathbf{I} - \mathbf{K}(k + 1)\mathbf{H}(k + 1)]\mathbf{P}(k + 1 \mid k)$$

Updated State

$$\hat{\mathbf{x}}(k + 1 \mid k + 1) = \hat{\mathbf{x}}(k + 1 \mid k) + \mathbf{K}(k + 1)\Delta\mathbf{z}(k + 1 \mid k)$$

$$\Delta\mathbf{z}(k + 1 \mid k) = \mathbf{z}(k + 1) - \mathbf{h}[\hat{\mathbf{x}}(k + 1 \mid k), k + 1]$$

are not determined until the state estimate at the present step has been computed, because the Kalman gain depends on the estimated state at each prior step. It is, therefore, usually necessary to compute the Kalman gain sequence on line as the measurement data and state estimates are processed.

If the approximation involved in the dynamic linearization of the equations is too crude because of large state changes between measurements, the filter behavior may become poor or even unstable.

8.7.8 Steady State Estimation

For a linear, step-invariant, stochastic system and constant $\mathbf{Q}$ and $\mathbf{R}$ matrices, the solutions for the prediction error covariance, estimation error covariance, and Kalman gain are bounded and converge to a unique steady state value for a sufficiently large number of steps k, provided that the system is completely observable. If $\mathbf{P}(k + 1 \mid k)$, $\mathbf{P}(k \mid k)$, and $\mathbf{K}(k + 1)$ reach limiting values for large k,

$$\lim_{k \to \infty} \mathbf{P}(k + 1 \mid k) = \mathbf{P}_1$$

$$\lim_{k \to \infty} \mathbf{P}(k \mid k) = \mathbf{P}$$

$$\lim_{k \to \infty} \mathbf{K}(k + 1) = \mathbf{K}$$

Then the gain and covariance equations become, for large k

$$\mathbf{K} = \mathbf{P}_1 \mathbf{H}^{\dagger} [\mathbf{H} \mathbf{P}_1 \mathbf{H}^{\dagger} + \mathbf{R}]^{-1}$$

$$\mathbf{P}_1 = \mathbf{F} \mathbf{P} \mathbf{F}^{\dagger} + \mathbf{Q}$$

$$\mathbf{P} = [\mathbf{I} - \mathbf{K} \mathbf{H}] \mathbf{P}_1$$

With substitution, the two covariance equations and the Kalman gain equation may be combined as follows:

$$\mathbf{P}_1 = \mathbf{F} \mathbf{P}_1 \mathbf{F}^{\dagger} - \mathbf{F} \mathbf{P}_1 \mathbf{H}^{\dagger} [\mathbf{H} \mathbf{P}_1 \mathbf{H}^{\dagger} + \mathbf{R}]^{-1} \mathbf{H} \mathbf{P}_1 \mathbf{F}^{\dagger} + \mathbf{Q}$$

This nonlinear matrix equation is known as the *algebraic Riccati equation* and has been studied extensively. This equation can be solved directly, or the method developed in Chapter 4 for the closed form solution of the optimal control problem can be adapted for calculating the steady state Kalman gain.

As a simple numerical example, consider the first-order system

$$x(k + 1) = 0.3x(k) + w(k)$$

$$z(k + 1) = 2x(k + 1) + v(k + 1)$$

with

$Q = 3$ and $R = 6$

The Riccati equation is

$$P_1 = 0.09P_1 - \frac{0.6P_1}{4P_1 + 6}(0.6P_1) + 3$$

$$= \frac{0.36P_1^2 + 0.54P_1 - 0.36P_1^2 + 12P_1 + 18}{(4P_1 + 6)}$$

$4P_1^2 - 6.54P_1 - 18 = 0$ $P_1 = -1.456, 3.091$

Therefore, the limiting values of P_1 and K are

$P_1 = 3.091$ $K = 0.336$

As another example, consider the same system but with $Q = 0$ and $R = 6$. The Riccati equation becomes

$$P_1 = 0.09P_1 - \frac{0.36P_1^2}{4P_1 + 6}$$

or

$P_1 = 0, -1.365$

The solution $P_1 = -1.365$ is rejected because $P_1 \geq 0$, and, therefore, the solution is $P_1 = 0$, $P = 0$, and $K = 0$.

This means that all future measurements will have no effect on the estimate. Were the system to undergo an unexpected disturbance, the filter would not respond to it. If there are slight modeling errors of the system, especially if the system is unstable, huge errors can occur as the Kalman gain goes to zero.

In practical filters, even when there is negligible state noise, $\mathbf{Q}$ is taken to be nonzero so that the Kalman gain does not go to zero. Alternatively, the Kalman gain may not be allowed to drop below a threshold value.

The steady state Kalman gain $\mathbf{K}$ is sometimes used for placing step-invariant observer eigenvalues.

8.8 Stochastic Optimal Control

In this final section, the stochastic optimal control problem is briefly discussed. Because the plant is modeled as a deterministic system with additional, inaccessible stochastic disturbance inputs, the Kalman filter, which produces a generalized weighted, least squares solution for the state estimate, gives a state estimate with minimum expected square error. When the plant state is inaccessible, the optimal feedback system, in the expected mean square sense, for the stochastic plant consists of the Kalman filter estimate of the plant state fed back through the optimal feedback gains for the deterministic linear regulator with state feedback.

8.8.1 Kalman Filter–Observer

When the state of a stochastic system is not available for feedback, the state estimate as generated by the Kalman filter can be used in place of the state itself. The Kalman filter for an nth-order, linear, discrete-time, stochastic system is another nth-order, linear, discrete-time system. The filter inputs are the plant outputs z and deterministic inputs u, and the filter output is the optimal plant state estimate, as indicated in Figure 8-22.

The previous Kalman filter equations involve step $k + 1$ measurements. In feedback control, however, it may not be possible to use the step $k + 1$ measurement to help in estimating the state at step $k + 1$ because of the time needed to perform the calculations. If it is possible, then so much the better.

Suppose that the Kalman filter must use only the plant measurements through step k to estimate the plant state at step $k + 1$. In this estimation problem, we assume that the first measurement is $z(0)$. The state and output equations of the stochastic system then have the form

$$\mathbf{x}(k + 1) = \mathbf{F}(k)\mathbf{x}(k) + \mathbf{B}(k)\mathbf{u}(k) + \mathbf{w}(k)$$
$$\mathbf{z}(k) = \mathbf{H}(k)\mathbf{x}(k) + \mathbf{D}(k)\mathbf{u}(k) + \mathbf{v}(k) \tag{8-72}$$

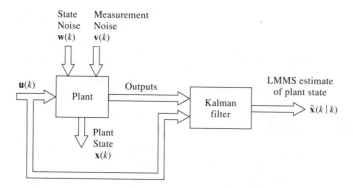

Figure 8-22 A Kalman filter as an observer of a plant.

As before, it is assumed that $\mathbf{w}(k)$ and $\mathbf{v}(k)$ are each zero-mean, white noise sequences with known covariances, and uncorrelated with one another. That is,

$$E[\mathbf{w}(k)] = \mathbf{0}$$

$$E[\mathbf{v}(k)] = \mathbf{0}$$

$$E[\mathbf{w}(j)\mathbf{w}^\dagger(k)] = \begin{cases} \mathbf{0} & j \neq k \\ \mathbf{Q}(k) & j = k \end{cases}$$

$$E[\mathbf{v}(j)\mathbf{v}^\dagger(k)] = \begin{cases} \mathbf{0} & j \neq k \\ \mathbf{R}(k) & j = k \end{cases}$$

and

$$E[\mathbf{v}(j)\mathbf{w}^\dagger(k)] = \mathbf{0} \quad \text{for all } j \text{ and } k$$

where $\mathbf{Q}(k)$ and $\mathbf{R}(k)$ are symmetric and positive semidefinite matrices.

Let the initial system state $\mathbf{x}(0)$ be probabilistic, with the zero mean, known positive-semidefinite covariance matrix $\mathbf{P}(0)$ and uncorrelated with $\mathbf{w}(k)$ and $\mathbf{v}(k)$.

The Kalman filter equations for the system in equation (8-72), using measurements from step 0 to step k, are collected in Table 8–8. Starting with a known intial state estimate $\hat{\mathbf{x}}(0 \mid -1)$ and a known covariance matrix $\mathbf{P}(0 \mid -1)$, the Kalman filter equations are executed in the order shown in the table.

Substituting the corrector equation into the predictor equation gives

$$\begin{aligned}
\hat{\mathbf{x}}(k + 1 \mid k) &= \mathbf{F}(k)\{\hat{\mathbf{x}}(k \mid k - 1) + \mathbf{K}(k)[\mathbf{z}(k) - \mathbf{H}(k)\hat{\mathbf{x}}(k \mid k - 1) \\
&\quad - \mathbf{D}(k)\mathbf{u}(k)]\} + \mathbf{B}(k)\mathbf{u}(k) \\
&= \mathbf{F}(k)\hat{\mathbf{x}}(k \mid k - 1) + \mathbf{B}(k)\mathbf{u}(k) + \mathbf{F}(k)\mathbf{K}(k)[\mathbf{z}(k) \\
&\quad - \hat{\mathbf{z}}(k \mid k - 1)]
\end{aligned}$$

The structure of this solution is identical to that for a full-order state observer. That is, if the observer's gain sequence is chosen to be

$$\mathbf{G}(k) = \mathbf{F}(k)\mathbf{K}(k)$$

then the observer is a Kalman filter.

It should not be inferred that the Kalman filter approach to determining observer gains for state estimation is necessarily superior or inferior to determining the gain by selecting the desired estimator convergence properties. For some kinds of state and measurement errors (white noise), the Kalman filter is the best we can do using linear operations. In many applications, it is a good approach even when the

Table 8-8 Kalman Filter Equations Using Measurements Only to Step k

---

System Model

$$\mathbf{x}(k + 1) = \mathbf{F}(k)\mathbf{x}(k) + \mathbf{B}(k)\mathbf{u}(k) + \mathbf{w}(k)$$
$$\mathbf{z}(k) = \mathbf{H}(k)\mathbf{x}(k) + \mathbf{D}(k)\mathbf{u}(k) + \mathbf{v}(k)$$

Initialization

Given

$$\hat{\mathbf{x}}(0 \mid -1) = E[\mathbf{x}(0)]$$
$$\mathbf{P}(0 \mid -1) = \mathbf{P}(0)$$

Kalman Filter Gain

(1) $\mathbf{K}(k) = \mathbf{P}(k \mid k - 1)\mathbf{H}^{\dagger}(k)[\mathbf{H}(k)\mathbf{P}(k \mid k - 1)\mathbf{H}^{\dagger}(k) + \mathbf{R}(k)]^{-1}$

(4) $\mathbf{P}(k \mid k) = [\mathbf{I} - \mathbf{K}(k)\mathbf{H}(k)]\mathbf{P}(k \mid k - 1)$

(6) $\mathbf{P}(k + 1 \mid k) = \mathbf{F}(k)\mathbf{P}(k \mid k)\mathbf{F}^{\dagger}(k) + \mathbf{Q}(k)$

Corrector Equations

(2) $\Delta\mathbf{z}(k \mid k - 1) = \mathbf{z}(k) - \hat{\mathbf{z}}(k \mid k - 1)$
$$= \mathbf{z}(k) - [\mathbf{H}(k)\hat{\mathbf{x}}(k \mid k - 1) + \mathbf{D}(k)\mathbf{u}(k)]$$

(3) $\hat{\mathbf{x}}(k \mid k) = \hat{\mathbf{x}}(k \mid k - 1) + \mathbf{K}(k)\,\Delta\mathbf{z}(k \mid k - 1)$

Predictor Equations

(5) $\hat{\mathbf{x}}(k + 1 \mid k) = \mathbf{F}(k)\hat{\mathbf{x}}(k \mid k) + \mathbf{B}(k)\mathbf{u}(k)$

---

error model is imprecise. In other situations, it is more sensible to specify observer convergence properties directly.

Any full-order state observer is a Kalman filter or a recursive least squares state estimator with some set of covariance matrices $\mathbf{Q}(k)$ and $\mathbf{R}(k)$. A Kalman filter is also a full-order state observer with some set of state transition matrices $\mathbf{\Phi}(k)$. When the covariance matrices can be sensibly chosen,[§] the Kalman filter approach to designing the observer gains is excellent. When it is more reasonable to select observer convergence properties directly, the least squares approach is better.

[§] If the covariance matrices $\mathbf{Q}(k)$ and $\mathbf{R}(k)$ are accurately known, the filter error covariance accurately assesses filter performance. If, however, $\mathbf{Q}(k)$ and $\mathbf{R}(k)$ are inaccurate, the error covariance may not be correct.

8.8.2 Optimal Linear Regulation

In the Kalman filter (or least squares) design of the previous section, the state estimate $\hat{\mathbf{x}}(k)$ that results in the minimum sum of squares of errors

$$J(k) = \sum_{i=0}^{k-1} \mathbf{w}^{\dagger}(i)\mathbf{Q}^{-1}\mathbf{w}(i) + \sum_{i=0}^{k} \mathbf{v}^{\dagger}(i)\mathbf{R}^{-1}(i)\mathbf{v}(i) \tag{8-73}$$

can be calculated as indicated in Table 8-8. The symmetric, positive-definite weighting matrices $\mathbf{Q}^{-1}$ and $\mathbf{R}^{-1}$ are expressed in J as inverses, for convenience. That way, $\mathbf{Q}$ and $\mathbf{R}$, rather than their inverses, appear in the corrector gain equations.

The minimization of J in equation (8-73) is the solution of a related linear–quadratic optimal control problem in which the performance measure to be minimized is

$$J = \mathbf{x}^{\dagger}(N)\mathbf{P}(N)\mathbf{x}(N) + \sum_{k=0}^{N-1} [\mathbf{x}^{\dagger}(k)\mathbf{Q}(k)\mathbf{x}(k) + \mathbf{u}^{\dagger}(k)\mathbf{R}(k)\mathbf{u}(k)]$$

which can be written as

$$J = \sum_{i=0}^{N-1} \mathbf{u}^{\dagger}(i)\mathbf{R}(i)\mathbf{u}(i) + \sum_{i=0}^{N} \mathbf{x}^{\dagger}(i)\mathbf{Q}(i)\mathbf{x}(i)$$

with $\mathbf{Q}(N) = \mathbf{P}(N)$.

Kalman, by applying this duality to the discrete-time Kalman filter, was the first to obtain the solution.

When the plant model is stochastic as in equations (8-72), the state is not accessible for feedback, and the cost functional is of the form

$$J = E\left\{\sum_{k=0}^{N} [\mathbf{x}^{\dagger}(k)\mathbf{Q}_0(k)\mathbf{x}(k) + \mathbf{u}^{\dagger}(k)\mathbf{R}_0(k)\mathbf{u}(k)]\right\}$$

then a separation principle applies to the solution. The control sequence that minimizes J consists of the MMS estimate of the plant state produced by the Kalman filter fed back through the optimal gains $\mathbf{E}(k)$ in Table 4-2 for the deterministic linear regulator with state feedback

$$\mathbf{u}(k) = \mathbf{E}(k)\hat{\mathbf{x}}(k \mid k)$$

as shown in Figure 8-23. The proof of the separation theorem for stochastic systems is very similar to the one presented in Chapter 5 for the observer feedback. The eigenvalues of the plant with Kalman filter

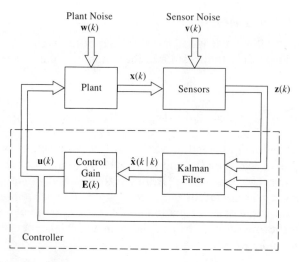

Figure 8-23 A Kalman filter estimate of the plant state fed back in place of the state.

state feedback are those of the state feedback system

$$|\lambda \mathbf{I} - (\mathbf{F} + \mathbf{BE})| = 0$$

and those of the Kalman filter

$$|\lambda \mathbf{I} - \mathbf{F} - \mathbf{FKH}| = 0$$

To sum up, the control system design for linear systems driven by random inputs can be accomplished in two parts:

1. Design the control system using state feedback as if the plant is entirely deterministic and the state is available for feedback.
2. Replace the state with a Kalman filter estimate of the state feedback.

One can always ignore the random inputs and design the control system as if the plant is strictly deterministic. Then one can only hope that the effects of the random inputs on the system performance will be slight. Simulation studies can be used to evaluate the wisdom of this choice.

8.9 Summary

In this final chapter, an introduction to stochastic systems and recursive estimation was presented. We first developed general solution methods for linear, step-invariant, single-input/single-output systems that are driven by random inputs. Then state variable stochastic models for systems with random inputs were introduced, and recursive algorithms for the propagation of the mean and the covariance were derived and demonstrated.

If more than the minimal number of measurements necessary to determine the state of a plant are available, a least squares state estimate can be used. The least squares estimation algorithm was derived and illustrated. Then the least squares calculations were arranged recursively so that the estimate was not completely recomputed whenever new, additional data was available. Instead, the previous least squares estimate was corrected to produce the new least squares estimate based on the previous estimate and the new data. The result for the case of scalar measurements is summarized in Table 8-2. Generalizations of the basic least squares problem and probabilistic interpretations of the results were discussed.

The basic LMMS estimation problem, which can be viewed as a generalization of least squares, was then formulated. When the present state of a plant was estimated with recursive least squares, with possible weighted state error and weighted measurement errors, the Kalman filter equations (Table 8-3) were derived, and computer programming considerations were discussed. Several extensions to the basic Kalman filter were developed, including the extended Kalman filter (Table 8-7).

The Kalman filter is a full-order state observer (Table 8-8) with a gain sequence that results in minimum weighted sum of squares of state and measurement errors at each step. When the errors are uncorrelated from step to step and when enough is known about the errors to choose the weighting matrices, the Kalman gain sequence is an attractive alternative to deriving observer gains from specific desired convergence properties. The chapter concluded with a discussion of optimal linear regulation of stochastic systems. Using the separation principle, the control system for linear systems driven by random inputs can be designed using state feedback, assuming the plant is entirely deterministic and the state is available for feedback. Then the state is replaced with a Kalman filter estimate of the state feedback.

This ends our tour of digital control system design, consisting of eight chapters and three appendices. We began with an introduction to discrete-time systems and ended with stochastic systems and recursive estimation. We sincerely hope that you enjoyed the tour as much as we have and that you will find the design philosophy and methods useful now and in the future.

References

The following is a list of references for those who wish to pursue these ideas further. We have made no attempt to be comprehensive or to attribute credit for various developments. The references listed here are chosen based on their suitability for extending this discussion to greater depth and to related topics.

A historical perspective of least squares estimation and Kalman filtering can be found in

K. F. Gauss, *Theory of Motion of the Heavenly Bodies about the Sun in Conic Section.* Dover, New York: 1963;

H. W. Sorenson, "Least-Squares Estimation: From Gauss to Kalman," *IEEE Spectrum*, July 1970, pp. 63–68.

The paper

T. Kailath, "A View of Three Decades of Linear Filtering Theory," *IEEE Tran. Info. Theory*, Vol. IT-20, March 1974, pp. 146–181

is a good survey of the subject.

More about least squares and recursive least squares can be found in the following books:

R. C. K. Lee, *Optimal Estimation, Identification and Control.* Cambridge, MA: M.I.T. Press, 1964;

H. W. Sorenson, *Parameter Estimation.* New York: Marcel Dekker, 1980;

J. M. Mendel, *Discrete Techniques of Parameter Estimation.* New York: Marcel Dekker, 1973;

C. L. Lawson and R. J. Hanson, *Solving Least Squares Problems.* Englewood Cliffs, NJ: Prentice-Hall, 1974.

The original papers on what is now known as Kalman filtering are

R. E. Kalman, "A New Approach to Linear Filtering and Prediction Problems," *Trans. ASME J. Basic Engr.*, Ser. D, 82, March 1960, pp. 35–45;

R. E. Kalman and R. S. Bucy, "New Results in Linear Filtering and Prediction Theory," *Trans. ASME J. Basic Engr.*, Ser. D, 83, December 1961, pp. 95–107.

The Kalman filter is covered in depth in such texts as

J. S. Meditch, *Stochastic Optimal Linear Estimation and Control.* New York: McGraw-Hill, 1969;

B. D. O. Anderson and J. B. Moore, *Optimal Filtering.* Englewood Cliffs, NJ: Prentice-Hall, 1973;

A. Gelb, Editor, *Applied Optimal Estimation.* Cambridge, MA: M.I.T. Press, 1974;

H. W. Sorenson, "Kalman Filtering Techniques." In *Advances in Control Systems*, Vol. 3, C. T. Leondes (Ed.), New York: Academic Press, 1966, pp. 219–289;

R. G. Brown, *Introduction to Random Signal Analysis and Kalman Filtering*. New York: Wiley, 1983.

Additional references on estimation theory and Kalman filtering are

N. Wiener, *The Extrapolation, Interpolation, and Smoothing of Stationary Time Series*, New York: Wiley, 1949;

N. Nahi, *Estimation Theory and Applications*. New York: Wiley, 1969;

T. P. McGarthy, *Stochastic Systems and State Estimation*. New York: Wiley, 1974;

F. G. C. Schweppe, *Uncertain Dynamic Systems*. Englewood Cliffs, NJ: Prentice-Hall, 1973;

A. P. Sage and J. L. Melsa, *Estimation Theory with Applications to Communications and Control*. New York: McGraw-Hill, 1971;

H. W. Sorenson (Ed.), *Kalman Filtering: Theory and Application*. New York: IEEE Press, 1985;

R. Deutch, *Estimation Theory*. Englewood Cliffs, NJ: Prentice-Hall, 1965;

J. Candy, *Signal Processing: The Model-Based Approach*. New York: McGraw-Hill, 1986;

P. Maybeck, *Stochastic Models, Estimation and Control* (Vol. 1). New York: Academic Press, 1979;

S. M. Bozic, *Digital and Kalman Filtering*. London: E. Arnold, Publisher, 1979;

M. Schwartz and L. Shaw, *Signal Processing: Discrete Analysis, Detection and Estimation*. New York: McGraw-Hill, 1975;

C. T. Leondes (Ed.), *Theory and Applications of Kalman Filtering*. North Atlantic Treaty Organization AGARD Report No. 139, February 1970.

Both continuous-time and discrete-time stochastic control are covered in

J. S. Melsa, *Stochastic Optimal Linear Estimation and Control*. New York: McGraw-Hill, 1969;

A. E. Bryson and Y. C. Ho, *Applied Optimal Control*. Waltham, MA: Ginn, 1969;

K. Åström, *Introduction to Stochastic Control Theory*. New York: Academic Press, 1970;

A. P. Sage and C. C. White, *Optimum Systems Control*, 2nd edition. Englewood Cliffs, NJ: Prentice-Hall, 1977.

The book

G. J. Bierman, *Factorization Methods for Discrete Sequential Estimation*. New York: Academic Press, 1979

and the articles

J. M. Mendel, "Computational Requirements for a Discrete Kalman Filter," *IEEE Trans. Automatic Control,* AC-16, December 1971, pp. 748–758;

A. Gura and G. J. Bierman, "On Computational Efficiency of Linear Filtering Algorithms," *Automatica*, 7, 1971, pp. 299–314;

R. A. Singer and R. G. Sea, "Increasing the Computational Efficiency of Discrete Kalman Filters," *IEEE Trans. Automatic Control*, AC-16, June 1971, pp. 254–257;

P. G. Kaminski, A. E. Bryson, Jr., and S. F. Schmidt, "Discrete Square Root Filtering: A Survey of Current Techniques," *IEEE Trans. Automatic Control*, Dec. 1971.

give important computational considerations.

Application examples of Kalman filtering can be found in

J. B. Pearson, "Kalman Filter Applications in Airborne Radar Tracking," *IEEE Trans. Aerospace Electronic Systems,* AES-10, May 1974, pp. 369–373;

B. D. Tapley, "Orbit Determination in the Presence of Unmodeled Accelerations," *IEEE Trans. Automatic Control*, AC-18, August 1973, pp. 369–373;

C. R. Szelag, "A Short-Term Forecasting Algorithm for Trunk Demand Servicing," *Bell System Tech. J.*, **61**, January 1982, pp. 67–96;

Y. Sawaragi, et al., "The Prediction of Air Pollution Levels by Nonphysical Models Based on Kalman Filtering Method," *J. Dynamic Systems, Measurement and Control*, **98**, December 1976, pp. 375–386;

"Special Issue on Applications of Kalman Filtering," *IEEE Trans. Automatic Control*, Vol. AC-28, March 1983, pp. 254–434;

P. K. Tam and J. B. Moore, "Improved Demodulation of Sampled-FM Signals in High Noise," *IEEE Trans. Communications*, COM-25, September 1977, pp. 1052–1053.

Kalman filter divergence is discussed in

R. J. Fitzgerald, "Divergence of the Kalman Filter," *IEEE Trans. Automatic Control*, AC-16, December 1971, pp. 736–747;

F. H. Schlee, C. J. Standish, and N. F. Toda, "Divergence in the Kalman Filter," *AIAA J.*, **5**, June 1967, pp. 1114–1120;

C. F. Price, "An Analysis of the Divergence Problem in the Kalman Filter," *IEEE Trans. Automatic Control*, AC-13, December 1968, pp. 699–702.

Additional important references on stochastic systems and stochastic processes are given in Appendix B.

Chapter Eight Problems

8-1. Find the power spectral density for each of the following auto-correlation functions. For each case, draw a pole-zero plot of the resulting PSD.

a. $R(k) = \delta(k)$

b. $R(k) = \begin{cases} 1 - \dfrac{|k|}{c} & |k| \le c \\ 0 & |k| > c \end{cases}$

c. $R(k) = \cos bk$

d. $R(k) = a^{-|k|} \cos bk$

e. $R(k) = 10 + 2e^{-|k|}$

8-2. Find and sketch the autocorrelation function for each of the following PSDs.

a. $S(z) = \dfrac{-4.8z}{z^2 - 5.2z + 1}$

b. $S(z) = \dfrac{12.5z^2}{(z^2 - 5.2z + 1)(z^2 - 2.9z + 1)}$

c. $S(z) = \dfrac{0.2(z^2 + 5z + 1)}{z}$

d. $S(\omega) = 0.3\cos \omega T + 1$

e. $S(\omega) = \dfrac{6 - 2 \cos \omega T}{8 - 4 \cos \omega T}$

8-3. For each of the following systems, find the mean, mean square, and PSD of the output for the indicated random input.

a. $w(k) \longrightarrow \boxed{\dfrac{z}{z - \frac{1}{2}}} \longrightarrow y(k)$

$E[w(k)] = 1$

$R_w(k) = (0.9)^{|k|}$

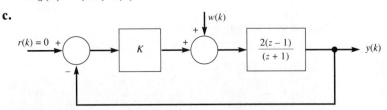

b. $w(k) \longrightarrow \boxed{\dfrac{(z-1)(z+1)}{z^2 - \frac{1}{4}}} \longrightarrow y(k)$

$$E[w(k)] = 0.3$$
$$R_w(k) = (0.1)\,\delta(k)$$

c.

$r(k) = 0 \xrightarrow{\ +\ } \bigcirc \xrightarrow{\ -\ } \boxed{K} \xrightarrow{\ +\ } \bigcirc \xrightarrow{\ +\ } \boxed{\dfrac{2(z-1)}{(z+1)}} \longrightarrow y(k)$

with $w(k)$ entering the second summing junction.

$$E[w(k)] = 0$$
$$S_w(z) = 0.01$$

d.

$w(k) \xrightarrow{\ +\ } \bigcirc \xrightarrow{\ -\ } \boxed{\dfrac{3}{z^2 + z - 2.76}} \longrightarrow y(k)$

$$E[w(k)] = 0.1$$
$$S_w(z) = 1$$

8-4. Consider the following system with random input $w(k)$

$w(k) \longrightarrow \boxed{H(z)} \longrightarrow y(k)$

a. Show that

$$S_{wy}(z) = H(z)\,S_w(z)$$

where $S_w(z)$ is the PSD of the input and $S_{wy}(z)$ is the cross PSD between the input and the output.

b. Show that

$$S_y(z) = H(z)\,S_{yw}(z)$$

8-5. For each of the following PSD functions, design a shaping filter which, when driven by a zero-mean, unit variance, white noise produces an output sequence that has the required PSD.

a. $S(z) = \dfrac{0.75(z^2 + \frac{5}{2}z + 1)}{z^2 + (13/6)z + 1}$

b. $S(z) = \dfrac{4z}{z^2 + (17/4)z + 1}$

c. $S(z) = \dfrac{4z^2}{z^2 + (17/4)z + 1}$

d. $S(z) = \dfrac{10z(z + 0.5)(z + 2)}{(z^2 + 0.45z + 0.05)(z^2 + 9z + 20)}$

8-6. If a linear, time-invariant, continuous-time system described by the transfer function $H(s)$ is driven by a stationary input with known spectral density function $S_{xx}(j\omega)$ show that

a. The mean of the output $y(t)$ is, in terms of the mean of the input $x(t)$ and the transfer function

$E[y(t)] = H(0)E[x(t)]$

b. $S_{yy}(j\omega) = H(j\omega)H(-j\omega)S_{xx}(j\omega)$

c. $S_{xy}(j\omega) = S_{yx}(-j\omega) = H(j\omega)S_{xx}(j\omega)$

8-7. Find and sketch the PSD for each of the following correlation functions:

a. $R(\tau) = A\,\delta(\tau)$

b. $R(\tau) = 1$

c. $R(\tau) = \sigma^2 e^{-\alpha|\tau|}$ σ, α are constants

d. $R(\tau) = \begin{cases} 1 - \dfrac{|\tau|}{T} & |\tau| \le T \\ 0 & |\tau| > T \end{cases}$

e. $R(\tau) = A\cos\omega_0\tau$

f. $R(\tau) = e^{-\alpha|\tau|}\cos\omega_0\tau$

8-8. For each of the following systems, the input $w(t)$ is a zero-mean, unit variance, white noise sequence. Find the mean, mean square, and PSD of the output.

a.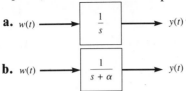

b. $w(t) \longrightarrow \boxed{\dfrac{1}{s + \alpha}} \longrightarrow y(t)$

c.

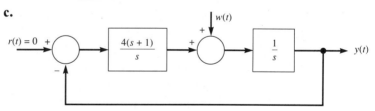

d.

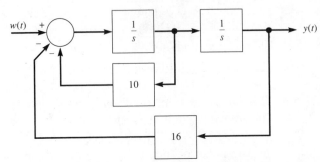

8-9. For each of the following systems, it is assumed that all initial conditions are zero and that the zero-mean, unit variance, white noise input $w(t)$ is applied at $t = 0$. In each case, find the mean square value of the output using the convolution integral of Table 8-1.

a. $G(s) = \dfrac{1}{s}$

b. $G(s) = \dfrac{10}{s + 10}$

c. $G(s) = \dfrac{1}{s^2}$

d. $G(s) = \dfrac{100}{s^2 + 100}$

e. $G(s) = \dfrac{10}{s(s + 10)}$

8-10. For each of the following PSD functions, design a shaping filter which, when driven by a zero-mean, unit variance, white noise produces an output sequence that has the required PSD.

a. $S(\omega) = \dfrac{4}{\omega^2 + 1}$

b. $S(\omega) = \dfrac{10}{\omega^4 + 6\omega^2 + 5}$

c. $S(\omega) = \dfrac{\omega^2 + 4}{\omega^4 + 20\omega^2 + 100}$

d. $S(\omega) = \dfrac{\omega^2 + 9}{\omega^4 + 625}$

8-11. Most computer systems have built-in random number generators that return independent sample values that are uniform in the range [0, 1]. Write, debug, and test a computer program that generates a Gaussian random number v with user-specified mean η and standard deviation σ. Check your program by calling it 500 times.

8-12. For each of the following cases, generate and plot a Gaussian noise sequence with the prescribed mean and standard deviation.

    a. zero mean, unit variance

    b. zero mean, 0.1 variance

    c. 0.3 mean, unit variance

    d. 0.6 mean, 0.1 variance

8-13. For each of the following systems, the input $w(k)$ is a zero-mean, white noise sequence with unit variance. Develop a computer program to generate and plot the output sequence $y(k)$ for 200 points.

    a.

    b.

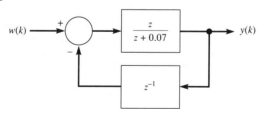

    c.

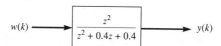

    d.

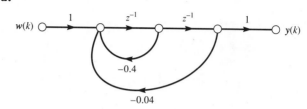

8-14. It is desirable to generate random time signals that have specific PSD functions. Using straight line approximations, derive a transfer function for each of the following magnitude squared functions. Note that for each plot the scale is 10 Log_{10}.

a.

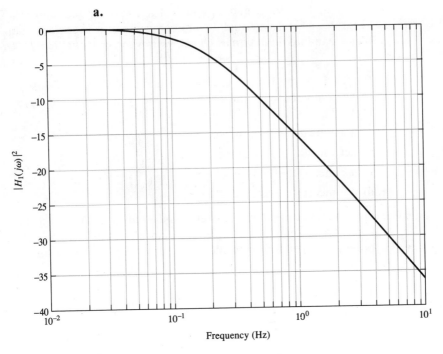

b.

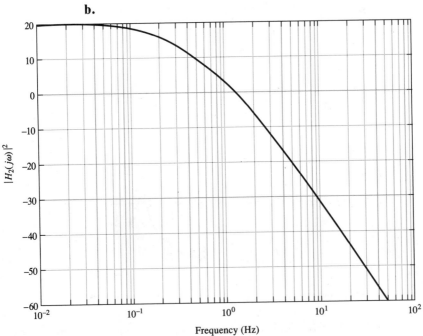

c.

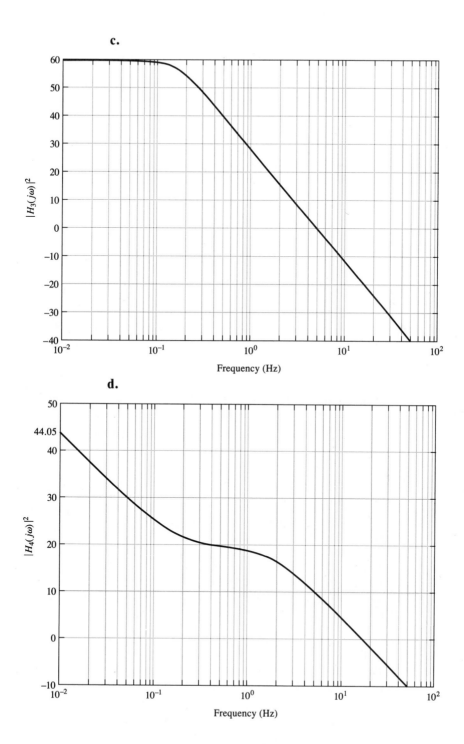

d.

e.

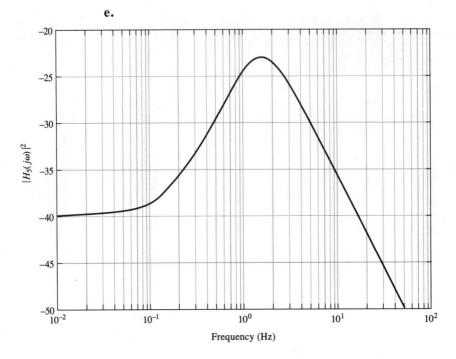

8-15. If the input to each of the shaping filters derived in problem 8-14 is a zero-mean, unit variance, white noise sequence

  a. Generate, using a computer program, an output $y(t)$ of each shaping filter

  b. Plot the spectral density of each output $y(t)$ generated in part (a) and compare with Problem 8-14

8-16. For the system

$$x(k + 1) = \frac{1}{4} x(k) + w(k)$$

where $w(k)$ is a zero-mean, unit variance, white noise sequence, and the initial condition $x(0)$ is a random variable independent of $w(k)$ with $\bar{x}(0) = \frac{1}{2}$ and $E\{[x(0) - \bar{x}(0)]^2\} = 1$, find and plot the mean $\bar{x}(k)$ and the autocovariance $P(k)$ of the state.

8-17. For each of the following systems with zero mean unit variance white noise inputs, find the steady state covariance matrix, if it exists, and determine its sign definiteness.

  a. $\mathbf{x}(k + 1) = \begin{bmatrix} -\frac{5}{6} & 1 \\ -6 & \frac{5}{2} \end{bmatrix} \mathbf{x}(k) + \begin{bmatrix} 1 \\ 0 \end{bmatrix} w(k)$

b. $\mathbf{x}(k + 1) = \begin{bmatrix} 10 & -4 \\ 24 & -10 \end{bmatrix} \mathbf{x}(k) + \begin{bmatrix} 0 \\ -1 \end{bmatrix} w(k)$

c. $\mathbf{x}(k + 1) = \begin{bmatrix} 4 & -\frac{3}{2} \\ 9 & -\frac{7}{2} \end{bmatrix} \mathbf{x}(k) + \begin{bmatrix} 2 \\ 1 \end{bmatrix} w(k)$

8-18. For the system

$$\dot{x}(t) = -3x(t) + 2w(t)$$

where $w(t)$ is a zero-mean, unit variance, white noise process, and the initial condition $x(0)$ is a random variable independent of $w(t)$ with $\bar{x}(0) = 1$ and $E\{[x(0) - \bar{x}(0)]^2\} = 2$, find and plot the mean $\bar{x}(t)$ and the covariance $P(t)$ of the state.

8-19. For each of the following systems with zero mean unit variance white noise inputs, find the steady state covariance matrix, if it exists, and determine its sign definiteness.

a. $\dot{\mathbf{x}}(t) = \begin{bmatrix} 6 & -3 \\ 18 & -9 \end{bmatrix} \mathbf{x}(t) + \begin{bmatrix} \frac{1}{6} \\ \frac{1}{2} \end{bmatrix} w(t)$

b. $\dot{\mathbf{x}}(t) = \begin{bmatrix} 10 & -5 \\ 30 & -15 \end{bmatrix} \mathbf{x}(t) + \begin{bmatrix} 1 \\ 1 \end{bmatrix} w(t)$

c. $\dot{\mathbf{x}}(t) = \begin{bmatrix} 1 & -1 \\ 6 & -4 \end{bmatrix} \mathbf{x}(t) + \begin{bmatrix} 2 \\ 1 \end{bmatrix} w(t)$

8-20. For the system

$$\begin{bmatrix} \dot{x}_1(t) \\ \dot{x}_2(t) \end{bmatrix} = \begin{bmatrix} 0 & 1 \\ -2 & -2 \end{bmatrix} \begin{bmatrix} x_1(t) \\ x_2(t) \end{bmatrix} + \begin{bmatrix} 0 \\ 1 \end{bmatrix} w(t)$$

the input $w(t)$ is a zero-mean, unit variance, white noise. Find the mean square values of $x_1(t)$ and $x_2(t)$, assuming zero initial conditions.

8-21. Consider the nth-order, linear, continuous-time system

$$\dot{\mathbf{x}}(t) = \mathbf{F}(t)\mathbf{x}(t) + \mathbf{L}(t)\mathbf{w}(t)$$

where the input vector $\mathbf{w}(t)$ is white noise with known mean $\bar{\mathbf{w}}(t)$ and known covariance

$$E\{[\mathbf{w}(t_1) - \bar{\mathbf{w}}(t_1)][\mathbf{w}(t_2) - \bar{\mathbf{w}}(t_2)]^\dagger\} = \mathbf{Q}(t_1)\, \delta(t_1 - t_2)$$

Assuming that the initial state vector $\mathbf{x}(0)$ is probabilistic with known mean $\bar{\mathbf{x}}(0)$ and known covariance

$$E\{[\mathbf{x}(0) - \bar{\mathbf{x}}(0)][\mathbf{x}(0) - \bar{\mathbf{x}}(0)]^\dagger\} = \mathbf{P}(0)$$

and that the input noise $\mathbf{w}(t)$ and the initial state $\mathbf{x}(0)$ are uncorrelated for all t,

a. Derive an expression for the mean $\bar{\mathbf{x}}(t)$ of the state vector

b. Derive an expression for the covariance $\mathbf{P}(t)$ of the state vector

8-22. For the system

$$\dot{x}(t) = -3x(t) + w(t)$$

where $w(t)$ is a white noise process with $\bar{w}(t) = 2$ and $Q(t) = 1$, find and plot the mean and covariance of the state assuming zero initial conditions. Also find the mean square value of the state.

8-23. Consider the system

$$\mathbf{x}(k + 1) = \mathbf{F}(k)\mathbf{x}(k) + \mathbf{L}(k)\mathbf{w}(k)$$
$$\mathbf{y}(k) = \mathbf{H}(k)\mathbf{x}(k) + \mathbf{v}(k)$$

where $\mathbf{w}(k)$ and $\mathbf{v}(k)$ are uncorrelated, Gaussian, white noise sequences with zero means and variances $\mathbf{Q}_w(k)$ and $\mathbf{R}_v(k)$, respectively.

a. Derive an expression for the mean $\bar{\mathbf{y}}(k)$

b. Derive an expression for the covariance of the output $\mathbf{y}(k)$

8-24. If the system in problem 8-23 is step-invariant,

a. Derive an expression for the PSD of the state vector $\mathbf{x}(k)$

b. Derive an expression for the PSD of the output vector $\mathbf{y}(k)$

8-25. For the system

$$x(k + 1) = 0.7x(k) + 0.2w(k)$$
$$y(k) = 2x(k) + v(k)$$

where $w(k)$ and $v(k)$ are uncorrelated, Gaussian, white noise sequences with zero means and variances $Q_w(k)$ and $R_v(k)$, respectively

a. Find the PSD of the state and the output

b. Find and plot the state and output covariances

c. Plot the output $y(k)$ if the initial condition $x(0)$ is a Gaussian random variable with unit mean and unit variance

8-26. Find least squares estimates for the following sets of equations:

a. $2x = 3.1$
$3x = 4.8$

$$-x = -0.2$$
$$4x = 8.6$$

b. $3x_1 - x_2 = 3.6$
$2x_1 - 3x_2 = 2.6$
$4x_1 + 2x_2 = -3.1$
$-x_1 - 2x_2 = -6$

8-27. Write a computer program to perform least squares estimation. Then find the least squares estimate of the solution to the following equations:

a. $2x_1 - x_2 + 3x_3 = -1$
$6x_1 - x_2 - 4x_3 = 0$
$11x_1 - 21x_2 + 11x_3 = 7$
$-x_1 + 3x_2 + x_3 = 9$
$9x_1 + 4x_2 - 7x_3 = 4$

b.
$$\begin{bmatrix} 2 & -2 & 0 & 4 \\ 6 & 4 & -2 & 0 \\ 2 & 2 & -2 & 2 \\ 4 & 2 & 4 & -2 \\ -2 & 4 & 2 & 0 \\ 0 & 0 & -4 & 6 \end{bmatrix} \mathbf{x} = \begin{bmatrix} 6 \\ 5 \\ 8 \\ -8 \\ 1 \\ 15 \end{bmatrix}$$

8-28. Find least squares curve fits for curves of the form indicated, with the data points given.

a. $y = \alpha_1 + \alpha_2 x + \alpha_3 x^2$
$(1, -1), (2, -3), (3, -4), (6.1, -3.2), (2.6, -4.3), (-1, 0.7)$

b. $y = \alpha_1 + \alpha_2 \sin 2t + \alpha_3 \cos 3t$
$(0, -2), (-1, 2), (3, 2), (6, -7), (2, -1), (4, -5)$

c. $z = \alpha_1 + \alpha_2 x + \alpha_3 y$
$(0, 0, 0), (1, 0, -1), (2, 1, 1)$

d. $y = \alpha_1 + \alpha_2 x + \alpha_3 x^2$
$(1, 1), (2, 2), (3, 0)$

8-29. In digital signal processing, the procedure for calculating the Fourier coefficients from the samples of a discrete-time signal $x(k)$, $k = 0, 1 \ldots, N - 1$, is termed the discrete Fourier transform (DFT)

$$d_m = \frac{1}{N} \sum_{k=0}^{N-1} x(k) \exp\left(-j\frac{2\pi m}{N} k\right) \quad m = 0, 1, 2, \ldots, N - 1$$

where the d_m's are the Fourier coefficients. On the other hand, the procedure for recovering the samples of the signal $x(k)$ from the Fourier coefficients is termed the inverse discrete Fourier transform (IDFT)

$$x(k) = \sum_{n=0}^{N-1} d_n \exp\left(j\frac{2\pi k}{N} n\right) \quad k = 0, 1, 2, \ldots, N - 1$$

Show, using the method of least squares, that the set d_m is the least squares estimate of the set d_n.

8-30. Find weighted least squares estimates for the following sets of equations, using the given weighting matrix $\mathbf{W}$.

a.
$$\begin{bmatrix} 2 \\ 2 \\ -1 \\ -2 \end{bmatrix} x = \begin{bmatrix} -2 \\ 1 \\ -3 \\ 1 \end{bmatrix} \quad \mathbf{W} = \begin{bmatrix} 3 & 1 & 0 & 0 \\ 1 & 2 & 0 & 0 \\ 0 & 0 & 4 & 1 \\ 0 & 0 & 1 & 1 \end{bmatrix}$$

b.
$$\begin{bmatrix} 1 & 1 \\ 2 & -4 \\ -1 & 2 \end{bmatrix} \begin{bmatrix} x_1 \\ x_2 \end{bmatrix} = \begin{bmatrix} 1 \\ -1 \\ 2 \end{bmatrix} \quad \mathbf{W} = \begin{bmatrix} 2 & 0 & 0 \\ 0 & 2 & 0 \\ 0 & 0 & 4 \end{bmatrix}$$

8-31. Modify the least squares estimation program in Problem 8-27 to perform weighted least squares estimation; then find the weighted least squares estimate of the solution to the following equations with the weighting matrix given:

a. Same as part (a) of Problem 8-27 with

$$\mathbf{W} = \begin{bmatrix} 2 & 1 & -1 & 1 & 0 \\ 1 & 3 & 0 & 0 & 0 \\ -1 & 0 & 4 & 0 & 0 \\ 1 & 0 & 0 & 1 & 0 \\ 0 & 0 & 0 & 0 & 1 \end{bmatrix}$$

b. Same as part (b) of Problem 8-27 with

$$W = \begin{bmatrix} 2 & 0 & 0 & 0 & 0 & 0 \\ 0 & 4 & 0 & 0 & 0 & 0 \\ 0 & 0 & 6 & 0 & 0 & 0 \\ 0 & 0 & 0 & 8 & 0 & 0 \\ 0 & 0 & 0 & 0 & 10 & 0 \\ 0 & 0 & 0 & 0 & 0 & 20 \end{bmatrix}$$

8-32. Use recursive least squares to find least squares estimates for the equations

a.
$$\begin{bmatrix} 3 \\ -2 \\ 2 \\ -1 \end{bmatrix} x = \begin{bmatrix} 6 \\ -1 \\ 1 \\ -2 \end{bmatrix}$$

b.
$$\begin{bmatrix} 3 & 1 \\ 0 & -2 \\ 2 & 2 \\ 1 & -1 \end{bmatrix} \begin{bmatrix} x_1 \\ x_2 \end{bmatrix} = \begin{bmatrix} -6 \\ 2 \\ 3 \\ -1 \end{bmatrix}$$

8-33. Write a computer program to perform recursive least squares estimation. Then find a sequence of least squares estimates for the following equations:

a.
$$\begin{bmatrix} 8.2 & -2.6 \\ -1.3 & 3.4 \\ 6.1 & 0.8 \\ 0.6 & 1 \\ 1.3 & -2 \end{bmatrix} \begin{bmatrix} x_1 \\ x_2 \end{bmatrix} = \begin{bmatrix} -11.3 \\ 2.6 \\ 11.3 \\ -8.6 \\ 2 \end{bmatrix}$$

b.
$$\begin{bmatrix} 2 & -1 & 3 \\ 0 & 1 & 6 \\ 4 & -2 & 3 \\ 1 & 1 & 1 \\ 3 & 0 & -2 \\ 11 & -2 & 6 \end{bmatrix} \begin{bmatrix} x_1 \\ x_2 \\ x_3 \end{bmatrix} = \begin{bmatrix} 1 \\ 3 \\ -2 \\ -1 \\ 1 \\ 6 \end{bmatrix}$$

8-34. Find the gain sequence for the first five steps of the Kalman filter for state estimation for the following systems:

a. $x(k + 1) = \dfrac{1}{2} x(k) + w(k)$

$z(k + 1) = 3x(k + 1) + v(k + 1)$

$Q = 0.01 \qquad R = 1$

b. $\begin{bmatrix} x_1(k + 1) \\ x_2(k + 1) \end{bmatrix} = \begin{bmatrix} -\frac{1}{3} & 1 \\ \frac{1}{3} & 0 \end{bmatrix} \begin{bmatrix} x_1(k) \\ x_2(k) \end{bmatrix} + \begin{bmatrix} 0 \\ 1 \end{bmatrix} w(k)$

$z(k + 1) = [2 \quad 2] \begin{bmatrix} x_1(k + 1) \\ x_2(k + 1) \end{bmatrix} + v(k + 1)$

$Q = 1 \qquad R = 2$

c. $\begin{bmatrix} x_1(k + 1) \\ x_2(k + 1) \\ x_3(k + 1) \end{bmatrix} = \begin{bmatrix} 0 & 1 & 0 \\ \frac{1}{2} & 0 & 1 \\ 0 & 0 & 0 \end{bmatrix} \begin{bmatrix} x_1(k) \\ x_2(k) \\ x_3(k) \end{bmatrix} + \begin{bmatrix} 1 \\ 2 \\ -1 \end{bmatrix} w(k)$

$\begin{bmatrix} z_1(k + 1) \\ z_2(k + 1) \end{bmatrix} = \begin{bmatrix} 1 & 0 & 0 \\ 1 & -1 & 2 \end{bmatrix} \begin{bmatrix} x_1(k + 1) \\ x_2(k + 1) \\ x_3(k + 1) \end{bmatrix} + \begin{bmatrix} 1 \\ 0 \end{bmatrix} v(k + 1)$

$Q = 3 \qquad R = 1$

8-35. For the first-order system

$x(k + 1) = 0.9x(k) + w(k)$

$z(k + 1) = 3x(k + 1) + v(k + 1)$

where

$Q = 1 \qquad R = 2 \qquad x(0) \text{ is } N(0, 0.01)$

$z(1) = 0.2 \qquad z(2) = 0.4$

calculate, using the Kalman filter algorithm, $\hat{x}(2 \mid 2)$.

8-36. For the first-order system

$x(k + 1) = 0.4x(k) + w(k)$

$z(k + 1) = 2x(k + 1) + v(k + 1)$

where

$Q = 2 \qquad R = 3 \qquad x(0) \text{ is } N(0, 1)$

$z(1) = 0.8 \qquad z(2) = 2.1 \text{ and } z(3) = -1.3$

calculate, using the Kalman filter algorithm, $\hat{x}(3 \mid 3)$.

8-37. For the second-order system

$\begin{bmatrix} x_1(k + 1) \\ x_2(k + 1) \end{bmatrix} = \begin{bmatrix} 0 & 1 \\ \frac{1}{6} & \frac{1}{6} \end{bmatrix} \begin{bmatrix} x_1(k) \\ x_2(k) \end{bmatrix} + \begin{bmatrix} 0 \\ 1 \end{bmatrix} w(k)$

$$z(k + 1) = \begin{bmatrix} 1 & 1 \end{bmatrix} \begin{bmatrix} x_1(k + 1) \\ x_2(k + 1) \end{bmatrix} + v(k + 1)$$

where

$$Q = 8 \qquad R = 0.01 \qquad x_1(0) = N(0, 1) \qquad x_2(0) = N(0, 1)$$

calculate, using the Kalman filter algorithm, $\hat{x}(2 \mid 2)$, in terms of the measurements $z(1)$ and $z(2)$.

8-38. For the second-order system

$$\begin{bmatrix} x_1(k + 1) \\ x_2(k + 1) \end{bmatrix} = \begin{bmatrix} 0 & 1 \\ \frac{1}{4} & 0 \end{bmatrix} \begin{bmatrix} x_1(k) \\ x_2(k) \end{bmatrix} + \begin{bmatrix} -1 \\ 2 \end{bmatrix} w(k)$$

$$z(k + 1) = \begin{bmatrix} -1 & 2 \end{bmatrix} \begin{bmatrix} x_1(k + 1) \\ x_2(k + 1) \end{bmatrix} + 3v(k + 1)$$

where

$$Q = 0.01 \qquad R = 1 \qquad x_i(0) = N(0, 0.01) \qquad i = 1, 2$$

$$z(1) = -\frac{1}{2} \qquad z(2) = 0.1$$

calculate, using the Kalman filter algorithm, $\hat{x}(2 \mid 2)$.

8-39. For the system

$$x(k + 1) = 0.8x(k) + 2w(k) + 3u(k)$$
$$z(k + 1) = 3x(k + 1) + 4v(k + 1)$$

where

$$Q = 0.01 \qquad R = 1 \qquad x(0) = N(0, 1) \qquad u(k) = 1 \quad \text{for } k \geq 0$$

a. Simulate the system for 100 points

b. Using a Kalman filter program, plot the corrected state estimate, state error covariance, and measurement residuals.

8-40. Using a Kalman filter program, simulate the following systems for 200 points and plot the corrected state estimate, state error covariance, and measurement residuals:

a. System of Problem 8-36

b. System of Problem 8-37

8-41. For the system

$$x(k + 1) = 0.6x(k) + w(k)$$
$$z(k + 1) = 2x(k + 1) + v(k + 1)$$

where

$$Q = 2 \quad R = 1 \quad x(0) = N(0.1, 1)$$
$$E[w(k)] = 0.1 \quad E[v(k + 1)] = 0.4$$

using a Kalman filter program, simulate the system for 200 points and plot the corrected state estimate, state error covariance, and measurement residuals.

8-42. For the system

$$x(k + 1) = \frac{1}{2} x(k) + w(k)$$

$$z(k) = x(k) + v(k)$$

$$Q = 2 \quad R = 1 \quad \hat{x}(0 \mid 0) = 0 \quad P(0 \mid 0) = 0$$

calculate $\hat{x}(2 \mid 2)$, using the Kalman filter algorithm in Table 8-8, in terms of $z(0)$, $z(1)$, and $z(2)$.

8-43. Find the steady state error covariance, estimation error covariance, and Kalman gain for the following systems:
  a. Problem 8-34(a), with $Q = 10$, $R = 2$
  b. Problem 8-35, with $Q = 1$, $R = 10$
  c. Problem 8-36, with $Q = 1$, $R = 0$

8-44. Consider the system described in Section 8.6.3. Suppose that the measurements are available only to step k instead of $k + 1$, that is,

$$z(k) = 2x(k) + v(k)$$

If $\hat{x}(0 \mid -1) = 0$ and $P(0 \mid -1) = 1$, calculate $\hat{x}(2 \mid 2)$, using the Kalman filter algorithm in Table 8-8, in terms of $z(0)$, $z(1)$, and $z(2)$.

8-45. Modify the Kalman filter program in Section 8.6.4 so that the measurements are available only to step k.

8-46. Using the computer program in Problem 8-45, simulate the system in Problem 8-41 for 200 points and plot the corrected state estimate, state error covariance, and measurement residuals.

8-47. Consider the system described in Section 8.8.1.
  a. Analogous to the eigenvalue placement with observer feedback, prove the separation theorem using Kalman filter feedback.

b. Find the transfer function matrix $\mathbf{H}_f(z)$ for the Kalman filter (assume $\mathbf{u} = \mathbf{0}$) relating the state estimate vector to the output vector.

8-48. For the plant

$$x(k + 1) = x(k) + 2u(k) + w(k)$$
$$z(k) = 3x(k) + v(k)$$

a. Find the transfer function of the Kalman filter in terms of Q and R. What is the effect of increasing Q on the filter poles and zeros for a fixed R? What is the effect of increasing R on the filter poles and zeros for a fixed Q?

b. If $Q = 0.0001$, $R = 0.01$, and $P = 1$, design a complete state feedback control system including a Kalman filter to place the eigenvalue of the system at $\lambda = 0.1$. Are the locations of the filter pole and zero satisfactory? If not, what do you suggest?

8-49. Consider the continuous-time system

$$\dot{\mathbf{x}}(t) = \mathbf{F}_c\mathbf{x}(t) + \mathbf{w}(t)$$
$$\mathbf{z}(t) = \mathbf{H}_c\mathbf{x}(t) + \mathbf{v}(t) \tag{1}$$

where $\mathbf{w}(t)$ and $\mathbf{v}(t)$ are zero-mean, continuous-time, noise processes with covariance matrices $\mathbf{Q}_c \, \delta(t - \tau)$ and $\mathbf{R}_c \, \delta(t - \tau)$, respectively. Consider also the discrete-time system

$$\mathbf{x}(k + 1) = \mathbf{F}_d\mathbf{x}(k) + \mathbf{w}(k)$$
$$\mathbf{z}(k + 1) = \mathbf{H}_d\mathbf{x}(k + 1) + \mathbf{v}(k + 1) \tag{2}$$

where $\mathbf{w}(k)$ and $\mathbf{v}(k)$ are zero-mean, white noise sequences with covariance matrices $\mathbf{Q}_d \, \delta(k)$ and $\mathbf{R}_d \, \delta(k)$, respectively. Derive expressions for $\mathbf{F}_d$, $\mathbf{H}_d$, $\mathbf{Q}_d$, and $\mathbf{R}_d$ in terms of $\mathbf{F}_c$, $\mathbf{H}_c$, $\mathbf{Q}_c$, and $\mathbf{R}_c$ such that system 2 is a discrete approximation of system 1.

Elements of Linear Algebra

Linear algebra is the language of modern control system design. In this appendix, we present topics in linear algebra that have immediate application to control system analysis and design. Our emphasis is on summarizing needed concepts and results and on establishing terminology and notation that are used throughout this book.

A.1 Matrix Methods

A *matrix* is a rectangular array of elements, most often numbers. Matrices are symbolized by letters, and the elements that comprise a matrix are denoted by a doubly subscripted letter, usually the lowercase of the symbol for the matrix itself. The first subscript denotes the row number of the element, and the second subscript denotes the column number of the element. A matrix with m rows and n columns is of dimension $m \times n$. A *square* matrix has the same number of rows as columns. A *column vector* (or just "vector") is a matrix with just one column. The elements of a column vector, called an *m-vector*, where m is the number of its elements, are labeled with a single subscript.

A matrix with all elements zero is a *zero* (or *null*) *matrix*. An $m \times n$ zero matrix is denoted by just $\mathbf{0}$ or by $\mathbf{0}_{mn}$ when its dimensions are not obvious. The main diagonal of a square matrix is the diagonal from the upper left-hand corner to the lower right-hand corner. A square matrix is also a *diagonal* matrix if all the elements that are off the main diagonal are zero. A diagonal matrix with all diagonal elements unity is an *identity* matrix. The $n \times n$ identity matrix is denoted by $\mathbf{I}$ or by $\mathbf{I}_n$.

A.1.1 Basic Matrix Operations

Two matrices are equal,

$$\mathbf{A} = \mathbf{B}$$

if and only if they have the same dimensions (they are then said to be *conformable*) and all corresponding elements in the two matrices are equal. The product of a matrix $\mathbf{A}$ with a scalar k is another matrix $\mathbf{B} = k\mathbf{A}$ of the same dimension as $\mathbf{A}$ but with corresponding elements k times those of $\mathbf{A}$.

$$b_{ij} = ka_{ij}$$

The sum or difference of two matrices, $\mathbf{C} = \mathbf{A} \pm \mathbf{B}$, makes sense only if $\mathbf{A}$ and $\mathbf{B}$ are conformable. The sum $\mathbf{C}$ has the same dimension as $\mathbf{A}$ and $\mathbf{B}$ and has elements that are the sums or differences of the corresponding elements of $\mathbf{A}$ and $\mathbf{B}$.

$$c_{ij} = a_{ij} \pm b_{ij}$$

The order in which several conformable matrices are added or subtracted is of no consequence (*associativity*).

A square matrix $\mathbf{A}$ for which $a_{ij} = a_{ji}$ is called *symmetric*. The definition does not restrict the diagonal elements. A square matrix $\mathbf{A}$ for which $a_{ij} = -a_{ji}$ is called *skew-symmetric*. The skew-symmetric definition requires that the diagonal elements be negatives of themselves, which means they must be zero. Any square matrix can be expressed as the sum of a symmetric matrix and a skew-symmetric matrix.

A *dagger* symbol, as in $\mathbf{B} = \mathbf{A}^\dagger$, denotes an interchange of the rows and columns of a matrix. The matrix $\mathbf{B}$ is the *transpose* of the matrix $\mathbf{A}$. The transpose of the transpose is the matrix itself:

$$(\mathbf{A}^\dagger)^\dagger = \mathbf{A}$$

The transpose of a column vector is a *row vector*

$$\mathbf{b}^\dagger = \begin{bmatrix} b_1 \\ b_2 \\ b_3 \\ \vdots \end{bmatrix}^\dagger = [b_1 \quad b_2 \quad b_3 \ldots]$$

The notation used here, which is common, is that lowercase symbols represent column vectors. A row vector is then represented as the transpose of the corresponding column vector.

The *product* of two matrices, $\mathbf{C} = \mathbf{AB}$, makes sense only if

(Number of columns of $\mathbf{A}$) = (Number of rows of $\mathbf{B}$)

The matrices are then said to be *conformable for multiplication*. If $\mathbf{A}$ is $m \times n$ and $\mathbf{B}$ is $n \times p$, the product is $m \times p$.

$$\underset{m \times p}{\mathbf{C}} = \underset{m \times n}{\mathbf{A}} \quad \underset{n \times p}{\mathbf{B}}$$

For conformable matrices,

$$c_{ij} = a_{i1}b_{1j} + a_{i2}b_{2j} + \cdots + a_{in}b_{nj} = \sum_{k=1}^{n} a_{ik}b_{kj}$$

Multiplication of a matrix on the left (*premultiplication*) or on the right (*postmultiplication*) by an identity matrix of proper dimension for conformability has no effect.

$$\mathbf{IA} = \mathbf{A} \quad \text{and} \quad \mathbf{AI} = \mathbf{A}$$

In general, however, products of conformable matrices have these differences from scalar products:

$\mathbf{AB}$ does *not* necessarily equal $\mathbf{BA}$

$\mathbf{AB} = \mathbf{AC}$ does *not* generally imply that $\mathbf{B} = \mathbf{C}$

$\mathbf{AB} = \mathbf{0}$ does *not* generally imply that $\mathbf{B} = \mathbf{0}$ or $\mathbf{A} = \mathbf{0}$

For multiple products, the ordering of the product is important, but the order in which the operations are carried out is of no consequence.

$$\mathbf{ABC} = \mathbf{A(BC)} = \mathbf{(AB)C}$$

Premultiplication and postmultiplication have the distributive property

$$\mathbf{A(B + C)} = \mathbf{AB} + \mathbf{AC}$$
$$\mathbf{(A + B)C} = \mathbf{AC} + \mathbf{BC}$$

and

$$\mathbf{A0} = \mathbf{0A} = \mathbf{0}$$

Partitioning of a matrix is indicated by dashed lines and is a way of indicating how a matrix is composed of smaller *submatrices*, that is, how matrices are joined to form larger matrices. For example,

$$\mathbf{A} = \left[\begin{array}{ccc:ccc} a_{11} & a_{12} & a_{13} & a_{14} & a_{15} \\ a_{21} & a_{22} & a_{23} & a_{24} & a_{25} \\ \hdashline a_{31} & a_{32} & a_{33} & a_{34} & a_{35} \\ a_{41} & a_{42} & a_{43} & a_{44} & a_{45} \end{array}\right] = \left[\begin{array}{c:c} \mathbf{A}_{11} & \mathbf{A}_{12} \\ \hdashline \mathbf{A}_{21} & \mathbf{A}_{22} \end{array}\right]$$

indicates that the matrix $\mathbf{A}$ is considered to be composed of the submatrices $\mathbf{A}_{11}$, $\mathbf{A}_{12}$, $\mathbf{A}_{21}$, and $\mathbf{A}_{22}$.

The partitioning

$$\mathbf{A} = [\mathbf{a} \mid \mathbf{b} \mid \cdots]$$

indicates that the columns of $\mathbf{A}$ are the vectors $\mathbf{a}$, $\mathbf{b}$, ... The partitioning

$$\mathbf{A} = \begin{bmatrix} \mathbf{a}^\dagger \\ \hline \mathbf{b}^\dagger \\ \hline \vdots \end{bmatrix}$$

indicates that the rows of $\mathbf{A}$ are the row vectors $\mathbf{a}^\dagger$, $\mathbf{b}^\dagger$, ... Matrix products, when the matrices involved are partitioned so that the submatrices are conformable for multiplication, have multiplication relations as if the submatrices were numbers, paying due respect to the ordering of the products.

The transpose of a matrix product is the product of the transposed matrices in reverse order

$$(\mathbf{AB})^\dagger = \mathbf{B}^\dagger \mathbf{A}^\dagger$$

Applying this result successively gives

$$(\mathbf{AB} \ldots \mathbf{PQ})^\dagger = \mathbf{Q}^\dagger \mathbf{P}^\dagger \ldots \mathbf{B}^\dagger \mathbf{A}^\dagger$$

The trace of a square matrix $\mathbf{A}$ is denoted by trace $(\mathbf{A})$ and is the sum of the diagonal elements

$$\text{trace } (\mathbf{A}) = a_{11} + a_{22} + \cdots + a_{nn} = \sum_{i=1}^{n} a_{ii}$$

The complex conjugate $\mathbf{A}^*$ of a matrix $\mathbf{A}$ is simply the matrix composed of the complex conjugates of the elements of $\mathbf{A}$.

The derivative of a matrix $\mathbf{A}$ with respect to a scalar is the matrix composed of elements that are the derivatives of the corresponding elements of $\mathbf{A}$. Higher derivatives, partial derivatives, and integrals are defined similarly, as matrices of the same dimension, composed of the corresponding derivatives or integrals of the original elements.

The partial derivative of a scalar function f with respect to a vector $\mathbf{x}$ is denoted by

$$\frac{\partial f}{\partial \mathbf{x}} = \begin{bmatrix} \dfrac{\partial f}{\partial x_1} & \dfrac{\partial f}{\partial x_2} & \cdots & \dfrac{\partial f}{\partial x_n} \end{bmatrix}$$

and is a *row* vector composed of the derivatives of f with respect to each of the components of $\mathbf{x}$. The derivative of a scalar function with respect to a vector is also termed the *gradient operation*. The partial derivative of a vector $\mathbf{y}$ with respect to a vector $\mathbf{x}$ is the matrix

$$
\frac{\partial \mathbf{y}}{\partial \mathbf{x}} =
\begin{bmatrix}
\dfrac{\partial y_1}{\partial \mathbf{x}} \\[2mm]
\dfrac{\partial y_2}{\partial \mathbf{x}} \\[2mm]
\vdots \\[2mm]
\dfrac{\partial y_m}{\partial \mathbf{x}}
\end{bmatrix}
=
\begin{bmatrix}
\dfrac{\partial y_1}{\partial x_1} & \dfrac{\partial y_1}{\partial x_2} & \cdots & \dfrac{\partial y_1}{\partial x_n} \\[2mm]
\dfrac{\partial y_2}{\partial x_1} & \dfrac{\partial y_2}{\partial x_2} & \cdots & \dfrac{\partial y_2}{\partial x_n} \\[2mm]
\vdots & & & \\[2mm]
\dfrac{\partial y_m}{\partial x_1} & \dfrac{\partial y_m}{\partial x_2} & \cdots & \dfrac{\partial y_m}{\partial x_n}
\end{bmatrix}
$$

The partial derivative of a scalar function f with respect to an $n \times m$ matrix

$$
\mathbf{A} = [\mathbf{a}_1 \mid \mathbf{a}_2 \mid \cdots \mid \mathbf{a}_m]
$$

is

$$
\frac{\partial f}{\partial \mathbf{A}} =
\begin{bmatrix}
\dfrac{\partial f}{\partial \mathbf{a}_1} \\[2mm]
\dfrac{\partial f}{\partial \mathbf{a}_2} \\[2mm]
\vdots \\[2mm]
\dfrac{\partial f}{\partial \mathbf{a}_m}
\end{bmatrix}
=
\begin{bmatrix}
\dfrac{\partial f}{\partial a_{11}} & \dfrac{\partial f}{\partial a_{21}} & \cdots & \dfrac{\partial f}{\partial a_{n1}} \\[2mm]
\dfrac{\partial f}{\partial a_{12}} & \dfrac{\partial f}{\partial a_{22}} & \cdots & \dfrac{\partial f}{\partial a_{n2}} \\[2mm]
\vdots & & & \\[2mm]
\dfrac{\partial f}{\partial a_{1m}} & \dfrac{\partial f}{\partial a_{2m}} & \cdots & \dfrac{\partial f}{\partial a_{nm}}
\end{bmatrix}
$$

Some important properties of derivatives involving matrices are listed in Table A-1. It is assumed that each of the variables involved in the partial derivatives is independent.

A.1.2 Determinants and Ranks

The determinant of a square matrix $\mathbf{A}$, denoted by $|\mathbf{A}|$, is uniquely defined by the three properties listed in Table A-2. Important properties that follow from the defining ones are also listed in the table. The determinant of a 2×2 matrix is

$$
\begin{vmatrix}
a_{11} & a_{12} \\
a_{21} & a_{22}
\end{vmatrix}
= a_{11}a_{22} - a_{12}a_{21}
$$

Table A-1 Some Properties of Matrix Derivatives

Derivative of a Matrix with Respect to a Scalar

A. $\dfrac{d}{dt}(k\mathbf{A}) = k\dfrac{d\mathbf{A}}{dt} + \dfrac{dk}{dt}\mathbf{A}$ if k is a scalar

$\dfrac{\partial}{\partial t}(k\mathbf{A}) = k\dfrac{\partial\mathbf{A}}{\partial t} + \dfrac{\partial k}{\partial t}\mathbf{A}$ if k is a scalar

B. $\dfrac{d}{dt}(\mathbf{A} + \mathbf{B}) = \dfrac{d\mathbf{A}}{dt} + \dfrac{d\mathbf{B}}{dt}$

$\dfrac{\partial}{\partial t}(\mathbf{A} + \mathbf{B}) = \dfrac{\partial\mathbf{A}}{\partial t} + \dfrac{\partial\mathbf{B}}{\partial t}$

C. $\dfrac{d}{dt}(\mathbf{AB}) = \dfrac{d\mathbf{A}}{dt}\mathbf{B} + \mathbf{A}\dfrac{d\mathbf{B}}{dt}$

$\dfrac{\partial}{\partial t}(\mathbf{AB}) = \dfrac{\partial\mathbf{A}}{\partial t}\mathbf{B} + \mathbf{A}\dfrac{\partial\mathbf{B}}{\partial t}$

Derivative of a Scalar with Respect to a Vector

D. $\dfrac{\partial}{\partial\mathbf{x}}(\mathbf{x}^\dagger\mathbf{y}) = \dfrac{\partial}{\partial\mathbf{x}}(\mathbf{y}^\dagger\mathbf{x}) = \mathbf{y}^\dagger$

E. $\dfrac{\partial}{\partial\mathbf{x}}(\mathbf{x}^\dagger\mathbf{Ax}) = \mathbf{x}^\dagger\mathbf{A}^\dagger + \mathbf{x}^\dagger\mathbf{A}$

Derivative of a Vector with Respect to a Vector

F. $\dfrac{\partial}{\partial\mathbf{x}}(\mathbf{Ax}) = \mathbf{A}$

Derivative of a Scalar with Respect to a Matrix

G. $\dfrac{\partial}{\partial\mathbf{A}}[\text{trace }(\mathbf{BA})] = \mathbf{B}^\dagger$

Determinants of square matrices of larger dimension can be found by manipulation, by using these properties, or by expanding the determinant of a larger matrix into determinants of smaller matrices.

The ith row, jth column *minor* $|\mathbf{M}_{ij}|$ of a matrix is the determinant of the matrix formed by deleting the ith row and the jth column of the original matrix. The ith row, jth column *cofactor* $|\mathbf{A}_{ij}|$ of a matrix is the same as the minor $|\mathbf{M}_{ij}|$ except for a possible difference in algebraic sign.

$$|\mathbf{A}_{ij}| = (-1)^{i+j}|\mathbf{M}_{ij}|$$

Cofactors are used in the Laplace expansion, which is a systematic method of finding the value of a determinant. In the Laplace expan-

Table A-2 Properties of Determinants

Defining Properties

A. $|\mathbf{A}|$ is unchanged if the elements of any row (or column) are replaced by the sums of the elements of that row (column) and the corresponding elements of another row (column).

B. The value of the determinant is multiplied by k if all of the elements of any row or column are each multiplied by a scalar k.

C. The determinant of an identity matrix is unity.

Further Properties

These properties of determinants follow from the defining properties:

D. $|\mathbf{A}|$ is unchanged if the elements of any row (or column) are replaced by the sums of each of these elements with any number times the corresponding elements of another row (column).

E. The algebraic sign of $|\mathbf{A}|$ is reversed if any two rows or columns are interchanged.

F. $|\mathbf{A}| = 0$ if all elements of a row or column are zero or if the corresponding elements of any two rows or columns are identical or have a common ratio.

G. The determinant of a diagonal matrix is the product of the diagonal elements.

H. $|\mathbf{A}^\dagger| = |\mathbf{A}|$

---

sion, a determinant is expressed in terms of cofactors. The cofactors involve determinants of smaller dimension than the original matrix, and their values in turn can be expressed in terms of *their* cofactors, and so on until the result is in terms of determinants of small matrices. A determinant can be Laplace-expanded along any row or any column. If it is expanded along the ith row, then

$$|\mathbf{A}| = a_{i1}|\mathbf{A}_{i1}| + a_{i2}|\mathbf{A}_{i2}| + \cdots + a_{in}|\mathbf{A}_{in}| = \sum_{j=1}^{n} a_{ij}|\mathbf{A}_{ij}|$$

Expansion along the jth column is similar

$$|\mathbf{A}| = a_{1j}|\mathbf{A}_{1j}| + a_{2j}|\mathbf{A}_{2j}| + \cdots + a_{nj}|\mathbf{A}_{nj}| = \sum_{i=1}^{n} a_{ij}|\mathbf{A}_{ij}|$$

If $\mathbf{A}$ and $\mathbf{B}$ are each square and of the same dimension, then

$$|\mathbf{AB}| = |\mathbf{A}||\mathbf{B}|$$

as can be verified by comparing the Laplace expansions of each side. In general, for $\mathbf{A}$, $\mathbf{B}$, . . . , $\mathbf{P}$, $\mathbf{Q}$, all $n \times n$,

$$|\mathbf{AB} \ldots \mathbf{PQ}| = |\mathbf{A}||\mathbf{B}| \ldots |\mathbf{P}||\mathbf{Q}|$$

The rank of a matrix, denoted by rank($\mathbf{A}$), is the dimension of the largest square array having nonzero determinants within the matrix $\mathbf{A}$ formed by deleting rows or columns or both as necessary. A matrix of the largest possible rank for its dimension is said to be of *full rank*. A matrix $\mathbf{A}$ and its transpose have the same rank.

$$\text{rank}(\mathbf{A}^\dagger) = \text{rank}(\mathbf{A})$$

A square matrix $\mathbf{A}$ that is not of full rank has $|\mathbf{A}| = 0$ and is said to be *singular*.

A.2 Linear Algebraic Equations

We now review the properties of linear algebraic equations and the use of matrices to represent them. A set of m simultaneous linear algebraic equations in the n variables $x_1, x_2, \ldots, x_n$ has the form

$$\begin{cases} a_{11}x_1 + a_{12}x_2 + \cdots + a_{1n}x_n = y_1 \\ a_{21}x_1 + a_{22}x_2 + \cdots + a_{2n}x_n = y_2 \\ \vdots \\ a_{m1}x_1 + a_{m2}x_2 + \cdots + a_{mn}x_n = y_m \end{cases} \tag{A-1}$$

where the a's are the *coefficients* of the equations. The left side of each equation is referred to as the "unknowns" side; it involves the x's, which are the "unknowns." The right side is the "knowns" side and consists of the y's, which are the "knowns." A set of variables ($x_1, x_2, \ldots, x_n$) satisfying all of the equations is a *solution* of the equations. A set of equations might have no solution, as is the case for the set

$$\begin{cases} x_1 + 2x_2 = 4 \\ 3x_1 + 6x_2 = -5 \end{cases}$$

It might have a unique solution, as does

$$\begin{cases} x_1 + x_2 = 1 \\ 2x_1 - x_2 = 8 \end{cases}$$

Or a set may have numerous solutions, as does

$$\begin{cases} x_1 + 3x_2 = 4 \\ -2x_1 - 6x_2 = -8 \end{cases}$$

One method of solving a set of linear algebraic equations is to replace the original set of equations with a new set that has the same solution but is simpler than the original set. Any equation in a set can be replaced by the sum of that equation and a nonzero constant times any other equation in the set. The new set of equations is equivalent to the original set; that is, it has the same solution(s).

Gauss-Jordan pivoting is a systematic procedure for obtaining the simplest set of equivalent algebraic equations. Multiples of one equation in the set are added to each other equation in the set to make the coefficients of the first variable in the other equations zero. The equation used to eliminate terms in the other equations is called the *pivot equation* for this *pivot cycle*. Then another equation (the next pivot equation) is selected, and multiples of that equation are added to each other equation to make the coefficients of the second variable in all other equations zero. The second pivot cycle is then complete. The process continues until all equations have been used as pivots, or until each variable has been eliminated from all but one equation, or both.

An equation

$$a_{11}x_1 + a_{12}x_2 + \cdots + a_{1n}x_n = y_1 \qquad \textbf{(A-2)}$$

is *linearly dependent* on other equations

$$a_{21}x_1 + a_{22}x_2 + \cdots + a_{2n}x_n = y_2$$
$$a_{31}x_1 + a_{32}x_2 + \cdots + a_{3n}x_n = y_3$$
$$\vdots$$
$$a_{m1}x_1 + a_{m2}x_2 + \cdots + a_{mn}x_n = y_m \qquad \textbf{(A-3)}$$

if it can be expressed as a linear combination of the other equations

$$a_{11}x_1 + a_{12}x_2 + \cdots + a_{1n}x_n = k_1(a_{21}x_1 + a_{22}x_2 + \cdots + a_{2n}x_n)$$
$$+ k_2(a_{31}x_1 + a_{32}x_2 + \cdots + a_{3n}x_n)$$
$$+ \cdots + k_{m-1}(a_{m1}x_1 + a_{m2}x_2 + \cdots + a_{mn}x_n) \quad \textbf{(A-4)}$$

and

$$y_1 = k_1 y_2 + k_2 y_3 + \cdots + k_{m-1}y_m \qquad \textbf{(A-5)}$$

for some constants $k_1, k_2, \ldots, k_{m-1}$. A linearly dependent equation is redundant and can be deleted from a set of equations without affecting any solution. Indeed, it is highly desirable to make such deletions to simplify the set of equations and to place the set in a nonredundant form. A linearly dependent equation is indicated by an equation with an entire row of zero coefficients and a zero ''known'' after Gauss-Jordan pivoting.

Equation (A-2) is said to be *inconsistent* with other equations (A-3) if the "unknowns" side of that equation can be expressed as a linear combination of the "unknowns" sides of the other equations, as in equation (A-4), but the same linear combination on the "knowns" side (A-5) is *not* equal.

$$y_1 \neq k_1 y_2 + k_2 y_3 + \cdots + k_{m-1} y_m$$

The presence of one or more inconsistent equations in a set means that the set has no solution. Inconsistent equations are indicated by an equation with an entire row of zero coefficients and a nonzero "known" after Gauss-Jordan pivoting.

A set of *homogeneous equations,* a set with all the "knowns" zero,

$$\begin{cases} a_{11}x_1 + a_{12}x_2 + \cdots + a_{1n}x_n = 0 \\ a_{21}x_1 + a_{22}x_2 + \cdots + a_{2n}x_n = 0 \\ \vdots \\ a_{m1}x_1 + a_{m2}x_2 + \cdots + a_{mn}x_n = 0 \end{cases}$$

cannot be inconsistent because any linear combination of elements on the "knowns" side always gives zero. Another way of viewing this result is that for homogeneous equations there is always the *trivial solution*

$$\begin{cases} x_1 = 0 \\ x_2 = 0 \\ \vdots \\ x_n = 0 \end{cases}$$

There might be other solutions that are nontrivial.

Suppose a set of linear algebraic equations is consistent and suppose that any linearly dependent equations have been deleted so that the set consists entirely of linearly independent equations. If there are n variables, the set cannot consist of more than n linearly independent equations because those equations in excess of n equations can be shown, using pivoting, to be linearly dependent on the others. If there are exactly n equations in the set, there is a *unique* solution to the set. If, instead, there are fewer than n equations, the set has a whole family of different solutions. If there are $(n - p)$ linearly independent equations in the set, the solutions can be found by assigning arbitrary values to p of the variables and solving the resulting $(n - p)$ linearly independent equations in $(n - p)$ variables. For example, in the set of equations

$$\begin{cases} x_1 + x_2 & = & 2 \\ x_1 - x_2 & = & -3 \\ x_2 - x_3 + x_4 = & 0 \end{cases}$$

the variables x_1 and x_2 are determined by the first two equations, but either of the variables x_3 or x_4 can be assigned an arbitrary value.

A set of m equations in n variables, as in equation (A-2), is written in matrix notation as

$$\mathbf{Ax} = \mathbf{y}$$

Matrix notation is useful because it saves a lot of writing and because it greatly simplifies descriptions of equation manipulations.

When a unique solution exists to a set of n linear algebraic equations in n variables

$$\mathbf{Ax} = \mathbf{y}$$

where $\mathbf{A}$ is $n \times n$, the solution can be given by *Cramer's rule*. The solution for the ith variable is

$$x_i = \frac{|\mathbf{A}_i|}{|\mathbf{A}|}$$

where the ith row *cofactor* $|\mathbf{A}_i|$ is the same as $|\mathbf{A}|$, except that the ith column is replaced by the column of "knowns."

Because some pairs of matrices are not conformable for multiplication and because, when they are conformable, the order in which two matrices are multiplied generally makes a difference to the product, matrix division is not defined. Instead, for square matrices with non-zero determinants, the *inverse* matrix is used. The product of a matrix with its inverse is an identity matrix

$$\mathbf{AA}^{-1} = \mathbf{I} \tag{A-6}$$

The inverse of a matrix can be found by using Cramer's rule to solve equation (A-6), obtaining

$$\mathbf{A}^{-1} = \frac{1}{|\mathbf{A}|} \begin{bmatrix} |\mathbf{A}_{11}| & |\mathbf{A}_{21}| & \cdots & |\mathbf{A}_{n1}| \\ |\mathbf{A}_{12}| & |\mathbf{A}_{22}| & \cdots & |\mathbf{A}_{n2}| \\ \vdots & & & \\ |\mathbf{A}_{1n}| & |\mathbf{A}_{2n}| & \cdots & |\mathbf{A}_{nn}| \end{bmatrix} = \frac{\text{adj } \mathbf{A}}{|\mathbf{A}|}$$

The *adjugate* (or *adjoint*) of a square matrix $\mathbf{A}$, denoted by adj $\mathbf{A}$, is the transpose of the matrix in which each element of $\mathbf{A}$ is replaced by its cofactor. If $|\mathbf{A}| = 0$, $\mathbf{A}^{-1}$ does not exist, and the matrix $\mathbf{A}$ is *singular*. Some properties of inverse matrices are listed in Table A-3.

Integer powers of a square matrix are defined according to

$$\mathbf{A}^i = \underbrace{\mathbf{A} \cdot \cdot \cdot \mathbf{A}}_{i \text{ terms}}$$

The additional definitions

$$\mathbf{A}^0 = \mathbf{I}$$

and (providing $\mathbf{A}$ is nonsingular)

$$\mathbf{A}^{-i} = \underbrace{(\mathbf{A}^{-1})(\mathbf{A}^{-1}) \cdot \cdot \cdot (\mathbf{A}^{-1})}_{i \text{ terms}}$$

are useful to make, so that

$$\mathbf{A}^{i_1}\mathbf{A}^{i_2} = \mathbf{A}^{(i_1 + i_2)}$$

Table A-3 Properties of Inverse Matrices

Defining Property

A. $\mathbf{A}\mathbf{A}^{-1} = \mathbf{I}$

Further Properties

These properties follow from the defining property:

B. $\mathbf{A}^{-1}\mathbf{A} = \mathbf{I}$

C. $\mathbf{I}^{-1} = \mathbf{I}$

D. $[\mathbf{A}^{-1}]^{-1} = \mathbf{A}$

E. $[\mathbf{A}^{-1}]^{\dagger} = [\mathbf{A}^{\dagger}]^{-1}$

F. The inverse of a nonsingular diagonal matrix $\mathbf{D}$ is diagonal. It has diagonal elements that are the reciprocals of the diagonal elements of $\mathbf{D}$.

G. $(k\mathbf{A})^{-1} = \dfrac{1}{k}\mathbf{A}^{-1}$ for k a nonzero scalar

H. $(\mathbf{A}\mathbf{B})^{-1} = \mathbf{B}^{-1}\mathbf{A}^{-1}$, provided that $\mathbf{A}$ and $\mathbf{B}$ are each square and nonsingular. Applying this result repeatedly

$$(\mathbf{A}\mathbf{B} \ldots \mathbf{P}\mathbf{Q})^{-1} = \mathbf{Q}^{-1}\mathbf{P}^{-1} \ldots \mathbf{B}^{-1}\mathbf{A}^{-1}$$

I. $\dfrac{d}{dt}(\mathbf{A}^{-1}) = -\mathbf{A}^{-1}\dfrac{d\mathbf{A}}{dt}\mathbf{A}^{-1}$

for all integers i_1 and i_2. A matrix polynomial is a function of a square matrix $\mathbf{A}$ of the form

$$\mathbf{F}(\mathbf{A}) = c_0\mathbf{I} + c_1\mathbf{A} + c_2\mathbf{A}^2 + c_3\mathbf{A}^3 + \cdots$$

where $c_1, c_2, c_3, \ldots,$ are scalars. If there is a finite number of terms in the polynomial, the polynomial is said to be finite. Otherwise, it is an infinite polynomial.

A.3 Vectors and Transformations

Solutions to linear equations are represented by vectors, as are many other quantities of special importance in linear algebra. Assignment of scalar size to each vector $\mathbf{x}$ in a set is termed a vector *norm*, denoted by $\|\mathbf{x}\|$, and required to have the four properties listed in Table A-4. The most commonly used norm is the *Euclidean norm*, which for real vector elements $x_1, x_2, \ldots, x_n$ is the vector magnitude or length

$$\|\mathbf{x}\| = \sqrt{x_1^2 + x_2^2 + \cdots + x_n^2} = \sqrt{\mathbf{x}^\dagger\mathbf{x}}$$

Norms other than the Euclidean norm find occasional uses. One possibility is the sum of the magnitudes of the elements of $\mathbf{x}$. Another is the maximum of the magnitudes of the elements of $\mathbf{x}$. Hereafter, the term *norm* means Euclidean norm unless otherwise noted. If the norm of a vector is unity, the vector is called a *unit vector*.

An *inner product* assigns a scalar measure of closeness in the direction of two vectors of the same dimension. For two vectors $\mathbf{x}$ and $\mathbf{y}$, the inner product is denoted by $\langle \mathbf{x}, \mathbf{y} \rangle$ and is a scalar function of the elements of the two vectors. It is required to have the four properties listed in Table A-5. The most commonly used inner product is the Euclidean inner product, which for real vector elements is

$$\langle \mathbf{x}, \mathbf{y} \rangle = \mathbf{x}^\dagger\mathbf{y} = x_1 y_1 + x_2 y_2 + \cdots + x_n y_n$$

Although other inner products are sometimes useful, the term *inner product* means the Euclidean inner product unless otherwise noted.

Table A-4 Requirements of a Vector Norm

A. $\|\mathbf{x}\| > 0$ for $\mathbf{x} \neq 0$
B. $\|\mathbf{x}\| = 0$ for $\mathbf{x} = 0$
C. $\|k\mathbf{x}\| = |k|\,\|\mathbf{x}\|$ for any scalar k
D. $\|\mathbf{x}\| + \|\mathbf{y}\| \geq \|\mathbf{x} + \mathbf{y}\|$ (triangle inequality)

Table A-5 Requirements of a Vector Inner Product

A. $\langle \mathbf{y}, \mathbf{x} \rangle = \langle \mathbf{x}, \mathbf{y} \rangle^*$ where * denotes the complex conjugate
B. $\langle k\mathbf{x}, \mathbf{y} \rangle = k\langle \mathbf{x}, \mathbf{y} \rangle = \langle \mathbf{x}, k^*\mathbf{y} \rangle$ where k is a scalar
C. $\langle \mathbf{x}, \mathbf{x} \rangle > 0$ for $\mathbf{x} \neq 0$
D. $\langle \mathbf{x} + \mathbf{u}, \mathbf{y} + \mathbf{v} \rangle = \langle \mathbf{x}, \mathbf{y} \rangle + \langle \mathbf{x}, \mathbf{v} \rangle + \langle \mathbf{u}, \mathbf{y} \rangle + \langle \mathbf{u}, \mathbf{v} \rangle$

Two vectors of nonzero norm $\mathbf{x}$ and $\mathbf{y}$ are *orthogonal* if

$$\langle \mathbf{x}, \mathbf{y} \rangle = 0$$

Orthogonality is the same as perpendicularity for vectors with real elements when the inner product used is the usual Euclidean one. For example, the 3-vectors

$$\mathbf{x} = \begin{bmatrix} 1 \\ -1 \\ 0 \end{bmatrix} \qquad \mathbf{y} = \begin{bmatrix} 1 \\ 1 \\ 0 \end{bmatrix} \qquad \text{and} \qquad \mathbf{z} = \begin{bmatrix} 0 \\ 0 \\ 3 \end{bmatrix}$$

are *mutually orthogonal* because

$$\begin{cases} \langle \mathbf{x}, \mathbf{y} \rangle = \mathbf{x}^\dagger \mathbf{y} = 0 \\ \langle \mathbf{x}, \mathbf{z} \rangle = \mathbf{x}^\dagger \mathbf{z} = 0 \\ \langle \mathbf{y}, \mathbf{z} \rangle = \mathbf{y}^\dagger \mathbf{z} = 0 \end{cases}$$

It is occasionally convenient to introduce superscript indices on vector symbols. Because x_i is used to denote the ith element of the vector $\mathbf{x}$, a superscript, for example $\mathbf{x}^j$, is used to indicate the jth vector in some set of vectors. There is no conflict with the notation for an exponent here because a vector is not conformable for multiplication with itself. A set of m vectors $\mathbf{x}^1, \mathbf{x}^2, \ldots, \mathbf{x}^m$ is said to be *linearly independent* if there are no scalars $k_1, k_2, \ldots, k_m$ except all k's zero for which

$$k_1\mathbf{x}^1 + k_2\mathbf{x}^2 + \cdots + k_m\mathbf{x}^m = \mathbf{0}$$

Otherwise, the set of vectors is *linearly dependent*. Linear independence means that no vector in the set is a linear combination of the other vectors in the set.

For example, the 3-vectors

$$\begin{bmatrix} 1 \\ 0 \\ 1 \end{bmatrix} \qquad \begin{bmatrix} 0 \\ 2 \\ -3 \end{bmatrix} \qquad \text{and} \qquad \begin{bmatrix} 1 \\ 3 \\ 2 \end{bmatrix}$$

are linearly independent because the equations

$$k_1 \begin{bmatrix} 1 \\ 0 \\ 1 \end{bmatrix} + k_2 \begin{bmatrix} 0 \\ 2 \\ -3 \end{bmatrix} + k_3 \begin{bmatrix} 1 \\ 3 \\ 2 \end{bmatrix} = \begin{bmatrix} 0 \\ 0 \\ 0 \end{bmatrix}$$

or

$$\begin{cases} k_1 + 0k_2 + k_3 = 0 \\ 0k_1 + 2k_2 + 3k_3 = 0 \\ k_1 - 3k_2 + 2k_3 = 0 \end{cases}$$

have no solution for the k's except all k's are zero.

No more than n n-vectors can be linearly independent of one another. A set of n linearly independent n-vectors can always be found; for example, the *unit coordinate vectors*

$$\mathbf{i}_1 = \begin{bmatrix} 1 \\ 0 \\ 0 \\ \vdots \\ 0 \\ 0 \end{bmatrix} \quad \mathbf{i}_2 = \begin{bmatrix} 0 \\ 1 \\ 0 \\ \vdots \\ 0 \\ 0 \end{bmatrix} \cdots \mathbf{i}_n = \begin{bmatrix} 0 \\ 0 \\ 0 \\ \vdots \\ 0 \\ 1 \end{bmatrix}$$

are obviously linearly independent. Any set of n linearly independent n-vectors is said to form a *basis* for the n-dimensional space of all n-vectors, and any n-vector can be expressed as a unique linear combination of a set of basis vectors.

A.3.1 Vector Interpretation of Linear Equations

In a set of linear algebraic equations

$$\mathbf{Ax} = \mathbf{y}$$

one may think of the rows of $\mathbf{A}$ as vectors

$$\mathbf{A} = \begin{bmatrix} \mathbf{a}_1^\dagger \\ \hline \mathbf{a}_2^\dagger \\ \hline \vdots \\ \hline \mathbf{a}_m^\dagger \end{bmatrix}$$

A solution $\mathbf{x}$ then has the property

$$\begin{cases} \langle \mathbf{a}_1, \mathbf{x} \rangle = \mathbf{a}_1^\dagger \mathbf{x} = y_1 \\ \langle \mathbf{a}_2, \mathbf{x} \rangle = \mathbf{a}_2^\dagger \mathbf{x} = y_2 \\ \quad \vdots \\ \langle \mathbf{a}_m, \mathbf{x} \rangle = \mathbf{a}_m^\dagger \mathbf{x} = y_m \end{cases} \qquad \text{(A-7)}$$

That is, the inner products of $\mathbf{x}$ with each of the vectors $\mathbf{a}_1, \mathbf{a}_2, \ldots, \mathbf{a}_m$ are $y_1, y_2, \ldots, y_m$.

For a set of homogeneous equations

$$\mathbf{Ax} = \mathbf{0} \qquad \text{(A-8)}$$

relation (A-7) with $\mathbf{y} = \mathbf{0}$ shows that a nontrivial solution $\mathbf{x}$, if it exists, is orthogonal to each of the vectors $\mathbf{a}_1, \mathbf{a}_2, \ldots, \mathbf{a}_m$. As no more than n n-vectors can be linearly independent, if there are n linearly independent n-vectors among the $\mathbf{a}$'s, that is, if the rank of $\mathbf{A}$ is n, the trivial solution

$$\mathbf{x} = \mathbf{0}$$

is the only possible solution of equation (A-8). If there are not n linearly independent n-vectors among the $\mathbf{a}$'s, then a nontrivial solution for $\mathbf{x}$ exists because it is always possible to find an n-vector that is orthogonal to less than n different n-vectors. The norm of any nontrivial solution is arbitrary because its only requirement is that it must be orthogonal to each of the $\mathbf{a}$'s. If only $n - p$ of the rows of $\mathbf{A}$ are linearly independent, that is, if $\mathbf{A}$ is of rank $n - p$, then p linearly independent n-vectors can be found that are orthogonal to each of the rows of $\mathbf{A}$.

As a numerical example of finding nontrivial solutions to homogeneous equations, consider the set

$$\begin{bmatrix} 1 & 2 & 3 \\ -1 & -2 & -3 \\ 2 & 4 & 6 \end{bmatrix} \begin{bmatrix} x_1 \\ x_2 \\ x_3 \end{bmatrix} = \mathbf{Ax} = \mathbf{0}$$

The matrix $\mathbf{A}$ has rank 1, the second and third equations being multiples of the first. Deleting the second and third equations (which are automatically satisfied if the first equation is) gives

$$x_1 + 2x_2 + 3x_3 = 0$$

Choosing $x_2 = 1$ and $x_3 = 0$ gives

$$\mathbf{x}^1 = \begin{bmatrix} -2 \\ 1 \\ 0 \end{bmatrix}$$

which is a nontrivial solution. Another solution is obtained by choosing $x_2 = 0$ and $x_3 = 1$

$$\mathbf{x}^2 = \begin{bmatrix} -3 \\ 0 \\ 1 \end{bmatrix}$$

and this solution is linearly independent of $\mathbf{x}^1$. The vectors

$$\begin{bmatrix} 1 \\ 2 \\ 3 \end{bmatrix} \qquad \mathbf{x}^1 = \begin{bmatrix} -2 \\ 1 \\ 0 \end{bmatrix} \qquad \text{and} \qquad \mathbf{x}^2 = \begin{bmatrix} -3 \\ 0 \\ 1 \end{bmatrix}$$

are linearly independent, and $\mathbf{x}^1$ and $\mathbf{x}^2$ are each orthogonal to the first vector. Any solution of the original homogeneous equations is a linear combination of $\mathbf{x}^1$ and $\mathbf{x}^2$.

A.3.2 Linear Transformations

Linear transformations are of special importance in the analysis of many problems because the description of system properties is often greatly simplified by judicious choice of a coordinate system. The multiplication of a vector $\mathbf{x}$ by a square matrix to form another vector $\mathbf{y}$

$$\mathbf{Ax} = \mathbf{y}$$

can be thought of as a transformation that converts the vector $\mathbf{x}$ to the vector $\mathbf{y}$. The transformation is linear because it always maps any scaled vector $\mathbf{x}$ to the $\mathbf{y}$ vector scaled by the same factor

$$\mathbf{A}(k\mathbf{x}) = (k\mathbf{y})$$

and sums of vectors map to the corresponding sums of individual results. If

$$\mathbf{Ax}^1 = \mathbf{y}^1 \qquad \text{and} \qquad \mathbf{Ax}^2 = \mathbf{y}^2$$

then

$$\mathbf{A}(\mathbf{x}^1 + \mathbf{x}^2) = \mathbf{y}^1 + \mathbf{y}^2$$

If the matrix $\mathbf{A}$ is square and nonsingular, the vector $\mathbf{x}$ can be recovered from $\mathbf{y}$ through

$$\mathbf{x} = \mathbf{A}^{-1}\mathbf{y}$$

Each possible vector $\mathbf{x}$ is transformed to a unique vector $\mathbf{y}$ by a nonsingular $n \times n$ transformation $\mathbf{A}$, and each $\mathbf{y}$ corresponds to a unique $\mathbf{x}$.

If, instead, the rank of A is $m < n$ so that just m of the rows of A are linearly independent, then the space of possible vectors y is m-dimensional. This space is called the *range space* of A. A basis for the range space of A is the largest set of linearly independent *columns* of A

$$A = [c_1 \mid c_2 \mid \cdots \mid c_n]$$

because a general vector x maps to

$$y = x_1 c_1 + x_2 c_2 + \cdots + x_n c_n$$

If the matrix A of a linear transformation is not of full rank, different vectors x are transformed to the same vector y. The space of vectors x for which

$$Ax = 0 \qquad\qquad (A-9)$$

is called the *null space* of the matrix A. Any vector x^0 in this space can be added to any vector x with no effect on the resulting y

$$y = Ax = A(x + x^0)$$

A basis for the null space of A is the largest set of linearly independent vectors x satisfying equation (A-9).

For example, for

$$y = \begin{bmatrix} 1 & 0 & 1 \\ 1 & 1 & 0 \\ 0 & -1 & 1 \end{bmatrix} x = Ax$$

the matrix A has rank 2. The pairs of vectors

$$\begin{bmatrix} 1 \\ 1 \\ 0 \end{bmatrix} \quad \text{and} \quad \begin{bmatrix} 0 \\ 1 \\ -1 \end{bmatrix}$$

$$\begin{bmatrix} 1 \\ 1 \\ 0 \end{bmatrix} \quad \text{and} \quad \begin{bmatrix} 1 \\ 0 \\ 1 \end{bmatrix}$$

$$\begin{bmatrix} 0 \\ 1 \\ -1 \end{bmatrix} \quad \text{and} \quad \begin{bmatrix} 1 \\ 0 \\ 1 \end{bmatrix}$$

each form a basis for the two-dimensional range space of A.

A nontrivial solution to

$$\begin{bmatrix} 1 & 0 & 1 \\ 1 & 1 & 0 \\ 0 & -1 & 1 \end{bmatrix} \mathbf{x} = \mathbf{0}$$

is

$$\mathbf{x} = \begin{bmatrix} -1 \\ 1 \\ 1 \end{bmatrix}$$

and this is a basis vector for the null space of **A**.

If a matrix **A**, which need not be square, is premultiplied or post-multiplied by a square nonsingular matrix **Q**, the rank of the product is the same as the rank of **A**. If the matrix **A** is taken to be the coefficient matrix for a set of homogeneous equations

$$\mathbf{Ax} = \mathbf{0}$$

the rank of **A** is the number of linearly independent equations in the set. Postmultiplication of **A** by a conformable, nonsingular matrix **Q** can be interpreted as a change of variables

$$\mathbf{x} = \mathbf{Qx}'$$

$$\mathbf{Ax} = (\mathbf{AQ})\mathbf{x}' = \mathbf{0}$$

in the equations. Because the old variables can be recovered from the new ones through

$$\mathbf{x}' = \mathbf{Q}^{-1}\mathbf{x}$$

the change of variables cannot change the intrinsic number of linearly independent equations in the set. Premultiplication by a nonsingular matrix also leaves the rank unchanged.

A.3.3 Orthogonal Matrices

If a nonsingular matrix **A** satisfies

$$\mathbf{A}^{-1} = \mathbf{A}^{\dagger}$$

the matrix **A** is termed an *orthogonal matrix*. Some properties of orthogonal matrices are listed in Table A-6.

Table A-6 Properties of Orthogonal Matrices

Defining Property

A. $\mathbf{A}^{-1} = \mathbf{A}^{\dagger}$

Further Properties

These properties follow from the defining propety

B. $|\mathbf{A}| = \pm 1$
C. The rows of an $n \times n$ orthogonal matrix $\mathbf{A}$ are an orthonormal set of n-vectors.
D. The columns of an $n \times n$ orthogonal matrix $\mathbf{A}$ are an orthonormal set of n-vectors.

For example, the matrix

$$\mathbf{A} = \begin{bmatrix} \dfrac{\sqrt{2}}{2} & 0 & -\dfrac{\sqrt{2}}{2} \\ 0 & 1 & 0 \\ \dfrac{\sqrt{2}}{2} & 0 & \dfrac{\sqrt{2}}{2} \end{bmatrix}$$

is an orthogonal matrix because it satisfies all the properties of orthogonal matrices.

A.4 The Characteristic Value Problem

One of the most useful changes of state variables is the one that results in a diagonal state coupling matrix. We now discuss the characteristic value problem. It is the solution to the problem of transforming linear, step-invariant state equations to diagonal form. Because of its close relation to diagonalization and its importance to topics covered in this book, the Cayley-Hamilton theorem is also covered in this section.

A.4.1 Eigenvalues and Eigenvectors

The characteristic value problem of linear algebra addresses the following question: For the set of linear equations

$$\mathbf{Ax} = \mathbf{y}$$

where $\mathbf{A}$ is square, is there a nonsingular transformation of the vectors

x and **y**

$$\mathbf{x} = \mathbf{P}\mathbf{x}' \qquad \mathbf{x}' = \mathbf{P}^{-1}\mathbf{x}$$

$$\mathbf{y} = \mathbf{P}\mathbf{y}' \qquad \mathbf{y}' = \mathbf{P}^{-1}\mathbf{y}$$

such that, in terms of the new coordinates, the equations are

$$\Lambda\mathbf{x}' = \mathbf{y}'$$

where Λ is a diagonal matrix? It is more convenient to define **P** as the transformation from **x'** to **x**, as has been done here, rather than as the transformation from **x** to **x'**. A transformation matrix **P** with this property is called a *modal matrix*. Substituting and premultiplying gives

$$\mathbf{A}\mathbf{x} = \mathbf{A}\mathbf{P}\mathbf{x}' = \mathbf{P}\mathbf{y}'$$

$$(\mathbf{P}^{-1}\mathbf{A}\mathbf{P})\mathbf{x}' = \mathbf{y}'$$

so that the diagonal matrix, called a *spectral matrix*, is related to **A** by

$$\Lambda = \mathbf{P}^{-1}\mathbf{A}\mathbf{P}$$

In terms of the spectral matrix Λ,

$$\mathbf{A}\mathbf{P} = \mathbf{P}\Lambda = [\mathbf{p}^1 \mid \mathbf{p}^2 \mid \cdots \mid \mathbf{p}^n] \begin{bmatrix} \lambda_1 & 0 & 0 & \cdots & 0 \\ 0 & \lambda_2 & 0 & \cdots & 0 \\ \vdots & & & & \\ 0 & 0 & 0 & \cdots & \lambda_n \end{bmatrix}$$

where the columns of the modal matrix **P** are denoted by the vectors $\mathbf{p}^1$, $\mathbf{p}^2, \ldots, \mathbf{p}^n$, and the elements along the diagonal of the spectral matrix Λ (which is a diagonal matrix) are $\lambda_1, \lambda_2, \ldots, \lambda_n$. Each column of **P** satisfies an equation of the form

$$\mathbf{A}\mathbf{p} = \lambda\mathbf{p}$$

or

$$(\lambda\mathbf{I} - \mathbf{A})\mathbf{p} = \mathbf{0} \tag{A-10}$$

Of course, the trivial solution $\mathbf{p} = \mathbf{0}$ is of no help because a column of zeros makes **P** singular.

The set of n individual homogeneous equations (A-10) in the n elements of **p** has a nontrivial solution only if their determinant is zero

$$|\lambda\mathbf{I} - \mathbf{A}| = 0$$

This is an nth-degree polynomial equation in λ, termed the *characteristic equation* of the matrix $\mathbf{A}$

$$\lambda^n + \alpha_{n-1}\lambda^{n-1} + \cdots + \alpha_1\lambda + \alpha_0 = 0$$

The n roots of the characteristic equation, $\lambda_1, \lambda_2, \ldots, \lambda_n$ are the *eigenvalues* of the matrix $\mathbf{A}$. A simple example of finding the three eigenvalues of the 3×3 matrix

$$\mathbf{A} = \begin{bmatrix} -1 & -2 & 0 \\ 1 & 2 & 0 \\ -2 & -1 & -3 \end{bmatrix}$$

is as follows:

$$\lambda\mathbf{I} - \mathbf{A} = \begin{bmatrix} (\lambda + 1) & 2 & 0 \\ -1 & (\lambda - 2) & 0 \\ 2 & 1 & (\lambda + 3) \end{bmatrix}$$

$$|\lambda\mathbf{I} - \mathbf{A}| = \lambda^3 + 2\lambda^2 - 3\lambda = \lambda(\lambda - 1)(\lambda + 3) = 0$$

$$\lambda_1 = 0 \quad \lambda_2 = 1 \quad \lambda_3 = -3$$

An $n \times n$ matrix always has an nth-degree characteristic equation and thus n eigenvalues. If the elements of $\mathbf{A}$ are real numbers, the coefficients of the characteristic equation are real numbers, and complex roots always occur in conjugate pairs. Table A-7 lists some other important properties of the eigenvalues of a matrix.

Corresponding to each of the n eigenvalues λ_i of a matrix $\mathbf{A}$ is an *eigenvector* $\mathbf{p}^i$, satisfying

$$\mathbf{A}\mathbf{p}^i = \lambda_i\mathbf{p}^i$$

The eigenvectors form the columns of the desired transformation matrix $\mathbf{P}$. If the eigenvalues λ_i are distinct, the eigenvectors can be found by determining nontrivial solutions $\mathbf{p}^i$ to the n homogeneous equations of the form

$$(\lambda_i\mathbf{I} - \mathbf{A})\mathbf{p}^i = \mathbf{0}$$

each involving a different eigenvalue λ_i. The sets of homogeneous equations each have a nontrivial solution, because the λ_i's are precisely the numbers necessary to make $(\lambda_i\mathbf{I} - \mathbf{A})$ singular. As with any solution of a set of homogeneous equations, a nonzero constant times

Table A-7 Properties of the Eigenvalues of a Matrix

Defining Property

A. $|\lambda \mathbf{I} - \mathbf{A}| = 0$

Further Properties

These properties follow from the defining property:

B. $\lambda = 0$ is an eigenvalue of $\mathbf{A}$ if and only if $\mathbf{A}$ is singular. Because for $\lambda = 0$

$$|\lambda \mathbf{I} - \mathbf{A}| = |-\mathbf{A}| = \pm|\mathbf{A}| = 0$$

C. The eigenvalues of the matrix $(k\mathbf{A})$ are k times the eigenvalues of $\mathbf{A}$, for any scalar k because

$$|k\lambda \mathbf{I} - k\mathbf{A}| = 0$$

for each value of λ for which

$$|\lambda \mathbf{I} - \mathbf{A}| = 0$$

D. The eigenvalues of $\mathbf{A}^t$ are the same as the eigenvalues of $\mathbf{A}$.

E. The eigenvalues of $\mathbf{A}^{-1}$, provided $\mathbf{A}^{-1}$ exists, are the inverses of the eigenvalues of $\mathbf{A}$

$$|\lambda \mathbf{I} - \mathbf{A}| = |\lambda \mathbf{A}\mathbf{A}^{-1} - \mathbf{A}| = |\mathbf{A}(\lambda \mathbf{A}^{-1} - \mathbf{I})| = |-\lambda \mathbf{A}|\left|\left(\frac{1}{\lambda}\right)\mathbf{I} - \mathbf{A}^{-1}\right|$$

F. The eigenvalues of $\mathbf{A}^k$ (k an integer) are the eigenvalues of $\mathbf{A}$ raised to the kth power

$$0 = |\lambda \mathbf{I} - \mathbf{A}| = |\lambda \mathbf{I} - \mathbf{A}||\lambda \mathbf{I} + \mathbf{A}|$$
$$= |(\lambda \mathbf{I} - \mathbf{A})(\lambda \mathbf{I} + \mathbf{A})| = |\lambda^2 \mathbf{I} - \mathbf{A}^2|$$

G. The eigenvalues of a diagonal matrix are the diagonal elements.

H. The sum of the eigenvalues of an $n \times n$ matrix $\mathbf{A}$ with the characteristic equation

$$\lambda^n + \alpha_{n-1}\lambda^{n-1} + \cdots + \alpha_1 \lambda + \alpha_0 = 0$$

is

$$\lambda_1 + \lambda_2 + \cdots + \lambda_n = -\alpha_{n-1} = \text{ trace } (\mathbf{A})$$

Further Properties

I. The product of the eigenvalues of an $n \times n$ matrix $\mathbf{A}$ with the characteristic equation

$$\lambda^n + \alpha_{n-1}\lambda^{n-1} + \cdots + \alpha_1 \lambda + \alpha_0 = (\lambda - \lambda_1)(\lambda - \lambda_2) \cdots (\lambda - \lambda_n) = 0$$

is

$$\lambda_1 \lambda_2 \cdots \lambda_n = (-1)^n \alpha_0 = |\mathbf{A}|$$

an eigenvector is also an eigenvector. For example, for the matrix

$$\mathbf{A} = \begin{bmatrix} -1 & -2 & 0 \\ 1 & 2 & 0 \\ -2 & -1 & -3 \end{bmatrix}$$

it was previously found that the eigenvalues were

$$\lambda_1 = 0 \qquad \lambda_2 = 1 \qquad \lambda_3 = -3$$

The eigenvector $\mathbf{p}^1$ corresponding to λ_1 has components that satisfy

$$(\lambda_1 \mathbf{I} - \mathbf{A})\mathbf{p}^1 = \begin{bmatrix} 1 & 2 & 0 \\ -1 & -2 & 0 \\ 2 & 1 & 3 \end{bmatrix} \begin{bmatrix} p_{11} \\ p_{21} \\ p_{31} \end{bmatrix} = \mathbf{0}$$

or

$$\begin{cases} p_{11} + 2p_{21} & = 0 \\ -p_{11} - 2p_{21} & = 0 \\ 2p_{11} + p_{21} + 3p_{31} = 0 \end{cases}$$

Deleting the first of these equations, because it is obviously linearly dependent, results in

$$\begin{cases} -p_{11} - 2p_{21} & = 0 \\ 2p_{11} + p_{21} + 3p_{31} = 0 \end{cases}$$

Choosing $p_{11} = 2$ for convenience gives

$$\mathbf{p}^1 = \begin{bmatrix} 2 \\ -1 \\ -1 \end{bmatrix}$$

The eigenvector $\mathbf{p}^2$ satisfies

$$(\lambda_2 \mathbf{I} - \mathbf{A})\mathbf{p}^2 = \begin{bmatrix} 2 & 2 & 0 \\ -1 & -1 & 0 \\ 2 & 1 & 4 \end{bmatrix} \begin{bmatrix} p_{12} \\ p_{22} \\ p_{32} \end{bmatrix} = \mathbf{0}$$

or

$$\begin{cases} 2p_{12} + 2p_{22} & = 0 \\ -p_{12} - p_{22} & = 0 \\ 2p_{12} + p_{22} + 4p_{32} = 0 \end{cases}$$

Deleting the first equation as being linearly dependent results in

$$\begin{cases} p_{12} + p_{22} & = 0 \\ 2p_{12} + p_{22} + 4p_{32} = 0 \end{cases}$$

Choosing $p_{12} = 4$ gives

$$\mathbf{p}^2 = \begin{bmatrix} 4 \\ -4 \\ -1 \end{bmatrix}$$

The eigenvector $\mathbf{p}^3$ satisfies

$$(\lambda_3 \mathbf{I} - \mathbf{A})\mathbf{p}^3 = \begin{bmatrix} -2 & 2 & 0 \\ -1 & -5 & 0 \\ 2 & 1 & 0 \end{bmatrix} \begin{bmatrix} p_{13} \\ p_{23} \\ p_{33} \end{bmatrix} = \mathbf{0}$$

or

$$\begin{cases} -2p_{13} + 2p_{23} = 0 \\ -p_{13} - 5p_{23} = 0 \\ 2p_{13} + p_{23} = 0 \end{cases}$$

The only solution for p_{13} and p_{23} is

$$\begin{cases} p_{13} = 0 \\ p_{23} = 0 \end{cases}$$

but p_{33} can be anything. Choosing $p_{33} = 1$ gives

$$\mathbf{p}^3 = \begin{bmatrix} 0 \\ 0 \\ 1 \end{bmatrix}$$

Important properties of the eigenvectors of a matrix are summarized in Table A-8. That the eigenvectors corresponding to distinct eigenvalues are linearly independent can be shown in the following way. Suppose that two eigenvectors $\mathbf{p}^i$ and $\mathbf{p}^j$, corresponding to two distinct eigenvalues λ_i and λ_j, are linearly dependent. This is equivalent to supposing $\mathbf{p}^i = \mathbf{p}^j$ because the scaling of an eigenvector is arbitrary. Then

$$(\lambda_i \mathbf{I} - \mathbf{A})\mathbf{p}^i = (\lambda_j \mathbf{I} - \mathbf{A})\mathbf{p}^j = (\lambda_j \mathbf{I} - \mathbf{A})\mathbf{p}^i = \mathbf{0}$$

Subtracting results in

$$(\lambda_i - \lambda_j)\mathbf{p}^i = \mathbf{0}$$

Table A-8 Properties of the Eigenvectors of a Matrix

Defining Property

A. $\mathbf{Ap} = \lambda\mathbf{p}$

Further Properties

These properties follow from the defining property and the properties of eigenvalues:

B. If two eigenvalues of a matrix are distinct, the corresponding eigenvectors are linearly independent.

C. For a matrix $\mathbf{A}$ with real elements, the eigenvectors, if complex, can be expressed as complex conjugate pairs.

D. The eigenvectors of a matrix $(k\mathbf{A})$ are identical to the eigenvectors of $\mathbf{A}$, for any scalar k. Because the eigenvalues of $(k\mathbf{A})$ are k times the eigenvalues of $\mathbf{A}$, if

$$\mathbf{Ap} = \lambda\mathbf{p}$$

then

$$(k\mathbf{A})\mathbf{p} = (k\lambda)\mathbf{p}$$

E. The eigenvectors of $\mathbf{A}^{-1}$ are the same as the eigenvectors of $\mathbf{A}$. Eigenvalues of $\mathbf{A}^{-1}$ are the inverses of the eigenvalues of $\mathbf{A}$ and if

$$\mathbf{Ap} = \lambda\mathbf{p}$$

then

$$\left(\frac{1}{\lambda}\right)\mathbf{A}^{-1}\mathbf{Ap} = \mathbf{A}^{-1}\mathbf{p}$$

or

$$\mathbf{A}^{-1}\mathbf{p} = \left(\frac{1}{\lambda}\right)\mathbf{p}$$

which can be true only for $\mathbf{p}^i = \mathbf{0}$ or $\lambda_i = \lambda_j$. Thus, $\mathbf{p}^i$ and $\mathbf{p}^j$ must be linearly independent. In a similar way, every pair of the eigenvectors can be shown to be independent of one another; hence, the eigenvectors are mutually independent.

A.4.2 Diagonalizing Transformations

When a matrix $\mathbf{A}$ has distinct eigenvalues, the eigenvectors are linearly independent and so the modal matrix

$$\mathbf{P} = [\mathbf{p}^1 \;\vdots\; \mathbf{p}^2 \;\vdots\; \cdots \;\vdots\; \mathbf{p}^n]$$

is nonsingular. The transformation

$$\mathbf{x} = \mathbf{P}\mathbf{x}' \qquad \mathbf{x}' = \mathbf{P}^{-1}\mathbf{x}$$

$$\mathbf{y} = \mathbf{P}\mathbf{y}' \qquad \mathbf{y}' = \mathbf{P}^{-1}\mathbf{y}$$

substituted into the equation

$$\mathbf{A}\mathbf{x} = \mathbf{y}$$

gives

$$\mathbf{A}\mathbf{P}\mathbf{x}' = \mathbf{P}\mathbf{y}'$$

$$(\mathbf{P}^{-1}\mathbf{A}\mathbf{P})\mathbf{x}' = \Lambda\mathbf{x}' = \mathbf{y}'$$

where the spectral matrix Λ is diagonal:

$$\Lambda = \begin{bmatrix} \lambda_1 & 0 & \cdots & 0 \\ 0 & \lambda_2 & \cdots & 0 \\ \vdots & & & \\ 0 & 0 & \cdots & \lambda_n \end{bmatrix}$$

The order in which the eigenvalues appear along the diagonal of the spectral matrix is the order in which the eigenvectors are placed as columns of the modal matrix $\mathbf{P}$. Thus, there are a number of different modal matrices, each yielding a different but closely related spectral matrix. Any column of a modal matrix can be multiplied by a nonzero constant, giving a new modal matrix, because the eigenvectors, being solutions of homogeneous equations, are of arbitrary norm. Eigenvector scaling changes in the modal matrix do not affect the spectral matrix.

Continuing the example with

$$\mathbf{A} = \begin{bmatrix} -1 & -2 & 0 \\ 1 & 2 & 0 \\ -2 & -1 & -3 \end{bmatrix}$$

and

$$\mathbf{p}^1 = \begin{bmatrix} 2 \\ -1 \\ -1 \end{bmatrix} \qquad \mathbf{p}^2 = \begin{bmatrix} 4 \\ -4 \\ -1 \end{bmatrix} \qquad \mathbf{p}^3 = \begin{bmatrix} 0 \\ 0 \\ 1 \end{bmatrix}$$

gives a modal matrix of

$$\mathbf{P} = \begin{bmatrix} 2 & 4 & 0 \\ -1 & -4 & 0 \\ -1 & -1 & 1 \end{bmatrix}$$

The inverse of this modal matrix is

$$\mathbf{P}^{-1} = \begin{bmatrix} 1 & 1 & 0 \\ -\frac{1}{4} & -\frac{1}{2} & 0 \\ \frac{3}{4} & \frac{1}{2} & 1 \end{bmatrix}$$

and

$$\mathbf{P}^{-1}\mathbf{A}\mathbf{P} = \begin{bmatrix} 0 & 0 & 0 \\ 0 & 1 & 0 \\ 0 & 0 & -3 \end{bmatrix} = \Lambda$$

As another numerical example of solving a characteristic value problem, this one having a matrix with complex eigenvalues, consider

$$\mathbf{A} = \begin{bmatrix} -2 & 1 \\ -5 & 0 \end{bmatrix}$$

Its characteristic equation is

$$|\lambda\mathbf{I} - \mathbf{A}| = \begin{vmatrix} (\lambda + 2) & -1 \\ 5 & \lambda \end{vmatrix} = \lambda^2 + 2\lambda + 5 = 0$$

and the eigenvalues are

$$\lambda_1, \lambda_2 = -1 \pm j2$$

The eigenvector $\mathbf{p}^1$ corresponding to the root $\lambda_1 = -1 + j2$ satisfies

$$(\lambda_1\mathbf{I} - \mathbf{A})\mathbf{p}^1 = \begin{bmatrix} (1 + j2) & -1 \\ 5 & (-1 + j2) \end{bmatrix}\begin{bmatrix} p_{11} \\ p_{21} \end{bmatrix} = \mathbf{0}$$

or

$$\begin{cases} (1 + j2)p_{11} & - p_{21} = 0 \\ 5 \quad p_{11} + (-1 + j2) & p_{21} = 0 \end{cases}$$

Because the determinant of these two equations is zero, the two equations are linearly dependent. The first equation multiplied by $(1 - j2)$

equals the second. Deleting the second equation and choosing $p_{11} = 1$ gives

$$p_{21} = 1 + j2$$

and

$$\mathbf{p}^1 = \begin{bmatrix} 1 \\ (1 + j2) \end{bmatrix}$$

The eigenvector corresponding to the complex conjugate root $\lambda_2 = \lambda_1^*$ is $\mathbf{p}^2 = \mathbf{p}^{1*}$, so

$$\mathbf{P} = [\mathbf{p}^1 \,\vdots\, \mathbf{p}^2] = \begin{bmatrix} 1 & 1 \\ (1 + j2) & (1 - j2) \end{bmatrix}$$

and

$$\mathbf{P}^{-1} = \begin{bmatrix} (\tfrac{1}{2} + j\tfrac{1}{4}) & -j\tfrac{1}{4} \\ (\tfrac{1}{2} - j\tfrac{1}{4}) & j\tfrac{1}{4} \end{bmatrix}$$

giving

$$\mathbf{P}^{-1}\mathbf{A}\mathbf{P} = \begin{bmatrix} (-1 + j2) & 0 \\ 0 & (-1 - j2) \end{bmatrix} = \begin{bmatrix} \lambda_1 & 0 \\ 0 & \lambda_2 \end{bmatrix}$$

Any nonsingular transformation that diagonalizes a matrix $\mathbf{A}$ also diagonalizes $\mathbf{A}^2$, $\mathbf{A}^3$, ... If

$$\mathbf{P}^{-1}\mathbf{A}\mathbf{P} = \Lambda$$

then

$$\mathbf{P}^{-1}\mathbf{A}^2\mathbf{P} = \mathbf{P}^{-1}\mathbf{A}\mathbf{I}\mathbf{A}\mathbf{P} = \mathbf{P}^{-1}\mathbf{A}\mathbf{P}\mathbf{P}^{-1}\mathbf{A}\mathbf{P} = (\mathbf{P}^{-1}\mathbf{A}\mathbf{P})(\mathbf{P}^{-1}\mathbf{A}\mathbf{P}) = \Lambda^2$$

Similarly,

$$\mathbf{P}^{-1}\mathbf{A}^3\mathbf{P} = \Lambda^3$$

and so on. A transformation that diagonalizes $\mathbf{A}$ also diagonalizes $\mathbf{A}^{-1}$.

$$(\mathbf{P}^{-1}\mathbf{A}\mathbf{P})^{-1} = \mathbf{P}^{-1}\mathbf{A}^{-1}\mathbf{P} = \Lambda^{-1}$$

Because Λ is diagonal, Λ^{-1} is diagonal, with diagonal elements that are the reciprocals of the diagonal elements of Λ.

If the matrix $\mathbf{A}$ has repeated eigenvalues, linearly independent eigenvectors corresponding to the same root must be found for each eigenvalue repetition, or else a nonsingular modal matrix $\mathbf{P}$ will not exist. There must be an additional linearly independent solution of

$$(\lambda_i\mathbf{I} - \mathbf{A})\mathbf{p}^i = \mathbf{0}$$

for each repetition of λ_i. This occurs only if

rank$(\lambda_i \mathbf{I} - \mathbf{A}) = n - r$

where r is the number of repetitions of the root λ_i, which is not likely to be the case.

For example, the matrix

$$\mathbf{A} = \begin{bmatrix} 1 & 0 & -4 \\ 0 & 1 & 2 \\ 0 & 0 & 3 \end{bmatrix}$$

has eigenvalues

$$\lambda_1 = \lambda_2 = 1 \qquad \lambda_3 = 3$$

For the repeated root,

$$(\lambda_i \mathbf{I} - \mathbf{A}) = \begin{bmatrix} 0 & 0 & 4 \\ 0 & 0 & -2 \\ 0 & 0 & -2 \end{bmatrix}$$

which has the rank 1. Eigenvectors corresponding to the repeated root then satisfy

$$(\lambda_i \mathbf{I} - \mathbf{A})\mathbf{p}^1 = \mathbf{0}$$

or

$$\mathbf{p}^1 = \begin{bmatrix} 1 \\ 0 \\ 0 \end{bmatrix}$$

Another linearly independent solution is

$$\mathbf{p}^2 = \begin{bmatrix} 0 \\ 1 \\ 0 \end{bmatrix}$$

The third eigenvector is

$$\mathbf{p}^3 = \begin{bmatrix} -2 \\ 1 \\ 1 \end{bmatrix}$$

A modal matrix is then

$$\mathbf{P} = [\mathbf{p}^1 \mathrel{\vdots} \mathbf{p}^2 \mathrel{\vdots} \mathbf{p}^3] = \begin{bmatrix} 1 & 0 & -2 \\ 0 & 1 & 1 \\ 0 & 0 & 1 \end{bmatrix}$$

On the other hand, the matrix

$$\mathbf{A} = \begin{bmatrix} 1 & 1 & 0 \\ 0 & 1 & 1 \\ 0 & 0 & 1 \end{bmatrix}$$

has the eigenvalues $\lambda_1 = \lambda_2 = \lambda_3 = 1$, but

$$(\lambda_i \mathbf{I} - \mathbf{A}) = \begin{bmatrix} 0 & -1 & 0 \\ 0 & 0 & -1 \\ 0 & 0 & 0 \end{bmatrix}$$

is of rank 2. Three, or even two, linearly independent eigenvectors cannot be found, so a modal matrix does not exist.

A.4.3 Similarity Transformation

Any nonsingular transformation $\mathbf{P}$ of a matrix $\mathbf{A}$ of the form

$$\mathbf{A}' = \mathbf{P}^{-1}\mathbf{A}\mathbf{P}$$

is termed a *similarity transformation*. The diagonalizing transformation by a modal matrix is a special kind of similarity transformation. Important properties of similarity transformation are listed in Table A-9. To

Table A-9 Properties of Similarity Transformations

A nonsingular transformation $\mathbf{P}$ of a matrix $\mathbf{A}$ of the form

$$\mathbf{B} = \mathbf{P}^{-1}\mathbf{A}\mathbf{P}$$

has the following properties:

A. $\mathbf{P}^{-1}(\mathbf{A}^2)\mathbf{P} = \mathbf{B}^2$

    $\mathbf{P}^{-1}(\mathbf{A}^3)\mathbf{P} = \mathbf{B}^3$
    and so on.

B. Provided $\mathbf{A}^{-1}$ exists,

    $\mathbf{P}^{-1}(\mathbf{A}^{-1})\mathbf{P} = \mathbf{B}^{-1}$

C. trace $(\mathbf{B})$ = trace $(\mathbf{A})$

D. $|\mathbf{B}| = |\mathbf{A}|$

E. The eigenvalues of $\mathbf{B}$ are identical to those of $\mathbf{A}$; they are unchanged by a similarity transformation.

show that a similarity transformation does not change the eigenvalues of a matrix, consider

$$|\lambda\mathbf{I} - \mathbf{A}'| = |\lambda\mathbf{P}^{-1}\mathbf{P} - \mathbf{P}^{-1}\mathbf{A}\mathbf{P}| = |\mathbf{P}^{-1}(\lambda\mathbf{I} - \mathbf{A})\mathbf{P}|$$
$$= |\mathbf{P}^{-1}||\lambda\mathbf{I} - \mathbf{A}||\mathbf{P}| = |\lambda\mathbf{I} - \mathbf{A}|$$

The determinant

$$|\lambda\mathbf{I} - \mathbf{A}'| = 0$$

if and only if

$$|\lambda\mathbf{I} - \mathbf{A}| = 0$$

A.4.4 The Cayley-Hamilton Theorem

The Cayley-Hamilton theorem describes a remarkable property of every square matrix $\mathbf{A}$. If the characteristic equation of $\mathbf{A}$ is

$$\lambda^n + \alpha_{n-1}\lambda^{n-1} + \cdots + \alpha_1\lambda + \alpha_0 = 0$$

the matrix itself satisfies the same equation, namely,

$$\mathbf{A}^n + \alpha_{n-1}\mathbf{A}^{n-1} + \cdots + \alpha_1\mathbf{A} + \alpha_0\mathbf{I} = \mathbf{0}$$

The result holds in general, but it will be shown now for matrices that can be diagonalized, that is, matrices with distinct eigenvalues. For a matrix with distinct eigenvalues,

$$\mathbf{A} = \mathbf{P}\boldsymbol{\Lambda}\mathbf{P}^{-1}$$

where $\mathbf{P}$ is a modal matrix and

$$\boldsymbol{\Lambda} = \begin{bmatrix} \lambda_1 & 0 & \cdots & 0 \\ 0 & \lambda_2 & \cdots & 0 \\ \vdots & & & \\ 0 & 0 & \cdots & \lambda_n \end{bmatrix}$$

is a spectral matrix for $\mathbf{A}$. And

$$\mathbf{A}^2 = \mathbf{P}\boldsymbol{\Lambda}\mathbf{P}^{-1}\mathbf{P}\boldsymbol{\Lambda}\mathbf{P}^{-1} = \mathbf{P}\boldsymbol{\Lambda}^2\mathbf{P}^{-1}$$

where

$$\boldsymbol{\Lambda}^2 = \begin{bmatrix} \lambda_1^2 & 0 & \cdots & 0 \\ 0 & \lambda_2^2 & \cdots & 0 \\ \vdots & & & \\ 0 & 0 & \cdots & \lambda_n^2 \end{bmatrix}$$

Similarly,

$$\mathbf{A}^m = \mathbf{P}\mathbf{\Lambda}^m\mathbf{P}^{-1}$$

Then

$$\mathbf{A}^n + \alpha_{n-1}\mathbf{A}^{n-1} + \cdots + \alpha_1\mathbf{A} + \alpha_0\mathbf{I}$$

$$= \mathbf{P}(\mathbf{\Lambda}^n + \alpha_{n-1}\mathbf{\Lambda}^{n-1} + \cdots + \alpha_1\mathbf{\Lambda} + \alpha_0\mathbf{I})\mathbf{P}^{-1}$$

$$= \mathbf{P}\begin{bmatrix} (\lambda_1^n + \alpha_{n-1}\lambda_1^{n-1} + \cdots + \alpha_1\lambda_1 + \alpha_0) & 0 & \cdots \\ 0 & (\lambda_2^n + \alpha_{n-1}\lambda_2^{n-1} + \cdots + \alpha_1\lambda_2 + \alpha_0) & \cdots \\ \vdots & & \\ 0 & 0 & \cdots \end{bmatrix}$$

$$= \mathbf{0}$$

For example, the matrix

$$\mathbf{A} = \begin{bmatrix} -1 & 1 & 0 \\ -3 & 0 & 1 \\ -2 & 0 & 0 \end{bmatrix}$$

has the characteristic equation

$$\lambda^3 + \lambda^2 + 3\lambda + 2 = 0$$

The matrix polynomial

$$\mathbf{A}^3 + \mathbf{A}^2 + 3\mathbf{A} + 2\mathbf{I} = \mathbf{0}$$

as is easily verified

$$\mathbf{A}^2 = \begin{bmatrix} -2 & -1 & 1 \\ 1 & -3 & 0 \\ 2 & -2 & 0 \end{bmatrix} \quad \mathbf{A}^3 = \begin{bmatrix} 3 & -2 & -1 \\ 8 & 1 & -3 \\ 4 & 2 & -2 \end{bmatrix}$$

$$\mathbf{A}^3 + \mathbf{A}^2 + 3\mathbf{A} + 2\mathbf{I} = \begin{bmatrix} 0 & 0 & 0 \\ 0 & 0 & 0 \\ 0 & 0 & 0 \end{bmatrix}$$

When the Cayley-Hamilton theorem is used, the nth power of $\mathbf{A}$ can be expressed in terms of lesser powers of $\mathbf{A}$

$$\mathbf{A}^n = -\alpha_{n-1}\mathbf{A}^{n-1} - \cdots - \alpha_1\mathbf{A} - \alpha_0\mathbf{I}$$

In fact, by repeatedly substituting for $\mathbf{A}^n$, any power of $\mathbf{A}$ can be expressed in terms of the $(n-1)$th power of $\mathbf{A}$ and lower powers, down to and including the zero power of $\mathbf{A}$, the identity matrix.

For example, the characteristic equation of

$$\mathbf{A} = \begin{bmatrix} 1 & -2 \\ 3 & 0 \end{bmatrix}$$

is

$$\lambda^2 - \lambda + 6 = 0$$

Thus,

$$\mathbf{A}^2 = \mathbf{A} - 6\mathbf{I}$$

so the polynomial

$$\mathbf{A}^4 + 2\mathbf{A}^3 + 3\mathbf{A} = (\mathbf{A} - 6\mathbf{I})^2 + 2\mathbf{A}(\mathbf{A} - 6\mathbf{I}) + 3\mathbf{A}$$

$$= 3\mathbf{A}^2 - 21\mathbf{A} + 36\mathbf{I} = 3(\mathbf{A} - 6\mathbf{I}) - 21\mathbf{A} + 36\mathbf{I}$$

$$= -18\mathbf{A} + 18\mathbf{I}$$

The inverse of a nonsingular matrix can be expressed as a matrix polynomial. Solving the following equations for the following equation for the identifying matrix

$$\mathbf{A}^n + \alpha_{n-1}\mathbf{A}^{n-1} + \cdots + \alpha_1\mathbf{A} + \alpha_0\mathbf{I} = 0$$

we obtain

$$\mathbf{I} = -\left(\frac{1}{\alpha_0}\right)(\mathbf{A}^n + \alpha_{n-1}\mathbf{A}^{n-1} + \cdots + \alpha_1\mathbf{A})$$

Multiplying this last equation by $\mathbf{A}^{-1}$ gives

$$\mathbf{A}^{-1} = \left(\frac{-1}{\alpha_0}\right)(\mathbf{A}^{n-1} + \alpha_{n-1}\mathbf{A}^{n-2} + \cdots + \alpha_1\mathbf{I})$$

Thus, a matrix inverse may be computed from the coefficients of the characteristic equation of the matrix and the powers of the matrix up through the $(n - 1)$th power.

For example,

$$\mathbf{A} = \begin{bmatrix} -2 & 1 & 0 \\ -3 & 0 & 1 \\ 4 & 0 & 0 \end{bmatrix}$$

has the characteristic equation

$$\lambda^3 + 2\lambda^2 + 3\lambda - 4 = 0$$

so

$$\mathbf{A}^{-1} = \left(\frac{1}{4}\right)(\mathbf{A}^2 + 2\mathbf{A} + 3\mathbf{I}) = \begin{bmatrix} 0 & 0 & \frac{1}{4} \\ 1 & 0 & \frac{1}{2} \\ 0 & 1 & \frac{3}{4} \end{bmatrix}$$

A.5 Quadratic Forms

A quadratic form is a scalar function of the form

$$f(x_1, x_2, \ldots, x_n) = f(\mathbf{x}) = \mathbf{x}^t \mathbf{S} \mathbf{x}$$

$$= [x_1 \; x_2 \cdots x_n] \begin{bmatrix} s_{11} & s_{12} & \cdots & s_{1n} \\ s_{21} & s_{22} & \cdots & s_{2n} \\ \vdots & & & \\ s_{n1} & s_{n2} & \cdots & s_{nn} \end{bmatrix} \begin{bmatrix} x_1 \\ x_2 \\ \vdots \\ x_n \end{bmatrix}$$

$$= s_{11}x_1^2 + s_{12}x_1x_2 + s_{13}x_1x_3 + \cdots + s_{1n}x_1x_n$$
$$+ s_{21}x_2x_1 + s_{22}x_2^2 + s_{23}x_2x_3 + \cdots + s_{2n}x_2x_n + \cdots$$
$$+ s_{n1}x_nx_1 + s_{n2}x_nx_2 + \cdots + s_{nn}x_n^2$$

A quadratic form is a linear combination of the products $x_i x_j$, including the squares of each variable. For example,

$$f(\mathbf{x}) = x_1^2 + 2x_1x_2 - 8x_1x_3 + 4x_2^2 + 4x_2x_3$$

$$= [x_1 \quad x_2 \quad x_3] \begin{bmatrix} 1 & 1 & -4 \\ 1 & 4 & 2 \\ -4 & 2 & 0 \end{bmatrix} \begin{bmatrix} x_1 \\ x_2 \\ x_3 \end{bmatrix}$$

is a three-variable quadratic form. A quadratic form can be expressed in terms of many different matrices $\mathbf{S}$. It is always possible, however, to choose $\mathbf{S}$ to be a symmetric matrix.

Symmetric matrices have three key properties that are of special importance for quadratic forms:

1. The eigenvalues of a symmetric matrix with real elements are real numbers. Consequently, the corresponding eigenvectors are always real.
2. Any two eigenvectors corresponding to different eigenvalues are orthogonal to one another.
3. If, for a real, symmetric matrix $\mathbf{S}$, an eigenvalue λ_i is repeated, the rank of $\lambda_i \mathbf{I} - \mathbf{S}$ is always reduced by the number of occurrences of the root λ_i. Thus, it is always possible to find n linearly independent eigenvectors and, hence, a nonsingular modal matrix. Even with repeated roots, the eigenvectors can always be chosen to be orthogonal.

As a numerical example, consider the symmetric matrix

$$\mathbf{S} = \begin{bmatrix} 2 & -36 & 0 \\ -36 & 23 & 0 \\ 0 & 0 & -75 \end{bmatrix}$$ **(A-11)**

which has the characteristic equation

$$\begin{vmatrix} (\lambda - 2) & 36 & 0 \\ 36 & (\lambda - 23) & 0 \\ 0 & 0 & (\lambda + 75) \end{vmatrix} = (\lambda + 75)(\lambda - 50)(\lambda + 25) = 0$$

The eigenvalues of $\mathbf{S}$ are

$$\lambda_1 = -25 \qquad \lambda_2 = 50 \qquad \lambda_3 = -75$$

which are real, as they should be. The corresponding eigenvectors (with convenient scaling) are

$$\mathbf{p}^1 = \begin{bmatrix} 4 \\ 3 \\ 0 \end{bmatrix} \qquad \mathbf{p}^2 = \begin{bmatrix} 3 \\ -4 \\ 0 \end{bmatrix} \qquad \mathbf{p}^3 = \begin{bmatrix} 0 \\ 0 \\ 1 \end{bmatrix}$$

and these are mutually orthogonal. Normalizing them (so that each eigenvector has unit norm) and forming a modal matrix results in

$$\mathbf{P}_0 = \begin{bmatrix} \frac{4}{5} & \frac{3}{5} & 0 \\ \frac{3}{5} & -\frac{4}{5} & 0 \\ 0 & 0 & 1 \end{bmatrix}$$ **(A-12)**

which is an orthogonal matrix

$$\mathbf{P}_0^{-1} = \mathbf{P}_0^{\dagger}$$

and for which

$$\mathbf{P}_0^{-1}\mathbf{S}\mathbf{P}_0 = \mathbf{P}_0^{\dagger}\mathbf{S}\mathbf{P}_0 = \begin{bmatrix} -25 & 0 & 0 \\ 0 & 50 & 0 \\ 0 & 0 & -75 \end{bmatrix} = \mathbf{S}_0$$

For a quadratic form

$$f(\mathbf{x}) = \mathbf{x}^{\dagger}\mathbf{S}\mathbf{x}$$

expressed in terms of a symmetric matrix $\mathbf{S}$, it is always possible to find an orthogonal change of variables

$$\mathbf{x} = \mathbf{P}_0\mathbf{x}' \qquad \mathbf{x}' = \mathbf{P}_0^{\dagger}\mathbf{x}$$

such that

$$f(\mathbf{x}) = \mathbf{x}^t \mathbf{S} \mathbf{x} = \mathbf{x}'^t (\mathbf{P}_0^t \mathbf{S} \mathbf{P}_0) \mathbf{x}' = \mathbf{x}'^t (\mathbf{P}_0^{-1} \mathbf{S} \mathbf{P}_0) \mathbf{x}' = \mathbf{x}'^t \mathbf{S}_0 \mathbf{x}'$$
$$= \lambda_1 x_1'^2 + \lambda_2 x_2'^2 + \cdots + \lambda_n x_n'^2$$

where $\mathbf{S}_0$ is diagonal. In terms of the new variables, the quadratic form is a sum of squares, with no cross-product terms. The eigenvalues of $\mathbf{S}$ and thus of $\mathbf{S}_0$ are coefficients of the squares of the new variables. For the quadratic form defined by the symmetric matrix in equation (A-12),

$$f(\mathbf{x}) = [x_1 \quad x_2 \quad x_3] \begin{bmatrix} 2 & -36 & 0 \\ -36 & 23 & 0 \\ 0 & 0 & -75 \end{bmatrix} \begin{bmatrix} x_1 \\ x_2 \\ x_3 \end{bmatrix}$$
$$= 2x_1^2 - 72x_1 x_2 + 23x_2^2 - 75x_3^2$$

the orthogonal change of variables,

$$\mathbf{x} = \mathbf{P}_0 \mathbf{x}' \qquad \mathbf{x}' = \mathbf{P}_0^{-1} \mathbf{x} = \mathbf{P}_0^t \mathbf{x} \qquad \qquad \text{(A-13)}$$

using equation (A-13) gives the following quadratic form

$$f(\mathbf{x}') = (\mathbf{P}_0 \mathbf{x}')^t \mathbf{S} (\mathbf{P}_0 \mathbf{x}') = \mathbf{x}'^t (\mathbf{P}_0^t \mathbf{S} \mathbf{P}_0) \mathbf{x}'$$

$$= \mathbf{x}'^t \mathbf{S}_0 \mathbf{x}' = [x_1' \quad x_2' \quad x_3'] \begin{bmatrix} -25 & 0 & 0 \\ 0 & 50 & 0 \\ 0 & 0 & -75 \end{bmatrix} \begin{bmatrix} x_1' \\ x_2' \\ x_3' \end{bmatrix}$$

$$= -25x_1'^2 + 50x_2'^2 - 75x_3'^2$$

which is a pure sum of squares.

A quadratic form

$$f(\mathbf{x}) = \mathbf{x}^t \mathbf{S} \mathbf{x}$$

and its associated symmetric matrix $\mathbf{S}$ are classified as sign definite as follows:

1. If for all nonzero real vectors $\mathbf{x}$

    $$\mathbf{x}^t \mathbf{S} \mathbf{x} > 0$$

    the quadratic form is *positive definite*.
2. If for all nonzero real vectors $\mathbf{x}$

    $$\mathbf{x}^t \mathbf{S} \mathbf{x} \geqslant 0$$

    the quadratic form is *positive semidefinite*.

3. If for all nonzero real vectors $\mathbf{x}$

$$\mathbf{x}^\dagger \mathbf{S} \mathbf{x} < 0$$

the quadratic form is *negative definite*.

4. If for all nonzero real vectors $\mathbf{x}$

$$\mathbf{x}^\dagger \mathbf{S} \mathbf{x} \leq 0$$

the quadratic form is *negative semidefinite*.

Otherwise, the quadratic form is not sign definite.

Sign definiteness of a quadratic form is unchanged by nonsingular transformations of variables, as noted in equation (A-13). For $\mathbf{S}$ to be positive definite, it is necessary and sufficient for all of the eigenvalues $\lambda_1, \lambda_2, \ldots, \lambda_n$ of $\mathbf{S}$ (which are real for $\mathbf{S}$ symmetric) to be positive. If all of the eigenvalues of $\mathbf{S}$ are nonnegative, with one or more roots equal to zero, the quadratic form is positive semidefinite. If all of the characteristic roots of $\mathbf{S}$ are negative, f is negative definite, and so forth. The quadratic form of equation (A-12) is not sign definite because it has both positive and negative eigenvalues.

A.5.1 Matrix Factorization

Every symmetric matrix $\mathbf{S}$ can be factored in a form

$$\mathbf{S} = \mathbf{\Psi}^\dagger \mathbf{\Psi}$$

where $\mathbf{\Psi}$ is termed a *square root* of $\mathbf{S}$. Generally, square roots of symmetric matrices are not unique and they can involve complex elements. A positive definite symmetric matrix $\mathbf{S}$ has a square root involving real numbers that is especially simple to visualize. If, when diagonalized, $\mathbf{S}$ becomes

$$\mathbf{S}_0 = \begin{bmatrix} s_1 & 0 & \cdots & \\ 0 & s_2 & \cdots & \\ \vdots & & & \\ 0 & 0 & \cdots & s_n \end{bmatrix}$$

then

$$\mathbf{S} = \mathbf{P}_0 \mathbf{S}_0 \mathbf{P}_0^{-1} = \mathbf{P}_0 \mathbf{S}_0 \mathbf{P}_0^\dagger$$

Defining the real, diagonal matrix

$$\Psi_0 = \Psi_0^\dagger = \begin{bmatrix} \sqrt{s_1} & 0 & \cdots & 0 \\ 0 & \sqrt{s_2} & \cdots & 0 \\ 0 & 0 & \cdots & \sqrt{s_n} \end{bmatrix}$$

so that

$$S_0 = \Psi_0^\dagger \Psi_0$$

gives

$$S = P_0 \Psi_0^\dagger \Psi_0 P_0^\dagger = \Psi^\dagger \Psi$$

where

$$\Psi = \Psi_0 P_0^\dagger$$

An efficient numeric algorithm for determining the square root of a symmetric positive definite matrix S is based on the lower-upper (LU) decomposition method. If the matrix Ψ is chosen to be upper triangular of the form

$$\Psi = \begin{bmatrix} \Psi_{11} & \Psi_{12} & \Psi_{13} & \cdots & \Psi_{1n} \\ 0 & \Psi_{22} & \Psi_{23} & \cdots & \Psi_{2n} \\ 0 & 0 & \Psi_{33} & \cdots & \Psi_{3n} \\ \vdots & & & & \\ 0 & 0 & 0 & \cdots & \Psi_{nn} \end{bmatrix}$$

its elements can be computed by matrix multiplication of $\Psi^\dagger$ by Ψ and by equating the product to the corresponding elements of the matrix S. It can easily be verified that the elements of the matrix Ψ can be determined using the following algorithm:

$$\Psi_{11} = \sqrt{s_{11}}$$

$$\Psi_{1j} = \frac{s_{1j}}{\Psi_{11}} \quad (j = 2, 3, \ldots, n)$$

$$\Psi_{ii} = \sqrt{s_{ii} - \sum_{k=1}^{i-1} \Psi_{ki}^2} \quad (i = 2, 3, \ldots, n)$$

$$\Psi_{ij} = \frac{1}{\Psi_{ii}} \left[s_{ij} - \sum_{k=1}^{i-1} \Psi_{ki} \Psi_{kj} \right] \quad \begin{cases} (j = i+1, i+2, \ldots, n) \\ \text{incrementing } j \text{ for every } i \end{cases}$$

$$= 0 \qquad\qquad i > j$$

This method of finding the square root of a positive definite symmetric matrix is also termed the *Cholesky decomposition method*.

For example, the square root of the symmetric positive definite matrix

$$S = \begin{bmatrix} 4 & 0 & 2 \\ 0 & 9 & 3 \\ 2 & 3 & 3 \end{bmatrix}$$

is determined using the previous algorithm as follows:

$$\Psi_{11} = \sqrt{4} = 2$$

$$\Psi_{12} = \frac{s_{12}}{2} = 0 \qquad \Psi_{13} = \frac{s_{13}}{2} = 1$$

$$\Psi_{22} = \sqrt{9 - \Psi_{12}^2} = \sqrt{9 - 0} = 3$$

$$\Psi_{23} = \frac{1}{3}[s_{23} - \Psi_{12}\Psi_{13}] = 1$$

and

$$\Psi_{33} = \sqrt{s_{33} - (\Psi_{13}^2 + \Psi_{23}^2)} = \sqrt{3 - (1+1)} = 1$$

Therefore,

$$\Psi = \begin{bmatrix} 2 & 0 & 1 \\ 0 & 3 & 1 \\ 0 & 0 & 1 \end{bmatrix}$$

is a square root of S.

A.6 Computation of e^{At}

The state transition matrix of a linear, time-invariant, continuous-time system is the matrix exponential function

$$\Phi(t) = \exp(At) = I + At + \frac{1}{2}A^2t^2 + \cdots + \frac{1}{i!}A^it^i + \cdots \qquad \text{(A-14)}$$

In this section, methods for determining the state transition matrix are discussed.

A.6.1 Evaluation by Diagonalization

One method of determining the state transition matrix is to use diagonalization. If a matrix A involved in the series given by equation

(A-14) can be diagonalized,

$$\mathbf{P}^{-1}\mathbf{A}\mathbf{P} = \Lambda$$

$$\mathbf{P}^{-1}\mathbf{A}^2\mathbf{P} = \Lambda^2$$

$$\vdots$$

$$\mathbf{P}^{-1}\mathbf{A}^m\mathbf{P} = \Lambda^m$$

then premultiplying equation (A-14) by $\mathbf{P}^{-1}$ and postmultiplying it by $\mathbf{P}$ gives

$$\mathbf{P}^{-1}e^{\mathbf{A}t}\mathbf{P} = \mathbf{P}^{-1}\left[\mathbf{I} + \mathbf{A}t + \left(\frac{1}{2}\right)\mathbf{A}^2t^2 + \cdots + \left(\frac{1}{i!}\right)\mathbf{A}^it^i + \cdots\right]\mathbf{P}$$

$$= \mathbf{I} + \Lambda t + \left(\frac{1}{2}\right)\Lambda^2t^2 + \cdots + \left(\frac{1}{i!}\right)\Lambda^it^i + \cdots$$

where

$$\Lambda = \begin{bmatrix} \lambda_1 & 0 & \cdots & 0 \\ 0 & \lambda_2 & \cdots & 0 \\ \vdots & & & \\ 0 & 0 & \cdots & \lambda_n \end{bmatrix}$$

Then

$$\mathbf{P}^{-1}e^{\mathbf{A}t}\mathbf{P} = \begin{bmatrix} \left(1 + \lambda_1 t + \frac{1}{2}\lambda_1^2t^2 + \cdots + \frac{1}{i!}\lambda_1^it^i + \cdots\right) & \cdots & 0 & 0 \\ 0 & \left(1 + \lambda_2 t + \frac{1}{2}\lambda_2^2t^2 + \cdots + \frac{1}{i!}\lambda_2^it^i + \cdots\right) & \cdots & 0 \\ \vdots & & & \\ 0 & \cdots & 0 & \left(1 + \lambda_n t + \frac{1}{2}\lambda_n^2t^2 + \cdots + \frac{1}{i!}\lambda_n^it^i + \cdots\right) \end{bmatrix}$$

$$= \begin{bmatrix} e^{\lambda_1 t} & 0 & \cdots & 0 \\ 0 & e^{\lambda_2 t} & \cdots & 0 \\ \vdots & & & \\ 0 & 0 & \cdots & e^{\lambda_n t} \end{bmatrix}$$

$$= e^{\Lambda t}$$

and, therefore,

$$e^{\Lambda t} = \mathbf{P}e^{\Lambda t}\mathbf{P}^{-1}$$

For example, the matrix

$$\mathbf{A} = \begin{bmatrix} 0 & 1 \\ -3 & -4 \end{bmatrix}$$

has eigenvalues

$$\lambda_1 = -1 \quad \text{and} \quad \lambda_2 = -3$$

and corresponding eigenvectors

$$\mathbf{p}^1 = \begin{bmatrix} 1 \\ -1 \end{bmatrix} \quad \text{and} \quad \mathbf{p}^2 = \begin{bmatrix} 1 \\ -3 \end{bmatrix}$$

Hence,

$$\mathbf{P} = \begin{bmatrix} 1 & 1 \\ -1 & -3 \end{bmatrix} \quad \text{and} \quad \mathbf{P}^{-1} = \frac{1}{2}\begin{bmatrix} 3 & 1 \\ -1 & -1 \end{bmatrix}$$

and

$$e^{\Lambda t} = \mathbf{P}e^{\Lambda t}\mathbf{P}^{-1}$$

$$= \begin{bmatrix} 1 & 1 \\ -1 & -3 \end{bmatrix}\begin{bmatrix} e^{-t} & 0 \\ 0 & e^{-3t} \end{bmatrix}\begin{bmatrix} \frac{3}{2} & \frac{1}{2} \\ -\frac{1}{2} & -\frac{1}{2} \end{bmatrix}$$

$$= \frac{1}{2}\begin{bmatrix} (3e^{-t} - e^{-3t}) & (e^{-t} - e^{-3t}) \\ (-3e^{-t} + 3e^{-3t}) & (-e^{-t} + 3e^{-3t}) \end{bmatrix}$$

A.6.2 Cayley-Hamilton Method

Another method of determining the exponential function $\exp(\mathbf{A}t)$ is to use the Cayley-Hamilton theorem.

The Cayley-Hamilton theorem states that every square matrix $\mathbf{A}$ satisfies its own characteristic equation. That is

$$\mathbf{A}^n + \alpha_{n-1}\mathbf{A}^{n-1} + \cdots + \alpha_1\mathbf{A} + \alpha_0\mathbf{I} = 0$$

where the α's are the coefficients of the characteristic equation of the matrix $\mathbf{A}$,

$$\lambda^n + \alpha_{n-1}\lambda^{n-1} + \cdots + \alpha_1\lambda + \alpha_0 = 0$$

Therefore all powers of $\mathbf{A}$ higher than $\mathbf{A}^{n-1}$ can be expressed in terms of lower powers of $\mathbf{A}$, giving an alternative finite series expansion for the function given by equation (A-14)

$$e^{\mathbf{A}t} = R(\mathbf{A}t) = d_{n-1}(t) \mathbf{A}^{n-1} + d_{n-2}(t) \mathbf{A}^{n-2} + \cdots + d_1(t) \mathbf{A} + d_0(t) \mathbf{I}$$

The infinite series $\exp(\mathbf{A}t)$ is said to *collapse* to the finite series $R(\mathbf{A}t)$.
The corresponding scalar series

$$e^{\lambda t} = 1 + \lambda t + \frac{1}{2} \lambda^2 t^2 + \cdots$$

does not in general collapse to a finite series. It is only for values of λ
that are eigenvalues of $\mathbf{A}$ for which the series collapses. Then it collapses with exactly the same coefficients as the matrix series

$$e^{\lambda_i t} = R(\lambda_i t) = d_{n-1}(t) \lambda_i^{n-1} + d_{n-2}(t) \lambda_i^{n-2} + \cdots + d_1(t) \lambda_i + d_0(t)$$

Distinct Eigenvalues

If the matrix $\mathbf{A}$ has distinct eigenvalues, the following n linearly independent equations can be solved for the n unknowns $d_{n-1}(t)$, $d_{n-2}(t)$,
$\ldots$, $d_1(t)$, and $d_0(t)$.

$$e^{\lambda_1 t} = d_{n-1}(t) \lambda_1^{n-1} + d_{n-2}(t) \lambda_1^{n-2} + \cdots + d_1(t) \lambda_1 + d_0(t)$$

$$e^{\lambda_2 t} = d_{n-1}(t) \lambda_2^{n-1} + d_{n-2}(t) \lambda_2^{n-2} + \cdots + d_1(t) \lambda_2 + d_0(t)$$

$$\vdots$$

$$e^{\lambda_n t} = d_{n-1}(t) \lambda_n^{n-1} + d_{n-2}(t) \lambda_n^{n-2} + \cdots + d_1(t) \lambda_n + d_0(t) \qquad \textbf{(A-15)}$$

Having obtained $d_{n-1}(t)$, $d_{n-2}(t)$, $\ldots$, $d_1(t)$, and $d_0(t)$,

$$e^{\mathbf{A}t} = R(\mathbf{A}t) = d_{n-1}(t) \mathbf{A}^{n-1} + d_{n-2}(t) \mathbf{A}^{n-2} + \cdots + d_1(t) \mathbf{A} + d_0(t) \mathbf{I}$$

is then found.

As an example of finding $\exp(\mathbf{A}t)$ using the Cayley-Hamilton method, consider again the matrix

$$\mathbf{A} = \begin{bmatrix} 0 & 1 \\ -3 & -4 \end{bmatrix}$$

The characteristic equation for the matrix $\mathbf{A}$ is

$$|\lambda \mathbf{I} - \mathbf{A}| = \lambda^2 + 4\lambda + 3 = 0$$

so the eigenvalues of $\mathbf{A}$ are $\lambda_1 = -1$ and $\lambda_2 = -3$. Because $\mathbf{A}$ is 2×2

$$R(\lambda t) = d_1(t) \lambda + d_0(t)$$

and

$$e^{\lambda_1 t} = e^{-t} = -d_1(t) + d_0(t)$$

$$e^{\lambda_2 t} = e^{-3t} = -3d_1(t) + d_0(t)$$

which has the solutions

$$d_1(t) = \frac{1}{2}(e^{-t} - e^{-3t})$$

and

$$d_0(t) = \frac{1}{2}(3e^{-t} - e^{-3t})$$

Then

$$e^{At} = R(At) = \frac{1}{2}(e^{-t} - e^{-3t})A + \frac{1}{2}(3e^{-t} - e^{-3t})I$$

$$= \frac{1}{2}\begin{bmatrix} (3e^{-t} - e^{-3t}) & (e^{-t} - e^{-3t}) \\ (-3e^{-t} + 3e^{-3t}) & (-e^{-t} + 3e^{-3t}) \end{bmatrix}$$

which is the same as the solution obtained with diagonalization.

Repeated Eigenvalues

If the matrix A has repeated eigenvalues, the set of equations (A-15) should be modified to account for the eigenvalue repetition. If the matrix A has an eigenvalue λ_i repeated r times, additional linearly independent equations of the form

$$\left.\frac{d(e^{\lambda t})}{d\lambda}\right|_{\lambda=\lambda_i} = \left.\frac{dR}{d\lambda}\right|_{\lambda=\lambda_i}$$

$$\left.\frac{d^2(e^{\lambda t})}{d\lambda^2}\right|_{\lambda=\lambda_i} = \left.\frac{d^2R}{d\lambda^2}\right|_{\lambda=\lambda_i}$$

$$\vdots$$

$$\left.\frac{d^{r-1}(e^{\lambda t})}{d\lambda^{r-1}}\right|_{\lambda=\lambda_i} = \left.\frac{d^{r-1}R}{d\lambda^{r-1}}\right|_{\lambda=\lambda_i}$$

are used to obtain the solution.

For example, for

$$A = \begin{bmatrix} -4 & 1 \\ -4 & 0 \end{bmatrix}$$

the characteristic equation for the matrix $\mathbf{A}$ is

$$|\lambda \mathbf{I} - \mathbf{A}| = \lambda^2 + 4\lambda + 4 = 0$$

so the eigenvalues of $\mathbf{A}$ are repeated

$$\lambda_1 = -2 \qquad \lambda_2 = -2$$

Because $\mathbf{A}$ is matrix then 2×2

$$R(\lambda t) = d_1(t)\, \lambda + d_0(t)$$

and for $\lambda = -2$,

$$e^{-2t} = -2d_1(t) + d_0(t)$$

Taking the derivative of this last equation with respect to $\lambda = -2$ gives

$$\left. \frac{d(e^{\lambda t})}{d\lambda} \right|_{\lambda=-2} = \left. \frac{dR(\lambda t)}{d\lambda} \right|_{\lambda=-2} = d_1(t)$$

Solving for $d_1(t)$ and $d_0(t)$ we obtain

$$d_1(t) = te^{-2t}$$

and

$$d_0(t) = 2te^{-2t} + e^{-2t}$$

Therefore,

$$e^{\mathbf{A}t} = R(\mathbf{A}t) = te^{-2t}\mathbf{A} + (e^{-2t} + 2te^{-2t})\mathbf{I}$$

$$= \begin{bmatrix} (e^{-2t} - 2te^{-2t}) & te^{-2t} \\ -4te^{-2t} & (e^{-2t} + 2te^{-2t}) \end{bmatrix}$$

A.6.3 Sylvester's Theorem

If a square matrix $\mathbf{A}$ has distinct eigenvalues $\lambda_1, \lambda_2, \ldots, \lambda_n$, then the state transition matrix is

$$e^{\mathbf{A}t} = \sum_{i=1}^{n} e^{\lambda_i t} \mathbf{Z}_i$$

where

$$\mathbf{Z}_i = \frac{\displaystyle\prod_{\substack{j=1 \\ j \neq i}}^{n} (\mathbf{A} - \lambda_j \mathbf{I})}{\displaystyle\prod_{\substack{j=1 \\ j \neq i}}^{n} (\lambda_i - \lambda_j)}$$

The matrix

$$\mathbf{A} = \begin{bmatrix} 0 & 1 & 0 \\ 7 & 0 & 1 \\ 6 & 0 & 0 \end{bmatrix}$$

for example, has the characteristic equation

$$\lambda^3 - 7\lambda - 6 = (\lambda + 1)(\lambda + 2)(\lambda - 3) = 0$$

The eigenvalues of $\mathbf{A}$ are

$$\lambda_1 = -1 \qquad \lambda_2 = -2 \qquad \text{and} \qquad \lambda_3 = 3$$

Using Sylvester's theorem gives

$$e^{\mathbf{A}t} = e^{-t} \frac{(\mathbf{A} + 2\mathbf{I})(\mathbf{A} - 3\mathbf{I})}{(-1 + 2)(-1 - 3)} + e^{-2t} \frac{(\mathbf{A} + \mathbf{I})(\mathbf{A} - 3\mathbf{I})}{(-2 + 1)(-2 - 3)}$$

$$+ e^{3t} \frac{(\mathbf{A} + \mathbf{I})(\mathbf{A} + 2\mathbf{I})}{(3 + 1)(3 + 2)}$$

$$= -\frac{1}{4} e^{-t} \begin{bmatrix} 1 & -1 & 1 \\ -1 & 1 & -1 \\ -6 & 6 & -6 \end{bmatrix} + \frac{1}{5} e^{-2t} \begin{bmatrix} 4 & -2 & 1 \\ -8 & 4 & -2 \\ -12 & 6 & -3 \end{bmatrix}$$

$$+ \frac{1}{20} e^{3t} \begin{bmatrix} 9 & 3 & 1 \\ 27 & 9 & 3 \\ 18 & 6 & 2 \end{bmatrix}$$

For matrices with repeated eigenvalues, the *confluent* form of Sylvester's theorem applies. If the root λ_m is repeated r times, the corresponding $\mathbf{Z}$ matrix can be shown to have the form

$$\mathbf{Z}_m = \frac{1}{r!} \frac{d^r}{d\lambda^r} \left(\frac{\text{adj}(\lambda\mathbf{I} - \mathbf{A})}{\displaystyle\prod_{\substack{j=1 \\ j \neq m}}^{n} (\lambda - \lambda_j)^r} \right) \Bigg|_{\lambda = \lambda_m}$$

A.6.4 Evaluation Using Residue Matrices

An efficient numerical procedure for determining the resolvent matrix

$$\mathbf{R}(s) = (s\mathbf{I} - \mathbf{A})^{-1}$$

is given by the Faddeev-Leverrier algorithm shown in Table A-10, in which

$$\mathbf{R}(s) = (s\mathbf{I} - \mathbf{A})^{-1} = \frac{\mathbf{D}_1 s^{n-1} + \mathbf{D}_2 s^{n-2} + \cdots + \mathbf{D}_{n-1} s + \mathbf{D}_n}{s^n + \alpha_{n-1} s^{n-1} + \alpha_{n-2} s^{n-2} + \cdots + \alpha_1 s + \alpha_0}$$

Table A-10 The Faddeev-Leverrier Algorithm

$\mathbf{D}_1 = \mathbf{I}$	$\alpha_{n-1} = -\text{trace} (\mathbf{AD}_1)$
$\mathbf{D}_2 = \mathbf{AD}_1 + \alpha_{n-1}\mathbf{I}$	$\alpha_{n-2} = -\dfrac{1}{2} \text{trace} (\mathbf{AD}_2)$
$\mathbf{D}_3 = \mathbf{AD}_2 + \alpha_{n-2}\mathbf{I}$	$\alpha_{n-3} = -\dfrac{1}{3} \text{trace} (\mathbf{AD}_3)$
$\vdots$	$\vdots$
$\mathbf{D}_k = \mathbf{AD}_{k-1} + \alpha_{n-k+1}\mathbf{I}$	$\alpha_{n-k} = -\dfrac{1}{k} \text{trace} (\mathbf{AD}_k)$
$\vdots$	$\vdots$
$\mathbf{D}_n = \mathbf{AD}_{n-1} + \alpha_1\mathbf{I}$	$\alpha_0 = -\dfrac{1}{n} \text{trace} (\mathbf{AD}_n)$

$\mathbf{D}_{n+1} = \mathbf{AD}_n + \alpha_0\mathbf{I}$ ($\mathbf{D}_{n+1}$ is used as an error check.)

$$\mathbf{A}^{-1} = -\frac{1}{\alpha_0} \mathbf{D}_n$$

$$|s\mathbf{I} - \mathbf{A}| = s^n + \alpha_{n-1}s^{n-1} + \alpha_{n-2}s^{n-2} + \cdots + \alpha_1 s + \alpha_0$$

$$\text{adj}(s\mathbf{I} - \mathbf{A}) = \mathbf{D}_1 s^{n-1} + \mathbf{D}_2 s^{n-2} + \cdots + \mathbf{D}_{n-1}s + \mathbf{D}_n$$

$$(s\mathbf{I} - \mathbf{A})^{-1} = \frac{\text{adj}(s\mathbf{I} - \mathbf{A})}{|s\mathbf{I} - \mathbf{A}|}$$

Distinct Eigenvalues

If the matrix $\mathbf{A}$ has distinct eigenvalues $\lambda_1, \lambda_2, \ldots, \lambda_n$, then expanding $\mathbf{R}(s)$ in matrix partial fractions gives

$$\mathbf{R}(s) = \frac{1}{s + \lambda_1} \mathbf{R}_1 + \frac{1}{s + \lambda_2} \mathbf{R}_2 + \cdots + \frac{1}{s + \lambda_n} \mathbf{R}_n \qquad \text{(A-16)}$$

where the matrices $\mathbf{R}_1, \mathbf{R}_2, \ldots, \mathbf{R}_n$ are termed *residue matrices*. Residue matrices are determined as follows:

$$\mathbf{R}_i = (s + \lambda_i)\mathbf{R}(s)\Big|_{s = -\lambda_i}$$

Having obtained the residue matrices, the state transition matrix is determined using the inverse Laplace transform of equation (A-16)

$$e^{\mathbf{A}t} = e^{-\lambda_1 t}\mathbf{R}_1 + e^{-\lambda_2 t}\mathbf{R}_2 + \cdots + e^{-\lambda_n t}\mathbf{R}_n \qquad \text{(A-17)}$$

An application of the Faddeev-Leverrier algorithm to the matrix

$$\mathbf{A} = \begin{bmatrix} 0 & 1 \\ -3 & -4 \end{bmatrix}$$

follows:

$$\mathbf{D}_1 = \mathbf{I} \qquad \alpha_1 = -\text{trace}\,(\mathbf{AD}_1) = 4$$

$$\mathbf{D}_2 = \mathbf{AD}_1 + \alpha_1\mathbf{I}$$

$$= \begin{bmatrix} 0 & 1 \\ -3 & -4 \end{bmatrix} + \begin{bmatrix} 4 & 0 \\ 0 & 4 \end{bmatrix} = \begin{bmatrix} 4 & 1 \\ -3 & 0 \end{bmatrix}$$

$$\alpha_0 = -\frac{1}{2}\,\text{trace}\,(\mathbf{AD}_2)$$

$$= -\frac{1}{2}\,\text{trace}\left(\begin{bmatrix} 0 & 1 \\ -3 & -4 \end{bmatrix}\begin{bmatrix} 4 & 1 \\ -3 & 0 \end{bmatrix}\right)$$

$$= 3$$

Then

$$\mathbf{R}(s) = (s\mathbf{I} - \mathbf{A})^{-1} = \frac{\begin{bmatrix} 1 & 0 \\ 0 & 1 \end{bmatrix}s + \begin{bmatrix} 4 & 1 \\ -3 & 0 \end{bmatrix}}{s^2 + 4s + 3}$$

$$= \frac{\begin{bmatrix} (s+4) & 1 \\ -3 & s \end{bmatrix}}{(s+1)(s+3)}$$

$$= \frac{1}{s+1}\mathbf{R}_1 + \frac{1}{s+3}\mathbf{R}_2$$

where

$$\mathbf{R}_1 = \left.\frac{\begin{bmatrix} (s+4) & 1 \\ -3 & s \end{bmatrix}}{s+3}\right|_{s=-1} = \frac{1}{2}\begin{bmatrix} 3 & 1 \\ -3 & -1 \end{bmatrix}$$

$$\mathbf{R}_2 = \left.\frac{\begin{bmatrix} (s+4) & 1 \\ -3 & s \end{bmatrix}}{s+1}\right|_{s=-3} = \frac{1}{2}\begin{bmatrix} -1 & -1 \\ 3 & 3 \end{bmatrix}$$

Therefore,

$$e^{\mathbf{A}t} = \frac{1}{2} e^{-t} \begin{bmatrix} 3 & 1 \\ -3 & -1 \end{bmatrix} - \frac{1}{2} e^{-3t} \begin{bmatrix} 1 & 1 \\ -3 & -3 \end{bmatrix}$$

$$= \frac{1}{2} \begin{bmatrix} (3e^{-t} - e^{-3t}) & (e^{-t} - e^{-3t}) \\ (-3e^{-t} + 3e^{-3t}) & (-e^{-t} + 3e^{-3t}) \end{bmatrix}$$

which can be easily verified.

Repeated Eigenvalues

If the matrix $\mathbf{A}$ has an eigenvalue repeated r times, the resolvent matrix is of the form

$$\mathbf{R}(s) = \frac{1}{s + \lambda_1} \mathbf{R}_1 + \frac{1}{(s + \lambda_1)^2} \mathbf{R}_2 + \cdots + \frac{1}{(s + \lambda_1)^r} \mathbf{R}_r + \cdots$$

$$+ \{\text{other terms that correspond to distinct eigenvalues}\}$$

The residue matrices corresponding to the repeated eigenvalue λ_1 may be found by evaluation according to

$$\mathbf{R}_i = \frac{1}{(r - i)!} \left\{ \frac{d^{r-i}}{ds^{r-i}} [(s + \lambda_1)^r \mathbf{R}(s)] \right\} \Bigg|_{s = -\lambda_1} \qquad i = 1, 2, \ldots, r$$

For example, for

$$\mathbf{A} = \begin{bmatrix} -4 & 1 \\ -4 & 0 \end{bmatrix}$$

the Faddeev-Leverrier algorithm gives

$$\mathbf{D}_1 = \mathbf{I} \qquad \alpha_1 = -\text{trace} (\mathbf{AD}_1) = 4$$

$$\mathbf{D}_2 = \mathbf{AD}_1 + \alpha_1 \mathbf{I}$$

$$= \begin{bmatrix} -4 & 1 \\ -4 & 0 \end{bmatrix} + \begin{bmatrix} 4 & 0 \\ 0 & 4 \end{bmatrix} = \begin{bmatrix} 0 & 1 \\ -4 & 4 \end{bmatrix}$$

$$\alpha_0 = -\frac{1}{2} \text{trace} (\mathbf{AD}_2) = 4$$

Then

$$\mathbf{R}(s) = (s\mathbf{I} - \mathbf{A})^{-1} = \frac{s \begin{bmatrix} 1 & 0 \\ 0 & 1 \end{bmatrix} + \begin{bmatrix} 0 & 1 \\ -4 & 4 \end{bmatrix}}{s^2 + 4s + 4}$$

$$= \frac{\begin{bmatrix} s & 1 \\ -4 & (s+4) \end{bmatrix}}{(s+2)^2} = \frac{1}{s+2} \mathbf{R}_1 + \frac{1}{(s+2)^2} \mathbf{R}_2$$

and

$$\mathbf{R}_1 = \frac{1}{1!} \left\{ \frac{d}{ds} \begin{bmatrix} s & 1 \\ -4 & (s+4) \end{bmatrix} \right\} \Bigg|_{s=-2}$$

$$= \begin{bmatrix} 1 & 0 \\ 0 & 1 \end{bmatrix}$$

Similarly,

$$\mathbf{R}_2 = \frac{1}{0!} \left\{ \begin{bmatrix} s & 1 \\ -4 & (s+4) \end{bmatrix} \right\} \Bigg|_{s=-2}$$

$$= \begin{bmatrix} -2 & 1 \\ -4 & 2 \end{bmatrix}$$

Hence,

$$\mathbf{R}(s) = \frac{1}{s+2} \mathbf{R}_1 + \frac{1}{(s+2)^2} \mathbf{R}_2$$

and

$$e^{At} = e^{-2t} \begin{bmatrix} 1 & 0 \\ 0 & 1 \end{bmatrix} + te^{-2t} \begin{bmatrix} -2 & 1 \\ -4 & 2 \end{bmatrix}$$

$$= \begin{bmatrix} (e^{-2t} - 2te^{-2t}) & te^{-2t} \\ -4te^{-2t} & (e^{-2t} + 2te^{-2t}) \end{bmatrix}$$

A.6.5 Evaluation Using Reciprocal Basis

Another method of determining the state transition matrix when $\mathbf{A}$ has distinct eigenvalues is to use the concept of reciprocal basis.

For a given set of basis vectors $\mathbf{p}^1$, $\mathbf{p}^2$, . . . , $\mathbf{p}^n$, any arbitrary vector $\mathbf{x}$ in the same vector space can be uniquely expressed as a linear combination of the basis vectors as

$$\mathbf{x} = \sum_{i=1}^{n} \xi_i \mathbf{p}^i \tag{A-18}$$

$$= [\mathbf{p}^1 \quad \mathbf{p}^2 \quad \cdots \quad \mathbf{p}^n] \begin{bmatrix} \xi_1 \\ \xi_2 \\ \vdots \\ \xi_n \end{bmatrix}$$

$$= \mathbf{P}\xi$$

A reciprocal basis is a set of vectors $\mathbf{q}_1^\dagger, \mathbf{q}_2^\dagger, \cdots, \mathbf{q}_n^\dagger$, such that

$$\mathbf{q}_i^\dagger \mathbf{p}^j = \begin{cases} 1 & i = j \\ 0 & i \neq j \end{cases}$$

Specifically, if the eigenvectors of the matrix $\mathbf{A}$ are chosen as the basis vectors, and if we define

$$\mathbf{Q} = \mathbf{P}^{-1}$$

and partition $\mathbf{Q}$ into n rows

$$\mathbf{Q} = \begin{bmatrix} \mathbf{q}_1^\dagger \\ \mathbf{q}_2^\dagger \\ \vdots \\ \mathbf{q}_n^\dagger \end{bmatrix}$$

then

$$\mathbf{QP} = \mathbf{I}$$

or

$$\begin{bmatrix} \mathbf{q}_1^\dagger \\ \mathbf{q}_2^\dagger \\ \vdots \\ \mathbf{q}_n^\dagger \end{bmatrix} [\mathbf{p}^1 \quad \mathbf{p}^2 \quad \cdots \quad \mathbf{p}^n] = \mathbf{I}$$

and, therefore,

$$\mathbf{q}_i^\dagger \mathbf{p}^j = \begin{cases} 1 & i = j \\ 0 & i \neq j \end{cases}$$

indicating that the reciprocal basis can be determined simply as a set of row vectors of $\mathbf{P}^{-1}$.

The row vector $\mathbf{q}_i^\dagger$ is termed the ith *left eigenvector* of the matrix $\mathbf{A}$ because it satisfies

$$\mathbf{q}_i^\dagger \mathbf{A} = \lambda_i \mathbf{q}_i^\dagger$$

To show this, recall

$$\Lambda = \mathbf{P}^{-1}\mathbf{A}\mathbf{P}$$
$$= \mathbf{Q}\mathbf{A}\mathbf{P}$$

Then

$$\Lambda\mathbf{Q} = \mathbf{P}^{-1}\mathbf{A} = \mathbf{Q}\mathbf{A}$$

or

$$\begin{bmatrix} \lambda_1 & 0 & 0 & \cdots & 0 \\ 0 & \lambda_2 & 0 & \cdots & 0 \\ \vdots & & & & \\ 0 & 0 & 0 & \cdots & \lambda_n \end{bmatrix}\begin{bmatrix} \mathbf{q}_1^\dagger \\ \mathbf{q}_2^\dagger \\ \vdots \\ \mathbf{q}_n^\dagger \end{bmatrix} = \begin{bmatrix} \lambda_1\mathbf{q}_1^\dagger \\ \lambda_2\mathbf{q}_2^\dagger \\ \vdots \\ \lambda_n\mathbf{q}_n^\dagger \end{bmatrix} = \mathbf{Q}\mathbf{A}$$

Therefore,

$$\begin{bmatrix} \mathbf{q}_1^\dagger \\ \mathbf{q}_2^\dagger \\ \vdots \\ \mathbf{q}_n^\dagger \end{bmatrix}\mathbf{A} = \begin{bmatrix} \lambda_1\mathbf{q}_1^\dagger \\ \lambda_2\mathbf{q}_2^\dagger \\ \vdots \\ \lambda_n\mathbf{q}_n^\dagger \end{bmatrix}$$

the stated result.

For example, the matrix

$$\mathbf{A} = \begin{bmatrix} 0 & 1 \\ -3 & -4 \end{bmatrix}$$

has the eigenvalues

$$\lambda_1 = -1 \quad \text{and} \quad \lambda_2 = -3$$

and corresponding eigenvectors

$$\mathbf{p}^1 = \begin{bmatrix} 1 \\ -1 \end{bmatrix} \quad \text{and} \quad \mathbf{p}^2 = \begin{bmatrix} 1 \\ -3 \end{bmatrix}$$

Hence,

$$\mathbf{P} = \begin{bmatrix} 1 & 1 \\ -1 & -3 \end{bmatrix}$$

$$\mathbf{Q} = \mathbf{P}^{-1} = -\frac{1}{2}\begin{bmatrix} -3 & -1 \\ 1 & 1 \end{bmatrix}$$

Thus,

$$\mathbf{q}_1^\dagger = -\frac{1}{2}[-3 \quad -1]$$

$$\mathbf{q}_2^\dagger = -\frac{1}{2}[1 \quad 1]$$

Therefore,

$$\mathbf{q}_1^\dagger \mathbf{A} = -\frac{1}{2}[-3 \quad -1]\begin{bmatrix} 0 & 1 \\ -3 & -4 \end{bmatrix}$$

$$= -\frac{1}{2}[3 \quad 1] = \lambda_1 \mathbf{q}_1^\dagger$$

and

$$\mathbf{q}_2^\dagger \mathbf{A} = -\frac{1}{2}[1 \quad 1]\begin{bmatrix} 0 & 1 \\ -3 & -4 \end{bmatrix}$$

$$= -\frac{1}{2}[-3 \quad -3] = \lambda_2 \mathbf{q}_2^\dagger$$

Continuous-Time Systems

Consider the state equation

$$\dot{\mathbf{x}}(t) = \mathbf{A}\mathbf{x}(t)$$

If the state vector is written as

$$\mathbf{x}(t) = \sum_{i=1}^{n} \xi_i(t)\mathbf{p}^i \qquad\qquad\text{(A-19)}$$

where ξ_i is a scalar, then

$$\dot{\mathbf{x}}(t) = \sum_{i=1}^{n} \dot{\xi}_i(t)\mathbf{p}^i = \mathbf{A}\mathbf{x}(t)$$

$$= \mathbf{A}\sum_{i=1}^{n} \xi_i(t)\mathbf{p}^i = \sum_{i=1}^{n} \lambda_i\xi_i(t)\mathbf{p}^i$$

because

$$\mathbf{A}\mathbf{p}^i = \lambda_i \mathbf{p}^i$$

Hence,

$$\dot{\xi}_i(t) = \lambda_i \xi_i(t)$$

and

$$\xi_i(t) = e^{\lambda_i t}\xi_i(0)$$

Substituting into equation (A-19) gives

$$\mathbf{x}(t) = \sum_{i=1}^{n} \xi_i(0)(e^{\lambda_i t}\mathbf{p}^i) \qquad \text{(A-20)}$$

where the term in parenthesis is called the ith *natural mode* of the system described by

$$\dot{\mathbf{x}}(t) = \mathbf{A}\mathbf{x}(t)$$

The initial conditions $\xi_i(0)$ in equation (A-20) are determined by setting $t = 0$ in equation (A-20)

$$\mathbf{x}(0) = \sum_{i=1}^{n} \xi_i(0)\mathbf{p}^i$$

and multiplying both sides by the ith left eigenvector

$$\mathbf{q}_i^{\dagger}\mathbf{x}(0) = \sum_{i=1}^{n} \xi_i(0)\mathbf{q}_i^{\dagger}\mathbf{p}^i = \xi_i(0)$$

Therefore,

$$\mathbf{x}(t) = \sum_{i=1}^{n} [\mathbf{q}_i^{\dagger}\mathbf{x}(0)]e^{\lambda_i t}\mathbf{p}^i$$

$$= \sum_{i=1}^{n} [e^{\lambda_i t}\mathbf{p}^i\mathbf{q}_i^{\dagger}]\mathbf{x}(0) = e^{\mathbf{A}t}\mathbf{x}(0)$$

which implies

$$e^{\mathbf{A}t} = \sum_{i=1}^{n} e^{\lambda_i t}\mathbf{p}^i\mathbf{q}_i^{\dagger} \qquad \text{(A-21)}$$

For example, the matrix

$$\mathbf{A} = \begin{bmatrix} 0 & 1 \\ -3 & -4 \end{bmatrix}$$

given previously, has eigenvalues

$$\lambda_1 = -1 \quad \text{and} \quad \lambda_2 = -3$$

and corresponding right and left eigenvectors

$$\mathbf{p}^1 = \begin{bmatrix} 1 \\ -1 \end{bmatrix} \quad \text{and} \quad \mathbf{p}^2 = \begin{bmatrix} 1 \\ -3 \end{bmatrix}$$

$$\mathbf{q}_1^\dagger = -\frac{1}{2}[-3 \quad -1] \quad \text{and} \quad \mathbf{q}_2^\dagger = -\frac{1}{2}[1 \quad 1]$$

Therefore,

$$
\begin{aligned}
e^{\mathbf{A}t} &= e^{-t}\mathbf{p}^1\mathbf{q}_1^\dagger + e^{-3t}\mathbf{p}^2\mathbf{q}_2^\dagger \\
&= \frac{1}{2}e^{-t}\begin{bmatrix} 3 & 1 \\ -3 & -1 \end{bmatrix} + \frac{1}{2}e^{-3t}\begin{bmatrix} -1 & -1 \\ 3 & 3 \end{bmatrix} \\
&= \frac{1}{2}\begin{bmatrix} (3e^{-t} - e^{-3t}) & (e^{-t} - e^{-3t}) \\ (-3e^{-t} + 3e^{-3t}) & (-e^{-t} + 3e^{-3t}) \end{bmatrix}
\end{aligned}
$$

which can be easily verified.

Incidently, the comparison of equations (A-21) and (A-17) shows that

$$\mathbf{p}^i\mathbf{q}_i^\dagger = \mathbf{R}_i$$

the residue matrix can be determined using the right and left eigenvalues of the matrix $\mathbf{A}$.

Discrete-Time Systems

Consider the linear, step-invariant, discrete-time system

$$\mathbf{x}(k + 1) = \mathbf{A}\mathbf{x}(k) \tag{A-22}$$

If the matrix $\mathbf{A}$ has distinct eigenvalues, the state vector can be expressed as

$$\mathbf{x}(k) = \sum_{i=1}^{n} \xi_i(k)\mathbf{p}^i \tag{A-23}$$

Substituting equation (A-23) into equation (A-22) gives

$$
\begin{aligned}
\mathbf{x}(k + 1) &= \mathbf{A}\sum_{i=1}^{n} \xi_i(k)\mathbf{p}^i \\
&= \sum_{i=1}^{n} \xi_i(k)\lambda_i\mathbf{p}^i
\end{aligned}
$$

but

$$\mathbf{x}(k + 1) = \sum_{i=1}^{n} \xi_i(k + 1)\mathbf{p}^i$$

so

$$\xi_i(k + 1) = \lambda_i \xi_i(k)$$

and

$$\xi_i(k) = \lambda_i^k \xi_i(0)$$

Therefore,

$$\mathbf{x}(k) = \sum_{i=1}^{n} \xi_i(0)\lambda_i^k \mathbf{p}^i \qquad \text{(A-24)}$$

The initial conditions $\xi_i(0)$ in equation (A-24) are determined by setting $k = 0$ in equation (A-24)

$$\mathbf{x}(0) = \sum_{i=1}^{n} \xi_i(0)\mathbf{p}^i \qquad \text{(A-25)}$$

and multiplying both sides of equation (A-25) by the ith left eigenvector

$$\mathbf{q}_i^\dagger \mathbf{x}(0) = \sum_{i=1}^{n} \xi_i(0)\mathbf{q}_i^\dagger \mathbf{p}^i = \xi_i(0)$$

therefore,

$$\mathbf{x}(k) = \sum_{i=1}^{n} \lambda_i^k \mathbf{p}^i \mathbf{q}_i^\dagger \mathbf{x}(0)$$

Hence,

$$\mathbf{A}^k = \sum_{i=1}^{n} \lambda_i^k \mathbf{p}^i \mathbf{q}_i^\dagger$$

References

There are many fine texts on matrix methods. Among them are

L. Mirsky, *An Introduction to Linear Algebra*. New York: Dover, 1982. Originally, Oxford: Clarendon Press, 1955;

R. Bellman, *Introduction to Matrix Analysis*. New York: McGraw-Hill, 1960;

L. A. Pipes, *Matrix Methods for Engineering*. Englewood Cliffs, NJ: Prentice-Hall, 1963;

G. W. Stewart, *Introduction to Matrix Computation*. New York: Academic Press, 1973;

G. Strang, *Linear Algebra and Its Applications,* 2nd edition. New York: Academic Press, 1980;

J. N. Franklin, *Matrix Theory.* Englewood Cliffs, NJ: Prentice-Hall, 1968;

B. Noble, *Applied Linear Algebra,* 2nd edition. Englewood Cliffs, NJ: Prentice-Hall, 1977;

F. R. Gantmakher, *Theory of Matrices* (2 Volumes). New York: Chelsea Publishing Co., 1959;

S. Barnett, *Matrices in Control Theory.* New York: Van Nostrand Reinhold, 1972;

G. Forsythe, M. Malcolm, and C. Moler, *Computer Methods for Mathematical Computations.* Englewood Cliffs, NJ: Prentice-Hall, 1977;

D. K. Faddeev and V. N. Faddeeva, *Computational Methods of Linear Algebra.* San Franciso: W. H. Freeman & Co., 1963;

G. H. Golub and C. F. Van Loan, *Matrix Computations.* Baltimore, MD: The Johns Hopkins University Press, 1983;

A. S. Householder, *The Theory of Matrices in Numerical Analysis.* Waltham, MA: Ginn-Blaisdell, 1964.

Problems for Appendix A

A-1. Show that for square matrices

  a. $B = A + A^\dagger$ is symmetric.

  b. $D = A^\dagger A$ is symmetric.

  c. trace (AB) = trace (BA).

A-2. Show that

$$\frac{d}{dt}(Ab) = \frac{dA}{dt}b + A\frac{db}{dt}$$

where b is a vector. From this result,

$$\frac{d}{dt}(AB) = \frac{dA}{dt}B + A\frac{dB}{dt}$$

follows easily because each column of the general relation is of the form of the vector relation. Using the above, find

$$\frac{d}{dt}(A^3)$$

in terms of A and dA/dt.

A-3. Indicate two different ways that the following matrices might be partitioned so that the submatrices are conformable for multiplication. Find the product using each of the partitions.

$$\begin{bmatrix} 0 & 3 & -2 \\ 1 & -1 & 4 \\ -2 & 3 & 8 \end{bmatrix} \begin{bmatrix} -4 & -4 & 4 & 1 \\ -5 & -2 & 2 & -1 \\ 7 & 0 & -3 & -5 \end{bmatrix}$$

A-4. Find the ranks of the following matrices:

a. $\begin{bmatrix} 1 & -3 & 2 & 4 \\ -2 & -1 & 3 & 0 \\ -3 & -5 & 8 & 4 \end{bmatrix}$

b. $\begin{bmatrix} 1 & 0 & 1 & 0 & 0 \\ -1 & 1 & 0 & 1 & -1 \\ 0 & 0 & -1 & 0 & 0 \\ 1 & 0 & -1 & 0 & 0 \end{bmatrix}$

A-5. Find the following:

a. Two square matrices A and B such that $AB \neq BA$.

b. Two *different* nonzero square matrices A and B such that $AB = BA$.

c. A set of nonzero matrices A, B, and C such that $AB = AC$ but $B \neq C$.

d. Two nonzero matrices A and B such that $AB = 0$.

e. A nonzero 2×2 matrix A such that $A^2 = 0$.

f. A 2×2 matrix A such that $A^2 = -I$. This matrix is analogous to $\sqrt{-1}$.

A-6. If possible, find the solutions for the following sets of equations:

a. $\quad 2x_1 + 3x_2 = 0$
$\quad -3x_1 - x_2 = 1$

b. $0x_1 + x_2 + x_3 = 2$
$\quad x_1 + 0x_2 + x_3 = 3$
$\quad x_1 + x_2 + 0x_3 = -4$

c. $2x_1 - x_2 + 3x_3 = 4$
$\quad 4x_1 - 2x_2 + 0x_3 = 6$
$\quad 0x_1 - 3x_2 + x_3 = 0$

A-7. Find the inverses, if they exist, of the following matrices, using

$$A^{-1} = \frac{\text{adj } (A)}{|A|}$$

a. $\mathbf{A} = \begin{bmatrix} 1 & -3 \\ 2 & 1 \end{bmatrix}$ b. $\mathbf{A} = \begin{bmatrix} 2 & 4 & 4 \\ 0 & 0 & 1 \\ -1 & 2 & -3 \end{bmatrix}$

c. $\mathbf{A} = \begin{bmatrix} 1 & 3 & -2 \\ 0 & 4 & 5 \\ -3 & -1 & 2 \end{bmatrix}$

A-8. Show that

a. $(\mathbf{A}^{-1})^{\dagger} = (\mathbf{A}^{\dagger})^{-1}$

b. $(\mathbf{A}^{-1})^{-1} = \mathbf{A}$

A-9. Show that if $\mathbf{A}$ and $\mathbf{B}$ are each nonsingular,

$$\begin{bmatrix} \mathbf{A} & \mathbf{0} \\ \mathbf{0} & \mathbf{B} \end{bmatrix}^{-1} = \begin{bmatrix} \mathbf{A}^{-1} & \mathbf{0} \\ \mathbf{0} & \mathbf{B}^{-1} \end{bmatrix}$$

A-10. Determine how many of the following vectors are linearly independent:

$$\begin{bmatrix} 1 \\ 2 \\ 1 \end{bmatrix} \quad \begin{bmatrix} 1 \\ 1 \\ 0 \end{bmatrix} \quad \begin{bmatrix} 0 \\ 1 \\ 1 \end{bmatrix} \quad \begin{bmatrix} 3 \\ 6 \\ 3 \end{bmatrix} \quad \begin{bmatrix} -1 \\ -2 \\ -1 \end{bmatrix}$$

Specify a subset of these vectors that contains the largest number of linearly independent vectors.

A-11. Find a basis for the range space and a basis for the null space for the following matrices:

a. $\begin{bmatrix} 2 & 0 & 3 \\ 1 & 0 & 1 \\ 1 & 0 & 2 \end{bmatrix}$ b. $\begin{bmatrix} 1 & -2 & 3 \\ 2 & 0 & 0 \\ -1 & 1 & 0 \end{bmatrix}$

c. $\begin{bmatrix} 2 & -2 & 1 \\ -2 & 2 & -1 \\ 4 & -4 & 2 \end{bmatrix}$

A-12. For the vectors

$$\mathbf{x} = \begin{bmatrix} 2 \\ 0 \\ 3 \end{bmatrix} \quad \text{and} \quad \mathbf{y} = \begin{bmatrix} -1 \\ 2 \\ 4 \end{bmatrix}$$

find the following:

a. $\|\mathbf{x}\|$

b. $\|\mathbf{y}\|$

c. $\langle \mathbf{x}, \mathbf{y} \rangle$

d. A unit vector in the direction of $\mathbf{x}$.

e. A vector in the direction of $\mathbf{y}$ but with norm of 2.

f. A vector that is orthogonal to $\mathbf{x}$.

g. A unit vector that is orthogonal to both $\mathbf{x}$ and $\mathbf{y}$.

A-13. Express the vector

$$\begin{bmatrix} 2 \\ 0 \\ -1 \\ 3 \end{bmatrix}$$

as a linear combination of the basis vectors

$$\begin{bmatrix} 1 \\ 0 \\ 1 \\ 0 \end{bmatrix} \quad \begin{bmatrix} 0 \\ 3 \\ 2 \\ 0 \end{bmatrix} \quad \begin{bmatrix} 0 \\ 2 \\ -1 \\ 0 \end{bmatrix} \quad \text{and} \quad \begin{bmatrix} 2 \\ -2 \\ 0 \\ -1 \end{bmatrix}$$

A-14. Verify that the matrix

$$\mathbf{A} = \begin{bmatrix} \dfrac{\sqrt{2}}{2} & \dfrac{\sqrt{2}}{2} & 0 \\ -\dfrac{\sqrt{2}}{2} & \dfrac{\sqrt{2}}{2} & 0 \\ 0 & 0 & 1 \end{bmatrix}$$

is orthogonal and that it satisfies the properties listed in Table A-6.

A-15. Find the eigenvalues of the following matrices. For which of these matrices does an inverse exist?

a. $\begin{bmatrix} -5 & 1 \\ -6 & 0 \end{bmatrix}$

b. $\begin{bmatrix} -1 & 2 & 0 \\ 1 & 5 & 6 \\ -1 & 9 & 6 \end{bmatrix}$

c. $\begin{bmatrix} 0 & 1 & 0 \\ 0 & 0 & 1 \\ -2 & -5 & -4 \end{bmatrix}$

A-16. Find the characteristic equation, eigenvalues, eigenvectors, and a modal matrix for each of the following matrices:

a. $\begin{bmatrix} 0 & 1 \\ -1 & 0 \end{bmatrix}$

b. $\begin{bmatrix} -5 & 1 & 0 \\ -4 & 0 & 1 \\ 0 & 0 & 0 \end{bmatrix}$

c. $\begin{bmatrix} 1 & 0 & -5 \\ 0 & -1 & -6 \\ 0 & 0 & 2 \end{bmatrix}$

d. $\begin{bmatrix} -6 & 3 \\ -8 & 4 \end{bmatrix}$

e. $\begin{bmatrix} 0 & 0 & 0 \\ 0 & 0 & 1 \\ 0 & -1 & 0 \end{bmatrix}$

A-17. For the following matrices, find a modal matrix **P** and the corresponding spectral matrix Λ. Verify that $\mathbf{P}^{-1}\mathbf{AP} = \Lambda$.

a. $\begin{bmatrix} -1 & 0 \\ -1 & 0 \end{bmatrix}$

b. $\begin{bmatrix} 1 & 0 \\ -1 & 4 \end{bmatrix}$

c. $\begin{bmatrix} 1 & 4 & 6 \\ 2 & 8 & 12 \\ -1 & -4 & -6 \end{bmatrix}$

d. $\begin{bmatrix} 0 & 1 & 5 \\ 1 & 0 & -1 \\ 0 & 0 & 2 \end{bmatrix}$

e. $\begin{bmatrix} 1 & 1 \\ -1 & 1 \end{bmatrix}$

A-18. Verify that the matrix

$$\mathbf{A} = \begin{bmatrix} 3 & 0 & 1 \\ -2 & 1 & 0 \\ 4 & -3 & 2 \end{bmatrix}$$

satisfies its own characteristic equation.

A-19. For the matrix

$$\mathbf{A} = \begin{bmatrix} 1 & 2 \\ -3 & 4 \end{bmatrix}$$

find $\mathbf{A}^5$ by first expressing it in terms of $\mathbf{A}$ and $\mathbf{I}$.

A-20. For the matrix

$$\mathbf{A} = \begin{bmatrix} 0 & 1 & 0 \\ 0 & 0 & 1 \\ 2 & -1 & 4 \end{bmatrix}$$

express the matrix polynomial

$$\mathbf{A}^4 + 2\mathbf{A}^3 + \mathbf{A}^2 - \mathbf{A} + 3\mathbf{I}$$

in terms of $\mathbf{A}^2$, $\mathbf{A}$, and $\mathbf{I}$.

A-21. Although every matrix satisfies its own characteristic equation, some matrices satisfy polynomial equations of lower degree than the characteristic equation. The polynomial equation of lowest degree that is satisfied by a matrix is called a *minimal polynomial*. Verify that the 3×3 matrix

$$\mathbf{A} = \begin{bmatrix} 6 & 0 & 0 \\ 0 & 2 & 0 \\ 0 & 0 & 6 \end{bmatrix}$$

satisfies the second-degree polynomial equation

$$\mathbf{A}^2 - 8\mathbf{A} + 12\mathbf{I} = \mathbf{0}$$

A-22. For any nonsingular matrix $\mathbf{A}$ with distinct eigenvalues, one can find $\mathbf{A}^{-1}$ by diagonalizing $\mathbf{A}$

$$\mathbf{P}^{-1}\mathbf{A}\mathbf{P} = \begin{bmatrix} \lambda_1 & 0 & \cdots \\ 0 & \lambda_2 & \\ \vdots & & \ddots \end{bmatrix}$$

using

$$\mathbf{P}^{-1}(\mathbf{A}^{-1})\mathbf{P} = \mathbf{R} = \begin{bmatrix} \left(\dfrac{1}{\lambda_1}\right) & 0 & \cdots \\ 0 & \left(\dfrac{1}{\lambda_2}\right) & \\ \vdots & & \ddots \end{bmatrix}$$

and forming

$$\mathbf{A}^{-1} = \mathbf{PRP}^{-1}$$

Use this method to find the inverse of

$$\mathbf{A} = \begin{bmatrix} 0 & 1 & 0 \\ 0 & 0 & 1 \\ -27 & -39 & -13 \end{bmatrix}$$

A-23. Express the following quadratic forms in terms of symmetric matrices $\mathbf{S}$ in $f(\mathbf{x}) = \mathbf{x}^t \mathbf{Sx}$.

a. $f(x_1, x_2, x_3) = 3x_1^2 - 4x_1x_2 + 4x_1x_3 + 5x_2^2 + 2x_2x_3 - x_3^2$

b. $f(x_1, x_2, x_3, x_4, x_5) = 6x_1^2 + 3x_1x_2 - 2x_1x_3 + 4x_1x_4$
$+ 16x_2^2 + 8x_2x_3 - x_2x_4 + x_2x_5 + 10x_3^2 - 4x_3x_4 - 8x_3x_5$
$- 12x_4^2 + 5x_4x_5 - x_5^2$

A-24. For the following symmetric matrices, find an orthogonal modal matrix $\mathbf{P}_0$ and the corresponding spectral matrix $\mathbf{S}_0$. Verify that $\mathbf{P}_0^t \mathbf{SP}_0 = \mathbf{S}_0$.

a. $\begin{bmatrix} 2 & -1 \\ -1 & 2 \end{bmatrix}$

b. $\begin{bmatrix} 2 & 0 & 0 \\ 0 & 2 & -1 \\ 0 & -1 & 2 \end{bmatrix}$

c. $\begin{bmatrix} -1 & 0 & -3 \\ 0 & 3 & 0 \\ -3 & 0 & -1 \end{bmatrix}$

A-25. For each of the following quadratic forms, find orthogonal matrices $\mathbf{P}_0$ that transform the function to a sum of squares through

$$\mathbf{x} = \mathbf{P}_0\mathbf{x}' \qquad \mathbf{x}' = \mathbf{P}_0^{-1}\mathbf{x}$$

a. $f(\mathbf{x}) = \mathbf{x}^t \begin{bmatrix} 3 & 2 \\ 2 & 3 \end{bmatrix} \mathbf{x}$

b. $f(\mathbf{x}) = \mathbf{x}^t \begin{bmatrix} 2 & -1 & 0 \\ -1 & 3 & -1 \\ 0 & -1 & 2 \end{bmatrix} \mathbf{x}$

A-26. Determine the sign definiteness, if any, of the following symmetric matrices:

a. $A_1 = \begin{bmatrix} 5 & -2 \\ -2 & 3 \end{bmatrix}$

b. $A_2 = \begin{bmatrix} 0 & 1 & 1 \\ 1 & 2 & -1 \\ 1 & -1 & 1 \end{bmatrix}$

c. $A_3 = \begin{bmatrix} -3 & 1 & 0 \\ 1 & 4 & -2 \\ 0 & -2 & 5 \end{bmatrix}$

A-27. Find the square root of each of the following matrices:

a. $S = \begin{bmatrix} 3 & -1 \\ -1 & 3 \end{bmatrix}$

b. $S = \begin{bmatrix} 1 & 0 & 0 \\ 0 & 2 & 1 \\ 0 & 1 & 1 \end{bmatrix}$

A-28. Find the square root of the matrix

$$S = \begin{bmatrix} 36 & -6 & -12 & 0 \\ -6 & 5 & 10 & 2 \\ -12 & 10 & 24 & 20 \\ 0 & 2 & 20 & 74 \end{bmatrix}$$

using the Cholesky decomposition method. Verify by direct multiplication that it is a square root.

A-29. Use the diagonalization method to find $\exp(At)$ for the following matrices:

a. $A = \begin{bmatrix} -2 & 1 \\ 0 & -4 \end{bmatrix}$ **b.** $A = \begin{bmatrix} 0 & 1 & 0 \\ 0 & 0 & 1 \\ -6 & -11 & -6 \end{bmatrix}$

A-30. Use the Cayley-Hamilton technique to find $\exp(At)$ for the following matrices:

a. $A = \begin{bmatrix} -1 & 2 \\ 0 & -1 \end{bmatrix}$

b. $A = \begin{bmatrix} 0 & 1 & 0 \\ 0 & 0 & 1 \\ -24 & -26 & -9 \end{bmatrix}$

A-31. Use Sylvester's theorem to find $\exp(\mathbf{A}t)$ for the following matrices:

a. $\mathbf{A} = \begin{bmatrix} -1 & 0 \\ 1 & -2 \end{bmatrix}$

b. $\mathbf{A} = \begin{bmatrix} 0 & 1 & 0 \\ 0 & 0 & 1 \\ 0 & -3 & -4 \end{bmatrix}$

A-32. Use the method of residues to find $\exp(\mathbf{A}t)$ for the following matrices:

a. $\mathbf{A} = \begin{bmatrix} -1 & 0 \\ -2 & -3 \end{bmatrix}$

b. $\mathbf{A} = \begin{bmatrix} 0 & 1 & 0 \\ 0 & 0 & 1 \\ -1 & -3 & -3 \end{bmatrix}$

A-33. Use the method of reciprocal basis to find $\exp(\mathbf{A}t)$ for the following matrices:

a. $\mathbf{A} = \begin{bmatrix} 0 & 1 \\ -6 & -5 \end{bmatrix}$

b. $\mathbf{A} = \begin{bmatrix} 0 & 1 & 0 \\ 0 & 0 & 1 \\ -8 & -14 & -7 \end{bmatrix}$

A-34. Use the method of reciprocal basis to find $\mathbf{A}^k$ for the matrices given in problem A-33.

A-35. Let $\lambda_1, \lambda_2, \ldots, \lambda_n$ be the eigenvalues and $\mathbf{p}^1, \mathbf{p}^2, \ldots, \mathbf{p}^n$ be a corresponding set of linearly independent eigenvectors of an $n \times n$ matrix $\mathbf{A}$; that is,

$$\mathbf{A}\mathbf{p}^i = \lambda_i \mathbf{p}^i \quad i = 1, 2, \ldots, n$$

a. Show that the corresponding residue matrices as given by

$$\mathbf{R}_i = \mathbf{p}^i \mathbf{q}_i^\dagger \quad i = 1, 2, \ldots, n$$

satisfy the following properties:

(1) $\mathbf{R}_i^m = \mathbf{R}_i$ for any positive integer m

(2) $\mathbf{R}_i \mathbf{R}_j = \mathbf{0}$ if $i \neq j$

(3) $\displaystyle\sum_{i=1}^{n} \mathbf{R}_i = \mathbf{I}$ the identity matrix

(4) $\mathbf{A} = \displaystyle\sum_{i=1}^{n} \lambda_i \mathbf{R}_i$

and

$$\mathbf{A}^k = \sum_{i=1}^{n} \lambda_i^k \mathbf{R}_i$$

b. Show that, for a nonsingular matrix $\mathbf{A}$,

$$\mathbf{A}^{-1} = \sum_{i=1}^{n} (\lambda_i)^{-1} \mathbf{R}_i$$

c. Using property (4) in part (**a**), find the matrix $\mathbf{A}$ that has the following set of eigenvalues and corresponding eigenvectors:

$$\lambda_1 = -4 \qquad \lambda_2 = -5$$

$$\mathbf{p}^1 = \begin{bmatrix} 0 \\ -1 \end{bmatrix} \qquad \mathbf{p}^2 = \begin{bmatrix} 1 \\ 3 \end{bmatrix}$$

Review of Selected Topics from the Theory of Probability and Stochastic Processes

The material presented in this appendix is intended to serve as a brief review of selected topics from the theory of probability and stochastic processes. The coverage is by no means complete, and only those topics relevant to our work are discussed. Those who wish to pursue a more detailed treatment of probability and stochastic processes are referred to the many well-written textbooks on the subject.

B.1 Probability Fundamentals

An experiment whose *outcome* is not known prior to the experiment being performed is called a *random experiment*. Tossing a coin, rolling a die, and drawing a card from a deck of cards are all examples of random experiments. An individual outcome ω of a random experiment is called a *sample point*, and the set of all possible outcomes of a random experiment is called the *sample space* Ω of the experiment. A *random event A* is an outcome or collection of outcomes of a random experiment.

For example, for flipping a coin once, the sample space is

$$\Omega = \{H, T\}$$

and for flipping a coin twice in succession, the sample space is

$$\Omega = \{HH, HT, TH, TT\}$$

where H stands for heads and T stands for tails. For flipping a coin twice in succession, the subsets

$$A_1 = \{HH,\ TT\} \quad \text{and} \quad A_2 = \{HT,\ TH\}$$

are random events. A nonempty set F of subsets of the sample space Ω for which

1. $A_1, A_2 \in F$ implies $A_1 \cup A_2 \in F$
2. $A \in F$ implies $\bar{A} \in F$

is called a *field of events*. The symbol $\in$ means "belongs to," the bar over A means the *complement* of A or "not A," and the symbol $\cup$ means *union* of sets.

If item 1 is changed to 1', for every countable collection of events, that is,

$$A_1, A_2, \ldots, A_i, \ldots \in F, \text{ implies that } \cup_{i=1}^{\infty} A_i \in F$$

then F is called a *sigma field* (σ-*field*) of events.

A real-valued set function Pr defined on the field of events such that

1. $Pr(A) \geq 0$ for any $A \in F$
2. $Pr(\Omega) = 1$
3. For every countable collection of events $A_1, A_2, \ldots$, finite or infinite, for which

$$A_i \cap A_j = \phi \quad i \neq j$$

$$Pr(\cup_{i=1}^{\infty} A_i) = \sum_{i=1}^{\infty} Pr(A_i)$$

is called a *probability measure* defined on the field F. Here ϕ denotes the *empty set*, and the symbol $\cap$ means *intersection* of sets.

B.1.1 Independence of Events

The *joint probability* of two events A and B is written as $Pr(A \cap B)$ or $Pr(AB)$. Events A and B are *independent* if and only if

$$Pr(A \cap B) = Pr(A)Pr(B)$$

The events $A_1, A_2, \ldots, A_n$ are said to be *mutually independent* if and only if

$$Pr(A_i \cap A_j \cap \cdots \cap A_k) = Pr(A_i)Pr(A_j) \cdots Pr(A_k)$$

for *every* combination of events, two at a time, three at a time, and so on. Pairwise independence is *not* sufficient for the entire set to be independent.

B.1.2 Conditional Probability

Consider two events A and B. The probability that event A will occur *given* that the event B has already occurred is termed *conditional probability* and is defined as

$$Pr(A \mid B) = \frac{Pr(A \cap B)}{Pr(B)}$$

whenever

$$Pr(B) \neq 0$$

For example, if a fair die is tossed once, what is the probability that the outcome is a 1 or a 2, given that the outcome is an even number?

According to our notation,

$$B = \{2, 4, 6\} \qquad A = \{1, 2\}$$

$$Pr(B) = \frac{1}{2}$$

$$Pr(A \cap B) = Pr\{\text{outcome is a 2}\} = \frac{1}{6}$$

Then

$$Pr(A \mid B) = \frac{Pr(A \cap B)}{Pr(B)} = \frac{\frac{1}{6}}{\frac{1}{2}} = \frac{1}{3}$$

Some properties of conditional probability are

1. $0 \leq Pr(A \mid B) \leq 1$
2. $Pr(\Omega \mid A) = 1$
3. If A and B are independent, then

   $$Pr(A \mid B) = Pr(A)$$

4. If $A_i \cap A_j = \phi$, $i \neq j$, then

   $$Pr[(\cup_i A_i) \mid B] = \sum_i Pr(A_i \mid B)$$

B.1.3 Bayes' Rule

Bayes' rule states that

$$Pr(A \mid B) = \frac{Pr(B \mid A)Pr(A)}{Pr(B)}$$

Because

$$Pr(A \mid B) = \frac{Pr(A \cap B)}{Pr(B)}$$

and

$$Pr(B \mid A) = \frac{Pr(B \cap A)}{Pr(A)}$$

combining these two relationships and noting that

$$Pr(A \cap B) = Pr(B \cap A)$$

completes the proof.

B.2 Random Variables

Consider a random experiment that may have a finite or infinite number of random outcomes. A function $x(\omega)$ that assigns a number to each sample point ω in the sample space Ω is called a *random variable*. The probability that a random variable has value $x \leq X$ as a function of X is termed the *probability distribution function* of x and is denoted by

$$F_x(X) = Pr[x \leq X]$$

The probability distribution function has the properties listed in Table B-1. If the probability distribution function is differentiable everywhere, the *probability density function* associated with the random variable x is

$$f_x(X) = \frac{dF_x(X)}{dX}$$

consequently,

$$F_x(X) = \int_{-\infty}^{X} f_x(X') \, dX'$$

Table B-1 Properties of Probability Distribution Functions

1. $0 \leq F_x(X) \leq 1$.
2. $F_x(-\infty) = 0$ and $F_x(\infty) = 1$.
3. $F_x(X)$ is a nondecreasing function of X.
4. $Pr\{X_1 < x \leq X_2\} = F_x(X_2) - F_x(X_1)$.
5. $Pr\{x > X\} = 1 - F_x(X)$.

Table B-2 Properties of Probability Density Functions

1. $f_x(X) \geq 0$, probability density is a nonnegative function.

2. $\int_{-\infty}^{\infty} f_x(X) \, dX = 1$.

3. $Pr[X_1 < x \leq X_2] = \int_{X_1}^{X_2} f_x(X) \, dX$.

The probability density function has the properties listed in Table B-2.

For example, the probability distribution function shown in Figure B-1(a) has a corresponding density function

$$f_x(X) = \begin{cases} \dfrac{6}{70} & -3 < X < 4 \\ 0.4\delta(X - 4) & X = 4 \\ 0 & \text{otherwise} \end{cases}$$

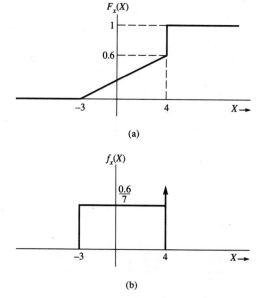

(a)

(b)

Figure B-1 An example of a probability distribution function and its corresponding density function. (a) Probability distribution function. (b) Corresponding probability density function.

as shown in Figure B-1(b). Clearly, the probability that the random variable x is between -2 and $+2$ is

$$Pr\{-2 \le x \le +2\} = \frac{30}{70} - \frac{6}{70} = 0.34$$

Random variables are either continuous, discrete, or mixed. A random variable is *continuous* if its probability distribution function is continuous everywhere and is differentiable except at a countable number of points. On the other hand, a random variable is said to be *discrete* if the derivative of the probability distribution function is zero everywhere except at a countable number of jump points. If a random variable is neither continuous nor discrete, it is termed *mixed*.

B.2.1 Function of a Random Variable

Let $Y = g(X)$ be a single-valued function of X. If x is a random variable, then

$$y = g(x)$$

is another random variable. If the random variable x is continuous and $g(X)$ is differentiable and monotonic, then

$$f_y(Y) = f_x(X) \left| \frac{dX}{dY} \right|$$

For example, consider the system shown in Figure B-2. If

$$Y = g(X) = 2X + 4$$

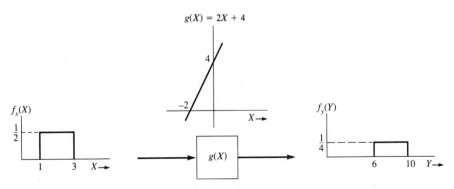

Figure B-2 An example of a density of a function of a random variable.

then

$$y = 2x + 4$$

and the density function of y is

$$f_y(Y) = f_x(X) \left| \frac{dX}{dY} \right|$$

where

$$\left| \frac{dX}{dY} \right| = \frac{1}{2}$$

In terms of Y

$$X = \frac{1}{2} Y - 2$$

so

$$f_y(Y) = \frac{1}{2} f_x \left(\frac{1}{2} Y - 2 \right)$$

as shown in the figure.

Two probability density functions that are useful in this book are the *uniform probability density* and the *Gaussian probability density*. These are shown in Figure B-3. Of course, there are many other density functions that are also useful; however, in control systems, these two are the most important.

Mathematically, the uniform density function is defined by

$$f_x(X) = \begin{cases} \dfrac{1}{X_2 - X_1} & X_1 \leq x \leq X_2 \\ 0 & \text{otherwise} \end{cases}$$

whereas the Gaussian density function is defined by

$$f_x(X) = \frac{1}{\sigma \sqrt{2\pi}} e^{-[(X-\eta)^2/2\sigma^2]} \quad -\infty \leq X \leq \infty$$

The parameters σ and η are constants. These two densities will be analyzed later.

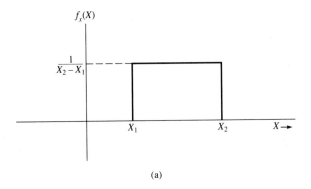

(a)

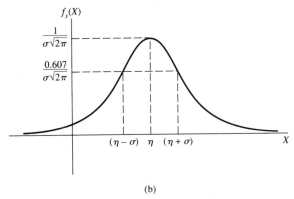

(b)

Figure B-3 Common probability density functions.
(a) Uniform density function. (b) Gaussian density
function.

B.2.2 Moments of a Random Variable

The *mean,* or *expected value,* of a random variable x, denoted by $E[x]$
or $\bar{x}$, is given by

$$E[x] = \int_{-\infty}^{\infty} X f_x(X) \, dX$$

For example, the mean of the random variable x that has the probability density function shown in Figure B-4 is

$$E[x] = \int_0^2 X \left(1 - \frac{1}{2} X\right) dX = \frac{2}{3}$$

Obviously,

$$E[C] = C$$

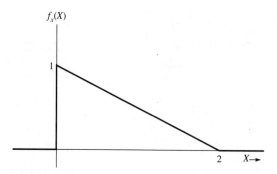

Figure B-4

where C is a known constant and

$$E[Cx] = CE[x]$$

where x is a random variable.

If $y = g(x)$ is a function of a random variable x, then

$$E[g(x)] = \int_{-\infty}^{\infty} g(X)f_x(X) \, dX$$

B.2.3 Higher Moments of a Random Variable

The nth *moment* of a random variable x is

$$E[x^n] = \int_{-\infty}^{\infty} X^n f_x(X) \, dX$$

Obviously, the mean of the random variable is the *first moment*, and the *mean square*

$$E[x^2] = \int_{-\infty}^{\infty} X^2 f_x(X) \, dX$$

is the *second moment*.

For the previous example, the mean square of the random variable is

$$E[x^2] = \int_{0}^{2} X^2 \left(1 - \frac{1}{2}X\right) dX = \frac{2}{3}$$

B.2.4. Central Moments

The nth *central moment* of a random variable x is

$$E[x - \bar{x})^n] = \int_{-\infty}^{\infty} (X - \bar{x})^n f_x(X) \, dX$$

The *second central moment*, denoted by σ^2, is

$$\sigma^2 = E[(x - \bar{x})^2] = \int_{-\infty}^{\infty} (X - \bar{x})^2 f_x(X) \, dX$$

and is called the *variance* of the random variable. The positive square root of the variance is called the *standard deviation* σ.

The variance of a random variable x can be expressed in terms of the mean and the mean square of the random variable as follows:

$$\sigma^2 = E[(x - \bar{x})^2] = E[x^2 - 2x\bar{x} + \bar{x}^2] = E[x^2] - \bar{x}^2$$

For the uniform density function shown in Figure B-3(a), the mean is given by

$$E[x] = \int_{X_1}^{X_2} \frac{X}{X_2 - X_1} \, dX = \frac{1}{X_2 - X_1} \left(\frac{X^2}{2} \right) \Big|_{X_1}^{X_2}$$

$$= \frac{1}{2} (X_1 + X_2)$$

and the variance is given by

$$\sigma^2 = \int_{X_1}^{X_2} (X - \bar{x})^2 \cdot \frac{1}{X_2 - X_1} \, dX$$

$$= \frac{(X_2 - X_1)^2}{12}$$

The mean of a Gaussian random variable is

$$E[x] = \frac{1}{\sigma \sqrt{2\pi}} \int_{-\infty}^{\infty} X e^{-[(X-\eta)^2/2\sigma^2]} \, dX$$

If

$$X' = X - \eta \qquad X = X' + \eta \qquad dX' = dX$$

$$E[x] = \frac{1}{\sigma \sqrt{2\pi}} \int_{-\infty}^{\infty} (X' + \eta) e^{-(X'^2/2\sigma^2)} \, dX'$$

$$= \frac{1}{\sigma \sqrt{2\pi}} \int_{-\infty}^{\infty} X' e^{-(X'^2/2\sigma^2)} \, dX' + \frac{1}{\sigma \sqrt{2\pi}} \int_{-\infty}^{\infty} \eta e^{-(X'^2/2\sigma^2)} \, dX'$$

because the integrand in the first integral is an odd function, it integrates to zero. Hence,

$$E[x] = \eta \left[\frac{1}{\sigma \sqrt{2\pi}} \int_{-\infty}^{\infty} e^{-(X'^2/2\sigma^2)} \, dX' \right] = \eta F_x(\infty)$$

and, therefore,

$$E[x] = \eta$$

Also note that

$$\int_{-\infty}^{\infty} e^{-(X^2/2)} \, dX = \sqrt{2\pi}$$

Using an argument similar to that above, we find that any random variable that has a density function with even symmetry about a point $X = m$ will have mean m.

The variance of a Gaussian random variable is

$$E[(x - \eta)^2] = \frac{1}{\sigma \sqrt{2\pi}} \int_{-\infty}^{\infty} (X - \eta)^2 e^{-[X-\eta^2/2\sigma^2]} \, dX$$

If

$$X' = \frac{X - \eta}{\sigma} \qquad X = \sigma X' + \eta \cdot \qquad dX = \sigma \, dX'$$

then

$$E[(x - \eta)^2] = \frac{\sigma^2}{\sqrt{2\pi}} \int_{-\infty}^{\infty} X'^2 e^{-(X'^2/2)} \, dX'$$

Integrating by parts gives

$$E[(x - \eta)^2] = \frac{\sigma^2}{\sqrt{2\pi}} \int_{-\infty}^{\infty} \underbrace{X'}_{u} \underbrace{X' e^{-(X'^2/2)} \, dX'}_{dv}$$

$$= \frac{\sigma^2}{\sqrt{2\pi}} \left[\underbrace{X'}_{u} \underbrace{(-e^{-(X'^2/2)})}_{v} \Big|_{-\infty}^{\infty} - \int_{-\infty}^{\infty} \underbrace{(-e^{-(X'^2/2)})}_{v} \underbrace{dX'}_{du} \right]$$

$$= \frac{\sigma^2}{\sqrt{2\pi}} \int_{-\infty}^{\infty} e^{-(X'^2/2)} \, dX' = \left(\frac{\sigma^2}{\sqrt{2\pi}} \right) (\sqrt{2\pi})$$

Hence,

$$E[(x - \eta^2] = \sigma^2$$

B.3 Joint Random Variables

For two random variables x and y, the *joint probability distribution* is

$$F_{xy}(X, Y) = Pr\{x \leq X, y \leq Y\} = F_{yx}(Y, X)$$

The properties of the joint distribution function are listed in Table B-3. The *joint probability density function* is defined as

$$f_{xy}(X, Y) = \frac{\partial^2 F_{xy}(X, Y)}{\partial X \, \partial Y} = \frac{\partial^2 F_{yx}(Y, X)}{\partial Y \, \partial X}$$

provided the derivative exists. Hence,

$$F_{xy}(X, Y) = \int_{-\infty}^{X} \int_{-\infty}^{Y} f_{xy}(X', Y') \, dX' \, dY'$$

The joint probability density function has properties that are listed in Table B-4.

Table B-3 Properties of Joint Probability Distribution

1. $0 \leq F_{xy}(X) \leq 1$.
2. $F_{xy}(-\infty, Y) = F_{xy}(X, -\infty) = 0$.
3. $F_{xy}(\infty, \infty) = 1$.
4. $F_{xy}(X, Y)$ is nondecreasing in X and in Y.
5. $Pr\{x \leq X, Y_1 < y \leq Y_2\} = F_{xy}(X, Y_2) - F_{xy}(X, Y_1)$.

Table B-4 Properties of Joint Probability Densities

1. $f_{xy}(X, Y) \geq 0$

   Density function is always nonnegative.

2. $\int_{-\infty}^{\infty} \int_{-\infty}^{\infty} f_{xy}(X, Y) \, dX \, dY = 1$.

3. $f_{xy}(X, Y) \, dX \, dY = Pr\{X < x \leq X + dX, Y < y \leq Y + dY\}$.

4. $Pr\{X_1 < x \leq X_2, Y_1 < y \leq Y_2\} = \int_{X_1}^{X_2} \int_{Y_1}^{Y_2} f_{xy}(X, Y) \, dX \, dY$.

For example, for the function

$$f_{xy}(X, Y) = \begin{cases} \alpha XY & 0 < X < 1 \quad 2 < Y < 4 \\ 0 & \text{otherwise} \end{cases}$$

to be a valid probability density function, it should satisfy the property

$$\int_{-\infty}^{\infty} \int_{-\infty}^{\infty} f_{xy}(X, Y) \, dX \, dY = 1$$

Hence,

$$\int_0^1 \int_2^4 \alpha XY \, dX \, dY = \alpha \int_0^1 \left[\int_2^4 XY \, dY \right] dX$$

$$= \alpha \int_0^1 \frac{XY^2}{2} \Big|_2^4 \, dX$$

$$= 6\alpha \int_0^1 X \, dX$$

$$= 3\alpha$$

Therefore,

$$\alpha = \frac{1}{3}$$

B.3.1 Marginal Probability Distributions and Densities

The *marginal probability distributions* of the random variables x and y are

$$F_x(X) = Pr\{x \le X, y \le \infty\} = F_{xy}(X, \infty)$$

$$= \int_{-\infty}^{X} \int_{-\infty}^{\infty} f_{xy}(X', Y) \, dX' \, dY$$

and

$$F_y(Y) = F_{xy}(\infty, Y) = \int_{-\infty}^{\infty} \int_{-\infty}^{Y} f_{xy}(X, Y') \, dX \, dY'$$

respectively. It follows that the *marginal probability density* of the random variables x and y are

$$f_x(X) = \int_{-\infty}^{\infty} f_{xy}(X, Y) \, dY$$

$$f_y(Y) = \int_{-\infty}^{\infty} f_{xy}(X, Y) \, dX$$

respectively.

If we continue with the previous example, the marginal probability density of the random variable x is

$$f_x(X) = \int_{-\infty}^{\infty} f_{xy}(X,\ Y)\ dY$$

$$= \int_{2}^{4} \frac{1}{3} XY\ dY$$

$$= 2X$$

Similarly,

$$f_y(Y) = \int_{0}^{1} \frac{1}{3} XY\ dX$$

$$= \frac{1}{6} Y$$

B.3.2 Moments of Joint Random Variables

For two random variables x and y that have a joint probability density function $f_{xy}(X,\ Y)$, the mean of x and the mean of y are

$$E[x] = \int_{-\infty}^{\infty} \int_{-\infty}^{\infty} Xf_{xy}(X,\ Y)\ dX\ dY$$

and

$$E[y] = \int_{-\infty}^{\infty} \int_{-\infty}^{\infty} Yf_{xy}(X,\ Y)\ dX\ dY$$

respectively. Furthermore, the variance of x and the variance of y are given by

$$\sigma_x^2 = E[(x - \bar{x})^2] = \int_{-\infty}^{\infty} \int_{-\infty}^{\infty} (X - \bar{x})^2 f_{xy}(X,\ Y)\ dX\ dY$$

and

$$\sigma_y^2 = E[(y - \bar{y})^2] = \int_{-\infty}^{\infty} \int_{-\infty}^{\infty} (Y - \bar{y})^2 f_{xy}(X,\ Y)\ dX\ dY$$

respectively.

For the previous example, where

$$f_{xy}(X,\ Y) = \begin{cases} \frac{1}{3}XY & 0 < X < 1 \quad 2 < Y < 4 \\ 0 & \text{otherwise} \end{cases}$$

$$E[x] = \int_0^1 \int_2^4 \frac{1}{3} X^2 Y \, dX \, dY$$

$$= \frac{2}{3}$$

and

$$E[y] = \int_0^1 \int_2^4 \frac{1}{3} XY^2 \, dX \, dY$$

$$= \frac{56}{18}$$

B.3.3 Correlation Between Random Variables

The *correlation* between two random variables x and y that have a joint density function $f_{xy}(X, Y)$ is defined as

$$R_{xy} = E[xy] = \int_{-\infty}^{\infty} \int_{-\infty}^{\infty} XYf_{xy}(X, Y) \, dX \, dY$$

Continuing with the previous example gives

$$E[xy] = \int_0^1 \int_2^4 \frac{1}{3} X^2 Y^2 \, dX \, dY$$

$$= \frac{56}{27}$$

B.3.4 Covariance of Random Variables

The *covariance* of two random variables x and y that have a joint density function $f_{xy}(X, Y)$ is defined as

$$\Gamma_{xy} = E[(x - \bar{x})(y - \bar{y})] = \int_{-\infty}^{\infty} \int_{-\infty}^{\infty} [(X - \bar{x})(Y - \bar{y})] f_{xy}(X, Y) \, dX \, dY$$

Higher-order moments are similarly defined.

B.3.5 Uncorrelated Random Variables

Two random variables x and y are said to be *uncorrelated* if

$$R_{xy} = E[xy] = E[x]E[y]$$

that is,

$\Gamma_{xy} = 0$

The two random variables x and y of the previous example are uncorrelated because

$$E[xy] = \frac{56}{27} = \left(\frac{2}{3}\right)\left(\frac{56}{18}\right)$$

B.3.6 Orthogonal Random Variables

Two random variables x and y are said to be *orthogonal* if

$R_{xy} = E[xy] = 0$

B.4 Conditional Distributions and Densities

For two random variables x and y, the *conditional probability distribution function* is defined as

$F_{x|y}(X|Y) = Pr\{x \leq X \mid y = Y\}$

If x and y have a joint density function $f_{xy}(X, Y)$, then the conditional distribution function is expressed as

$$F_{x|y}(X|Y) = \frac{\int_{-\infty}^{X} f_{xy}(X', Y) \, dX'}{f_y(Y)}$$

provided $f_y(Y) \neq 0$.

The density function, on the other hand, is given by

$$f_{x|y}(X|Y) = \frac{\partial F_{x|y}(X|Y)}{\partial X} = \frac{f_{xy}(X, Y)}{f_y(Y)}$$

$$= \frac{f_{xy}(X, Y)}{\int_{-\infty}^{\infty} f_{xy}(X, Y) \, dX}$$

provided $f_y(Y) \neq 0$.

For the previous example, the conditional probability density function is

$$f_{x|y}(X|Y) = \frac{\frac{1}{3}XY}{\frac{1}{6}Y} = 2X \quad 0 < X < 1 \qquad 2 < Y < 4$$

B.4.1 Conditional Mean

The *conditional mean* of the random variable x given the random variable y is, in terms of the conditional density,

$$E[x|y] = \int_{-\infty}^{\infty} X f_{x|y}(X|Y) \, dX$$

B.4.2 Conditional Correlation

The *conditional correlation* of x given y is

$$R_{x|y} = E[x^2|y] = \int_{-\infty}^{\infty} X^2 f_{x|y}(X|Y) \, dX$$

B.4.3 Conditional Covariance

For the two random variables x and y that have a conditional density function $f_{x|y}(X|Y)$, the *conditional covariance* of x given y is

$$\Gamma_{x|y} = E[(x - E[x \mid y])^2 \mid y]$$

$$= \int_{-\infty}^{\infty} (X^2 - 2E[x \mid y]X + E^2[x \mid y]) f_{x|y}(X \mid Y) \, dX$$

$$= \int_{-\infty}^{\infty} X^2 f_{x|y}(X \mid Y) \, dX - E^2[x \mid y]$$

$$= R_{x|y} - E^2[x \mid y]$$

If we continue with the previous example, the conditional mean of the random variable x is

$$E[x \mid y] = \int_{0}^{1} 2X^2 \, dX = \frac{2}{3}$$

and the conditional correlation of x given y is

$$R_{x|y} = \int_{0}^{1} 2X^3 \, dX = \frac{2X^4}{4}\Big|_{0}^{1} = \frac{1}{2}$$

Hence,

$$\Gamma_{x|y} = \frac{1}{2} - \frac{4}{9} = \frac{1}{18}$$

B.4.4 Independence of Joint Random Variables

The random variables x and y with joint density function $f_{xy}(X, Y)$ are *independent* if and only if

$$Pr[x \leq X, y \leq Y] = Pr(x \leq X)Pr(y \leq Y)$$

In terms of the probability distribution function,

$$F_{xy}(X, Y) = F_x(X)F_y(Y)$$

and in terms of the probability density function,

$$f_{xy}(X, Y) = f_x(X)f_y(Y)$$

In the previous example, the two random variables x and y are independent because

$$f_{xy}(X, Y) = \frac{1}{3} XY = (2X) \left(\frac{1}{6} Y \right)$$

B.4.5 Correlation of Independent Random Variables

Independent random variables x and y are always *uncorrelated*, but uncorrelated random variables are *not* necessarily independent.

$$R_{xy} = E[xy] = \int_{-\infty}^{\infty} \int_{-\infty}^{\infty} XYf_{xy}(X, Y) \, dX \, dY$$

$$= \int_{-\infty}^{\infty} Xf_x(X) \, dX \int_{-\infty}^{\infty} Yf_y(Y) \, dY$$

$$= E[x]E[y]$$

If the random variables x and y are uncorrelated and Gaussian, they are independent.

B.4.6 Conditional Probability Density of Independent Random Variables

If the random variables x and y are independent with probability density function $f_{xy}(X, Y)$, then

$$E[x \mid y] = \int_{-\infty}^{\infty} Xf_{x|y}(X \mid Y) \, dX = \int_{-\infty}^{\infty} Xf_x(X) \, dX$$

$$= E[x]$$

Similarly,

$$R_{x|y} = R_x = E[x^2] = \int_{-\infty}^{\infty} X^2 f_x(X)\, dX$$

and

$$\Gamma_{x|y} = \Gamma_x = E[(x - \bar{x})^2] = R_x - (\bar{x})^2$$

B.4.7 The Central Limit Theorem

The *central limit theorem* states that if n random variables $x_1, x_2, \ldots,$ x_n are independent, identically distributed with mean $\bar{x}_i$ and variance $\sigma_{x_i}^2$, then the random variable

$$y = \frac{\sum_{i=1}^{n}(x_i - \bar{x}_i)}{\sigma_{x_i}^2 \sqrt{n}}$$

is asymptomatically Gaussian with zero mean and unit variance, written as $N(0, 1)$. In particular, if the random variables are uniformly distributed between 0 and 1 (set $X_1 = 0$ and $X_2 = 1$ in Figure B-3), then

$$\bar{x}_i = \frac{1}{2}$$

$$\sigma_{x_i}^2 = \frac{1}{12} \quad i = 1, 2, \ldots, n$$

Hence, for $n = 12$,

$$y = \sum_{i=1}^{12} x_i - 6$$

is approximately Gaussian, $N(0, 1)$. As mentioned in Chapter 8, this relationship is very helpful for generating white Gaussian noise from a zero-mean and unit variance random variable. This result can also be used to generate samples from a Gaussian distribution $N(0, 1)$ with user-specified mean η and standard deviation σ as follows:

$$z = \left[\sum_{i=1}^{12} x_i - 6\right]\sigma + \eta$$

Thus, the random variable z is approximately Gaussian $N(\eta, \sigma)$. The central limit theorem is also true for independent random variables with the same mean and the same variance, regardless of the density of the individual variables.

B.4.8 Characteristic Function

The *characteristic function* of a random variable that has a probability density function $f_x(X)$ is

$$\phi_x(u) = E[e^{jux}] = \int_{-\infty}^{\infty} f_x(X)e^{juX}\, dX$$

where j is the complex operator $j = \sqrt{-1}$. Obviously, the characteristic function is the Fourier transform of the probability density function with the transform variable u replaced by $-u$ for historical reasons.

The characteristic function of a Gaussian random variable is

$$\phi_x(u) = \int_{-\infty}^{\infty} f_x(X)e^{juX}\, dX$$

$$= \int_{-\infty}^{\infty} \frac{1}{\sigma\sqrt{2\pi}}\, e^{-(x-\eta)^2/2\sigma^2} e^{juX}\, dX$$

$$= e^{ju\eta} e^{-\sigma^2 u^2/2}$$

For a given characteristic function, the probability density function can be recovered via the inverse transform

$$f_x(X) = \frac{1}{2\pi} \int_{-\infty}^{\infty} \phi_x(u)e^{-juX}\, du$$

The characteristic function may be used to determine the moments of a random variable. Differentiating the characteristic function,

$$\frac{d\phi_x(u)}{du} = \int_{-\infty}^{\infty} f_x(X)(jX)e^{juX}\, dX$$

and setting $u = 0$ gives

$$\left.\frac{d\phi_x(u)}{du}\right|_{u=0} = j \int_{-\infty}^{\infty} X f_x(X)\, dX = jE[x]$$

Thus, in general,

$$E[x^n] = \frac{1}{j^n}\left[\frac{d^n\phi_x(u)}{du^n}\right]\bigg|_{u=0}$$

B.4.9 Joint Characteristic Function

The *joint characteristic function* of two random variables x and y that have a joint probability density function $f_{xy}(X, Y)$ is

$$\phi_{xy}(u, v) = \int_{-\infty}^{\infty}\int_{-\infty}^{\infty} f_{xy}(X, Y)e^{j(uX+vY)}\, dX\, dY$$

Then

$$f_{xy}(X, Y) = \frac{1}{(2\pi)^2} \int_{-\infty}^{\infty} \int_{-\infty}^{\infty} \phi_{xy}(u, v) e^{-j(uX+vY)} \, du \, dv$$

The *marginal characteristic functions* are

$$\phi_x(u) = \int_{-\infty}^{\infty} f_x(X) e^{juX} \, dX$$

and

$$\phi_y(v) = \int_{-\infty}^{\infty} f_y(Y) e^{jvY} \, dY$$

B.4.10 Independence in Terms of the Characteristic Function

If the random variables x and y are independent, then

$$\phi_{xy}(u, v) = \int_{-\infty}^{\infty} \int_{-\infty}^{\infty} f_{xy}(X, Y) e^{j(uX+vY)} \, dX \, dY$$

$$= \int_{-\infty}^{\infty} f_x(X) e^{juX} \, dX \int_{-\infty}^{\infty} f_y(Y) e^{jvY} \, dY$$

$$= \phi_x(u)\phi_y(v)$$

B.5 Stochastic Processes

The concept of a stochastic process is a generalization of the concept of a random variable. By definition, a *real stochastic process*, or *random process*, is a family of functions $x(t, \omega)$, which, for every outcome ω in the sample space Ω, assigns a function of the parameter t ranging over an index set T. The dependence on ω is often suppressed, with the process denoted by $x(t)$.

A *complex stochastic process* is a complex combination of two real stochastic processes defined on the same probability space. In this book, unless otherwise specified, we assume that all random processes are real. If the parameter t (here interpreted as time) is from a discrete set, the process is termed *discrete-time*. If it is from a continuum, the process is *continuous-time*.

A specific possible function $x(t)$ is termed a *sample function* of the stochastic process. The set of all possible functions $x(t)$ is termed the *ensemble* of the stochastic process. For example, let the underlying experiment be the flip of a coin, thus,

$$\Omega = \{\text{heads, tails}\}$$

and let the field of events be

$F = \{\phi, \{\text{heads}\}, \{\text{tails}\}, \Omega\}$

and let the probability measure be defined by

$$Pr(\phi) = 0 \qquad Pr(\Omega) = 1$$

$$Pr(\{\text{heads}\}) = \frac{1}{2} \qquad Pr(\{\text{tails}\}) = \frac{1}{2}$$

Let the parameter set be

$T = \{t: 0 \le t \le 1\}$

Now we define the continuous stochastic process as

$$x(t, \omega) = \begin{cases} 1 - t & \omega = \text{heads} \\ \sin \dfrac{\pi}{2} t & \omega = \text{tails} \end{cases}$$

If ω is fixed, $x(t, \omega)$ is a deterministic function, and if t is fixed, $x(t, \omega)$ is a random variable. For fixed ω and t, $x(t, \omega)$ is a fixed number.

As another example, let the underlying experiment be the same as in the previous example and let the parameter set be

$T = \{0, 1, 2\}$

Now let the stochastic sequence be

$$x(t, \omega) = \begin{cases} 1 - t & \omega = \text{heads} \\ \sin \dfrac{\pi}{2} t & \omega = \text{tails} \end{cases}$$

More specifically, let the stochastic sequence be

$$x(t, \omega) = \begin{cases} 1, 0, -1 & \omega = \text{heads} \\ 0, 1, \quad 0 & \omega = \text{tails} \end{cases}$$

That is,

$$x(0, \omega) = \begin{cases} 1 & \omega = \text{heads} \\ 0 & \omega = \text{tails} \end{cases}$$

$$x(1, \omega) = \begin{cases} 0 & \omega = \text{heads} \\ 1 & \omega = \text{tails} \end{cases}$$

$$x(2, \omega) = \begin{cases} -1 & \omega = \text{heads} \\ 0 & \omega = \text{tails} \end{cases}$$

B.5.1 Distribution and Density Functions of Stochastic Processes

If $x(t)$ is a stochastic process, then for every fixed $t \in T$, $x(t)$ is a random variable, and thus $x(t)$ has a probability distribution that, in general, depends on t. The probability distribution function of $x(t)$ is denoted by

$$F_x(X, t) = Pr\{x(t) \le X\}$$

A random process also has a probability density function

$$f_x(X, t) = \frac{dF_x(X, t)}{dX}$$

if the derivative exists.

B.5.2 Expectations and Moments

The *mean* of a random process $x(t)$ is given by

$$E[x(t)] = m(t) = \int_{-\infty}^{\infty} X f_x(X, t) \, dX$$

In general, the mean is a function of time. If the mean

$$E[x(t)] = 0$$

for all t, the stochastic process is said to be *zero mean*. As an example, the function

$$f_x(X, t) = \begin{cases} (1 + t^2)e^{-(1+t^2)X} & X \ge 0 \\ 0 & X < 0 \end{cases}$$

is a valid density function because it is nonnegative and

$$\int_{-\infty}^{\infty} f_x(X, t) \, dX = \int_{0}^{\infty} (1 + t^2)e^{-(1+t^2)X} \, dX$$

$$= \frac{(1 + t^2)e^{-(1+t^2)X}}{-(1 + t^2)} \bigg|_{0}^{\infty} = 1$$

for all t. The mean is

$$E[x(t)] = \int_{-\infty}^{\infty} X f_x(X, t) \, dX = \int_{0}^{\infty} X(1 + t^2)e^{-(1+t^2)X} \, dX$$

If we change variables,

$$X' = (1 + t^2)X \qquad dX' = (1 + t^2) \, dX$$

then

$$E[x(t)] = \frac{1}{1 + t^2} \int_0^\infty X' e^{-X'} \, dX' = \frac{1}{1 + t^2}$$

The mean square of a stochastic process is given by

$$E[x^2(t)] = \int_{-\infty}^\infty X^2 f_x(X, t) \, dX$$

The variance of a stochastic process is given by

$$E[(x(t) - E[x(t)])^2] = \sigma^2(t) = \int_{-\infty}^\infty (X - E[x(t)])^2 f_x(X, t) \, dX$$

The mean and mean square of a discrete-time stochastic process are given by

$$E[x(k)] = \int_{-\infty}^\infty X f_x(X, k) \, dX$$

$$E[x^2(k)] = \int_{-\infty}^\infty X^2 f_x(X, k) \, dX$$

respectively.

B.5.3 Joint Distribution and Density Functions

Consider the two stochastic processes $x(t)$ and $y(t)$ and the associated random variables $x(t_1)$ and $y(t_2)$, where t_1 and t_2 are fixed. The *joint probability distribution function* of these two random variables is

$$F_{xy}(X, Y; t_1, t_2) = Pr\{x(t_1) \le X, y(t_2) \le Y\}$$
$$= F_{yx}(Y, X; t_2, t_1)$$

and the *joint probability density function* is

$$f_{xy}(X, Y; t_1, t_2) = \frac{\partial^2 F_{xy}(X, Y; t_1, t_2)}{\partial X \, \partial Y}$$
$$= f_{yx}(Y, X; t_2, t_1)$$

The usual properties of joint distributions apply, including relations with the marginal distributions and densities.

B.5.4 Cross Correlation

The cross correlation between two stochastic processes is defined as

$$R_{xy}(t_1, t_2) = E[x(t_1)y(t_2)] = R_{yx}(t_2, t_1)$$

$$= \int_{-\infty}^{\infty} \int_{-\infty}^{\infty} XY f_{xy}(X, Y; t_1, t_2) \, dX \, dY$$

The processes are said to be *uncorrelated* if

$$R_{xy}(t_1, t_2) = E[x(t_1)y(t_2)] = E[x(t_1)]E[y(t_2)]$$

for all t_1 and t_2.

The processes are said to *orthogonal* if

$$R_{xy}(t_1, t_2) = 0$$

for all t_1 and t_2.

For example, the processes with joint probability density

$$f_{xy}(X, Y; t_1, t_2) = \begin{cases} \dfrac{(1 + t_1^2)e^{-(1+t_1^2)X}}{1 + t_2^2} & X \geq 0 \quad 0 \leq Y \leq (1 + t_2^2) \\ 0 & \text{otherwise} \end{cases}$$

have the cross correlation

$$R_{xy}(t_1, t_2) = \int_{-\infty}^{\infty} \int_{-\infty}^{\infty} XY f_{xy}(X, Y; t_1, t_2) \, dX \, dY$$

$$= \int_0^{\infty} X(1 + t_1^2)e^{-(1+t_1^2)X} \left[\int_0^{1+t_2^2} \frac{Y}{1 + t_2^2} \, dY \right] dX$$

$$= \int_0^{\infty} X(1 + t_1^2)e^{-(1+t_1^2)X} \, dX \left[\frac{1 + t_2^2}{2} \right]$$

$$= \frac{1 + t_2^2}{2(1 + t_1^2)}$$

The means of the processes are

$$E[x(t_1)] = \int_{-\infty}^{\infty} X f_x(X, t_1) \, dX = \int_0^{\infty} X(1 + t_1^2)e^{-(1+t_1^2)X} \, dX$$

$$= \frac{1}{1 + t_1^2}$$

and

$$E[y(t_2)] = \int_{-\infty}^{\infty} Y f_y(Y, t_2) \, dY = \int_0^{1+t_2^2} \frac{Y}{1 + t_2^2} \, dY$$

$$= \frac{1 + t_2^2}{2}$$

Because

$$R_{xy}(t_1, t_2) = E[x(t_1)] E[y(t_2)]$$

these processes are uncorrelated.

Processes with a separable probability density function, such as in this example, are always uncorrelated, but the reverse is not necessarily true.

B.5.5 Cross Covariance

The cross covariance between two stochastic processes is defined as

$$\Gamma_{xy}(t_1, t_2) = E[(x(t_1) - E[x(t_1)])(y(t_2) - E[y(t_2)])]$$

$$= \Gamma_{yx}(t_2, t_1)$$

$$= \int_{-\infty}^{\infty} \int_{-\infty}^{\infty} (X - E[x(t_1)])(Y - E[y(t_2)]) f_{xy}(X, Y; t_1, t_2) \, dX \, dY$$

$$= R_{xy}(t_1, t_2) - E[x(t_1)] E[y(t_2)]$$

For two uncorrelated stochastic processes

$$R_{xy}(t_1, t_2) = E[x(t_1)] E[y(t_2)]$$

thus,

$$\Gamma_{xy}(t_1, t_2) = 0$$

for all t_1 and t_2.

B.5.6 Higher-Order Distributions and Densities

For the stochastic process $x(t)$, $x(t_1)$ and $x(t_2)$ are distinct random variables with joint (second-order) distribution given by

$$F_{xx}(X, \zeta; t_1, t_2) = Pr\{x(t_1) \le X, x(t_2) \le \zeta\}$$

and joint probability density function

$$f_{xx}(X, \zeta; t_1, t_2) = \frac{\partial^2 F_{xx}(X, \zeta; t_1, t_2)}{\partial X \, \partial \zeta}$$

if the derivative exists. If we use this definition, the joint moments between $x(t_1)$ and $x(t_2)$ are defined as discussed in the following sections.

B.5.7 Autocorrelation

The *autocorrelation function* of a random process at two distinct instants of time t_1 and t_2 is defined as

$$R_{xx}(t_1, t_2) = E[x(t_1)x(t_2)] = \int_{-\infty}^{\infty} \int_{-\infty}^{\infty} X \, \zeta f_{xx}(X, \zeta; t_1, t_2) \, dX \, d\zeta$$

Because of the symmetry of the definition

$$R_{xx}(t_1, t_2) = R_{xx}(t_2, t_1)$$

If $t_1 = t_2 = t$, then

$$R_{xx}(t, t) = E[x^2(t)]$$

B.5.8 Covariance

The *covariance* of a random process at two instants of time is defined as

$$\Gamma_{xx}(t_1, t_2) = E[(x(t_1) - E[x(t_1)])(x(t_2) - E[x(t_2)])]$$
$$= R_{xx}(t_1, t_2) - E[x(t_1)]E[x(t_2)]$$

Due to the symmetry in the definition

$$\Gamma_{xx}(t_1, t_2) = \Gamma_{xx}(t_2, t_1)$$

B.5.9 Stationary Processes

A stochastic process $x(t)$ is said to be *stationary in the strict sense* if and only if every probability distribution function

$$F_x(X, t)$$
$$F_{xx}(X, \zeta; t_1, t_2)$$
$$F_{xxx}(X, \zeta, \alpha; t_1, t_2, t_3)$$

$$\vdots$$

is unchanged by a shift in the time origin. This means that F_x must be a function of X alone; F_{xx} must depend on X, ζ, and the time difference $(t_1 - t_2)$; F_{xxx} must depend only on X, ζ, α, $(t_1 - t_2)$ and $(t_1 - t_3)$; and so on.

If every distribution function through the Kth order has this property, the process is said to be *stationary of order K*.

Two processes $x(t)$ and $y(t)$ are *jointly stationary in the strict sense* if and only if each is individually stationary and if *every* joint probability distribution function

$$F_{xy}(X, Y; t_1, t_2)$$
$$F_{xxy}(X, \zeta, Y; t_1, t_2, t_3)$$
$$F_{xyy}(X, Y, \alpha; t_1, t_2, t_3)$$

$$\vdots$$
$$\vdots$$

is also unchanged by a shift in the time origin.

A stochastic process is called *wide-sense stationary* if it has the following properties:

1. $|E[x(t)]| < +\infty$ for all $t \in T$
2. $|E[x(t_1)x(t_2)]| < +\infty$ for all t_1, $t_2 \in T$
3. $E[x(t)] = c$ a constant
4. $R_{xx}(t_1, t_2) = R_{xx}(t_1 - t_2)$

Obviously, from item 4

$$E[x^2(t)] = R_{xx}(t, t) = R_{xx}(0)$$

For example, consider the random process

$$x(t) = A \cos \omega_0 t - B \sin \omega_0 t$$

where the random variables A and B are independent, and each has zero mean and unity variance. Then

$$E[x(t)] = E[A] \cos \omega_0 t - E[B] \sin \omega_0 t$$
$$= 0$$

and

$$E[x(t_1)x(t_2)] = E[(A \cos \omega_0 t_1 - B \sin \omega_0 t_1)(A \cos \omega_0 t_2 - B \sin \omega_0 t_2)]$$
$$= E[A^2] \cos \omega_0 t_1 \cos \omega_0 t_2 - E[AB] \cos \omega_0 t_1 \sin \omega_0 t_2$$
$$- E[AB] \sin \omega_0 t_1 \cos \omega_0 t_2 + E[B^2] \sin \omega_0 t_1 \sin \omega_0 t_2$$

Because A and B are independent and each has unity variance

$$E[x(t_1)x(t_2)] = \cos \omega_0 t_1 \cos \omega_0 t_2 + \sin \omega_0 t_1 \sin \omega_0 t_2$$
$$= \cos \omega_0(t_1 - t_2)$$

Hence, the random process $x(t)$ is wide-sense stationary.

A wide-sense stationary process $x(t)$ is *even* because

$$R_{xx}(\tau) = E[x(t + \tau)x(t)] = E[x(t)x(t + \tau)]$$
$$= R_{xx}(-\tau)$$

That the autocorrelation function is maximum at the origin can be shown as follows:

$$E[(x(t + \tau) \pm x(t))^2] = R_{xx}(0) \pm 2R_{xx}(\tau) + R_{xx}(0)$$

Because the left side of the equality is always nonnegative, then

$$2[R_{xx}(0) \pm R_{xx}(\tau)] \geq 0$$

which gives the stated result.

Note that stationarity to the second order is *not* the same as wide-sense stationary because the moments in the former need not exist.

Any strict sense stationary process is also stationary in the wide sense (assuming the first and second moments exist), but, of course, the reverse is not true. Wide-sense stationary does not necessarily imply that all distributions are unchanged by a shift in the time origin.

Two processes $x(t)$ and $y(t)$ are *jointly wide-sense stationary* if each is individually wide-sense stationary and

$$E[x(t_1)y(t_2)] = R_{xy}(t_1, t_2)$$

may be expressed as a function of the single variable $(t_1 - t_2)$. That is,

$$R_{xy}(\tau) = E[x(t + \tau)y(t)] = E[y(t)x(t + \tau)]$$

and, therefore,

$$R_{xy}(\tau) = R_{yx}(-\tau)$$

B.5.10 Power Spectral Density

If a discrete-time stochastic process $x(k)$ is wide-sense stationary with autocorrelation sequence $R_{xx}(n)$, then the *power spectral density* (PSD) S_{xx} of the autocorrelation sequence is defined as

$$S_{xx}(e^{j\omega T}) = \sum_{n=-\infty}^{\infty} R_{xx}(n)e^{-jn\omega T}$$

which is periodic in ω with period $2\pi/T$. It is the discrete Fourier transform of the autocorrelation sequence. Note that the autocorrelation sequence $R_{xx}(n)$ of a wide-sense stationary process is defined for positive as well as negative n.

For a given PSD, the autocorrelation sequence may be recovered as

$$R_{xx}(n) = \frac{1}{2\pi} \int_{-\pi}^{\pi} S_{xx}(e^{j\omega T}) e^{jn\omega T} \, d\omega$$

The mean square (average power) of the stochastic process $x(k)$ is

$$E[x^2(k)] = R_{xx}(0) = \frac{1}{2\pi} \int_{-\pi}^{\pi} S_{xx}(e^{j\omega T}) \, d\omega$$

It can be shown that the PSD has the properties of being real, even, and nonnegative.

In the z-domain, the PSD and autocorrelation sequence are related by

$$S_{xx}(z) = \sum_{n=-\infty}^{\infty} R_{xx}(n) z^{-n}$$

and

$$R_{xx}(n) = \frac{1}{2\pi j} \oint_c S_{xx}(z) z^{n-1} \, dz$$

where the contour of integration is chosen to be the unit circle $|z| = 1$.

The mean square may be expressed in the z-domain as

$$E[x^2(k)] = R_{xx}(0) = \frac{1}{2\pi j} \oint_c S_{xx}(z) z^{-1} \, dz$$

which can be calculated using residue theory or expansion methods for partial fractions.

The PSD of a continuous-time random process $x(t)$ is defined as the Fourier transform of the autocorrelation function of the random process

$$S_{xx}(j\omega) = \int_{-\infty}^{\infty} R_{xx}(\tau) e^{-j\omega\tau} \, d\tau$$

Consequently, the PSD of a continuous-time random process can be expressed in terms of the two-sided Laplace transform as

$$S_{xx}(s) = \int_{-\infty}^{\infty} R_{xx}(\tau) e^{-s\tau} \, d\tau$$

The autocorrelation function may be recovered via the inverse Fourier transform of the corresponding PSD

$$R_{xx}(\tau) = \mathcal{F}^{-1}[S_{xx}(j\omega)] = \frac{1}{2\pi} \int_{-\infty}^{\infty} S_{xx}(j\omega)e^{j\omega\tau}\, d\omega$$

Similar to discrete-time processes, the PSD of a continuous-time process is real, even, and nonnegative.

B.5.11 Cross Power Spectral Density

The cross power spectral density (cross PSD) of two discrete-time random processes $x(k)$ and $y(k)$ which are jointly wide-sense stationary is

$$S_{xy}(e^{j\omega T}) = \sum_{n=-\infty}^{\infty} R_{xy}(n)e^{-jn\omega T}$$

which may not be real, even, or nonnegative.

B.5.12 Discrete-Time White Noise

A discrete-time random process $w(k)$ is called *white noise* if its value at one step in time provides no information about its values at other steps.
    Mathematically, whiteness of the stochastic sequence $w(k)$ means that

$$E[w(n_1)w(n_2)] = \begin{cases} q(n_1) & n_1 = n_2 \\ 0 & n_1 \neq n_2 \end{cases}$$

where the mean square $q(n_1)$ is nonnegative. That is, $w(k)$ is uncorrelated with itself at any other time step. If the values of a random process $w(k)$ are correlated in time, the process is called *colored noise*.
    If a discrete-time white noise is wide-sense stationary, then

1. $E[w(k)] = m_w$ a constant
2. $E[w(k + n)w(k)] = R_{ww}(n)$

Hence, using the whiteness property above,

$$R_{ww}(k) = q\, \delta(k);\ m_w = 0$$

where $\delta(k)$ is the pulse sequence defined as

$$\delta(k) = \begin{cases} 1 & k = 0 \\ 0 & \text{otherwise} \end{cases}$$

If a discrete-time white noise is wide-sense stationary and has a zero mean, then

$$R_{ww}(k) = \sigma^2 \, \delta(k)$$

where σ is the standard deviation of the noise sequence. And, therefore, the PSD

$$S_{ww}(e^{j\omega T}) = \sigma^2$$

is a constant for all ω.

B.6 Vector Stochastic Processes

The definitions presented thus far can be easily extended to stochastic vectors.

A *vector* stochastic process is expressed as

$$\mathbf{x}(t) = \begin{bmatrix} x_1(t) \\ x_2(t) \\ \vdots \\ x_n(t) \end{bmatrix}$$

where n is the dimension of the vector, and each component of the vector is a stochastic process.

The probability distribution function of the vector $\mathbf{x}(t)$ is

$$F_{\mathbf{x}}(\mathbf{X}; t) = F_{x_1 \dots x_n}(X_1, X_2, \dots, X_n; t)$$
$$= Pr[x_1(t) \le X_1, x_2(t) \le X_2, \dots, x_n(t) \le X_n]$$

and thus the joint probability density of $\mathbf{x}(t)$ is

$$f_{\mathbf{x}}(\mathbf{X}; t) = \frac{\partial^n F_{\mathbf{x}}(X_1, X_2, \dots, X_n; t)}{\partial X_1 \, \partial X_2 \dots \partial X_n}$$

The joint probability distribution function of the random vectors $\mathbf{x}(t_1)$ and $\mathbf{x}(t_2)$ is

$$F_{\mathbf{xx}}(\mathbf{X}_1, \mathbf{X}_2; t_1, t_2) = F_{x_1 \dots x_n}(X_{11}, X_{12}, \dots, X_{1n}, X_{21}, X_{22}, \dots,$$
$$X_{2n}; t_1, t_2)$$
$$= Pr[x_1(t_1) \le X_{11}, \dots, x_n(t_1) \le X_{1n}, x_1(t_2) \le$$
$$X_{21}, \dots, x_n(t_2) \le X_{2n}]$$
$$= Pr[\mathbf{x}(t_1) \le \mathbf{X}_1, \mathbf{x}(t_2) \le \mathbf{X}_2]$$

where the inequality is on a component-by-component basis, and the joint probability density function is

$$f_{\mathbf{xx}}(\mathbf{X}_1, \mathbf{X}_2; t_1, t_2) = \frac{\partial^{2n} F(X_{11}, \ldots, X_{1n}, X_{21}, \ldots, X_{2n}; t_1, t_2)}{\partial X_{11} \ldots \partial X_{1n} \partial X_{21} \ldots \partial X_{2n}}$$

Now we define the following set of moments:

1. The *mean*, or expected value, of a stochastic vector

$$\mathbf{m}(t) = E[\mathbf{x}(t)] = \begin{bmatrix} E[x_1(t)] \\ E[x_2(t)] \\ \vdots \\ E[x_n(t)] \end{bmatrix} = \begin{bmatrix} m_1(t) \\ m_2(t) \\ \vdots \\ m_n(t) \end{bmatrix}$$

2. The *variance* matrix

$$\mathbf{Q}(t) = E[(\mathbf{x}(t) - \mathbf{m}(t))(\mathbf{x}(t) - \mathbf{m}(t))^{\dagger}]$$
$$= [E[(x_i(t) - m_i(t))(x_j(t) - m_j(t))]]$$

3. The *covariance* matrix

$$\mathbf{P}(t) = E[\mathbf{x}(t)\mathbf{x}^{\dagger}(t)] = [E[x_i(t)x_j(t)]]$$

4. The *autocorrelation* matrix

$$\mathbf{R}_{\mathbf{xx}}(t_1, t_2) = E[\mathbf{x}(t_1)\mathbf{x}^{\dagger}(t_2)] = [E[x_i(t_1)x_j(t_2)]]$$

When $t_1 = t_2 = t$,

$$\mathbf{R}_{\mathbf{xx}}(t, t) = \mathbf{P}(t)$$

5. The *autocovariance* matrix

$$\mathbf{\Gamma}_{\mathbf{xx}}(t_1, t_2) = E[(\mathbf{x}(t_1) - \mathbf{m}(t_1))(\mathbf{x}(t_2) - \mathbf{m}(t_2))^{\dagger}]$$
$$= [E[(x_i(t_1) - m_i(t_1))(x_j(t_2) - m_j(t_2))]]$$

6. The *cross-correlation* matrix

$$\mathbf{R}_{\mathbf{xy}}(t_1, t_2) = E[\mathbf{x}(t_1)\mathbf{y}^{\dagger}(t_2)] = [E[x_i(t_1)y_j(t_2)]]$$

7. The *cross-covariance* matrix

$$\mathbf{\Gamma}_{\mathbf{xy}}(t_1, t_2) = E[(\mathbf{x}(t_1) - \mathbf{m}_x(t_1))(\mathbf{y}(t_2) - \mathbf{m}_y(t_2))^{\dagger}]$$
$$= E[(x_i(t_1) - m_{x_i}(t_1))(y_j(t_2) - m_{y_j}(t_2))]$$

B.6.1 Gaussian Vector Stochastic Processes

Probably the most important probability density function is the Gaussian, or normal, density. Some of the reasons for its importance are

1. It is a good mathematical model for many physical situations.
2. It is completely specified by its mean and variance. Often it is the only case for which a problem may be solved analytically.
3. The central limit theorem states that if the n random variables x_1, $x_2, \ldots, x_n$ are independent and have identical density functions, then the sum

$$y = \frac{1}{\sqrt{n}} [x_1 + x_2 + \cdots + x_n]$$

   will approach the Gaussian density for large n.
4. It is one of the few density functions that can easily be extended to handle random functions of several variables.
5. Linear combinations of Gaussian random variables are also Gaussian.

In the remainder of this section, we will derive formulas that are very useful for solving the estimation problem presented in Chapter 8.

The probability density function of a Gaussian vector stochastic process $\mathbf{x}(t)$ is given by

$$f_{\mathbf{x}}(\mathbf{X}; t) = \frac{1}{(2\pi)^{n/2} |\mathbf{Q}(t)|^{1/2}} \exp\left\{ -\frac{1}{2} (\mathbf{X} - \mathbf{m}(t))^{\dagger} \mathbf{Q}^{-1}(t)(\mathbf{X} - \mathbf{m}(t)) \right\}$$

where $\mathbf{m}(t)$ is the expected value of $\mathbf{x}(t)$ and $\mathbf{Q}(t)$ is the variance matrix of $\mathbf{x}(t)$.

The joint probability density function of the two random vectors $\mathbf{x}(t_1)$ and $\mathbf{x}(t_2)$ is

$$f_{\mathbf{xx}}(\mathbf{X}_1, \mathbf{X}_2; t_1, t_2)$$

$$= \frac{1}{(2\pi)^n |\Sigma(t_1, t_2)|^{1/2}} \exp\left\{ -\frac{1}{2} \begin{bmatrix} (\mathbf{X}_1 - \mathbf{m}(t_1)) \\ (\mathbf{X}_2 - \mathbf{m}(t_2)) \end{bmatrix}^{\dagger} \Sigma^{-1}(t_1, t_2) \begin{bmatrix} (\mathbf{X}_1 - \mathbf{m}(t_1)) \\ (\mathbf{X}_2 - \mathbf{m}(t_2)) \end{bmatrix} \right\}$$

where

$$\Sigma(t_1, t_2) = \begin{bmatrix} \mathbf{Q}(t_1) & \Gamma_{\mathbf{xx}}(t_1, t_2) \\ \Gamma_{\mathbf{xx}}(t_2, t_1) & \mathbf{Q}(t_2) \end{bmatrix}$$

and where $\Gamma_{\mathbf{xx}}(t_1, t_2) = \Gamma_{\mathbf{xx}}^{\dagger}(t_2, t_1)$. Of course, $\Sigma(t_1, t_2)$ is symmetric. Higher-order joint probability density functions for several random variables $\mathbf{x}(t_1), \mathbf{x}(t_2), \ldots, \mathbf{x}(t_p)$ are similarly defined.

B.6.2 Conditional Gaussian Probability Densities

Consider two jointly Gaussian vector random variables $\mathbf{x}_1$ and $\mathbf{x}_2$ with the statistics

$$E[\mathbf{x}_1] = \mathbf{m}_1 \qquad E[\mathbf{x}_2] = \mathbf{m}_2$$

$$E[\mathbf{x}_1\mathbf{x}_1^\dagger] = \mathbf{R}_{11} \qquad E[\mathbf{x}_1\mathbf{x}_2^\dagger] = \mathbf{R}_{12}$$

$$E[\mathbf{x}_2\mathbf{x}_1^\dagger] = \mathbf{R}_{21} \qquad E[\mathbf{x}_2\mathbf{x}_2^\dagger] = \mathbf{R}_{22}$$

$$= \mathbf{R}_{12}^\dagger$$

If

$$\mathbf{y}_1 = \mathbf{x}_1 - \mathbf{m}_1 \qquad \mathbf{y}_2 = \mathbf{x}_2 - \mathbf{m}_2$$

and

$$\Sigma = \left[\begin{bmatrix} \mathbf{y}_1 \\ \mathbf{y}_2 \end{bmatrix} \begin{bmatrix} \mathbf{y}_1 \\ \mathbf{y}_2 \end{bmatrix}^\dagger \right] = \begin{bmatrix} E[\mathbf{y}_1\mathbf{y}_1^\dagger] & E[\mathbf{y}_1\mathbf{y}_2^\dagger] \\ E[\mathbf{y}_2\mathbf{y}_1^\dagger] & E[\mathbf{y}_2\mathbf{y}_2^\dagger] \end{bmatrix} = \begin{bmatrix} \mathbf{\Gamma}_{11} & \mathbf{\Gamma}_{12} \\ \mathbf{\Gamma}_{21} & \mathbf{\Gamma}_{22} \end{bmatrix}$$

and if

$$\Sigma^{-1} \triangleq \mathbf{\Lambda} = \begin{bmatrix} \mathbf{\Lambda}_{11} & \mathbf{\Lambda}_{12} \\ \mathbf{\Lambda}_{21} & \mathbf{\Lambda}_{22} \end{bmatrix}$$

then the joint probability density of $\mathbf{y}_1$ and $\mathbf{y}_2$ can be written as

$$f_{\mathbf{y}_1\mathbf{y}_2}(\mathbf{Y}_1, \mathbf{Y}_2) = \frac{1}{(2\pi)^{(n+m)/2}|\mathbf{\Lambda}|^{-1/2}} \exp\left\{ -\frac{1}{2} \begin{bmatrix} \mathbf{Y}_1 \\ \mathbf{Y}_2 \end{bmatrix}^\dagger \begin{bmatrix} \mathbf{\Lambda}_{11} & \mathbf{\Lambda}_{12} \\ \mathbf{\Lambda}_{21} & \mathbf{\Lambda}_{22} \end{bmatrix} \begin{bmatrix} \mathbf{Y}_1 \\ \mathbf{Y}_2 \end{bmatrix} \right\}$$

where it is assumed that $\mathbf{y}_1$ is an n-dimensional vector and $\mathbf{y}_2$ is an m-dimensional vector. Hence,

$$f_{\mathbf{y}_1\mathbf{y}_2}(\mathbf{Y}_1, \mathbf{Y}_2) = \frac{1}{(2\pi)^{(n+m)/2}\,|\mathbf{\Lambda}|^{-1/2}}$$

$$\times \exp\left\{ -\frac{1}{2} [\mathbf{Y}_1^\dagger\mathbf{\Lambda}_{11}\mathbf{Y}_1 + 2\mathbf{Y}_1^\dagger\mathbf{\Lambda}_{12}\mathbf{Y}_2 + \mathbf{Y}_2^\dagger\mathbf{\Lambda}_{22}\mathbf{Y}_2] \right\}$$

From this we can generate

$$f_{\mathbf{y}_2}(\mathbf{Y}_2) = \int_{-\infty}^{\infty} f_{\mathbf{y}_1\mathbf{y}_2}(\mathbf{Y}_1, \mathbf{Y}_2)\, d\mathbf{Y}_1 = \frac{1}{(2\pi)^{(n+m)/2}\,|\mathbf{\Lambda}|^{-1/2}}$$

$$\times \int_{-\infty}^{\infty} \exp\left\{ -\frac{1}{2} [\mathbf{Y}_1^\dagger\mathbf{\Lambda}_{11}\mathbf{Y}_1 + 2\mathbf{Y}_1^\dagger\mathbf{\Lambda}_{12}\mathbf{Y}_2] \right\} d\mathbf{Y}_1$$

$$\times \exp\left\{ -\frac{1}{2} \mathbf{Y}_2^\dagger\mathbf{\Lambda}_{22}\mathbf{Y}_2 \right\}$$

But because

$$f_{y_1|y_2}(\mathbf{Y}_1 \mid \mathbf{Y}_2) = \frac{f_{y_1y_2}(\mathbf{Y}_1, \mathbf{Y}_2)}{f_{y_2}(\mathbf{Y}_2)}$$

then

$$f_{y_1|y_2}(\mathbf{Y}_1 \mid \mathbf{Y}_2) = \frac{\exp\{-\frac{1}{2}[\mathbf{Y}_1^\dagger \Lambda_{11}\mathbf{Y}_1 + 2\mathbf{Y}_1^\dagger \Lambda_{12}\mathbf{Y}_2]\}}{\int_{-\infty}^{\infty} \exp\{-\frac{1}{2}[\mathbf{Y}_1^\dagger \Lambda_{11}\mathbf{Y}_1 + 2\mathbf{Y}_1^\dagger \Lambda_{12}\mathbf{Y}_2]\} \, d\mathbf{Y}_1}$$

If we let

$$l(\mathbf{Y}_2) \triangleq \frac{1}{\int_{-\infty}^{\infty} \exp\{-\frac{1}{2}[\mathbf{Y}_1^\dagger \Lambda_{11}\mathbf{Y}_1 + 2\mathbf{Y}_1^\dagger \Lambda_{12}\mathbf{Y}_2]\} \, d\mathbf{Y}_1}$$

then

$$f_{y_1|y_2}(\mathbf{Y}_1 \mid \mathbf{Y}_2) = l(\mathbf{Y}_2) \exp\left\{-\frac{1}{2}[\mathbf{Y}_1^\dagger \Lambda_{11}\mathbf{Y}_1 + 2\mathbf{Y}_1^\dagger \Lambda_{12}\mathbf{Y}_2]\right\}$$

Completing the square on the exponent as follows gives

$$f_{y_1|y_2}(\mathbf{Y}_1 \mid \mathbf{Y}_2) = l(\mathbf{Y}_2) \exp\left\{-\frac{1}{2}[\mathbf{Y}_2^\dagger \Lambda_{12}^\dagger \Lambda_{11}^{-1}\Lambda_{12}\mathbf{Y}_2]\right\}$$

$$\times \exp\left\{-\frac{1}{2}[\mathbf{Y}_1 + \Lambda_{11}^{-1}\Lambda_{12}\mathbf{Y}_2]^\dagger \Lambda_{11}[\mathbf{Y}_1 + \Lambda_{11}^{-1}\Lambda_{12}\mathbf{Y}_2]\right\}$$

$$= l'(\mathbf{Y}_2) \exp\left\{-\frac{1}{2}[\mathbf{Y}_1 + \Lambda_{11}^{-1}\Lambda_{12}\mathbf{Y}_2]^\dagger \Lambda_{11}[\mathbf{Y}_1 + \Lambda_{11}^{-1}\Lambda_{12}\mathbf{Y}_2]\right\}$$

Because this is a Gaussian conditional probability density, then

$$E[\mathbf{y}_1 \mid \mathbf{Y}_2] = -\Lambda_{11}^{-1}\Lambda_{12}\mathbf{Y}_2$$

and

$$\text{Var}[\mathbf{y}_1 \mid \mathbf{Y}_2] = \Lambda_{11}^{-1}$$

Note that $E[\mathbf{y}_1 \mid \mathbf{Y}_2]$ is linear in $\mathbf{Y}_2$ and $\text{Var}[\mathbf{y}_1 \mid \mathbf{Y}_2]$ is independent of $\mathbf{Y}_2$.

Because

$$\Sigma = \begin{bmatrix} \Gamma_{11} & \Gamma_{12} \\ \Gamma_{21} & \Gamma_{22} \end{bmatrix}$$

we can find the matrices Λ_{11}, Λ_{12}, Λ_{22} by noting that

$$\begin{bmatrix} \Gamma_{11} & \Gamma_{12} \\ \Gamma_{21} & \Gamma_{22} \end{bmatrix} \begin{bmatrix} \Lambda_{11} & \Lambda_{12} \\ \Lambda_{21} & \Lambda_{22} \end{bmatrix} = \begin{bmatrix} \mathbf{I} & \mathbf{0} \\ \mathbf{0} & \mathbf{I} \end{bmatrix}$$

We form the two matrix equations

$$\Gamma_{11}\Lambda_{11} + \Gamma_{12}\Lambda_{21} = \mathbf{I}$$
$$\Gamma_{21}\Lambda_{11} + \Gamma_{22}\Lambda_{21} = \mathbf{0}$$

which can be easily solved to obtain

$$\Lambda_{21} = -\Gamma_{22}^{-1}\Gamma_{21}\Lambda_{11}$$
$$(\Gamma_{11} - \Gamma_{12}\Gamma_{22}^{-1}\Gamma_{21})\Lambda_{11} = \mathbf{I}$$

or

$$\Lambda_{11} = (\Gamma_{11} - \Gamma_{12}\Gamma_{22}^{-1}\Gamma_{21})^{-1}$$
$$\Lambda_{21} = -\Gamma_{22}^{-1}\Gamma_{21}(\Gamma_{11} - \Gamma_{12}\Gamma_{22}^{-1}\Gamma_{21})^{-1} = \Lambda_{12}^{\dagger}$$

Now we can form

$$
\begin{aligned}
E[\mathbf{y}_1 \mid \mathbf{Y}_2] &= -\Lambda_{11}^{-1}\Lambda_{12}\mathbf{Y}_2 \\
&= (\Gamma_{11} - \Gamma_{12}\Gamma_{22}^{-1}\Gamma_{21})(\Gamma_{11} - \Gamma_{12}\Gamma_{22}^{-1}\Gamma_{21})^{-1}\Gamma_{12}\Gamma_{22}^{-1}\mathbf{Y}_2 \\
&= \Gamma_{12}\Gamma_{22}^{-1}\mathbf{Y}_2
\end{aligned}
$$

and

$$\mathrm{Var}[\mathbf{y}_1 \mid \mathbf{Y}_2] = \Lambda_{11}^{-1} = (\Gamma_{11} - \Gamma_{12}\Gamma_{22}^{-1}\Gamma_{21})$$

Again, note that $E[\mathbf{y}_1 \mid \mathbf{Y}_2]$ is linear in $\mathbf{Y}_2$ and that $\mathrm{var}[\mathbf{y}_1 \mid \mathbf{Y}_2]$ is independent of $\mathbf{Y}_2$. This second result is most useful in estimation theory.

We now show that the *unconditional* expected value of $(\mathbf{y}_1 - E[\mathbf{y}_1 \mid \mathbf{Y}_2])(\mathbf{y}_1 - E[\mathbf{y}_1 \mid \mathbf{Y}_2])^{\dagger}$ is the same as the variance of $\mathbf{y}_1$ conditioned on $\mathbf{Y}_2$. We form

$$
\begin{aligned}
E[(\mathbf{y}_1 &- E[\mathbf{y}_1 \mid \mathbf{Y}_2])(\mathbf{y}_1 - E[\mathbf{y}_1 \mid \mathbf{Y}_2])^{\dagger}] \\
&= E[(\mathbf{y}_1 - \Gamma_{12}\Gamma_{22}^{-1}\mathbf{Y}_2)(\mathbf{y}_1 - \Gamma_{12}\Gamma_{22}^{-1}\mathbf{Y}_2)^{\dagger}] \\
&= E[\mathbf{y}_1\mathbf{y}_1^{\dagger}] - \Gamma_{12}\Gamma_{22}^{-1}E[\mathbf{Y}_2\mathbf{y}_1^{\dagger}] - E[\mathbf{y}_1\mathbf{Y}_2^{\dagger}]\Gamma_{22}^{-1}\Gamma_{12}^{\dagger} \\
&\quad + \Gamma_{12}\Gamma_{22}^{-1}E[\mathbf{Y}_2\mathbf{Y}_2^{\dagger}]\Gamma_{22}^{-1}\Gamma_{12}^{\dagger} \\
&= \Gamma_{11} - \Gamma_{12}\Gamma_{22}^{-1}\Gamma_{12}^{\dagger} - \Gamma_{12}\Gamma_{22}^{-1}\Gamma_{12}^{\dagger} + \Gamma_{12}\Gamma_{22}^{-1}\Gamma_{12}^{\dagger} \\
&= \Gamma_{11} - \Gamma_{12}\Gamma_{22}^{-1}\Gamma_{12}^{\dagger}
\end{aligned}
$$

where $\mathbf{Y}_2$ is treated as a random vector.

For the nonzero-mean random vectors $\mathbf{x}_1$ and $\mathbf{x}_2$,

$$
\begin{aligned}
E[\mathbf{y}_1 \mid \mathbf{Y}_2] &= E[\mathbf{y}_1 \mid \mathbf{y}_2 = \mathbf{Y}_2] \\
&= E[\mathbf{y}_1 \mid \mathbf{x}_2 - \mathbf{m}_2 = \mathbf{Y}_2] \\
&= E[\mathbf{y}_1 \mid \mathbf{x}_2 = \mathbf{Y}_2 + \mathbf{m}_2] = E[\mathbf{y}_1 \mid \mathbf{X}_2] \\
&= E[(\mathbf{x}_1 - \mathbf{m}_1) \mid \mathbf{X}_2] \\
&= E[\mathbf{x}_1 \mid \mathbf{X}_2] - E[\mathbf{m}_1 \mid \mathbf{X}_2] \\
&= E[\mathbf{x}_1 \mid \mathbf{X}_2] - \mathbf{m}_1
\end{aligned}
$$

Thus,

$$E[\mathbf{x}_1 \mid \mathbf{X}_2] = E[\mathbf{y}_1 \mid \mathbf{Y}_2] + \mathbf{m}_1 = \Gamma_{12}\Gamma_{22}^{-1}(\mathbf{X}_2 - \mathbf{m}_2) + \mathbf{m}_1$$

where

$$\mathbf{X}_2 = \mathbf{Y}_2 + \mathbf{m}_2$$

Furthermore,

$$\text{Var}[\mathbf{y}_1 \mid \mathbf{Y}_2]$$
$$= \text{Var}[\mathbf{y}_1 \mid \mathbf{y}_2 = \mathbf{Y}_2]$$
$$= E[(\mathbf{y}_1 - E[\mathbf{y}_1 \mid \mathbf{Y}_2])(\mathbf{y}_1 - E[\mathbf{y}_1 \mid \mathbf{Y}_2])^\dagger \mid \mathbf{y}_2 = \mathbf{Y}_2]$$
$$= E[(\mathbf{y}_1 + \mathbf{m}_1 - E[\mathbf{y}_1 \mid \mathbf{Y}_2] - \mathbf{m}_1)(\mathbf{y}_1 + \mathbf{m}_1 - E[\mathbf{y}_1 \mid \mathbf{Y}_2] - \mathbf{m}_1)^\dagger$$
$$\mid \mathbf{x}_2 - \mathbf{m}_2 = \mathbf{Y}_2]$$
$$= E[(\mathbf{x}_1 - E[\mathbf{x}_1 \mid \mathbf{X}_2])(\mathbf{x}_1 - E[\mathbf{x}_1 \mid \mathbf{X}_2])^\dagger \mid \mathbf{x}_2 = \mathbf{X}_2]$$
$$= \text{Var}[\mathbf{x}_1 \mid \mathbf{X}_2]$$

Finally, we can write

1. $E[\mathbf{x}_1 \mid \mathbf{X}_2] = \Gamma_{12}\Gamma_{22}^{-1}(\mathbf{X}_2 - \mathbf{m}_2) + \mathbf{m}_1$
2. $\text{Var}[\mathbf{x}_1 \mid \mathbf{X}_2] = \Gamma_{11} - \Gamma_{12}\Gamma_{22}^{-1}\Gamma_{12}^\dagger$
3. $\text{Var}[\mathbf{x}_1 \mid \mathbf{X}_2] = E[(\mathbf{x}_1 - E[\mathbf{x}_1 \mid \mathbf{X}_2])(\mathbf{x}_1 - E[\mathbf{x}_1 \mid \mathbf{X}_2])^\dagger]$

References

A. Papoulis, *Probability, Random Variables, and Stochastic Processes,* 2nd edition. New York: McGraw-Hill, 1984.

J. L. Melsa and A. P. Sage, *An Introduction to Probability and Stochastic Processes.* Englewood Cliffs, N.J.: Prentice-Hall, 1973.

G. R. Cooper and C. D. McGillem, *Probabilistic Methods of Signal and System Analysis*, 2nd edition. New York: Holt, Rinehart and Winston, 1986.

W. B. Davenport, Jr., and W. L. Root, *Introduction to Random Signals and Noise.* New York: McGraw-Hill, 1958.

C. W. Helstrom, *Probability and Stochastic Processes for Engineers.* New York: Macmillan, 1984.

E. Parzen, *Modern Probability Theory and Its Applications.* New York: John Wiley and Sons, 1960.

P. Z. Peebles, *Probability, Random Variables, and Random Signal Principles.* New York: McGraw-Hill, 1980.

H. J. Larson and B. O. Shubert, *Probabilistic Models in Engineering Sciences* (Vols. 1 & 2). New York: Wiley, 1979.

P. M. Breipohl, *Probabilistic Systems Analysis*. New York: Wiley, 1970.

J. H. Lanning, Jr., and R. H. Battin, *Random Processes in Automatic Control*. New York: McGraw-Hill, 1956.

A. H. Jazwinski, *Stochastic Processes and Filtering Theory*. New York: Academic Press, 1970.

J. S. Bendat and A. G. Piersol, *Random Data: Analysis and Measurement Procedures*. New York: Wiley-Interscience, 1971.

R. Hogg and A. Craig, *Introduction to Mathematical Statistics*. New York: Macmillan, 1970.

R. B. Blackman and J. W. Tukey, *The Measurement of Power Spectra*. New York: Dover Publications, 1958.

G. M. Jenkins and D. G. Watts, *Spectral Analysis and Its Applications*. San Francisco: Holden-Day, 1968.

S. L. Maples, *Digital Spectral Analysis with Applications*. Englewood Cliffs, N.J.: Prentice-Hall, Signal Processing Series, 1987.

Basic Results for Linear Minimum Mean Square Error Estimation

The fundamental problem of estimation is that of generating a value for a parameter when a direct and error-free measurement of the parameter cannot be obtained. A (probably) noisy measurement is made of a set of related variables, and their values are used to generate an estimate of the unknown parameter. Of course, in any sensible estimation problem, the unknown parameter and the measured variables are related in some sense. More formally, a generic linear estimation problem can be formulated as follows:

Given

1. An unknown parameter x
2. A measurable set of variables z that are related in some sense to x
3. A priori information about x and the relationship between x and z
4. A performance index that allows a comparison of different estimates of x

It is desirable to find a linear estimation rule (algorithm), called a *linear estimator*, which maps the measured value of z into a *linear estimate* (in z) of x so that the performance index is optimized.

We assume that the unknown parameter is a random vector $\mathbf{x}$ of dimension n, and the measurable variable is a random vector $\mathbf{z}$ of dimension m. We also assume that $\mathbf{x}$ and $\mathbf{z}$ are related by a joint probability density function $f(\mathbf{x}, \mathbf{z})$, but that any a priori information about $\mathbf{x}$ and $\mathbf{z}$ is in the form of first and second moments and joint moments $E[\mathbf{x}]$, $E[\mathbf{z}]$, $E[\mathbf{xz}^{\dagger}]$, $E[\mathbf{xx}^{\dagger}]$, and $E[\mathbf{zz}^{\dagger}]$. Although many different performance indices could be used, unless otherwise specified, we will consider the following performance index

$$J = E[(\mathbf{x} - \hat{\mathbf{x}})^\dagger(\mathbf{x} - \hat{\mathbf{x}})] \tag{C-1}$$

where $\hat{\mathbf{x}}$ is a linear function of $\mathbf{z}$ that minimizes J. We let

$$\hat{\mathbf{x}} = \hat{\mathbf{K}}\mathbf{z} + \hat{\mathbf{b}} \tag{C-2}$$

and let $\mathbf{x}_a$ be an arbitrary linear estimate

$$\mathbf{x}_a = \mathbf{K}_a\mathbf{z} + \mathbf{b}_a \tag{C-3}$$

where $\hat{\mathbf{K}}$ and $\mathbf{K}_a$ are constant $n \times m$ matrices and $\hat{\mathbf{b}}$ and $\mathbf{b}_a$ are constant n-dimensional vectors.

In the material to follow, we prove four basic results for linear minimum mean square (LMMS) estimation that are useful in the derivation of the Kalman filter equations.

C.1 Minimum Mean Square Estimate

If $\mathbf{x}$ and $\mathbf{z}$ are zero-mean random vectors, an LMMS estimate of $\mathbf{x}$ based on $\mathbf{z}$ is

$$\hat{\mathbf{x}} = E[\mathbf{x}\mathbf{z}^\dagger]\{E[\mathbf{z}\mathbf{z}^\dagger]\}^{-1}\mathbf{z} \tag{C-4}$$

C.1.1 Proof

Because $\hat{\mathbf{x}}$ is an LMMS estimate of $\mathbf{x}$, the performance index

$$J = E[(\mathbf{x} - \hat{\mathbf{x}})^\dagger(\mathbf{x} - \hat{\mathbf{x}})]$$

is a minimum. If we introduce the arbitrary linear estimate $\mathbf{x}_a$, the performance measure in equation (C-1) can be rewritten as

$$J = E[(\mathbf{x} - \mathbf{x}_a + \mathbf{x}_a - \hat{\mathbf{x}})^\dagger(\mathbf{x} - \mathbf{x}_a + \mathbf{x}_a - \hat{\mathbf{x}})]$$
$$= E[(\mathbf{x} - \mathbf{x}_a)^\dagger(\mathbf{x} - \mathbf{x}_a)] + 2E[(\mathbf{x} - \mathbf{x}_a)^\dagger(\mathbf{x}_a - \hat{\mathbf{x}})]$$
$$+ E[(\mathbf{x}_a - \hat{\mathbf{x}})^\dagger(\mathbf{x}_a - \hat{\mathbf{x}})] \tag{C-5}$$

Because $\mathbf{x}_a$ is an arbitrary estimate, we choose it so that

$$E[(\mathbf{x} - \mathbf{x}_a)^\dagger(\mathbf{x}_a - \hat{\mathbf{x}})] = 0 \tag{C-6}$$

independent of the specific values of $\hat{\mathbf{K}}$, $\mathbf{K}_a$, $\hat{\mathbf{b}}$, and $\mathbf{b}_a$.

Then equation (C-5) becomes

$$J = E[(\mathbf{x} - \hat{\mathbf{x}})^\dagger(\mathbf{x} - \hat{\mathbf{x}})]$$

$$= E[(\mathbf{x} - \mathbf{x}_a)^\dagger(\mathbf{x} - \mathbf{x}_a)] + E[(\mathbf{x}_a - \hat{\mathbf{x}})^\dagger(\mathbf{x}_a - \hat{\mathbf{x}})]$$

$$\geq E[(\mathbf{x} - \mathbf{x}_a)^\dagger(\mathbf{x} - \mathbf{x}_a)] \qquad \text{(C-7)}$$

Because $\hat{\mathbf{x}}$ minimizes J, $\mathbf{x}_a$ cannot be a better estimate than $\hat{\mathbf{x}}$. The inequality in equation (C-7) implies that $\mathbf{x}_a$ is as good as $\hat{\mathbf{x}}$, however; that is, $\mathbf{x}_a$ is also an optimum estimate of $\mathbf{x}$. Then

$$E[(\mathbf{x}_a - \hat{\mathbf{x}})^\dagger(\mathbf{x}_a - \hat{\mathbf{x}})] = 0$$

which implies that $\mathbf{x}_a = \hat{\mathbf{x}}$ in the mean square sense. (We use $\mathbf{x}_a$ and $\hat{\mathbf{x}}$ interchangeably.)

Now it is desirable to find the LMMS estimate $\hat{\mathbf{x}}$. Replacing $\mathbf{x}_a$ with $\hat{\mathbf{x}}$, equation (C-6) becomes

$$E[(\mathbf{x} - \hat{\mathbf{x}})^\dagger(\mathbf{x}_a - \hat{\mathbf{x}})] = 0 \qquad \text{(C-8)}$$

independent of the specific estimators $\mathbf{x}_a$ and $\hat{\mathbf{x}}$. Subtracting equation (C-2) from equation (C-3) gives

$$\mathbf{x}_a - \hat{\mathbf{x}} = (\mathbf{K}_a - \hat{\mathbf{K}})\mathbf{z} + (\mathbf{b}_a - \hat{\mathbf{b}})$$

$$= \mathbf{K}\mathbf{z} + \mathbf{b} \qquad \text{(C-9)}$$

where $\mathbf{K}$ is apparently an arbitrary constant $n \times m$ matrix and $\mathbf{b}$ is an arbitrary constant n-dimensional vector.

Substituting equation (C-9) into equation (C-8) yields

$$E[(\mathbf{x} - \hat{\mathbf{x}})^\dagger(\mathbf{x}_a - \hat{\mathbf{x}})] = E[(\mathbf{x} - \hat{\mathbf{x}})^\dagger\mathbf{K}\mathbf{z}] + E[(\mathbf{x} - \hat{\mathbf{x}})^\dagger\mathbf{b}] = 0 \qquad \text{(C-10)}$$

Equation (C-10) must hold independent of the values of $\mathbf{K}$ and $\mathbf{b}$.

Suppose that $\mathbf{b}$ is any nonzero vector, then for any choice of $\mathbf{K}$, in particular for $\mathbf{K} = \mathbf{0}$, this equation must hold. But this cannot be true unless $\hat{\mathbf{x}} = \mathbf{x}$ (which by our definition of the estimation problem cannot be true except in the trivial case). Thus, it is necessary that $\mathbf{b} = \mathbf{0}$, and, therefore, equation (C-10) becomes

$$E[(\mathbf{x} - \hat{\mathbf{x}})^\dagger\mathbf{K}\mathbf{z}] = 0$$

for all constant matrices $\mathbf{K}$. This implies that each element of $\mathbf{z}$ must be orthogonal to every element of $\mathbf{x} - \hat{\mathbf{x}}$, that is,

$$E[(\mathbf{x} - \hat{\mathbf{x}})\mathbf{z}^\dagger] = \mathbf{0} \qquad \text{(C-11)}$$

Because $\mathbf{b}$ is zero, and $\mathbf{b}_a$ and $\hat{\mathbf{b}}$ are arbitrary, equation (C-2) yields

$$\hat{\mathbf{x}} = \hat{\mathbf{K}}\mathbf{z} \qquad \text{(C-12)}$$

Substituting equation (C-12) into equation (C-11) gives

$$E[(\mathbf{x} - \hat{\mathbf{K}}\mathbf{z})\mathbf{z}^\dagger] = E[\mathbf{x}\mathbf{z}^\dagger] - \hat{\mathbf{K}}E[\mathbf{z}\mathbf{z}^\dagger] = 0 \qquad \text{(C-13)}$$

Solving equation (C-13) for $\hat{\mathbf{K}}$ gives

$$\hat{\mathbf{K}} = E[\mathbf{xz}^\dagger]\{E[\mathbf{zz}^\dagger]\}^{-1} \tag{C-14}$$

For this value of $\hat{\mathbf{K}}$ and $\hat{\mathbf{b}} = \mathbf{0}$, equation (C-8) is satisfied independently of $(\hat{\mathbf{x}} - \mathbf{x}_a)$, and, therefore, the LLMS estimate given by equation (C-12) is

$$\hat{\mathbf{x}} = E[\mathbf{xz}^\dagger]\{E[\mathbf{zz}^\dagger]\}^{-1}\mathbf{z} \tag{C-15}$$

which completes the proof.

C.2 Orthogonality of Estimation Error and Data

A linear estimator $\hat{\mathbf{x}}$ is a minimum mean square estimator if and only if the estimation error $\mathbf{x} - \hat{\mathbf{x}}$ is orthogonal to (uncorrelated with) the measurement vector $\mathbf{z}$. That is,

$$E[(\mathbf{x} - \hat{\mathbf{x}})\mathbf{z}^\dagger] = \mathbf{0}$$

C.2.1 Proof of Necessity

$$E[(\mathbf{x} - \hat{\mathbf{x}})\mathbf{z}^\dagger] = E[\mathbf{xz}^\dagger] - E[\hat{\mathbf{x}}\mathbf{z}^\dagger] \tag{C-16}$$

Substituting equation (C-4) into equation (C-16) gives

$$\begin{aligned}
E[(\mathbf{x} - \hat{\mathbf{x}})\mathbf{z}^\dagger] &= E[\mathbf{xz}^\dagger] - E\{E[\mathbf{xz}^\dagger]\{E[\mathbf{zz}^\dagger]\}^{-1}\mathbf{zz}^\dagger\} \\
&= E[\mathbf{xz}^\dagger] - E[\mathbf{xz}^\dagger]\{E[\mathbf{zz}^\dagger]\}^{-1}E[\mathbf{zz}^\dagger] \\
&= E[\mathbf{xz}^\dagger] - E[\mathbf{xz}^\dagger] = \mathbf{0}
\end{aligned}$$

C.2.2 Proof of Sufficiency

Substituting equation (C-12)

$$\hat{\mathbf{x}} = \hat{\mathbf{K}}\mathbf{z}$$

into equation (C-11) gives

$$E[(\mathbf{x} - \hat{\mathbf{x}})\mathbf{z}^\dagger] = E[\mathbf{xz}^\dagger] - E[\hat{\mathbf{K}}\mathbf{zz}^\dagger] = E[\mathbf{xz}^\dagger] - \hat{\mathbf{K}}E[\mathbf{zz}^\dagger] = \mathbf{0}$$

Hence,

$$\hat{\mathbf{K}} = E[\mathbf{xz}^\dagger]\{E[\mathbf{zz}^\dagger]\}^{-1}$$

and, therefore,

$$\hat{\mathbf{x}} = E[\mathbf{xz}^\dagger]\{E[\mathbf{zz}^\dagger]\}^{-1}\mathbf{z}$$

C.3 Estimation of a Linear Composition

For the random vectors $\mathbf{x}$, $\mathbf{y}$, $\mathbf{z}$, and $\mathbf{w}$, if

$$\mathbf{x} = \mathbf{A}\mathbf{y} + \mathbf{B}\mathbf{w}$$

the LMMS estimate of $\mathbf{x}$ based on $\mathbf{z}$ is given by

$$\hat{\mathbf{x}} = \mathbf{A}\hat{\mathbf{y}} + \mathbf{B}\hat{\mathbf{w}} \tag{C-17}$$

where $\hat{\mathbf{y}}$ is the LMMS estimate of $\mathbf{y}$ and $\hat{\mathbf{w}}$ is the LMMS estimate of $\mathbf{w}$.

C.3.1 Proof That the Estimate is Minimum Mean Square

$$E[(\mathbf{x} - \hat{\mathbf{x}})\mathbf{z}^{\dagger}] = \mathbf{A}E[(\mathbf{y} - \hat{\mathbf{y}})\mathbf{z}^{\dagger}] + \mathbf{B}E[(\mathbf{w} - \hat{\mathbf{w}})\mathbf{z}^{\dagger}] = \mathbf{0}$$

C.4 Incorporation of Orthogonal Data

If $\mathbf{x}$, $\mathbf{z}_1$, and $\mathbf{z}_2$ are zero-mean random vectors, with $\mathbf{z}_1$ and $\mathbf{z}_2$ orthogonal,

$$E[\mathbf{z}_1\mathbf{z}_2^{\dagger}] = \mathbf{0}$$

and the LMMS estimate of $\mathbf{x}$ based on $\mathbf{z}_1$ and $\mathbf{z}_2$ is given by

$$\hat{\mathbf{x}} = \hat{\mathbf{x}}_1 + \hat{\mathbf{x}}_2$$

where $\hat{\mathbf{x}}_1$ is the LMMS estimate of $\mathbf{x}$ based on $\mathbf{z}_1$ and $\mathbf{x}_2$ is the LMMS estimate of $\mathbf{x}$ based on $\mathbf{z}_2$.

C.4.1 Proof

Let

$$\mathbf{z} = \begin{bmatrix} \mathbf{z}_1 \\ \hline \mathbf{z}_2 \end{bmatrix}$$

denote the combined set of measurements. Because $\mathbf{z}_1$ and $\mathbf{z}_2$ are orthogonal

$$E[\mathbf{z}\mathbf{z}^{\dagger}] = \begin{bmatrix} E[\mathbf{z}_1\mathbf{z}_1^{\dagger}] & \mathbf{0} \\ \hline \mathbf{0} & E[\mathbf{z}_2\mathbf{z}_2^{\dagger}] \end{bmatrix}$$

and

$$\{E[\mathbf{z}\mathbf{z}^{\dagger}]\}^{-1} = \begin{bmatrix} \{E[\mathbf{z}_1\mathbf{z}_1^{\dagger}]\}^{-1} & \mathbf{0} \\ \hline \mathbf{0} & \{E[\mathbf{z}_2\mathbf{z}_2^{\dagger}]\}^{-1} \end{bmatrix}$$

Because

$$\hat{\mathbf{x}} = E[\mathbf{x}\mathbf{z}^{\dagger}]\{E[\mathbf{z}\mathbf{z}^{\dagger}]\}^{-1}\mathbf{z}$$

then

$$\hat{\mathbf{x}} = E[\mathbf{x}\mathbf{z}_1^{\dagger}]\{E[\mathbf{z}_1\mathbf{z}_1^{\dagger}]\}^{-1}\mathbf{z}_1 + E[\mathbf{x}\mathbf{z}_2^{\dagger}]\{E[\mathbf{z}_2\mathbf{z}_2^{\dagger}]\}^{-1}\mathbf{z}_2$$

$$= \hat{\mathbf{x}}_1 + \hat{\mathbf{x}}_2$$

which completes the proof.

Index

A

Actuators, 382, 533
Adaptive control 215, 551
Aliasing, 70 (*see also* foldover)
Analog-to-digital (A/D) conversion, 4, 56
   for high-speed smoothing, 524–528
   in data acquisition, 533, 536
Analog plant input filtering, 520–524
Analog system, 2 (*see also* continuous-time systems)
Antialiasing filter, 74
Antiwindup reset, 574
Assembly language, 543
Autocorrelation, 582, 772, 778
Autocovariance, 778

B

Bandlimiting, 65
Basis, 694
Bayes' rule, 748
Bertram, John, 98
Between–sample response, 512–515
   design of, 520–521
   with analog plant input filtering, 520–524
   with high–rate digital filtering, 524–528
Bias (unbias), 620, 649
Bilinear transformation, 50, 496, 502
Binary code, 3, 472, 541
Bootstrap instructions, 529
Bumpless transfer, 578

C

Causal system, 32, 585
Cayley–Hamilton 207, 711–713, 721–723
Central limit theorem, 764
Central processor unit (CPU), 529
Characteristic equation, 119, 701
Characteristic function, 765–766
Characteristic value problem, 699–710
Cholesky decomposition method, 719
Colored noise, 593, 603, 649, 776
Compensator or controller, 1, 7–9, 91–94, 261–266, 358–360, 491–512
Compiler, 543
Computer
   general–purpose, 529–533
   multiprocessor architectures, 531–533
   software, 543–551
   special purpose, 539–543
Conditional
   correlation, 762
   covariance, 762
   Gaussian probability, 780–783
   mean, 762
   probability, 748
   probability density, 761, 763
   probability distribution, 761
Constant–parameter, 4
Constructability, 208
Continuous–time systems, 3–4
   from discrete–time models, 571
   response of, 116–128
   sampling of, 143–149
   state equations for, 110–123

Control
 classical, 5–9
 hardware, 528–543
 optimal, 267–297
 state, 225–229
 step-varying, 229, 250–253
 terminology, 1–5
Controllability
 definition of, 170–171, 222
 index, 223
 matrix, 222–224
 of a system's state, 225–229
 of continuous-time systems, 224
 of repeated modes, 173–176
Controllable form, 149–152
 for eigenvalue placement, 236–238
 transformation to, 176–179, 361
Convolution, 37, 41–43, 122–123, 134–136
Coprocessor, 531
Correlation, 760, 762–763
Covariance, 772–778
Cramer's rule, 690
Cross–correlation, 770, 778
Cross–covariance, 772, 778

D

Data acquisition and distribution, 533–538 (see also analog–to–digital conversion and digital–to–analog conversion)
Decoupling state equations, 158–170
Delay diagram, 129–134, 141
Determinant
 definition and properties, 684–687
 of a signal flow graph, 150
Difference equations, 31–33
Digital control system, 2
Digital filter, 492, 502, 524–526, 568–571
Digital–to–analog (D/A) conversion, 61–62, 533–538
Direct memory access (DMA), 529
Discrete–time signals, 16–31
Discrete–time systems, 31–55
 digitized analog controllers, 491–512
 equivalents, 76–80
 frequency response of, 51–55
 from continuous–time systems, 143–149, 512–519

Discrete–time systems (continued)
 response of, 31–33, 129–143
 state equations for, 129–134, 141
Disturbance rejection and cancellation, 455–465, 476–477
Duality, 223–224

E

Eigenvalues, 119, 139, 699–702
Eigenvalue placement
 with observer feedback, 355–367
 with output feedback, 253–261
 with state feedback, 235–250
 with feedback compensation, 261–266
Eigenvectors, 699–705
Elevator controller, 552–557
Error covariances, 628–637
Estimation
 bias, 649
 least squares, 605–607
 linear minimum mean square, 617–624
 of a linear composition, 624
 recursive least squares, 609–615
 state, 608–609
 steady state, 652–653
Equations, 687–690

F

Faddeev–Leverrier algorithm, 725
Feedback, 2, 205, 235–236, 250–253
Field of events, 747
Final value theorem, 8, 25–27
Finite impulse response (FIR), 397
First–order hold, 64–65
Foldover, 70
Fourier transform, 65–66
Franklin, Gene, 98
Friedland, Bernard, 98

G

Gain margin, 6
Gaussian random variable
 conditional density of, 780
 joint probability density function, 779
 mean of, 755
 probability density of, 752
 variance of, 756
 vector stochastic process, 779

Gauss–Jordan pivoting, 688–689
Geometric sequence, 20
Gradient, 684

H

Hidden oscillations, 514–515
Higher–rate digital filtering, 524–528
Holds, 62–65, 527–528
Homogeneous equations, 689
Housekeeping tasks, 545
Hybrid system simulation, 515–519

I

Ideal tracking design, 405–421
    with multiple plant inputs, 414–417
    with multiple reference inputs, 417–421
    single–input, single–output, 409–414
Identification, 11, 215–220, 303–304
Impulse train, 62–65
Inconsistent equations, 689
Independence of events, 747
Inner product of vectors, 692–693
Integration diagrams, 111–115
Integrals, approximations to, 143–149, 492–501
Interrupts, 546–549
Inverse filters, 409

J

Joint random variables, 757, 759, 760
Jordan forms of a matrix, 165–170
Jury, Eli, 46, 98
Jury stability test, 45–50

K

Kalman filter, 625–653
    algorithm for multivariable, 641–643
    equations, 632
    extended, 650–652
    extensions, 642
    linearized, 650
    -observer, 654–656
    program for first order, 637–642
Kalman gain, 627–637
Kalman, Rudolph, 581

L

Laplace expansion of a determinant, 685–686
Laplace transform, 58, 116–121
Least squares estimation, 605–617
    batch, 605–607
    initialization for recursion, 611–614
    of initial state, 608–609
    recursive, 609–615
    weighted, 615–617
    probabilistic interpretation of, 615
Left eigenvector, 731
Linear algebra, 680–745
Linear independence
    of equations, 688–689
    of inputs, 227
    of outputs, 212
    of vectors, 693–694
Linear transformation, 696–697
Linearization about an operating point, 13–14
Linear estimator, 785
Linear point connector, 65
Local area network (LAN), 533
Luenberger, David, 312

M

Marginal probability, 758
Mason's gain rule, 121, 150
Matrix
    adjugate or adjoint, 691
    algebra, 680–745
    cofactor, 685
    complex conjugate, 683
    derivatives, 683–685
    determinant, 684–687
    diagonal, 680
    identity, 680
    inverse, 690–691
    inversion lemma, 610
    modal, 700
    minimal, 741
    minor, 685
    null, 680
    null space, 697
    orthogonal, 698–699
    partitioning, 682–683
    polynomial, 691–692
    range space, 697
    rank, 687

Matrix (*continued*)
  singular, 691
  skew-symmetric, 681
  square root, 717–718 (*see also* Cholesky decomposition method)
  symmetric, 681, 714–717
  trace, 683
  transpose, 681
Measurement interruption, 471–472
Measurement residuals (innovations), 626
Memory controller, 529
Memory mapped input/output, 530
Minimum mean square estimate, 786–788
Mnemonic machine code, 543
Model algorithmic control, 487
Modes, 171–176, 203, 382, 733
Monorail system, 182–185
Multiprocessor architectures, 531–533
Multiplier-accumulator, 539–543

N

Natural mode, 733
Negative definite, 717
Negative semidefinite, 717
Nonlinear system, 4, 13–14, 650–651
Norm of a vector, 692
Null space, 697
Nyquist criterion, 6

O

Observability, 170–176
  definition of, 171, 206–208
  index, 209
  loss of due to sampling, 514
  matrices, 206–209, 220–221
  of a system's state, 210–214
  of repeated modes, 173–176
  step-varying, 220–221
Observable form, 152–156, 314–315
  transformation to, 179–181, 321–322, 362–363
Observation matrix, 331–332
Observer, 312
  as error feedback system, 330, 359, 369
  deadbeat, 316–317, 322–323, 371, 397

Observer (*continued*)
  design of, 319–323, 333–355, 370–380
  feedback, 355–367, 379–380
  for multiple-output systems, 325–330, 376–379
  full-order state, 312–325, 360–364, 367–369
  minimal order, 349–355
  of a linear state transformation, 330–332, 379–380
  ongoing design, 371–376
  orders for feedback eigenvalue placement, 354–355
  reduced order, 344–349, 364–367
  step-varying, 367–380
Operating point, 13
Optimal control, 267–297
  examples, 267–268, 274–282, 291–297
  state feedback solution, 268–278
  steady state, 294–297
Optimal linear regulation, 657–658
Orthogonal data, 624
Orthogonal random variable, 761
Orthogonal stochastic processes, 770
Orthogonal transformations, 715–716
Orthogonality principle, 624, 788
Overflow, 541

P

Parameter estimation, 215–220
Partial derivative of a scalar function, 683
Peripheral interface circuits, 529
Performance measure or index, 268–269, 274, 280, 283, 292, 605, 616, 620, 657
Persistent excitation, 218
Phase-locked loop, 465–472, 488
Phase margin, 6
Plant, 1
Plant input filtering, 520, 559–560
Polling, 549
Portability of software, 544
Positioning mechanism, 557–562
Positive definite, 716
Positive semidefinite, 716
Power spectral density, 582, 774–776

Prefiltering, 74, 538
Probability measure, 747
Processing delay, 539
Program structures, 543–551

Q

Quadratic forms, 714–717
Quadratic optimal regulation, 267–297
Quantization, 2–3, 56–57, 471

R

Ragazzini, John, 98
Ramp–invariant approximation, 508
Random event, 746
Random experiment, 746
Random variable
  continuous, 751
  discrete, 751
  expected value, 753
  function of, 751
  mean of, 753
  mean square of, 754
  mixed, 751
  moments of, 753–755
  probability density function of, 749
  probability distribution function of, 749
  standard deviation of, 755
  variance of, 755
Random access memory (RAM), 529
Range space, 697
Rank test, 206, 222
Reachability, 224
Read–only memory (ROM), 529
Real–time, 544
Reciprocal basis, 730
Reconstruction
  bandlimited, 74
  definition of, 55
  step, 61–65
  with significant rise time, 64
Reference input filters, 409, 422
Reference model tracking design, 438–455, 473–476
  higher–order controllers for, 446–450
  of multiple reference inputs, 450–455

Reference model tracking design (*continued*)
  reference signal models for, 438–439
  single–input, single–output, 439–450
  step–varying, 472–476
Regulation, 312, 355–367
  optimal, 267–297
Residue matrix, 726–729
Response
  between–sample, 512–515
  of continuous–time systems, 116–119, 121–128
  of step–invariant systems, 134–140
  of step–varying systems, 141–143
  transient, 6
Response model tracking design, 421–437
  using feedback observer inputs, 430–434
  with multiple plant inputs, 427–429
  of multiple reference inputs, 427–429
  single–input, single–output, 421–427
  tracking error feedback, 434–437
Root locus, 5, 91–95, 258
Routh–Hurwitz test, 6, 46, 51

S

Sample–and–hold (S/H), 4–5, 62
Sample point, 746
Sample space, 746
Sampling, 56, 143–149
  interval, 4, 18
  loss of observability, 514
  of a continuous–time signal, 55–60
  theorem, 65–70
Saturation nonlinearity, 552–553
Separation theorem, 356–367
Set
  empty, 747
  intersection of, 747
  union of, 747
Set point, 5
Servosystem, 5 (*see also* tracking system)
Sequence, 17–31

Sequencer, 540
Shaping filter, 593–594, 603–604
Sigma field of events, 747
Signal Flow graph, 150–156
Sign definiteness, 716–717
Similarity transformation, 156–157,
    710–711
Simulation, 95–96, 515–519
Simulation diagram, 111–115
Sklansky, Jack, 98
Software, 543–551
    for adaptation to failures, 551
    design of, 549–551
    housekeeping tasks, 545
    interrupts, 546–549
    portability, 544
    subroutines, 546–549
Spacecraft roll attitude control,
    278–282, 294–297
Spectral factorization, 592–594
Spectral matrix, 700
State
    control, 225–229
    determination, 206–209, 220–221
    equations, 110–120, 125–126,
        129–134, 141–143
    feedback, 235–253
    space, 9
    transition matrix, 122–128, 186–
        187, 199, 719–735
State variables
    change of 156–157
    controllable form for, 149–152
    decoupling of, 158–170
    observable form for, 152–156,
        314
Steady state error, 8, 90–94
Steady State response, 6
Stochastic optimal control, 653–658
Stochastic process, 582–585
    autocorrelation of, 582
    colored, 593
    complex, 766
    continuous–time, 766
    correlated, 603–604, 649–650
    discrete–time, 766
    ensemble of, 766
    mean of, 582
    mean square value of, 582, 769
    power spectral density of, 582

Stochastic process (continued)
    random, 766
    real, 766
    response of system to, 585–604
    strict sense stationary, 772–773
    variance of, 769
    wide–sense stationary, 582, 773
Subroutines, 546–549
Symmetric matrices, 681, 714–717

T

Temperature control system, 9–14
Timer, 547
Time–invariant, 4
Tracking system, 5, 7, 405–406, 472
    error feedback form, 434–437
    ideal system design, 406–421
    reference model design, 438–455,
        473–476
    response model design, 421–437
    using feedback observer inputs,
        430–434
Tracking outputs, 406–409
Transfer functions 6, 33–45, 119–
    121, 139–141
Tustin's approximation, 496, 498–
    501

U

Uniform random variable
    mean of, 755
    probability density of, 752
    variance of, 755

V

Vector
    basis, 694
    column, 680
    inner product, 692–693
    linear independence of, 693–694
    norm, 692
    orthogonality, 693
    row, 681
    unit, 692
    unit coordinate, 694
Videotape drive control, 88–96

W

Weighted least squares, 615–617
White noise, 590, 641–642

Z

Zadeh, Lotfi, 98
Zero–input and zero–state response, 6–9, 117–118, 135–136

Zero–order hold, 62–63, 527–528
Z–transfer function, 33–45, 139–140
Z–transform
    convergence of, 20–21
    definition of, 18
    inversion of, 20, 27–31
    pairs, 19–21
    properties of, 21–27
    of state variable equations, 129–130, 136–140